# Electron and Magnetization Densities in Molecules and Crystals

# NATO ADVANCED STUDY INSTITUTES SERIES

**A series of edited volumes comprising multifaceted studies of contemporary scientific issues by some of the best scientific minds in the world, assembled in cooperation with NATO Scientific Affairs Division.**

**Series B: Physics**

## *RECENT VOLUMES IN THIS SERIES*

*Volume 41* – Fiber and Integrated Optics
edited by D. B. Ostrowsky

*Volume 42* – Electrons in Disordered Metals and at Metallic Surfaces
edited by P. Phariseau, B. L. Györffy, and L. Scheire

*Volume 43* – Recent Advances in Group Theory and Their Application to Spectroscopy
edited by John C. Donini

*Volume 44* – Recent Developments in Gravitation – *Cargèse* 1978
edited by Maurice Lévy and S. Deser

*Volume 45* – Common Problems in Low- and Medium-Energy Nuclear Physics
edited by B. Castel, B. Goulard, and F. C. Khanna

*Volume 46* – Nondestructive Evaluation of Semiconductor Materials and Devices
edited by Jay N. Zemel

*Volume 47* – Site Characterization and Aggregation of Implanted Atoms in Materials
edited by A. Perez and R. Coussement

*Volume 48* – Electron and Magnetization Densities in Molecules and Crystals
edited by P. Becker

*Volume 49* – New Phenomena in Lepton-Hadron Physics
edited by Dietrich E. C. Fries and Julius Wess

*Volume 50* – Ordering in Strongly Fluctuating Condensed Matter Systems
edited by Tormod Riste

*Volume 51* – Phase Transitions in Surface Films
edited by J. G. Dash and J. Ruvalds

This series is published by an international board of publishers in conjunction with NATO Scientific Affairs Division

| | |
|---|---|
| A Life Sciences<br>B Physics | Plenum Publishing Corporation<br>London and New York |
| C Mathematical and Physical Sciences | D. Reidel Publishing Company<br>Dordrecht and Boston |
| D Behavioral and Social Sciences | Sijthoff International Publishing Company<br>Leiden |
| E Applied Sciences | Noordhoff International Publishing<br>Leiden |

# Electron and Magnetization Densities in Molecules and Crystals

Edited by

P. Becker

*CMOA, Centre National de la Recherche Scientifique*
*Paris, France*
*and Physics Department, University of Nancy I*

PLENUM PRESS • NEW YORK AND LONDON
Published in cooperation with NATO Scientific Affairs Division

Library of Congress Cataloging in Publication Data

Nato Advanced Study Institute on Electron and Magnetization Densities in Molecules and Crystals, Arles, France, 1978.
Electron and magnetization densities in molecules and crystals.

(Nato advanced study institutes series: Series B, Physics; v. 48)
"Lectures presented at the NATO Advanced Study Institute on Electron and Magnetization Densities in Molecules and Crystals, held in Arles, France, August 16–31, 1978."
Includes index.
1. Molecular structure–Congresses. 2. Crystals–Electric properties–Congresses. 3. Crystals–Magnetic properties–Congresses. I. Becker, Pierre. II. Title. III. Series.
QD461.N35 1978 530.4'1 79-19022
ISBN-13: 978-1-4684-1020-4 e-ISBN-13: 978-1-4684-1018-1
DOI: 10.1007/978-1-4684-1018-1

Lectures presented at the NATO Advanced Study Institute on Electron and Magnetization Densities in Molecules and Crystals, held in Arles, France, August 16–31, 1978.


Softcover reprint of the hardcover 1st edition 1980
A Division of Plenum Publishing Corporation
227 West 17th Street, New York, N.Y. 10011

# Preface

The interest of describing the ground state properties of a system in terms of one electron density (or its two spin components) is obvious, in particular due to the simple physical significance of this function. Recent experimental progress in diffraction made the measurement of charge and magnetization densities in crystalline solids possible, with an accuracy at least as good as theoretical accuracy. Theoretical developments of the many-body problem have proved the extreme importance of the one electron density function and presently, accurate methods of band structure determination become available.

Parallel to the diffraction techniques, other domains of research (inelastic scattering, resonance, molecular spectroscopy) deal with quantities directly related to the one particle density. But the two types of studies do not interfere enough and one should obviously gain more information by interpreting all experiments that are related to the density together.

It became necessary to have an International School that reviews the status of the art in the domain of "ELECTRON AND MAGNETIZATION DENSITIES IN MOLECULES AND CRYSTALS". This was made possible through the generous effort of N.A.T.O.'s Scientific Affairs Division, and I would specially thank Dr. T. KESTER, the head of this Division, for his help and competence. An Advanced Study Institute was thus held in ARLES, south France, from the 16th to the 31st of August 1978.

Over 180 applications were received and 105 persons could be accepted as participants, coming from 23 countries: Canada, 3; Denmark, 3; France, 24; Germany, 10; Italy, 2; Netherlands, 7; Norway, 2; Portugal, 2; Turkey, 1; United Kingdom, 5; United States, 13; Australia, 5; Brazil, 2; Egypt, 1; Finland, 3; India, 1; Israel, 2; Japan, 2; Poland, 2; South Africa, 1; Sweden, 6; Switzerland, 1; Venezuela, 1. They included crystallographers, physical chemists, solid state physicists, theorists. There were 19 lecturers, and among them were great specialists of the field.
Besides 3 to 4 lectures per day, each of 1-1/2 hour, 2 hours were

devoted every day to exercises of application. Participants insisted on having the major exercises and their solution included in the book of the proceedings. I do hope that this will make concepts easier to understand and to practically apply.

It would not have been possible to organize such a meeting without the NATO help and competence. The financial support of the Centre National de la Recherche Scientifique, of the Commissariat a l'Energie Atomique, of the European Research Office of the US Army, and of the City of Arles made possible to grant many participants from non-NATO countries and I am very grateful to those organizations.

During the preparation of the Institute, I was greatly helped by Madame Claire DESAUBLIAUX, who did the secretariat. I would like especially to thank Dr. Michel BONNET who accepted to help me in the scientific preparation of the school. The Palais des Congres d'Arles offered us, at very interesting conditions, the beautiful Conference Hall and I am grateful to Monsieur LIN for his efforts.

I would like to acknowledge the fantastic work accomplished by my wife, Dr. Monique BECKER, who took care of the whole organization during the Institute.

The recreational part of the meeting was made very pleasant through the efforts of the city of Arles. I would also like to thank the pianist Deniz ARMEN GELENBE who accepted to give a recital for us and Madame Nicole d'AGGAGIO who presented a painting exhibit: it was very nice to have those two artists with us during several days.

Of course, I want to thank all the participants, and especially the lecturers, chairmen, exercise instructors, for their active participation and I do hope that this book will help to the development of the field and to a stronger cooperation between researchers from various disciplines. It would be nice to feel the necessity for a School on the applications of one particle densities in a couple of years.

The book is divided into five parts, corresponding to the goals I had in mind when preparing the School:

1. A critical review of the description of ground state systems in terms of one particle densities, both in molecules and solids: the relationship between density and energy, and the role of density in chemical binding.

2. A large part is devoted to the physical and experimental problems related to the determination of electron and spin densities: mainly, how is a structure factor related to the measured intensity?

3. The representation of a charge or spin density in terms of hopefully physically interpretable quantities is a fundamental task: when the answer is not unique, how can one compare several me methods? The importance of a good description of the coupling between vibrations and electronic deformation is emphasized.

4. How are other techniques involving one particle densities related to this quantity ? What is the compatibility between charge and momentum densities ? A discussion of high energy electron scattering is included, with a detailed application to nitrogen. Magnetic resonance technique is also presented, at a fundamental level.

5. A final part is devoted to the applications of charge or spin densities : their relationship with macroscopic physics, with intermolecular forces, chemical reactivity, magnetism...
It is time to use experimental densities to predict all kinds of properties.

Practically many lecture notes include aspects that are relevant to several chapters and the separation outlined above was perhaps more pronounced in the oral presentations.

Pierre BECKER
Director of the Institute

## Contents

PART I : FUNDAMENTAL CONCEPTS AND THEORY

Electron Densities and Reduced Density Matrices ...... 3
V.H. Smith, Jr.

On the Calculation and Accuracy of Theoretical Electron Densities ........................ 27
V.H. Smith, Jr.

An Electrostatic Description of Chemical Binding ..... 47
F.L. Hirshfeld

Charge and Spin Densities in Solids .................. 63
N.H. March

Local Density Approach to Bulk and Surface Charge Densities .................................. 83
A.J. Freeman

Calculation of One-Electron Charge, Momentum and Spin Densities : The Statistical Exchange Approximation ..................... 107
D.E. Ellis

PART II : DIFFRACTION PHYSICS AND EXPERIMENTAL PROBLEMS

Introduction to Wave Scattering by Atoms and Molecules in the Free State ................ 137
R.A. Bonham

Diffraction by Real Crystals : I. Kinematical Theory and Time Averaging .................. 173
P. Becker

Diffraction by Real Crystals : II. Extinction ........ 213
P. Becker

Exercise : Extinction and Ferroelectric Transition in $RbH_2PO_4$ .................................. 231
P. Becker, P. Bastie , and J. Lajzerowicz

Exercise : Extinction Treatment in Polarized Neutron Experiments ....................... 235
A. Delapalme

On Extinction (with an exercise) .................... 237
N. Kato

Magnetic Neutron Scattering .......................... 255
P.J. Brown

Exercises : Magnetic Neutron Scattering ............. 271
P.J. Brown

Optimal Experimental Conditions for Charge Density Determination ...................... 287
M.S. Lehmann

Exercises : Experimental Problems ...................
M.S. Lehmann

The Measurement of Magnetic Structure Factors ........ 323
J.B. Forsyth

Error Analysis in Experimental Density Determination .............................. 355
M.S. Lehmann

PART III : ANALYSIS OF EXPERIMENTAL DENSITIES

Analysis and Partitioning of Electronic Densities .... 375
P. Becker

Multipolar Expansion of One-Electron Densities ....... 405
R.F. Stewart

Partitioning of Hartree-Fock Atomic Form Factors into Core and Valence Shells ....... 427
R.F. Stewart

Restricted Radial Functions for Analysis of Molecular Form Factors .................. 433
C. Ceccarelli and R.F. Stewart

Algorithms for Fourier Transforms of Analytical Density Functions ..................... 439
R.F. Stewart

Exercises on Form Factor Models ....................... 443
J.A. Avery

Real vs. Reciprocal Space Analysis of Electron Densities .......................... 447
P. Coppens and E.D. Stevens

Imposing Non-Crystallographic Symmetry Constraints on Multipole Populations and Defining Local Atomic Coordinate Systems .......................... 473
E.D. Stevens

Interpretation of the Spin Densities in Metals and Alloys ........................... 479
J. Schweizer

Interpretation of Magnetization Densities ............ 501
J. Schweizer

Thermal Smearing and Chemical Bonding ................ 521
P. Coppens

Elastic X-Ray Scattering from Solids Containing Non-Rigid Pseudoatoms .................. 545
S.W. Wilkins

Vibrational Studies of X-Ray Molecular Form Factors and Intensities .................... 549
J. Epstein and R.F. Stewart

Force Constants in Diatomic Molecules, from Charge Densities ........................... 569
P. Becker

PART IV : RELATED TECHNIQUES

Compton Scattering (with exercises) .................. 573
R.J. Weiss

Determination of Charge Densities and Related Quantities by use of High Energy Electron Scattering ................ 581
R.A. Bonham and M. Fink

Exercise : Properties of Charge Densities Obtained from Electron Scattering from Nitrogen .............................. 601
R.A. Bonham

Magnetic Resonance and Related Techniques ........... 633
J. Maruani

PART V : GOING TO THE REAL WORLD

Electronic Densities in Molecules, Chemical Bonds and Chemical Reactions ............... 695
R. Daudel

Spin, Charge, Momentum Densities : Their Relationships Amongst Themselves and With Other Physical Measurements ...................... 723
P.J. Brown

Electron Distributions and Thermodynamics ........... 735
R.J. Weiss

Exercise : Form Factors and their Application to Calculating Properties .................. 743
R.A. Bonham

Chemical Interpretation of Deformation Densities ..... 757
F.L. Hirshfeld

Exercise : Density Analysis for the Cubic Modification of $C_2H_2$ ...................... 769
A. Vos

Applications of Electron Density Studies to Complexes of the Transition Metals ......... 779
E.N. Maslen

The Chemical Interpretation of Magnetization Density Distributions ...................... 791
J.B. Forsyth

Exercise : Relationship of Multipole Populations to Orbital Occupancies of a Transition Metal Atom in a Tetragonally Distorted Octahedral Field ........................................ 823
E.D. Stevens

Exercise : Calculation of the Magnetic Form Factor for the Spin Density of a 3d-shell in a Given Environment ............ 827
J.X. Boucherle

Effect of Crystal Forces and Hydrogen Bonding on Charge Density........................... 831
I. Olovsson

Index ............................................. 895

# Part I
# Fundamental Concepts and Theory

# ELECTRON DENSITIES AND REDUCED DENSITY MATRICES

Vedene H. Smith, Jr.

Department of Chemistry, Queen's University, Kingston

Ontario, K7L 3N6, Canada

## I. INTRODUCTION

In the present lectures we are concerned with the use of the bound-state solutions of the time-independent Schrödinger equation

$$H\Psi = E\Psi \qquad (1)$$

to provide a conceptual and mathematical framework for the analysis and interpretation of the experimentally observed electron densities (charge, spin, and momentum) which are the subject of this Advanced Study Institute.

We consider the electronic wave function $\Psi(1,2,\ldots,N)$ for an N-electron system which was obtained by solution of (1) or by a variational or other approximation to that problem. We use the notation $J = (r_j,s_j)$ to denote the spin $(s_j)$ and space coordinates $(r_j)$ of electron $j$ and assume that $\Psi$ is antisymmetric, i.e. that for any permutation P of the coordinates $(1,2,\ldots,N)$

$$\Psi(1,2,\ldots,N) = (-1)^p P\Psi(1,2,\ldots,N) \qquad (2)$$

where p is the parity of P and that $\Psi$ is normalized, i.e. the normalization integral

$$\langle\Psi|\Psi\rangle = \int|\Psi(1,2,\ldots,N)|^2 d1d2\ldots dN \qquad (3)$$

exists and equals one.

We are interested in the expectation value

$$\langle F\rangle = \langle\Psi|\hat{F}|\Psi\rangle/\langle\Psi|\Psi\rangle \tag{4}$$

of a physical quantity F represented by an Hermitian operator

$$\hat{F} = \hat{F}_o + \sum_i \hat{F}_i + \frac{1}{2!}\sum_{i,j}{}' \hat{F}_{ij} + \frac{1}{3!}\sum_{i,j,k} \hat{F}_{ijk} + \ldots \tag{5}$$

where each term is symmetrical in the electron indices and the primes denote the omission of all terms with two or more equal indices. We recognize that the Hamiltonian operator for a molecule with frozen nuclei may be written using the first three terms in the above equation, i.e.,

$$\hat{H} = \hat{H}_o + \sum_i \hat{H}_i + \frac{1}{2}\sum_{i,j}{}' \hat{H}_{ij} \tag{6}$$

where

$$\hat{H}_o = \sum_{\alpha<\beta} \frac{Z_\alpha Z_\beta}{R_{\alpha\beta}} \tag{7}$$

$$\hat{H}_i = -1/2\,\nabla_i^2 - \sum_\alpha Z_\alpha/r_{i\alpha} \tag{8}$$

$$\hat{H}_{ij} = 1/r_{ij} \tag{9}$$

Here the index i denotes the spatial coordinates of electron i, $Z_\alpha$ is the atomic number of nucleus $\alpha$, $r_{i\alpha}$ is the distance between electron i and nucleus $\alpha$, $r_{ij}$ is the distance of separation between electrons i and j, and $R_{\alpha\beta}$ is the separation between nuclei $\alpha$ and $\beta$. We use atomic units (a.u.) where $\hbar = h/2\pi = 1$, $e = 1$, $m_e = 1$. Thus the a.u. of length (bohr) = $a_o = 4\pi\varepsilon_o\hbar^2/m_e e^2 \approx 0.529177 \times 10^{-10}$ m, and the a.u. of energy (hartree) $\approx 4.359814 \times 10^{-18}$ J.

The evaluation of $\langle\Psi|\hat{F}|\Psi\rangle$ involves the 4N-dimensional integral

$$\langle\hat{F}\rangle = \langle\Psi|\hat{F}|\Psi\rangle = \int\Psi^*(1,2,\ldots,N)\hat{F}\Psi(1,2,\ldots,N)d1d2\ldots dN \tag{10}$$

which can be simplified by observing that most of the integrations in (10) are independent of the particular form of F. We rewrite (10) in the form

$$\langle F\rangle = \int\hat{F}\Psi(1,2,\ldots,N)\Psi^*(1',2',\ldots,N')d1d2\ldots dN \tag{11}$$

where $\int dJ$ now means to let $\hat{F}$ operate on $\Psi(1,2,\ldots,N)$, set $J = J'$ and integrate with respect to dJ. We introduce the Nth order density matrix

$$\Gamma^N(1,2,\ldots,N;1',2',\ldots,N') = \Psi(1,2,\ldots,N)\Psi^*(1',2',\ldots,N') \quad (12)$$

as a convenient notation in order to write (11) in a more compact form:

$$\langle\hat{F}\rangle = \int \hat{F}\Gamma^N(1,2,\ldots,N;1',2',\ldots,N')d1d2\ldots dN \quad (13)$$

As before $\int dJ$, etc. have the same meaning. Operationally the evaluation of (13) proceeds in three stages

a) $\hat{F}\Gamma^N(\overbrace{1,2,\ldots,N};1',2',\ldots,N')$

b) Set $1 = 1'$, $2 = 2'$, $\ldots\ldots, N = N'$

c) Integrate over $d1, d2, \ldots, dN$.

We now write (13) in terms of the expansion (5) for $\hat{F}$, i.e.,

$$\langle\hat{F}\rangle = \int \hat{F}_o\Gamma^N(1,2,\ldots,N;1',2',\ldots,N')d1d2\ldots dN$$

$$+\sum_i \int \hat{F}_i\Gamma^N(1,2,\ldots,N;1',2',\ldots N')d1d2\ldots dN$$

$$+\frac{1}{2!}\sum_{i,j}{}' \int \hat{F}_{ij}\,\Gamma^N(1,2,\ldots,N;1',2',\ldots;N')d1d2\ldots dN$$

$$+\ldots \quad (14)$$

Since F is symmetric and $\Psi$ is antisymmetric in the electron coordinates, it follows that each term in any one sum has the same value and therefore

$$\langle\hat{F}\rangle = \hat{F}_o + N\int \hat{F}_1\Gamma^N(1,2,\ldots,N;1',2',\ldots,N')d1d2\ldots dN$$

$$+\binom{N}{2}\int \hat{F}_{12}\Gamma^N(1,2,\ldots,N;1',2',\ldots,N')d1\ldots dN$$

$$+\ldots \quad (15)$$

We can now eliminate the coordinates which are not operated on by

$\hat{F}_1$, $\hat{F}_{12}$, $\hat{F}_{123}$, etc. To do this, we define a sequence of reductions of $\Gamma^N$, i.e. for p = 1,...,N, we define the pth reduced density matrix

$$\Gamma^p(1,2,\ldots,N;1',2',\ldots,P')$$

$$= \binom{N}{p} \int \Gamma^N(1,2,\ldots,N;1',2',\ldots,N')d(P+1)\ldots dN \tag{16}$$

i.e. N - p coordinates P + 1,...,N have been eliminated or reduced from $\Gamma^N$. Alternately $\Gamma^p$ can be obtained from $\Gamma^{p+1}$ by a single reduction

$$\Gamma^p(1,2,\ldots,P;1',2',\ldots,P') \tag{17}$$

$$= \frac{(p+1)}{(N-p)} \int \Gamma^{p+1}(1,2,\ldots,(P+1);1',2',\ldots,(P+1)')d(P+1)$$

Then

$$\langle\hat{F}\rangle = \hat{F}_o + \int \hat{F}_1 \Gamma^1(1;1')d1$$

$$+\int \hat{F}_{12}\Gamma^2(1,2;1',2')d1d2$$

$$+\int F_{123}\Gamma^3(1,2,3;1',2',3')d1d2d3+\ldots \tag{18}$$

In the specific case of the electronic Hamiltonian (6) we may write (18) in the form

$$\langle H\rangle = \int [\hat{K}^2(1,2)\Gamma^2(1,2;1',2')]d1d2 \tag{19}$$

where

$$\hat{K}^2(1,2) = 2N^{-1}(N-1)^{-1}H_o + (N-1)^{-1}(H_1 + H_2) + H_{12} \tag{20}$$

is called the reduced Hamiltonian.

Inspection of (19) shows that the 4N dimensional integral required for $\langle H\rangle$ in the variational principle

$$E \leq \frac{\langle\Psi|H|\Psi\rangle}{\langle\Psi|\Psi\rangle} \tag{21}$$

has been reduced to an 18 dimensional one. However one can not vary (19) simply with respect to any $\Gamma^2$. If one does so, it is possible to obtain an energy lower than the ground state energy E /1,2/.

The ground state energy E corresponds to the minimum of $\int \hat{K}^2\Gamma^2(1,2;1',2')d1d2$ such that $\Gamma^2$ is expressible in the form of (16), i.e. that $\Gamma^2$ may be obtained from an antisymmetric wave function by means of (16). Such a $\Gamma^2$ is said to be N-representable and hence, (16) gives a necessary and sufficient criterion for N-representability. However, a simple mathematical test equivalent to (16) is not known. The subject of N-representability is of much current interest /3,4/.

Some obvious necessary criteria for N-representability include the normalization, hermiticity and anti-symmetry of the density matrices. The $\Gamma^p$ defined above are normalized according to the Löwdin convention /5/

$$\int \Gamma^p(1,2,\ldots,P;1',2',\ldots,P')d1d2\ldots dP = \binom{N}{p} \tag{22}$$

and are hermitian

$$\Gamma^p(1,2,\ldots,P;1',2',\ldots,P') = [\Gamma^p(1',2',\ldots,P';1,2,\ldots,P)]^* \tag{23}$$

as well as antisymmetric with respect to the interchange of electrons, e.g., for p = 2

$$\begin{aligned}\Gamma^2(1,2;1',2') &= -\Gamma^2(2,1;1',2') = \Gamma^2(2,1;2',1')\\ &= -\Gamma^2(1,2;1',2')\end{aligned} \tag{24}$$

Other conventions for the normalization of the reduced density matrices are used in the literature. For example, Coleman /1/ replaces the factor $\binom{N}{p}$ in the defining equation (16) by unity: We shall refer to density matrices so defined by the symbol $D^p(1,2,\ldots,P;1',2',\ldots P')$.

The Löwdin normalization will be convenient for our purpose of relating the reduced density matrices to the various experimentally observable electron densities. We note that in this normalization $\Gamma^1(1;1)d1$ is the number of electrons times the probability of finding an electron with spin s, in $d\vec{r}_1$ at $\vec{r}_1$ while $\Gamma^2(1,2;1,2)d1d2$ is the number of pairs times the probability of finding one electron with spin $s_1$ in $d\vec{r}_1$ at $\vec{r}_1$ and another with spin $s_2$ in $d\vec{r}_2$ at $\vec{r}_2$.

## II. DENSITY MATRICES AND THE HARTREE-FOCK MODEL

We observe that given a complete set of orthonormal spin orbitals $\{\Psi_1\}$, $\Gamma^1(1,1')$ may be written as an expansion

$$\Gamma^1(1;1') = \sum_{i,j} \gamma_{ij} \Psi_i(1)\Psi_j^*(1') \tag{25}$$

Since $\Gamma^1(1;1')$ represents the matrix corresponding to the operator $\Gamma^1$, we may write

$$\int\Gamma^1(1;1')d1 = \sum_{i,j} \gamma_{ij}\int\Psi_i(1)\Psi_j^*(1') = \sum_i\gamma_{ii} = N \tag{26}$$

This expansion of the Hermitian operator $\Gamma^1$ may be diagonalized, i.e.

$$\Gamma^1(1,1') = \sum_i \nu_i\chi_i(1)\chi_i^*(1') \tag{27}$$

in terms of the eigenfunctions of $\Gamma^1,\{\chi_i\}$ which are usually called natural spin orbitals. The eigenvalues (or occupation numbers) $\gamma_i$ satisfy

$$0 \leq \nu_i \leq 1, \tag{28}$$

$$\sum_i \nu_i = N. \tag{29}$$

In the Hartree-Fock or independent particle model, the wave function is a Slater determinant

$$\Phi = \frac{1}{\sqrt{N!}} \det|\phi_1(1)\phi_2(2)\ldots\phi_N(N)| \tag{30}$$

built up from a set of N orthonormal spin orbitals $\Phi_i$. The one-particle density matrix corresponding to $\Phi$, $\Gamma^1_\Phi$ (1,1') is called the Fock-Dirac density matrix and may be written in terms of the N occupied spin-orbitals, as

$$\Gamma^1_\Phi(1;1') = \sum_{i=1}^{N} \phi_i(1)\phi_i^*(1') \tag{31}$$

i.e. the occupied spin orbitals have occupation numbers of unity while the unoccupied have zero occupation numbers. It is idempotent, i.e.

$$\Gamma^1_\Phi(1,1') = \int\Gamma^1_\Phi(1,2)\Gamma^1_\Phi(2,1')d2 \tag{32}$$

Since it may be shown that

$$\Gamma_{\Phi}^{2}\ (1,2;1',2') = \frac{1}{2}\{\Gamma_{\Phi}(1,1')\Gamma_{\Phi}(2,2') - \Gamma_{\Phi}(1,2')\Gamma_{\Phi}(2,1')\} \quad (33)$$

the energy and all one- and two-particle properties are functionals of $\Gamma_{\Phi}^{1}$. As a result $\Gamma_{\Phi}^{1}$ is called the fundamental invariant of the HF scheme /5/. Furthermore, it is invariant to unitary transformations among the N occupied $\phi_i$. One can develop the HF (SCF) scheme to solve directly for $\Gamma_{\Phi}^{1}$.

We note that the first term in (33) gives rise to the so-called Coulomb contribution to the energy while the second is responsible for the exchange term. When $\Gamma_{\Phi}(1,1')$ is expanded in terms of a set of atomic orbitals $\omega_i$

$$\Gamma_{\Phi}(1,1') = \sum_{i,j} \rho_{ij}\ \omega_i(1)\ \omega_j^*(1') \quad (34)$$

it is often called the charge and bond order matrix.

## III. ONE AND TWO-ELECTRON DENSITY FUNCTIONS

The familiar one-electron density $\rho(\vec{r})$ is obtained by the spin-trace of $\Gamma^1(1,1')$ i.e.

$$\rho(\vec{r}) = \rho(\vec{r},\vec{r}) \quad (35)$$

where

$$\rho(\vec{r},\vec{r}\,') = \int \Gamma^1(1,1')ds_1 \quad (36)$$

We recall the radial electron density D(r) which is defined in terms of the spherical average of $\rho(\vec{r})$

$$D(r) = 4\pi r^2 \rho_o(r) \quad (37)$$

$$\rho_o(r) = (4\pi)^{-1} \int_o^{2\pi} \int_o^{\pi} \rho(\vec{r})\ d\Omega_r \quad (38)$$

In a similar manner, we introduce two-electron density functions which are useful for the interpretation of $\Gamma^2(1,2;1',2')$ and expectation values of two-particle operators.

We define the spin-traced two-matrix by

$$\Gamma(\vec{r}_1,\vec{r}_2;\vec{r}_1{}',\vec{r}_2{}') = \int \Gamma^2(1,2;1',2')ds_1\ ds_2 \quad (39)$$

The intracule matrix /6-8/ is obtained from $\Gamma$ as follows. Define the extracular coordinates

$$\vec{R}_o = (\vec{r}_1 + \vec{r}_2)/2, \ \vec{R}_o{}' = (\vec{r}_1{}' + \vec{r}_2{}')/2 \tag{40}$$

and the intracular coordinates

$$\vec{r}_{12} = \vec{r}_1 - \vec{r}_2, \ \vec{r}_{12}{}' = \vec{r}_1{}' - \vec{r}_2{}'. \tag{41}$$

Then we change coordinates to obtain

$$\tilde{\Gamma}(\vec{R}_o,\vec{r}_{12};\vec{R}_o{}',\vec{r}_{12}{}') = \Gamma(\vec{r}_1,\vec{r}_2;\vec{r}_1{}',\vec{r}_2{}'). \tag{42}$$

After integration over the extracular coordinates, we obtain the intracule matrix

$$J(\vec{r}_{12},\vec{r}_{12}{}') = \int \tilde{\Gamma}(\vec{R}_o,\vec{r}_{12};\vec{R}_o,\vec{r}_{12}{}')d\vec{R}_o. \tag{43}$$

The spherical average of a diagonal element of J is then given by

$$h(r_{12}) = (4\pi)^{-1} \int_0^{2\pi} \int_0^{\pi} J(\vec{r}_{12},\vec{r}_{12})\sin\alpha d\alpha d\beta, \tag{44}$$

where $r_{12} = (r_{12},\alpha,\beta)$.

We observe that $h(r_{12})$ is related to the electron-electron distribution function /9/ $P_o(r_{12})$ by

$$h(r_{12}) = P_o(r_{12})/4\pi r_{12}^2 \tag{45}$$

and the weighted pair distribution function /10,11/ $g(r_{12})$ by

$$h(r_{12}) = g(r_{12})/4\pi r_{12}. \tag{46}$$

We observe that there is a close parallelism between $h(r_{12})$ and $\rho_o(r)$ expressed by the following set of equations:

$$\rho_o(0) = \sum_i \langle\delta(\vec{r}_i)\rangle, \ h(0) = \sum_{i<j} \langle\delta(\vec{r}_{ij})\rangle, \tag{47}$$

$$4\int_0^{\infty} r^{k+2} \rho_o(r)dr = \sum_i \langle r_i^k\rangle, \ 4\pi \int_0^{\infty} r_{12}^{k+2} h(r_{12})dr_{12} = \sum_{i<j}\langle r_{ij}^k\rangle, \tag{48}$$

$$4\pi\int_0^\infty r^2\rho_o(r)dr = N, \quad 4\pi\int_0^\infty r_{12}^2\, h(r_{12})dr_{12} = \binom{N}{2}. \tag{49}$$

Thakkar and Smith /7/ have shown that the exact h satisfies an electron-electron cusp condition

$$\lim_{r_{12}\to 0} [1/2h'(r_{12})/h(r_{12})] = 1/2 \tag{50}$$

which is analogous to the electron-nuclear cusp condition /12/

$$\lim_{r_\alpha \to 0} [1/2\ \rho'_{o\alpha}\ (r_\alpha)/\ \rho_{o\alpha}\ (r_\alpha)] = -Z_\alpha \tag{51}$$

satisfied by $\rho$ at each of the nuclei $\alpha$ where

$$\rho_{o\alpha}(r_\alpha) = (4\pi)^{-1} \int \rho(\vec{r})d\Omega_\alpha \tag{52}$$

Since the left-hand-side of (50) vanishes for a HF function,

$$\Delta h(r_{12}) = h_{CORR}(r_{12})-h_{HF}(r_{12}) \tag{53}$$

should provide /7/ an effective measure of the so-called Coulomb hole especially for smaller values of $r_{12}$.

We present in Fig. 1, $\rho_o(r)$, $D(r)$, $h(r_{12})$ and $P_o(r_{12})$ calculated /8/ from a very accurate correlated wavefunction /13/for the ground state of atomic helium. By inspection of this figure one may observe the electron-nuclear cusp for $\rho_o(r)$ and the electron-electron cusp for $h(r_{12})$.

The intracule matrix $J(\vec{r}_{12},\vec{r}_{12})$ is accessible experimentally by means of total X-ray scattering /17,11,14-16/

$$I_{tot}(\vec{S})/I_{cl} = N + 2 \int J(\vec{r}_{12},\vec{r}_{12})e^{i\vec{s}\cdot\vec{r}_{12}}d\vec{r}_{12} \tag{54}$$

In the gas phase, this becomes

$$I_{tot}(S)/I_{cl} = N + 2 \int h(r_{12})j_o(Sr_{12})d\vec{r}_{12} \tag{55}$$

where $j_o(x) = \sin x/x$ is a zeroth-order spherical Bessel function. Thus, the inverse Fourier-Bessel transform of $(I_{tot}(s)/I_{cl}-N)$ will yield $h(r_{12})$ and consequently by means of (48), expectation values of $r_{12}$,$\langle r_{12}^k\rangle$, can be obtained. This includes in particular $\langle 1/r_{12}\rangle$ which is the important electron-electron repulsion

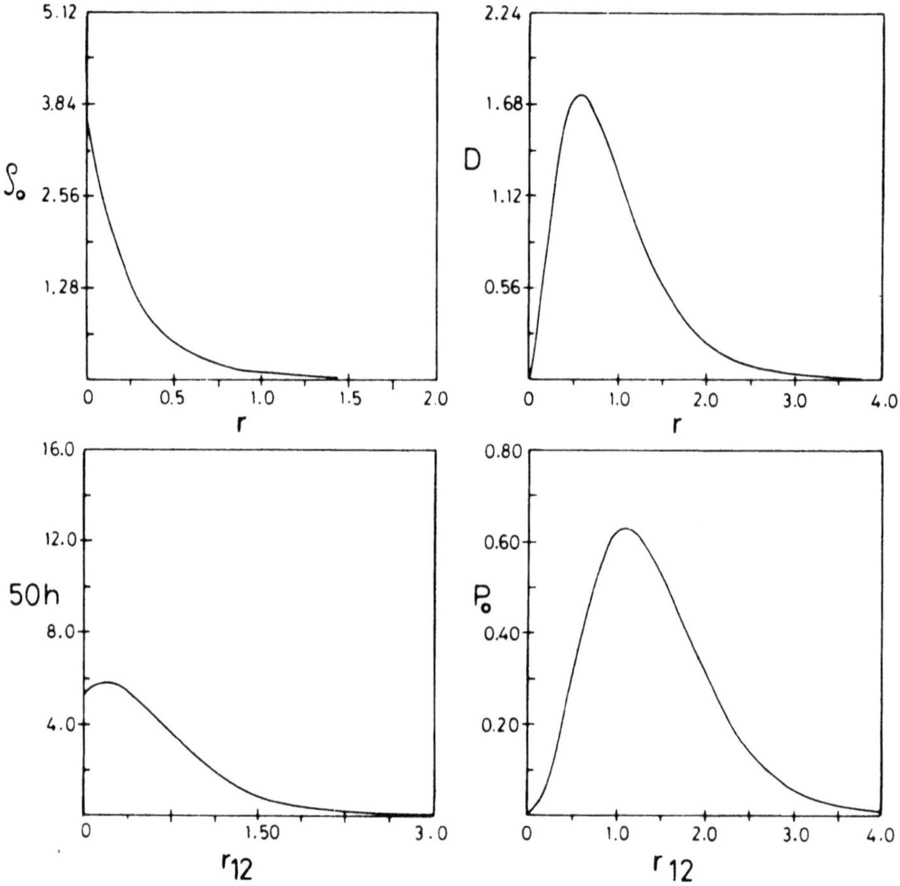


Figure 1 One- and two-electron density functions for the ground ($^1S$) state of atomic helium.

contribution (see (9) and (19)) to $\langle H \rangle$.

It will be shown below that $\Gamma^1(1,1')$ provides access to other types of one-electron densities, namely the momentum density and the spin density.

## IV. MOMENTUM DENSITIES

One can take the Dirac-Fourier transform of $\rho(\vec{r},\vec{r}')$ to obtain /15,17-19/ $\hat{\rho}(\vec{p},\vec{p}')$, i.e.

$$\hat{\rho}(\vec{p},\vec{p}') = (2\pi)^{-3}\iint \rho(\vec{r},\vec{r}')e^{-i\vec{p}\cdot\vec{r}+i\vec{p}'\cdot\vec{r}'}d\vec{r}d\vec{r}', \tag{56}$$

and then the momentum density $\hat{\rho}(\vec{p})$ is simply

$$\hat{\rho}(\vec{p}) = \hat{\rho}(\vec{p},\vec{p}) \tag{57}$$

Alternatively one can perform the Dirac-Fourier transform of the wave function $\Psi$:

$$\hat{\Psi}(\hat{1},\hat{2},\ldots,\hat{N}) = (2\pi)^{\frac{-3N}{2}} \int \Psi(1,2,\ldots N)e^{-i\sum_j \vec{p}_j\cdot\vec{r}_j} d\vec{r}_1 d\vec{r}_2\ldots d\vec{r}_N \tag{58}$$

where $\hat{J} = (p_j,s_j)$, and then calculate

$$\hat{\Gamma}^1(\hat{1},\hat{1}') = N\int\hat{\Psi}(\hat{1},\hat{2},\ldots,\hat{N})\Psi^*(\hat{1}',\hat{2},\ldots,\hat{N})d\hat{2}\ldots d\hat{N} \tag{59}$$

followed by the spin-trace

$$\hat{\rho}(\vec{p},\vec{p}') = \int\hat{\Gamma}^1(\hat{1},\hat{1}')ds_1 \tag{60}$$

Since the one-particle charge density matrix $\rho(\vec{r},\vec{r}')$ may be written in terms of its eigenfunctions,

$$\rho(\vec{r},\vec{r}') = \sum_i \lambda_i\tau_i(\vec{r})\ \tau_i^*(\vec{r}') \tag{61}$$

an equivalent procedure (except possibly for a set of measure zero) is to transform each of these eigenfunctions individually

$$\hat{\tau}_i(\vec{p}) = (2\pi)^{-3/2} \int e^{-i\vec{p}\cdot\vec{r}}\ \tau_i(\vec{r})d\vec{r} \tag{62}$$

to obtain

$$\rho(\vec{p},\vec{p}\,') = \sum_i \lambda_i \hat{\tau}_i(\vec{p}) \hat{\tau}_i^*(\vec{p}\,') \tag{63}$$

It should be emphasized that $\hat{\rho}(\vec{p})$ is not the Fourier transform of $\rho(r)$. The Fourier transform of the latter is the form factor /14,15/

$$F(\underset{\sim}{k}) = \int e^{i\underset{\sim}{k}\underset{\sim}{r}} \rho(\underset{\sim}{r})d\underset{\sim}{r} \tag{64}$$

which we pointed out /19/ may be written as an "auto-correlation" of $\rho(\vec{p},\vec{p}\,')$:

$$F(\vec{k}) = \int \rho(\vec{p},\vec{p} + \vec{k})d\vec{p} \tag{65}$$

Similarly one may define /19/ a function (now called $B(\vec{r})$ in the literature)

$$B(\vec{r}) = \int \hat{\rho}(\vec{p})e^{-i\vec{p}\cdot\vec{r}}d\vec{p} \tag{66}$$

which is an "auto-correlation" of $\rho(\vec{r},\vec{r}\,')$ i.e.

$$B(\vec{r}) = \int \rho(\vec{r}\,',\vec{r}\,' + \vec{r})d\vec{r}\,' \tag{67}$$

The description of the integral of $\rho$ in (65) and (67) as an "auto-correlation" of $\rho$ is by analogy with the usual definition. In fact, if we write $\rho$ in terms of its diagonal expansions (61) and (63), (65) and (67) become respectively

$$F(\vec{k}) = \sum_i \lambda_i \int \hat{\tau}_i(\vec{p}) \hat{\tau}_i^*(\vec{p} + \vec{k})d\vec{p} \tag{68}$$

$$B(\vec{r}) = \sum_i \lambda_i \int \tau_i(\vec{r}\,') \tau_i^*(\vec{r}\,' + \vec{r})d\vec{r} \tag{69}$$

and the individual orbital contributions may be regarded as auto-correlations. The relationships among these quantities are illustrated in Figure 2 where the Patterson function or density auto-correlation function $P(\vec{r})$

$$P(\vec{r}) = \int \rho(\vec{r}\,') \rho(\vec{r}\,' + \vec{r})d\vec{r}\,' \tag{70}$$

$$= \int |F(\vec{k})|^2 e^{i\vec{k}\cdot\vec{r}}d\vec{k} \tag{71}$$

and a momentum space analogue $\hat{P}(\vec{p})$

$$\hat{P}(\vec{p}) = \int \hat{\rho}(\vec{p}\,') \hat{\rho}(\vec{p}\,' + \vec{p})d\vec{p}\,' \tag{72}$$

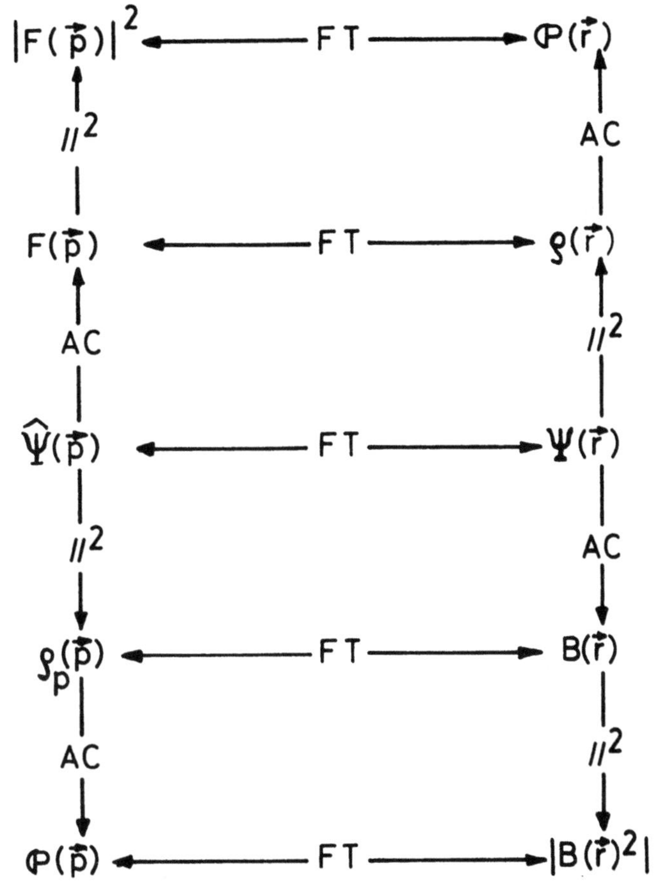


Figure 2 Relationships of various position and momentum space functions illustrated in the case of a one-electron wavefunction (orbital). FT, $//^2$, and AC denote three-dimensional Dirac-Fourier transformation, modulus squared, and "auto-correlation", respectively.

$$= \int |B(\vec{r})|^2 e^{-i\vec{p}\cdot\vec{r}} d\vec{r} \qquad (73)$$

are included for completeness. The $B(\vec{r})$ functions have recently become of practical interest because Weyrich, Pattison and Williams /20-22/ (see also Thulstrup /23/) realized that they were related in a simple way to the experimentally measureable Compton profile function /18/.

$$J(\vec{q}) = \int \hat{\rho}(\vec{p}) \, \delta(\vec{p}\cdot\hat{\vec{q}}-q)d\vec{p} \qquad (74)$$

where $\hat{q} = \frac{\vec{q}}{q}$ and $q = |q|$. $J(\vec{q})$ may be interpreted as an integration of $\rho(\vec{p})$ over a plane in $\vec{p}$-space perpendicular to the scattering vector $\vec{q}$ at a distance $q$ from the origin. The isotropic Compton profile $\overline{J(q)}$ is given by

$$\overline{J(q)} = \int J(\vec{q})d\Omega q = 2\pi \int_{|q|}^{\infty} p\overline{\rho(p)}dp. \qquad (75)$$

By choosing the $p_z$ axis in the $\hat{q}$ direction we may rewrite (74)

$$J(p_z) = \int \rho(\vec{p})dp_x dp_y \qquad (76)$$

and

$$B(0,0,z) = \int J(p_z) \, e^{-ip_z z} \, dp_z \qquad (77)$$

Similarly

$$\overline{B(\vec{r})} = \int \overline{J(q)} \, e^{-iqr} dq \qquad (78)$$

Therefore by one-dimensional Fourier transformation of the various $J(\vec{q})$ we can obtain $B(\vec{r})$ which in turn is an "auto correlation" of $\rho(\vec{r},\vec{r}')$.

## V. SPIN DENSITIES

As we remarked above, the spin density distribution may be obtained from $\Gamma^1(1,1')$. We first state the operational definition of the spin density $\gamma_s(\vec{r})$

$$\gamma_s(\vec{r}) = \gamma_s(\vec{r},\vec{r}) \qquad (79)$$

where

$$\gamma_s(\vec{r},\vec{r}') = (2M)^{-1} \int s_z(1)\, \Gamma^1(1,1')ds_1, \tag{80}$$

$s_z(1)$ is the operator for the Z-component of the spin angular momentum, and M is the eigenvalue of the total $S_z$ operator

$$S_z\Psi = M\Psi \tag{81}$$

where

$$S_z = \sum_i S_z(i). \tag{82}$$

Similarly we can define the spin density in momentum space $\hat{\gamma}_s(\vec{p})$ by

$$\hat{\gamma}_s(\vec{p}) = \gamma_s(\vec{p},\vec{p}) \tag{83}$$

where

$$\gamma_s(\vec{p},\vec{p}') = (2M)^{-1} \int s_z(1)\, \Gamma^1(\hat{1},\hat{1}')ds_1 \tag{84}$$

and may be obtained by the analogous Dirac-Fourier transformation of $\gamma_s(\vec{r},\vec{r}')$ as was used in (56).

We now consider the decomposition of $\Gamma^1(1,1')$ in terms of $\rho(\vec{r},\vec{r}')$ and $\gamma_S(\vec{r},\vec{r}')$. We first note that if $\Psi$ is an eigenfunction of an Hermitian operator of the form $\sum_i \hat{F}_i$, then $\hat{F}_1$ commutes with $\Gamma^1$. This means that $\Gamma^1$ is block-diagonal in any basis of eigenfunctions of F. Since $S_z$ is such an operator (see (81) and (82)), $\Gamma^1$ must be block diagonal in $\alpha$ and $\beta$, i.e.

$$\Gamma^1(1,1') = \rho_{\alpha\alpha}(\vec{r},\vec{r}')\alpha\alpha'^* + \rho_{\beta\beta}(\vec{r},\vec{r}')\beta\beta'^* \tag{85}$$

This may be rewritten as

$$\Gamma^1(1,1') = \rho(\vec{r},\vec{r}')\ [1/2(\alpha\alpha'^* + \beta\beta'^*)]$$
$$+2M\gamma_s(\vec{r},\vec{r}')\ [1/2(\alpha\alpha'^* - \beta\beta'^*)] \tag{86}$$

where $\rho(\vec{r},\vec{r}')$ and $\gamma_s(\vec{r},\vec{r}')$ were defined previously (36) and (80). In terms of $\rho_{\alpha\alpha}$ and $\rho_{\beta\beta}$ they are

$$\rho(\vec{r},\vec{r}') = \rho_{\alpha\alpha}(\vec{r},\vec{r}') + \rho_{\beta\beta}(\vec{r},\vec{r}') \tag{87}$$

$$2M\gamma_s(\vec{r},\vec{r}') = \rho_{\alpha\alpha}(\vec{r},\vec{r}')-\rho_{\beta\beta}(\vec{r},\vec{r}') \quad (88)$$

The latter is just the net density of unpaired spin. If we define $N_\alpha, N_\beta$ by

$$\int \rho_{\alpha\alpha}(\vec{r},\vec{r})d\vec{r} = N_\alpha \quad (89)$$

$$\int \rho_{\beta\beta}(\vec{r},\vec{r})d\vec{r} = N_\beta \quad (90)$$

then

$$\int 2M\gamma_s(1,1')d1 = 2M\int s_z(1)\Gamma'(1,1')d1$$

$$= M = 1/2(N_\alpha-N_\beta) \quad (91)$$

## VI. CUSP AND ASYMPTOTIC CONDITIONS ON $\rho(r)$

Knowledge of the short and long-range behaviour of the exact electron density $\rho(r)$ is useful for assessing the quality of experimental and calculated electron densities.

Both the exact $\rho(r)$ and the exact $\rho_{HF}(r)$ satisfy a Kato-type electron-nuclear cusp condition

$$\lim_{r_\alpha \to 0} \left(\frac{\partial}{\partial r_\alpha} + 2Z\right)\rho_{o\alpha}(r_\alpha) = 0 \quad (92)$$

at each of the nuclei where

$$\rho_{o\alpha}(r_\alpha) = (4\pi)^{-1} \int \rho(\vec{r})d\Omega_\alpha \quad (93)$$

The long-range form of the electron density in the HF model is

$$\rho_{HF}(r) \sim \exp[-2(-2\varepsilon_{max})^{1/2}r] \quad (94)$$

where $\varepsilon_{max}$ is the least negative of all the occupied orbital energies $\varepsilon_i$ (and by Koopmans' theorem /24/-$\varepsilon_{max}$ is identified with the first ionization potential).

In the case of the exact density, a similar relation holds /26/

$$\rho(r) \sim \exp[-2(-2\mu_{max})^{1/2}r] \quad (95)$$

where $\mu_{max}$ is the least negative eigenvalue of a certain matrix /26-28/, $\mu_{k\ell}$, which depends on $\Gamma^1$ and $\Gamma^2$ and is defined by

$$\mu_{k\ell} = (\nu_k\nu_\ell)^{-1/2}\{\nu_\ell\langle\chi(1)|\hat{h}_\ell|\chi(1)\rangle$$

$$+2\int\frac{1}{r_{12}}\chi_k^*(1)\chi_1(3)\;\Gamma^2(3,2,1,2)d\vec{r}_3d\vec{r}_2d\vec{r}_1\} \tag{96}$$

for k and $\ell$ such that $\nu_k \neq 0$ and $\nu_\ell \neq 0$. $\chi_k(1)$ and $\nu_k$ are the eigenfunctions and eigenvalues, respectively, of $\Gamma^1(1,1')$ defined by (27). $-\mu_{max}$ is similar to the first ionization potential I and is bounded by it:

$$-\mu_{max} \geq I. \tag{97}$$

We note that Ahlrichs /29/ has recently criticized this result.

Other bounds to the long-range behaviour of the exact density were recently given by Hoffmann-Ostenhof and Hoffmann-Ostenhof /30/. In the case of an atom of nuclear charge Z, they established that for $|r| \geq Z/I$,

$$\rho(r) \geq kr^{\beta}e^{-\alpha r} \tag{98}$$

where $\beta = Z/\alpha-1$ and $\alpha = (2I)^{1/2}$. Their conjecture that this result could be improved by replacing Z in the definition of $\beta$ by $Z^* = Z-N+1$ has been established recently by Tal /31/.

## VII. THE FUNDAMENTAL RÔLE OF THE ELECTRON DENSITY

It was shown above that the exact energy and all one- and two-particle expectation values are functionals of $\Gamma^2(1,2;1',2')$ (see (19)) while in the HF scheme they are functionals of the Fock-Dirac density matrix $\Gamma^1_\Phi(1,1')$ or charge and bond-order matrix (see (32) and (33)). Since much of our computational knowledge of atoms and molecules is at the HF (SCF) level, attention has been directed toward an understanding and analysis of $\Gamma^1_\Phi$. However, many have realized that the electron density $\rho(\vec{r})$ itself plays a very fundamental role and have directed their attention to it as the major source of information about electron systems.

This idea has been made rigorous by Hohenberg and Kohn /32,33/ who showed that a <u>non-degenerate</u> (electronic) ground state is a unique functional of $\rho(\vec{r})$. Their proof is based upon the assumption that the electronic Hamiltonian

$$\hat{H} = -1/2\sum_i \nabla_i^2 + \sum_i V(r_i) + \sum_{i<j} 1/r_{ij} \tag{99}$$

has an exact non-degenerate ground state $\Psi$ with energy E. Here $\hat{V}(\vec{r})$ is a local potential (usually the electron-nuclear attraction). If $\Psi'$ is the ground-state wavefunction with energy $E'$ of another Hamiltonian $\hat{H}'$ which differs from $\hat{H}$ by a different local potential $\hat{V}'$, then $\Psi'$ is not an eigenfunction of $\hat{H}$ provided $\hat{V}-\hat{V}'$ is not a constant. Application of the variation principle to $\hat{H}$ with $\Psi'$ as a trial function yields

$$E<E' + \int(V-V')\rho'(\vec{r})d\vec{r} \tag{100}$$

But the same procedure with $\hat{H}'$ and $\Psi$ results in

$$E'<E + \int(V'-V)\rho(\vec{r})d\vec{r} \tag{101}$$

However, if $\rho = \rho'$, then the sum of these two equations gives

$$E + E' < E + E' \tag{102}$$

This contradiction implies that $\rho \neq \rho'$ and therefore there is a one-to-one correspondence between the local potential $\hat{V}$ and $\rho(\vec{r})$. This means that E, $\hat{V}$, and $\Psi$ itself are all functionals of $\rho$ under the above hypothesis.

Hohenberg and Kohn /32/ only established the existence of such functionals. The problem as to their construction remains /34-37/. We can express E as a functional of $\rho$:

$$E[\rho] = \int\hat{V}(\vec{r})\rho(\vec{r})d\vec{r} + G[\rho] \tag{103}$$

where $G[\rho]$ is a functional to be determined. If another density $\rho'$ corresponding to $\hat{V}'$ and $\Psi'$ is used

$$E[\rho'] = \int V(\vec{r})\rho'(\vec{r})d\vec{r} + G[\rho'] = \langle\Psi'|H|\Psi'\rangle \geq E[\rho] \tag{104}$$

Thus the minimum is obtained for $\rho$ and one has a variational principle for determining $\rho$.

Although not mentioned explicitly above, the proof of the Hohenberg-Kohn theorem has required that $\rho(\vec{r})$ be derivable from an admissible anti-symmetric wavefunction $\Psi$, i.e. be N-representable. Similarly, the application of the Hohenberg-Kohn variation principle for E as a functional of $\rho$ would require that any admissible trial $\rho$ be N-representable.

We note that Gilbert /38,40,3/ has recently shown that any finite, non-negative, differentiable function $\rho(r)$ is N-representable provided that

$$\int \rho(\vec{r})d\vec{r} = N \tag{105}$$

The proof involves an algorithm for the construction of the wavefunction from a given $\rho(r)$. The algorithm brings out clearly the fact that there are many possible wavefunctions corresponding to a given electron density including Slater determinants.

This fact is important not only for the foundations of the density-functional approach to electronic structure but also for the approach of Massa and collaborators /40/ for the analysis of experimental $\rho(\vec{r})$ which assumes that such electron densities may be obtained from a Slater determinant.

We pointed out above that there is a one-to-one correspondence between the local potential $\hat{V}$ and $\rho(\vec{r})$. Given $\rho(\vec{r})$ for an electronic system, one can construct $\hat{V}$ in the form

$$\hat{V} = \sum_{\alpha} Z_{\alpha}/r_{\alpha} \tag{106}$$

by searching for the cusps of $\rho(\vec{r})$. Equations (91) and (92) would yield the nuclear locations $\vec{r}_{\alpha}$ and the atomic numbers $Z_{\alpha}$.

## VIII. TOPOGRAPHICAL FEATURES OF THE ELECTRON DENSITY

As discussed above, the charge density $\rho(\vec{r})$ occupies the central position in the understanding of electronic systems. As a result, it is natural to analyze $\rho(\vec{r})$ itself and focus attention on its topographical features /41/. The most interesting of these for any scalar function are those where the various-order derivatives vanish or are discontinuous. As noted above (section (VI)), the electron density ("at rest") is not everywhere smooth ($C^{\infty}$) as it has cusps at the various nuclei. These electron-nuclear cusps are maxima, but the gradient of $\rho(\vec{r})$ is discontinuous there and thus the second derivatives as well as the gradient itself are not defined there. At non-nuclear points, it is believed that $\rho(\vec{r})$ is a smooth function (or at least $C^2$) and thus can be characterized by its critical points where $\nabla\rho(\vec{r})$ vanishes. Critical points $\bar{x} = (\bar{x}_1,\ldots,\bar{x}_n)$ can be classified in terms of the <u>invariants</u> of the Hessian or curvature matrix $(h_{ij})$ where

$$h_{ij} = \left.\frac{\partial^2 f}{\partial x_i \partial x_j}\right|_{\bar{x}} \tag{107}$$

These are its <u>rank</u> r (the number of non-zero eigenvalues), <u>index</u> i(the number of negative eigenvalues) and <u>signature</u> p = r-2i (the difference between the numbers of positive and negative eigenvalues). Non-degenerate critical points have rank n. We list such critical points in Table 1 for functions of 1,2 and 3 variables.

Table 1. Non-degenerate Critical Points for Functions of 1,2, and 3 Variables /41/

| Dimension | (r,i) | (r,p) | Type | Name/42,43/ |
|---|---|---|---|---|
| 1 | (1,0) | (1,1) | minimum | pit |
| | (1,1) | (1,-1) | maximum | peak |
| 2 | (2,0) | (2,2) | minimum | pit |
| | (2,1) | (2,0) | saddle point | pass |
| | (2,2) | (2,-2) | maximum | peak |
| 3 | (3,0) | (3,3) | minimum | pit |
| | (3,1) | (3,1) | saddle point | pale |
| | (3,2) | (3,-1) | saddle point | pass |
| | (3,3) | (3,-3) | maximum | peak |

If we assume that there are no degenerate critical points, then

$$\chi_f(m) = \sum_{i=o}^{n} (-1)^i \nu_i \tag{108}$$

where $\chi_f(m)$ is the homological Euler characteristic of the domain m (of dimension n) of f and $\nu_i$ is the number of critical points of index i. For three-dimensional real space, $\chi = -1$, and for the infinite periodic lattice, $\chi = 0$.

The requirement of non-degenerate critical points is not practically restricting, as degenerate ones can be removed by infinitesimal changes of the function. In the case of a molecular electron density for a given nuclear configuration, an infinitesimal charge in the coordinates of any one of the nuclei would be expected to remove a degeneracy. This does not mean that they can be ignored as they can be vital in some applications such as the representation /44/ of bonding in terms of the catastrophe points /45/ of the electron density. The seminal paper of Collard and Hall /44/ and the extensive studies of Bader and collaborators /46/ discuss the relevance of the critical points of the electron density. The number of maxima (they are not true critical points since $\nabla\rho$ is not defined there) of $\rho(\vec{r})$ is equivalent to the number of nuclei, the number of passes with the number of bond paths or bonds, the number of pales with the number of rings and the number of pits with the number of cages. The extensive studies of Bader and co-workers /47/ on the virial partitioning of $\rho(\vec{r})$ are based on the topological properties of $\nabla\rho(\vec{r})$ and lead to definitions of concepts such as atoms in a molecule and bond paths.

Support of this research by the Natural Sciences and Engineering Research Council of Canada is gratefully acknowledged.

REFERENCES

1. A.J. Coleman, Rev. Mod. Phys. 35, 668 (1963).

2. C. Elson and V.H. Smith, Jr. (unpublished).

3. V.H. Smith and I. Absar, Isr. J. Chem. 16, 87 (1977).

4. A.J. Coleman, Int. J. Quantum Chem. 13, 67 (1978).

5. P.-O. Löwdin, Phys. Rev. 97, 1474 (1955).

6. A.J. Coleman, Int. J. Quantum Chem. 1S, 457 (1967).

7. A.J. Thakkar and V.H. Smith, Jr., Chem. Phys. Lett. 42, 476 (1976).

8. A.J. Thakkar and V.H. Smith, Jr., J. Chem. Phys. 67, 1191 (1977).

9. R. Benesch and V.H. Smith, Jr., J. Chem. Phys. 55, 482 (1971).

10. T.L. Gilbert, Rev. Mod. Phys. 35, 491 (1963).

11. V.H. Smith, Jr., Chem. Phys. Letters 7, 226 (1970); 11, 152 (1971).

12. E. Steiner, J. Chem. Phys. 39, 2365 (1961).

13. A.J. Thakkar and V.H. Smith, Jr., Phys. Rev. A 15, 1 (1977).

14. R. Benesch and V.H. Smith, Jr. Acta Crystallogr. A26, 579 (1970).

15. R. Benesch and V.H. Smith, Jr. in Wave Mechanics-The First Fifty Years edited by W.C. Price, S.S. Chissick and T. Ravers-dale (Butterworths, London, 1973).

16. R.F. Stewart, Isr. J. Chem. 16, 111 (1977).

17. R. Benesch and V.H. Smith, Jr., Chem. Phys. Letters 5, 601 (1970).

18. P. Kaijser and V.H. Smith, Jr., Adv. Quantum Chem. 10, 37 (1977).

19. R. Benesch, S.R. Singh and V.H. Smith, Jr., Chem. Phys. Letters 10, 151 (1971).

20. W. Weyrich, Habilitationsschrift, TH Darmstadt (1978).

21. P. Pattison and B. Williams, Sol. State Commun. 20, 585 (1976).

22. P. Pattison, W. Weyrich and B. Williams, Sol. State Commun. 21, 967 (1977).

23. P. Thalstrup, J. Chem. Phys 65, 3386 (1976).

24. T. Koopmans, Physica 1, 104 (1933).

25. N.C. Handy, M.T. Marron, and H.J. Silverstone, Phys. Rev. 180, 45 (1969).

26. M.M. Morrell, R.G. Parr and M. Levy, J. Chem. Phys. 62, 549 (1975).

27. D.W. Smith and O.W. Day, J. Chem. Phys. 62, 113 (1975).

28. V.H. Smith, Jr. and Y. Öhrn, Queen's Papers Pure Appl. Math. 40, 139 (1974).

29. R. Ahlrichs, J. Chem. Phys. 64, 2706 (1976).

30. M. Hoffmann-Ostenhof and T. Hoffmann-Ostenhof, Phys. Rev. A 16, 1782 (1977).

31. Y. Tal, Phys. Rev. A 18, 1781 (1978).

32. P. Hohenberg and W. Kohn, Phys. Rev. 136B, 864 (1964).

33. W. Kohn and L.J. Sham, Phys. Rev. 140 A, 1133 (1965).

34. S. Lundqvist in Quantum Science: Methods and Structure, edited by J.-L. Calais, O. Goscinski, J. Linderberg, and Y. Öhrn (Plenum Press, New York, 1976).

35. H. Primas, Int. J. Quantum Chem. 1, 493 (1967).

36. H. Nakatsuji and R.G. Parr, J. Chem. Phys. 63, 1112 (1975).

37. S.T. Epstein and C.M. Rosenthal, J. Chem. Phys. 64, 247 (1976).

38. T.L. Gilbert, Phys. Rev. B 12, 2111 (1975).

39. K. Kurkio-Suonio, Isr. J. Chem. 16, 132 (1977).

40. W.L. Clinton, C.A. Frishberg, L.J. Massa and P.A. Oldfield, Int. J. Quantum Chem. 7S, 505 (1973).

41. V.H. Smith, Jr., P.F. Price and I. Absar, Isr. J. Chem. 16, 187 (1977).

42. M. Morse, Pits, Peaks and Passes, Math. Assoc. Amer. (1966).

43. C.K. Johnson, A.C.A. Winter Meeting (Asilomar, 1977), communicated abstracts, p. 30.

44. K. Collard and G.G. Hall, Int. J. Quantum Chem. 12, 623 (1977).

45. R. Thom, Structural Stability and Morphogenesis, (Benjamin, Reading, Mass, 1975).

46. R.F.W. Bader (private communication).

47. S. Srebrenik, R.F.W. Bader, and T. Tung Nguyen-Dang, J. Chem. Phys. 68, 3667 (1978) and references therein.

# ON THE CALCULATION AND ACCURACY OF THEORETICAL ELECTRON DENSITIES

Vedene H. Smith, Jr.

Department of Chemistry, Queen's University, Kingston

Ontario, K7L 3N6, Canada

## I. INDEPENDENT PARTICLE MODEL

The basic method for the calculation of the electronic structure of atoms and molecules is the Independent Particle Model wherein there is a one-to-one correspondence between electrons and spin-orbitals and therefore each electron moves in the average field of the other N-1 electrons /1/. The wave function in this model is the Slater determinant

$$\Phi(1,2,\ldots,N) = \frac{1}{\sqrt{N!}} \det |\phi_1(1)\phi_2(2)\ldots\phi_N(N)| \tag{1}$$

where the $\{\phi_i(J)\}$ are a set of N-spin-orbitals and J is the combined space-spin coordinate of electron j. As we discussed elsewhere /2/, the one-particle density matrix in this model, $\Gamma^1_\Phi(1,1')$, is called the Fock-Dirac density matrix and may be written in the form

$$\Gamma^1_\Phi(1,1') = \sum_{i=1}^{N} \phi_i(1)\phi_i^*(1') \tag{2}$$

Use of (1) (or equivalently (2)) in the variation principle leads to the Hartree-Fock equations which are a set of integrodifferential equations for the set of orthonormal spin orbitals $\{\phi_i(J)\}$ which are optimum in the sense of the energetic criterion of the variation principle. They may be written in the form

$$\hat{F}(1)\phi_i(1) = \sum_{j=1}^{N} \lambda_{ij}\, \phi_j(1);\ i = 1,\ldots,N, \tag{3}$$

where $\hat{F}(1)$, called the Fock operator, is a functional of the set

$\{\phi_i(J)\}$ and will be defined below. It is a one-electron operator which is the effective Hamiltonian operator for an electron (described by spin-orbital $\phi_i$) in the attractive field of the nuclei and the average replusive field of the other electrons.

The exact solutions of these equations will henceforth be called the Hartree-Fock spin-orbitals. The practical method of solution is the so-called Self-Consistent-Field (SCF) approach /3/ and we shall designate such solutions as SCF spin-orbitals, etc.

This method was developed by Roothaan /4/ and Hall /5/ as a computational scheme with the spin-orbitals expanded in terms of a known set (basis) of one-electron functions. We shall discuss the computational aspects in more detail in section II below.

We consider a system of N electrons in the field of $\eta$ fixed nuclei. The electronic Hamiltonian $\hat{H}$ can be written as

$$\hat{H} = \sum_{i=1}^{N} \hat{h}(i) + \sum_{i<j} \hat{g}(i,j) \tag{4}$$

where

$$\hat{h}(i) = -\frac{1}{2}\nabla_i^2 - \sum_{\alpha=1}^{\eta} Z_\alpha / r_{i\alpha} \tag{5}$$

$$\hat{g}(i,j) = 1/r_{ij} \tag{6}$$

Here the variable i represents the spatial coordinates of the electron i, $Z_\alpha$ is the atomic number of nucleus $\alpha$, $r_{i\alpha}$ is the distance between electron i and nucleus $\alpha$, and $r_{ij}$ is the interelectronic separation between i and j.

Then the matrix element $F_{ij}$ between $\phi_i$ and $\phi_j$ of the Fock one-electron effective Hamiltonian operator, $\hat{F}(1)$, may be written in the form

$$\begin{aligned} F_{ij} &= \int \phi_i^*(1)\hat{F}(1)\phi_j(1)d1 \\ &= H_{ij} + \sum_{k=1}^{N} \{(ik|\hat{g}|jk)-(ik|\hat{g}|kj)\} \end{aligned} \tag{7}$$

where the one-electron integral $H_{ij}$ is defined by

$$H_{ij} = \int \phi_i^*(1)\hat{h}(1)\phi_j(1)d1 \tag{8}$$

and the two-electron integrals are defined as follows:

$$(pq|\hat{g}|rs) = \int \phi_p^*(1)\phi_q^*(2)r_{12}^{-1}\phi_r(1)\phi_s(2)d1d2. \tag{9}$$

It may be shown that the Slater determinant $\Phi$ is determined only within a unitary transformation amongst the N- members of the set $\{\phi_i(1)\}$. Thus the spin orbitals themselves are not unique and one may obtain different sets related by unitary transformations of the set $\{\phi_i(1)\}$. Quite often, the unitary transformation is chosen such that the matrix of the elements $F_{ij}$ is diagonal, i.e.

$$F_{ij} = \delta_{ij}\varepsilon_i \tag{10}$$

where $\varepsilon_i$ are called the orbital energies of the HF spin-orbitals and are given by

$$\varepsilon_i = H_{ii} + \sum_{k=1}^{N} (J_{ik}-K_{ik}) \tag{11}$$

where J and K are the two-electron Coulomb and exchange integrals respectively and are defined as

$$J_{ik} = (ik|\hat{g}|ik) \tag{12}$$

$$K_{ik} = (ik|\hat{g}|ki) \tag{13}$$

The $\phi_i$ under the requirement (10) are usually called canonical (or spectroscopic) HF spin-orbitals. The designation spectroscopic for the canonical spin-orbitals is explained by Koopmans' theorem /6/ that the ionization energy of an electron in $\phi_i$ is just $-\varepsilon_i$. This theorem is based upon the assumption that both the ion and its parent can be described by Slater determinants built up from the HF spin-orbitals for the ground state of the parent.

The set of equations (3) may be written for the canonical HF spin-orbitals in the form

$$\hat{F}(1)\phi_i(1) = \varepsilon_i\phi_i(1); \; i = 1,2,\ldots,N; \tag{14}$$

which justifies the earlier description of $\hat{F}(1)$ as a one-electron effective Hamiltonian operator.

Often the unitary transformation is chosen to localize the spin-orbitals according to some criterion. Such localized orbitals correspond more closely to familiar concepts such as bond and lone pairs than do the canonical orbitals. It should be emphasized that the total HF wavefunction and its expectation values (including the electron density) are unchanged by such transformations. Only the individual spin-orbitals, -orbital densities and -orbital expectation values are changed.

We mention an additional theorem (Brillouin's /7/) which will be required later in our discussion of the accuracy of the HF method. If $\phi_a$ designates a spin-orbital not occupied in the Slater determinant $\Phi$, we introduce the notation $\Phi_i^a$ for a determinant in which spin-orbital i is replaced by a. Then the theorem states that

$$<\Phi_i^a|H|\Phi> = 0. \tag{15}$$

So far the discussion has involved spin-orbitals without specifying their form. As a result, $\Phi$ does not necessarily obey the symmetry of the problem. Often restrictions are introduced due to the symmetry, intuition about the physical situation or because of computational convenience. Typical restrictions of a general spin-orbital

$$\Psi(1) = A^{\uparrow}(\vec{r})\alpha + B^{\downarrow}(\vec{r})\beta \tag{16}$$

include

$$\Psi(1) = A(\vec{r})\alpha \text{ or } B(\vec{r})\beta, \tag{17}$$

the usual restriction of a definite spin for the spin-orbital and (in the atomic care)

$$A(\vec{r}) = A(r,\theta,\phi) = a(r)Y_{\ell m}(\theta,\phi) \tag{18}$$

$$a(r) = a_{\ell m_\ell}(r) = a_\ell(r) \tag{19}$$

and

$$a_{ms}(r) = a(r) \tag{20}$$

The first of these (18) is the separation of the radial and angular coordinates (central-field model), (19), is the constraint that $2p_+$, $2p_0$ and $2p_-$ orbitals, for example, all have the same radial part, and (20) is the familiar restriction to double occupancy of the orbitals.

We shall continue to designate the spin-orbitals with no restrictions imposed as Hartree-Fock. Some authors would call them Unrestricted Hartree-Fock (UHF) -- although other schemes are also called UHF. The restrictions become especially important in the discussion of systems which are not of closed-shell type and for which in particular, an accurate calculation of the spin density is desired /8,9/.

## II. SOME COMPUTATIONAL ASPECTS OF THE SCF METHOD

The most commonly used procedure for obtaining solutions of the Hartree-Fock problem is the Self-Consistent-Field (SCF) procedure together with the assumption that the required one-electron orbitals are a linear combination of a set of basis functions.

In theory the use of a complete set of linearly independent continuous functions, should lead to the least (best) possible value of energy obtainable in this approximation, which is called the Hartree-Fock limit. However, a judicious choice of the basis functions can lead to values close to the Hartree-Fock limit with a finite-sized basis set. Such basis functions are usually obtained by appropriate modifications of the analytic solutions of the one-electron atomic problem.

The latter can be obtained in spherical polar coordinates $(r,\theta,\phi)$ and are classified according to their symmetries using a notation based on the natural quantum numbers which result from the classical degrees of freedom of the problem. Therefore, these solutions are designated by $n\ell_m$ (e.g. $1s_0$, $2p_{+1}$, $2p_0$, $2p_{-1}$, $3d_{+2}$, $3d_{+1}$, $3d_0$, etc.), where n is the principal quantum number and is restricted to positive integers, $\ell$ is the azimuthal quantum number and for any given n takes integer values from zero to $(n-\ell)$, and m is the magnetic quantum number which for a specified $\ell$ takes integer values from $-\ell$ to $+\ell$. The letters s, p, d, f, g, h, etc., designate $\ell$ = 0, 1, 2, 3, 4, 5, etc., respectively. The degenerate sets of orbitals, for example the p type, are more commonly used in their Cartesian form i.e. $2p_x$, $2p_y$, $2p_z$, which are simply appropriately directed linear combinations of the functions $2p_{+1}$, $2p_0$, $2p_{-1}$. The total electronic wavefunction for a state of the electron in the hydrogen atom is a product of a radial function R(r) which depends on n, an angular function $Y(\theta,\phi)$ (spherical harmonic) which depends on $\ell$ and m, and a spin function w which depends on the spin quantum number s, with values $+\frac{1}{2}$ or $-\frac{1}{2}$ depending on the direction of spin, i.e.

$$\Psi(1) = R_n(r)Y_{\ell m}(\theta,\phi)w(s) \tag{21}$$

and

$$w(1/2) = \alpha, \quad w(-1/2) = \beta \tag{22}$$

In two of the more commonly used families of basis functions (Slater-type and Gaussian-type) the angular function is held fixed and the radial function varied. Slater-type radial functions are of the form $r^{n-1}e^{-\zeta r}$, where r is the radial distance from the nucleus, n is the principal quantum number and $\zeta$ (zeta) is an orbital exponent parameter. Slater-type functions have the advantage that they have the correct form for the atomic one-electron case and are therefore expected to give a better description of the electron density especially in the near-nuclear region. Gaussian-type functions, on the other hand, have computational advantages, because integrations over these functions are much faster.

Gaussian functions are most commonly used in the so-called Cartesian form. They are:

$$e^{-\alpha r^2} \qquad \text{(s-type)} \tag{23}$$

$$xe^{-\alpha r^2},\ ye^{-\alpha r^2},\ ze^{-\alpha r^2} \qquad \text{(p-type)} \tag{24}$$

$$x^2e^{-\alpha r^2},\ xye^{-\alpha r^2},\ xze^{-\alpha r^2},$$

$$y^2e^{-\alpha r^2},\ yze^{-\alpha r^2},\ z^2e^{-\alpha r^2} \qquad \text{(d-type)} \tag{25}$$

etc.

It should be noted that there are six functions in the list (25) of d's. One can choose five linear combinations which correspond to the familiar d-orbitals and one combination

$$(x^2 + y^2 + z^2)e^{-\alpha r^2} = r^2e^{-\alpha r^2} \tag{26}$$

which is of s-type. It has been demonstrated that the Hartree-Fock limit can be approached with Gaussian functions; however, the basis-set size required is larger than those presently in common use/9/.

With the angular functions fixed, a finite basis set of r (Slater-type of Gaussian-type) functions is specified by the r exponent parameters ($\zeta$ for Slater functions, $\alpha$ for Gaussian functions). Given such a basis set, one obtains the expansion coefficients of the required atomic or molecular functions which give the least energy for the specified exponent parameters. For atoms, these linear combinations of basis functions are called atomic orbitals (AO) and for molecules they are called molecular orbitals (MO). Usually the MO's are constructed as linear combinations of atomic orbitals (LCAO) of the appropriate symmetry on each atomic center and are called LCAO-MO.

It is possible to improve the given basis set by variation of the exponent parameters. The least energy thus obtained is called the SCF limit for that basis set and the basis set is said to be atom- (or molecule-) optimized. When limited basis sets are used, it is the usual practice to refer to a basis set with one Slater-type radial function for each occupied orbital in the constituent atoms as the minimum Slater (or single-zeta) basis set. If two or more Slater-type functions are employed per occupied atomic orbital, we speak of double- (or higher) zeta basis sets. If each Slater function is expanded as a fixed linear combination of Gaussian functions, we say that contracted Gaussian functions are used and the number of Gaussians per Slater of each symmetry type is specified. Uncontracted Gaussian basis sets use Gaussian functions of appropriate symmetry type, and no Slater exponents need to be considered, e.g. a (5,2) Gaussian basis set for carbon means 5 s-type Gaussian and 2 p-type Gaussian functions are used to describe the carbon atom. When the basis set is augmented with orbitals of higher azimuthal quantum numbers than the ones occupied in the atomic ground states, these orbitals are called polarization functions, e.g., only s- and p-type orbitals are occupied in carbon or silicon, so the d-type functions are polarization functions. If the basis set uses too few s and p functions, the d-functions may simply be repairing the defects in the basis set rather than contributing to a physical phenomenon. This can be seen, for example, in an instructive study by Clementi and Popkie /10/ using a number of basis sets for $H_2O$. In order to really distinguish between these two phenomena, one should study the addition of d functions at the sp limit. The sp limit is the least energy that may be obtained by the use of s and p functions alone. spd and spdf etc. limits can be similarly defined.

The inclusion of functions of higher azimuthal quantum numbers can be a qualitative necessity due to molecular symmetry. If an occupied MO transforms according to an irreducible representation of the point group of the molecule for which s and p functions on first- and second-row atoms (s functions on hydrogen) do not provide a basis, then higher functions must be included in the basis. Examples of such cases have been given in the literature /11,12/.

When finite basis sets are used, it is important to consider the problem of how many functions of each symmetry type should be used to describe each orbital with the same degree of accuracy. This is called balancing of the basis set. For example, to describe the ground state of the carbon atom with two electrons in each of the 1s, 2s and 2p orbitals, using sixteen functions, one could use 15 s-type functions and 1 p-type function or 2 s-type function and 14 p-type functions or any intermediate composition. To determine the preferred composition the minimum in total energy fails to remain a good criterion because the 1s (or core) orbitals have considerably lower energy than the 2s or 2p orbitals. One way to decide upon a balanced basis set is to use the criterion that the energy of each orbital should be the same percentage of its energy at the Hartree-Fock limit.

Such balancing of the basis set is of even more importance in molecules where electronegativity differences of the atoms should be taken into account.

For electron-density studies balanced basis sets are of utmost importance, because a good approximate wavefunction for such purposes would be one where the density in all interesting regions is approximated equally well. Examples are well-known /13/ where wavefunctions of lower or comparable quality give quite inadequate descriptions of the electron density. For example, a one-centre calculation for $H_2O$ yields a much lower energy (-75.922 a.u.) than a minimal Slater LCAO (-75.703 a.u.) and comparable to that of an extended Slater LCAO (-76.005 a.u.). Comparison of the electron densities shows that the one-centre calculation was, as expected, extremely poor in the bonding regions and the region of the hydrogen nuclei, while the other calculations described the bonding more or less adequately.

Minimum-basis-set calculations give a correct qualitative variation of electron density with distance and in general predict the bonding in molecules; however, they are inadequate for quantitative description of the charge distribution in the important regions, namely the bonding regions and the near-nuclear regions. Cade /14/ has compared the electron density at the Hartree-Fock limit for $O_2$ and FH with results for a variety of basis sets and has pointed out that the minimal basis set places too little electronic charge in the bonding region, has the expected inadequacies in the nuclear region, and overestimates the density in the "lone pair" regions.

Such errors can become very serious on the scale of the deformation density $\Delta\rho_m$:

$$\Delta\rho_m = \rho_{molecule} - \rho_{SFA} \quad (27)$$

where $\rho_{SFA}$ is the density of the spherically-averaged free atoms superimposed at the molecular geometry. In Fig. 1, $\Delta\rho_m$ is shown /15/ along the molecular axis for CO for three different SCF wavefunctions: with an extended basis (EB) including polarization functions, a double-zeta (DZ) basis, and a minimal (or single-zeta, SZ) basis. In contrast to the larger bases, the SZ density shows an absence of charge in most of the bonding region. Comparison of the EB and DZ curves indicates the importance of the polarization functions in the description of the bonding charge density.

An alternative to adding higher azimuthal quantum numbers to

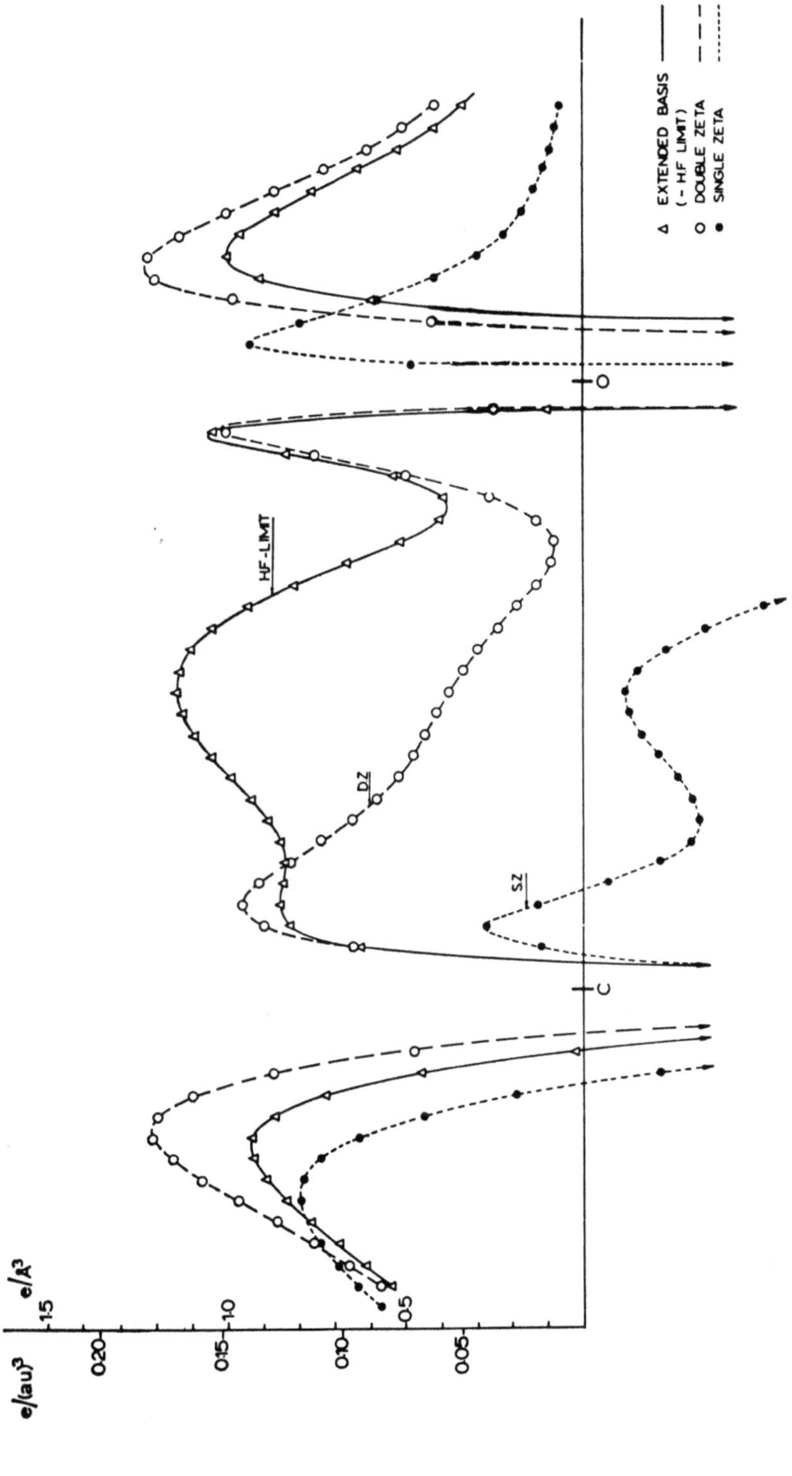


Figure 1 Comparison of $\Delta\rho_m$ along the molecular axis of CO for SCF wavefunctions constructed with basis sets of different quality /15/.

a basis set is to use s- and p- type functions at off-nuclear points in the bonding regions /16,17/ such as the mid-point of the bond.

Molecular calculations may be carried out using atom-optimized basis sets of basis sets where exponent parameters are optimized for the molecular Hamiltonian. Polarization functions, of course, should be molecularly optimized. Computer program packages such as GAUSSIAN 70, POLYATOM, IBMOL, and others are now available to conduct fairly extensive *ab initio* calculations with relatively large basis sets for smaller systems and in moderately limited basis sets for larger systems. The sizes of the basis sets used are limited by the present computer capabilities, because the computational time involved in an SCF calculation is proportional to the fourth power of the number of basis functions used. The data below shows the rapid increase in the number of one- and two-electron integrals as a function of the number of basis orbitals employed.

| Number of basis orbitals | Number of integrals | |
|---|---|---|
| | one-electron ($H_{ij}$) | two-electron ($\langle pq\|\hat{g}\|rs\rangle$) |
| 10 | 55 | 1,540 |
| 20 | 210 | 22,155 |
| 50 | 1,275 | 814,725 |
| 100 | 5,050 | 12,751,250 |
| 200 | 20,100 | 202,015,050 |
| 300 | 45,150 | 1,019,261,250 |

These integrals must be stored and recalled in each iteration of the SCF cycle. Therefore, an *ab initio* calculation using, say, 100 basis functions requires not only relatively large amounts of computer time but also a large amount of computer core-space. The core-space problem is reduced to some extent by not storing those integrals that are identically or essentially zero. Semi-empirical methods attempt to reduce the number of integrals by introducing approximations for many of the integrals.

## III. ERRORS IN EXPECTATION VALUES

We would like to analyze the error in the expectation value of an operator $\hat{F}$ when it is calculated using an approximate wavefunction $\Phi$ instead of the exact wavefunction $\Psi$. We assume that one formula for the expectation value $\langle F\rangle$ is used for all the wavefunctions considered in this discussion. Thus we do not consider the problem due to the fact that for many F, there are several formulae that are equivalent for the exact $\Psi$ but yield different results for an approximate $\Phi$. As one example, there are length, velocity, and acceleration formulae used for the calculation of transition probabilities.

We write $\Psi$ in terms of $\Phi$ and its orthogonal correction function $\omega$, i.e.

$$\Psi = a(\Phi + \varepsilon\omega) \tag{28}$$

where

$$\langle\omega|\Phi\rangle = 0 \tag{29}$$

and $\varepsilon$ is presumably a small, real number. In order that all three functions be normalized

$$\langle\Psi|\Psi\rangle = \langle\Phi|\Phi\rangle = \langle\omega|\omega\rangle = 1 \tag{30}$$

a becomes

$$a = (1 + \varepsilon^2)^{-1/2} \tag{31}$$

The expectation value of $\hat{F}$ is

$$F_{\Psi\Psi} = \langle\Psi|\hat{F}|\Psi\rangle = (1 + \varepsilon^2)^{-1}\{F_{\Phi\Phi} + 2\varepsilon F_{\Phi\omega} + \varepsilon^2 F_{\omega\omega}\} \tag{32}$$

where

$$F_{\Phi\Phi} = \langle\Phi|\hat{F}|\Phi\rangle \tag{33}$$

$$F_{\Phi\omega} = \langle\Phi|\hat{F}|\omega\rangle \tag{34}$$

$$F_{\omega\omega} = \langle\omega|\hat{F}|\omega\rangle \tag{35}$$

The error in $\langle F\rangle$, $\Delta F = F_{\Psi\Psi} - F_{\Phi\Phi}$ is

$$\Delta F = (1 + \varepsilon^2)^{-1}\{2\varepsilon F_{\Phi\omega} + \varepsilon^2(F_{\omega\omega} - F_{\Phi\Phi})\} \tag{36}$$

or in terms of the overlap S of $\Phi$ with $\Psi$

$$S = \langle\Phi|\Psi\rangle = (1 + \varepsilon^2)^{-1/2} \tag{37}$$

$$\Delta F = 2S(1 - S^2)^{1/2}F_{\Phi\omega} + (1 - S^2)(F_{\omega\omega} - F_{\Phi\Phi}) \tag{38}$$

$$= AF_{\Phi\omega} + B(F_{\omega\omega} - F_{\Phi\Phi}) \tag{39}$$

We can not evaluate S or for that matter any of the above quantities (except for $F_{\Phi\Phi}$) without knowing the exact $\Psi$. However, there do exist techniques for bounding S. We suppose that $\Phi$ is sufficiently accurate so that S is nearly unity and write /18/ $S = (1 - 10^{-n})$ and study the two coefficients, A and B, in the last equation.

| | | | | |
|---|---|---|---|---|
| n | 1 | 2 | 4 | 6 |
| S | 0.90 | 0.990 | 0.9999 | 0.999999 |
| $\varepsilon$ | 0.48 | 0.14 | 0.014 | 0.0014 |
| A | 0.785 | 0.279 | 0.028 | $2.8 \times 10^{-3}$ |
| B | 0.19 | 0.002 | 0.0002 | $2 \times 10^{-6}$ |

Although $F_{\Phi\omega}$ and $F_{\omega\omega}$, are not known the coefficient multiplying the former does not become very small until S has attained a very high degree of overlap. As a result, it would be helpful if one could work with $\Phi$'s such that $F_{\Phi\omega}$ is as small as possible while S is as close to unity as possible.

Examination of the second term in (39), namely $B(F_{\omega\omega} - F_{\Phi\Phi})$ leads to the hope that its importance may be reduced when F is either a positive-definite or negative definite operator, since cancellation in $(F_{\omega\omega} - F_{\Phi\Phi})$ will occur.

Since our primary interest is in one-particle operators and the electron density, three methods to achieve the above stated goal exist within the framework of the Independent Particle Model /1/. From this point on we therefore restrict $\Phi$ to be a single Slater determinant (1). We expand $\Psi$ in terms of a set of normalized Slater determinants over a complete set of orthonormal spin orbitals whose first N members are the spin orbitals occupied in the Slater determinant $\Phi$,

$$\Psi = a\{\Phi + C_S\Phi_S + C_D\Phi_D + \ldots\} \tag{40}$$

where $C_S\Phi_S$, $C_D\Phi_D$, etc. represent respectively the sums over the singly, doubly, etc. substituted determinants $\Phi_i^a$, $\Phi_{aj}^{ab}$, etc. with respect to $\Phi$ and a is the normalization factor.

One can choose $\Phi$ such that, S, the overlap of $\Phi$ with $\Psi$, is a maximum. The resulting spin orbitals are called the maximum overlap or Brueckner spin orbitals /1,19/. They have the interesting property that the singly substituted determinants do not appear in the expansion (40), i.e. $C_S = 0$. The result is that $F_{\Phi\omega}$ vanishes for one-electron operators and $\Delta F$ depends only on the second term $B(F_{\omega\omega} - F_{\Phi\Phi})$. Of course, we need to know $\Psi$ to determine these spin orbitals exactly. Approximate methods for their determination have been suggested /20/.

Another procedure for the choice of $\Phi$ would be to construct it from the first N natural spin orbitals or eigenfunctions of $\Gamma^1_\Psi$. These "best-density" spin orbitals /1/ are optimum in the sense that they provide a best least-squares approximation of $\Gamma^1_\Psi$ by a Fock-Dirac density matrix $\Gamma^1_\Phi$. In this model, the singly substituted determinents are not missing in general from the expansion (40). But $F_{\Phi\omega}$ vanishes and $\Delta F$ again depends only on the second term $B(F_{\omega\omega} - F_{\Phi\Phi})$. However, $\varepsilon$, S, $F_{\omega\omega}$ and $F_{\Phi\Phi}$ are different for each of these models except in the case N = 2 where the best-overlap and best density spin orbitals are the same /21/.

The third and most practical model to consider is the HF(SCF) one. By its very definition, in terms of the variation principle, it is the Independent Particle Model which is the most accurate from the energy point of view. Since it is often said to offer the same degree of accuracy for the energy,the electron density and expectation values of single particle operators it is important to examine this statement.

Since Brillouin's theorem (15) states that the matrix element of H between $\Phi_{HF}$ and $\phi_S$ vanishes i.e.

$$\langle\Phi_{HF}|H|\Phi_S\rangle = 0 \tag{41}$$

it follows from perturbation theory that only the determinants $\Phi_D$, doubly substituted with respect to $\Phi_{HF}$ contribute in first order to the expansion

$$\Psi = a\{\Phi_{HF} + C_S\Phi_S + C_D\Phi_D + C_T\Phi_T + \ldots\} \tag{42}$$

of the exact wavefunction. Here $C_S\Phi_S$, $C_D\Phi_D$, etc. have the same meaning as in (40) but represent substitutions with respect to $\Phi_{HF}$. The single, triple and quadruple substitutions appear in second order through their coupling with the doubles. In terms of the perturbation theory parameter $\lambda$, this means that

$$C_D = O(\lambda),\ C_S = O(\lambda^2),\ C_T = O(\lambda^2),\ C_Q = O(\lambda^2) \tag{43}$$

As a result, for a one-electron operator, $F_{\Phi\omega}$ vanishes to first order (Moller-Plesset Theorem /22/) and we are left with corrections of second and higher order to $F_{\Psi\Psi}$ (from both terms in (36)). Since the HF energy is also correct to second order, this gives rise to the hope that the relative errors should be similar, i.e.

$$\frac{\Delta E}{E_{HF}} \sim O(\lambda^2) \tag{44}$$

$$\frac{\Delta F^1}{F^1_{HF}} \sim O(\lambda^2). \tag{45}$$

This discussion of the accuracy of HF expectation values of one-particle operators and $\rho(r)$* has left the important question of the magnitude of the second-order correction, to which both $\Phi_S$ and $\Phi_D$ contribute. This is in contrast to the energy, where only the latter contributes in the second and third orders. Consequently, for a long time the singly substituted terms were neglected because they were not very important from an energetic point of view. Difficulties encountered in the prediction by CI of the dipole moments of CO and LiH led to the discovery that the singly substituted terms are actually more important than the doubles in the second-order correction to the density. This should not be construed to mean that the doubles should be omitted in correlated studies of the electron density. The coupling of the singles with $\Phi_{HF}$ through the doubles is quite important. Thus, in a study of the dipole moment of LiH the inclusion of only singles led to essentially the HF value. Hence, one must include the singles and the doubles simultaneously /23-25/.

Consider the second-order contributions to $\rho(r)$ in the expansion:

$$\rho(\vec{r}) = (1 + |C_S|^2 + |C_D|^2 + \ldots)^{-1}\{\rho_{oo} + 2R_e(C_S\rho_{oS}) + |C_S|^2\rho_{SS} + |C_D|^2\rho_{DD} + \ldots\} \tag{46}$$

where

$$\rho_{oS}(\vec{r}) = N\int\Phi^*_{HF}(1,2,\ldots,N)\Phi_S(1,2,\ldots,N)ds_1d2\ldots dN \tag{47}$$

$$\rho_{SS}(\vec{r}) = N\int\Phi^*_S(1,2,\ldots,N)\Phi_S(1,2,\ldots,N)ds_1d2\ldots dN \tag{48}$$

$$\rho_{DD}(\vec{r}) = N\int\Phi^*_D(1,2,\ldots,N)\Phi_D(1,2,\ldots,N)ds_1d2\ldots dN \tag{49}$$

etc,

*Since the charge density $\rho(r)$ can be written as an expectation value of the one-electron operator

$$\hat{\rho}(\vec{r}) = \sum_{i=1}^{N} \delta(\vec{r} - \vec{r}_i)$$

the preceeding analysis would apply to the charge density as well.

namely

$$\rho(\vec{r})^{(2)} = 2R_e(C_s\rho_{os}) + |C_D|^2(\rho_{DD} - \rho_{oo}). \tag{50}$$

A typical term $\rho_{DD}(\vec{r})$ may be written in the form

$$\rho_{DD} = \rho_{oo} + \rho_{aa} + \rho_{bb} - \rho_{ii} - \rho_{jj} \tag{51}$$

where $\rho_{aa}$ denotes an orbital density,

$$\rho_{aa}(\vec{r}) = \phi_a^*(\vec{r})\phi_a(\vec{r}) \tag{52}$$

with a similar meaning for $\rho_{bb}$, $\rho_{ii}$ and $\rho_{jj}$. Then elimination of $\rho_{oo}$ yields

$$\rho(\vec{r})^{(2)} = 2R_e(C_s\rho_{os}) + |C_D|^2(\rho_{aa} + \rho_{bb} - \rho_{ii} - \rho_{jj}) \tag{53}$$

Since the most important double excitations are expected to involve orbitals localized in the same region as the orbitals for which they are substituted, the factor $(\rho_{aa} + \rho_{bb} - \rho_{ii} - \rho_{jj})$ should be comparatively small and the first term in Eq. (53) should dominate /24/. In the case of the dipole moment operator, the evidence is that one single substitution is the most important /25/. For LiH, HF, HCl, and ClF the most important single substitution is that of the lowest unoccupied sigma molecular orbital (LUσMO) for the highest occupied sigma molecular orbital (HOσMO). For CO and CS, it is the corresponding LUπMO substitution for the HOπMO.

In recent years, a number of calculations have been made of CI wavefunctions which include all possible singly and doubly substituted configurations from the SCF determinant as well as the latter. Such wavefunctions are called SDCI in contrast to DCI wavefunctions which include only double substitutions. Since the virtual orbitals are generated from the same basis set as the SCF determinant, Brillouin's theorem and all the above discussion applies. As a result, we should expect to be able to determine $\rho$ rather accurately for a given basis set with such SDCI wavefunctions. The problem would remain of choosing a good basis set for a given system.

How large are these correlation corrections to $\rho$ and do we need to worry about their effect on $\Delta\rho$ in the chemically interesting regions of the molecule?

Duben and Lowe have made a very instructive study of $H_3^+$. They constructed SCF, DCI and SDCI wavefunctions for each of two AO basis sets. Since $H_3^+$ is a two-electron system, SDCI would be a full CI for the given basis set. $\Delta\rho_c = \rho_{CI} - \rho_{SCF}$ for all four of their wavefunctions shows the same qualitative behaviour, i.e. charge accumulation around and behind the nuclei with removal from the central region, such that

$$\int \Delta\rho_c(\vec{r})d\vec{r} = 0 \tag{54}$$

is satisfied. Comparison of $\Delta\rho_c(\vec{r})$ for their two DCI wavefunctions or their two SDCI wavefunctions (36) shows the comparatively small effect of basis set size on $\Delta\rho_c$. On the other hand, comparison of $\Delta\rho_c$ for their SDCI and DCI wavefunctions shows that the effect of including the single substitutions is a much larger one /26/.

2D contour maps /27/ of $\Delta\rho_c$ and the relative density difference function $\Delta\rho_c/\rho_{CI}$ are shown in Fig. 2 from our recent calculation /28-30/ in the case of the water molecule where SDCI and SCF wavefunctions constructed from the same basis set were used. Examination of $\Delta\rho_c$ shows that there is a piling up of charge in regions at and behind each of the nuclei, charge removal from the OH bond regions, and a small depression in the lone-pair region behind the oxygen. These results are consistent with the dipole-moment reduction upon correlation mentioned above. We calculated $\Delta\rho_c$ and $\Delta\rho_c/\rho_{CI}$ for a number of other molecules $(H_2O)_2$, $CH_4$, $C_2H_2$, $C_2H_4$ wavefunctions of the SDCI type. Becker /31/ et al have recently calculated $\Delta\rho_c$ for the CO and $N_2$ molecules while Lipscomb et al /32/ have studied $H_2O$, $H_2S$, and BH, with conclusions similar to ours.

Two questions which are often raised are whether one can correct basis-set inadequacy by means of CI or if one could add a basis-set "error" correction calculated at the SCF level to the CI result. Green /25/ has considered this problem for the expectation values $(F^1)$ of one-particle operators. He has shown that if a small change in the basis set leads to a change in $(F^1)$ at the SCF level, then it will lead to essentially the same change in $(F^1)$ in a CI calculation constructed from the same basis. This means that the CI calculation cannot correct the error due to basis-set quality in the SCF calculation of $(F^1)$.

It should be pointed out that some of the features of the $\Delta\rho_c$ contour maps reflect the effects of electron correlation at the atomic level /27/. It would be useful, therefore, to calculate second difference densities $\Delta\Delta\rho_c$ defined by

$$\Delta\Delta\rho_c = \Delta\rho_m^{CI} - \Delta\rho_m^{SCF} \tag{55}$$

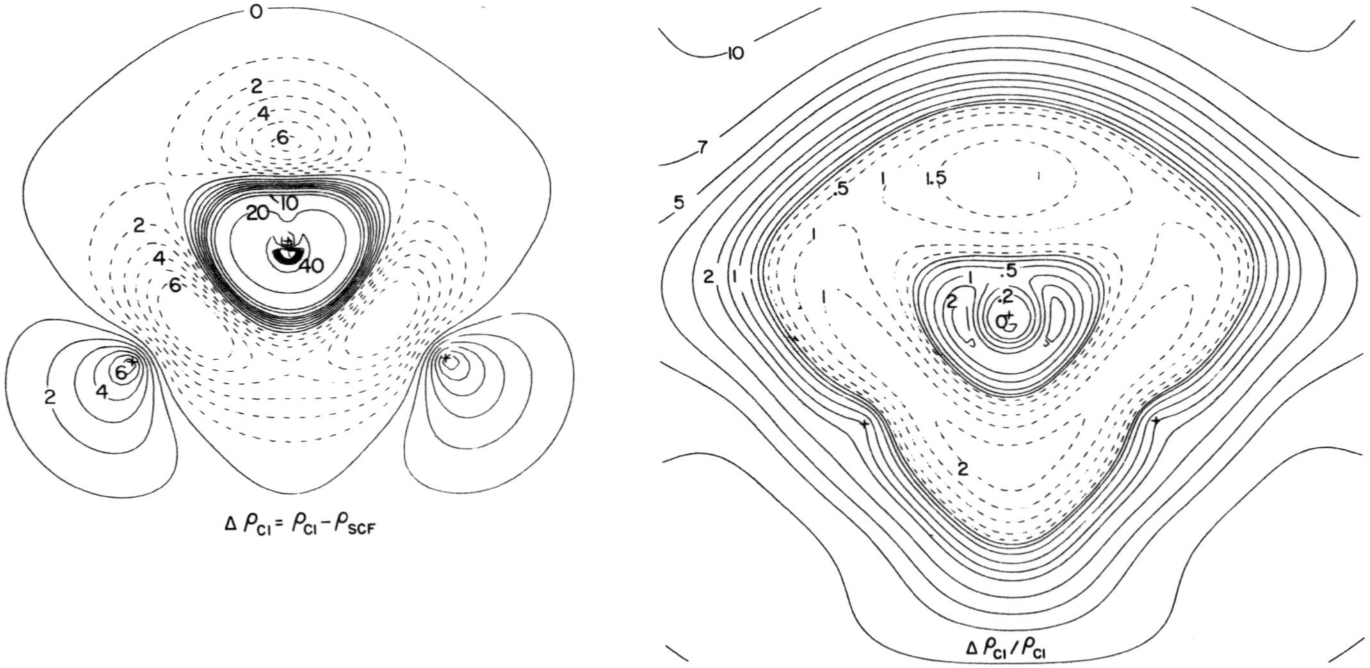


Figure 2 2D contourmaps of $\Delta\rho_{CI}$ and $\Delta\rho_{CI}/\rho_{CI}$ for $H_2O$ in the molecular plane /28-30/. Positive and negative contours are drawn with solid and dashed lines, repectively. The units are: $\Delta\rho_{CI}$ ($10^{-3}$ e/a.u$^3$) and $\Delta\rho_{CI}/\rho_{CI}$ (%).

$$= (\rho^{CI}_{molecule} - \rho^{CI}_{SFA}) - (\rho^{SCF}_{molecule} - \rho^{SCF}_{SFA}) \tag{56}$$

$$= \Delta\rho_{c,molecule} - \Delta\rho_{c,SFA} \tag{57}$$

where it is presumed that the same atomis basis sets were used in the construction of all of the densities involved.

It should not be forgotten that this discussion of the accuracy of the Hartree-Fock model has been with reference to that model without any restrictions, i.e. the Slater determinant $\Phi$ is constructed from N orthonormal spin orbitals. For closed shells, restrictions are not involved, since the Fock operator leads to doubly occupied molecular orbitals and the above discussion is fully applicable. For other systems, the discussion would not hold for the restricted Hartree-Fock (RHF) determinant since $C_s$ is no longer $O(\lambda^2)$ but of $O(\lambda)$. For example in the case of the $^2S$ ground state of atomic lithium, the RHF determinant would be

$$\Phi_{RHF} = \frac{1}{\sqrt{3!}} \det|1s^2 2s| = \frac{1}{\sqrt{3!}} \det|1s\alpha 1s\beta 2s\alpha| \tag{58}$$

i.e. with a doubly occupied K-shell while the HF determinant is

$$\Phi_{HF} = \frac{1}{\sqrt{3!}} \det|1s\alpha 1s'\beta 2s\alpha| \tag{59}$$

One describes the latter as exhibiting a spin-polarized core. Although the Brillouin and Moller-Plesset Theorems hold, $\Phi_{HF}$ is not an eigenfunction of $S^2$. These are typical of the problems that arise in the calculation of spin densities /8,19/.

Support of this research by the National Research Council of Canada is gratefully acknowledged.

REFERENCES

1. W. Kutzelnigg and V.H. Smith, Jr., J. Chem. Phys. 41, 896 (1964).

2. V.H. Smith, Jr., "Electron Densities and Reduced Density Matrices", this volume, pp. 1-23.

3. D.R. Hartree, Calculation of Atomic Structures, (Wiley, New York, 1957).

4. C.C.J. Roothaan, Rev. Mod. Phys. 23, 69 (1951); 32, 179 (1960).

5. G.G. Hall, Proc. Roy. Soc. A205, 541 (1951).

6. T. Koopmans, Physica 1, 104 (1933).

7. L. Brillouin, Actual. Sci. Ind. 71, 1 (1933); 159, 1 (1934); 160, 1 (1934).

8. R.E. Brown, S. Larsson and V.H. Smith, Jr. Phys. Rev. A 2, 593 (1970); 6, 1375 (1972); 8, 2765E (1973).

9. W. Von Niessen, G.H.F. Diercksen and W.P. Kraemer in Quantum Chemistry: The State of the Art, edited by V.R. Saunders and S. Brown (Atlas Laboratory, 1974).

10. E. Clementi and H. Popkie, J. Chem. Phys. 57, 1077 (1972).

11. N. Rösch, V.H. Smith, Jr. and M.H. Whangbo, J. Amer. Chem. Soc. 96, 5984 (1974).

12. M.A. Ratner and J.R. Sabin, J. Amer. Chem. Soc. 99, 3954 (1977).

13. J. Van Wazer and I. Absar, Electron Densities in Molecules and Molecular Orbitals (Academic Press, New York, 1975).

14. P.E. Cade, Trans. Amer. Crystallogr. Assoc. 8, 1 (1972).

15. G. DeWith and D. Feil, Chem. Phys. Lett. 30, 279 (1975) and private communication.

16. S. Rothenberg and H.F. Schaefer, III, J. Chem. Phys. 54, 2764 (1971).

17. H.L. Hase and A. Schweig, Angew. Chem. 89, 264 (1977).

18. B.T. Sutcliffe in Computational Techniques in Quantum Chemistry and Molecular Crystals edited by G.H.F. Diercksen, B.T. Sutcliffe, and A. Viellard (D. Reidel, Dordrechl, 1974).

19. S. Larsson and V.H. Smith, Jr., Phys. Rev. 178, 137 (1969); Intern. J. Quantum Chem. 6, 1019 (1972).

20. S. Larsson, Chem. Phys. Letts. 7, 165 (1970).

21. V.H. Smith, Jr. and W. Kutzelnigg, Ark. Fysik 38, 309 (1968).

22. C. Moller and M.S. Plesset, Phys. Rev. 46, 618 (1934); W. Kutzelnigg and V.H. Smith, Jr., J. Chem. Phys. 42, 2791 (1965); J. Goodisman and W.J. Klemperer, J. Chem. Phys. 38, 721 (1963). I. Shavitt in Modern Theoretical Chemistry, Vol. II, edited by H.F. Schaefer III (Plenum, New York 1976).

23. S. Green, J. Chem. Phys. 54, 827 (1971); Adv. Chem. Phys. 25, 179 (1974).

24. C.F. Bender and E. Davidson, J. Chem. Phys. 49, 4222 (1968).

25. S. Green, Adv. Chem. Phys. 25, 179 (1974).

26. A.J. Duben and J.P. Lowe, J. Chem. Phys. 55, 4276 (1971).

27. V.H. Smith, Jr., P.F. Price, and I. Absar, Isr. J. Chem. 16, 187 (1977).

28. V.H. Smith, Jr., Phys. Scr. 15, 147 (1977).

29. V.H. Smith, Jr. and I. Absar, Isr. J. Chem. 16, 87 (1977).

30. V.H. Smith, Jr., P.F. Price, T. Winsor and G.H.F. Diercksen, (to be published).

31. P. Becker, M.E. Stephens and F. Grimaldi, (to be published).

32. J. Bicerano, D.S. Marynick and W.N. Lipscomb, J. Amer. Chem. Soc. 100, 732 (1978).

AN ELECTROSTATIC DESCRIPTION OF CHEMICAL BINDING

F.L. Hirshfeld

Department of Structural Chemistry

The Weizmann Institute of Science, Rehovot, Israel

The reasons for our interest in charge-density maps are probably extremely varied. But surely one important reason is the hope of learning something about chemical binding. For this purpose it is useful to think of the chemical bond in electrostatic terms.

## IMPLICATIONS OF HELLMANN-FEYNMAN THEOREM

The basic justification for an electrostatic interpretation of the chemical bond lies in the Hellmann-Feynman electrostatic theorem[1]. This tells us that if we know the charge distribution in a molecule, i.e. that computed from the *exact* electronic wavefunction (in the presence of a collection of fixed nuclear charges), the forces on the nuclei, in the chosen configuration, can be evaluated exactly by a classical electrostatic calculation, assuming ordinary Coulomb repulsion between the nuclei and attraction between nuclei and electrons.

A corresponding theorem[2] is valid for the exact Hartree-Fock approximation (and some others) to the true electronic wavefunction. This generalization says that the force on each nucleus derived electrostatically from the Hartree-Fock charge density is the same as we would obtain by displacing the nucleus a small distance in one direction or another, calculating the modified Hartree-Fock energy, and so differentiating the calculated energy with respect to each of the nuclear coordinates.

In the equilibrium molecular configuration the true forces on all nuclei must vanish. So if a given wavefunction fails to predict vanishing forces on all nuclei we can conclude that either

a. the postulated configuration does not correspond to equilibrium, or

b. the electronic wavefunction is inexact.

But if the *vector sum* of the calculated forces on all the nuclei does not vanish either, the only explanation is b (whether or not a is also to blame). For whatever the assumed molecular geometry, there can be no net force on the molecule as a whole if we have not applied an external field. This turns out, in fact, to be one of the most useful applications of the Hellmann-Feynman theorem - testing whether a particular approximation is close to the exact, or the exact Hartree-Fock, wavefunction. The Hellmann-Feynman forces are found to be highly sensitive to small inaccuracies in the wavefunction, especially near the nuclear positions. So if a trial wavefunction is almost, but not quite, at the Hartree-Fock limit the error will usually reveal itself in an appreciably non-vanishing sum of the Hellmann-Feynman forces on the nuclei.

## ELECTROSTATIC ROLE OF DEFORMATION DENSITY

But this lengthy introduction has not yet taught us anything about chemical binding. Let us now adopt the familiar crystallographic approach and write the true molecular charge density as

$$\rho^{\mathrm{mol}}(r) = \rho^{\mathrm{pro}}(r) + \delta\rho(r) \qquad (1)$$

The first term on the right is the promolecule density[3], given by

$$\rho^{\mathrm{pro}}(r) = \sum_{\alpha}\rho_{\alpha}^{\mathrm{at}}(r), \qquad (2)$$

where $\rho_{\alpha}^{\mathrm{at}}$ is the spherically symmetric free-atom density of atom $\alpha$ centered at the appropriate atomic position in the molecule. The promolecule thus corresponds to an assemblage of free atoms that have been brought together from infinite separation but not allowed to interact. The final term in equation (1), long slighted by crystallographers as a negligible correction to their deceptively successful free-atom model, is where all the chemistry lies. This is the deformation density $\delta\rho$ and it represents the charge migration whereby the collection of non-interacting atoms is converted into a chemically bound molecule.

If the molecule is in its equilibrium geometry and the charge density $\rho^{mol}(r)$ is exact, the electrostatic field computed from this density (with the nuclear charges included) must vanish at each nuclear position. Now this electrostatic field may be evaluated as a sum of two terms, according to the decomposition of $\rho^{mol}$ given in equation (1). Thus we have the field $E_p$ due to the promolecule (including nuclei) plus a part $E_\Delta$ due to the deformation density $\delta\rho$, and the sum of these must vanish at every nucleus, i.e.

$$E_p + E_\Delta = 0 \qquad (3)$$

We must, however, remember that the promolecule does not exist, except on paper. And if it did, we should still have no justification for applying the Hellmann-Feynman theorem to it. Nevertheless we can formally equate the field at a nucleus in the actual molecule, where the Hellmann-Feynman theorem does apply, with the sum of two hypothetical fields attributable, respectively, to the promolecule and to the deformation density. Equation (3) refers to these two fictitious fields.

Having adopted such a viewpoint, we can argue that one necessary function of the deformation density $\delta\rho$ is to produce an electrostatic field at each nucleus just sufficient to offset the opposite field due to the promolecule and so permit the molecule to attain electrostatic equilibrium. In justification of this argument we note that the fields $E_p$ are generally repulsive and accordingly are counteracted by attractive fields $E_\Delta$ arising from the charge deformation. This is most simply seen in a diatomic promolecule, where each nucleus is repelled by the incompletely shielded nuclear charge of the other atom (Fig. 1). We can thus call this repulsive field $E_p$ the penetration field since it arises from the penetration of the nucleus inside the electronic envelope of its neighbor. It is balanced by an attractive migration field $E_\Delta$ due to the deformation density. This argument clearly supports our contention that the deformation density $\delta\rho$ is the source of chemical binding. It also explains why the most prominent features in many deformation density maps are compact peaks in the bonds, which can act to pull the two atoms towards one another.

Whatever meaning we may attach to its separate terms, equation (3) has one important consequence. The penetration fields $E_p$ are derived from the free-atom densities, which constitute the promolecule, and can be simply evaluated provided we

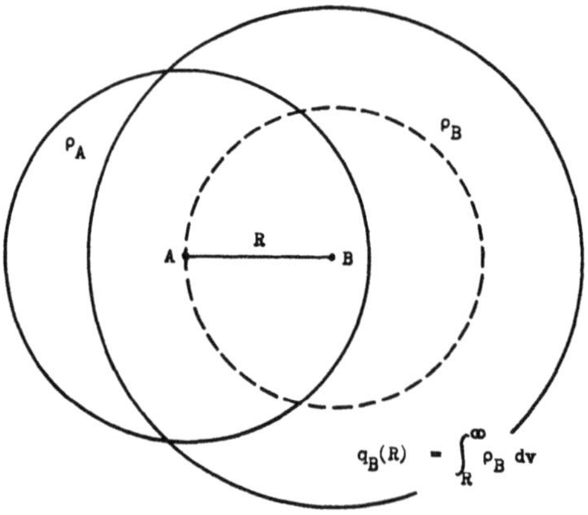


Figure 1. Hellmann–Feynman fields in diatomic promolecule AB. Field at nucleus A : $E_{p,A} = q_B(R)/R^2$ (repulsive).

know the molecular dimensions. So even before we have measured or calculated the deformation density, we know in advance the values of the fields $E_\Delta$ it must produce at the several nuclei in order to satisfy equation (3).

In some cases we can regard the magnitude of the migration field $E_\Delta$ (or of $E_p$) on a nucleus as a rough measure of the strength of the bond joining that atom to the rest of the molecule. This view is supported, for example, by an excellent correlation (Fig. 2) between the migration fields $E_\Delta$ at the two nuclei and the bond dissociation energy in two families of diatomic molecules[3]. In polyatomic molecules such a correlation cannot be so straightforward since the net migration field $E_\Delta$ at a particular nucleus cannot be uniquely divided into contributions from the several bonds in which the atom participates. Nevertheless it is not hard to imagine that if such a division were possible it would show, for example, that among aromatic C-C bonds of varying length the shorter and stronger bonds correspond to the larger penetration fields, associated with the deeper penetration of each nucleus inside the charge cloud of its ligand. A more direct indication of this is seen in the clear inverse correlation found experimentally between the length of a C-C bond, in an assortment of organic molecules, and the amount of charge in the bond peak[4]. Evidently, the more excess charge is localized near the bond center the greater the attractive field it will exert on the two nuclei.

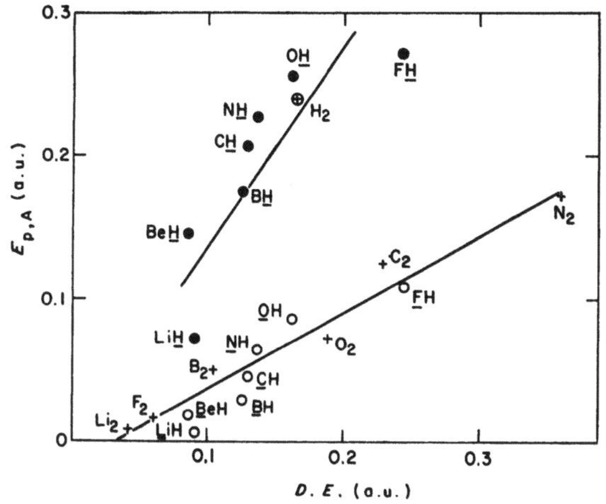


Figure 2. Penetration fields versus dissociation energies in $A_2$ and AH molecules. + $E_{p,A}$ in $A_2$. ○ $E_{p,A}$ in AH. ● $E_{p,H}$ in AH.

In the molecule HF the penetration field at the proton $E_{p,H}$ is 2½ times greater than that at the fluorine nucleus $E_{p,F}$ (Table 1). The corresponding repulsive *force* is almost four times as large on the fluorine nucleus, because of its greater nuclear charge, as on the proton. Thus we require of the deformation density $\delta\rho$ that it act very asymmetrically on the two nuclei, producing a stronger attractive field, but a much weaker force, on the hydrogen than on the fluorine nucleus.

In the water molecule each proton encounters a penetration field made up of two contributions -- a strong repulsion from the oxygen atom and a much weaker repulsion from the other hydrogen atom. The resultant field $E_{p,H}$ is slightly inclined to the direction of the O-H vector. Consequently the migration field $E_{\Delta,H}$ must also have a small component perpendicular to the O-H line. A reasonable deduction is that the deformation density near the proton is not symmetric about the O-H line, i.e. the O-H bond is bent.

These few examples illustrate how the magnitude and directions of the penetration fields at the several nuclei may be qualitatively interpreted to give a fair indication of the relative strengths of different chemical bonds, the asymmetry of a given bond with respect to the two bonded atoms, or the bending of a bond by a transverse components of the penetration field.

## ORBITAL DECOMPOSITION OF MIGRATION FIELD

But simply knowing the values of the penetration fields at the several nuclear positions is far from enough to predict what the deformation density will look like. We must still examine actual deformation maps for a variety of chemical systems to see what can be learned about the detailed manner in which the charge migration around the several kinds of atoms satisfies the constraints imposed by equation (3).

We have chosen to examine[3] the two families of diatomic molecules $A_2$ and AH, where A is a first-row atom. These offer several advantages for such a survey of bonding patterns:

- They form a set of closely related molecules and so permit direct correlations among similar chemical systems;
- Near Hartree-Fock wavefunctions are available for all of them[5];
- Each atom is bonded to only one neighboring atom;
- The high molecular symmetry facilitates a partial decomposition by orbitals of the deformation density and, thus, of the migration fields.

Taking the last point first, we have, in each molecule as well as in the separate atoms, an unambiguous separation of the orbitals into $\sigma$ and $\pi$ symmetries. Since the total electron density is a sum of contributions from the several orbitals, we can thus divide the deformation density $\delta\rho$, and the corresponding migration fields $E_\Delta$, into $\sigma$ and $\pi$ contributions. However, there is a small complication. In many cases the number of $\sigma$, or $\pi$, electrons in the molecule is different from that in the separate atoms, when these are chosen to be spherically symmetric. For example, in the $O_2$ promolecule each O atom has four electrons in its $1s$ and $2s$ orbitals and four in $2p$ orbitals. The latter must be shared equally among the three equivalent $2p$ orbitals, to assure spherical symmetry, so we assign 4/3 e to the $p_\sigma$ orbital and 4/3 e to each of two $p_\pi$ orbitals. This gives a total of $5\frac{1}{3}$ $\sigma$ electrons *vs* $2\frac{2}{3}$ $\pi$ electrons per atom, i.e. 10 $\frac{2}{3}$ *vs* $5\frac{1}{3}$ for the pair of atoms. But the $O_2$ molecule has 10 $\sigma$ electrons and 6 $\pi$ electrons. We can handle such a situation in either of two ways:

- We can define, say, the $\sigma$ deformation density as the difference between the molecular and the atomic $\sigma$ densities even though this may have a non-vanishing integral, i.e. it may represent not a pure migration but also a net creation or annihilation of electronic charge.

- We can define three components of the deformation density: a $\sigma$ and a $\pi$ deformation, each integrating to zero, plus an additional term that adjusts the numbers of $\sigma$ and $\pi$ electrons in each atom to match the corresponding numbers in the molecule.

We have selected the latter alternative so as to avoid the conceptual difficulty of having to create and annihilate electrons unnecessarily.

TABLE 1. Penetration fields $E_{p,A} = q_B/R_{AB}^2$ and migration fields $E_{\Delta,A}$ in fourteen AH and $A_2$ molecules at their equilibrium bond lengths $R_{AB}$. Molecular wavefunctions from Cade *et al*[5]. All quantities in atomic units.

| Molecule | $R$(bohr) | $q_A$(e) | $q_H$(e) | $E_{p,A}$ | $E_{\Delta,A}$ | $E_{p,H}$ | $E_{\Delta,H}$ |
|---|---|---|---|---|---|---|---|
| $H_2$ | 1.400 | | 0.469 | | | 0.240 | -0.245 |
| LiH | 3.015 | 0.664 | 0.061 | 0.007 | -0.009 | 0.073 | -0.073 |
| BeH | 2.538 | 0.941 | 0.118 | 0.018 | -0.016 | 0.146 | -0.145 |
| BH | 2.336 | 0.958 | 0.155 | 0.028 | -0.034 | 0.175 | -0.184 |
| CH | 2.124 | 0.934 | 0.204 | 0.045 | -0.054 | 0.207 | -0.221 |
| NH | 1.9614 | 0.874 | 0.250 | 0.065 | -0.074 | 0.227 | -0.246 |
| OH | 1.8342 | 0.857 | 0.291 | 0.086 | -0.095 | 0.255 | -0.278 |
| FH | 1.7328 | 0.812 | 0.327 | 0.109 | -0.117 | 0.270 | -0.296 |
| $Li_2$ | 5.051 | 0.210 | | 0.008 | -0.007 | | |
| $B_2$ | 3.005 | 0.453 | | 0.050 | -0.045 | | |
| $C_2$ | 2.3481 | 0.689 | | 0.125 | -0.123 | | |
| $N_2$ | 2.068 | 0.734 | | 0.172 | -0.183 | | |
| $O_2$ | 2.282 | 0.376 | | 0.072 | -0.070 | | |
| $F_2$ | 2.68 | 0.115 | | 0.016 | -0.007 | | |

But we can also decompose the $\sigma$ deformation density into core and valence contributions, identifying the lowest-energy molecular orbitals as core orbitals that correspond to the similarly low lying 1*s* orbitals of the first-row atoms. In this way we divide the deformation density in each of the diatomic molecules into, at most, four terms:

the core contribution $\delta\rho_k$, always present (except in $H_2$);

the $\sigma$ valence term $\delta\rho_\sigma$, always present,

a $\pi$ term $\delta\rho_\pi$, present if the molecule has at least one occupied $\pi$ orbital;

an intraatomic polarization term $\delta\rho_I$, present if one of the atoms has a ground state of $^1P$ symmetry (atoms B, C, O, or F).

Table 1 confirms that, for all molecules listed, the migration fields $E_\Delta$ derived, by numerical integration, from the deformation densities corresponding to the molecular wavefunctions of Cade *et al*[5] agree well with the values of $-E_p$, as required by equation (3). Other evidence implies that the small discrepancies found, especially at the hydrogen positions, are due primarily to the Hartree-Fock approximation itself, which leads to predicted equilibrium bond lengths somewhat shorter than the observed distances, while the larger discrepancies generally found at the first-row nuclei reflect inaccuracies of the wavefunctions, due to basis-set inadequacy, in the immediate vicinity of these nuclei.

## CORE POLARIZATION

Examining the orbital breakdown of the calculated migration fields $E_{\Delta,A}$ at the first-row nuclei (Table 2), we find appreciable contributions $E_k$ from the core deformation density $\delta\rho_k$ in almost all cases. These contributions evidently arise from a slight, but highly effective, polarization of the core density very close to the nuclear position. Such polarization is almost imperceptible in contour maps, where the region around the nucleus often cannot be clearly mapped at all because of excessive crowding of the contours. Moreover, crystallographic studies, for a variety of experimental reasons, as well as most theoretical studies, especially those using Gaussian basis sets, are both incapable of providing sufficiently precise information on the details of the charge density in the nuclear regions to establish reliable estimates of this inner polarization. Fortunately, the wavefunctions of Cade *et al*.[5], calculated with a Slater-type

TABLE 2. Decomposition of calculated migration fields $E_\Delta$ at first-row nuclei into contributions from intraatomic $\sigma$-$\pi$ polarization ($E_I$), core polarization ($E_k$), $\sigma$ ($E_\sigma$) and $\pi$ ($E_\pi$) valence deformations. Atomic units x $10^{-4}$.

| Molecule | $E_I$ | $E_k$ | $E_\sigma$ | $E_\pi$ | $E_\Delta$ |
|---|---|---|---|---|---|
| LiH | | 537 | −625 | | −89 |
| BeH | | 197 | −357 | | −161 |
| BH | | −447 | 104 | | −342 |
| CH | | −579 | 218 | −179 | −536 |
| NH | | −665 | 335 | −412 | −743 |
| OH | | −732 | 456 | −673 | −949 |
| FH | | −758 | 573 | −982 | −1167 |
| $Li_2$ | | 249 | −315 | | −66 |
| $B_2$ | 154 | 56 | −67 | −594 | −450 |
| $C_2$ | 500 | 139 | −168 | −1697 | −1226 |
| $N_2$ | | −570 | 1204 | −2465 | −1831 |
| $O_2$ | 263 | −709 | 1027 | −1281 | −700 |
| $F_2$ | 302 | −510 | 526 | −389 | −71 |

basis, appear accurate enough in the nuclear regions to justify reasonable confidence in the calculated values of $E_k$. This confidence is supported by the rather close agreement, noted above, between $E_\Delta$ and $E_p$ in all fourteen molecules as well as by the systematic nature of the variation from molecule to molecule in the quantities $E_k$. This systematic variation also provides a secure basis for generalizing the observed trends to other classes of molecules.

Two immediately obvious regularities in the tabulated values of $E_k$ are:-

- they vary systematically, in both AH and $A_2$ series of molecules, from positive (anti-binding) in Li to negative (binding) in N, O, and F;

- they are consistently opposite in sign and comparable in magnitude to the $\sigma$ valence term $E_\sigma$.

The latter observation suggests that it may be profitable to inquire first into the significance of the pattern presented by the calculated values of $E_\sigma$ in these two families of molecules. And indeed we can understand the general trend of these quantities if we attribute the positive values found in the heavier atoms in both series to the anti-binding effect of lone-pair density localized in compact peaks close behind the atomic nuclei on the molecular axis. Such lone-pair peaks, whenever they are present, consistently outweigh the binding effect of the somewhat broader bond peaks found on the bond axis between the two atoms.

## IMPORTANCE OF RADIAL NODES

But closer examination of the $\sigma$ valence density in these molecules reveals another highly significant feature. Being derived from the atomic $2s$ orbitals, or from $sp$ hybrid orbitals, the molecular $\sigma$ valence orbitals retain the nodal surfaces of the parent atomic $2s$ orbitals surrounding the first-row nuclei. And just as the orbital amplitudes reverse sign on these nodal surfaces, so does the net charge polarization seem to have opposite directions inside and outside the radial nodes. Thus, those first-row atoms that, in these molecules, have no $\sigma$ lone-pair orbitals show a net forward polarization of their $\sigma$ valence density into the bond region between the two atoms. But this is partly offset by a slight backward polarization of the inner portion of this same valence density close to the atomic nucleus. In the heavier atoms, in both series, the picture is dominated by the lone-pair density behind the atomic nucleus and the overall situation is reversed. Here the net polarization of the outer portion of the $\sigma$ valence density is backward, towards the lone-pair peak, while the inner region is polarized in the forward direction.

The polarization of the inner portion of the $\sigma$ valence density is an exceedingly small effect. Consequently, even though its electrostatic effect on the nucleus is highly enhanced by proximity to the nucleus, it never succeeds in overriding the net effect, whether binding or anti-binding, of the more prominent, though more distant, outer features of the $\sigma$ deformation density. Thus the net effect of this density, measured by the quantity $E_\sigma$, is binding in the first members of each series, anti-binding in the later members which have lone-pair peaks behind the first-row nuclei.

But now we return to the core deformation $\delta\rho_k$. In every case examined this has the same sense of polarization as the inner portion of the valence density $\delta\rho_\sigma$. The combined effect of this core polarization and the polarization, in the same direction, of the inner part of $\delta\rho_\sigma$ is generally large enough to offset most of the field at the nucleus produced by the outer portions of $\delta\rho_\sigma$. In several cases it even succeeds in reversing the sign of the net σ contribution, given by $E_k + E_\sigma$. In all cases this net σ contribution, whether binding or anti-binding, is too small to constitute a major factor unless the binding is altogether rather weak. Generalizing this observation, we conclude that the σ density as a whole *never plays a prominent role* in the electrostatic binding of a first-row nucleus.

This relative ineffectiveness of the σ orbitals in first-row atoms is closely related to the presence of radial nodes in these orbitals. By contrast, no such disability weakens the electrostatic binding capacity of the π orbitals. These are consistently polarized in the forward direction and so make substantial contributions $E_\pi$ to the net binding in all those molecules that have occupied π orbitals. Thus we find that even in a molecule like HF, where there is no formal π bond between the atoms, the binding of the fluorine nucleus comes almost entirely from a uniform forward polarization of the π deformation density, which provides almost 85% of the net migration field $E_\Delta$ at the fluorine nucleus.

Similarly, the unique position of hydrogen may be associated with the fact that it is the only element whose valence σ orbital lacks a radial node. Thus polarization of this orbital, when it is involved in chemical bonding, is entirely free of the complicating effects of oppositely behaving inner and outer densities. In fact, binding of the hydrogen atom by forward polarization of the σ deformation density is consistently strong in all the hydrides examined.

From the present electrostatic viewpoint, the most relevant classification of molecular orbital types thus appears to be one that sharply distinguishes between orbitals that have a nodal surface about a given nucleus and those that do not. In such a classification, σ orbitals are strongly binding for hydrogen as are π orbitals for first-row atoms, while σ binding, or anti-binding, of first-row atoms is uniformly weak. Similarly, we expect that if this generalization extends to heavier elements, second-row atoms should display relatively weak electrostatic binding because of the radial nodes in their valence orbitals, two in the 3*s* orbitals and one in the 3*p* orbitals. This simple classification can thus help us to understand why hydrogen is uniquely capable of forming strong σ bonds, why lithium and beryllium show little aptitude for covalent binding, and why the first-row elements C, N, and O form much stronger π bonds than their second-row homologs Si, P, and S.

## INTERPRETATION DIFFICULTIES

In this manner, our interpretation of chemical binding in terms of the electrostatic role of the deformation density seems to throw new light on well known regularities in the chemical properties of the elements. At the same time it emphasizes the important conclusion that a chemical bond is far from the simple phenomenon that we naively represent by a straight line drawn on paper between two atomic symbols. Rather the net binding typically arises as a delicate balance between competing tendencies among the several groups of molecular orbitals, each governed by its peculiar set of rules and circumstances. The hope is that if we can uncover these rules with sufficient precision, we may arrive at a clearer understanding of the ways atoms bind together into molecules.

But the particular electrostatic approach described above has led us into a number of disturbing paradoxes. Among these anomalies is the predominant binding role we have assigned to the $\pi$ orbitals in molecules that have no formal $\pi$ bonds, such as $F_2$ and the hydrides of N, O, and F. Another is the conclusion that the two atoms in a heteronuclear diatomic molecule may be bound unequally, if we accept the value of the migration field $E_\Delta$ at each nucleus as a measure of the strength of its binding. This particular paradox has its origin in our attempt to apply electrostatic reasoning to the hypothetical promolecule, where such a viewpoint requires us to conclude that there may be a net force acting between the electronic cloud as a whole and the assemblage of nuclei. It then follows that among the functions of the deformation density, apart from counteracting the net repulsion between the nuclei, is the restoration of electrostatic equilibrium between the electronic envelope and the nuclear framework.

One lesson from all this is that classical electrostatics may be applied legitimately to the actual molecule, where the Hellmann-Feynman theorem will protect us from going too far astray in our conclusions, but when we apply similar arguments to such fictitious constructions as the promolecule or the deformation density we have no such insurance policy to save us from absurdity.

## INTERACTION OF RIGID WHOLE ATOMS

For this reason it may be instructive to consider an alternative interpretation of the electrostatic forces in the promolecule. This time we imagine that each atom consists of a frozen spherical cloud of electrons with the nucleus rigidly attached at its center. Rather than consider the nuclei alone as subject to electrostatic forces exerted by the electrons and the other nuclei, we now take the total atom, nucleus plus electrons, as a rigid unit subject to electrostatic interactions with other similarly rigid atomic units. An immediate advantage of such a picture is that, in a diatomic promolecule AB for example, the total force on atom A, due to its interaction with atom B, is exactly equal and opposite to the force on atom B exerted by the electrostatic field of A.

It should be emphasized that this model of the promolecule is no more valid than that given above and, in fact, it may lead us into far greater absurdities than the previous approach. The main reason for introducing it is that by balancing one viewpoint against the other we may attain a clearer understanding of the limits of validity of each of them.

The most glaring defect in this second interpretation of the promolecule is that it assigns distinguishable electrons to each atom and so flagrantly violates the Pauli principle. Having explicitly defined a pair density for the electronic system - our former interpretation left the pair density quite undefined - that is inconsistent with the requirement of antisymmetry, we cannot even rely on the variational theorem to set a lower bound to the calculated energy of our system. Indeed when we evaluate the total energy of this "classical" promolecule we find that this often lies well below that of the actual molecule.

But the most striking contrast with the previous picture is that, in almost every case, the net force between the atoms in our diatomic promolecule is now found to be attractive rather than repulsive. This means that the electronic envelope of each atom is attracted more strongly to the other atom than its nucleus is repelled by it. The role of the deformation density $\delta\rho$, on this interpretation, is no longer primarily that of binding the nuclei together against their mutual repulsion. Rather it has the primary function of restoring antisymmetry while keeping the total energy as low as possible. In this process the net change in energy may be positive or negative.

The crucial significance of antisymmetry in this picture is underlined by the exceptional behavior of $H_2$. This is the only molecule in our list for which the classical electrostatic force between atoms in the promolecule is repulsive. Is this simply another manifestation of the usual perversity of hydrogen?

Two hydrogen atoms, like any pair of rigid spherical atoms, experience a mutual electrostatic attraction as soon as they come close enough so that their electronic envelopes begin to overlap. As they approach more closely this attraction increases, but eventually it gives way to a net repulsion as the repulsive force between the two nuclei dominates the other interactions. At some intermediate distance the attractive and repulsive forces just balance and this is the separation at which the classical electrostatic energy of the system is a minimum. But in most molecules the two atoms never, in fact, come this close together. The requirement of antisymmetry modifies the molecular electronic distribution and shifts the actual equilibrium distance to a value well beyond the classical electrostatic minimum. As a result, the observed bond distance occurs in the region where the classical interatomic force is still attractive. This is just what our calculations show. The net classical force on each atom, resulting from the repulsive force on the nucleus, given by

$$F_A^n = Z_A \; E_{p,A},$$

acting against the attractive force on the spherical electron cloud $F_A^e$, is negative (attractive) in thirteen of the molecules listed in Table 3.

What makes $H_2$ exceptional is that its two atoms are able to come even closer together ($R$ = 1.40 bohr) than the classical minimum in energy (1.87 bohr). And the explanation for this is that in this two-electron system the constraints of the Pauli exclusion principle may be satisfied simply by the pairing of the electron spins. The spatial part of the electronic wave-function is thus allowed to be totally symmetric, i.e. the electronic clouds of the two atoms can overlap completely without hindrance from the Pauli principle. In this way the atoms are able to come more closely together than would be possible if they retained their separate spherical electron clouds and interacted classically. This is the meaning of the repulsive classical force $F$ in the $H_2$ promolecule (Table 3). What actually happens is that the charge cloud around each nucleus is polarized strongly in the forward direction and contracted inwards. Both effects lower the energy of the molecule, leading to a Hartree-Fock

TABLE 3. Classical electrostatic force $F$ on each of two spherical atoms in the promolecule, decomposed into forces $F^n$ on the nucleus and $F^e$ on the electron cloud. Also listed are the classical binding energy $V_c$ in the promolecule and the Hartree-Fock molecular binding energy $\Delta E_{HF}$. Atomic units x $10^{-3}$.

| Molecule | $F_A{}^n$ | $F_A{}^e$ | $F_H{}^n$ | $F_H{}^e$ | $F$ | $V_c$ | $\Delta E_{HF}$ |
|---|---|---|---|---|---|---|---|
| $H_2$ | | | 240 | −139 | +100 | −2 | −134 |
| LiH | 20 | −23 | 73 | −76 | −3 | −12 | −55 |
| BeH | 74 | −85 | 146 | −157 | −11 | −35 | −80 |
| BH | 142 | −175 | 175 | −209 | −33 | −59 | −102 |
| CH | 271 | −329 | 207 | −265 | −57 | −85 | −91 |
| NH | 454 | −539 | 227 | −312 | −85 | −111 | −77 |
| OH | 692 | −798 | 255 | −361 | −107 | −132 | −111 |
| FH | 981 | −1111 | 270 | −400 | −130 | −152 | −161 |
| $Li_2$ | 25 | −31 | | | −6 | −14 | −6 |
| $B_2$ | 251 | −384 | | | −133 | −148 | −33 |
| $C_2$ | 749 | −1084 | | | −335 | −336 | −29 |
| $N_2$ | 1202 | −1890 | | | −688 | −532 | −191 |
| $O_2$ | 577 | −1265 | | | −688 | −372 | −47 |
| $F_2$ | 145 | −444 | | | −299 | −133 | +49 |

binding energy $\Delta E_{HF}$ (0.134 hartree) almost seven times greater than would be possible with undeformed atoms.

The more typical situation, in which the requirement of antisymmetry plays a decisive role, is illustrated by the many-electron systems $N_2$, $O_2$, and $F_2$, in which doubly occupied valence orbitals on the two atoms overlap severely in the promolecule. The molecule responds to this violation of the exclusion principle by polarizing the $\sigma$ density away from the region of overlap (e.g. by *s-p* hybridization) into a lone-pair peak behind each atomic nucleus. Such backward polarization is clearly destabilizing and, indeed, the net binding energy $\Delta E_{HF}$ in these three molecules is much smaller (less negative) than the classical promolecule energy $V_c$.

An intermediate situation arises when a doubly occupied valence orbital on one atom overlaps with a singly occupied orbital on the other, as in most of the hydrides. The answer in such a case is to polarize backwards the $\sigma$ density of the first-row atom, but the resulting destabilization is offset by the forward polarization of its $\pi$ density and of the hydrogen $\sigma$ density as well as by the customary pronounced contraction of the hydrogen density. The net effect in most cases is a final energy not too different from that of the classical promolecule.

## CONCLUSIONS

In summary, it appears that by combining two different views of the electrostatic forces in the hypothetical promolecule we can begin to understand the meaning of certain tendencies that are apparent in the properties of chemical bonds between different kinds of atoms. The presence or absence of radial nodes in the valence orbitals of an atom is seen to be a major factor in determining whether or not it will form strong covalent bonds. While the balance of opposing influences leading to electrostatic equilibrium at a particular molecular geometry is extremely sensitive to a nearly undetectable polarization of the core density around each nucleus, the nature of this polarization is sufficiently systematic to be approximately predictable in simple molecules. Finally, it appears that the $\sigma$ orbitals of first-row atoms are, rather surprisingly, less effective electrostatically than the $\pi$ orbitals. This is due both to their radial nodes and to their tendency to overlap on the bond axis in violation of the Pauli exclusion principle.

## REFERENCES

1. H. Hellmann, *Einführung in den Quantenchemie*, Franz Deuticke, Leipzig, 1937; R.P. Feynman, Phys. Rev. 56, 340 (1939).

2. A.C. Hurley in *Molecular Orbitals in Chemistry, Physics, and Biology*, Löwdin and Pullman, eds., Academic Press, N.Y., 1964, p.161; C.W. Kern and M. Karplus, J. Chem. Phys. 40, 1374, (1964).

3. F.L. Hirshfeld and S. Rzotkiewicz, Mol. Phys. 27, 1319 (1974)

4. Z. Berkovitch-Yellin and L. Leiserowitz, J. Amer. Chem. Soc. 99, 6106 (1977).

5. P.E. Cade, K.D. Sales, and A.C. Wahl, J.Chem.Phys. 44,1973(1966); P.E. Cade and W.M. Huo, J. Chem. Phys. 47, 614 (1967); P.E. Cade, private communication, 1968.

# CHARGE AND SPIN DENSITIES IN SOLIDS

N.H. March

Theoretical Chemistry Department

University of Oxford, Oxford OX1 3TG, England

CONTENTS

1. Introduction
2. Localized versus delocalized description of $\rho(\underset{\sim}{r})$ in crystals
   2.1. Examples of localized descriptions of $\rho(\underset{\sim}{r})$
      (a) Rare gas crystals
      (b) Body-centred cubic lithium metal
      (c) Classical ionic crystals, e.g. NaCl
      (d) Covalently bonded semi-conductors, e.g. silicon
3. Density calculated from one-body theory
   3.1. Approximate one-body potential in slowly varying electron cloud
   3.2. One-body potential calculation of electron density in lithium metal
      Comparison with superposition of screened ions
4. Phonons and linear response
   4.1. Rigid screened ion model
5. Spin density description by one-body potential
   5.1. Hartree-Fock theory of uniform electron assembly
   5.2. Spin-dependent one-body potentials
6. Directional bonding in solids
   6.1. Bond electron density in Si from LCAO method
   6.2. Scattering from amorphous Si
   6.3. Relation between bonding and energy bands
   6.4. Scattering factor in terms of Wannier functions
7. Summary

## 1. INTRODUCTION

In these lectures we shall develop basically the theme that the ground state electron density $\rho(\underset{\sim}{r})$ has a truly major role to play in the many-electron description of solids. Following the pioneering work of Thomas (1926) and Fermi (1928), in which the electronic properties of heavy atoms were described by the electron density, two parallel descriptions of electron states have been employed :

(i) Using one-body wave functions $\Psi_i(\underset{\sim}{r})$, the Hartree and Hartree-Fock self-consistent field methods being the most famous example. Here the ground state density in the solid is constructed as

$$\rho(\underset{\sim}{r}) = \sum_{\substack{\text{occupied}\\ \text{states}}} \Psi_i^*(\underset{\sim}{r})\Psi_i(\underset{\sim}{r}) \qquad (1.1)$$

from the one-electron wave functions, which, in turn are constructed from an appropriate one-electron Schrödinger equation.

(ii) By avoiding explicit use of wave functions, either one-body or many-body, and attempting to work directly with the electron density $\rho(\underset{\sim}{r})$. This is the basis of the modern density functional approach.

We shall demonstrate here that in spite of the fact that electron-electron interactions often play a major role in solids, in fact the exact many-body form of the electron density $\rho(\underset{\sim}{r})$ can indeed be built up exactly as in eq. (1.1). This is tantamount to the claim that in perfect crystalline solids, where the one-electron wavefunctions $\Psi_i(\underset{\sim}{r})$ take Bloch form

$$\Psi_{\underset{\sim}{k}}(\underset{\sim}{r}) = \exp(i\underset{\sim}{k}.\underset{\sim}{r})\; u_{\underset{\sim}{k}}(\underset{\sim}{r}) \qquad (1.2)$$

where $u_k(\underset{\sim}{r})$ is periodic with the period of the lattice, the electronic density $\rho(\underset{\sim}{r})$ can, in principle, be calculated exactly from energy band theory. Of course, in practice, the difficulty which remains is to construct the one-body potential energy $V(\underset{\sim}{r})$, periodic with the period of the lattice, which will allow the Bloch wave functions $\Psi_k(\underset{\sim}{r})$ to be generated, and hence $\rho(\underset{\sim}{r})$ calculated from eqn (1.1). One of the major tasks of the theory is to construct $V(\underset{\sim}{r})$, which, as we shall see below, incorporates in an essential way the exchange and correlation interactions between electrons. However, before going on to discuss the calculation of the one-body potential $V(\underset{\sim}{r})$ we want to stress two important properties of the electron density in a perfect crystalline solid :

(i) $\rho(\underset{\sim}{r})$ is an observable, in contrast, say to a many-body ground-state wave function, since it can in principle be determined from measuring X-Ray scattering intensity at the Bragg reflections.

(ii) Once $\rho(\underset{\sim}{r})$ is known, the ground-state energy is a unique functional of the electron density. This important result was assumed in the pioneering works of Thomas (1926), Fermi (1928) and Dirac (1930) on the density description of atomic systems. It was proved for a non-degenerate many-electron ground state by Hohenberg and Kohn (1964), whose work therefore formally completes the Thomas-Fermi theory.

## 2. LOCALIZED VERSUS DELOCALIZED DESCRIPTION OF $\rho(\underset{\sim}{r})$ IN CRYSTALS

The classic crystallographic approach to the representation of the periodic electron density $\rho(\underset{\sim}{r})$ in perfect crystals is to expand $\rho(\underset{\sim}{r})$ in a Fourier series using the reciprocal lattice. Thus, we write:

$$\rho(\underset{\sim}{r}) = \sum_{\underset{\sim}{K}} \rho_{\underset{\sim}{K}} \exp(i\underset{\sim}{K}.\underset{\sim}{r}) \tag{2.1}$$

where the $\underset{\sim}{K}$'s denote the reciprocal lattice vectors.

Such a delocalized picture of the electron density, it has to be said immediately, is an unescapable consequence of solving the Schrödinger equation for the many-body wave function $\Phi(\underset{\sim}{r}_1 \ldots \underset{\sim}{r}_N)$ for the ground state of the crystal with N electrons, and forming the density from :

$$\rho(\underset{\sim}{r}_1) = N \int \Phi^*(\underset{\sim}{r}_1 \ldots \underset{\sim}{r}_N)\, \Phi(\underset{\sim}{r}_1 \ldots \underset{\sim}{r}_N)\, d\underset{\sim}{r}_2 \ldots d\underset{\sim}{r}_N . \tag{2.2}$$

As remarked, such a $\rho(\underset{\sim}{r})$ will belong to the crystal as a whole and, except in such an extreme case as a classic ionic crystal like NaCl, to be referred to briefly below, it would not be possible from the solution of the wave equation alone, to divide $\rho(\underset{\sim}{r})$ into localized densities.

Nevertheless, as will be stressed throughout these lectures, it will often be of prime importance for chemistry and for physics to attempt such a decomposition. But then we must be quite clear that any such localized description in a crystal must :

(a) Be in keeping with the knowledge we have, from both experiment and from theory, of the periodic electron density $\rho(\underset{\sim}{r})$

(b) Be consistent with our chemical and physical intuition

(c) Be useful in describing other properties of crystals than merely the periodic ground state density (for example phonons and simple defects).

## 2.1. Examples of Localized descriptions of $\rho(\underset{\sim}{r})$

Below we shall take a number of examples to illustrate how we can use intuition to give a localized description of the electron density $\rho(\underset{\sim}{r})$ in a wide variety of different types of crystals.

(a) Rare gas crystals. We start with the most elementary type, the rare gas solids, held together predominantly by van der Waals forces. In crystalline argon, say it is then intuitively plausible to write :

$$\rho(\underset{\sim}{r}) = \sum_{\underset{\sim}{R}} \sigma(\underset{\sim}{r}-\underset{\sim}{R}) \tag{2.3}$$

where the $\underset{\sim}{R}$'s denote the direct lattice vectors. Not only is $\sigma(\underset{\sim}{r})$ a localized density centred on each nucleus in the crystal, but in crystalline argon all available evidence points to the fact that it is well represented by the free atom density, which, because of the closed shell atoms involved, can be taken as spherical.

One piece of diffraction evidence which can be cited in support of this point of view has been put forward by Egelstaff, March and McGill (1974) on liquid argon. Thus, one compares X-ray and neutron diffraction patterns from the liquid, the neutron diffraction yielding directly the liquid structure factor S(k), which is the Fourier transform of the radial distribution function g(r) through

$$S(k) = 1 + \rho_0 \int [g(r)-1] \exp(i\underset{\sim}{k}.\underset{\sim}{r})\, d\underset{\sim}{r} \tag{2.4}$$

$\rho_0$ being the number of atoms per unit volume. This is a formula to which we shall return below when we discuss non-crystalline solid silicon.

If we have a liquid scatterer of N atoms, and we represent the atomic scattering factor by f(k), then the intensity I(k) of X-rays scattered through an angle $\theta$, with $k=4\pi\sin\theta/\lambda$, $\lambda$ being the X-ray wavelength, is given by :

$$I(k) = N f^2(k)\, S(k) \tag{2.5}$$

By comparing X-ray and neutron scattering at the principle peak of S(k), Egelstaff et al (1974) confirm that f(k) is correctly represented by the neutral atomic scattering factor. We do not know of Bragg reflection experiments on crystalline rare gases, but, we stress, that even if these could be carried out, they would only determine the Fourier transform $\sigma(k)$ of the localized electron density $\sigma(r)$ at the reciprocal lattice vectors K.

Thus we stress that it is not possible, from appeal to crystalline data alone, to uniquely determine $\sigma(r)$. One must appeal to models based on physical or chemical intuition. Nevertheless, it is of the greatest interest to explore such localized descriptions, which are surely intimately related to the role of electron-electron interactions in the crystal under discussion.

(b) Body-centred cubic lithium metal. We shall refer in some detail below to the case of lithium metal. It emerges that again a localized description of the form (2.3) is appropriate, to a useful approximation. But now $\sigma(\underset{\sim}{r})$ is not correctly given by a neutral free atom density. Rather, it is characteristic of a screened ion, with the associated, and characteristic Friedel oscillations.

What we shall show below, is that a potential $V(\underset{\sim}{r})$ can be set up, including electron-electron interactions, which can be used in a band theory calculation to determine the periodic charge density $\rho(\underset{\sim}{r})$. This will then be compared with a summation of screened ion densities, following eqn (2.3).

One of the important consequences of the form (2.3) is that it allows a description of the force fields in terms of a pair potential. For crystalline argon, this is most basically taken as the vacuum interaction between two argon atoms in free space. One must add three-body interactions as significant corrections arising in the condensed phase, though these are usually small corrections. Likewise, in lithium metal, the pair interaction is that between the screened ions, but the total energy as a function of nuclear configuration is a sum of pair potentials plus a volume-dependent energy to allow fully for the presence of the conduction electron gas. We return to this example briefly below.

(c) Classical ionic crystals, e.g. NaCl. In the case of NaCl, we write the crystal density as the superposition of $Na^+$ and $Cl^-$ ions, namely

$$\rho(\underset{\sim}{r}) = \sum_{\underset{\sim}{R}_+} \sigma_+(\underset{\sim}{r}-\underset{\sim}{R}_+) + \sum_{\underset{\sim}{R}_-} \sigma_-(\underset{\sim}{r}-\underset{\sim}{R}_-) \tag{2.6}$$

where the cation densities $\sigma_+$ are centred on sites $\underset{\sim}{R}_+$, etc. Again, there is experimental evidence supporting the superposition assumption (2.6) through the work of Castmann et al (1971).

(d) Covalently bonded semiconductors, e.g. silicon. Our fourth and final example is that of the covalently bonded semiconductors in the diamond lattice structure. Here, because of the pronounced chemical bonding, it is not reasonable to approximate $\rho(\underset{\sim}{r})$ as a superposition of localized distributions centred on the nuclei. Rather, the basic building block is the chemical bond, and we therefor write :

$$\rho(\underset{\sim}{r}) = \Sigma\, \sigma_{bond} \tag{2.7}$$

We shall see below that, for a perfect crystal, the electron density $\rho(\underset{\sim}{r})$ can be calculated from either band theory or from a chemical approach based on $sp^3$ hybrids, and the results are in good agreement. But the merit of the chemical approach is that it can also be applied to the amorphous (i.e. non crystalline) solid phase of silicon, whereas band theory is no longer applicable when the long range order of the perfect solid is lost. Furthermore, just as the form (2.3) led to a natural zeroth order description of the force field in terms of pair potentials, so the form (2.7) leads rather naturally to the calence force field, discussed for example by Musgrave and Pople (1962).

The conclusion from these examples is that while $\rho(\underset{\sim}{r})$ is usually delocalized, and no immediate nor unique decomposition is possible into localized building blocks, nevertheless such localized descriptions are of the greatest physical and chemical interest, as we have illustrated with reference to the description of force fields.

## 3. DENSITY CALCULATED FROM ONE-BODY POTENTIAL THEORY

This is the point to return to the result (1.1) for the electron density in terms of one-body wave functions. These, as already remarked, will satisfy an appropriate one-electron Schrödinger equation

$$\nabla^2\Psi_i + \frac{2m}{\hbar^2}[\varepsilon_i - V(\underset{\sim}{r})]\,\Psi_i = 0 \qquad (3.1)$$

The basic question then concerns the form of $V(\underset{\sim}{r})$.

From our knowledge of $V(\underset{\sim}{r})$, we expect that we must calculate a major contribution from (a) the potential energy $V_n$ of all nuclei and (b) the potential $V_e$ generated according to electrostatics by the ground state electron density $\rho(\underset{\sim}{r})$. Calling $V_n+V_e$ the Hartree potential $V_H$, we therefore write :

$$V(\underset{\sim}{r}) = V_H(\underset{\sim}{r}) + V_{xc}(\underset{\sim}{r}) \qquad (3.2)$$

where $V_{xc}(\underset{\sim}{r})$ is written for that contribution from the many-body exchange and correlation interactions. Naturally, exact knowledge of $V_{xc}(\underset{\sim}{r})$ would be equivalent to exact solution of the many-body problem : an impossible task. Thus, the essence of the one-body potential approach to the calculation of the ground state electron density $\rho(\underset{\sim}{r})$ lies in making a judicious approximation to $V_{xc}(\underset{\sim}{r})$ in eqn (3.2). Since we can only solve the many-electron problem at all accurately for a uniform assembly of interacting electrons, it is essential that we attempt to utilize the knowledge we have of this problem in approximating to $V_{xc}(\underset{\sim}{r})$.

### 3.1. Approximate one-body potential in slowly varying electron cloud

Instead of assuming a constant electron density, we shall below have in mind the situation in which $\rho(\underset{\sim}{r})$ varies only slowly in space. Unfortunately, though this affords a useful starting point, it is only realistic for such special cases as the conduction electron density in metallic Na, in which there is ample evidence that the electron-ion interaction is exceptionally weak.

Nevertheless, motivated by the Thomas-Fermi-Dirac theory, let us write the ground-state energy E in the form :

$$E = \int t_r[\rho]d\underset{\sim}{r} + \int \rho V_n d\underset{\sim}{r} + \frac{1}{2}\int \rho V_e d\underset{\sim}{r} + \int \varepsilon_{xc}[\rho]d\underset{\sim}{r} \qquad (3.3)$$

In the T.F.D. theory, the single particle kinetic energy $t_r[\rho]$ is :

$$t_r[\rho] = C_k[\rho(\underset{\sim}{r})]^{5/3}$$

while the exchange and correlation energy density $\varepsilon_{xc}[\rho]$ is approximated by the exchange energy density of a uniform electron gas, used locally at $\underset{\sim}{r}$. As first shown by Dirac (1930) this implies :

$$\varepsilon_{xc}[\rho] = - C_e\, [\rho(\underset{\sim}{r})]^{4/3} \qquad (3.4)$$

where $C_e = \frac{3}{4}\, e^2 (\frac{3}{\pi})^{1/3}$. If we perform the minimization of the energy E with respect to $\rho(\underset{\sim}{r})$, subject to the normalisation requirement

$$\int \rho(\underset{\sim}{r})\, d\underset{\sim}{r} = N \qquad (3.5)$$

with N the given total number of electrons, then we find :

$$\mu = \frac{\delta t_r[\rho]}{\delta\rho} + V_H + \frac{\delta\varepsilon_{xc}[\rho]}{\delta\rho} \qquad (3.6)$$

$\mu$ being the chemical potential, which comes in as the Lagrange multiplier taking account of the normalisation condition (3.5).

Since $t_r[\rho]$ is a single-particle kinetic energy-density, it is clear that eqn 53.6) has the form of a one electron problem with potential (c.f. Kohn and Sham, 1965)

$$V(\underset{\sim}{r}) = V_H(\underset{\sim}{r}) + \frac{\delta\varepsilon_{xc}[\rho]}{\delta\rho} \qquad (3.7)$$

If we use the approximate form (3.4) in eqn (3.7) then we obtain the potential

$$V(\underset{\sim}{r}) = V_H(\underset{\sim}{r}) - \frac{4}{3} C_e[\rho(\underset{\sim}{r})]^{1/3} \qquad (3.8)$$

which is the Dirac-Slater exchange potential approximation to eqn (3.7). In general, we ought to incorporate an approximate account of correlation by utilizing locally the correlation energy density of a uniform electron gas.

Many examples exist of the use of the potential (3.8), sometimes supplemented by approximate inclusion of correlation. We shall briefly discuss below one example, that of the electron density in body centred cubic lithium metal (Perrin, Taylor, March, 1975), the second example considered in section 2.1 above.

### 3.2. One-body potential calculation of electron density in lithium metal

The band theory calculation of Perrin et al used the Korringa Kohn Rostoker (KKR) method. Of most interest to us here is how the potential V(r) was constructed within the Wigner-Seitz cell. Core-core and core-conduction effects were approximated by the use of the semi-empirical Seitz potential for $Li^+$. The conduction-conduction potential was added, as calculated from the basic potential eqn (3.7) this potential being cut off at the muffin tin radius. The input used to compute $V_H$ was the s component of the crystal density generated by the screened ion superposition referred to a little further. The model potential used to calculate the screened ion density took into account non linearities in the conduction electron response to a single ion scattering event (Rasolt, Taylor, 1975-Dagens et al, 1975). The KKR and screened ion superposition s components of the density $\rho(\underset{\sim}{r})$ agree well enough that the potential of Perrin et al can be regarded as almost self-consistent. For $\varepsilon_{xc}[\rho]$, Perrin et al employed the Nozières and Pines (1958) expression to the exchange-correlation energy of the uniform electron assembly. Explicitely, the form is :

$$\frac{\delta\varepsilon_{xc}}{\delta\rho} = -\frac{2}{\pi}\left\{[3\pi^2\rho(\underset{\sim}{r})]^{1/3} - [3\pi^2\bar{\rho}]^{1/3}\right\} - \frac{.031}{3}\ln\frac{\rho(\underset{\sim}{r})}{\bar{\rho}} \qquad (3.9)$$

where $\rho(\underset{\sim}{r})$ and $\bar{\rho}$, the uniform electron density are expressed in atomic units and the energy unit is the Rydberg.

Actually, Perrin et al (1975) calculated the conduction electron density $\rho(\underset{\sim}{r},E)$, giving the number of electrons per unit volume lying below energy E for values of E equal to $1/4E_f$, $1/2E_f$, $3/4E_f$ as well as the electron density

$\rho(\underset{\sim}{r}) \equiv \rho(\underset{\sim}{r}E_f)$, $E_f$ being the Fermi energy. Fig.1 shows the s-term of $\rho(\underset{\sim}{r}E)$, $\rho_o(rE)$, the distance r from the center of the cell being measured in units of the radius of the inscribed sphere,

$r_i$ say, in the Wigner-Seitz cell. Curves A-D are the KKR results for $\rho_c(rE)$, for E = $1/4E_f$, $1/2E_f$, $3/4E_f$, and $E_f$. Curve E represents the superposition of screened ions density, which is seen to be in excellent agreement with the KKR density (curve D)

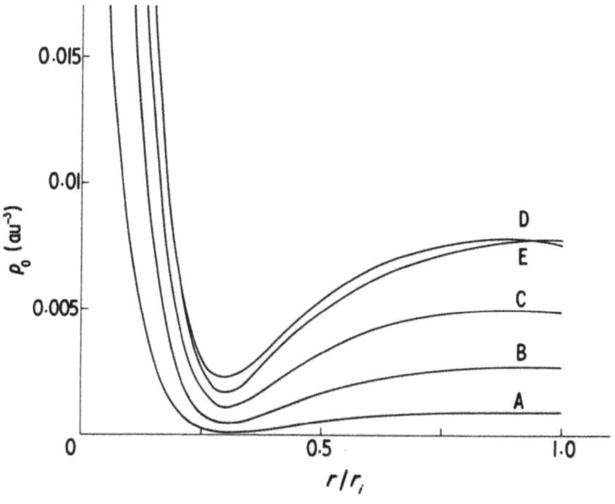


Fig.1. Density $\rho$(rE) of electrons in conduction band of body-centred cubic Lithium below energy E

Curves A to D correspond to $E/E_f$ = 1/4, 1/2, 3/4, 1, $E_f$ being the Fermi energy.

Curve E is the superposition of screened ions densities.

The localized screened ion density used in constructing curve E was obtained by calculating the screening charge which is piled up round a single $Li^+$ ion embedded in a Fermi gas of electrons having the average density appropriate to the conduction band in bcc Li metal.

(a) Comparison with superposition of screened ions.

One center self-consistent non linear calculations, using the one-body potential theory referred to above, have been carried out for for this system by Dagens et al (1975). In this calculation, the wave number dependent dielectric function of the interacting uniform electron assembly was used, in the form given by Geldart and Taylor (1970).

The scattering factor for the valence electrons in metallic Li was also calculated by Perrin et al (1975) from the screened ion density. It turns out that at the first two reciprocal lattice vectors the scattering factor is, in absolute value, larger for the metal conduction electrons than for the 2s atomic wave function. Though Bragg reflection intensities are by now available for a number of pure metals (eg Be,Al,Cu,Cr and Fe), to our knowledge no such data is available for lithium. It would be of interest if such experiments could be carried out.

The results of Brown (1972) on Be show clearly in that metal that directional effects in the charge density are important, and one must not therefore expect that a superposition of spherical screened ions will be an adequate approximation for Be metal, even though it works quite nicely for Li as we have seen.

The very recent work of Léonard (1978) also demonstrates that KKR calculations of the periodic density of the conduction electrons in Al are well represented by a similar screened ion model.

## 4. PHONONS AND LINEAR RESPONSE

Having established a formally exact one-body potential method of calculating the density $\rho(\underset{\sim}{r})$, let us briefly consider how to treat phonons in the Born Oppenheimer and harmonic approximations. Here, we wish to find the density change, $\Delta\rho(\underset{\sim}{r})$ say, as nuclei are displaced from lattice sites $\underset{\sim}{\ell}$ by amounts $\underset{\sim}{u}_\ell$. Suppose the one-body potential change is $\Delta V(\underset{\sim}{r})$. Then, linear response theory allows one to write :

$$\Delta\rho(\underset{\sim}{r}) = \int F(\underset{\sim}{r}\,\underset{\sim}{r}')\,\Delta V(\underset{\sim}{r}')\,d\underset{\sim}{r}' \qquad (4.1)$$

where F is determined by the Bloch wave functions and energies generated by the one-body periodic potential V. Secondly, one may express the potential energy change as :

$$\Delta V(\underset{\sim}{r}) = \Delta V(\underset{\sim}{r})_{\text{electrostatic}} + \Delta V_{xc}(\underset{\sim}{r}) \qquad (4.2)$$

and since $\varepsilon_{xc}[\rho]$ generates $V_{xc}(\underset{\sim}{r})$, it is clear that in a linear theory,

$$\Delta V_{xc}(\underset{\sim}{r}) = \int U(\underset{\sim}{r}\,\underset{\sim}{r}')\,\Delta\rho(\underset{\sim}{r}')\,d\underset{\sim}{r}' \qquad (4.3)$$

U involves, of course, the exchange and correlation interactions in the periodic lattice.

### 4.1. Rigid screened ion model.

At this point, it is helpful to consider $\Delta\rho(\underset{\sim}{r})$ in the rigid screened ion model. One simply writes :

$$\rho(\underset{\sim}{r}) = \sum_{\underset{\sim}{\ell}} \sigma(\underset{\sim}{r}-\underset{\sim}{\ell})_{\text{screened ion}} \qquad (4.4)$$

where $\sigma$ is the localized density discussed above. For the density change, in this model, one simply moves $\underset{\sim}{\ell}$ by $\underset{\sim}{u}_\ell$ and to first order in the displacement, the result is :

$$\Delta\rho(\underset{\sim}{r}) = \sum_{\underset{\sim}{\ell}} \underset{\sim}{u}_\ell . \underset{\sim}{\nabla}\sigma(\underset{\sim}{r}-\underset{\sim}{\ell}) \qquad (4.5)$$

In fact this rigid screened ion model for the phonons has been used by Dagens et al (1975) for Li. The agreement with experiment is quite good. The computed phonon curves were about 5% higher than the measured frequencies and the cross-over between the two (100) branches was correctly reproduced. Thus the rigid screened ion model gives a good description of the lattice dynamics of Li, as well as the spherical component of the charge distribution. (cf Fig. 1)

Jones and March (1970) have shown that the generalisation of the screened rigid ion model resulting from eqns (4.2) and (4.3) can be written

$$\Delta\rho(\underset{\sim}{r}) = \sum_{\underset{\sim}{\ell}} \underset{\sim}{u}_\ell . \underset{\sim}{R}(\underset{\sim}{r}-\underset{\sim}{\ell}) \qquad (4.6)$$

where the localized vector quantity $\underset{\sim}{R}(\underset{\sim}{r})$, determined by F and U defined above, cannot in general be written as the gradient of a scalar density . For all $\underset{\sim}{u}_\ell$ equal, that is simply translation of the crystal, we have :

$$\underset{\sim}{\nabla}\rho(\underset{\sim}{r}) = \sum_{\underset{\sim}{\ell}} \underset{\sim}{R}(\underset{\sim}{r}-\underset{\sim}{\ell}) \qquad (4.7)$$

which is a formally exact vector rigid ion model. Fortunately, for Li, it appears useful to approximate $\underset{\sim}{R}$ as $\underset{\sim}{\nabla}\sigma$. The resulting rigid spherical screened ion model is equivalent to the central pair forces. In other cases, deviations of $\underset{\sim}{R}(\underset{\sim}{r})$ from $\underset{\sim}{\nabla}\sigma(\underset{\sim}{r})$ reflect the need for introducing many-body forces.

For plane waves, the linear response function F in eqn (4.1) can be calculated explicitely as :

$$F(\underset{\sim}{r}\,\underset{\sim}{r}') = -\frac{k_f^3}{2\pi^3}\,\frac{j_1(2k_f|\underset{\sim}{r}-\underset{\sim}{r}'|)}{|\underset{\sim}{r}-\underset{\sim}{r}'|^2} \qquad (4.8)$$

$$j_1(x) = \frac{\sin x - x\cos x}{x^2}$$

where $k_f$ is the radius of the Fermi surface ; that is the Fermi wave number. If this result for F is inserted in eqn (4.1) with a localized potential change $\Delta V$, one is led to a linear screened ion density which at large r varies as :

$$\Delta\rho \simeq A\cos(2k_f r)/r^3 \qquad (4.9)$$

exhibiting the so-called Friedel oscillations having wavelength $\pi/k_f$ ; ie determined by the de Broglie wavelength of electrons at the Fermi surface. $\Delta V$ when calculated self-consistently from $\Delta\rho$ in eqn (4.9) also behaves in precisely the same way as eqn (4.9) at large r. The Hellmann-Feynman theorem, discussed by Professor Hirshfeld in his lectures, then leads to a central pair potential $\Phi(r)$ between screened ions, having the asymptotic form

$$\Phi(r) \quad \text{constant} \frac{\cos 2k_f r}{r^3} \tag{4.10}$$

Such central pair potentials, added up over all pairs of screened ions, need supplementing by a volume dependent but structure independent energy, $E(\Omega)$ say, due to the conduction electrons. Thus, these central pair potentials between screened ions are most suitable for describing nuclear rearrangement at constant volume $\Omega$. The form (4.10) is, it must be stressed, a characteristically metallic interaction.

Very recently Flores et al (1978) have discussed the asymptotic behaviour of the linear response function F as influenced by the shape of the Fermi surface of metals and hence the result (4.9) for the density change. The answer depends in detail on the principal curvatures of the Fermi surface at a point of stationary phase. Anisotropy in $\Delta\rho(\underset{\sim}{r})$ results and it is shown by Flores et al that under certain conditions F can decay as $r^{-1}$ or $r^{-2}$, ie can be longer range than for a spherical Fermi surface (cf eqn (4.8)) along certain specific directions. This will naturally imply non-central force fields for particular Fermi surface topologies.

In concluding this section, it is relevant to remark that an attempt has been made by March and Wilkins (1978) to provide a method of treating elastic X-Ray scattering from solids containing non rigid pseudoatoms but it is too early to assess the utility of such a treatment of dynamic deformability.

## 5. SPIN DENSITY DESCRIPTION BY ONE-BODY POTENTIALS

### 5.1. Hartree-Fock theory of uniform electron assembly

As the most elementary example of the charge density description developed above generalized to treat spin density, consider the generalisation of the Thomas-Fermi-Dirac theory for uniform electrons, decribed by plane waves. If $\rho_\uparrow$ and $\rho_\downarrow$ denote the densities (assumed constant) for the upward and downward spins, the single-particle kinetic energy density of the Thomas-Fermi theory is readily generalized to read :

$$t[\rho_\uparrow,\rho_\downarrow] = \text{constant}[\rho_\uparrow^{5/3} + \rho_\downarrow^{5/3}] \tag{5.1}$$

Similarly the exchange energy density takes the form

$$\varepsilon_x[\rho_\uparrow,\rho_\downarrow] = \text{constant}[\rho_\uparrow^{4/3} + \rho_\downarrow^{4/3}] \tag{5.2}$$

As Bloch was the first to show, by minimizing $t+\varepsilon_x$ with respect to $\rho_\uparrow$ and $\rho_\downarrow$, subject to $\rho_\uparrow+\rho_\downarrow=N/\Omega$ for N electrons in volume $\Omega$, the ferromagnetic state becomes stable at sufficiently low densities, relative to the paramagnetic state. In terms of $\rho = \rho_\uparrow + \rho_\downarrow$ and the exchange constant $c_e$ in eqn (3.4), the condition for ferromagnetism in the ground state in this model is found to be :

$$c_e \, \rho^{1/3} / E_f \; > \; 4\,[2^{1/3} + 1]\,/5 \tag{5.3}$$

where $E_f$ is the Fermi energy.

This example is merely illustrative, and we caution the reader that the Hartree-Fock model overestimates the tendency to ferromagnetism, because in this theory there are Fermi statistical correlations between all parallel spin electrons but no correlation between antiparallel spins. In fact, in the uniform electron assembly the introduction of correlation stabilizes the antiferromagnetic phase relative to the paramagnetic phase at sufficiently low densities, and ground state ferromagnetism most probably never occurs for any density in this system.

### 5.2. Spin-dependent one body potentials

Below, a sketch of the generalization of one-body potential theory to describe spin densities in inhomogeneous electron clouds will be given, leading to spin dependent potentials (Stoddart and March, 1971). Suppose a spin dependent external potential energy $V^\sigma_{ext}$ is switched on, $\sigma$ being either $\uparrow$ or $\downarrow$. The ground state energy can now be written formally as

$$E[\rho_\uparrow(\underset{\sim}{r}),\rho_\downarrow(\underset{\sim}{r})] \equiv E[\rho(\underset{\sim}{r}),m(\underset{\sim}{r})] \tag{5.4}$$

where m(r) is the resultant spin density $\rho_\uparrow - \rho_\downarrow$. This can be written alittle more explicitely as

$$E = G[\rho(\underset{\sim}{r}),m(\underset{\sim}{r})] + \frac{1}{2}\int \rho V_e d\underset{\sim}{r} + \int \rho V_{ext}(\underset{\sim}{r})d\underset{\sim}{r} + E_{field} \tag{5.5}$$

where G subsumes kinetic, exchange and correlation energy densities as a function of $\rho$ and m. $E_{field}$ has been added to represent the interaction with a magnetic field.

Performing the minimization of E with respect to the spin densities, as a generalization of the charge density argument of section 3 leads to two spin dependent potentials :

$$V_\sigma(\underset{\sim}{r}) = V_{ext}(\underset{\sim}{r}) + \int \frac{\rho(\underset{\sim}{r}')}{|\underset{\sim}{r} - \underset{\sim}{r}'|} d\underset{\sim}{r}' + V^\sigma_{xc}(\underset{\sim}{r}) \tag{5.6}$$

Again, local density approximations are usually made for $V_{xc}(\underset{\sim}{r})$ and Professors Freeman and Ellis give examples of explicit calculation of spin densities from such one-body potentials in their lectures. Reference should also be made here to the work of von Barth and Hedin (1972), Rajagopal and Callaway (1973) and Langlinais and Callaway (1972).

## 6. DIRECTIONAL BONDING IN SOLIDS

As the final topic of these lectures, directional bonding in solids will be treated. As a suitable starting point, chemical arguments will be employed to discuss the anisotropic charge density in Si. Following that, we shall return to the discussion of anisotropic metals like Be, Fe and Cr and examine in these metallic crystals how chemical bonding may be related to one-body potential theory, ie to energy bands.

For solid Si, the electron density will be considered both in the crystalline and in the amorphous phases. In the latter case, the long range order is lost and $\rho(\underset{\sim}{r})$ cannot presently be calculated by energy band theory because Bloch's theorem no longer applies and in particular the wave vector k no longer constitutes a useful set of quantum numbers. We shall describe below the calculations of Stenhouse et al (1977) using the most elementary form of the classical bonding approach.

### 6.1. Bond Electron Density in Si from LCAO Method

The usual $sp^3$ atomic hybrid orbitals were employed by Stenhouse et al. The density associated with one Si-Si bond was then calculated using the linear combination of atomic orbitals method. The contours of constant electron density thereby obtained are shown in Figure 2.

Using this density in the superposition from (2.7) the X-Ray scattering amplitude f at the Bragg reflections for crystalline silicon was obtained and the results are shown by the circles in Fig.3, the crosses denoting the experimental values. The agreement between theory and experiment is good, at least comparable with the best band theory calculation. In this latter connection, we should mention the discussion of Bennett and Inkson (1977) of exchange and correlation in crystalline silicon. Such an approach ought to be incorporated, via eqn (3.7) in subsequent band calculations in Si.

We note that the so called 'forbidden' reflection (222), which would be absent if we added up spherical neutral atom densities, though of small amplitude, is very well given by this chemical bonding approach.

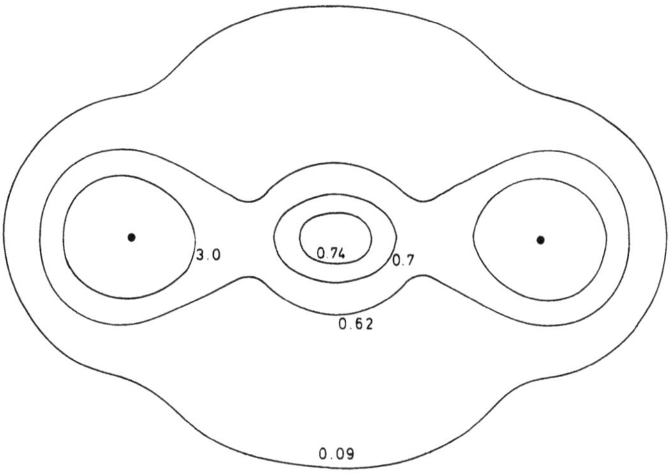


Fig.2. Density contours in Si-Si bond using $sp^3$ hybrid atomic orbitals and LCAO method. Note that in addition to the two valence electrons per bond, one quarter of the core charge density has been assigned to this basic Si-Si scattering unit.

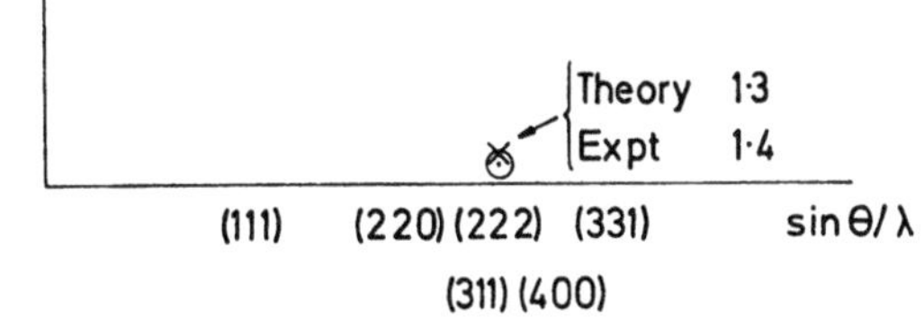

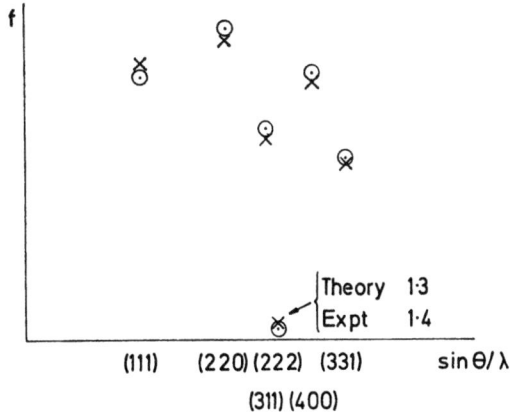

Fig. 3. X-Ray scattering amplitude f at Bragg reflections for crystalline silicon. Crosses show the experimental results while circles are results of the localized bond density model.

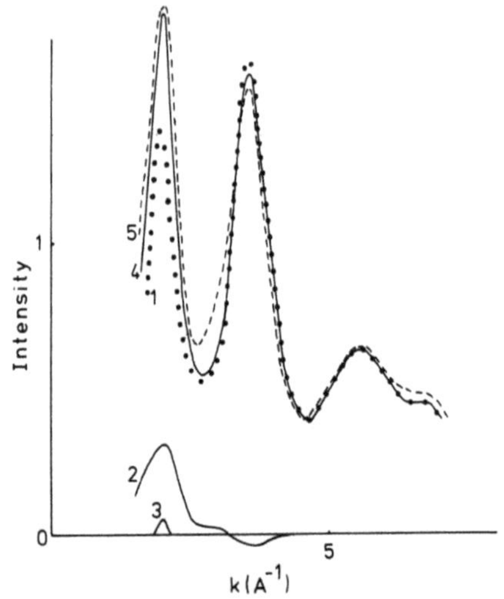


Fig 4. X-Ray diffraction from amorphous silicon.
curve 1. Contribution from Si-Si correlations
curve 2. Contributions from Si-bond centre correlations
curve 3. Contribution from centre-bond centre correlations
curve 4. Total intensity from chemical bond method plus continuous random network description of structure
curve 5. Experimental intensity as determined by Richter and Breitling (1958)

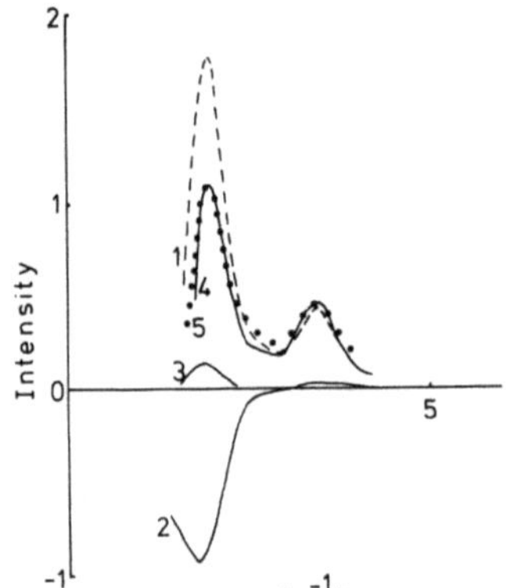


Fig. 5. Electron scattering intensity for amorphous silicon. Labelling (1)-(4) as in Fig.4. Experimental points labelled 5 are from the work of Moss and Graczyk (1969)

### 6.2. Scattering from Amorphous Silicon

As a further test of the electron density obtained from this chemical bond approach, Stenhouse et al calculated the X-Ray and electron scattering from amorphous silicon. Of course, to do so requires not only this knowledge of the electron density but also a suitable description of the structure. Suffice it to say that the so called random network model is the best available description of the structure of such amorphous covalently bonded semiconductors. Stenhouse et al thereby obtained the structure factor S(k) defined in eqn (2.4) for the Si nuclei. In order to allow for chemical bonding, described via the concept of bond charge centred at the middle of the chemical bond, Stenhouse et al also calculated from the random network model the bond-centre-bond-centre structure factor and the bond-centre-nuclei structure factor.

Using these partial structure factors and the above chemical bonding approach to the electron density the X-Ray diffraction results of the theory are compared with experiment in Figure 4. The agreement is excellent, the bonding charge being required to obtain quantitative agreement with experiment regarding the height of the first two peaks. An equally good account of the electron diffraction is given as can be seen from Figure 5.

### 6.3. Relation between bonding and energy bands

Finally, let us return to the anisotropic electron density in crystalline metals. Energy band theory, based on the periodic one-body potential of the form (3.7) can be employed. Below, this Bloch wave description will be recast into a localized form, using Wannier functions. For metals, these specific, localized functions appear to afford one route to link energy band calculations with chemical bonding ideas, as emphasized by Matthai et al (1978).

Below, their argument will be developed for the very simplest case of a non-degenerate band with a spherical Fermi surface. However, to be completely consistent, these restrictions will have to be relaxed eventually (eg the linear response arguments in section 4 showed that Fermi surface shape can introduce anisotropy into the charge density).

For the crystalline density in terms of Bloch waves $\Psi_{\underset{\sim}{k}}$, one has

$$\rho(\underset{\sim}{r}) = \sum_{\substack{\text{occupied}\\ \text{states}}} \Psi^{*}_{\underset{\sim}{k}}(\underset{\sim}{r})\Psi_{\underset{\sim}{k}}(\underset{\sim}{r}) \equiv \sum_{\substack{\text{occupied}\\ \text{states}}} \rho_{\underset{\sim}{k}}(\underset{\sim}{r}) \qquad (6.1)$$

The localized Wannier functions $a(\underset{\sim}{r})$ are now introduced through :

$$\Psi_{\underset{\sim}{k}}(\underset{\sim}{r}) = \sum_{j} \exp(i\underset{\sim}{k}.\underset{\sim}{R}_j)\ a(\underset{\sim}{r}-\underset{\sim}{R}_j) \qquad (6.2)$$

where the vectors $\underset{\sim}{R}_i$ define the crystal lattice sites. Then $\rho_{\underset{\sim}{k}}(\underset{\sim}{r})$ in eqn (6.1) is given by :

$$\rho_{\underset{\sim}{k}}(\underset{\sim}{r}) = \sum_{m\,n} \exp[i\underset{\sim}{k}.(\underset{\sim}{R}_m-\underset{\sim}{R}_n)]\ a(\underset{\sim}{r}-\underset{\sim}{R}_m)a^*(\underset{\sim}{r}-\underset{\sim}{R}_n) \qquad (6.3)$$

For the simple example of a spherical Fermi surface of radius $k_f$, the sum over k can be carried out to obtain for such a model of a metal :

$$\rho(\underset{\sim}{r}) = \sum_{|k|<k_f} \rho_{\underset{\sim}{k}}(\underset{\sim}{r}) = \sum_{m\,n} \frac{3j_1(k_f|\underset{\sim}{R}_m-\underset{\sim}{R}_n|)}{k_f|\underset{\sim}{R}_m-\underset{\sim}{R}_n|}\ a(\underset{\sim}{r}-\underset{\sim}{R}_m)a^*(\underset{\sim}{r}-\underset{\sim}{R}_n) \qquad (6.4)$$

where as usual $j_1(x) = (\sin x - x\cos x)/x^2$. The 'overlap charge density' determined from eqn (6.4) by $a(\underset{\sim}{r}-\underset{\sim}{R}_m)a(\underset{\sim}{r}-\underset{\sim}{R}_n)$ will, in fact, be largest for hybrid bands, where the Wannier functions can decay as an inverse power of distance, in contrast to the exponential decay for a simple band.

### 6.4. Scattering factor in terms of Wannier functions

Following Matthai et al (1978), one next calculates the Fourier transform $f(\underset{\sim}{k})$ of $\rho(\underset{\sim}{r})$ given by the model form (6.4). Then one sees that the form factor of the bonding charge involves the quantity

$$I_R(\underset{\sim}{k}) = \int a(\underset{\sim}{r}-\underset{\sim}{R}_m)a^*(\underset{\sim}{r}-\underset{\sim}{R}_n)\ \exp(i\underset{\sim}{k}.\underset{\sim}{r})\ d\underset{\sim}{r} \qquad (6.5)$$

where R is the bond length $|\underset{\sim}{R}_m-\underset{\sim}{R}_n|$. Defining

$$f_R(\underset{\sim}{k}) = I_R(\underset{\sim}{k})\ 3j_1(k_fR)\ /\ k_fR \qquad (6.6)$$

as the form factor of a charge between nuclei, distance R apart, one can separate $f(\underset{\sim}{k})$ into a sum over the crystal lattice sites $\underset{\sim}{R}_m$, plus a sum over stes of superlattices $\underset{\sim}{L}_n$ corresponding respectively to near-neighbour bond centres, next near neighbour bond centres, etc, namely

$$f(\underset{\sim}{k}) = f_o(k)\ \sum_m \exp(-i\underset{\sim}{k}.\underset{\sim}{R}_m) + \sum_{\substack{\text{cell}\\ \text{superlattices}}} f_n(k)\ \exp(-i\underset{\sim}{k}.\underset{\sim}{L}_n) \qquad (6.7)$$

If the spherical rigid ion model (4.4) holds, then only $\sum_m \exp(-i\underset{\sim}{k}.\underset{\sim}{R}_m)$ enters $f(\underset{\sim}{k})$, but otherwise structure factors of superlattices of sites on which the bonding charges are centred also contribute.

As examples, we shall consider briefly bcc Fe and Cr. On rather general grounds, one expects near neighbour and next near neighbour 'bonds' to be important. The scattering factor for which the coefficient of $f_1(k)$ is zero, 1 denoting the near neighbour bond centre lattice, becomes $(f_o-f_2)$. Hence, until Wannier functions become available for these transition metals, one might approximately identify $f_2$ at these reflections with the difference $\Delta F$ say, between the Hartree-Fock free atom and experimental form factors. Though $\Delta F$ has large error bars, these reflections plotted against $\sin\theta/\lambda$ lie on smooth curves for both Fe and Cr. This is encouraging but final results must await Wannier functions calculations.

Brown's (1972) results for Be may be analyzed along somewhat similar lines but the reader should refer to Matthai et al for further details.

Various other examples of the relation between chemical descriptions and energy band theory may be found, especially in the work of Freeman and Zunger, some of which is discussed in Professor Freeman's lecture.

## 7. SUMMARY

The main points covered in these lectures can be summarized as follows :

(i) The charge density $\rho(\underset{\sim}{r})$ can be calculated, in principle exactly, from the one-boby potential $V(\underset{\sim}{r})$ in eqn (3.7). In practice, local density approximations, plus non local corrections are employed. For Be metal, for example, the non local correction to $V_{xc}(\underset{\sim}{r})$ are at most 30% of the local $V_{xc}(\underset{\sim}{r})$ and usually much less, within the unit cell of this metal.

(ii) Splitting up the delocalized density $\rho(\underset{\sim}{r})$ into localized distributions is important for :

(a) Calculating phonon properties

(b) Relating to (a), describing interionic and interatomic force fields

(c) Treatment of disorder (eg amorphous phases)

(iii) In metals, with directional bonding, it is shown that Wannier decomposition is a way to bring together chemical bonding and energy band theory. In the model worked out by Matthai et al (1978), the total bond charge is zero however. This may eventually require the relaxation of the orthogonality of the localized orbitals.

Acknowledgement

Part of the research work reported here was made possible through the contractual support of the European Research Office of the U.S. Army.

REFERENCES

Bennett M., Inkson J.C., 1977, J. Phys. C, 10, 987
Brown P.J., 1972, Phil. Mag., 26,1377
Castmann B., Pettersson G., Vallin J., 1971, Physica Scripta, 3, 35
Dagens L., Rasolt M., Taylor R., 1975, Phys. Rev., B11, 2726
Dirac P.A.M.,1930, Proc. Camb. Phil. Soc., 26, 376
Egelstaff P.A., March N.H., McGill N.C., 1974, Can. J. Phys., 52, 1651
Fermi E., 1928, Z. Phys., 48, 73
Flores F., March N.H., Ohmura Y., Stoneham A.M., 1978, to be published
Geldart D.J.W., Taylor R., 1970, Can. J. Phys., 48, 167
Hohenberg P.C., Kohn W., 1964, Phys. Rev.,B136, 864
Jones W., March N.H., 1970, Proc. Roy. Soc. A317, 359 ; see also Theoretical Solid State Physics, Wiley-Interscience-London, 1973
Kohn W., Sham L.J., 1965, Phys. Rev., A140, 1133
Langlinais J., Callaway J., 1972, Phys. Rev.,B124
Léonard P., 1978, J. Phys. F, 8, 467
March N.H., Wilkins S.W., 1978, Acta Cryst A34, 19
Matthai C.C., Grout P.J., March N.H., 1978, to be published
Moss S.C., Graczyk J.F., 1969, Phys. Rev. Lett., 23, 1167
Musgrave M.J.P., Pople J.A., 1962, Proc. Roy. Soc. A268, 474
Nozières P., Pines D., 1958, Phys. Rev., 111, 442
Perrin, R.C., Taylor R., March N.H., 1975, J. Phys. F, 5, 1490
Rajagopal A.K., Callaway J., 1973, Phys. Rev., B7, 1912
Rasolt M., Taylor R., 1975, Phys. Rev., B11, 2717
Richter H., Breitling G., 1958, Z. Naturf., a13, 988
Stenhouse B., Grout P.J., March N.H., Wenzel J., 1977, Phil. Mag, 36, 129
Stoddart J.C., March N.H., 1971, Annals of Physics, 64, 174
Thomas L.H., 1926, Proc. Camb. Phil. Soc., 23, 542
von Barth U., Hedin L., 1972, J. Phys. C, 5, 1629

# LOCAL DENSITY APPROACH TO BULK AND SURFACE CHARGE DENSITIES*

A. J. Freeman
Physics Department
Northwestern University
Evanston, IL 60201

## I. INTRODUCTION

As is clear from some of the other lectures given at this Institute, the current popularity of energy band theory stems from its successful application to the study of increasingly diverse problems in solid state physics. Recent new sophisticated experiments on both traditional materials and those having complex crystallographic structures have demanded, however, not only theoretical descriptions of eigenvalue phenomena but also detailed and precise wave functions with which to determine not only charge and spin densities but also the expectation values of different observable operators. Such a demanding test of the predictions of one-electron theory has the additional virtue of permitting, by their comparison with experiment, accurate determinations of the relative magnitude and importance of many-body effects in real solids. Thus, as emphasized by Norman March, there has developed considerable interest in applying the Hohenberg-Kohn-Sham (1,2) local density functional (LDF) formalism, and its recent extension as a local spin density functional (LSDF) formalism (3), to the investigation of various ground state properties of solids despite the usual difficulties of solving the associated one-particle equation characterized by a multi-center non-spherical potential.

Applications of the LDF formalism to atoms (4,5) and molecules (6) have yielded encouraging results. Similar applications for solids are complicated by (i) the need to consider both the short range and the long range multicenter crystal potential having non-spherical components, (ii) the difficulties in obtaining full self-consistency in a periodic system and (iii) the need to provide a basis set with sufficient variational flexibility. Hence, theoretical studies of ground state electronic properties of solids in the LDF formalism have been mainly limited to muffin-tin models for the potential, non-self-consistent schemes, treatments of simplified jellium models or spherical cellular schemes.

Zunger and I recently proposed (7,8) a general self-consistent method for solving the LDF formalism one-particle equation for realistic solids using a numerical basis set LCAO (Linear Combination of Atomic Orbitals) expansion and retaining all non-spherical parts of the crystal potential. We have demonstrated a rapid convergence of the self-consistent (SC) cycle when the treatment of the full crystal charge density is suitably apportioned between real-space and Fourier transformed reciprocal space parts and have indicated the large degree of variational flexibility offered by a non-linearly optimized (exact) numerical atomic-like basis set. We have shown that all multi-center interactions as well as the non-constant parts of the crystal potential are efficiently treated by a three dimensional Diophantine integration scheme.

The method has been applied to a number of different materials (9) and shown to describe successfully their electronic structure. More recently, C. S. Wang and I have extended this treatment to thin films in order to study surface properties. This lecture describes some results of these studies.

## II. METHOD

As described by Norman March, the Hohenberg-Kohn-Sham (1,2) local density formalism is based on the fundamental theorem that the ground state properties of an inhomogeneous interacting electron system are functionals of the electron density, $\rho(\underset{\sim}{r})$, and that in the presence of an external potential, $V_{ext}(r)$, the total ground state energy in its lowest variational state can be written as:

$$E[\rho(\underset{\sim}{r})] = \int V_{ext}(\underset{\sim}{r})\rho(\underset{\sim}{r})d\underset{\sim}{r} + G[\rho(\underset{\sim}{r})] \qquad (1)$$

where $G[\rho(\underset{\sim}{r})]$ is a universal functional of $\rho(\underset{\sim}{r})$ and is <u>independent of the external potential</u> $V_{ext}(\underset{\sim}{r})$. This theorem forms the basis of our approach to the electronic structure problem in that it provides an effective one-particle equation relating self-consistently the ground state wavefunctions to the energy functionals (i.e., potential) of the electronic system. Identifying the external potential for a polyatomic system as the electron-nuclear and internuclear interactions and varying $E[\rho(\underset{\sim}{r})]$ with respect to $\rho(\underset{\sim}{r})$, one obtains an effective one-particle equation of the form:

$$\{-\frac{1}{2}\nabla^2 + \sum_m \frac{Z_m}{|\underset{\sim}{R}_m-\underset{\sim}{r}|} + \int \frac{\rho(\underset{\sim}{r}')}{|\underset{\sim}{r}-\underset{\sim}{r}'|} d\underset{\sim}{r}' + \frac{\delta E_{xc}[\rho(\underset{\sim}{r})]}{\delta\rho(\underset{\sim}{r})}\}\psi_j(\underset{\sim}{r})=\epsilon_j\psi_j(\underset{\sim}{r}) \quad (2)$$

Here $Z_m$ denotes the nuclear charge of the particle at site $\underset{\sim}{R}_m$ and $E_{xc}[\rho(\underset{\sim}{r})]$ the total exchange and correlation energy of the interacting (inhomogeneous) electron system (square brackets are used to denote functional dependence). The eigenfunctions $\psi_j(\underset{\sim}{r})$ are simply related to the total ground state charge density of the $\sigma_{oc}$ occupied one-particle states by:

$$\rho(\underset{\sim}{r}) = \sum_{j=1}^{\sigma_{oc}} \psi_j^*(\underset{\sim}{r})\psi_j(\underset{\sim}{r}) \quad (3)$$

which, in turn, determines self-consistently the local density functional in Eq. (2).

No satisfactory formulation of $E_{xc}[\rho(\underset{\sim}{r})]$ has been obtained so far for a general $\rho(\underset{\sim}{r})$. In the limit of slowly varying density, gradient expansions of $E_{xc}[\rho(\underset{\sim}{r})]$ have been suggested. (2),(10) Although there seem to be no compelling evidence for the possible suitability of such expansions to realistic models of polyatomic systems, there still seems to be some interest in applying the LDF formalism with the presently available first term expansion of $E_{xc}[\rho(\underset{\sim}{r})]$ as a first step towards a more complete electronic structure theory based on accurate local density functionals. Note that the LDF formalism in the form described above makes no claim on the physical significance of the eigenvalues $\epsilon_j$ in Eq. (2).

Retaining only the non-gradient terms in the expansion of $E_{xc}[\rho(\underset{\sim}{r})]$, the exchange and correlation potential becomes:

$$\frac{\delta E_{xc}[\rho(\underset{\sim}{r})]}{\delta\rho(\underset{\sim}{r})} \cong F_{ex}[\rho(\underset{\sim}{r})] + F_{corr}[\rho(\underset{\sim}{r})] \quad (4)$$

where the exchange potential has the well known form:

$$F_{ex}[\rho(\underset{\sim}{r})] = \frac{4}{3}\epsilon_x[\rho(\underset{\sim}{r})] \equiv -(\frac{3}{\pi}(\rho(\underset{\sim}{r}))^{1/3} \quad (5)$$

The correlation energy of a uniform electron gas with local density $\rho(\underset{\sim}{r})$ has been calculated by many authors using different techniques. The agreement between their most recent results lies within 5-8 mRyd in the metallic density range. We use the results of Singwi et al. (11) fitted to a convenient analytical form: (12)

$$F_{corr}[\rho(\underset{\sim}{r})] \equiv A\ell n(1 + B\ \rho^{1/3}(\underset{\sim}{r})) \tag{6}$$

where A=0.0899 and B=33.8518, in a.u.

Our work has focused on obtaining fully self-consistent solutions of the one-particle equations in a periodic solid within the local density functional formalism without including approximations not inherent in LDF. The method is based on systematic extensions of non-self-consistent real space techniques of Ellis, Painter and collaborators and the self-consistent reciprocal space methodologies of Chaney, Lin, Lafon and co-workers. Thus, it is designed and developed to incorporate special features with which to overcome difficulties encountered by other methods. Specifically, the method combines a discrete variational treatment of all potential terms (Coulomb, exchange, and correlation) arising from the superposition of spherical atomic-like overlapping charge-densities, with a rapidly convergent three-dimensional Fourier series representation of all the multi-center potential terms that are not expressible by a superposition model.

The basis set consists of the exact numerical valence orbitals obtained from a direct solution of the local density atomic one-particle equations. To obtain increased variational freedom, this basis set is then augmented by virtual (numerical) atomic orbitals, charge-transfer (ion pair) orbitals, and 'free' Slater one-site functions. The initial crystal potential consists of a non-muffin-tin superposition potential, including non-gradient free-electron correlation terms calculated beyond the Random-Phase-Approximation. The Hamiltonian matrix elements between Bloch states are calculated by the three-dimensional Diophantine integration scheme of Haselgrove (13), and Ellis and Painter (14), thereby avoiding the usual multi-center integrations encountered in the LCAO tight-binding formalism.

Self-consistency is obtained in two stages: in the first stage ('charge and configuration self-consistency'), the atomic superposition potential and the corresponding numerical basis orbitals are modified simultaneously and non-linearly by varying (iteratively) the atomic occupation numbers (on the basis of the computed Brillouin-zone averaged band population) so as to minimize the deviation, $\Delta\rho(\underset{\sim}{r})$, between the band charge density and the superposition charge density. This step produces the 'best' atomic configuration (for the employed numerical basis orbitals) within the superposition model for the crystal charge density and tends to remove all the sharp 'localized' features in the function $\Delta\rho(\underset{\sim}{r})$ by allowing for intra-atomic charge redistribution to take place. Having obtained a low-amplitude smooth function $\Delta\rho(\underset{\sim}{r})$ that contains zero charge, we proceed in the second stage of self-consistency to solve the three-dimensional multi-center Poisson equation associated with $\Delta\rho(\underset{\sim}{r})$ through a Fourier series representation of $\Delta\rho(\underset{\sim}{r})$. The solution of the band problem is repeated until the changes in the Fourier coefficients of $\Delta\rho(\underset{\sim}{r})$ in successive iterations are lower than a prescribed tolerance. The calculated observables include the total crystal ground state energy, equilibrium lattice constants, electronic pressure, X-ray scattering factors, and directional Compton profile in addition to the one-electron band structure.

## III. ILLUSTRATIVE EXAMPLES - BULK PROPERTIES

We here illustrate some aspects of applications of the LDF method to the electronic structure of real systems.

### A. Diamond

1. Charge Densities and X-ray Scattering Factors. Diamond has been long considered as a prototype for covalently bonded insulators (15) and a great deal of experimental work has been done on its ground state properties, including cohesive energy (16), lattice constant studies, (17) X-ray scattering factors (18,19), charge density(18) and directional Compton profile (20,21). In addition, theoretical studies on its ground state properties within the Restricted Hartree Fock (RHF) model are available(22-25) so comparison with the predictions of the LDF formalism is possible.

In an extensive study, Zunger and I(8) have presented detailed results for the X-ray scattering factors, charge density, directional Compton profile, total energy and equilibrium lattice constant. In order to examine the effects of exchange and correlation on the ground state charge density in diamond we performed three fully self-consistent calculations; the first employed only the electrostatic electron-electron and electron-nuclear potential in the one-particle equation ("electrostatic model"), the second incorporated also the local exchange ("exchange model"), while in the third calculation, the correlation potential was also considered ("exchange and correlation model"). All three calculations used an extended numerical set (1s,2s,2p,3s and 3p orbitals per carbon) and all lattice sums were performed to convergence. The lattice constant was fixed at 6.740 au.

The main conclusions to be drawn from these comparisons are: (i) The exchange charge density alters the electrostatic charge density by as much as 10-40% in the region around the center of the

bond while the correlation charge density is responsible for only a 0.1-0.3% change. The exchange charge is about two orders of magnitude larger than the correlation charge in most of the bond region.
(ii) Both the exchange and correlation charge densities show substantial $\underset{\sim}{k}$-space dispersion. The relative changes in various $\underset{\sim}{k}$-space contributions are larger than the corresponding variations in the contributions to the total valence density. The directional anisotropy of the exchange and correlation charges in real space is quite small.
(iii) The overall effect of both exchange and correlation on the charge density is to enhance the charge build-up around the center of the bond and to substantially increase the charge localization on the atoms, at the expense of deleting some charge density from a doughnut-shaped region at about 0.2-0.4 a.u. from each atom. This observed effect in the solid compared to the free atom indicates an additional charge localization mechanism in the solid near the cores due to the enhancement of the overlap of wave functions from neighbouring sites. It thus seems that the charge density changes introduced by exchange and correlation, both close to the nuclei and around the bond center, act to increase the stability of the solid relative to the non-interacting atoms.

Table I shows the calculated X-ray scattering factors in diamond at the three levels of local-density approximations, together with the experimental results (19) and the canonical Hartree-Fock results of Euwema *et.al.* obtained using an s/p Gaussian basis set (23,25). It is apparent that exchange acts to increase the low-angle scattering factors quite dramatically, reflecting the increased localization of charge in the interatomic region, with the correlation effect being much smaller. In particular, the calculated (222) forbidden reflection (forbidden in the approximation in which the charge density is given as a superposition of spherically symmetric, possibly overlapping, atomic densities, last column in Table I) is increased by about a factor of 2 upon introducing exchange in the potential. The value of this scattering factor turned out to be particularly sensitive to the details of the self-consistency maintained in the calculation and to the quality of the basis set (e.g. a minimal basis set yielded a value of 0.082 while a non-self-consistent extended basis set yielded a value of 0.089). We note from Table I that a band model that allows for wave-function overlap produces markedly improved results over a spherical superposition model (last column).

The general agreement between our calculated results and experiment is reasonable. It is apparent from the comparison of the atomic and crystal values that, as expected, the scattering factors constitute a sensitive test for the details of the calculated charge density only for the first few reflections to which the density in the outer regions of the cell contribute. Similar conclusions can be drawn from comparing our exchange-model results with those obtained in the literature (26-30) by a number of approximations to

| $hk\ell$ | Electrostatic Model | Exchange Model | Exchange + Correlation Model | EXP[(1)] | Hartree Fock[(3)] | Superposition Model |
|---|---|---|---|---|---|---|
| 111 | 3.062 | 3.273 | 3.281 | 3.32 | 3.29 | 3.005 |
| 220 | 1.936 | 1.992 | 1.995 | 1.98 | 1.93 | 1.964 |
| 311 | 1.656 | 1.720 | 1.692 | 1.66 | 1.69 | 1.760 |
| 222 | 0.066 | 0.137 | 0.139 | 0.144[(2)] | 0.08 | 0.0 |
| 400 | 1.470 | 1.494 | 1.493 | 1.48 | 1.57 | 1.585 |
| 331 | 1.625 | 1.600 | 1.605 | 1.58 | 1.55 | 1.519 |
| 422 | 1.411 | 1.423 | 1.408 | 1.42 | 1.42 | 1.432 |
| 511 | 1.347 | 1.385 | 1.392 | 1.42 | ---- | 1.387 |
| 333 | 1.346 | 1.381 | 1.392 | 1.42 | ---- | 1.387 |

TABLE: I: Calculated and experimental X-ray scattering factors for diamond

$$f(\underset{\sim}{K}_s) = \frac{1}{\Omega}\int (i\underset{\sim}{K}_s\underset{\sim}{r})\rho_{cry}(\underset{\sim}{r})d\underset{\sim}{r} \quad . \quad f(0,0,0) \text{ equals } 6.0.$$

The 'Superposition Model' results are calculated in the exchange model assuming the $1s^2 2s^2 2p^2$ configuration for the atoms and a charge density given in Eq.(3).

(1) Ref. 18.

(2) Ref. 19.

(3) Ref. 22,25.

| hkℓ | SCOPW(1) | Pseudopotential t-matrix(2) | Pseudopotential OPW(3) | Equivalent Orbitals(4) | Present Study |
|---|---|---|---|---|---|
| 111 | 3.23 | 3.21 | 3.32 | 3.31 | 3.27 |
| 220 | 1.92 | 1.91 | 1.97 | 1.94 | 1.99 |
| 311 | 1.64 | 1.58 | 1.66 | 1.69 | 1.72 |
| 222 | 0.12 | 0.13 | 0.15 | 0.064 | 0.137 |
| 400 | 1.52 | 1.51 | 1.48 | 1.57 | 1.49 |
| 331 | 1.52 | ---- | ---- | 1.54 | 1.60 |
| 422 | 1.40 | ---- | ---- | ---- | 1.42 |
| 511 | 1.35 | ---- | ---- | ---- | 1.38 |
| 333 | 1.32 | ---- | ---- | ---- | 1.38 |

TABLE II: Comparison of several model calculations of the X-ray scattering factors in diamond.

(1) Self-consistent OPW calculation, Ref. (26) for an exchange coefficient of 2/3.
(2) Ref. 27.
(3) Ref. 28-29.
(4) Ref. 30.

the local exchange problem (Table II). Although these calculations employ various independent approximations, the results vary only in the first few reflections.

2. Anisotropy in Compton Profiles. Although the LDF formalism makes no definite predictions on observables that depend on the individual eigenfunctions of the one-particle equation solved here (since they are not actual one-particle states), its resemblance to an 'effective' Hartree-Fock type equation makes it interesting to examine the quality of these wave functions as 'pseudo'-one-particle states. Such a test is naturally provided by comparing the calculated directional Compton profile with experiment.

We give in Fig. 1 our calculated Compton profile differences for the [100]-[110] and [100]-[111] directions together with the experimental results and those yielded by the restricted HF model of Wepfer et.al. (24). Table III gives our calculated results for the [100] direction together with the experimental results of Reed and Eisenberger (20) and Weiss and Phillips (21) and for comparison, the calculated results of Seth (31) in the non-SC local-exchange model (with exchange coefficient of 0.70) and the restricted HF results of Wepfer et al. (24).

The main conclusions to be drawn from this comparison are:
(i) While the RHF results for the profile lack sufficient high momentum components (and hence produce too high a profile at low momentum transfer), our model predicts a slight excess of high momentum components. The non-self-consistent exchange model (31) tends to produce a profile that is even higher than the RHF profile at low q . This result is in line with previous calculations on atomic Compton profiles using the exchange model (32,33) and with our results for a non-SC superposition model. We thus conclude that both the inclusion of correlation and the iteration towards self-consistency tend to redistribute the charge so as to add some high momentum components to it.
(ii) While the RHF results agree with both experiment and with our results for the anisotropy along [100]-[110] (Fig. 1b), the RHF directional profile for [100]-[111] (Fig. 1a) is considerably more anisotropic than experiment and our data. We found that upon iterating our results to self-consistency (varying our basis set non-linearly so as to allow more variational participation of the virtual 3s and 3p atomic orbitals into the valence band) the calculated anisotropy decreased. It would thus seem that by allowing the formerly virtual states (having long-range tails) to participate in the bonding manifold, the localization of the valence band states decreases and a broader $\Delta J(\underset{\sim}{q})$ is produced (along with an increase in the binding energy). It is possible that incorporation of such 'plane wave' character would also diminish the anistropy in the RHF profile.

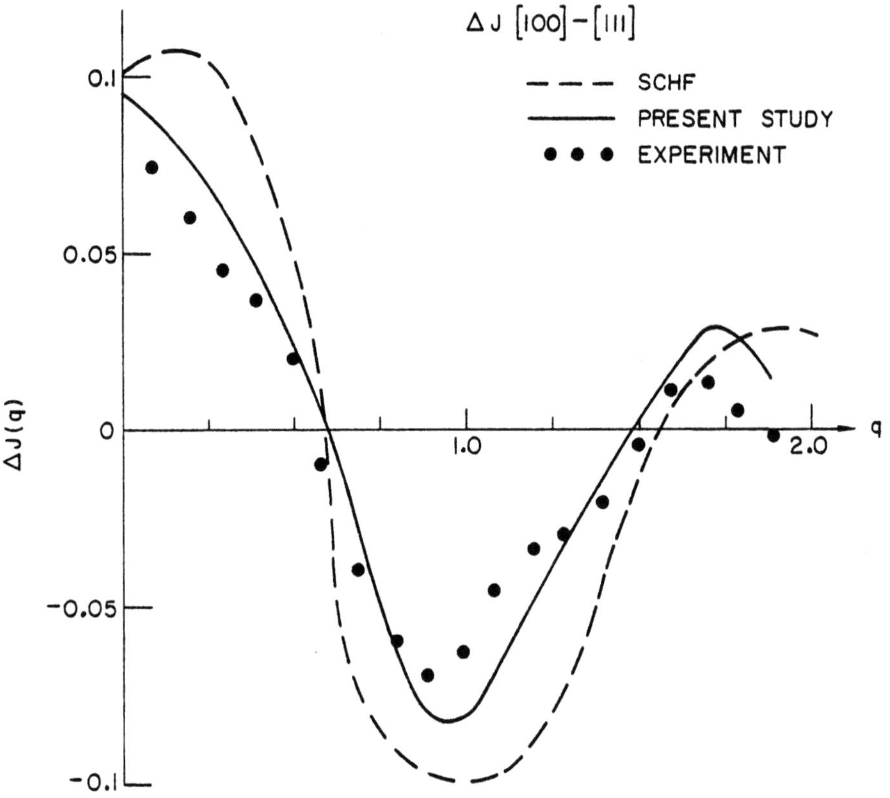


Fig. 1a: Directional Compton profile in diamond as obtained from the self-consistent exchange and correlation model (8): ΔJ[100]-[111].

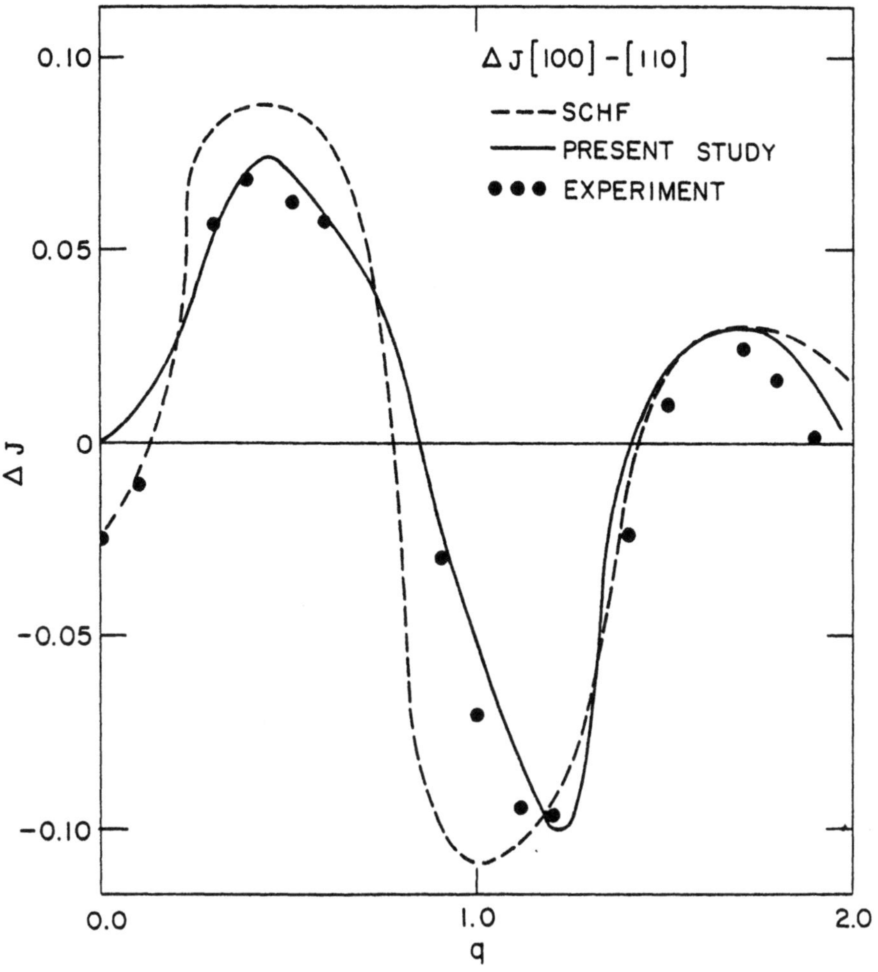


Fig. 1b: Directional Compton profile in diamond as obtained from the self-consistent exchange and correlation model (8): ΔJ[100]-[110].

| Momentum (a.u.) | Non-SC exchange Model(a) | SC exchange and correlation model | Exp(b) | SC HF model(c) |
|---|---|---|---|---|
| 0.0 | 2.230 | 2.05 | 2.09,2.08 | 2.180 |
| 0.4 | 2.075 | 1.93 | 1.91,1.94 | 2.046 |
| 0.8 | 1.575 | 1.52 | 1.46,1.55 | 1.548 |
| 1.2 | 0.890 | 0.96 | 0.86,0.94 | 0.884 |
| 1.6 | 0.455 | 0.48 | 0.47,0.45 | 0.460 |
| 2.0 | 0.300 | 0.33 | ---- 0.31 | 0.293 |
| 4.0 | 0.080 | 0.11 | ---- 0.10 | ----- |

TABLE III: Compton profile in diamond along the [100] direction.

(a) Non-SC results Ref. (31) using an exchange coefficient of 0.70 and a double-zeta Slater basis set.
(b) The first value refers to the experimental results of Reed and Eisenberger (20) while the second number gives the data of Weiss and Phillips (21).
(c) Ref. 24.

The overall agreement of our calculated Compton profile with experiment seems reasonable. It would thus seem that at least for diamond, the apparent anomaly (i.e., $J(\underset{\sim}{q})$ too high at low q) of the calculated Compton profile in the exchange model for atoms (with an exchange coefficient of 2/3) disappears in a SC exchange and correlation model. In such a covalently bonded system high momentum components are added due to correlation, localization of charge in the core regions and enhancement of the charge build-up in the bond center.

B. Ground and Excited State Properties of LiF

After diamond (8) and BN (9), we extended our study to ionic solids, for which LiF has been chosen as a prototype. We considered the description of ground state properties of the system, such as the band structure, charge density, X-ray scattering factors, cohesive energy, equilibrium lattice constant, and behavior under pressure, and compared the predictions of the LDF model with both experimental data and with available restricted Hartree-Fock (HF) results (34).

1. Charge Density and X-ray Scattering Factors. Figure 2 shows the total ground state charge density calculated in (i) the exchange model and (ii) in the exchange and correlation model, along the [100] direction in the unit cell. The position of their minima is given in Table IV, and compared with the relevant experimental determination (34,36). These quantities are given here as a percentage of the lattice constant (i.e., twice the nearest-neighbor distance), as is commonly done in the literature. Unlike the older measurements of Krug, et al. (35) (in which dispersion corrections were not applied and rather high fluorine Debye-Waller factors used) which show a substantially larger Li radius, more recent structure factor measurements (37-39) seem to agree better with the data of Merisalo and Inkinen (36) which exhibit systematically shorter metal radii. Our results agree better with the latter data and indicate that the effect of correlation is to expand the electropositive Li site at the expense of contracting the electronegative F. In this context it is interesting to note that Pauling's ionic radii (0.6Å for $Li^+$ and 1.36Å for $F^-$) predict a much larger disproportion between the size of the lattice ions (15.3% and 34.7% for $Li^+$ and $F^-$, respectively) than do both the observed and the calculated values in the crystal.

The precise value of the minimum charge density in the unit cell is difficult to evaluate accurately from the experimental data since small changes in the temperature parameters and structure factors introduce significant changes into this small quantity. The measurements of Krug, et al. indicate a minimum density of

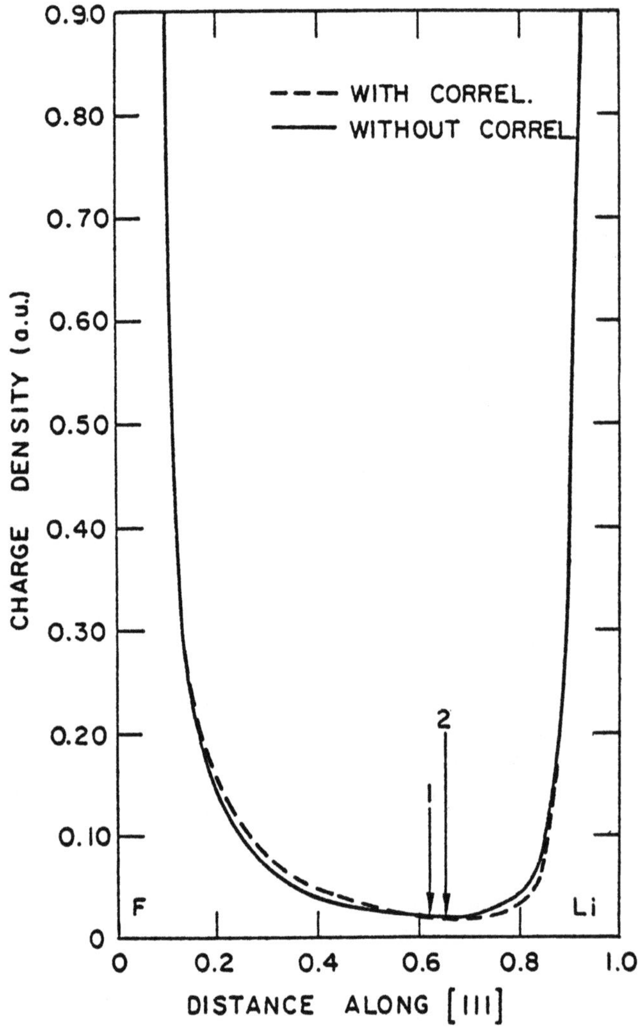


Fig. 2: Total ground-state charge density calculated (9) in (i) the exchange and correlation model, and (ii) the exchange model along the [100] direction in the unit cells. The arrows point to the positions of minimum density in the corresponding models.

| | Exchange and Correlation Model | Exchange Model | EXP[a] | EXP[b] |
|---|---|---|---|---|
| $R_{Li}$ | 19.2 | 17.5 | 19.4 | 22.9 |
| $R_F$ | 30.8 | 32.5 | 30.6 | 27.1 |

TABLE IV: The distances from Li and F sites, at which the charge density reaches a minimum, expressed as a percentage of the lattice constant a = 4.01852Å.

(a) Ref. (47).
(b) Ref. (46).

0.19 e/Å$^3$ (0.028 e/a.u.$^3$) while that of Merisalo and Inkinen show a minimum of approximately 0.15 e/Å$^3$ (0.022 e/a.u.$^3$). While our calculation shows a minimal density of 0.155 e/Å$^3$ (0.023 e/a.u.$^3$) in good agreement with both measurements, the large uncertainties in the observed values may make this agreement fortuitious. It is interesting to note that Hartree-Fock molecular calculations (40) predict a much higher minimum along the Li-F bond namely, 0.675 e/Å$^3$(0.1 e/a.u.$^3$).

Our calculation of the total (numerically integrated) electronic charge enclosed in spheres of varying sizes around the Li and F sites whose radii are chosen to form touching spheres gives results that are in remarkably good agreement with the measured data of Merisalo and Inkinen. We find in our exchange and correlation model that the radius at which the Li sphere enclosed exactly 2.0 electrons is 0.69Å (17.1% of the lattice parameter) which is only slightly smaller than the position of the minimum in the charge density (17.5% given in Table II), at which the Li sphere contains 2.05 electrons.

We have also compared our calculated X-ray structure factors with those calculated in the Hartree-Fock model (34,41) and with the observed data (36) with the temperature factors removed. Several conclusions can be drawn from this comparison: (i) The spherical free-ion HF model yields lower $F_{h,k,l}$ values than the $\alpha$=1 LDF model, but higher values than those obtained with the $\alpha$= 2/3 LDF model. The HF charge density is hence more spatially diffuse than that yielded by the $\alpha$=1 results but slightly more contracted than that predicted by the $\alpha$= 2/3 calculation. The exchange and correlation model with $\alpha$ = 2/3 produced values that are slightly larger than those produced by the $\alpha$ = 2/3 exchange-only model. (ii) Nonspherical corrections to the free-ion HF model produce a more localized charge density (i.e., have the effect of approaching the LDF spherical results with $\alpha$ greater than 2/3). (iii) The crystalline HF and LDF results show a general increase in the structure factors relative to the corresponding spherical free-ion results which is much larger in the LDF model than in the HF model. The increase in the HF crystalline structure factors relative to the spherical free-ion limit seems insufficient to account for the experimental crystal data and is significantly lower than that yielded by the nonspherical free-ion HF model. This might reflect both numerical inaccuracies in the crystal HF model, which is significantly more complex than the non-spherical free-ion model, and the superiority of the Slater basis set representation used in the latter as compared with a Gaussian set used in the crystalline HF model. The overall agreement of both the HF and the LDF crystal results with experiment is good, with the HF systematically lower and the LDF systematically higher than the observed results.

2. Excited States. We have also considered the excited state properties of LiF by first using the standard band approach to excitation energies (i.e., viewing them as differences in the band eigenvalues between unoccupied and occupied bands). Contrary to what has been previously suggested (42-44), we find that this approach fails completely in the case of a heteropolar wide gap material such as LiF due to the localized nature of many of the electron-hole states in the system. Thus, we are led to a generalized band model in which both the initial and the final crystal states are allowed to become localized to some extent and hence to exhibit relaxation, polarization, and electron-hole interaction effects. For this purpose, we use a defect superlattice representation (45,46) in which a locally excited species in the solid is viewed as a point-defect placed at the center of a large unit cell and the excitation energy is determined as the difference of total energies between independent self-consistent calculations ('ΔSCF') of the corresponding excited and ground state systems. The effects introduced by final state orbital relaxation and valence band polarization by the hole state are found to be very large and to account for most of the discrepancies between the unperturbed band model predictions and the observed data. In addition, the model is used to calculate core and valence exciton states in the system by a direct diagonalization of the locally perturbed crystal Hamiltonian rather than by conventional effective-mass or perturbational Frenkel models. Good agreement is found with the experimentally observed exciton energies and binding over a large spectral region. Electron-hole interaction energies are found to be very large (2-9 eV) and to vary considerably as one moves from a deep core to a valence state exciton. The exciton band width is calculated directly from the self-consistent perturbed crystal wave functions using standard techniques and is found to be rather large (0.3 eV) for valence excitons.

In the superlattice (or periodic cluster) representation, we construct a large crystallographic unit cell with a defect placed at its center and solve the associated Bloch Hamiltonian problem (with periodicity imposed with this large supercell) using band structure techniques. To the extent that the defect-defect interaction present in this superlattice model (monitored by the dispersion of the defects' one-electron band) can be kept small by choosing a sufficiently large SPC, the solution would form a good approximation to that of a single point defect in the lattice. For example, in the case of LiF, we use a basic unit cell containing 8 (fcc structure) or 16 (simple cubic structure) atoms, with one of them being locally excited (and the defect-defect distance is a and $\sqrt{2}a$, respectively, where a=4.02Å is the lattice constant). One obtains the ΔSCF estimate for the excitation energies simply by subtracting the ground state energy from that of the defect-containing crystal model.

## IV. LDF APPROACH FOR SURFACES

The study of transition metal surfaces has become a subject of intense interest and study. As in the case of bulk transition metal studies which have taken place over the last 40 years, the difficulty of treating localized d electrons along with the itinerant s-p electrons has provided the challenge and impetus for developing the sophisticated theoretical methods necessary for accurately determining the electronic structure of transition metal surfaces.

The superlattice representation described above also permits the treatment of the electronic structure of unsupported thin films (or slabs) by the same self-consistent LDF methods described above. Here I describe some aspects of the (mostly unpublished) work which C.S. Wang and I (46-49) have done on transition metals, notably Ni(001) and with oxygen chemisorbed on c(2X2). We consider a film of m layers with the origin of the system midway between the two surface layers and the z axis normal to the surfaces. The unit cell consists of a parallel-piped whose z dimension extends to $\pm\infty$. The Coulomb potential is formed as a non-muffin tin superposition of spherical atomic potentials which can contain long-range ionic components to account for charge transfer in the film. A superposition of overlapping spherical atomic charge densities is used to construct the local density Kohn-Sham ($\alpha = 2/3$)$\rho^{1/3}$ exchange potential. Atoms up to 25 a.u. away from the atomic site were included in the two-dimensional direct lattice sum to obtain the superposition potential and charge density. The long-range ionic component of the Coulomb potential accompanying the charge transfer near the surface is included through a generalized Ewald type procedure.

For these films, one is interested in both total density of states (DOS) and projected local DOS (by layer plane) for comparison with photo-emission experiments. For the case of Ni(001) distinct differences are obtained between the surface and central plane local DOS. The central plane DOS is found to converge rapidly to the DOS of bulk paramagnetic Ni obtained by Wang and Callaway(50). Only a very small surplus charge is found on the surface planes— in agreement with earlier jellium model calculations. For this Institute, the results of interest are the SC charge and spin densities. Thus we present in Fig. 3 the SC valence electronic charge density map of a 5-layer Ni(001) film on the (110) plane with the surface normal along the vertical section. It is interesting to note the way the charge density gradually smoothes out parallel to the surface as one enters into the vacuum region. Surprisingly, the charge density at one layer below the surface already appears to be bulk-like. Around each atom there is a fairly large region where the charge density is spherically symmetric. Therefore, the commonly

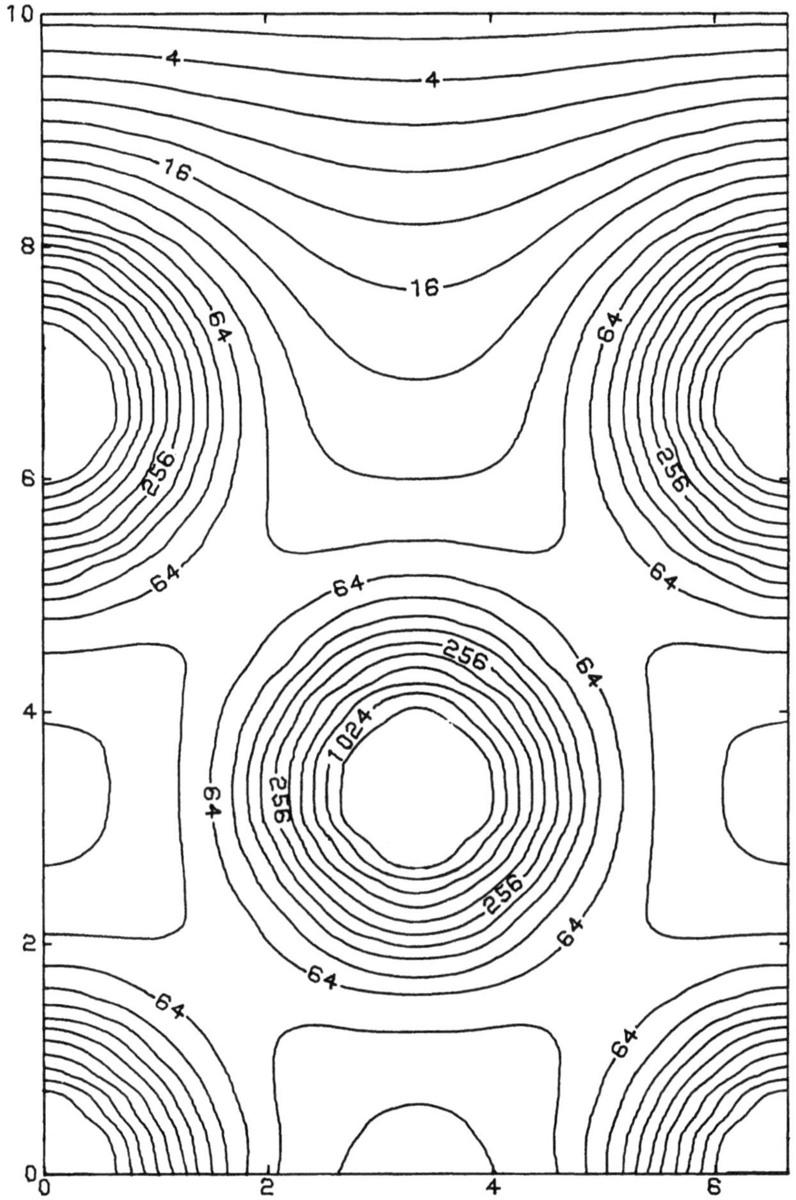


Fig. 3: Self-consistent conduction electron charge density map in units of 0.001 e shown in the [110] plane as obtained from the 5-layer Ni(001) film calculation (47). Each contour line differs by a factor of $\sqrt{2}$.

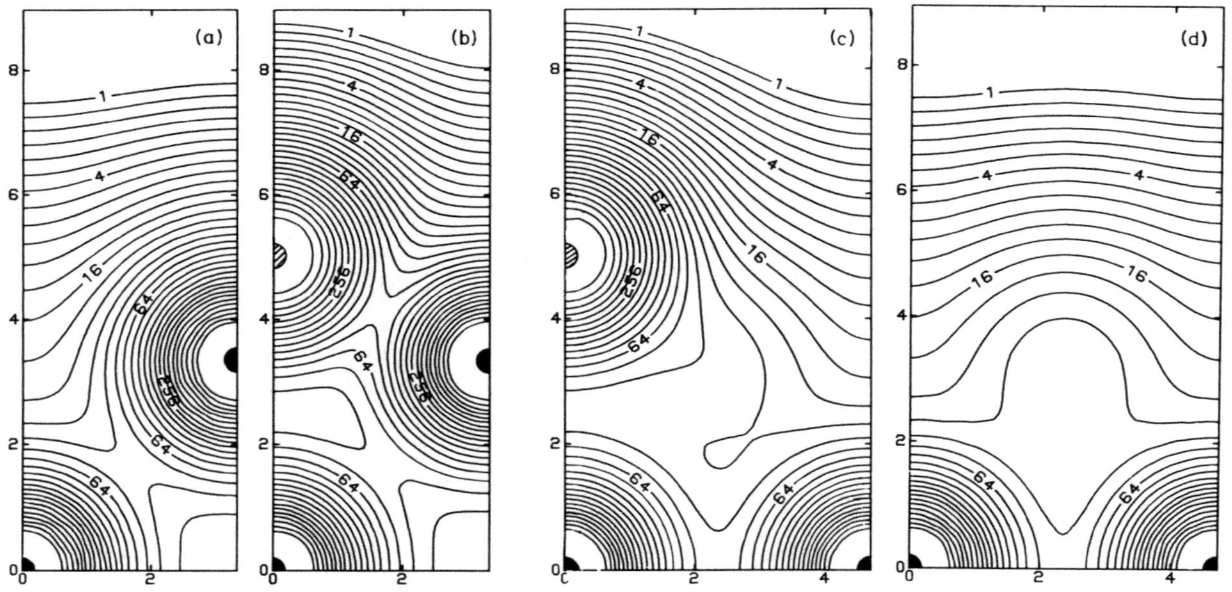


Fig. 4: Self-consistent conduction electron charge density map on the [100] plane in units of 0.001 e (a) before and (b) after oxygen chemisorption (49). Corresponding results on the [110] plane are shown in (d) and (c) respectively.

used muffin-tin approximation, which assumes spherical symmetry around each atom, z dependence outside the surface plane, and constant in the interstitial region is quite adequate except in the interstitial region on the surface plane where the charge density varies more rapidly due to the open surface structure.

It is instructive to inspect the charge density for the case of c(2X2) O on Ni(001). Several features of the chemisorption bond are shown by the charge density contour maps in Fig. 4 for (110) and (100) planes with the vertical along the surface normal and contours plotted at $\sqrt{2}$ times the previous value. Comparing the clean Ni and chemisorbed Ni map shows the buildup of charge on the O site and even in the bonding region between the Ni atoms. The (100) plane contours show more clearly the interaction of the O atoms with the surface plane Ni atoms in the formation of the adsorbate-substrate bond and the enhancement of the charge density in the interstitial region.

## REFERENCES

* Supported by the NSF and the AFOSR.

1. P. Hohenberg and W. Kohn, Phys. Rev. B 136, 864 (1964).

2. W. Kohn and L. J. Sham, Phys. Rev. A 140, 1133 (1965).

3. O. Gunnarsson and B. I. Lundqvist, Phys. Rev. B 13, 4274 (1976).

4. S. Lundqvist and C. W. Ufford, Phys. Rev. 144, A1 (1965).

6. O. Gunnarsson, et al., Int. J. of Quant. Chem. S9, 83 (1975).

7. A. Zunger and A. J. Freeman, Phys. Rev. B 15, 4716 (1977).

8. A. Zunger and A. J. Freeman, Phys. Rev. B 15, 5049 (1977).

9. For example see A. Zunger and A. J. Freeman, Phys. Rev. B 16, 906 (1977) for $TiS_2$; Phys. Rev. B 16, 2901 (1977) for LiF; Phys. Rev. B 17, 2030 (1978) for cubic BN; Phys. Rev. B 17, 1839 (1978) for $TiSe_2$; Phys. Rev. B 17, 4850 (1978) for cubic CdS.

10. M. Rasolt and D.J.W. Geldart, Phys. Rev. Lett. 35, 1234 (1975).

11. K. S. Singwi, et al., Phys. Rev. B 1, 1044 (1970).

12. L. Hedin and B. J. Lundqvist, J. Phys. C 4, 2064 (1971).

13. C. B. Hazelgrove, Math. Compt. 15, 373 (1961).

14. D. E. Ellis and G. S. Painter, Phys. Rev. B 2, 7887 (1970)

15. See for instance the review by G. S. Buberman in Soviet Phys. Uspekhi 14, 180 (1971).

16. H. D. Hagstrum, Phys. Rev. 72, 947 (1947) and "JANAF Tables of Thermochemical Data" edt. D. R. Stull, Midland, Michigan (1965).

17. J. Thewlis and A. R. Davey, Phil. Mag. 1, 409 (1958).

18. S. Götlicher and E. Wölfel, Z. Electrochem 63, 891 (1959).

19. M. Renninger, Acta Cryst. 8, 606 (1955).

20. W. A. Reed and P. Eisenberger, Phys. Rev. B 6, 4596 (1972).

21. R. J. Weiss and W. C. Phillips, Phys. Rev. 176, 900 (1968).

22. R. N. Euwema, D. L. Wilhite and G. T. Surratt, Phys. Rev. B 7, 818 (1973).

23. G. T. Surratt, R. N. Euwema and D. L. Wilhite, Phys. Rev. B 8, 4019 (1973).

24. G. G. Wepfer, R. N. Euwema, G. T. Surratt and D. L. Wilhite, Phys. Rev. B 9, 2670 (1974).

25. R. N. Euwema and R. L. Greene, J. Chem. Phys. 62, 4455 (1975).

26. P. M. Raccah, R. N. Euwema, D. J. Stukel and T. C. Collins, Phys. Rev. B 1, 756 (1970).

27. K. H. Bennemann, Phys. Rev. 133, A1045 (1964).

28. L. Kleinman and J. C. Phillips, Phys. Rev. 125, 819 (1962).

29. I. Goroff and L. Kleinman, Phys. Rev. 164, 1100 (1967).

30. H. Clark, Phys. Lett. 11, 41 (1964).

31. A. Seth and D. E. Ellis, J. Phys. C 10, 181 (1977).

32. R. N. Euwema and G. T. Surratt, J. Phys. C7, 3655 (1974).

33. J. R. Sabin and S. B. Trickey, J. Phys. B8, 7593 (1975).

34. R. N. Euwema et al., Phys. Rev. B 9, 5249 (1974).

35 J. Krug, H. Witte, and E. Wölfel, Z. Phys. Chem. 4, 36 (1955).

36. M. Merisalo and O. Inkinen, Phys. Fen. 207A, 3 (1966).

37. R.C.G. Killean, J. L. Lawrence, and V. C. Sharma, Acta Crystallogr. A 28, 405 (1972).

38. W. H. Zachariasen, Acta. Crystallogr. A 24, 324 (1968).

39. C. J. Howard and R. G. Khadake, Acta. Crystallogr. A 30, 296 (1974).

40. R.F.W. Bader and A. D. Bandrank, J. Chem. Phys. 49, 1653 (1968).

41. O. Aikala and K. Mansikka, Phys. Kondens. Mater. 11, 243 (1970); 13, 58 (1971); 14, 105 (1972).

42. W. P. Menzel, C. C. Lin, D. Fouquet, E. E. Lafon and R. C. Chaney, Phys. Rev. Lett. 30, 1313 (1973).

43. R. C. Chaney, E. E. Lafon, and C. C. Lin, Phys. Rev. B 4, 2734 (1971).

44. F. C. Brown, C. Gahwiller, A. B. Kunz, and N. O. Lipari, Phys. Rev. Lett. 25, 927 (1970).

45. A. Zunger, J. Chem Phys. 62, 1861 (1975); 63, 1713 (1975); A. Zunger and R. Englman, Phys. Rev. B 17, 641 (1978); B 17, 676 (1978)

46. C. S. Wang and A. J. Freeman, Phys. Rev. B 18, 1714 (1978).

47. C. S. Wang and A. J. Freeman, Phys. Rev. B (1978) to appear.

48. C. S. Wang and A. J. Freeman, J. Appl. Phys. March (1979) to appear.

49. C. S. Wang and A. J. Freeman, Phys. Rev. B (1979) to appear.

50. C. S. Wang and J. Callaway, Phys. Rev. B 7, 1096 (1973).

CALCULATION OF ONE-ELECTRON CHARGE, MOMENTUM, AND SPIN DENSITIES: THE STATISTICAL EXCHANGE APPROXIMATION

D. E. Ellis

Northwestern University

Evanston, Illinois 60201

## A. DENSITY MATRICES AND DISTRIBUTIONS

In these notes we will present a certain view of the status of the theory of one-electron densities, using concrete examples to illustrate the applicability of theoretical models to experimentally interesting systems. Our task is made easier by the excellent lectures on the fundamental theory given by Professor March and other participants of this school. Let us begin by remembering the connection between the basic N-electron quantum state $\Psi_\ell(1,2,.. N)$ and the density matrices whose properties we eventually measure: The $p^{th}$ order density matrix can be defined as /1,2/

$$\rho^p_\ell(1,2,...p|1'2'...p') = \binom{N}{p} \int \Psi^*_\ell(1,2,...N)\ \Psi_\ell(1'2'...p',p+1,...N)\ d(p+1)d(p+2)...d(N) \quad (1)$$

Evidently, integration over the coordinates (p+1)...N suppresses a great deal of information contained in the state $\Psi$ so that one eventually arrives at the simple one-particle and two-particle distributions or densities which are accessible to experiment. We can best display the hierarchy of density matrices and their final profection onto various "islands of experiment" in Fig. 1, borrowing from the Dick Weiss' delightful model.

In the following we shall be interested almost entirely in approximations to the one-electron matrix $\rho^1(1|1')$, and its manifestations in the charge density $\rho(\vec{r})$, the momentum density $\tilde{\rho}(\vec{p})$, and the spin density $m(\vec{r})$, but we must never lose sight of the framework in which these studies are made. Our contention is that measurements and models of, say, $\rho(\vec{r})$ become most valuable when they are correlated with data from other sources, and we are very grateful for the discussions and exchanges of information which this school has made possible.

In order to introduce the one-electron model, we may write down the natural spin-orbital expansion of the first order density matrix /3/:

$$\rho^1(\vec{r}\mu|\vec{r}'\mu') = \sum_i^{\infty} n_i \psi_i^*(\vec{r}\mu)\psi_i(\vec{r}'\mu') \qquad (2)$$

Here the spin-orbitals $\psi_i$ form a complete orthonormal set in the and spin coordinates $(\vec{r}\mu)$ of a single particle, and the $n_i$ play a role of occupation numbers. While the exact distribution $\rho^1$ can be written in this form, practical interest centers on approximate models which can give more or less accurate representations of Eqn. (2). The most famous example is of course the Hartree-Fock (HF) model, in which N of the $n_i$ are assumed to have value 1, and the remainder 0. This density matrix can be obtained (not uniquely!) from a many-electron wavefunction which is the antisymmetrized product (Slater determinant) formed from the N spin-orbitals.

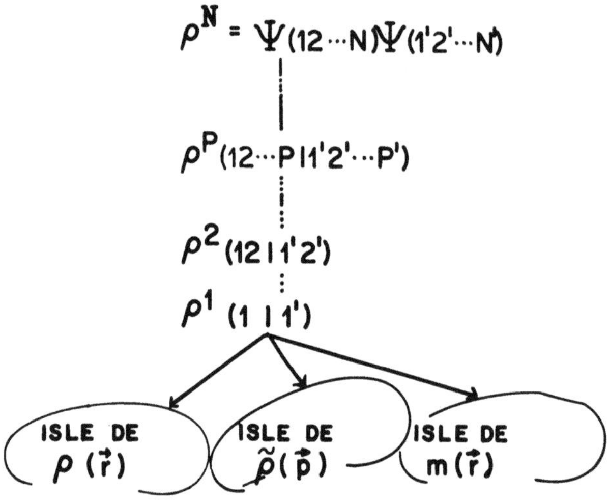


Fig. 1 Hierarchy of density matrices and their projection onto experimental "islands".

In the HF model the expectation value of the total energy is minimized with respect to variations in the spin-orbitals $\delta\Psi_i$, and with suitable constraints this provides a practical method for determining the $\Psi_i$/4/. It is important to note that this procedure "works" only because the second order density matrix, required to evaluate the electron-electron interaction energy, can be expressed directly in terms of the first order density:

$$\rho^2_{HF}(12|1'2') = \frac{1}{2}\{ \rho^1(1|1')\rho^1(2|2') - \rho^1(1|2')\rho^1(2|1')\} \quad (3)$$

Other more exact theories will not be so simple, and one is obliged then to consider electron-pair correlations in detail. Expansion of $\rho^2$ in a two-particle basis (natural geminals) is an attractive approach /1-3/, which has unfortunately proved difficult except for rather small few-electron systems. It would appear that the machinery of many-body perturbation theory is better suited for applications even to atoms, as well as molecules and solids.

Now let us return to some simple properties of the first order density, beginning with the kinetic energy density,

$$T(\vec{r}\mu) = \sum_i n_i \psi_i^*(\vec{r}\mu) t_{op} \psi_i(\vec{r}\mu) \quad (4)$$

The expectation value of T, averaged over both space and spin variables, gives the kinetic energy of the system characterized by orbitals $\psi_i$ and occupation numbers $n_i$. This provides a useful example of the procedure used to obtain expectation values of operators: the operator $t_{op}$ is sandwiched between primed and unprimed variables, allowed to act, and finally the "primes" are removed. This operation scheme is completely consistent with the ordinary rules of quantum mechanics, as a glance at Eqn. (1) will show, but lies at the root of difficulties encountered in trying to determine $\rho^1$ by direct methods. For example, in the Thomas-Fermi method or its modern elaborations one attempts to obtain a density function $\rho(\vec{r})$ which can be interpreted as the diagonal part of $\rho$, and hence the charge density of a system, Since $t_{op}\rho(\vec{r})$ is very different from $T(\vec{r})$ one is faced with formal as well as computational difficulties. For this and other reasons so-called one-electron methods, which take Eqn. (2) as an <u>ansatz</u> and attempt to determine optimum orbitals $\psi_i$ by various schemes have become widely used and form the basis for much of our present understanding of electronic structure in molecules and solids.

As a second exercise let us consider the representation of $\rho^1$ in momentum space, which we denote as $\tilde{\rho}(\vec{p}|\vec{p}')$. Suppressing the spin variables, this is given as the double Fourier transform

$$\tilde{\rho}(\vec{p}|\vec{p}') = (2\pi)^{-3} \iint \exp\{i(\vec{p}\cdot\vec{r}-\vec{p}'\cdot\vec{r}')\}\ \rho(\vec{r}|\vec{r}')\ d^3r d^3r' \tag{5a}$$

$$= \sum_i n_i \chi_i^*(\vec{p})\chi_i(\vec{p}') \tag{5b}$$

where

$$\chi(\vec{p}) = (2\pi)^{-3/2} \int \exp(-i\vec{p}\cdot\vec{r})\psi(\vec{r}) \tag{6}$$

is seen to be simply the p-space representation of the natural spin orbital. The diagonal part $\tilde{\rho}(\vec{p}) \equiv \tilde{\rho}(\vec{p}|\vec{p})$ is the momentum density of the system. It is quite apparent that $\tilde{\rho}(\vec{p})$ is <u>not</u> simply the Fourier transform of the charge density $\rho(\vec{r})$, so that experiments like Compton (inelastic photon) scattering which measure aspects of $\tilde{\rho}$ are complementary to techniques like elastic x-ray scattering which probe $\rho$. Dick Weiss has given a comprehensive survey of the Compton scattering technique in his lectures. Here we only wish to comment on some interesting possibilities for analyzing Compton data.

In the simple non-relativistic <u>impulse approximation</u> one can obtain from the data a directional Compton profile which is related to the momentum density by

$$J(\hat{k},q) = \int_{-\infty}^{\infty} \tilde{\rho}(\vec{p})\ \delta(q - \hat{k}\cdot\vec{p})d^3p \tag{7}$$

Here $\hat{k}$ is a unit vector in the direction of momentum transfer $\vec{K}$ (see Fig. 2) and q is a measure of the Doppler broadening of the Compton line due to the motion of the scattering electrons /5,6/. Now consider the one-dimensional Fourier transform of J, given as

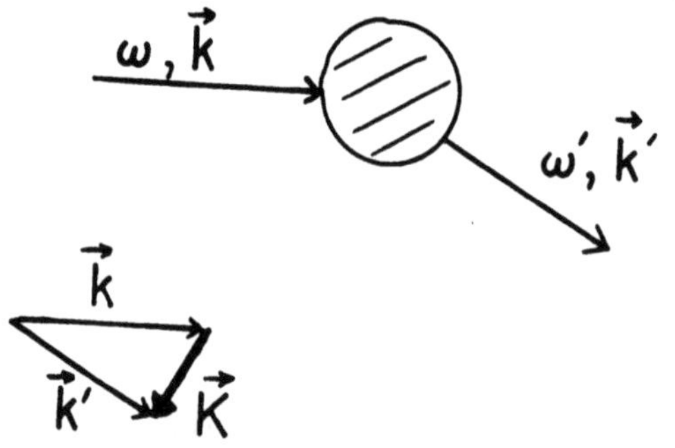


Fig. 2 The Compton inelastic photon scattering process; $\vec{K}$ is the momentum transfer.

$$B(\vec{t}) = \int_{-\infty}^{\infty} \exp(iqt)J(\hat{k},q)dq \quad (8)$$

with $\vec{t} = \hat{k}t$

If we ignore difficulties in extrapolating J beyond experimentally accessible q values, several interesting features emerge.

For example, in a free electron gas model for metals and alloys one can approximate the momentum density as

$$\tilde{\rho}(\vec{p}) = \rho_o \qquad 0 \leq p \leq p_F$$
$$= 0 \qquad p > p_F$$

where $p_F$ is the Fermi momentum. Direct substitution gives

$$B(\vec{t}) = 4\pi\rho_o \int_0^{p_F} p^2 j_o(t_p)dp \propto \frac{j_1(tp_F)}{tp_F} \quad (9)$$

This method has been recently used in efforts to make a precise determination of the Fermi momentum in both simple metals and alloys /7/. In principle, this scheme could be used to study anisotropy of Fermi surfaces in metals which are insufficiently pure or insufficiently perfect to perform conventional deHaas-vanAlphen resonance studies.

To emphasize another point of view, one can easily derive the identity

$$B(\vec{t}) = \int_{-\infty}^{\infty} \rho^1(\vec{r}|\vec{r}+\vec{t})d^3r \quad (10)$$

which reveals B as a kind of autocorrelation function for spin orbitals. Possibilities for extracting information about off-diagonal components of $\rho^1(\vec{r}|\vec{r}')$ need to be explored.

## B. STATISTICAL EXCHANGE AND LOCAL DENSITY METHODS

The Hartree-Fock one electron method was for many years the primary first principles method available for the study of atoms and small molecules. The technology of this method has continued to evolve so that with large scale computers calculations can be made on molecules like $Ni(CO)_6$ and on a few crystalline systems. The HF spin-orbitals are obtained by approximately solving the one-electron Schrodinger equation

$$(h - \epsilon_i)\psi_i(\vec{r}\mu) = 0 \quad (11)$$

where h is obtained in the process of minimizing the total

energy expression with respect to variations $\delta\psi_i$. The effective Hamiltonian has the form

$$h = t_{op} + V_{coulomb} + V_{ex} \tag{12}$$

consisting of kinetic energy, electron-nucleus and electron-electron Coulomb potentials, and the exchange operator $V_{ex}$. The non-local nature of the exchange operator, arising from the antisymmetry of the HF determinantal wavefunction has been the principal cause of difficulty in extending applications to larger and more complex systems. The basic problems of generating and manipulating the $\sim N^4$ two electron integrals required for the iterative solution of Eqn. (11) clearly become severe for $N \gtrsim 100$ with present day machines. In the usual variational procedures which expand the spin orbital in a basis the number of orbitals which can be treated with any precision is thus very limited /8,9/.

Of course there exist ways of escaping from the $N^4$ trap -- the rapid development of pseudopotential methods for suppressing calculation of unwanted core electron states is a good example /10,11/. Embedding techniques or molecular fragment models which focus attention on interesting portions of a molecule or solid while maintaining suitable boundary conditions have similar encouraging features. We will only be able to briefly discuss the simplest <u>local density</u> methods in which $V_{ex}$ is replaced by a potential which is a specific function of the charge (and spin) density.

One of the early justifications and derivations of a local density scheme was given in Slater's statistical exchange model. Here one thinks of averaging the exchange operator over the set of occupied orbitals, and finally approximating the average by the analytical result calculated for a free electron gas of the appropriate density. This argument and other derivations leads to the form

$$V_{ex}(\vec{r}\mu) = -3\alpha\left[\frac{3\rho(\vec{r}\mu)}{4\pi}\right]^{1/3} \quad \text{(Hartree units)} \tag{13}$$

where $2/3 \le \alpha \le 1$ is a constant whose value depends upon the particular derivation /12-14/. We should remark that the exchange potential in an open shell system will be different for spin $\uparrow$ and for spin $\downarrow$ electrons, leading to different spatial orbitals for either spin. Following the observation that the ground state energy must be a unique functional of the charge density, foundations for a more rigorous local density theory were laid. Recent years have seen many significant efforts to improve upon the description given by Eqn. (13), but the quality of results obtainable with the simplest form is quite sufficient for most spectroscopic investigations /15/. In fact the main limitations on accuracy of local density calculations often seem to arise from restrictions on the form or number of basis functions, and approximation errors in representing the Coulomb potential -- a remarkable state of affairs!

Variants of the scheme which we have been describing, called Hartree-Fock-Slater, $X\alpha$, Kohn-Sham, local-spin-density, etc. are now well established methods and their applications go far beyond the HF theory. The routine successful application of local density models in the interpretation of excited state properties has stimulated a search for justifiable approaches and rigorous theorems for treating excitations of many electron systems in the one-electron framework which is far from complete. In this connection we want to mention the transition state approach of Slater and coworkers which has been taken up and extended by Ziegler and others /16,17/. In this scheme many electron excitation energies are calculated as eigenvalues of optimized single particle Schrodinger equations with a rather high degree of precision; modest extensions of the procedure permit evaluation of multiplet splittings. Much less progress has been made in the area of describing properties which depend upon the excited state wavefunctions, such as photoionization cross-sections, although "practical results" have certainly begun to appear.

A consistent relativistic local density theory can be obtained as a straightforward extension of the models just described. We wish to solve an eigenvalue equation of the same form as Eqn. (11); however we will now adopt the Dirac Hamiltonian /18,19/:

$$h = c\,\underset{\approx}{\alpha} \cdot (\vec{p} - \frac{e}{c}\vec{A}) + eV + m_o c^2 \underset{\approx}{\beta} \tag{14}$$

Here $\vec{A}$ and V are the electromagnetic vector and scalar potentials respectively and and $m_0c^2$ is the electron rest mass. The 4x4 matrices $\underset{\approx}{\alpha}$ and $\underset{\approx}{\beta}$ can be conveniently written in terms of the 2x2 Pauli spin matrices as

$$\underset{\approx}{\alpha} = \begin{bmatrix} o & \vec{\sigma} \\ \vec{\sigma} & o \end{bmatrix} \qquad \underset{\approx}{\beta} = \begin{bmatrix} I & o \\ o & -I \end{bmatrix} \tag{15}$$

and the wave function is a column vector with components $\psi_i^1, \psi_i^2, \psi_i^3, \psi_i^4$. The four terms can be thought of as a pair of spin-orbitals coupled by the four first order differential equations, and for a spherical potential where the total angular momentum j is a good quantum number, the eigenfunctions can be written as

$$\Phi(\vec{r}\mu) = \begin{bmatrix} f_{n\ell j}(r)Y_{\ell jm}(\hat{r}\mu) \\ \\ ig_{n\bar{\ell}j}(r)Y_{\bar{\ell}jm}(\hat{r}\mu) \end{bmatrix} \tag{16}$$

The spherical spinors $Y_{\ell jm}$ contain both angular and spin coordinates, with allowed values of the orbital momentum, $\ell$ and $\bar{\ell}$, being j ± ½. In self-consistent field atomic calculations the principal quantum number n has its usual meaning in determining the number of nodes in the radial function. For light atoms $f_{n\ell j}$ can be identified with the usual nonrelativistic (NR) radial wave function.

In the relativistic Dirac-Fock counterpart of the Hartree-Fock theory, the N-electron wavefunction is given as an antisymmetrized product of four-component single particle states. For a closed shell system the vector potential A can be set zero and the electrostatic potential V treated as in the NR case. Explicit relativistic contributions to the potential, such as the Breit interaction, are usually treated by perturbation theory. Exactly as in the HF problem, the non-local exchange operator causes the greatest difficulties in performing calculations /19,20/. Again in analogy to NR models one can bypass these problems by making the transition to a local density model--variously known as Dirac-Fock-Slater or Dirac-Slater /21/. Thus, for example, if one wishes to perform relativistic molecular orbital calculations on heavy atom systems such as uranyl, $UO_2^{2+}$, it is relatively simple to set up a variational expansion in the free atom or ion four-component basis.

$$\psi_i(\vec{r}\mu) = \sum_n \Phi_n(\vec{r}-\vec{R}_n, \mu)\, C_{ni} \tag{17}$$

with atomic basis functions $\Phi_n$ centered on the nuclei /22,23/. The variational coefficients $C_{ni}$ can be found by solving the familiar matrix secular equation

$$(\underset{\approx}{H} - E\underset{\approx}{S})\underset{\approx}{C} = 0 \tag{18}$$

by standard techniques. A second feasible method, which we describe in a later section in connection with energy band models, arises from a scattered wave or Green's function description of $\psi$/24/.

## C. SOME ATOMIC AND MOLECULAR EXAMPLES

It might seem that every interesting atomic calculation would have been completed by now. On the contrary, not only do atomic problems serve as testing grounds for more sophisticated theories and methods, but there also remain basic questions of some practical importance. The example which we choose is concerned with the magnetization distribution in heavy atoms for which relativistic effects are not negligible. The magnetization consists in general of both orbital and spin contributions. For the present let us consider the case of partially filled shells for which the vector potential $\vec{A}$ is zero; this is analogous to those NR situations in which $m(\vec{r})$ is entirely due to the unpaired spins. As a simple alternative to the computationally demanding Unrestricted Dirac-Fock scheme developed by Desclaux /20/, Rosén /19/, and others we may generalize the relativistic local density scheme in a similar spirit to the NR spin-polarized method /25,26/.

Experiment shows that the $^8S$ ground state of the $Eu^{2+}$ ion arising from the $4f^7$ configuration has a moment consistent with seven unpaired f-electrons.

Mossbauer hyperfine measurements on solids like EuO show the existence of a magnetic field at the nucleus which is attributed to the polarization of inner shells by the 4f magnetization /27/. Let us compare in Figs. 3 and 4 nonrelativistic HFS and relativistic DS descriptions of the charge and magnetization density in the ion. First, in comparing $\rho(r)$ and the difference $\Delta\rho(r)$ we see the relativistic contraction of the core expected from analytic solutions for one-electron ions. Fig. 3 also reveals the more subtle variations of charge density in the valence region which are due to the self-consistent response of all electrons to changes in electrostatic shielding. Such indirect relativistic effects are of major importance for understanding "effects of relativity" in chemical bonding and reactions.

The magnetization density m(r) and the difference $\Delta m(r)$ are shown in Fig. 4, but this figure requires some further explanation. As we have already mentioned, the NR case corresponds to an orbital-filled, spin-half-filled situation, so the magnetization is simply proportional to the difference between spin ↑ and spin ↓ components of the charge density, i.e., the expectation value of $S_z$ at every point $\vec{r}$. However, in the Dirac theory spin and orbital momenta are coupled together and $S_z$ is not a diagonal operator; instead we can choose $J_z$. Thus we have performed a relativistic "moment polarized" calculation with a potential corresponding to the $(4f_{5/2})^3(4f_{7/2})^4$ configuration with all $m_j$ parallel (moment ↑). Core polarization is included self consistently by allowing moment ↑ and moment ↓ radial functions to differ. The difference between this relativistic density and the NR spin-polarized density is displayed as $\Delta m(r)$ in Fig. 4. We see the expected dominant $4f^7$ contribution to the magnetization m(r), and the relativistic enhancement of the magnetization in the core region is clearly evident. The moment polarized scheme is sufficiently simple to permit applications to both molecules and solids, allowing self-consistent treatment of polarization and relativistic effects on an equal footing.

As a second example we wish to discuss the comparison of molecular charge densities calculated by HF and local density theories and obtained by x-ray diffraction, the amplitude $F(\vec{K})$ for scattering vector $\vec{K}$ being simply related to the Fourier transform of the charge density:

$$F(\vec{K}) = \int \exp(i\vec{K}\cdot\vec{r})\rho(\vec{r})d^3r \tag{19}$$

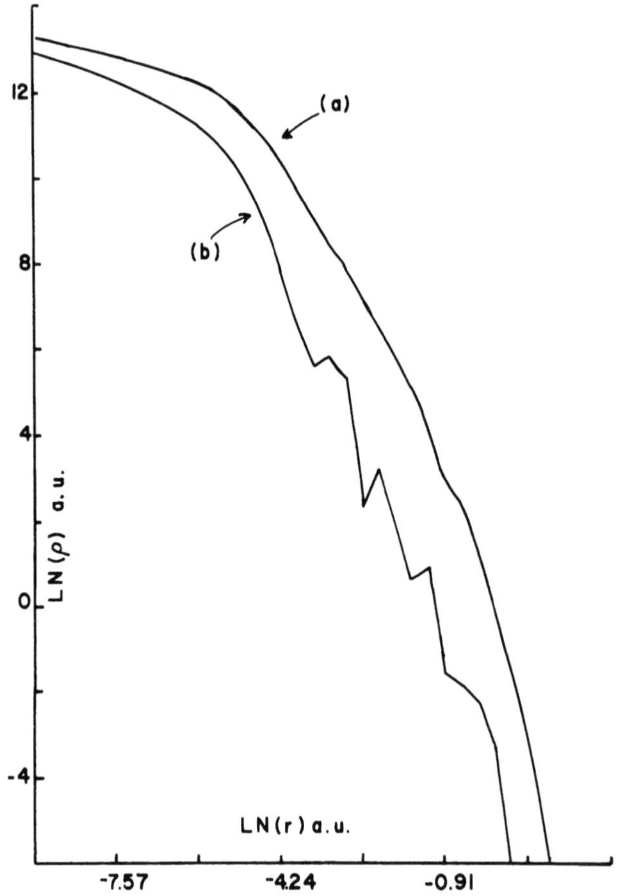


Fig. 3 Comparison of relativistic and non-relativistic (NR) charge densities for the $Eu^{2+}$ ion in the $(4f)^7$ configuration: (a) R-charge density, (b) |R-NR| charge density difference.

Since x-ray intensities $I \propto |F|^2$ can be measured with an absolute precision of ~1%, inversion procedures have been developed which produce rather good quality "experimental" maps of $\rho$ for a number of molecules and solids. As the experimental lectures in this school testify the x-ray technique, sometimes augmented by neutron

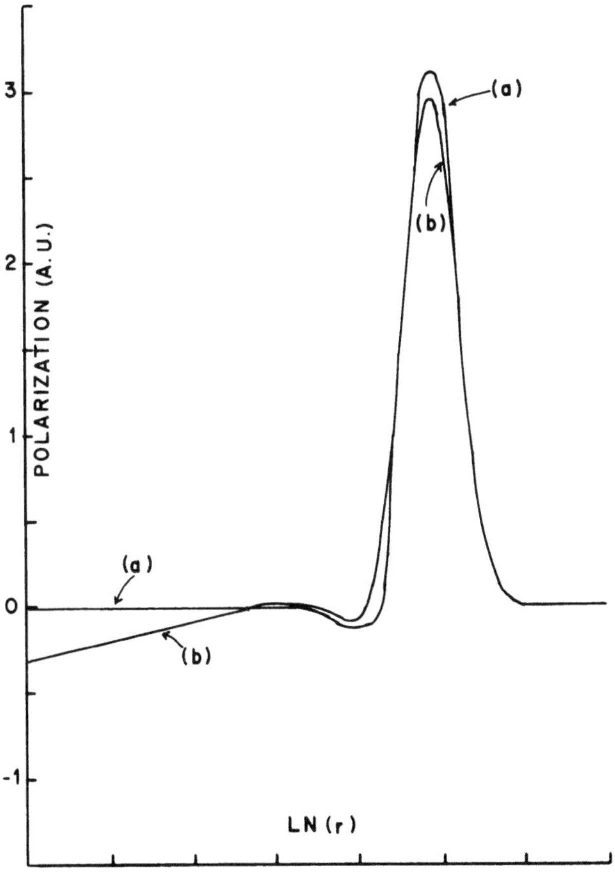


Fig. 4 Comparison of relativistic and non-relativistic magnetization densities for the $Eu^{2+}$ ion in the $(4f)^7$ configuration: (a) R-magnetization density, (b) (R-NR) magnetization density difference.

diffraction data, can be used to map the charge density even in complex metal-organic compounds.

Baerends and Ros have made a study of HFS versus HF theoretical densities for a number of small molecules /28,29/. Some of their results for CO are reproduced in Fig. 5. A definite conclusion of their work is that HF and HFS ground state bonding densities differ very little, providing that sufficient care is taken with basis set quality and in representation of the molecular potential and its matrix elements. Several variants of local density potentials were explored; there is as yet no clear-cut preference for one or another version of the exchange and correlation potential-differences in charge densities as well as energy levels tend to be rather small.

With confidence gained from small molecule studies, it becomes reasonable to attack molecular systems of greater complexity. Experimental density differences (molecule minus superposition of free atoms) have been obtained for $Cr(CO)_6$ recently, and HFS calculations have been made by Baerends and Ros /29/. Rather good quantitative agreement is found in comparing theory and experiment, so the way appears to be clear for extensive applications to transition metal complexes and metal-organic systems of interest.

As a third example we can discuss the description of localized states in solids. For many purposes it is convenient to describe properties of extended solids in terms of molecular cluster models.

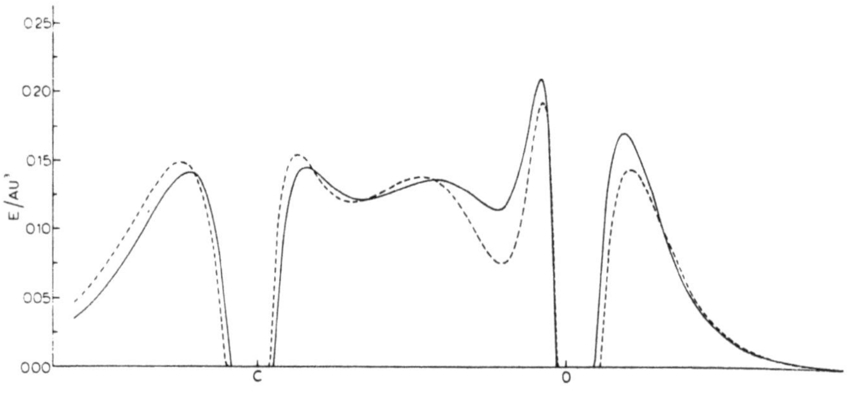


Fig. 5. Theoretical charge density differences in carbon monoxide, plotted along C-O bond axis (Ref. 29). Dashed line : Hartree Fock; solid line: Hartree-Fock-Slater; both calculations made with extended Slater-type orbital basis. Calculations made with smaller (double zeta) bases show a pronounced dip in the bonding region.

Not only are such calculations simpler and faster to perform, but it is also often easier to extract information about bonding and near-neighbor interactions than to go through the machinery of band theory. For models of defects and impurities where the periodicity of the crystal lattice is broken, advantages of a cluster approach are even more obvious.

It has become increasingly clear that an essential part of the cluster method involves the use of constraints which couple the cluster to the host medium. The alternative of treating larger and larger free clusters rapidly becomes unattractive. The development of molecular-field potentials and boundary conditions suitable for the cluster embedding problem is still taking place /30/. We will describe a simple self-consistent embedding scheme and its application to crystalline $NbO_2$.

$NbO_2$ is a member of the interesting group of transition metal oxides which exhibit metal-insulator or metal-semiconductor transitions as a function of temperature or applied pressure. In the high temperature rutile phase $NbO_2$ is metallic with conduction processes dominated by the Nb 4d band complex. Each niobium atom is coordinated to six oxygen atoms in distorted octahedral geometry, and Nb atoms form a pseudo-one dimensional chain along the crystal c-axis. In the low temperature semiconducting phase, Nb atoms move together to form pairs along the chain, accompanied by minor rearrangement of oxygen positions. This phenomenon reminds us of the famous Peierls transition which is often quoted as an example of electron correlation effects. Many experimental and theoretical efforts have been aimed toward understanding this phenomenon /31/. Experimental efforts to detect changes in the electron momentum distribution, by positron annihilation and Compton scattering, are particularly relevant to this meeting /32/.

Two sets of cluster calculations were made /33/: one with the nearest neighbors, $NbO_6$ and the second-neighbor cluster, $Nb_3O_6$. Although the formal valence of the atoms corresponds to $Nb^{4+}$ and $O^{2-}$, a mixture of covalent and ionic bonding is expected and so the number of electrons associated with a given cluster has to be determined as part of the self-consistent potential iterations. The cluster wavefunctions were expanded in a linear combination of atomic orbitals, with the basis set chosen to be numerical solutions of the HFS equations for free atoms, as in Eqn. (17). For a given cluster potential, matrix elements of the Hamiltonian and overlap were calculated by numerical integration, and the secular equation (18) was solved to obtain energies and eigenvectors.

The cluster potential which was adopted can be written as

$$V = V_{Coul} + V_{ex} + V_{PP} \tag{20}$$

where $V_{Coul}$ and $V_{ex}$ are the Coulomb and exchange potentials constructed from the charge density of the entire crystal (cluster plus surrounding atoms) and thus has the full perioaicity of the crystal. The repulsive pseudopotential term $V_{PP}$ is added to keep cluster electrons from populating core and valence levels of atoms exterior to the cluster. The physics of $V_{PP}$ is an expression of the Pauli exclusion principle; its form can be derived by arguments of pseudopotential theory, or by simply orthogonalizing the cluster basis to exterior atom states. To maintain the simplicity of the model we have parametrized $V_{PP}$, by truncating the potential at a value $V_F$ in regions exterior to the cluster /34/. This amounts to clipping off the deep potential wells of exterior ions (see Fig. 6); results are found to be rather insensitive to $V_F$. The iteration loop potential-eigenvector-potential is closed by projecting the eigenvectors onto a set of atomic occupation numbers, similar to the Mulliken polulation analysis, and substituting these atomic populations throughout the crystal for the subsequent calculation of potential. The iteration process is terminated when input and output populations agree to the required precision.

A crystal momentum density and directional Compton profiles can be synthesized from the cluster results, after the latter have been broken down into Nb-Nb, Nb-O, and O-O components. The smallest cluster will essentially inelude only nearest neighbor contributions, while larger clusters contain second neighbor and

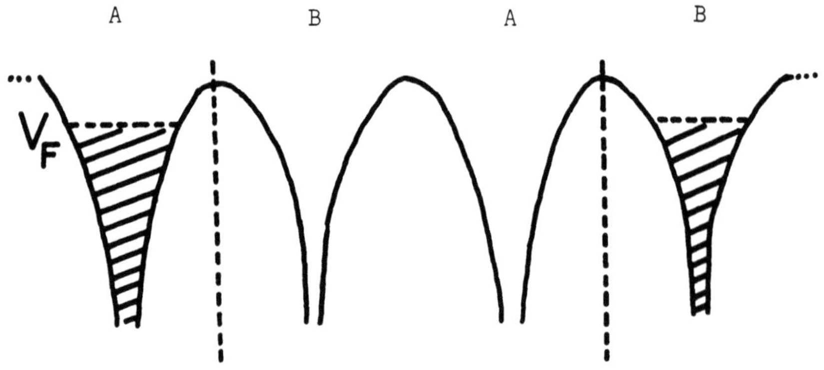


Fig. 6. Model potential for cluster embedded in solid (from Ref. 34) note truncation parameter $V_F$ for potential wells of exterior ions.

higher interactions as well. Thus in comparing the $NbO_6$ and $Nb_3O_6$ clusters one would be particularly interested in the role of the metal-metal interaction, as revealed by $\tilde{\rho}(\vec{p})$. The cluster (and crystal) momentum density can be expanded in spherical harmonics,

$$\tilde{\rho}(\vec{p}) = \sum_{\ell m} \tilde{\rho}_{\ell m}(p)\ Y_{\ell m}(\hat{p}) \tag{21}$$

with radial expansion functions $\tilde{\rho}_{\ell m}$ determined by numerical least squares procedures. In practice, the spherical harmonics $Y_{\ell m}$ are combined together to form symmetrized combinations characteristic of the point symmetry of the system. This expansion is useful in obtaining the corresponding expansion of the directional Compton profile

$$J(q, \hat{k}) = \sum_{\ell m} J_{\ell m}(q) Y_{\ell m}(\hat{k}) \tag{22}$$

since the expansion functions $J_{\ell m}(q)$ can be obtained by one-dimensional numerical integration:

$$J_{\ell m}(q) = 2\pi \int_{|q|}^{\infty} \tilde{\rho}_{\ell m}(p)\ P_{\ell}(q/p)p dp \tag{23}$$

where $P_{\ell}$ is a Legendre polynomial /35-37/.

In Fig. 7 the spherical, $\ell = 0$, part of the Compton profile (CP) of the rutile phase of $NbO_2$ is given, as inferred from the $NbO_6$ cluster results. The rather featureless total CP is shown in the first panel, followed by one-center Nb-Nb, bonding Nb-O, and oxygen-oxygen contributions. Contributions of the Nb-O bonding momentum density to the CP are seen to be sizable, with considerable variation in the range $0 \leq p \leq 2$ a.u. However, they only account for a few per cent of the scattering intensity in that momentum range. Changes in the anisotropy of the CP which would give desired information about electronic rearrangement in the phase transition represent a further challenge to experimentalists. As an example, we show the difference in CP's for nickel along two crystallographic directions, (100)-(110), in Fig.8. Since it appears that relative differences of $\sim$0.1% in CP can be determined we may hope to obtain contour maps of CP anisotropy in d-electron compounds in the near future.

## D. ENERGY BAND THEORY

Recall again that in local density theory we have a Hamiltonian which is a specified functional of the first order density $\rho^1$, where

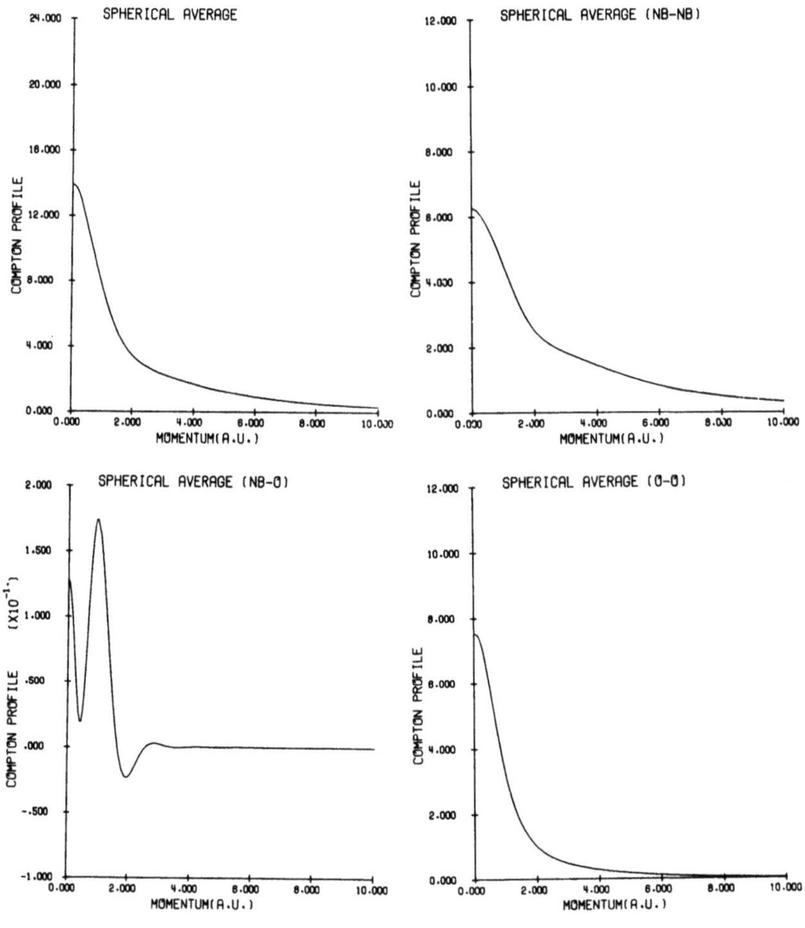


Fig. 7 Calculated $\ell = 0$ part of Compton profile for $NbO_2$ (Ref. 33): (a) total CP, (b) One-center Nb contribution, (c) Bonding Nb-O contribution, (d) Oxygen-oxygen contributions.

h can be written in the form of Eqn. (12) (nonrelativistic) or Eqn. (14) (relativistic). If $\rho^1$ is periodic $\rho^1(\vec{r} + \vec{R}_\mu | \vec{r}' + \vec{R}_\mu) = \rho^1(\vec{r}|\vec{r}')$ such that h commutes with the lattice translation operators $T(\vec{R}_\mu)$ we know that the stationary states can be obtained as Bloch functions,

$$[h - \epsilon_n(\vec{k})]\Psi_n(\vec{k}, \vec{r}) = 0 \tag{24}$$

where $T(\vec{R}_\mu)\Psi_n(\vec{k},\vec{r}) = e^{i\vec{k}\cdot\vec{R}_\mu}\,\Psi_n(\vec{k}, \vec{r})$. (25)

Similarly, the charge density can be written as

$$\rho(\vec{r}) = \sum_{n,\vec{k}} f_n(\vec{k}) |\Psi_n(\vec{k}, \vec{r})|^2 \tag{26}$$

where the occupation numbers $f_n(\vec{k})$ are unity for $\epsilon_n(\vec{k}) < E_F$ (Fermi energy) and zero otherwise, at a temperature $T = 0$. Fermi-Dirac statistics govern the occupation numbers at finite temperature. For a given crystal, only a finite number of bands, with band index n, will be occupied; however, the number of states in a given band, denoted by wave vector $\vec{k}$, is quasi-infinite. In practical applications a certain discrete set of states, $\{k_i\}$, will be sampled and the integration over $\vec{k}$ implied in Eqn. (26) will be replaced by a weighted sum.

Norman March and Art Freeman have already spoken about many properties of band electron states; our purpose here will be to provide a few additional details and elaborations. A great variety of computational methods have been developed to calculate band states, so that one really needs a glossary. Among the active methods are: Augmented Plane Wave (APW), Linear Combination of Atomic Orbitals (LCAO), Scattered Wave/Green's function (KKR), Orthogonalized Plane Wave (OPW), Pseudopotential (PP), Linear Muffin-Tin Orbital (LMTO), etc., but with rare exceptions these methods (and more!) will produce the same results, when applied c carefully to the same crystal potential, selection of method being a matter of efficiency and individual taste.

With modern developments in computational technique theoretical descriptions not only of the energy band spectrum and related properties, but also the spatial densities depending explicitly on the wave functions have become available. Let us consider the momentum distribution first. The momentum wavefunction corresponding to a particular Bloch state is

$$\chi_n(\vec{k}, \vec{p}) = (2\pi)^{-3/2} \int \exp(-i\vec{p}\cdot\vec{r}) \Psi_n(\vec{k},\vec{r}) d^3r \tag{27a}$$

$$= \delta(\vec{k}-\vec{p}+\vec{K}_\mu) b_n(\vec{k}, \vec{p}) \tag{27b}$$

where $\vec{K}_\mu$ is a reciprocal lattice vector.

We can interpret this result as the product of a lattice momentum selection rule and the Fourier transform of a function $a_n(\vec{k}, \vec{r})$ associated with a single cell of the crystal lattice. There are interesting connections between this function and the Wannier theory of localized states in crystals. Now we may construct the crystal momentum density,

$$\tilde{\rho}(\vec{p}) = \sum_{n,\vec{k}} f_n(\vec{k}) |\chi_n(\vec{k}, \vec{p})|^2 \tag{28}$$

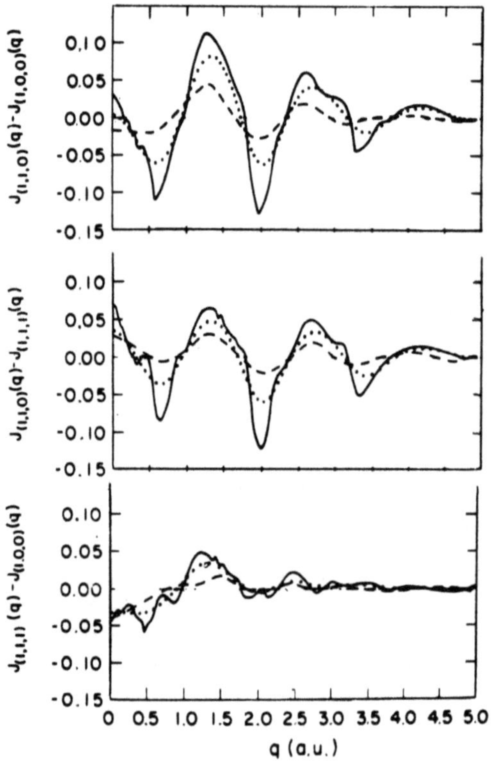


Fig. 8 Anisotropy of the Compton profile of nickel, from Ref. 39. Band theory results (solid line) are also shown broadened with the experimental resolution function (dotted line). Experimental results of Ref. 44 are given as the dashed line.

noting that the selection rule in Eqn. (27b) will reduce the k-integration to a simple sum. The calculation of Compton profiles is then straightforward, so that we can make some comparisons with experiment. However, it is important to reiterate a comment and warning by Professor March: We have taken an indirect route to obtain $\tilde{\rho}(\vec{p})$ via an approximate one-electron calculation of $\Psi_n(\vec{k}, \vec{r})$ using a potential optimized for a particular total energy expression. The "optimum" $\tilde{\rho}(\vec{p})$ may be determined by a rather different crystal potential; nevertheless available comparisons suggest that the differences are not very large.

The LCAO-Gaussian basis studies on the 3d transition metals by Callaway and coworkers have provided a great deal of insight into capabilities of local density models for describing charge, spin, and momentum densities /38-41/. Their work and that of Wakoh and Yamashita /42,42/ has been of key importance in understanding the 3d-metal Compton data. There appears to be rather good quantitative agreement between theory and experiment for the total, spherically averaged, profile. The anisotropy in $J(q, \hat{k})$ calculated for nickel and iron is at least in semi-quantitative agreement with experiment, as shown in for Ni in Fig. 8, suggesting that Fermi surface features can be detected /39,44/. For iron, the spin-dependence of the Compton profile was also successfully calculated /40/. The recent work on vanadium is interesting, since a detailed prediction of the anisotropy in J is given /41/, which is apparently in gross disagreement with experiment. (The predicted difference (111)-(100) appears to be ~4 times too large.) Dick Weiss has pointed out at this meeting and elsewhere that a simple atomic model, with two different radial 3d functions (for $e_g$ and $t_{2g}$ symmetries in crystal field notation) and empirically determined orbital populations can give a satisfactory fit to both x-ray and Compton data /45/. He further suggests that the band calculations could be improved by allowing greater radial freedom. However, this can hardly be the case since radial functions are determined independently at each $\vec{k}$ point in quantitative band calculations, which have much greater variational freedom than two or three parameter models. So the discrepancy, if it persists, is far more serious and points to some basic problem either in one-electron theory or in interpretation of the data.

It is worthwhile to consider the x-ray form factors for vanadium for further clues as to whether there are basic problems. In Table I we give a number of recent theoretical results and experimental data. The systematic deviation between theory and experiment for increasing $\vec{K}$ is due primarily to thermal smearing effects. The traditional scheme of applying a spherically symmetric Debye-Waller factor to correct the core form factors largely removes the discrepancy, as indicated in the table /46/. Some discussion has been given of the anisotropy of the charge distribution $\rho(\vec{r})$ indicated by the ratios of scattering factors for the paired reflections $(hk\ell)$ = (330)/(411) and (442)/(600). The experimental ratios are about 1.02 and 1.04 respectively, and are considerably larger than band-theoretic ratios. Thus theoretical anisotropies for vanadium appear to be too large in $\vec{p}$-space and too small in $\vec{r}$-space. While further studies on vanadium are clearly called for, it would also be extremely interesting to obtain good quality x-ray and Compton data for the 4d-metal niobium, where very intensive theoretical work has also been done/49/.

Table I. Comparison of theoretical and experimental x-ray scattering factors in vanadium.

| $(hk\ell)$ | DDK[a] | LWC[b] | BMK[c] | WKY[d] | D-W[e] | EXP[f] | EXP[g] |
|---|---|---|---|---|---|---|---|
| 110 | 15.72 | 15.75 | 15.82 | 15.84 | 0.29 | 15.62 | 15.90±0.18 |
| 200 | 13.06 | 13.11 | 13.13 | 13.15 | 0.49 | 12.64 | 13.22±0.17 |
| 211 | 11.40 | 11.43 | 11.41 | 11.36 | 0.65 | 10.67 | |
| 220 | 10.20 | 10.23 | 10.19 | 10.12 | 0.77 | 9.61 | |
| 310 | 9.30 | 9.32 | 9.32 | 9.25 | 0.88 | 8.48 | |
| 222 | 8.71 | 8.73 | 8.67 | 8.63 | 0.98 | 7.44 | |
| 321 | 8.20 | 8.22 | 8.17 | 8.16 | 1.08 | 7.16 | |
| 400 | 7.76 | 7.78 | 7.80 | 7.75 | 1.16 | 6.55 | |
| 330 | 7.50 | 7.52 | 7.49 | 7.48 | 1.25 | 6.31 | |
| 411 | 7.47 | 7.49 | 7.48 | 7.47 | 1.25 | 6.31 | |
| 420 | 7.23 | 7.24 | 7.23 | 7.22 | 1.33 | 5.84 | |

a) Self-consistent APW method with complete potential; D. D. Koelling (unpublished).

b) Self-consistent LCAO-Gaussian basis method, Reference 41.

c) Self-consistent APW method with muffin-tin potential; B. M. Klein (unpublished).

d) Self-consistent APW method; the D-W correction indicated in the following column has been subtracted from the published results (Ref. 43) for ease of comparison.

e) Debye-Waller correction for finite temperature experimental data.

f) Experimental data, Reference 47.

g) Experiment, Reference 48.

Freeman and coworkers have been very active in developing band techniques, usually using the APW method, capable of describing conduction electron polarization, spin densities, and the neutron magnetic form factors of heavy metals /50-52/. Magnetic properties of heavy metals are complicated by relativistic effects and the admixture of spin and orbital contributions to the magnetization density. In cases like the rare earths with a very localized 4f shell carrying the majority of the moment, it has even been unclear whether band theory was applicable, or whether intraatomic correlations had to be treated explicitly. Since the magnetic scattering amplitude and magnetization are related by

$$F(\vec{k}) = \int \exp(i\vec{k}\cdot\vec{r})m(\vec{r})d^3r \tag{29}$$

experimentalists have been able to invert their data and prepare detailed maps of $m(\vec{r})$ to bedevil their theorist friends. Parametrization of $m(\vec{r})$ usually proceeds by finding a localized "atomic moment" contribution and a delocalized "conduction band" part. Although they are typically not carried to self-consistency band-theory calculations have at least progressed so that calculated and experimental conduction band contributions in metals like gadolinium strongly resemble each other /50/.

For paramagnetic metals like Pd and Pt where interest centers on the magnetic-field induced conduction electron polarization, the theory seems to be quite accurate. For the case of palladium, excellent agreement was found with the magnetic form factor of Cable <u>et al</u>./53/. The total spherical part of the induced spin density of platinum calculated by the relativistic APW method is shown in Fig. 9. It is interesting that the magnetization density found here closely resembles that of the $Pt^{2+}$ ion, calculated in the same one-electron model. It is also an important prediction that the induced <u>spin</u> density is highly anisotropic, and almost purely $t_{2g}$ in character.

As a final example of the applications of band theory we return to the LCAO scheme and studies of the Compton profile of compounds. The tetrahedrally bonded semiconductors are important here because of interest in the covalent bond; diamond is unique in that it has been studied by a variety of theoretical methods /37/. By carrying out measurements and calculations on a series of related compounds we can hope to draw general conclusions about the momentum-space representation of the bonding density. This hope seems to be realized for the spherical part of the Compton profile and the corresponding momentum density; e.g., $J_{oo}(q)$ when plotted in the appropriate reduced units is virtually indistinguishable for diamond, silicon, and silicon carbide. Anisotropic contributions show tantalizing similarities and dissimilarities, since some features are clearly controlled by the

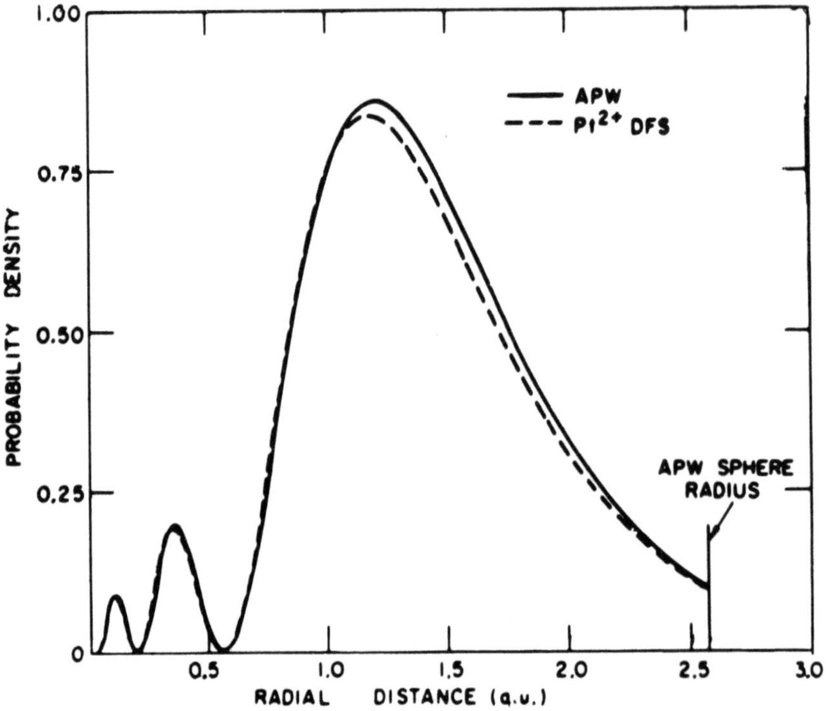


Fig. 9. The total spherical part of the induced spin density of platinum, calculated by the RAPW method (solid line), and the charge density of the $Pt^{2+}$ ion calculated by the DFS theory (dotted line). (From Ref. 52)

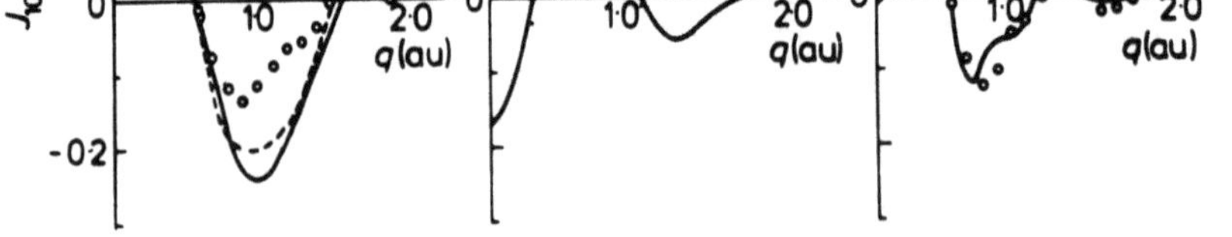

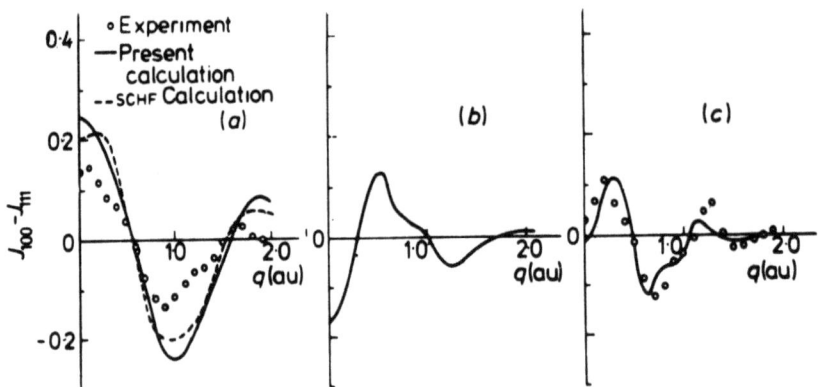

Fig. 10 The difference between Compton profile for directions <100> and <111> of (a) diamond, (b) SiC, and (c) silicon from Ref. 37.

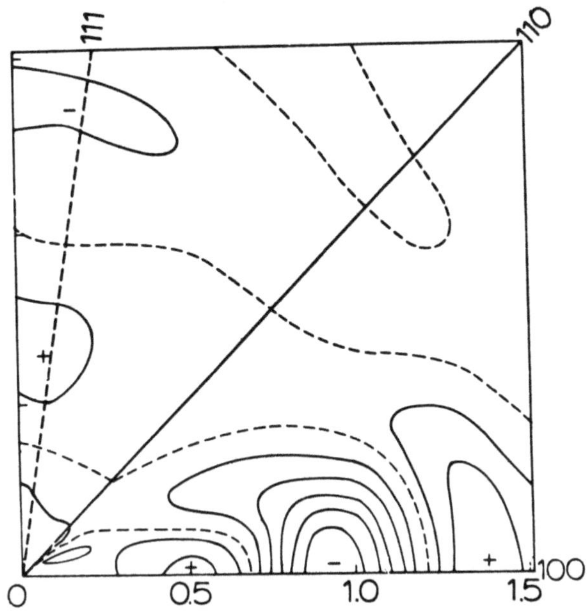


Fig. 11. Calculated anisotropy in J(q, $\hat{k}$) for TiC, from Ref. 54. Dotted line denotes zero anisotropy contour; contour level intervals are 0.1 a.u.

geometry of the bond, and others by the shape of the bonding density. In Fig. 10 we show anisotropy in J, (100) - (111), for the three crystals. The agreement between theory and available experimental data is generally of the order of 2-5%.

A very detailed comparison of experiment and band theory for TiC has been carried out by Seth et al., again achieving very satisfactory quantitative agreement for both spherical and anisotropic parts of J(q, $\hat{k}$) /54/. In this rock salt structure where each atom lies in six-fold cubic coordination to its neighbors, momentum density is found to pile up along the (100) bonding directions. A contour map of the resulting anisotropy in J, showing the corresponding increase in scattering intensity along (100), is shown in Fig. 11.

ACKNOWLEDGMENTS

Thanks are especially due to Art Freeman, Dale Koelling, Dick Weiss and Barry Klein for ideas, suggestions, and access to unpublished data. This work was supported in part by the National Science Foundation, Grant No. DMR 77-22644. We gratefully acknowledge use of the facilities of Northwestern's Materials Research Center.

## REFERENCES

1. R. McWeeny and B.T. Sutcliffe, Methods of Molecular Quantum Mechanics, (Academic, New York, 1969).

2. R. Benesch and V.H. Smith, Jr., in Wave Mechanics - The First Fifty Years, W.C. Price, S.S. Chissick and T. Ravensdale, editors (Butterworths, London, 1973); Acta Cryst. A26, 579 (1970).

3. P.O. Löwdin, Phys. Rev. 97, 1474 (1955).

4. J.C. Slater, Quantum Theory of Molecules and Solids, Vol. I (McGraw-Hill, New York, 1963) App. 4.

5. P. Eisenberger and P.M. Platzman, Phys. Rev. A2, 415 (1970).

6. B. Williams, editor, Compton Scattering (McGraw-Hill, New York, 1977).

7. P. Pattison and B. Williams, Sol. Sta. Commun. 20, 585 (1976); B. Kramer, P. Krusius, W. Schröder and W. Schülke, Phys. Rev. Letters 38, 1227 (1977).

8. J.C. Slater, Quantum Theory of Molecules and Solids, Vol. I (McGraw-Hill, New York, 1973) App. 7.

9. C.C.J. Roothaan, Rev. Mod. Phys. 32, 179 (1960).

10. J.C. Phillips and L. Kleinman, Phys. Rev. 116, 287 (1959).

11. W.A. Harrison, Pseudopotentials in the Theory of Metals (Benjamin, New York, 1966).

12. J.C. Slater, Phys. Rev. 81, 385 (1951).

13. R. Gaspar, Acta. Phys. Acad. Sci. Hung. 3, 263 (1954).

14. W. Kohn and L.J. Sham, Phys. Rev. 140, A1133 (1965).

15. For example, see U. von Barth and L. Hedin, J. Phys. C5, 1629 (1972); L. Hedin and B.I. Lundquist, J. Phys. C4, 2064 (1971).

16. J.C. Slater, The Self-Consistent Field for Molecules and Solids (McGraw-Hill, New York, 1974) p. 51.

17. T. Ziegler and A. Rauk, Theor. Chim. Acta 46, 1 (1977).

18. H.A. Bethe and R.W. Jackson, Intermediate Quantum Mechanics, 2nd edition (Benjamin, New York, 1968).

19. I. Lindgren and A. Rosen, Case Studies Atom. Phys. 4, 93 (1974); ibid, 197 (1974).

20. J.P. Desclaux, D.F. Mayers and F. O'Brien, J. Phys B: Atom. Molec. Phys. 4, 631 (1971); J.P. Desclaux and P. Pyykkö, Chem. Phys. Lett. 29, 534 (1974).

21. D.A. Liberman, J.T. Waber and D.T. Cromer, Phys. Rev. 137, A27 (1965); J. Chem. Phys. 51, 664 (1969).

22. A. Rosén and D.E. Ellis, Chem. Phys. Lett. 27, 595 (1974); J. Chem. Phys. 62, 3039 (1975).

23. P.F. Walch and D.E. Ellis, J. Chem. Phys. 65, 2387 (1976); D.D. Koelling, D.E. Ellis and R.J. Bartlett, J. Chem. Phys. 65, 3331 (1976).

24. C.Y. Yang, Chem. Phys. Lett. 41, 588 (1976); C.Y. Yang and S. Rabii, Phys. Rev. A12, 362 (1975); B.G. Cartling and D.M. Whitmore, Chem. Phys. Lett. 35, 51 (1975).

25. D.E. Ellis, J. Phys. B: Atom. Moelc. Phys. 10, 1 (1977).

26. D.E. Ellis, Intern. J. Quantum Chem. Symp. 11, 201 (1977); D.E. Ellis, V.A. Gubanov and A. Rosén, Proc. 3rd Intern. Conf. Elec. Struc. Actinides, Grenoble, France, Aug. 1978 (to be published).

27. U.F. Klein, G. Wortmann and G.M. Kalvius, Proc. ICM-73 (Moscow) IV, 149 (1974); G.M. Kalvius, U.F. Klein and G. Wortmann, J. Phys. (Paris) 35, C6-139 (1975).

28. For an introduction to the scattered wave technique for molecules, see K.H. Johnson, J. Chem. Phys. 45, 3085 (1966); for LCAO variational methods, see E.J. Baerends, D.E. Ellis and P. Ros, Chem. Phys. 2, 41 (1973); E.J. Baerends and P. Ros, Chem. Phys 2, 52 (1973); also H. Sambe and R.H. Felton, J. Chem. Phys. 62, 1122 (1975).

29. E.J. Baerends and P. Ros. Intern. J. Quantum Chem. Symp. (to be published). This review contains many useful references.

30. See for example, R.P. Messmer, C.W. Tucker, Jr. and K.H. Johnson, Chem. Phys. Lett, 36, 423 (1975); R.P. Messmer, S.K. Knudson, K.H. Johnson, J.B. Diamond and C.Y. Yang, Phys. Rev. B13, 1396 (1976); P.J. Jennings, G.S. Painter and R.O. Jones, Surface Sci. 60, 255 (1976); D.E. Ellis, G.A. Benesh and E. Byrom, Phys. Rev. B16, 3308 (1977); M. Gupta, V.A. Gubanov and D.E. Ellis, J. Phys. Chem. Solids 38, 499 (1977); V.A. Gubanov, D.E. Ellis and A.A. Fotiev, J. Sol. Sta. Chem. 21, 303 (1977).

31. M. Posternak, A.J. Freeman and D.E. Ellis, Phys. (submitted for publication).

32. But note lack of success in studies on $V_2O_3$ and $Fe_3O_4$: A. Greenberger and S. Berko, Bull. Amer. Phys. Soc. 17, 358 (1972); P.E. Mijnarends and R.M. Singru, Appl. Phys. 4, 303 (1974); R. Lässer, R.M. Singru and B. Lengeler, Sol. Sta. Commun. 25, 345 (1978).

33. C. Umrigar and D.E. Ellis (to be published).

34. D.E. Ellis, G.A. Benesh and E. Byrom, Phys. Rev. B16, 3308 (1977); J. Appl. Phys. 49, 1543 (1978).

35. P.E. Mijanarends, Phys. Rev. 160, 512 (1967); Physica 63, 235 (1973); Physica 63, 248 (1973).

36. R. M. Singru and P.E. Mijnarends, Phys. Rev. B9, 2372 (1974).

37. A. Seth and D.E. Ellis, J. Phys. C10, 181 (1977) and references therein.

38. J. Rath, C.S. Wang, R.A. Tawil and J. Callaway, Phys. Rev. B8, 5139 (1973).

39. C.S. Wang and J. Callaway, Phys. Rev. B11, 2417 (1975); Phys. Rev. B15, 298 (1977).

40. J. Callaway and C.S. Wang, Phys. Rev. B16, 2095 (1977).

41. D.G. Laurent, C.S. Wang and J. Callaway, Phys. Rev. B17, 455 (1978).

42. S. Wakoh and J. Yamashita, J. Phys. Sec. Japan 30, 422 (1971) J. Phys. Soc. Japan 21, 1712 (1966); 25, 1272 (1968); 35, 1394 (1973).

43. S. Wakoh, Y. Kubo and J. Yamashita, J. Phys. Soc. Japan 40, 1043 (1976).

44. P. Eisenberger and W.A. Reed, Phys. Rev. B9, 3242 (1974).

45. R.J. Weiss, Phil. Mag. (to be published) 1978.

46. We thank D. D. Koelling for assembling most of this information.

47. M.V. Linkeaho, Phys. Scripta 5, 271 (1972).

48. O. Terasaki, Y. Uchida and D. Watanabe, J. Phys. Soc. Japan

39, 1277 (1975).

49. N. Elyashar and D.D. Koelling, Phys. Rev. B15, 3620 (1977) and references therein.

50. B.N. Harmon and A.J. Freeman, Phys. Rev. B10, 1979 (1974).

51. A.J. Freeman, B.N. Harmon and T.J. Watson-Yang, Phys. Rev. Letters 34, 281 (1975); T.J. Watson-Yang, B.N. Harmon and A.J. Freeman, J. Mag. and Magn. Mat. 2, 334 (1976).

52. T.J. Watson-Yang, A.J. Freeman and D.D. Koelling, J. Mag. and Magn. Mat. 5, 277 (1977).

53. J.W. Cable, E.O. Wollan, G.P. Felcher, T.O. Brun and S.P. Hornfeldt, Phys. Rev. Letters 34, 278 (1975).

54. A. Seth, T. Paakkari, S. Manninen and A. Christensen, J. Phys. C10, 3127 (1977).

# Part II
# Diffraction Physics and Experimental Problems

INTRODUCTION TO WAVE SCATTERING BY ATOMS AND MOLECULES IN THE FREE STATE

R. A. Bonham

Chemistry Department, Indiana University

Bloomington, Indiana

## I. DERIVATION OF FUNDAMENTAL EQUATIONS BY CLASSICAL WAVE CONSTRUCTION

### A. Introduction and Mathematical Preliminaries

Let us imagine an incident plane wave source of radiation impinging on a fixed array of scattering centers. We graphically represent such a situation by drawing lines of maximum wave amplitude. The incident source originating to the readers left and traveling or propagating toward the readers right will be represented by a series of parallel vertical lines with the inter line spacing corresponding to the wavelength. As the incident wave arrives at a particular scattering center a disturbance or scattering occurs, and the part of the incident wave impinging on the scattering point is re-emitted in the form of an outgoing spherical wave. This outgoing spherical wave is represented by a series of concentric circles again with each circle representing a maxima of the wave amplitude. The inter circle spacings will be constant and will correspond to the wavelength of the diffracted wave. This wavelength of the scattered wave does not have to be the same as that for the incident plane wave. If it is the same the wave diffraction is said to be elastic, that is no energy transfer takes place between the incident wave and the diffraction object. If the diffracted wave length is different the incident wave is said to be scattered inelastically. In Fig. 1 a schematic picture of the situation for elastic scattering is shown.

In order to render such a simple picture into quantitative terms it is necessary to introduce a mathematical model. Let us first consider a solution of Maxwell's wave equation for field free space as

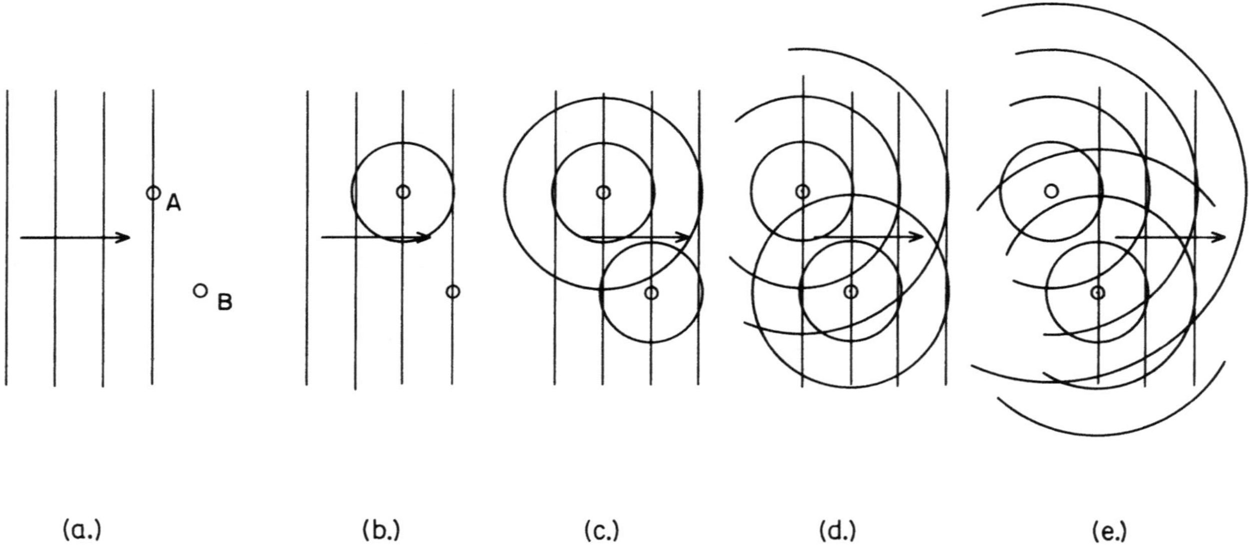


Fig. 1. An incident plane wave passing through two diffracting points. The progression of the wave through the two points is followed from left to right in Figures (a.) through (e). The diffracted waves are shown by circles. Note that for clarity only the same four maximum amplitude lines of the incident plane wave are shown.

$$[\nabla^2 - \frac{1}{c^2}(\frac{\partial}{\partial t})^2] \left\{ \begin{matrix} \vec{E} \\ \vec{H} \end{matrix} \right\} = 0 \qquad (1.0)$$

where $\vec{E}$ is the electric field vector and $\vec{H}$ is the magnetic field vector. The symbol { } indicates that $\vec{E}$ and $\vec{H}$ are separately solutions of the wave equation. The solution to Eq. (1.0) can be found to be

$$\left\{ \begin{matrix} \vec{E} \\ \vec{H} \end{matrix} \right\} = \left\{ \begin{matrix} \vec{E}_0 \\ \vec{H}_0 \end{matrix} \right\} e^{i(\vec{k}\cdot\vec{r} - \omega t)} \qquad (1.1)$$

where $k = \frac{2\pi}{\lambda}$ with $\lambda$ the wavelength and $\omega = 2\pi\nu$ with $\nu$ the frequency of the incident radiation. The vector $\vec{r}$ denotes a point in space and t is the time. For the plane wave shown in Fig. 1 we take $\vec{k}$ to be in the direction of propagation (parallel to the arrows) and we choose this direction (left to right) as the positive z direction in a cartesian coordinate system. The amplitude of an incident plane wave traveling in the z direction (left to right) will henceforth be written as

$$\Psi(r,t) = Ae^{i(kz - \omega t)} \qquad (1.2)$$

where A is a constant.

Problem 1.0 Show that Eq. (1.2) is a solution of Eq. (1.0).
Hint: Recall that $\nu\lambda = c$ for electromagnetic radiation.

What about the scattered waves? It is clear that the section of an incident wave which encounters a diffracting point will be scattered. Let us call the width of this section $\epsilon$. Along any line passing through the center of a scattering point the wave amplitude will clearly have the form

$$\Psi_{sc}(\vec{r},t) = A(r)e^{i(kr - \omega t)} \qquad (1.3)$$

That is along any radius from a scattering center the outgoing scattered wave looks like a plane wave. Note however that we have not assumed that A(r) is a constant!

In electromagnetic theory the energy stored in an electric field $\vec{E}$ is

$$W = \frac{1}{2}\int dV|\vec{E}|^2 \qquad (1.4)$$

If we consider the energy content of an electromagnetic wave in a shell of width $\epsilon$, $\epsilon \ll 1$, a radius $r_0$ from the point of a disturbance we have from Eq. (1.3) and (1.4)

$$W = \frac{1}{2}(4\pi)\int_{r_0 - \epsilon/2}^{r_0 + \epsilon/2} dr\, r^2|A(r)|^2 \qquad (1.5)$$

At some other distance $r_0'$ from the point of disturbance we can calculate the energy content also for a shell of width $\epsilon$. The energy calculated in spherical polar coordinates is again

$$W = \frac{1}{2}(4\pi)\int_{r_0' - \epsilon/2}^{r_0' + \epsilon/2} dr\ r^2|A(r)|^2 \tag{1.6}$$

where since the energy content in a portion of the wave of width $\epsilon$ along r must be a constant independent of r in order to conserve energy we must choose A(r) such that

$$\Psi_{sc}(r,t) = \frac{A(0)}{r} e^{i(kr - \omega t)} \tag{1.7}$$

We will later encounter cases where A(0) may depend on scattering angle but at this stage we assume that it is a constant. This will be approximately true for x-ray and neutron scattering but not electron scattering.

---

Problem 1.1 Show that Eq. (1.7) when used in Eq. (1.6) leads to a result independent of $r_0'$ for the energy content of the scattered wave. Also show that Eq. (1.7) is a solution of Eq. (1.0).

---

We are now in a position to consider the scattering from any array of point scatterers. First it is necessary to select a reference time from which the experiment can begin. We arbitrarily select the time at which the first wave amplitude maximum reaches the center of mass of a collection of diffracting mass points as our time origin. We note as shown in Fig. 2 that the first wave maximum may arrive at some diffracting points before reaching the mass center and in some cases after reaching the mass center. This means that a scattered wave will have an amplitude at the point of detection (P in Fig. 2) of

$$\Psi^j_{sc}(\vec{r},t) = \frac{A_j(0)}{|\vec{r}-\vec{r}_j|} e^{i(k|\vec{r}-\vec{r}_j| - \omega t + \omega t_j)} \tag{1.8}$$

where $|\vec{r}-\vec{r}_j|$ is the distance between the jth diffracting point and the detecting point P. The time t is the time the wave takes to travel from the center of mass point to the detector P and $t_j$ is the time required for the first incident plane wave maximum to reach the jth disturbing point after passing the center of mass. Note that if the disturbing point lies to the left of the center of mass that $t_j$ will be negative.

For electromagnetic radiation we have $\nu\lambda = c = \omega/k$ and $t_j = \frac{z_j}{c}$ so that the scattered wave can finally be written as

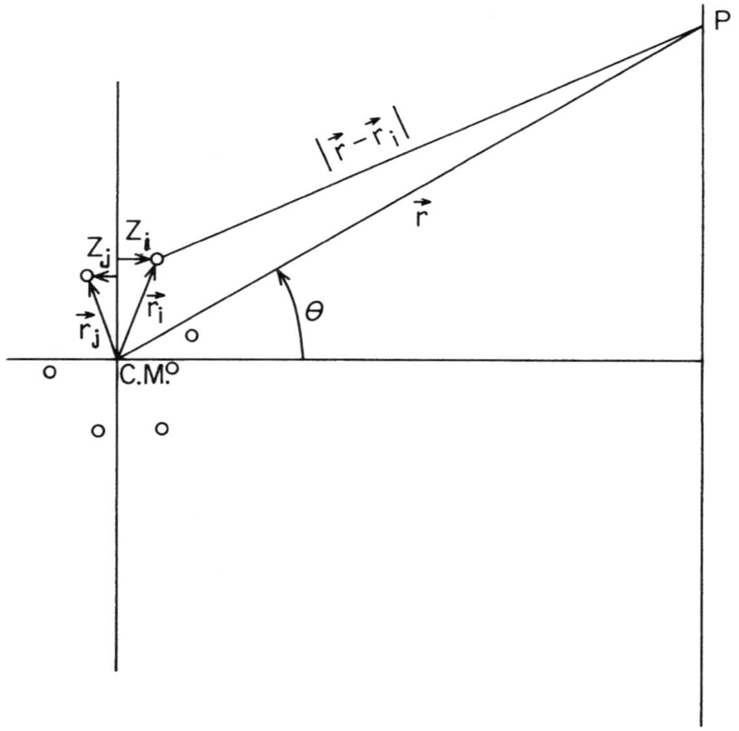


Fig. 2. Arrangement of scattering centers about the center of mass (C.M.) and the scattering angle θ for detection of scattered radiation from the ith scatterer at the point P.

$$\Psi^j_{sc}(\vec{r},t) = \frac{A_j(0)}{|\vec{r}-\vec{r}_j|} e^{i(k|\vec{r}-\vec{r}_j| + kz_j - \omega t)} \tag{1.9}$$

---

Problem 1.2 Derive Eq. (1.9) for matter waves where $E = \hbar^2k^2/2m$. Hint: The particle velocity is not given by $v = D/t$ but must be inferred from the phase of the particle wave packet.

---

We next assume that the total amplitude observed at the point P at a time t after the wave front reaches the mass center is simply the sum of the single diffracting point amplitudes given in Eq. (1.9) from all of the diffracting centers. In fact this is not strictly correct as a portion of the scattered wave from center j may re-scatter on encountering another center. One can imagine many such rescattering possibilities. These rescattering or non primary scattering processes are collectively referred to as multiple scattering. Here we shall focus only on the primary or single scattering. The assumption that the total amplitude is given by the sum of single scattering amplitudes is referred to as the principle of linear superposition.

Before proceeding it will prove helpful to take advantage of certain physical characteristics of all diffraction experiments on atoms and molecules. We will always be interested in diffracting objects (atoms and molecules) with a physical extent of the order of $10^{-8}$cm (Å's) while the wavelength of the probe radiation will be of a similar magnitude. The distance r between target and detector on the other hand will be of the order of 10's of centimeters. This means that in all cases that we shall study $|\vec{r}| >> |\vec{r}_j|$ or $z_j$ and hence we can write

$$|\vec{r}-\vec{r}_j| \cong r - \frac{\vec{r}}{r} \cdot \vec{r}_j \tag{1.10}$$

if $|\vec{r}-\vec{r}_j|$ occurs in the phase of a trigonometric function and

$$|\vec{r}-\vec{r}_j| \cong r \tag{1.11}$$

if $|\vec{r}-\vec{r}_j|$ does not occur in the phase of a trigonometric function.

---

Problem 1.3 Derive Eq. (1.10) and show that $\frac{1}{|\vec{r}-\vec{r}_j|} \cong \frac{1}{r}$ but that

$e^{ik|\vec{r}-\vec{r}_j|} \neq e^{ikr}$ or that $e^{ik|\vec{r}-\vec{r}_j|} \cong e^{ikr} \cdot e^{-ik\frac{\vec{r}}{r}\cdot r_j}$

---

Using Eqs. (1.10) and (1.11) and the principal of linear super position we can now write the total scattered amplitude as

$$\Psi_{sc}^{TOTAL}(\vec{r},t)=\sum_{j=1}^{N}\Psi_{sc}^{j}(\vec{r},t)=\frac{e^{i(kr-\omega t)}}{r}\sum_{j=1}^{N}A_j(0)e^{i(\vec{k}_i-\vec{k}_s)\cdot\vec{r}_j} \tag{1.12}$$

where $\vec{k}_i$ and $\vec{k}_s$ are vectors of magnitude k in the incident and scattered directions.

Problem 1.4 Derive (1.12) from Eqs. (1.10), (1.11) and the linear super position principal.

The scattered intensity at the point P in Fig. 2 can then be computed as

$$I=|\Psi_{sc}^{TOTAL}(\vec{r},t)|^2=\sum_{j}\frac{|A_j(0)|^2}{r^2}+2Rp\sum_{i>j}\sum\frac{A_i^*(0)\ A_j(0)}{r^2}e^{i(\vec{k}_i-\vec{k}_s)\cdot\vec{r}_{ij}} \tag{1.13}$$

where Rp signifies the real part of the complex quantity to the right.

Eqs. (1.12) and (1.13) will serve as the basis for analyzing all scattering from multi center systems. Before proceeding it is worthwhile to note that since the momentum of the radiation field is given by

$$\vec{p}=\frac{E}{c}\hat{k} \tag{1.14}$$

where E is the energy $\hbar ck$ and $\hat{k}$ is a unit vector in the direction of propagation of the field we may write $\vec{k}_i-\vec{k}_s$ as $(\vec{p}_i-\vec{p}_s)/\hbar$ or $\Delta p/\hbar$ where $\Delta\vec{p}$ is the momentum transferred to the target in the diffracting process. We will choose to label $|\Delta\vec{p}|$ by K although the symbols s and q are also often used (s is especially used in the case of x-ray scattering).

A second point is the meaning of the amplitudes at r = 0, the $A_j(0)$'s. Note in Eq. (1.13) that if only a single diffracting point exists, say the <u>jth</u> point, then only the single term $|A_j(0)|^2/r^2$ occurs. It is thus clear that $A_j(0)/r$ is the scattered amplitude for the incident radiation in question diffracted by the <u>jth</u> scattering point. The term furthest to the right in Eq. (1.13) arises from the interference of outgoing waves from two different centers. It is this term which contains the information concerning the detailed structure of the collection of diffracting points.

In order to apply Eq. (1.13) to actual situations of interest we need to consider how electromagnetic radiation, electrons and neutrons are scattered by planetary target electrons and nuclei in atoms and molecules. Consider first the case of scattering by an atom. In Table 1 we give the scattered intensity for each type of radiation from electrons and nuclei in terms of the effective cross sectional area possessed by the target. Roughly speaking x-rays are scattered only by the planetary electrons[1], neutrons by the nuclei[2] and electrons by both.[3] In the case of neutron scattering additional contributions can occur from spin interactions. Further the center of mass and laboratory coordinate systems for the neutron case will differ significantly and the cross section for scattering by a free atom is given as

$$\sigma_{\text{free atom}} = \frac{4\pi b^2}{(1 + \frac{m}{M})^2} \tag{1.15}$$

where m is the neutron rest mass, M is the mass of the target and b is the neutron scattering length.

In order to apply Eqs. (1.12) or (1.13) to scattering by an atom we can substitute the appropriate amplitude from Table 1 ($A_X$, $A_E$ or $A_N$) for $A_j(0)/r$. The individual particle cross sections $|A_X|^2$ for x-ray scattering and $|A_E|^2$ for electron scattering are simply the classical Thomson cross section, including polarization, and the Rutherford cross section respectively. Let us consider as a concrete example x-ray scattering. Because the x-ray scattering contribution from the nucleus is negligible we have

$$\Psi_{sc}^{TOTAL}(\vec{r},t) = A_X e^{i(kr - \omega t)} \sum_{j=1}^{N} e^{i\vec{K}\cdot\vec{r}_j} \tag{1.16}$$

or

$$I_{sc}^{TOTAL} = A_X^2 [N + 2 \sum_{i>j} e^{i\vec{K}\cdot\vec{r}_{ij}}] \tag{1.17}$$

where $|\vec{r}_j|$ is the distance from the mass center (nucleus) to the instantaneous position of the jth target electron and N is the total number of target electrons. Because the planetary electrons of the atom are in constant motion the x-rays will not see the same electronic configuration in two different atoms or even in the same atom if observed at different times. This means that we must carry out an average over all possible electronic configurations weighted by the probability that the electrons can be found in each possible configuration. For Eq. (1.16) it is clear that to carry out the required average we must know the probability of finding the jth electron at the point $\vec{r}_j$. This is equivalent to the probability of finding the electron in a volume element of vanishingly small size. We denote this probability or probability density by $\rho_j(\vec{r}_j)$ so that the average of the amplitude becomes

Table 1. The cross sections and scattered amplitudes for electromagnetic, electron and neutron scattering by both electrons and nuclei.

| Projectile | Usual Projectile Wave Length Range | Cross Section Electron | Cross Section Nucleus | Amplitude for Main Scattering Contribution |
|---|---|---|---|---|
| X Rays | .5 -1.5 x $10^{-8}$cm | 0.66 x $10^{-24}$cm$^2$ | ~ $10^{-31}$cm$^2$ | $A_X = A(0)[e^2/m_T c^2 r]^{\frac{1}{2}} \cos\chi$ + |
| Electrons | .05 -1 x $10^{-8}$cm | $10^{-15}$ - $10^{-22}$cm$^2$ | $10^{-14}$ - $10^{-22}$cm$^2$ | $A_E = \frac{\pm 4 Z_T m_e e^2}{\hbar^2 K^2 r}$ * |
| Neutrons | ~ 1 x $10^{-8}$cm | Magnetic scattering (can be significant for materials with unpaired electrons) | $2\times10^{-24}$-$2\times10^{-27}$cm$^2$ (may also have a magnetic scattering component) | $A_N = \frac{b}{r}$ ‡ |

Notes: + A(O) is the amplitude of the electric field.
e is the electrons charge.
$\chi$ is the angle between the direction of polarization of the scattered wave (radiation) and the direction of polarization of the incident field. See Section II for a more detailed treatment of the polarization.
$m_T$ is the mass of the target particle (either an electron or the total number of protons).
c is the velocity of light.
R is the distance between target and detector.

* $Z_T$ is the number of electronic charges contained in the target particle; one for electrons and Z, the atomic number, for nuclei.
$\hbar$ is Planck's constant divided by $2\pi$.
K is the momentum transfer on scattering.
$m_e$ is the electron mass.
$\pm$ The plus sign pertains to electron-electron scattering and the minus sign to electron-proton scattering.

‡ b is the neutron scattering length which is normally a real constant for thermal neutrons.

$$\langle \Psi_{sc}^{TOTAL}(\vec{r},t)\rangle_{AVE} = A_X e^{i(kr-\omega t)} \sum_{j=1}^{N} \int d\vec{r}_j \rho_j(\vec{r}_j) e^{i\vec{K}\cdot\vec{r}_j} \tag{1.18}$$

or by substituting $\vec{r} = \vec{r}_j$ for all j we have

$$\langle \Psi_{sc}^{TOTAL}(\vec{r},t)\rangle_{AVE} = A_X e^{i(kr-\omega t)} \int d\vec{r} \rho(\vec{r}) e^{i\vec{K}\cdot\vec{r}} \tag{1.19}$$

where $\rho(\vec{r}) = \sum_{j=1}^{N} \rho_j(\vec{r})$ is a total probability density function. The integral is a Fourier transform of the probability density $\rho(\vec{r})$ and is referred to as the x-ray coherent scattering factor designated as $F(\vec{K})$. In the case of Eq. (1.17) the average requires a different probability density function namely the probability of finding electron j at the point $\vec{r}_j$ when electron i is at the point $\vec{r}_i$. We call this probability density function the electron pair correlation density $\rho_c^{ij}(\vec{r}_{ij})$. The average of I then can be written as

$$\langle I_{sc}\rangle_{AVE} = A_X^2[N + \int d\vec{r}\ \rho_c(\vec{r}) e^{i\vec{K}\cdot\vec{r}}] \tag{1.20}$$

where $\rho_c(\vec{r}) = \sum_{i\neq j}^{N} \rho_c^{ij}(\vec{r})$ is the total electron pair correlation density. One important property of the total probability functions is that since the total probability of finding the ith electron or the i-jth electron pair somewhere in the atom (total volume of space) must be unity we have

$$\int d\vec{r}\ \rho_j(\vec{r}) = 1 \text{ and } \int d\vec{r}\ \rho_c^{ij}(\vec{r}) = 1$$

so that

$$\int d\vec{r}\ \rho(\vec{r}) = N \text{ and } \int d\vec{r}\ \rho_c(\vec{r}) = N(N-1)$$

where N is the total number of electrons in the atom. Note in Eq. (1.20) that the average intensity is also related to a Fourier transform of a density.

We are now faced with a delima. The intensity of scattering derived from Eq. (1.19) is not the same as that given by Eq. (1.20)! Which of the two possible ways of performing the average is the correct one?

B. The Question of the Order of Averaging

As observed in Section A it is not at all clear how averages over an assembly of diffracting objects in constant motion should be made. Should the average be carried out on the intensity or on the amplitude? The quantum theory provides an unequivocal answer to this question. The average is always carried out on the amplitude if one projectile particle is scattered by a single target particle. Having said this we must be quick to point out that in some cases an approximately correct result may appear as the average over an intensity expression. In other words both expressions (1.19) and (1.20) in the last section may occur.

Let us consider the case of x-ray scattering from molecules.[4] The molecule must initially be in some well defined quantum mechanical state described by a wave function $\Psi_i$ and after the scattering in a well defined state described by a wave function $\Psi_f$. The quantum theory tells us that the average scattering amplitude (or probability amplitude) for this case must be given as

$$\Psi_{sc}^{TOTAL}(\vec{r},t) = A_X e^{i(kr-\omega t)} \langle \Psi_f | \sum_{j=1}^{N} e^{i\vec{K}\cdot\vec{r}_j} | \Psi_i \rangle \tag{2.0}$$

where the intensity will be

$$I_{sc}^{TOTAL} = \left(\frac{k_f}{k_i}\right)^{\ell} |A_X|^2 |\langle \Psi_f | \sum_{j=1}^{N} e^{i\vec{K}\cdot\vec{r}_j} | \Psi_i \rangle|^2 \tag{2.1}$$

with $\ell = 2$ for x-ray scattering and $\ell = 1$ for electron and neutron scattering.

Problem 2.0 By consideration of the flux conservation or current density vector $\vec{J} = \frac{i\hbar}{2m}[\Psi^*(\vec{r},t)\vec{\nabla}\Psi(\vec{r},t)-\Psi(\vec{r},t)\vec{\nabla}\Psi^*(\vec{r},t)]$ show that the ratio of outgoing to incoming flux is $(k_f/k_i)\frac{1}{R^2}|\langle\Psi_f|e^{iK\cdot r}|\Psi_i\rangle|^2$.

Further the quantum theory tells us that if we scatter from an assembly of target species and only single scattering occurs then we must sum the intensity expression for a single scattering event over all final states f which are excited by the incident radiation and average over all populated initial states i. The population of initial states for a gas in thermal equilibrium obeys the Boltzman law hence

$$I_{sc}^{TOTAL} = \sum_f \sum_i \frac{e^{-E_i/k_B T}}{M_N} |A_X|^2 \left(\frac{k_f}{k_i}\right)^{\ell} |\langle \Psi_f | e^{i\vec{K}\cdot\vec{r}_j} | \Psi_i \rangle|^2 \tag{2.2}$$

where $M_N$ is the normalization constant, $M_N = \sum_i e^{-E_i/K_BT}$, $k_B$ is the Boltzman factor, T is the absolute temperature and $E_i$ is the energy of the initial state. We shall let i and j stand for all the quantum numbers needed to specify the molecular motions and sums are understood to be over all such quantum numbers unless otherwise stated.

For molecular scattering we will assume that the molecular wave function for any state n can be written as the product.[5]

$$\Psi_n = \phi_n^T \phi_n^R \phi_n^V \phi_n^E \tag{2.3}$$

where $\phi_n^T$ is the translational wave function, $\phi_n^R$ is the rotational wave function, $\phi_n^V$ is the vibrational wave function and $\phi_n^E$ is the electronic wave function. The separation of the true molecular wave function $\Psi_n$ into an electronic part and a remainder as $[\phi_n^T \phi_n^R \phi_n^V]\phi_n^E$ is referred to as the Born-Oppenheimer approximation. The factorization $\phi_n^T[\phi_n^R \phi_n^V \phi_n^E]$ is called the separation out of the center of mass motion. The general principle behind all these factorizations is that the energy spacing between adjacent translational energy levels is small compared to adjacent rotational energy levels compared to adjacent vibrational levels and so on. Since the energy separations are radically different for different types of molecular motion it is a good approximation to consider the motions as occurring independently from each other. One consequence of the assumed factorized form of the molecular wave function is that the total molecular energy will be the sum of the energies of the factored parts as $E_n = E_n^T + E_n^R + E_n^V + E_n^E$. Factorization on the other hand does not mean that the various $\varphi_n$ are necessarily independent of the coordinates of other $\varphi_n$'s.

It is of interest to carry out a series of hypothetical experiments assuming that we have in our possession an energy analyzer with continuously adjustable energy resolution ranging from zero to infinity. First let us set the energy resolution to ∞. This means we can resolve any two spectral lines no matter how close they can come to each other unless they are truly degenerate. The case of true degeneracy will be discussed at the end of this section. The intensity expression in this case will be

$$I_{sc}^{TOTAL} = \sum_i \frac{e^{-E_i/k_BT}}{M_N} \left(\frac{k_f}{k_i}\right) \ell |A_X|^2 |\langle \varphi_i^T \varphi_i^R \varphi_i^V \varphi_i^E | \sum_{j=1}^{N} e^{iK\cdot r_j} | \varphi_f^T \varphi_f^R \varphi_f^V \varphi_f^E \rangle|^2 \tag{2.4}$$

where if f = i we have the case of pure elastic scattering and for

$f \neq i$ and $E_i \neq E_f$ we have inelastic scattering.

Suppose we now detune our perfect energy analyzer to the point where we are still able to resolve transitions between rotational states but not between translational levels. Eq. (2.2) tells us that we must sum over all excited final states. If we cannot resolve any features of the translational spectrum but can resolve the closest rotational excitations (starting from the ground state) we are likely to be carrying out an experiment in which the incident energy is larger than the energy needed to excite very high translational levels. This means that to an excellent degree of approximation we can consider the sum over final translational states to include a complete set of such states. By use of the quantum mechanical closure theorem for the complete manifold of translational wave functions

$$\sum_f |\phi_f^T\rangle\langle\phi_f^T| = \sum_f \varphi_f^T(\vec{R})\varphi_f^{T^*}(\vec{R}') = \delta(\vec{R}-\vec{R}') \qquad (2.5)$$

the intensity expression for this case then reduces to

$$I_{sc}^{TOTAL} = \sum_i \frac{e^{-E_i/k_BT}}{M_N} \left(\frac{k_f}{k_i}\right)^{\ell} |A_X|^2 \langle\phi_i^T|\,|\langle\varphi_i^R\varphi_i^V\varphi_i^E| \sum_{j=1}^{N} e^{i\vec{K}\cdot\vec{r}_j} |\varphi_f^R\varphi_f^V\varphi_f^E\rangle|^2 |\varphi_i^T\rangle \qquad (2.6)$$

Note that in utilizing the closure argument two further approximations were necessary. First

$$k_f = \sqrt{k_i^2-(E_f-E_i)} \simeq k_i \qquad (2.7)$$

or we assume that $k_f$ is independent of the final translational energy. The momentum transfer K also depends on $E_f^T$ through $\mathbf{k}_f$ since $K^2 = k_i^2 + k_f^2 - 2k_ik_f\cos\theta$ with $\theta$ the scattering angle. The above approximation thus also renders K independent of the sum over f. The second point is that since not all excited translational states may be accesible to the incident energy the sum over f actually terminates and states with excitation energies greater than $k_i^2$ are not included. Hence we must assume that $k_i^2$ is large enough to guarantee that the closure relation is at least approximately correct.

Continuing in this same fashion we can show that for vibrational resolution

$$I_{sc}^{TOTAL} = \sum_i \frac{e^{-E_i/k_BT}}{M_N} \left(\frac{k_f}{k_i}\right)^{\ell} |A_X|^2 \langle \phi_i^T \phi_i^R | | \langle \varphi_i^V \varphi_i^E | \sum_{j=1}^{N} e^{i\vec{K}\cdot\vec{r}_j} | \varphi_f^V \varphi_f^E \rangle |^2 | \varphi_i^T \varphi_i^R \rangle \tag{2.8}$$

and for electronic resolution

$$I_{sc}^{TOTAL} = \sum_i \frac{e^{-E_i/k_BT}}{M_N} \left(\frac{k_f}{k_i}\right)^{\ell} |A_X|^2 \langle \phi_i^T \phi_i^R \phi_i^V | | \langle \phi_i^E | \sum_{j=1}^{N} e^{i\vec{K}\cdot\vec{r}_j} | \varphi_f^E \rangle |^2 | \varphi_i^T \varphi_i^R \varphi_i^V \rangle . \tag{2.9}$$

Finally with resolution so poor that no states can be resolved we have

$$I_{sc}^{TOTAL} = \sum_i \frac{e^{-E_i/k_BT}}{M_N} \left(\frac{k_f}{k_i}\right)^{\ell} |A_X|^2 \langle \varphi_i^T \varphi_i^R \varphi_i^V \varphi_i^E | N + \sum_{i \neq j}^{N} \sum^{N} e^{i\vec{K}\cdot\vec{r}_{ij}} | \varphi_i^T \varphi_i^R \varphi_i^V \varphi_i^E \rangle \tag{2.10}$$

---

Problem 2.1 Show that the total inelastic scattering (excluding elastic) can be written in terms of the x-ray incoherent scattering factor $S(K) = N + \langle \Psi_0 | e^{i\vec{K}\cdot\vec{r}_{12}} | \Psi_0 \rangle - |F(K)|^2$ if the energy resolution is just sufficient to resolve the electronically inelastic scattering from the elastic scattering.

---

We must keep in mind that in each case we are limited to experiments in which the incident energy is large compared to the largest excitation energy and the energy resolution is poor compared to the largest energy level spacing from the ground state for the unresolvable part of the spectrum.

It is now of interest to compare Eq. (2.4) for the elastic case with the intensity given by use of Eq. (1.19) and also to compare Eq. (2.10) with Eq. (1.20). We see that Eq. (1.19) is the correct result for pure elastic scattering while Eq. (1.20) is an approximately correct result for total scattering (elastic plus all inelastic scattering).

In actual practice the case of pure elastic scattering is not likely to be encountered because of experimental resolution problems. Hence the cases in Eq. (2.9) and (2.10) are the ones most likely to be observed.

One possibility which we have not discussed is what happens when one scatters elastically from a system in a state with degenerate levels. If we have scattering from degenerate levels then at high resolution we have

$$I_{sc}=|A_X|^2 \frac{e^{-E_i/k_BT}}{M_N} \left|\sum_{\ell}\sum_{\ell'}\langle\varphi_\ell^T\varphi_\ell^R\varphi_\ell^V\varphi_\ell^E|\sum_{j=1}^{N} e^{iK\cdot r_j}|\phi_{\ell'}^T\phi_{\ell'}^R\phi_{\ell'}^V\phi_{\ell'}^E\rangle\right|^2 \quad (2.11)$$

where the sums over $\ell$ and $\ell'$ are over all states which have the same total energy $E_i$. That is the amplitudes add coherently.

C. Some Examples of Averaging

Before proceeding it may be instructive to consider some concrete examples of the averages implied by Eqs. (2.6), (2.8) and (2.9). In Eq. (2.6) the amplitude $\langle\phi_i^R\phi_i^V\phi_i^E|\sum_{j=1}^{N} e^{i\vec{K}\cdot\vec{r}_j}|\phi_f^R\phi_f^V\phi_f^E\rangle$ can be approximately referenced to a fixed laboratory frame in the case of scattering by x-rays and fast electrons by noting that $\vec{r}_j = \vec{R}_{cm} + \vec{r}_j\,'$ where $\vec{r}_j$ is the position of the $j$th electron referenced to the lab frame, $\vec{R}_{cm}$ is the position of the center of mass with reference to the lab frame and $\vec{r}_j\,'$ is the position of the $j$th electron with reference to the center of mass coordinate. The squared matrix element in Eq. (2.6) is independent of

$R_{cm}$ (i.e. consider $|e^{i\vec{K}\cdot\vec{R}_{cm}}\langle\phi_i^R\phi_i^V\phi_i^E|\sum_{j=1}^{N} e^{i\vec{K}\cdot\vec{r}_j\,'}|\phi_f^R\phi_f^V\phi_f^E\rangle|^2$)

and since the $\phi_i^T$ are normalized we have

$$\sum_i \frac{e^{-E_i/k_BT}}{M_N} \langle\phi_i^T|\phi_i^T\rangle = 1 \quad (3.0)$$

so that no contribution from the translational motion occurs.[6]

In the case of Eq. (2.8) we must consider the average over the initial rotational state motion. To be specific assume we have a diatomic molecule in the gas phase. The rotational coordinates are the orientation angles of the internuclear axis $\theta$ and $\varphi$. The squared matrix element to be averaged is at most a function of $\theta,\varphi$ which we denote by $|g(\theta,\varphi)|^2$. The wave function for a rigid rotator is the spherical harmonic $Y_{J,m}(\theta,\varphi) = \sqrt{\frac{2J+1}{4\pi}}\, P_J^{|m|}(\cos\theta)e^{im\varphi}$. The rotational average is thus

$$\sum_i \frac{e^{-E_i/k_BT}}{M_N} \int d\Omega_{ROT}\, Y_{J,m}^*(\theta,\varphi)|g(\theta,\varphi)|^2 Y_{J,m}(\theta,\varphi) \tag{3.1}$$

where $\int d\Omega_{rot} = \int_0^{\pi} d\theta \sin\theta \int_0^{2\pi} d\varphi$ and $E_i^R = \frac{\hbar^2}{2\mu R^2} J(J+1)$. An important sum rule for the spherical harmonics is

$$\sum_{m=-J}^{J} Y_{J,m}^*(\theta,\varphi)\, Y_{J,m}(\theta,\varphi) = \frac{(2J+1)}{4\pi}\ . \tag{3.2}$$

Because $g(\theta,\varphi)$ will not depend on m and $E_i^R$ does not depend on m we can sum over m and utilize Eq. (3.2) to obtain

$$I_{sc}^{TOTAL} = \frac{1}{4\pi} \sum_i \frac{e^{-E_i/k_BT}(2J+1)}{M_N} \int d\Omega_{ROT} |g(\theta,\varphi)|^2 \tag{3.3}$$

where the sum over the quantum numbers i no longer includes a sum over the quantum number m which specifies the spatial degeneracy of the rotational wave function. Eq. (3.3) is simply the classical average over all angles of orientation of the bond axis with respect to a space fixed coordinate system. This result is quite general for molecules in the gas phase providing we replace (2J+1) by the actual spatial degeneracy of each rotational energy level possessing a different energy and understand $\frac{1}{4\pi}\int d\Omega_{ROT}$ as an average over the Euler angles $\alpha, \beta, \gamma$ as $\frac{1}{8\pi^2}\int_0^{2\pi} d\gamma \int_0^{2\pi} d\alpha \int_0^{\pi} d\beta \sin\beta$ where $\alpha, \beta, \gamma$ define the instantaneous orientation of the rigid molecular framework with respect to a space fixed axis.[7] Usually the z direction of the spaced fixed axis is chosen to be parallel to the momentum transfer $\vec{K}$. Note that for linear molecules the angle $\gamma$ is unnecessary and the average reduces to that given in Eq. (3.3). It is important to understand that the occurrence of the classical rotational average depends on the fact that the scattering includes all possible rotational excitations.

Problem 2.1 Using closure show that the result for elastic scattering from a molecule in a particular rotational level J yields the classical rotational average of $|g(\theta,\varphi)|^2$ (i.e. consider explicitly the spatial degeneracy of the Jth rotational state).

If one has instead the case of pure rotationally elastic scattering one has

$$I_{sc}^{TOTAL} = \sum_J \frac{e^{-E_J/k_BT}}{M_N} \left| \frac{(2J+1)}{4\pi} \int d\Omega_{ROT} g(\theta,\varphi) \right|^2 \qquad (3.4)$$

for the intensity. Again notice the occurrence of both the average of an amplitude (elastic case) and the average of an intensity (total inelastic case). Note that any cases between these extremes will require the actual rotational wave functions to carry out the needed averages. Remember that in x-ray scattering from oriented single crystals the rotational motion will be replaced by additional types of vibrational motion.

Finally we deal with the case where the energy resolution does not permit separation of the vibrational levels. For the case of a homonuclear diatomic molecule using Eq. (2.9) we have for elastic scattering[8]

$$I_{sc}^{EL} = |A_X|^2 \sum_i \frac{e^{-E_i/k_BT}}{M_N} \frac{1}{4\pi} \int d\Omega_{ROT} \langle \varphi_i^V | \, | \langle \varphi_i^E | \sum_{j=1}^{M} e^{iK\cdot r_j} | \phi_i^E \rangle |^2 | \phi_i^V \rangle \qquad (3.5)$$

In order to evaluate Eq. (3.5) we adopt the "perfectly following approximation" which assumes that the electronic charge of the molecule is partitioned into separate parts with each part belonging to a different atomic nucleus but contained in the same molecule and that all electrons assigned to a particular nucleus move in phase with that nucleus. This means that the x-ray coherent form factor can be written as

$$F(\vec{K}) = \int d\vec{r} \rho(\vec{r}, \vec{R}) e^{i\vec{K}\cdot\vec{r}} = \sum_{\nu=1}^{M} e^{i\vec{K}\cdot\vec{R}_\nu} \int d\vec{r} \rho_\nu(\vec{r}) e^{i\vec{K}\cdot\vec{r}} = \sum_{\nu=1}^{M} e^{i\vec{K}\cdot\vec{R}_\nu} F_\nu(\vec{K}) \qquad (3.6)$$

where $\vec{R}$ denotes a dependence on the nuclear positions, $\vec{R}_\nu$ is a vector from the center of mass to the $\nu$th nucleus, $\rho_\nu(\vec{r})$ is that part of the molecular density partitioned off and assigned as belonging to nucleus $\nu$ and $F_\nu(\vec{K})$ is the x-ray coherent scattering factor for scattering by the $\nu$th "atomic" charge $\rho_\nu(\vec{r})$. The essence of the "perfectly following approximation" is that the "atomic" charges $\rho_\nu(\vec{r})$ are assumed to be independent of the vibrational motion. The vibrational average in Eq. (3.5) now reduces to

$$I^{EL} = |A_X|^2 \frac{1}{4\pi} \int d\Omega_{ROT} \sum_i \frac{e^{-E_i/k_BT}}{M_N} \langle \varphi_i^V | \sum_\nu |F_\nu(\vec{K})|^2 + \sum_{\mu\neq\nu} e^{i\vec{K}\cdot\vec{R}_{\mu\nu}} F_\mu^*(\vec{K}) F_\nu(\vec{K}) | \phi_i^V \rangle \qquad (3.7)$$

where we have inverted the order of taking averages because it will be simpler to average over the vibrational motion first.

Let us consider a diatomic molecule with the coordinate system fixed to one of the nuclei and with the vibrational motion assumed to be harmonic. We can write the vibrational Hamiltonian in dimensionless coordinates as

$$H = \left(\frac{d}{dQ}\right)^2 - Q^2 \tag{3.8}$$

by use of the transformation

$$Q = \sqrt{\frac{\mu\omega}{\hbar}}\,(R-R_e) \tag{3.9}$$

with $\mu$ the rest mass, $\omega$ the frequency and R the equilibrium bond length. The ortho-normal vibrational wave functions for simple harmonic motion can be found by solving Eq. (3.8) and are

$$\phi_n^{HO}(Q) = \frac{(\hbar/\mu\omega)^{1/4}}{\sqrt{2^n n!\sqrt{\pi}}}\,H_n(Q)e^{-Q^2/2} \tag{3.10}$$

where the $H_n(Q)$ are the Hermite polynomials. For simple harmonic motion the eigenvalue is $\epsilon_n = \hbar\omega(n + 1/2)$. Next we may write $e^{i\vec{K}\cdot\vec{R}_v}$ using Eq. (3.9) as $e^{iKR_e\cos\theta}\, e^{iK\sqrt{\frac{\hbar}{\mu\omega}}\,Q\cos\theta}$ where $\theta$ is the orientation angle of the bond with respect to the fixed momentum transfer direction.

The required vibrational average can now be written as

$$I^{EL} = |A_X|^2 \frac{1}{4\pi}\int d\Omega_{ROT} \sum_{n=0}^{\infty} \frac{e^{-\frac{\hbar\omega(n+1/2)}{k_BT}}}{M_N 2^n n!\sqrt{\pi}} \int_{-\infty}^{\infty} dQ\, H_n^2(Q)e^{-Q^2}$$

$$\Big[|F_1(K)|^2+|F_2(K)|^2 + e^{iKR_e\cos\theta}\, F_1^*(K)F_2(K)e^{iK\ell(0)\cos\theta Q}$$

$$+ e^{-iKR_e\cos\theta}\, F_1(K)F_2^*(K)e^{-iK\ell(0)\cos\theta Q}\Big] \tag{3.11}$$

where $\ell(0) = \sqrt{\frac{\hbar}{\mu\omega}}$ and 1 and 2 refer to the two possibly different atoms in the molecule. One can carry out the sum over n first by using the useful identity

$$\sum_{n=0}^{\infty} \frac{e^{-\frac{\hbar\omega}{k_B T}(n+1/2)}}{M_N 2^n n! \sqrt{\pi}} e^{-Q^2} H_n^{\,2}(Q) = \sqrt{\frac{\tanh(\hbar\omega/2k_B T)}{\pi}}\, e^{-Q^2 \tanh(\hbar\omega/2k_B T)} \tag{3.12}$$

The average in Eq. (3.11) can now be carried out with the result

$$I^{EL} = |A_X|^2 \frac{1}{4\pi} \int d\Omega_{ROT} \Big\{ |F_1(K)|^2 + |F_2(K)|^2 + [e^{iKR_e\cos\theta} F_1^{*}(K)F_2(K) + e^{-iKR_e\cos\theta} F_1(K)F_2^{*}(K)] e^{-\frac{K^2}{2}\ell(T)^2\cos^2\theta} \Big\} \tag{3.13}$$

where, providing the atomic charge densities are spherically symmetric, steepest descent evaluation of the integral over $\theta$ yields the approximate result

$$I^{EL} = |A_X|^2 \left[ |F_1(K)|^2 + |F_2(K)|^2 + 2R_e F_1^{*}(K)F_2(K) e^{-K^2\ell^2(T)/2} \frac{\sin KR_e}{KR_e} \right] \tag{3.14}$$

with $\ell^2(T) = \frac{\hbar}{\mu\omega}\coth(\hbar\omega/2k_B T)$. Except for the term $e^{-K^2\ell^2(T)/2}$ we can recognize that Eq. (2.25) is identical to what we would expect if no vibration took place. It is interesting therefore to consider the significance of the term $\ell^2(T)$. To do this consider the virial theorem for simple harmonic motion as

$$\epsilon_n = \hbar\omega(n+1/2) = \left(\frac{k}{2}\right) \langle \phi_n^{V} | (R-R_e)^2 | \phi_n^{V} \rangle \tag{3.15}$$

where the force constant $k = \mu\omega^2$. Hence the mean square displacement of one atom with respect to its bonded neighbor for a diatomic molecule in the <u>nth</u> vibrational state is

$$\langle (R-R_e)^2 \rangle_{nn} = \frac{2\hbar}{\mu\omega}(n+1/2) \tag{3.16}$$

For n = 0, the ground state, we see that $\ell(0) = \hbar\omega$ is the mean square displacement from equilibrium.

Problem 3.1 Prove that $\ell^2(T)$ is the Boltzman average of Eq. (3.16). In other words the scattering is damped by a Gaussian weight factor which depends on the product of the momentum transfer and the root mean square amplitude of vibration $l(T)$, often called the amplitude of vibration. The factor $l^2(T)/2$ is often referred to as the Debye-Waller factor in the case of x-ray scattering and denoted as B(T).

### D. X-Ray Intensities and the Energy of the target

It can be observed from Eq.(2.9) for electronic elastic scattering (inelasticity included for all other motions) and Eq.(2.10) for electronically inelastic scattering that the scattered elastic and total scattered intensities can be written in terms of charge densities for the atomic case as

$$I^{EL} = |A_X|^2 \frac{1}{4\pi} \int d\Omega_{ROT} \int d\vec{r}\, \rho(\vec{r})\, e^{i\vec{K}.\vec{r}} = |A_Z|^2 \int d\vec{r}\, \rho(\vec{r}) \int d\vec{r}'\, \rho(\vec{r}')\, j_o(K|\vec{r}-\vec{r}'|) \tag{4.0}$$

with $I^{EL}/|A_X|^2 = N^2$ at K=0 and vanishing as $1/K^8$ as $K\to\infty$. Since $I^{EL}/A_X^2$ is well behaved as a function of K and localized to a region of K that is experimentally accessible we can consider the integral

$$\int_o^\infty dK\left(\frac{I^{EL}}{|A_X|^2}\right) = \int d\vec{r}\, \rho(\vec{r}) \int d\vec{r}'\, \rho(\vec{r}') \int_o^\infty dK\, j_o(K|\vec{r}-\vec{r}|) = \frac{\pi}{2} \int d\vec{r}\, \rho(\vec{r}) \int d\vec{r}'\, \rho(\vec{r}')\, \frac{1}{|\vec{r}-\vec{r}'|} = \frac{\pi}{2}\bar{V}_{ee}^{Coul} \tag{4.1}$$

where $\bar{V}_{ee}^{Coul}$ is the classical Coulomb repulsive potential energy between electrons. This relation was first dicovered by Silverman and Obata[9].

If we consider the total x-ray scattering in the same way then we have

$$\int_o^\infty dK\left[\frac{I^{TOTAL}}{|A_X|^2} - N\right] = \int d\vec{r}\, \rho_c(\vec{r}) \int_o^\infty dK\, j_o(Kr) = \frac{\pi}{2}\int d\vec{r}\, \frac{\rho_c(\vec{r})}{r} = \bar{V}_{ee} \tag{4.2}$$

where $\bar{V}_{ee}$ is the average electron-electron repulsive potential energy. Eq (4.2) was first discovered by Tavard, Roux and Rouault[10]. In addition if the total inelastic scattering is observed we can see that since

$$I^{INEL} = I^{TOTAL} - I^{EL}$$

we have

$$\int_0^\infty dK\left[\frac{I_{sc}^{TOTAL}}{|A_X|^2} - N - \frac{I_{sc}^{EL}}{|A_X|^2}\right] = \frac{\pi}{2}\left[\bar{V}_{ee} - \bar{V}_{ee}^{coul}\right] \tag{4.3}$$

which by definition must be the average electron-electron exchange potential energy of the atom. A third sum rule also obtained by Silverman and Obata[9] can be derived by considering the integral over K of the x-ray coherent scattering factor averaged over all orientations in space as

$$\int_0^\infty dKF(K) = \int dr\rho(r)\int_0^\infty dKj_o(Kr) = \frac{\pi}{2}\int dr\ \frac{\rho(r)}{r} = \frac{\pi}{2}\langle\frac{1}{r}\rangle = -\frac{\pi}{2Z}\bar{V}_{en} \tag{4.4}$$

where Z is the nuclear charge and $\bar{V}_{en}$ is the average electron-nuclear attractive energy. Similar relations can be derived for the molecular case and for electron scattering.[10, 11]

## II. X-RAY SCATTERING:SEMI CLASSICAL RADIATION THEORY

### A. Review of Classical Electricity and Magnetism

Let us consider the force of an electromagnetic field on a charged particle. This is given by the Lorentz force law in c.g.s. units (esu's and emu's) for a single electron in a vacuum as

$$\vec{F} = e\vec{E} + \frac{e}{c}\vec{v} \times \vec{H} \tag{5.0}$$

where E and H are the electric and magnetic field vectors and $\vec{v}$ is the velocity of the electron upon which the field acts. It is convenient for this problem to express both $\vec{E}$ and $\vec{H}$ in terms of the vector and scalar potentials $\vec{A}$ and $\varphi$ which are related to $\vec{E}$ and $\vec{H}$ by

$$\vec{E} = -\frac{1}{c}\frac{\partial\vec{A}}{\partial t} - \vec{\nabla}\varphi \tag{5.1}$$

and

$$\vec{H} = \vec{\nabla} \times \vec{A} \tag{5.2}$$

For the problem under consideration, a single charged particle in a vacuum with no other charges present, Maxwell's equations take on the particularly simple form

$$\vec{\nabla} \times \vec{E} = -\frac{1}{c}\frac{\partial \vec{H}}{\partial t}, \tag{5.3}$$

$$\vec{\nabla}\cdot H = 0, \tag{5.4}$$

$$\vec{\nabla} \times \vec{H} = \frac{1}{c}\frac{\partial \vec{E}}{\partial t}, \tag{5.5}$$

and

$$\vec{\nabla}\cdot\vec{E} = 0 \tag{5.6}$$

By substituting Eqs. (5.2) and (5.1) into Eq. (5.5) and Eq. (5.1) into Eq. (5.6) we obtain

$$\vec{\nabla} \times (\vec{\nabla}\times\vec{A}) = \vec{\nabla}(\vec{\nabla}\cdot\vec{A}) - \nabla^2\vec{A} = -\frac{1}{c^2}\left(\frac{\partial}{\partial t}\right)^2\vec{A} - \frac{1}{c}\frac{\partial}{\partial t}\vec{\nabla}\varphi \tag{5.7}$$

and

$$-\frac{1}{c}\frac{\partial}{\partial t}(\vec{\nabla}\cdot\vec{A}) - \nabla^2\varphi = 0 \tag{5.8}$$

If we next transform $\vec{A}$ and $\varphi$ to a new vector and scalar potential $\vec{A}_o$ and $\varphi_o$ by means of the Gauge transformation

$$\vec{A}_o = \vec{A} - \vec{\nabla}\Psi \tag{5.9}$$

and

$$\varphi_o = \varphi + \frac{1}{c}\frac{\partial \Psi}{\partial t} \tag{5.10}$$

and choose the scalar function $\Psi$ in such a way that $\vec{\nabla}\cdot\vec{A} = \nabla^2\Psi$ (i.e. the Coulomb Gauge) then Eq. (5.8) simplifies to

$$\nabla^2\varphi = 0 \tag{5.11}$$

which is just Laplaces equation $\nabla^2\varphi = -4\pi\rho$ with no charges present (i.e. $\rho = 0$). If $\varphi$ is continuous over the region in space and vanishes at the boundaries then Eq. (5.11) implies that $\varphi$ must be everywhere zero. Under these conditions Eq. (5.7) becomes the familiar wave equation

$$\nabla^2\vec{A} - \frac{1}{c^2}\left(\frac{\partial}{\partial t}\right)^2\vec{A} = 0 \tag{5.12}$$

with solutions of the form

$$\vec{A}(\vec{r},t) = \vec{A}(0)e^{i(\vec{k}\cdot\vec{r}-\omega t)} \tag{5.13}$$

The Lorentz force law can now be written as

$$\vec{F} = \frac{d\vec{p}}{dt} = -\frac{e}{c}\frac{\partial\vec{A}}{\partial t} + \frac{e\hbar i}{cm}\vec{\nabla}\times(\vec{\nabla}\times\vec{A}) \tag{5.14}$$

by use of the quantum mechanical operator representation of $\vec{v}$ in Eq. (5.0) and simplifies by use of Eqs. (5.7) and (5.12) to

$$\frac{d\vec{p}}{dt} = \frac{\partial}{\partial t}\left[-\frac{e}{c}\vec{A} - \frac{e}{c}\frac{\hbar i}{mc^2}\frac{\partial}{\partial t}\vec{A}\right] \tag{5.15}$$

which can be integrated to obtain

$$\vec{p}(t)-\vec{p}(0) = -\frac{e}{c}[A(\vec{r},t)-A(\vec{r},0)] + \frac{\hbar i}{mc^2}\left[\frac{\partial}{\partial t}A(\vec{r},t)\Big|_{t=t} - \frac{\partial}{\partial t}A(\vec{r},t)\Big|_{t=0}\right] \tag{5.16}$$

For the plane wave field in Eq. (5.13), Eq. (5.16) assumes the form

$$\vec{p}(t)-\vec{p}(0) = -\frac{e}{c}\left[1 + \frac{\hbar\omega}{mc^2}\right][\vec{A}(\vec{r},t) - \vec{A}(\vec{r},0)] \tag{5.17}$$

where the term proportional to $\hbar\omega/mc^2$ is the magnetic field contribution which is normally neglected. We deduce from Eq. (5.17) that if an electron starts from rest at time $t = 0$ it will attain a momentum $p(t)$ at the end of a time $t$ by the action of the vector potential at the point $\vec{r}$.

## B. Effect of a Radiation Field on the Motion of an Electron

How do we formulate a quantum mechanical treatment of the effect of electromagnetic radiation on electrons? First of all we must answer such questions as whether or not we need to worry about relativistic corrections[13] and whether or not it is necessary to quantize the radiation field. Here we will neglect relativistic effects and show that the concept of the photon will not be necessary in order to obtain the correct non relativistic treatment of the scattering problem. For a critique on the limitations of semi classical radiation theory (non quantized electro-magnetic field) the reader should see Ref. 14. Eq. (5.17) suggests that we modify the kinetic energy operator $\frac{p^2}{2m} = -\frac{\hbar^2}{2m}\nabla^2$ to include the effect of the electromagnetic field. The momentum operator for an electron is normally given by $\vec{p} = i\hbar\vec{\nabla}$ but in Eq. (5.17) we saw that the electron also undergoes a change in momentum from the action of the electromagnetic field. This suggests we write the total momentum of the electron as

$$\vec{p} = i\hbar\vec{\nabla} - \frac{e}{2c}\,[\vec{A}^{*}(\vec{r},t) + \vec{A}(\vec{r},t)] \qquad (6.0)$$

that is we must write the classical operator $\vec{A}(\vec{r},t)$ in an Hermitean form before adding it to the orbital momentum operator $i\hbar\vec{\nabla}$. The Hamiltonian can then be written as

$$H = \frac{1}{2m}[i\hbar\vec{\nabla} - \frac{e}{2c}(\vec{A}^{*} + \vec{A})]\cdot[i\hbar\vec{\nabla} - \frac{e}{2c}(\vec{A}^{*} + \vec{A})] + V(\vec{r}) \qquad (6.1)$$

where Eq. (6.1) can be generalized to the many electron case by assuming that since the vector potential will act equally on all electrons the total kinetic energy for a system containing N-electrons can be expressed as

$$\sum_{i=1}^{N} \left\{i\hbar\vec{\nabla}_i - \frac{e}{2c}[\vec{A}^{*}(\vec{r}_i,t)+\vec{A}(\vec{r}_i,t)]\right\}\cdot\left\{i\hbar\vec{\nabla}_i - \frac{e}{2c}[\vec{A}^{*}(\vec{r}_i,t)+\vec{A}(\vec{r}_i,t)]\right\} \qquad (6.2)$$

Before solving the Schrödinger equation with the Hamiltonian given in Eq. (6.1) we need to specify the explicit form of the vector potential $\vec{A}$. We know that the solution of Maxwell's wave equation for a plane wave is $\vec{A}(0)e^{i(\vec{k}\cdot\vec{r}-\omega t)}$ where $\vec{A}(0)$ is the electric field perpendicular to the direction of propagation $\vec{k}$. In a scattering experiment we actually have two fields to consider. The first is the incident field $\vec{A}_i(0)e^{i(\vec{k}_i\cdot\vec{r}-\omega t)}$ and the second is the field produced as the result of the scattering process $\vec{A}_s(0)e^{i(\vec{k}_s\cdot\vec{r}-\omega t)}$. The field acting on the electron is thus

$$A(\vec{r},t) = \left[\vec{A}_i(0)e^{i(\vec{k}_i\cdot\vec{r}-\omega t)} + \vec{A}_s(0)e^{i(\vec{k}_s\cdot\vec{r}-\omega t)}\right] \qquad (6.3)$$

A question which arises here is how does the form used in (Eq. 6.3) compare with the form for an outgoing spherical wave? We recall that if an outgoing spherical wave is observed at a point which is a distance R from the center of mass of an atom or molecule then the approximate form of the wave is $\frac{e^{ik_sR}}{R}e^{i\vec{k}_s\cdot\vec{r}}$ where $\vec{r}$ defines a point within the scatterer with $R \gg |\vec{r}|$ and $|\vec{R}+\vec{r}|$ is the distance between the point of observation and the position of the electron. Here we see that the part of the wave which depends on the field is in the form of a plane wave as we have assumed and that we can define

$$\vec{A}_i(0) = E^i_\alpha \hat{n}^i_\alpha e^{i\vec{k}_i \cdot \vec{R}} \tag{6.4}$$

and

$$\vec{A}_s(0) = E^s_\alpha \hat{n}^s_\alpha \frac{e^{ikR}}{R} \tag{6.5}$$

where $E^i_\alpha$ and $E^s_\alpha$ are the magnitudes of the electric fields in the incident (i) and scattered directions (s) with polarization direction $\hat{n}^i_\alpha$ or $\hat{n}^s_\alpha$. The term $\frac{e^{ikR}}{R}$ represents an outgoing spherical wave. The gauge guarantees that the field is polarized perpendicular to the direction of propagation of the field, that is transverse polarization. Without loss of generality we can subdivide the polarization into a component in the scattering plane (i.e. plane containing both $k_i$ and $k_s$) and a component perpendicular to the scattering plane.

Clearly two separate experiments corresponding to the two possible choices of initial polarization can be carried out with two possible choices for the polarization of the scattered wave. In addition we could imagine scattering by an unpolarized source which would be the sum over all final state polarizations and an average over the initial polarizations. Let us use 1 to label the polarization direction in the scattering plane and 2 to label the perpendicular polarization direction. To make things simple let us consider elastic scattering where $|k_i|=|k_s|$ and $\omega=\omega'$. The vector potential operator is then expressible as

$$\vec{A}(\vec{r},t)= \frac{1}{2}\left[A_i(0)\hat{n}_\alpha{}^i e^{i\vec{k}_i \cdot \vec{r}} + A_s(0)\hat{n}_\alpha{}^s e^{i\vec{k}_s \cdot \vec{r}}\right] \tag{6.6}$$

where $\alpha$ = 1 or 2.

## C. Time Dependent Perturbation Theory

The next step is to consider the solution to the time dependent Schrödinger equation

$$H\Psi(\vec{r},t) = [H_0 + V(\vec{r},t)]\Psi(\vec{r},t) = i\hbar \frac{\partial\Psi(\vec{r},t)}{\partial t} \tag{7.0}$$

with $H_0 = -\frac{\hbar^2}{2m}\nabla^2 + V(\vec{r})$ the usual Hamiltonian in the absence of the electromagnetic field and with the perturbation $V(\vec{r},t)$ taken as all those terms depending on $\vec{A}(\vec{r},t)$. From Eq. (6.6) we see that

$$V(\vec{r},t) = \frac{e}{2mc}[\vec{A}(r,0)\cdot\vec{P}e^{-i\omega t} + \vec{A}^*(\vec{r},0)\cdot\vec{P}e^{i\omega t}]$$
$$+ \frac{e^2\hbar^2}{8mc^2}[\vec{A}(\vec{r},0)\cdot\vec{A}^*(\vec{r},0) + \vec{A}^*(\vec{r},0)\cdot\vec{A}(\vec{r},0)$$
$$+ \vec{A}(\vec{r},0)\cdot\vec{A}(\vec{r},0)e^{-2i\omega t} + \vec{A}^*(\vec{r},0)\cdot\vec{A}^*(\vec{r},0)e^{2i\omega t}] \tag{7.1}$$

where terms such as $\vec{\nabla}\cdot\vec{A}\Psi$ have been written as $(\vec{\nabla}\cdot\vec{A})\Psi + \vec{A}\cdot\vec{\nabla}\Psi$ which since $\vec{\nabla}\cdot\vec{A}$ yields $i\vec{k}_i\cdot\vec{A}i$ and $i\vec{k}_s\cdot\vec{A}s$ will vanish because $\vec{k}_i \perp \vec{A}_i$ and $\vec{k}_s \perp \vec{A}_s$.

Eq. (7.0) must now be solved for the perturbation in Eq. (7.1). To do this we expand $\Psi(\vec{r},t)$ in eigenfunctions $\varphi_n(\vec{r})e^{-iE_nt/\hbar}$ of the operator $(H_0 - i\hbar\frac{\partial}{\partial t})$ as[15]

$$\Psi(\vec{r},t) = \sum_{n=0}^{\infty} C_n(t)\varphi_n(\vec{r})e^{-iE_nt/\hbar} \tag{7.2}$$

where substitution of Eq. (7.2) back into Eq. (7.0), multiplication by $\varphi_n(\vec{r})$ and integration over $\vec{r}$ yields the differential equation for the time dependent coefficient $C_n(t)$ as

$$\frac{dC_n(t)}{dt} = -\frac{i}{\hbar}\sum_{p=0}^{\infty} C_p(t)V_{np}(t)e^{i(E_n-E_p)t/\hbar} \tag{7.3}$$

where $V_{np}(t) = \langle\varphi_n(\vec{r})|V(\vec{r},t)|\varphi_p(\vec{r})\rangle$. Eq. (7.3) may be solved by iteration but to better understand how this should be done it is worthwhile to consider the physical significance of the coefficients $C_n(t)$. If we damand that the total probability $P(\vec{r},t)$, remain constant in time we have

$$\frac{d}{dt}\int d\vec{r}P(\vec{r},t) = \frac{d}{dt}\int d\vec{r}\Psi^*(\vec{r},t)\Psi(\vec{r},t) = \frac{d}{dt}\sum_{n=0}^{\infty}|C_n(t)|^2 = 0 \qquad (7.4)$$

hence the sum $\sum_{n=0}^{\infty}|C_n(t)|^2$ must be a constant. If the system is found in a particular initial state i at time t=0 then $|C_i(0)|^2 = 1$ and $|C_n(0)|^2 = 0$ for $n \neq i$. The system may then evolve in time but the sum of the $|C_n(t)|^2$ must remain equal to unity. Physically $|C_n(t)|^2$ represents the probability of finding the system in the state n at the time t. Let us solve Eq. (7.3) by iteration starting with the zeroth order solution $C_i^{(0)}(t) = 1$, $C_n^{(0)}(t) = 0$; $n \neq i$. Substitution of this solution in to the right hand side of Eq. (7.3) and integrating over t yields the first order solutions as

$$C_i^{(1)}(t) = 1 - \frac{i}{\hbar}\int_0^t dt' V_{ii}(t') \qquad (7.5)$$

and

$$C_{n\neq i}^{(1)}(t) = -\frac{i}{\hbar}\int_0^t dt' V_{ni}(t') e^{i(E_n - E_i)t'/\hbar} \qquad (7.6)$$

These solutions can then be reiterated to obtain the cofficients through second order as

$$C_i^{(2)}(t) = 1 - \frac{i}{\hbar}\int_0^t dt' V_{ii}(t') - \frac{1}{\hbar^2}\int_0^t dt' V_{ii}(t')\int_0^{t'} dt'' V_{ii}(t'') \qquad (7.7)$$

$$- \frac{1}{\hbar^2}\sum_{n\neq i}\int_0^t dt' V_{in}(t') e^{i(E_i - E_n)t'/\hbar}\int_0^{t'} dt'' V_{ni}(t'') e^{i(E_n - E_i)t''/\hbar}$$

and

$$C_n^{(2)}(t) = -\frac{i}{\hbar}\int_0^t dt' V_{ni}(t')e^{i(E_n-E_i)t'/\hbar} \tag{7.8}$$

$$-\frac{i}{\hbar^2}\int_0^t dt' V_{ni}(t')e^{i(E_n-E_i)t'/\hbar}\int_0^{t'} dt'' V_{ii}(t'')$$

$$-\frac{1}{\hbar^2}\sum_{\ell\neq i}\int_0^t dt' V_{n\ell}(t')e^{i(E_n-E_\ell)t'/\hbar}\int_0^{t'} dt'' V_{\ell i}(t'')e^{i(E_\ell-E_i)t''/\hbar}$$

Problem 7.0 Verify Eqs. (7.5) through (7.8).

We are interested in the second order solutions because for elastic scattering we must collect together all terms of the same order in the vector potential $\vec{A}$. This will become clearer as we proceed. The experiment we invision is the sudden turning on of an x-ray source, bathing the target with this continuous source and after some very long time suddenly turning the source off. This procedure is often referred to as the sudden approximation.

Let us organize the calculation by writing

$$V(\vec{r},t) = V_1(\vec{r},t) + V_2(\vec{r},t) \tag{7.9}$$

with

$$V_1(\vec{r},t) = -\frac{e}{2mc}\Big\{[e^{i\vec{k}_i\cdot\vec{r}}\vec{A}_i\cdot\vec{p} + e^{i\vec{k}_s\cdot\vec{r}}\vec{A}_s\cdot\vec{p}]e^{-i\omega t} \tag{7.10}$$

$$+ [e^{-i\vec{k}_i\cdot\vec{r}}\vec{A}_i{}^*\cdot\vec{p} + e^{-i\vec{k}_s\cdot\vec{r}}\vec{A}_s{}^*\cdot\vec{p}]e^{i\omega t}\Big\}$$

and

$$V_2(\vec{r},t) = \frac{\hbar^2e^2}{8mc^2}\Big\{[e^{i\vec{k}_i\cdot\vec{r}}\vec{A}_i + e^{i\vec{k}_s\cdot\vec{r}}\vec{A}_s]\cdot[e^{-i\vec{k}_i\cdot\vec{r}}\vec{A}_i{}^* + e^{-i\vec{k}_s\cdot\vec{r}}\vec{A}_s^*] \tag{7.11}$$

$$+[e^{-i\vec{k}_i\cdot\vec{r}}\vec{A}_i{}^* + e^{-i\vec{k}_s\cdot\vec{r}}\vec{A}_s^*]\cdot[e^{i\vec{k}_i\cdot\vec{r}}\vec{A}_i + e^{i\vec{k}_s\cdot\vec{r}}\vec{A}_s]$$

$$+[e^{i\vec{k}_i\cdot\vec{r}}\vec{A}_i + e^{i\vec{k}_s\cdot\vec{r}}\vec{A}_s]\cdot[e^{i\vec{k}_i\cdot\vec{r}}\vec{A}_i + e^{i\vec{k}_s\cdot\vec{r}}\vec{A}_s]e^{-2i\omega t}$$

$$+[e^{-i\vec{k}_i\cdot\vec{r}}\vec{A}_i{}^* + e^{-i\vec{k}_s\cdot\vec{r}}\vec{A}_s{}^*]\cdot[e^{-i\vec{k}_i\cdot\vec{r}}\vec{A}_i{}^* + e^{-i\vec{k}_s\cdot\vec{r}}\vec{A}_s{}^*]e^{2i\omega t}$$

$$= \alpha + \beta e^{-2i\omega t} + \beta^* e^{2i\omega t}$$

It is convenient to consider the contribution from $V_2(\vec{r},t)$ first. The result of a first order calculation is

$$C_i^{(1)}(t) = -\frac{it}{\hbar}\langle i|\alpha|i\rangle + \frac{\langle i|\beta|i\rangle[e^{-2i\omega t}-1]}{2\hbar\omega} - \frac{\langle i|\beta^*|i\rangle[e^{2i\omega t}-1]}{2\hbar\omega} \tag{7.12}$$

$$= -\frac{it}{\hbar}\langle i|\alpha|i\rangle - i\langle i|\beta|i\rangle\frac{e^{-i\omega t}\sin\omega t}{\hbar\omega} - i\langle i|\beta^*|i\rangle\frac{e^{i\omega t}\sin\omega t}{\hbar\omega}$$

Problem 7.1 Verify Eq. (7.12).

We see that the first term on the right becomes more important the longer the perturbation is left on. What about the other terms? We note that since $\lim_{t\to\infty}\frac{\sin\omega t}{\omega} = \pi\delta(\omega)$ the longer the perturbation is left on the more singular these contributions become. That is they only contribute at $\omega=0$ and are resonance terms of no interest to the present aims. Clearly they are of no interest to the scattering problem.

If we place $V_1(\vec{r},t)$, Eq. (7.10), into the first order calculation we get only resonance terms and hence no contribution to the scattering. In other words we must carry the $V_1(\vec{r},t)$ contribution to second order in the preturbation. After a lot of algebra one finds that the only non resonance type terms are

$$+(i\hbar t)\left(\frac{e^2}{4m^2c^2}\right)\left[\frac{F_{in}F_{ni}^*}{E_n-E_i+\hbar\omega} + \frac{F^*_{in}F_{ni}}{E_n-E_i-\hbar\omega}\right]$$

where

$$F_{in} = \langle i|\, e^{i\vec{k}_i\cdot\vec{r}}\,\vec{A}_i\cdot\vec{\mathbf{p}} + e^{i\vec{k}_s\cdot\vec{r}}\,\vec{A}_s\cdot\vec{\mathbf{p}}\,|n\rangle$$

Problem 7.2 Derive the non resonance type terms from $V_1(\vec{r},t)$ in the second order perturbation coefficient.

Recalling that $e^2/mc^2$ is the square root of the Thomson cross section, $\sqrt{\sigma_T}$, and retaining only the scattering terms (outgoing spherical waves) we have for non resonant (NR) contributions to $C_i(t)$

for all terms of order $A^2$

$$C_{NR}(t)=\left(\frac{\hbar t}{i}\right)\frac{\sqrt{\sigma_T}}{4}\left\{\vec{A}_i{}^*\cdot\vec{A}_s F(\vec{K})-\frac{1}{m}\frac{\langle i|e^{i\vec{k}_s\cdot\vec{r}}\vec{A}_s\cdot\vec{p}|n\rangle\langle n|e^{-i\vec{k}_i\cdot\vec{r}}\vec{A}_i{}^*\cdot\vec{p}|i\rangle}{E_n-E_i+\hbar\omega}\right.$$

$$\left.-\frac{1}{m}\frac{\langle i|e^{-i\vec{k}_i\cdot\vec{r}}\vec{A}_i{}^*\cdot\vec{p}|n\rangle\langle n|e^{i\vec{k}_s\cdot\vec{r}}\vec{A}_s\cdot\vec{p}|i\rangle}{E_n-E_i-\hbar\omega}\right\} \tag{7.13}$$

Problem 7.3 Derive Eq. (7.13). Hint: Neglect all self terms and keep only terms with outgoing scattered wave boundary conditions.

Eq. (7.13) using $|\vec{A}_i|=|E_0|$ and $|\vec{A}_s|=|E_0|\,\gamma$ with $|E_0|$ the magnitude of the electric field and $\gamma$ an unknown coefficient we can write

$$C^{\alpha}_{NR}(t)=\left(\frac{\hbar t}{i}\right)\frac{\sqrt{\sigma_T}|E_0|^2\gamma\, e^{i(kR-kR\cos\theta)}}{4R}\left\{(\hat{n}_\alpha{}^i\cdot\hat{n}_\alpha{}^s)F(\vec{K})\right.$$

$$-\frac{1}{m}\sum_n\frac{\langle i|e^{i\vec{k}_s\cdot\vec{r}}(\hat{n}_\alpha{}^s\cdot\vec{p})|n\rangle\langle n|e^{-i\vec{k}_i\cdot\vec{r}}(\hat{n}_\alpha{}^i\cdot\vec{p})|i\rangle}{(E_n-E_i+\hbar\omega)}$$

$$\left.-\frac{1}{m}\sum_n\frac{\langle i|e^{-i\vec{k}_i\cdot\vec{r}}(\hat{n}_\alpha{}^i\cdot\vec{p})|n\rangle\langle n|e^{i\vec{k}_s\cdot\vec{r}}(\hat{n}_\alpha{}^s\cdot\vec{p})|i\rangle}{(E_n-E_i-\hbar\omega)}\right\} \tag{7.14}$$

Eq. (7.14) with the neglect of the last two terms on the right is called the form factor approximation. While the term depending on $F(\vec{K})$ is normally the largest term we must always remember that all three terms on the right are of the same order in the electric field. In general the last two terms on the right will make important contributions to the amplitude only when $E_n-E_i$ is comparable to $\hbar\omega$. This happens for a fixed value of the x-ray frequency $\omega$ as the atomic size increases. We can safely assume for x-ray scattering from light atoms that the form factor approximation is adequate (more on this later). We see from Eq. (7.14) that except for the unknown constant factor $\gamma^2$ the probability $|C_i(t)|^2$ for scattering the field elastically with momentum transfer K is given for an unpolarized incident x-ray beam by

$$|C_i^{(1)}(t)|^2=\frac{\sigma_T(\hbar^2t^2)|E_0|^4\gamma^2}{16R^2}\left\{\frac{1}{2}(1+\cos^2\theta)|F(\vec{K})|^2\right\} \tag{7.15}$$

where $\frac{1}{2}(1+\cos^2\theta)$ is called the polarization correction.

Problem 7.4 Derive Eq. (7.15) starting from Eq. (7.14).

It can be argued that the scattered x-ray intensity must also be

proportional to $|C_i(t)|^2$ as given in Eq. (7.15).
The unknown factor $\gamma^2$ is clearly proportional to the Thomson cross section $\sigma_T$.

D. Analysis of the Form Factor Approximation

Florescu and Gavrila[17] have evaluated the relativistic form of Eq. (7.14) for K shell electrons in the high energy small angle limit. For Z = 29 the form factor is 2.15% too large, 6.8% too large at Z = 50 and 13.86% too large for Z = 80. These are the maximum corrections since the incident x-ray energy is assumed to be $\infty$.

Kissel and Pratt[18] have used the complete relativistic formalism for the elements Z = 30, Z = 73 and Z = 82 using Dirac-Hartree-Fock-Slater self consistent potentials for the neutral atom. They find for 60 keV incident photons that for Z = 30 (Zn) the form factor approximation is low by about 5-10% over the range $\theta$ = 60 -150°. For Z = 73 (Ta) the form factor approximation is too high by about a factor of two between 90° and 150°. Brown and Mayers[19] in a similar calculation found that for mercury (Z = 80) at 323 keV incident energy that the form factor approximation was too high by 8% at 0° and 38% at 60° for the scattered amplitude for no polarization change on scattering. The amplitude with polarization change appears to be rather well given by the form factor approximation.

Brown and Mayers[19] have suggested a form factor approximation of the form

$$F_c(\vec{K}) = \int d\vec{r}\rho(\vec{r})e^{i\vec{K}\cdot\vec{r}}f(r) \tag{8.0}$$

for the amplitude without polarization change where

$$f(r) = \frac{mc^2}{E + |V(r)|} \tag{8.1}$$

with V(r), the spherically symmetric potential an electron feels in the presence of the neutral atom and E is the total relativistic energy of the atom, written as

$$E = mc^2 - |E_0| \tag{8.2}$$

where $|E_0|$ is the magnitude of the total electronic energy of the atom in its ground state. The function f(r) can thus be written as

$$f(r) = \frac{1}{\left[1 + \frac{|V(r)| - |E_0|}{mc^2}\right]} \tag{8.3}$$

Eq. (8.0) has been found to give improved results for the form factor without change in polarization as long as the momentum transfer is less than about 250 $\text{Å}^{-1}$ which means that it should be useful for all x-ray studies.

In summary it appears that the classical form factor approximation as a concept for interpreting x-ray scattering in terms of charge densities should not be used for elements much heavier than argon without careful analysis of the corrections involved. Further it would seem that quoted accuracies for charge densities obtained for the heavier elements should be modified to account for deficiencies in the form factor concept. In the absence of rigorous relativistic calculations, it is recommended that the approximation given in Eq. (8.0) be used to compute form factor corrections and that these be applied to correct experimental data involving heavy atoms. Useful bibliographic tools for the literature on form factors can be found in Ref. 20.

## E. Alternative Expressions for the Electromagnetic Field Operator

Some authors have employed the operator $\vec{E}\cdot\vec{\mu}$ for the effect of the electromagnetic field on an electron where $\vec{E}$ is the electric field vector and $\vec{\mu}$ is the quantum mechanical dipole moment operator. The conditions under which this formulation is equivalent to that employed in the discussion here was first discussed by Goeppert-Mayer.[21]

We consider the Hamiltonian

$$H = \frac{P^2}{2m} + V(\vec{r}) - \frac{e}{mc}\vec{A}(\vec{r},t)\cdot\vec{P} + \frac{e^2}{2mc^2}\vec{A}(\vec{r},t)\cdot\vec{A}(\vec{r},t) \tag{9.0}$$

which we can treat as the classical mechanical Hamiltonian function. This can be transformed to the classical Lagrangian, $L = \dot{\vec{r}}\cdot\vec{p} - H$, by means of the relations $\dot{x} = \frac{\partial H}{\partial p_x}$, $\dot{y} = \frac{\partial H}{\partial p_y}$ and $\dot{z} = \frac{\partial H}{\partial p_z}$. We can see that $\dot{x} = \frac{p_x}{m} - \frac{e}{mc}A_x$, etc., and that $p_x = m\dot{x} + \frac{e}{c}A_x$, etc. These last relations allow us to obtain the result

$$L = \frac{m}{2}\dot{\vec{r}}\cdot\dot{\vec{r}} - V(\vec{r}) + \frac{e}{c}\dot{\vec{r}}\cdot\vec{A}(\vec{r},t) \tag{9.1}$$

The Langrangian has the property that

$$\delta \int_{t_1}^{t_2} dt\ L(\vec{r},\vec{p},t) = 0 \tag{9.2}$$

where $\delta$ signifies a variation of the integral. It is clear that any function which is a total derivative with respect to time can be added to L without alteration of Eq. (9.2). In particular we may add and subtract the total derivative term $\frac{e}{c}\frac{d}{dt}(\vec{r}\cdot\vec{A})$ and

Problem 9.0 Prove this last statement.

define a new Lagrangian

$$L' = L - \frac{e}{c}\frac{d}{dt}(\vec{r}\cdot\vec{A}) = \frac{m}{2}\dot{\vec{r}}\cdot\dot{\vec{r}} - V(\vec{r}) - \frac{e}{c}\vec{r}\cdot\frac{d\vec{A}}{dt} \tag{9.3}$$

The Hamiltonian function corresponding to this new Lagrangian is

$$H' = \frac{p^2}{2m} + V(\vec{r}) + \frac{e}{c}\vec{r}\cdot\frac{d\vec{A}}{dt} \tag{9.4}$$

Problem 9.1 Prove that $L' = \dot{\vec{r}}\cdot\vec{p} - H'$.

We could employ Eq. (9.4) rather than Eq. (9.0) since it is entirely equivalent. We may also recall that for our hypothetical single electron in a vacuum problem that the electric field vector $\vec{E}$ is given in terms of the vector potential as

$$\vec{E} = -\frac{1}{c}\frac{\partial\vec{A}}{\partial t} \tag{9.5}$$

which suggests substitution in Eq. (9.4). However we must be careful to recall that

$$\frac{d\vec{A}(\vec{r},t)}{dt} = \frac{\partial\vec{A}}{\partial x}\frac{\partial x}{\partial t} + \frac{\partial\vec{A}}{\partial y}\frac{dy}{dt} + \frac{\partial\vec{A}}{\partial z}\frac{\partial z}{\partial t} + \frac{\partial\vec{A}}{\partial t} \tag{9.6}$$

hence we can only replace $\frac{d\vec{A}}{dt}$ by $\frac{\partial\vec{A}}{\partial t}$ if the field $\vec{A}$ is very slowly varying (dipole approximation). Under these conditions we have

$$H' = \frac{p^2}{2m} + V(\vec{r}) - \vec{\mu}\cdot\vec{E} \tag{9.7}$$

where $\vec{\mu}$ is the quantum mechanical dipole moment operator $e\vec{r}$. Fiutak[22] has shown that using a higher order approximation the term $-\vec{\mu}\cdot\vec{E}$ is replaced by a multipolar expansion of the form

$$-\vec{P}\cdot\vec{E}-\vec{M}\cdot\vec{H}-\vec{Q}:\nabla\vec{E} + \ldots$$

where $\vec{P}$ is the polarization vector, $\vec{M}$ is the magnetization vector and $\vec{Q}$ is the quadrupole tensor. Obviously one does not want to use Eq. (9.7) in spite of its simplicity unless the dipole approximation can be expected to be valid. At first glance Eq. (9.4), which does have general validity, would seem to possess some advantages because the perturbation contains only a single term. This is illusary however since for a plane wave of the form $A_o e^{i(k_i \cdot r - \omega t)}$ we have

$$\frac{e}{c}\,\vec{r}\cdot\frac{d\vec{A}}{dt} = \frac{ie}{c}\,(\vec{r}\cdot\vec{A})[\vec{k}_i\cdot\dot{\vec{r}}-\omega] \tag{9.8}$$

which brings us back to two terms and includes an operator, $\dot{\vec{r}}$, which is not easily dealt with. One might guess that the quantum mechanical equivalent of this operator is

$$\frac{e}{c}\,\vec{r}\cdot\frac{d\vec{A}}{dt} = -ik_i[(\vec{\mu}\cdot\vec{A})-\frac{(\hat{k}_i\cdot\vec{P})}{2mc}(\vec{\mu}\cdot\vec{A})-(\vec{\mu}\cdot\vec{A})\frac{(\hat{k}_i\cdot\vec{P})}{2mc}] \tag{9.9}$$

but it does not seem to have been widely used previously. One problem is that since Eq. (9.9) is linear in $\vec{A}$ all Rayleigh scattering terms will occur in second order in the perturbation theory.

## References

1. A. H. Compton and S. K. Allison, "X-Rays in Theory and Experiment", Chapter III (D. Van Nostrand Co., Inc., New York, 1948).
2. G. E. Bacon, "Neutron Diffraction", Chapter 2 (Oxford University Press, London, 1962).
3. R. A. Bonham and M. Fink, "High Energy Electron Scattering", Chapter 1 (Van Nostrand Rrinhold Co., New York, 1974).
4. See Ref. 3, Chapter IV.
5. See for example F. L. Pilar, "Elementary Quantum Chemistry" (McGraw-Hill, Inc., New York, 1968).
6. In the case of thermal neutron scattering this conclusion does not apply. For a treatment of excitation of translational motion by slow neutrons see A. C. Zemach and R. J. Glauber, Phys. Rev. 101, 118; 129 (1956),
7. T. Iijima, R. A. Bonham and T. Ando, J. Phys. Chem. 67, 1472 (1963).
8. R. A. Bonham and J. Geiger, J. Chem. Phys. 51, 5246 (1969); 4586 (1975).
9. J. N. Silverman and Y. Obata, J. Chem. Phys. 38, 1254 (1963).
10. C. Tavard and M. Roux, Compt. Rend. 260, 4933 (1965). C. Tavard, M. Rouault and M. Roux, J. Chim. Phys. 62, 1410 (1965).
11. R. A. Bonham, J. Phys. Chem. 71, 856 (1967).

12. J. D. Jackson, "Classical Electrodynamics", (John Wiley and Sons, Inc., New York, 1962).
13. L. I. Schiff, "Quantum Mechanics", (McGraw Hill Book Co., New York, 1968) Third Edition, Page 472.
14. R. K. Nesbet, Phys. Rev. Letters 27, 553 (1971).
15. See Ref. 13, Page 280.
16. Derivation of this equation is generally ascribed to: I. Waller, Z. Phys. 58, 75 (1929).
17. V. Florescu and M. Gavrila, Phys. Rev. A14, 211 (1976).
18. L. Kissel and R. H. Pratt, Phys. Rev. Letters 40, 387 (1978).
19. G. E. Brown and D. F. Mayers, Proc. Roy. Soc. London, A242, 89 (1957).
20. J. H. Hubbell, J. De Phys. 32, C4 (1971). (Colloque C4, Supplement au n°10). J. H. Hubbell, Nat. Bur. Stand. (U.S.) 29 (1969). (NSRDS-NBS 29).
21. M. Göppert-Mayer, Ann. Phys. 9, 273 (1931).
22. J. Fiutak, Can. J. Phys. 41, 12 (1963) and E. A. Power and S. Zienau, Phil. Trans. Roy. Soc. 251A, 54 (1959).

DIFFRACTION BY REAL CRYSTALS :

I. KINEMATICAL THEORY AND TIME AVERAGING

Pierre BECKER

CMOA, CNRS and University Paris VI

23 rue du Maroc, 75019 Paris, France

CONTENTS

I. Introduction to lattice dynamics
II. Kinematical scattering
III. X Ray elastic scattering
IV. Thermal diffuse scattering

## I. INTRODUCTION TO LATTICE DYNAMICS

My intention is only to discuss a few basic concepts that are relevant to the physics of diffraction. There are many excellent books and reviews on this subject (1).

We consider a crystal composed with N unit cells (labelled $\ell$) and n atoms in each cell (labelled $\kappa$). $\underset{\sim}{R}_{\kappa\ell}$ is the equilibrium position of an atom :

$$\underset{\sim}{R}_{\kappa\ell} = \underset{\sim}{R}_{\kappa} + \underset{\sim}{\ell}$$

Its displacement is called $\underset{\sim}{u}_{\kappa\ell}$. The kinetic energy T of the crystal is :

$$T = \sum_{\kappa\ell} \frac{1}{2} m_{\kappa} \dot{\underset{\sim}{u}}_{\kappa\ell}^{2} \tag{1}$$

The potential energy V can be expressed in a Taylor series about equilibrium. Limiting presently ourselves to the harmonic terms, we can write :

$$V = V_0 + \frac{1}{2} \sum_{\kappa\ell} \sum_{\kappa'\ell'} (\underset{\sim}{u}_{\kappa\ell} \cdot \underset{\sim}{\nabla}_{\kappa\ell})(\underset{\sim}{u}_{\kappa'\ell'} \cdot \underset{\sim}{\nabla}_{\kappa'\ell'}) V_{eq} \tag{2}$$

The equation of motion for nucleus $\kappa\ell$ is therefore :

$$m_\kappa \ddot{\underset{\sim}{u}}_{\kappa\ell} = -\sum_{\kappa\ell\kappa'\ell'} G_{\kappa\ell,\kappa'\ell'} \underset{\sim}{u}_{\kappa'\ell'} \tag{3}$$

where the matrix $G_{\kappa\ell,\kappa'\ell'}$ can be written :

$$G_{\kappa\ell,\kappa'\ell'} = \underset{\sim}{\nabla}_{\kappa\ell} \underset{\sim}{\nabla}_{\kappa'\ell'} V = G_{\kappa\kappa'}(\underset{\sim}{\ell}' - \underset{\sim}{\ell}) \tag{4}$$

$$m_\kappa \ddot{\underset{\sim}{u}}_{\kappa\ell} = -\sum_{\kappa'\ell'} G_{\kappa\kappa'}(\underset{\sim}{\ell}') \, \underset{\sim}{u}_{\kappa',\ell+\ell'} \tag{5}$$

By virtue of Bloch's theorem, there must exist one vector $\underset{\sim}{q}$ such that :

$$\underset{\sim}{u}_{\kappa\ell} = \exp(i\underset{\sim}{q}\cdot\underset{\sim}{\ell}) \, \underset{\sim}{u}_{\kappa o} \tag{6}$$

$\underset{\sim}{q}$ is not uniquely defined, since (6) is unchanged if $\underset{\sim}{q}$ is replaced by $(\underset{\sim}{q}+2\pi\underset{\sim}{H})$, $\underset{\sim}{H}$ being a reciprocal vector. $\underset{\sim}{q}$ should be therefore limited to one reciprocal cell, namely the first Brillouin zone (BZ).

Boundary Conditions

We shall use the Born von Karman conditions. The easiest picture is to say that there exist N allowed $\underset{\sim}{q}$ vectors, equidistributed in the BZ. Suppose that there are $N_1$ cells in the $\underset{\sim}{a}_1$ direction, $N_2$ along $\underset{\sim}{a}_2$, $N_3$ along $\underset{\sim}{a}_3$, with $N=N_1N_2N_3$, the allowed values of $\underset{\sim}{q}$ are :

$$\underset{\sim}{q} = \sum_i^3 \frac{n_i}{N_i} \underset{\sim}{b}_i$$

where $\underset{\sim}{b}_i$ define the reciprocal lattice : $\underset{\sim}{b}_1 = 2\pi \frac{\underset{\sim}{a}_2 \times \underset{\sim}{a}_3}{v_c}$

$v_c$ will be the unit cell volume

$v$ is the crystal volume : $v = Nv_c$. The volume of the reciprocal cell is $v_c^* = (2\pi)^3/v_c$.

Each $\underset{\sim}{q}$ vector occupies a reciprocal volume $(2\pi)^3/v$. The density of $\underset{\sim}{q}$ vectors is thus : $(2\pi)^{-3}v$. Since N is very large, we shall replace the discrete sum

$$\sum_{\underset{\sim}{q}} \ldots\ldots \text{ by the integral } (2\pi)^{-3} v \int_{BZ} d\underset{\sim}{q} \ldots\ldots$$

I should add that this equidistribution of $\underset{\sim}{q}$ in the BZ is not obvious and there exist several difficulties in solid state physics that are related to the B.V.K. boundary conditions.

Vibrational modes of a crystal

In equation (6), $\mathbf{u}_{\kappa\varepsilon}$ depends only on q. Let us rewrite it as :

$$\underset{\sim}{u}_{\kappa o} = \underset{\sim}{U}_\kappa(\underset{\sim}{q}) \exp(i\underset{\sim}{q}\cdot\underset{\sim}{R}_\kappa) \tag{7}$$

(5) can then be transformed into :

$$m_\kappa \ddot{\underset{\sim}{U}}_\kappa(\underset{\sim}{q}) = - \sum_{\kappa'\ell'} [(G_{\kappa\kappa'}(\underset{\sim}{\ell}')\exp(i\underset{\sim}{q}.(\underset{\sim}{R}_{\kappa'}-\underset{\sim}{R}_\kappa))) \exp(i\underset{\sim}{q}.\underset{\sim}{\ell}')] \underset{\sim}{U}_{\kappa'}(\underset{\sim}{q}) \quad (8)$$

If one defines :

$$G_{\kappa\kappa'}(\underset{\sim}{q}) = \sum_{\ell'} G_{\kappa\kappa'}(\underset{\sim}{\ell}')\exp(i\underset{\sim}{q}.(\underset{\sim}{R}_{\kappa'}-\underset{\sim}{R}_\kappa))\exp(i\underset{\sim}{q}.\underset{\sim}{\ell}') \quad (9)$$

we are left with the fundamental equation :

$$m_\kappa \ddot{\underset{\sim}{U}}_\kappa(\underset{\sim}{q}) = - \sum_{\kappa'} G_{\kappa\kappa'}(\underset{\sim}{q}) \underset{\sim}{U}_{\kappa'}(\underset{\sim}{q}) \quad (10)$$

(10) is similar to the familiar equation of motion for a molecule. The 3nN dimensional problem is reduced to N problems of dimension 3n, each of the N "pseudo-molecular" problems being defined by a value of $\underset{\sim}{q}$. Each of the 3n modes varies with $\underset{\sim}{q}$, defining 3n branches. We write $\underset{\sim}{U}_1(\underset{\sim}{q}),\ldots\ldots, \underset{\sim}{U}_n(\underset{\sim}{q})$ as a column matrix $\mathbb{U}(\underset{\sim}{q})$. Let also $\mathbb{G}(\underset{\sim}{q})$ be the matrix composed of $G_{\kappa\kappa'}(\underset{\sim}{q})$. (10) becomes :

$$\mathbb{M}\, \ddot{\mathbb{U}}(\underset{\sim}{q}) = - \mathbb{G}(\underset{\sim}{q})\mathbb{U}(\underset{\sim}{q}) \quad (11)$$

where $\mathbb{M}$ is the diagonal matrix of masses.
Let now :

$$\mathbb{V}(\underset{\sim}{q}) = \mathbb{M}^{1/2}\, \mathbb{U}(\underset{\sim}{q})$$
$$\mathbb{D}(\underset{\sim}{q}) = \mathbb{M}^{-1/2}\, \mathbb{G}(\underset{\sim}{q})\, \mathbb{M}^{1/2} \quad (12)$$

we finally obtain :

$$\ddot{\mathbb{V}}(\underset{\sim}{q}) = - \mathbb{D}(\underset{\sim}{q})\mathbb{V}(\underset{\sim}{q}) \quad (13)$$

Assuming a solution :

$$\mathbb{V}(\underset{\sim}{q},t) = \mathbb{V}(\underset{\sim}{q})\exp(-i\omega(\underset{\sim}{q})t)$$

we get the fundamental dispersion relation :

$$\omega^2(\underset{\sim}{q})\, \mathbb{V}(\underset{\sim}{q}) = \mathbb{D}(\underset{\sim}{q})\, \mathbb{V}(\underset{\sim}{q}) \quad (14)$$

There are 3n solutions $[\omega_j(\underset{\sim}{q}), \mathbb{V}_j(\underset{\sim}{q})]$ .

The major problem is thus the calculation of $\mathbb{D}(\underset{\sim}{q})$, the dynamical matrix. Through the Hellman-Feynman theorem, one may show that the dynamical matrix depends only on the density $\rho(\underset{\sim}{r})$ and its derivatives with respect to the nuclear positions. This will be discussed by R.F. Stewart. I just want here to insist on the *fundamental relationship between lattice dynamics and the determination of charge densities*. In order to get reliable models for the density, use should be made of the results of lattice dynamics spectra.

It can be shown that 3 branches have frequencies $\omega_j(\underset{\sim}{q}) \Rightarrow 0$ when $\underset{\sim}{q} \Rightarrow 0$. Moreover, one may show that for small $\underset{\sim}{q}$, these 3 modes are equivalent to the elastic modes of the crystal :

$$\omega_j(\underset{\sim}{q}) = \underset{\sim}{v}_j \cdot \underset{\sim}{q}$$

where $\underset{\sim}{v}_j$ is the velocity of the elastic wave. These 3 modes are called acoustic modes, a generalisation of translational motion in molecular systems.

In a molecular system, 3 rotational modes have also a zero frequency. But in a crystal there is no more free rotation, and even for $q \Rightarrow 0$, these modes do not have a zero frequency. We may distinguish between molecular and non molecular systems, however. In the first case, we can still talk about rotational (librational) modes, the frequency of which is small compared to the (3n-6) optic modes. For non molecular crystals, the (3n-3) modes are equivalent on that respect. The frequency of optic modes does not vary very strongly with q, and for molecular crystals, one often admits that $\omega_j(\underset{\sim}{q}) = \omega_j(0)$ for the (3n-6) optic modes, the characteristics of which are those of individual molecules.

## Quantisation of normal modes. Phonons

Let us write :

$$\underset{\sim}{V}_j(\underset{\sim}{q}) = \xi_j(\underset{\sim}{q})\, \underset{\sim}{e}_j(\underset{\sim}{q}) \tag{15}$$

where

$$\underset{\sim}{e}^{\dagger}_{j'}(\underset{\sim}{q}) \cdot \underset{\sim}{e}_j(\underset{\sim}{q}) = \delta_{jj'} \tag{16}$$

$\xi_j(\underset{\sim}{q})$ is thus the normal coordinate of the mode. It can be quantized, in terms of creation anihilation operators. We define the phonon as the quasi-particle having energy $\hbar\omega_j(\underset{\sim}{q})$ and pseudo-momentum $\hbar\underset{\sim}{q}$. It is a quasi-particle since $\underset{\sim}{q}$ is not uniquely defined.

At a given temperature T, the energy of each mode is :

$$E_j(\underset{\sim}{q}) = \hbar\omega_j(\underset{\sim}{q})\, [1/2 + \bar{n}_j(\underset{\sim}{q})] \tag{17}$$

where $\bar{n}_j(\underset{\sim}{q})$ is the thermal average number of phonons $[\omega_j(\underset{\sim}{q}), \underset{\sim}{q}]$.

$$\bar{n}_j(\underset{\sim}{q}) = [\exp(\frac{\hbar\omega_j(\underset{\sim}{q})}{k\,T}) - 1]^{-1} \tag{18}$$

Furthermore,

$$E_j(\underset{\sim}{q}) = N\, \omega_j^2(\underset{\sim}{q}) \langle \xi_j \xi_j^{\dagger} \rangle \tag{19}$$

$\langle \xi_j \xi_j^{\dagger} \rangle^{1/2}$ being the amplitude of the mode.

We can consider two limiting cases :

a. High temperature limit.

$\hbar\omega_j(\underset{\sim}{q}) \ll kT$ (classical limit)

$\bar{n}_j(\underset{\sim}{q}) = \dfrac{kT}{\hbar\omega_j(\underset{\sim}{q})}$ $E_j(\underset{\sim}{q}) = kT$

$$< \xi_j^2(\underset{\sim}{q}) > = \frac{kT}{N\omega_j^2(\underset{\sim}{q})} \qquad (20)$$

b. Low temperature limit.

$\hbar\omega_j(\underset{\sim}{q}) \gg kT$

$\bar{n}_j(\underset{\sim}{q}) = 0$ $E_j(\underset{\sim}{q}) = 1/2\ \hbar\omega_j(\underset{\sim}{q})$

$$< \xi_j^2(\underset{\sim}{q}) > = \frac{kT}{2N\omega_j(\underset{\sim}{q})} \qquad (21)$$

We see that the amplitude of motion increases when the frequency diminishes. For our concerns, the main effects come from low frequency phonons : in non molecular crystals, the 3 acoustic branches, in molecular crystals the 6p translational and librational branches (p molecules per cell).

In most cases, one may use the classical limit for these modes, though the low temperature limit is valid for optic modes. However, when decreasing the temperature, the differenciation between low and high frequency modes becomes less pronounced.

COMMON APPROXIMATE MODELS

A. Einstein Model. It assumes that the frequency associated with a given branch is constant. It is often assumed to be valid for optic modes in molecular crystals.

B. Debye Model. This model is very popular and is only valid for acoustic modes. We replace the three acoustic branches by an average one, with :

$$\omega_j(\underset{\sim}{q}) = v_s q \qquad (22)$$

Obviously the frequency distribution function $g(\omega)$ is :

$$g(\omega) = A\,\omega^2$$

The BZ is approximated by a sphere of radius $q_D$ such that :

$$(2\pi)^3/v_c = 4/3\,\pi q_D^2 \qquad (23)$$

Therefore, $g(\omega)$ is :

$$g(\omega) = 9N\frac{\omega^2}{\omega_D^2} \qquad \omega < \omega_D$$
$$= 0 \qquad \omega > \omega_D \tag{24}$$

We define the Debye temperature $\Theta_D$ as :

$$\hbar\omega_D = k\Theta_D \tag{25}$$

Specific heats or mean square displacements can be calculated using this model. Conversely the exact values can be fitted to the Debye model, defining a temperature dependent Debye temperature $\Theta(T)$. For specific heats, we call it $\Theta_C(T)$, for the displacements $\Theta_M(T)$. The results for $CaF_2$ are shown on Fig.1, showing that the model is reasonable for representing the atomic displacements, though cruder in the case of specific heat.

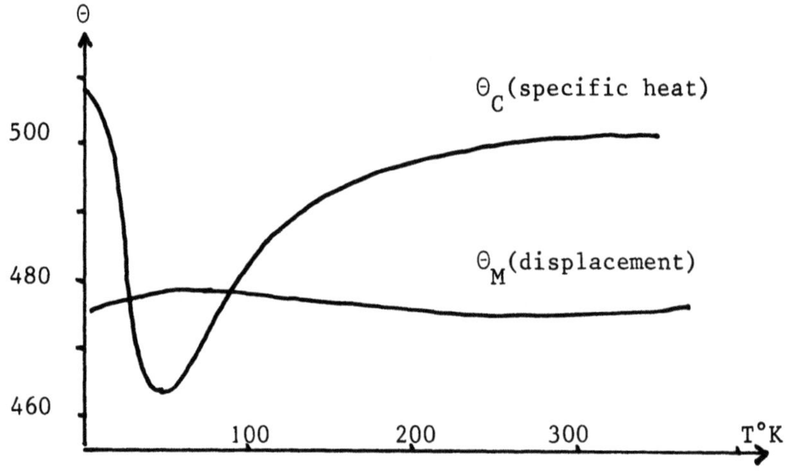


Figure 1

Debye Temperatures for displacement and specific heat for $CaF_2$, as a function of temperature.

C. Elastic wave approximation. The theory of elasticity is a macroscopic bulk theory and we shall only discuss its main features. Let u be the local displacement vector in the solid. The strain tensor $\eta$ may be defined as :

$$\eta_{ij} = \frac{\partial u_i}{\partial x_j} + \frac{\partial u_j}{\partial x_i} \qquad (i \neq j)$$
$$\eta_{ii} = \frac{\partial u_i}{\partial x_i} \tag{26}$$

If $\eta$ does not depend on the position, the strain is uniform.

We also define the stress tensor $\tau$ : $\tau_{ij}$ is the force applied in the direction i to a unit area whose normal is along j. It can be shown that $\tau$ is a symmetric tensor.

The application of Hooke's law leads to the existence of a linear relationship between $\eta$ and $\tau$ :

$$\tau_{ij} = \sum_k \sum_l C_{ijkl}\, \eta_{kl} \tag{27}$$

which defines the elastic constant tensor C.
Since $\tau$ and $\eta$ have only 6 independent coefficients, C has only 36 :

$$C_{ijkl} = C_{jikl} = C_{jilk} = C_{ijlk}$$

Furthermore one may show that : $C_{ijkl} = C_{klij}$. There are thus only 21 independent elastic constants. This number can be further reduced by symmetry (ref.2).

Let us now express the equation of motion of a small volume v, limited by a closed surface S. Let $\tau_{in}$ be the force, per unit area acting in the i direction, acting on a unit surface of normal $\underset{\sim}{n}$ :

$$\tau_{in} = \sum_j \tau_{ij}\, n_j$$

Therefore :

$$\iiint_v \rho \frac{\partial^2 u_i}{\partial t^2}\, dv = \iint_S \tau_{in}\, dS$$

We now use Green's theorem, getting :

$$\rho \frac{\partial^2 u_i}{\partial t^2} = \sum_j \frac{\partial \tau_{ij}}{\partial x_j} = \sum_k \sum_l \sum_j C_{ijkl} \frac{\partial \eta_{kl}}{\partial x_j} \tag{28}$$

If we assume the solution of (28) to be :

$$\underset{\sim}{u} = \underset{\sim}{\varepsilon}\, \exp(i(\omega t - \underset{\sim}{q}.\underset{\sim}{r})) \tag{29}$$

where $\underset{\sim}{\varepsilon}$ is a unit polarisation vector, and if we define A by :

$$A_{lm} = q^{-2} \sum_r \sum_s C_{rlsm}\, q_r q_s \tag{30}$$

(28) becomes :

$$\sum_j \left( A_{ij}\, q^2 - \omega^2 \rho\, \delta_{ij} \right) \varepsilon_j = 0 \tag{31}$$

(31) has 3 solutions. It is obvious that :

$$\omega_j(\underset{\sim}{q}) = v_j\, q \qquad j=1,2,3 \tag{32}$$

In reality, one must solve the dynamical matrix, which gives the elastic spectrum only asymptotically. For soft crystals (organic molecules) the elastic theory is only valid over a small range of q.

Let us finally derive a useful expression. For a mode $(j, \underset{\sim}{q})$ supposed to be elastic, the amplitude of displacement of all atoms should be the same. The vector $\underset{\sim}{e}_j(\underset{\sim}{q})$ defined in (15) is formed with n atomic components $\underset{\sim}{e}_{\kappa j}(\underset{\sim}{q})$. The displacement of atom $\kappa$ is proportional to :

$$\underset{\sim}{e}_{\kappa j}(\underset{\sim}{q}) \,/\, \sqrt{m_\kappa}\ .$$

The elastic approximation implies :

$$m_1^{-1/2}\, \underset{\sim}{e}_{1j}(\underset{\sim}{q}) = m_2^{-1/2}\, \underset{\sim}{e}_{2j}(\underset{\sim}{q}) = \ldots = c\, \underset{\sim}{\varepsilon}_j(\underset{\sim}{q})$$

where $\underset{\sim}{\varepsilon}_j(\underset{\sim}{q})$ is the unit polarisation vector. Applying (16) gives :

$$m_1^{-1/2}\, \underset{\sim}{e}_{1j}(\underset{\sim}{q}) = \ldots. = m^{-1/2}\, \underset{\sim}{e}_{j}(\underset{\sim}{q}) = m^{-1/2}\, \underset{\sim}{\varepsilon}_j(\underset{\sim}{q}) \tag{33}$$

$$m = m_1 + m_2 + \ldots + m_n$$

$$\kappa = 1,\ 2,\ \ldots,\ n$$

D. Molecular Crystals. In this case, the most common approximation is the separation into rigid and non rigid body motions. Each molecule is thus supposed to retain its individual optic vibrations. Only its 6 external coordinates are coupled through the crystal forces. This approximation has proved to be very useful. March and coworkers (ref.3) have proved that it is equivalent to a pairwise potential interaction model. The dynamical equations have to be modified in this case. We assume that we have n molecules in the unit cell (labelled ). $\underset{\sim}{u}_{\kappa\ell}$ and $\underset{\sim}{\theta}_{\kappa\ell}$ are the instantaneous translation and rotation of molecule $(\kappa\ell)$. Let $M_{\kappa,mol}$ be the 6x6 matrix :

$$M_{\kappa,mol} = \begin{pmatrix} M_\kappa & 0 \\ 0 & I_\kappa \end{pmatrix} \qquad M_\kappa = \begin{pmatrix} m_\kappa & 0 & 0 \\ 0 & m_\kappa & 0 \\ 0 & 0 & m_\kappa \end{pmatrix} \tag{34}$$

$I_\kappa$ : Inertia moments tensor

$$M_{\kappa,mol}\begin{pmatrix}\ddot{u}_{\kappa\ell}\\ \ddot{\theta}_{\kappa\ell}\end{pmatrix} = -\sum_{\kappa'\ell'} G_{\kappa\kappa'}(\ell'-\ell)\begin{pmatrix}u_{\kappa\ell}\\ \theta_{\kappa\ell}\end{pmatrix} \tag{35}$$

replaces equation (5). G are 6x6 matrices that are difficult to construct, unless one uses pairwise potential approximations (ref.4). From (35) the derivations are similar to those outlined previously. The only difficulty concerns the calculation of $M^{1/2}$, where M is formed with n block-diagonal matrices $M_{\kappa,mol}$ (ref.5). The method has been successfully applied to hexamethylenetetramine (ref.6) and to urea (ref.7). In the case of HMT, $I_\kappa$ is diagonal and there is one molecule per cell.

In the dynamical matrix, translation and rotation are no more separated. An example is given for the 6 modes travelling along the [100] direction of HMT (Fig.2). Branch 1 is purely translational. Branches 2 and 3, degenerate, are translational at the BZ center and become rotational at the BZ boundary : these branches are depicted in Fig.3. Branch 4 is purely rotational. 5 and 6, degenerate, are also of mixed type. The dispersion curves are shown in Fig.2.

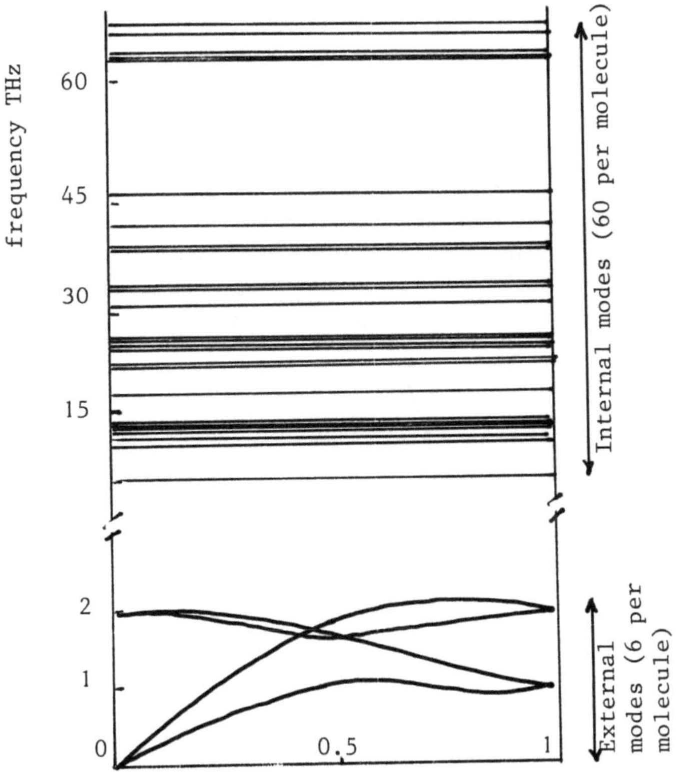


Figure 2a
Dispersion curves for HMT, in the (1,0,0) direction.

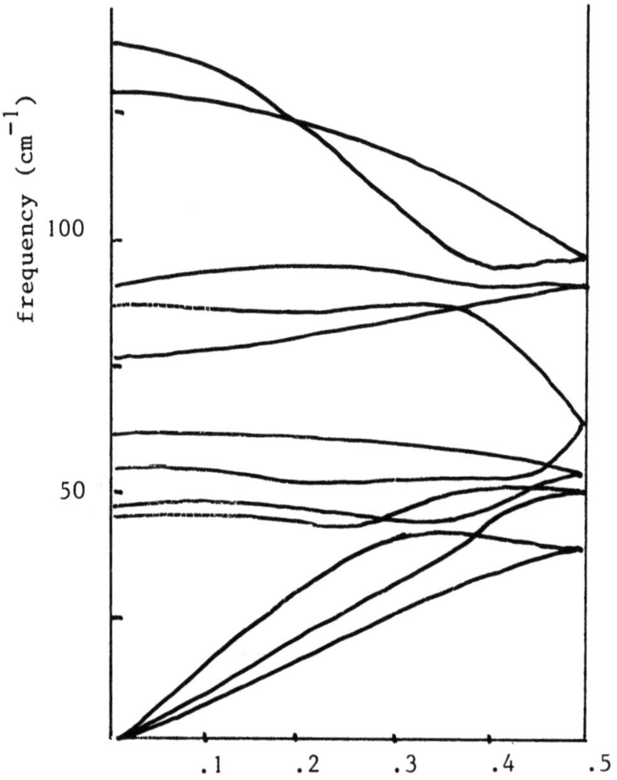


Figure 2b
Dispersion curves for naphtalene in the (0,1,0) direction : lattice modes

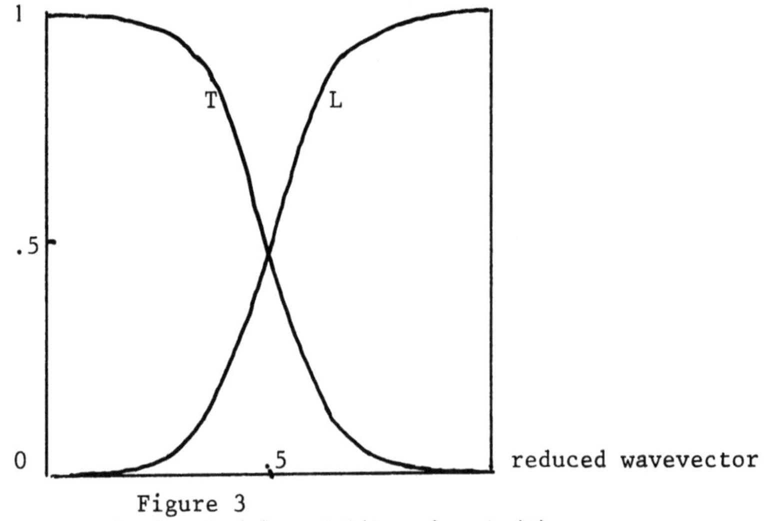


Figure 3
Translational (T) and librational (L) character for HMT in the (1,0,0) direction

## References

1.-Maradudin A.A., Montroll E.W., Weiss G.A., Ipatowa I.P. (1971) Lattice Dynamics in Harmonic Approximation. N.Y. Academic Press
-Cochran W. (1973). The Dynamics of Atoms in Crystals. Arnold
-Ghatak A.K., Kothari L.S. (1973). Lattice Dynamics. Addison Wesley
-Willis B.T.M., Pryor A.W. (1975). Vibrations in Crystals. Cambridge
-Cochran W., Cowley R.A. (1967). Handbuch der Physik 25,2a,59. Springer
-Reissland J.A. (1973). The Physics of Phonons. Wiley

2.-Nye J.F. (1957). Physical Properties of Crystals. Oxford
-Brillouin L. (1953). Wave Propagation in Periodic Structures

3. Jones W., March N.H. (1970). Proc. Roy. Soc. A317, 359

4. Pawley G.S. (1970). Phys. Stat. Sol. 20, 347

5. Venkataraman G., Sahni V.C. (1970). Rev. Mod. Phys. 42, 409

6. Cochran W., Pawley G.S. (1964). Proc. Roy. Soc. A280, 1

7. Mc Kenzie P.R., Pryor A.W. (1971). J. Phys. C., 4, 2304

## II. KINEMATICAL SCATTERING

Kinematical theory of scattering is based on the first Born approximation (see for example a paper by Feil, ref.1). We shall mainly discuss here elastic scattering, except for thermal diffuse component.

### A1. X-Ray Elastic Scattering.

$\underset{\sim}{k}_o$ and $\underset{\sim}{k}$ are the vectors in the incident and scattered directions. The scattering vector $\underset{\sim}{S}$ is defined as :

$$\underset{\sim}{S} = \underset{\sim}{k} - \underset{\sim}{k}_o \qquad (1)$$

In elastic scattering, $k_o=k=2\pi/\lambda$, and thus $S=4\pi\ \sin\theta/\lambda$.
We shall suppose that the nuclei are fixed. $|\Psi_o>$ is the electronic state vector. We shall adopt the following notations :

a : classical radius of the electron = $e^2/(4\pi\varepsilon_o mc^2)$

C : polarisation factor =1 for the perpendicular component of the electric field =cos2θ for the parallel component

$\hat{D}$ : scattering operator :

$$\hat{D} = \sum_j^{el} \exp(i\underset{\sim}{S}.\underset{\sim}{r}_j)$$

The Waller Hartree approximation tells that the intensity scattered at a distance $R_o$ from the sample, in the direction $\underset{\sim}{k}$, is :

$$I(\underset{\sim}{S}) = \left|\frac{aC}{R_o}\right|^2 |\langle\Psi_o|\hat{D}|\Psi_o\rangle|^2 \tag{2}$$

It is easy to show that :

$$\langle\Psi_o|\hat{D}|\Psi_o\rangle = A(\underset{\sim}{S}) = \int_V \rho(\underset{\sim}{r}) \exp(i\underset{\sim}{S}.\underset{\sim}{r})\, dv \tag{3}$$

where $\rho(\underset{\sim}{r})$ is the electron density and $A(\underset{\sim}{S})$ is therefore the amplitude of scattering relative to the free electron.

We now assume the system to be periodic :

$$\rho(\underset{\sim}{r}) = v_c^{-1} \sum_{\underset{\sim}{K}} F(\underset{\sim}{K}) \exp(-i\underset{\sim}{K}.\underset{\sim}{r}) \tag{4}$$

where the structure factor $F(\underset{\sim}{S})$ is defined as :

$$F(\underset{\sim}{S}) = \int_{v_c} \rho(\underset{\sim}{r}) \exp(i\underset{\sim}{S}.\underset{\sim}{r}) \tag{5}$$

Let :

$$\Delta(\underset{\sim}{S}) = \sum_{\ell} \exp(i\underset{\sim}{S}.\underset{\sim}{\ell}) \tag{6}$$

be the interference function :

$$A(\underset{\sim}{S}) = F(\underset{\sim}{S}).\ \Delta(\underset{\sim}{S}) \tag{7}$$

$F(\underset{\sim}{S})$ is a smoothly varying function of $\underset{\sim}{S}$, while $\Delta(\underset{\sim}{S})$ peaks only at reciprocal lattice vectors. If the crystal is finite, there is a little width of the reflection around $\underset{\sim}{K}$. We may write :

$$A(\underset{\sim}{S}) = F(\underset{\sim}{K}).\ \Delta(\underset{\sim}{S}) \tag{8}$$

The kinematical intensity becomes :

$$I_k(\underset{\sim}{S}) = \left|\frac{aF(K)C}{R_o}\right|^2 |\Delta(\underset{\sim}{S})|^2 \tag{9}$$

A2. Neutron Scattering

For neutron nuclear scattering, the theory is identical. But $F(\underset{\sim}{S})$ is the Fourier transform of a weighted nuclear distribution :

$$F(\underset{\sim}{S}) = \sum_j b_j \exp(i\underset{\sim}{S}.\underset{\sim}{R}_j) \tag{10}$$

$C = 1$ ; $a = 10^{-12}$ cm.

## B. RECORDED INTEGRATED POWER

We assume that the observations are made in the horizontal plane, the crystal being able to rotate around a vertical axis. Let $\underset{\sim}{u}_o$ and $\underset{\sim}{u}$ be the unit vectors along the incident and diffracted directions. $\underset{\sim}{u}_o^\circ$ and $\underset{\sim}{u}^\circ$ are the vectors when the Bragg condition is exactly fulfilled. We define the geometry in Fig.1. $\underset{\sim}{\tau}_1$ and $\underset{\sim}{\tau}_2$ are unit vectors in the plane of diffraction, $\tau_3$ is along the vertical axis. We can write :

$$\underset{\sim}{S} = \underset{\sim}{K} + 2\pi\lambda^{-1}\underset{\sim}{\varepsilon}$$
$$\underset{\sim}{\varepsilon} = \varepsilon_1\underset{\sim}{\tau}_1 + \varepsilon_2\underset{\sim}{\tau}_2 + \varepsilon_3\underset{\sim}{\tau}_3 \qquad (11)$$

The detector subtends a solid angle $\Omega$ (integration over $\varepsilon_2$ and $\varepsilon_3$) and the crystal is rotated around $\underset{\sim}{\tau}_3$ during a time t :

$$dt = \frac{d\varepsilon_1}{\omega_o},$$ $\omega_o$ being the angular velocity of scanning.

For a given position of the crystal ($\varepsilon_1$ constant), the power recorded in the counter is :

$$P_k(\varepsilon_1) = R_o^2 \iint I_k(\underset{\sim}{\varepsilon})\, d\varepsilon_2 d\varepsilon_3 = v\,\sigma(\varepsilon_1) \qquad (12)$$

$\sigma(\varepsilon_1)$ is the kinematical diffraction function and can be calculated for any crystal shape (ref.2). The integrated unit diffracting power $Q_k$ is obtained as :

$$Q_k = \int\sigma(\varepsilon_1)\, d\varepsilon_1 = \left|\frac{aFC}{v_c}\right|^2 \frac{\lambda^3}{\sin 2\theta} \qquad (13)$$

The total recorded power is :

$$E = \int I_k(\underset{\sim}{\varepsilon})\, d\Omega\, dt = \frac{v}{\omega_o} Q_k \qquad (14)$$

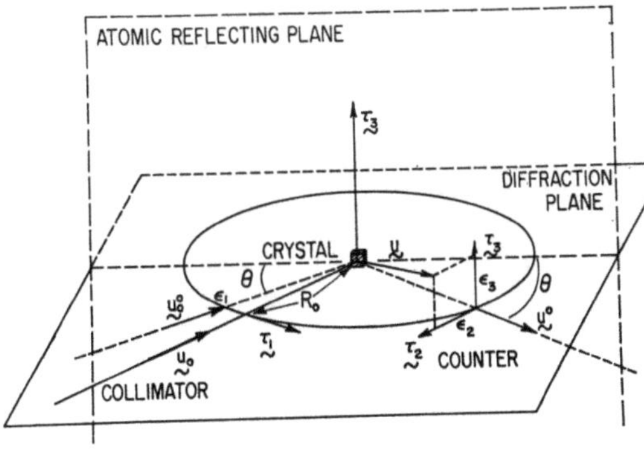


Figure 1
Definition of the directions of the incident beam $\underset{\sim}{u}_o$ and of the diffracted beam $\underset{\sim}{u}$ with respect to their ideal values $\underset{\sim}{u}_o^{\,o}$ and $\underset{\sim}{u}^o$, when the Bragg condition is fulfilled. $R_o$ is the distance between crystal and counter

C. THERMAL SCATTERING IN THE X-RAY CASE

We work within the Born Oppenheimer approximation. The state of the system is represented by :

$$|\psi_o(\underset{\sim}{r}_i, \underset{\sim}{Q})\ \phi_\nu(\underset{\sim}{Q})\rangle$$

where $\phi_\nu(\underset{\sim}{Q})$ is the nuclear wavefunction. By scattering , the system can be excited to

$$|\psi_o(\underset{\sim}{r}_i, \underset{\sim}{Q})\ \phi_{\nu'}(\underset{\sim}{Q})\rangle$$

implying creation or anihilation of a phonon. Since we are in the X-Ray case, the associated change in the wavelength of the photons is still negligible and we shall assume :

$$k \cong k_o$$

$|\langle\Psi_o|D|\Psi_o\rangle|^2$ in (2) is replaced by :

$$X = \sum_\nu \sum_{\nu'} p_\nu |\langle\psi_o\phi_\nu|D|\ \psi_o\phi_{\nu'}\rangle|^2 \tag{15}$$

We define :

$$A(\underset{\sim}{S}, \underset{\sim}{Q}) = \int_v \rho(\underset{\sim}{r}, \underset{\sim}{Q})\ \exp(i\underset{\sim}{S}.\underset{\sim}{r})\ dv = \langle\Psi_o|D|\Psi_o\rangle \tag{16}$$

$$X = \sum_\nu p_\nu \sum_{\nu'} \langle\phi_\nu|A|\phi_{\nu'}\rangle\langle\phi_{\nu'}|A^*|\phi_\nu\rangle$$

$$= \sum_\nu p_\nu \langle\phi_\nu|AA^*|\phi_\nu\rangle \tag{17}$$

The nuclear distribution function $P(\underset{\sim}{Q})$ is defined as :

$$P(\underset{\sim}{Q}) = \sum_\nu p_\nu \phi_\nu(\underset{\sim}{Q})\phi_\nu^*(\underset{\sim}{Q}) \tag{18}$$

Then :

$$X = \langle|A(\underset{\sim}{S},\ \underset{\sim}{Q})|^2\rangle = \int P(\underset{\sim}{Q})\ |A^2(\underset{\sim}{S}, \underset{\sim}{Q})|\ d\underset{\sim}{Q}$$

(2) is thus replaced by :

$$I_k(\underset{\sim}{S}) = \left|\frac{aC}{R_o}\right|^2 \langle|A(\underset{\sim}{S}, \underset{\sim}{Q})|^2\rangle \tag{19}$$

To go further, we have to assume a definite response of the electronic distribution to the vibrational excitations. This is a complicated subject which has not been considered seriously enough in the past. The unique approximation until now is the convolution approximation. One assumes that $\rho(\underset{\sim}{r}, \underset{\sim}{Q})$ can be partitioned into terms that rigidly follow the motion of a given nucleus. Recent attempts have been done to generalize this model (ref. 3).

$$\rho(\underset{\sim}{r}, \underset{\sim}{Q}) = \sum_{\kappa\ \ell} \rho_\kappa(\underset{\sim}{r} - \underset{\sim}{R}_{\kappa\ell} - \underset{\sim}{u}_{\kappa\ell}) \tag{20}$$

The generalized scattering factors are :

$$f_\kappa(\underset{\sim}{S}) = \int \rho_\kappa(\underset{\sim}{r})\ \exp(i\underset{\sim}{S}.\underset{\sim}{r})\ dv \tag{21}$$

so that we get :

$$A(\underset{\sim}{S},\underset{\sim}{Q}) = \sum_{\kappa\ell} f_{\kappa} \exp[i\underset{\sim}{S}.(\underset{\sim}{R}_{\kappa}+\underset{\sim}{\ell}+\underset{\sim}{u}_{\kappa\ell})] \tag{22}$$

$$|A^2| = \sum_{\kappa\ell} \sum_{\kappa'\ell'} f_{\kappa} f^*_{\kappa'} \exp i\underset{\sim}{S}.(\underset{\sim}{R}_{\kappa}-\underset{\sim}{R}_{\kappa'}) \exp[i\underset{\sim}{S}.(\underset{\sim}{\ell}-\underset{\sim}{\ell}')] \; <\exp[i\underset{\sim}{S}.(\underset{\sim}{u}_{\kappa\ell}-\underset{\sim}{u}_{\kappa'\ell'})]> \tag{23}$$

From this point, we can develop the theory in terms of the normal modes or in terms of the displacements (ref.4). We shall use a mixed treatment, and assume the harmonic approximation.

$$\begin{aligned} <\exp[i\underset{\sim}{S}.(\underset{\sim}{u}_{\kappa\ell}-\underset{\sim}{u}_{\kappa'\ell'})]> &= \exp[-\tfrac{1}{2} <\underset{\sim}{S}.(\underset{\sim}{u}_{\kappa\ell}-\underset{\sim}{u}_{\kappa'\ell'})^2>] \\ &= \exp -\tfrac{1}{2}<(S.u_{\kappa\ell})^2> \\ &\times \exp -\tfrac{1}{2}<(S.u_{\kappa'\ell'})^2> \\ &\times \exp <(\underset{\sim}{S}.\underset{\sim}{u}_{\kappa\ell})(\underset{\sim}{S}.\underset{\sim}{u}_{\kappa'\ell'})> \end{aligned} \tag{24}$$

If the third term would be one there should not exist any coupling between the motion of two different atoms. The effect of phonon waves should thus be negligible. This is the zero-phonon approximation. In the non harmonic approach, this would still mean :

$$<\exp[i\underset{\sim}{S}.(\underset{\sim}{u}_{\kappa\ell}-\underset{\sim}{u}_{\kappa'\ell'})]> \approx <\exp(i\underset{\sim}{S}.\underset{\sim}{u}_{\kappa\ell})><\exp(-i\underset{\sim}{S}.\underset{\sim}{u}_{\kappa'\ell'})> \tag{25}$$

Let us moreover assume that $<\exp(i\underset{\sim}{S}.\underset{\sim}{u}_{\kappa\ell})>$ does not depend on $\ell$. This is the Debye Waller factor $W_{\kappa}(\underset{\sim}{S})$. Using (25), (23) becomes :

$$X_o = |\sum_{\kappa} f_{\kappa} W_{\kappa} \exp(i\underset{\sim}{S}.\underset{\sim}{R}_{\kappa})|^2 |\Delta^2(\underset{\sim}{S})| \tag{26}$$

$$= |<F(\underset{\sim}{S},\underset{\sim}{Q})>|^2 |\Delta^2(\underset{\sim}{S})| \tag{27}$$

The zero-phonon approximation corresponds to scattering by the average structure, periodic :

$$\begin{aligned} \rho(\underset{\sim}{r}) &= \int P(\underset{\sim}{Q})\, \rho(\underset{\sim}{r},\underset{\sim}{Q})\, d\underset{\sim}{Q} \\ F(\underset{\sim}{S}) &= \int P(\underset{\sim}{Q})\, F(\underset{\sim}{S},\underset{\sim}{Q})\, d\underset{\sim}{Q} \end{aligned} \tag{28}$$

From that point, the kinematical theory developed above is unchanged. It is obvious that the concept of Debye-Waller factor in X-Ray scattering is associated with the convolution approximation. Beyond this approximation, the Debye-Waller factor does not strictly exist.

Coming back to (15), we see that :

$$X_o = |\sum_{\nu} p_{\nu} <\phi_{\nu}|A|\phi_{\nu}>|^2$$

though the elastic part of the scattering is given by :

$$X_{el} = \sum_{\nu} p_{\nu} |\langle\phi_{\nu}|A|\phi_{\nu}\rangle|^2 \tag{29}$$

The difference between $X_{el}$ and $X_o$ is discussed by Stewart (ref.5).

Using equation (24), we can write :

$$X = \langle A^2\rangle = \sum_{0}^{\infty} X_n$$

$$X_n = \sum_{\kappa\ell} \sum_{\kappa'\ell'} f_{\kappa} f_{\kappa'} W_{\kappa} W_{\kappa'} \exp[i\underset{\sim}{S}.(\underset{\sim}{R}_{\kappa\ell}-\underset{\sim}{R}_{\kappa'\ell'})] \frac{1}{n!} \langle(\underset{\sim}{S}.\underset{\sim}{u}_{\kappa\ell})(\underset{\sim}{S}.\underset{\sim}{u}_{\kappa'\ell'})\rangle^{n} \tag{30}$$

$X_n$ gives rise to the $n^{th}$ order diffuse scattering. There is only one basic coupling term :

$$\langle(\underset{\sim}{S}.\underset{\sim}{u}_{\kappa\ell})(\underset{\sim}{S}.\underset{\sim}{u}_{\kappa'\ell'})\rangle$$

Normal mode expansion. We have just seen that the important quantities to be evaluated are : $\langle(\underset{\sim}{S}.\underset{\sim}{u}_{\kappa\ell})^2\rangle$ and $\langle(\underset{\sim}{S}.\underset{\sim}{u}_{\kappa\ell})(\underset{\sim}{S}.\underset{\sim}{u}_{\kappa'\ell'})\rangle$
We first recall the expression for $\underset{\sim}{u}_{\kappa\ell}$ :

$$\underset{\sim}{u}_{\kappa\ell} = \sum_{j\underset{\sim}{q}} \xi_j(\underset{\sim}{q}) \frac{\underset{\sim}{e}_{\kappa j}(\underset{\sim}{q})}{\sqrt{m_{\kappa}}} \exp(i\underset{\sim}{q}.\underset{\sim}{R}_{\kappa\ell}) \tag{31}$$

$\underset{\sim}{e}_{\kappa j}(\underset{\sim}{q})$ is the $\kappa^{th}$ atom component of matrix vector $\underset{\sim}{e}_j(\underset{\sim}{q})$ (see I.15)
Therefore :

$$\langle(\underset{\sim}{S}.\underset{\sim}{u}_{\kappa\ell})^2\rangle = \frac{1}{m_{\kappa}} \sum_{j\underset{\sim}{q}} \langle\xi_j^2(\underset{\sim}{q})\rangle\, [\underset{\sim}{S}.\underset{\sim}{e}_{\kappa j}(\underset{\sim}{q})]^2 \tag{32}$$

We remember from (I.19) that :

$$\langle\xi_j^2(\underset{\sim}{q})\rangle = \frac{E_j(\underset{\sim}{q})}{N\omega_j^2(\underset{\sim}{q})}$$

Consider now $y = \langle(\underset{\sim}{S}.\underset{\sim}{u}_{\kappa\ell})(\underset{\sim}{S}.\underset{\sim}{u}_{\kappa'\ell'})\rangle$

$$y = \sum_{j\underset{\sim}{q}} \langle\xi_j^2(\underset{\sim}{q})\rangle[\underset{\sim}{S}.\frac{\underset{\sim}{e}_{\kappa j}(\underset{\sim}{q})}{\sqrt{m_{\kappa}}}]\ [\underset{\sim}{S}.\frac{\underset{\sim}{e}^{\dagger}_{\kappa' j}(\underset{\sim}{q})}{\sqrt{m_{\kappa'}}}]\exp[i\underset{\sim}{q}.(\underset{\sim}{R}_{\kappa\ell}-\underset{\sim}{R}_{\kappa'\ell'})] \tag{33}$$

The first order diffuse scattering $X_1$ can be written :

$$X_1 = \sum_{j\underset{\sim}{q}} I_1(j,\underset{\sim}{q})$$

$$I_1(j,\underset{\sim}{q}) = \langle\xi_j^2(\underset{\sim}{q})\rangle\ |[\underset{\sim}{S}.\frac{\underset{\sim}{e}_{\kappa j}(\underset{\sim}{q})}{\sqrt{m_{\kappa}}}]f_{\kappa} W_{\kappa} \exp[i(\underset{\sim}{q}+\underset{\sim}{S}).\underset{\sim}{R}_{\kappa}]|^2\ |\Delta^2(\underset{\sim}{q}+\underset{\sim}{S})| \tag{34}$$

Each term corresponds to some generalized structure factor. Because of the term $\Delta(\underset{\sim}{q}+\underset{\sim}{S})$, scattering only occurs for :

$$\underset{\sim}{q} + \underset{\sim}{S} = \underset{\sim}{K} \tag{35}$$

which is momentum conservation, if the scattering involves one phonon of momentum $\hbar \underset{\sim}{q}$. $X_1$ thus corresponds to the one phonon diffuse scattering. We can write :

$$I_1(j,\underset{\sim}{q}) = <\xi_j^2(\underset{\sim}{q})> S^2 |G_j(\underset{\sim}{q})|^2 |\Delta(\underset{\sim}{q}+\underset{\sim}{S})|^2$$

$$G_j(\underset{\sim}{q}) = \frac{\underset{\sim}{S}}{S} \cdot \sum_K f_K W_K \frac{\underset{\sim}{e}_{Kj}(\underset{\sim}{q})}{\sqrt{m_K}} \exp(i\underset{\sim}{K}.\underset{\sim}{R}_K) \tag{36}$$

It is easily seen that $X_2$, ..., $X_p$, ... involve 2, ..., p,... phonon scattering. We shall not consider these terms here. They do not involve more physics, but just mathematical complications.

## D. NEUTRON THERMAL SCATTERING

The major difference in the neutron case comes from the fact that the neutron energy is of the same order as the phonon energies. We can no longer assume that $k=k_o$, unless we consider the zero-phonon process, which has the same characteristics as X-Ray zero-phonon process, i.e. :

$$F(\underset{\sim}{S}) = \sum_k b_k W_k \exp(i\underset{\sim}{S}.\underset{\sim}{R}_k) \tag{37}$$

We now consider the one phonon process. Let us define the $< 0$ or $> 0$ frequency by :

$$\hbar\omega = \frac{\hbar^2}{2m_n}(k_o^2 - k^2)$$

$$\lambda = \frac{2\pi\hbar}{m_n v_n} \tag{38}$$

$\lambda$ is the wavelength of the neutron, $v_n$ its speed.

The energy conservation during the scattering process is :

$$\hbar\omega = \pm \hbar\omega_j(\underset{\sim}{q}) \qquad \begin{array}{l} + : \text{phonon creation} \\ - : \text{phonon anihilation} \end{array} \tag{39}$$

The momentum conservation is expressed by :

$$\underset{\sim}{S} + \underset{\sim}{q} = \underset{\sim}{K} \tag{40}$$

The advantage of neutron inelastic scattering comes from the fact that one can analyze the experiment both in energy and momentum. Let $dE = \hbar\, d\omega$ .

$\frac{dX_1}{dE}$ dE is the intensity corresponding to the energy band dE :

$$\frac{dX_1(j,\underset{\sim}{q})}{dE} = \frac{d^2\sigma_1}{d\Omega dE} \qquad \text{(double differential cross section)}$$

Phonon creation and phonon anihilation are equally probable but a change in the number of phonons will occur. It can be shown (ref.6) that :

$$\frac{d^2\sigma_1}{d\Omega dE} = \frac{1}{2N}\frac{k}{k_o}\frac{1}{\omega_j(\underset{\sim}{q})}\left|\sum_\kappa b_\kappa W_\kappa \frac{\underset{\sim}{S}.\underset{\sim}{e}_{\kappa j}(\underset{\sim}{q})}{\sqrt{m_\kappa}} \exp[i(\underset{\sim}{q}+\underset{\sim}{S}).\underset{\sim}{R}_\kappa]\right|^2$$
$$|\Delta(\underset{\sim}{q}+\underset{\sim}{S})|^2 \tag{41}$$
$$\{[\bar{n}_j(\underset{\sim}{q})+1]\ \delta(\omega-\omega_j(\underset{\sim}{q}))+\bar{n}_j(\underset{\sim}{q})\delta(\omega+\omega_j(\underset{\sim}{q}))\}$$

In the case where $k \to k_o$, i.e. $\omega_j(\underset{\sim}{q}) \to 0$

$$\int \frac{d^2\sigma_1}{d\Omega dE}\, dE \to I_1^{XR}(j,\underset{\sim}{q})$$

Obviously, neutron inelastic scattering is the dedicated method to determine lattice dynamical characteristics of a given crystal.

For neutrons, Debye-Waller factors are well defined. If we use the convolution approximation, D.W. X-Ray approximate factors should be the same.

It can be shown that the quantity corresponding to $X_1$ is :

$$X_1 = \sum_{j\underset{\sim}{q}} I_1(j,\underset{\sim}{q})$$
$$I_1(j,\underset{\sim}{q}) = \frac{k}{k_o} |\Delta(\underset{\sim}{q}+\underset{\sim}{S})|^2\ S^2\ |G_j(\underset{\sim}{q})|^2 \tag{42}$$
$$\times\ \frac{1}{2}\left[\frac{E_j(\underset{\sim}{q})+1/2\ \hbar\omega_j(\underset{\sim}{q})}{J_j^+(\underset{\sim}{q})\ \omega_j^2(\underset{\sim}{q})} + \frac{E_j(\underset{\sim}{q})-1/2\ \hbar\omega_j(\underset{\sim}{q})}{J_j^-(\underset{\sim}{q})\ \omega_j^2(\underset{\sim}{q})}\right]$$

with
$$J_j^\pm(\underset{\sim}{q}) = 1 \pm \frac{\underset{\sim}{v}_n.\underset{\sim}{V}_j(\underset{\sim}{q})}{v_n^2} \tag{43}$$

$\underset{\sim}{V}_j(\underset{\sim}{q})$ : group velocity of the phonon : $\nabla_{\underset{\sim}{q}}\ \omega_j(\underset{\sim}{q})$

$v_n$ : velocity of the neutron

## E. GENERAL EXPRESSION FOR THERMAL DIFFUSE SCATTERING CORRECTION

Let $E_o$ be the zero-phonon integrated power :

$$E_o = \frac{N\lambda^3}{v_c \omega \sin 2\theta} (aC)^2 |F(\underset{\sim}{K})|^2$$

A schematic case, including thermal diffuse scattering, is drawn in Fig.2. The total integrated power E is written as :

$$E = E_o ( 1 + \alpha ) \tag{44}$$

being the TDS correction that we want to evaluate. If this is done from E we get $E_o$, that is to say $F(\underset{\sim}{K})$ and thus $\rho(\underset{\sim}{r})$.

Calling $E^p$ and $E^b$ the recorded powers corresponding to the peak and background measurements, we get :

$$\alpha = \frac{E^p - E^b}{E_o} \tag{45}$$

We generalize the discussion of part B for first order scattering :

$$E = \frac{N\lambda^3}{v_c \omega \sin 2\theta} (aC)^2 \sum_{\underset{\sim}{q}}' \sum_{j}' s^2 |G_j(\underset{\sim}{q})|^2 \langle\xi_j^2(\underset{\sim}{q})\rangle \tag{46}$$

The summation is limited to those $\underset{\sim}{q}$ that are within the region of BZ spanned during recording. A similar expression can be obtained for neutrons, using the last term of (42) instead of $\langle\xi_j^2(\underset{\sim}{q})\rangle$ .

Let us define $B(\underset{\sim}{q})$ as :

$$B(\underset{\sim}{q}) = \frac{Nv_c}{|F(\underset{\sim}{K})|^2} \sum_j s^2 |G_j(\underset{\sim}{q})|^2 \langle\xi_j^2(\underset{\sim}{q})\rangle \tag{47}$$

and replace the sum by an integral

$$\alpha = \frac{1}{(2\pi)^3} \left[ \int_{peak} B(\underset{\sim}{q})\, d\underset{\sim}{q} - \int_{background} B(\underset{\sim}{q})\, d\underset{\sim}{q} \right] \tag{48}$$

### References


1. Feil D. (1977). Israel J. Chem. 16, 103
2. Becker P., Coppens P. (1974). Acta Cryst. A30, 129
3. -March N.H., Wilkins S.W. (1978) Acta Cryst. A34, 19
   -Jones W., March N.H. (1970). Proc. Roy. Soc. A317, 359
4. Borie B. (1970) Acta Cryst. A26, 533
5. Stewart R.F. (1977) Israel J. Chem. 16, 137
6. Marshall W., Lovesey S.W. (1971) Theory of Thermal Neutron Scattering. Oxford-Clarendon

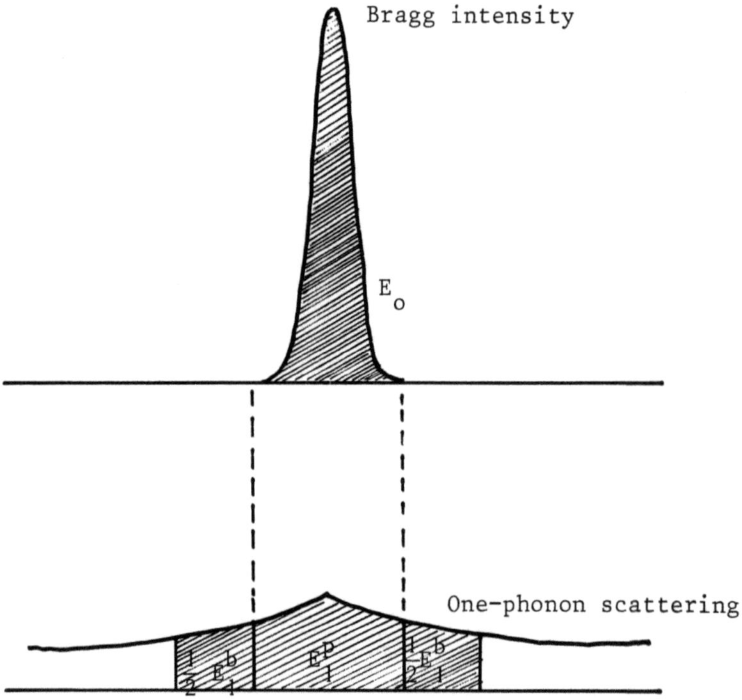


Figure 2

Experimental measurement of Bragg intensity : scan across Bragg peak

## III. X-RAY ELASTIC SCATTERING.

We have just seen in the preceding chapter that the elastic component of scattering is equivalent to scattering by the thermally averaged density. The structure factor $F(\underset{\sim}{S})$ is the Fourier transform of $\rho(\underset{\sim}{r})$

$$\rho(\underset{\sim}{r}) = \int \rho(\underset{\sim}{r},\underset{\sim}{Q})P(\underset{\sim}{Q})d\underset{\sim}{Q} \tag{1}$$

$$F(\underset{\sim}{S}) = \int \rho(\underset{\sim}{r}) \exp(i\underset{\sim}{S}.\underset{\sim}{r})d\underset{\sim}{r}$$

$$= \int F(\underset{\sim}{S},\underset{\sim}{Q})P(\underset{\sim}{Q})d\underset{\sim}{Q} \tag{2}$$

The optimal knowledge from an elastic scattering experiment is a set of $F(\underset{\sim}{S})$.

In order to calculate electronic properties, one should decorrelate electronic and nuclear motion contributions (ref. 1). This is a very difficult subject that is extensively discussed throughout this school.

If we assume the rigid pseudoatom model to be valid, one can measure Debye-Waller factors from neutron elastic scattering. Then, using a modelized shape for the pseudoatoms, one can obtain an approximate charge density at any geometry $\rho_m(\underset{\sim}{r},\underset{\sim}{Q})$.

Stewart originated a systematic method of generating pseudoatoms, expanding each pseudoatom in a spherical harmonic basis. The method has been extensively used by Stewart, Maslen, Coppens, Hirshfeld and coworkers. Coppens et al. (ref. 2) and Scheringer (ref. 3) developed models to calculate theoretical averaged densities. Becker (ref. 4) introduced a method of decorrelation of electronic properties, in the rigid body approximation.

We shall here only focus attention on some points that I believe to become more and more relevant to charge density work. The book of Pryor and Willis (ref. 5) extensively reviews the subject of Debye Waller factors.

### A. RIGID PSEUDOATOM APPROXIMATION

The validity of this approximation has been studied by March (ref. 6). He gives the following simple argument. Let us assume a simple crystal containing one atom per cell. We write :

$$\rho_o(\underset{\sim}{r}) = \sum_{\underset{\sim}{\ell}} \sigma(\underset{\sim}{r}-\underset{\sim}{\ell}) \tag{3}$$

for the at rest density, where $\sigma$ is the pseudoatomic continuous density, whose Fourier transform is $f(\underset{\sim}{S})$. We can only measure $F(\underset{\sim}{S})$ at lattice points and we have :

$$F(\underset{\sim}{S}) = f(\underset{\sim}{S})\ \delta(\underset{\sim}{S}-\underset{\sim}{K}) \qquad (4)$$

Obviously, $f(\underset{\sim}{S})$, thus $\sigma(\underset{\sim}{r})$, cannot be uniquely defined.

We now impose the rigid following approximation. If we displace atom at $\underset{\sim}{\ell}$ by $\underset{\sim}{u}_\ell$, the change in density is, to the first order :

$$\rho^{(1)}_{rb}(\underset{\sim}{r}) = \sum_\ell \underset{\sim}{u}_\ell . \underset{\sim}{\nabla}_\ell \sigma(\underset{\sim}{r}-\underset{\sim}{\ell}) \qquad (5)$$

If we do not impose the rigid body constraint :

$$\begin{aligned} \rho^{(1)}(\underset{\sim}{r}) &= \sum_\ell \underset{\sim}{u}_\ell . \underset{\sim}{\nabla}_\ell \rho \\ &= \sum_\ell \underset{\sim}{u}_\ell . \underset{\sim}{R}_\ell(\underset{\sim}{r}) \\ &= \sum_\ell \underset{\sim}{u}_\ell . \underset{\sim}{R}(\underset{\sim}{r}-\underset{\sim}{\ell}) \end{aligned} \qquad (6)$$

$\underset{\sim}{R}$ can, at least theoretically, be calculated from linear response theory, within the density functional approach.

Now take the particular case of a uniform translation $\underset{\sim}{u}_o$. (6) becomes :

$$\rho^{(1)}(\underset{\sim}{r}) = \underset{\sim}{u}_o . \underset{\sim}{\nabla}\rho_o = \underset{\sim}{u}_o . \sum_\ell \underset{\sim}{R}(\underset{\sim}{r}-\underset{\sim}{\ell}) \qquad (7)$$

and for (5)

$$\rho^{(1)}_{rb}(\underset{\sim}{r}) = \underset{\sim}{u}_o . \sum_\ell \underset{\sim}{\nabla}\sigma(\underset{\sim}{r}-\underset{\sim}{\ell}) \qquad (8)$$

From (7) we get the generally valid result :

$$\underset{\sim}{\nabla} \times \sum_\ell \underset{\sim}{R}(\underset{\sim}{r}-\underset{\sim}{\ell}) \qquad (9)$$

The rigid body approximation imposes the much stronger constraint :

$$\underset{\sim}{\nabla} \times \underset{\sim}{R} = 0 \qquad (10)$$

which is generally not fulfilled. It is possible to show (ref. 7) that the rigid pseudoatom approximation corresponds to the assumption of a potential limited to pairwise interactions (two body theory).

Note that the Fourier transform of $\rho^{(1)}$ and $\rho^{(1)}_{rb}$ are :

$$\begin{aligned} F^{(1)}(\underset{\sim}{S}) &= i\underset{\sim}{u}_o . \underset{\sim}{S}F(\underset{\sim}{S}) \\ F^{(1)}_{rb}(\underset{\sim}{S}) &= i\underset{\sim}{u}_o . \underset{\sim}{S}f(\underset{\sim}{S}) \end{aligned} \qquad (11)$$

and are identical for observable Bragg S vectors. Only scattering for $\underset{\sim}{S} \neq \underset{\sim}{K}$ can allow for a unique definition of a rigid pseudoatom. In the case of molecules, Stewart is able to obtain this unique solution because he deals with $F(\underset{\sim}{S})$ at any point in reciprocal space (see his lectures).

Parr and coworkers (ref. 8) have used the rigid body approximation to obtain universal approximate expressions for force constants in diatomics and recently polyatomics. See exercise by the author.

March and Wilkins (ref. 9) have recently relaxed this model, taking into account the mutual polarisability of pseudoatoms. The method follows an argument by Born (ref. 10) which has been discussed by the author (ref. 11). Each pseudoatom is written as :

$$\sigma_k(\underset{\sim}{r}-\underset{\sim}{R}_k-\underset{\sim}{u}_{k\ell},\{\underset{\sim}{u}_{k'\ell'}\}) = \sigma_k{}^{\circ}(\underset{\sim}{r}-\underset{\sim}{R}_k-\underset{\sim}{u}_{k\ell})$$
$$+\sum_{k'}\sum_{\ell'}\underset{\sim}{u}_{k'\ell'}\cdot\underset{\sim}{\nabla}_{k'\ell'}[\sigma_k(\underset{\sim}{r}-\underset{\sim}{R}_k-\underset{\sim}{u}_{k\ell},\{\underset{\sim}{u}_{k'\ell'}\})] \quad (12)$$

The major result is an apparent reduction of the Debye-Waller factor. For F in NaF :

$$\Delta B/B = -\ 0.036$$

The effect is the most pronounced for polarisable species. See the exercise by Wilkins.

There are cases where the rigid pseudoatom model fails :

. when collective electronic effects are present, as in metals, where conduction electrons do not follow any center.

. in spin density determination, one observes now spin density around the ligands (see for example ref. 11). Owing to the nature of the interactions involved, one can hardly believe that the ligand spin density is rigidly attached to this ligand. This is one of the reasons why such densities are too contracted around the ligand position. But the present resolution does not allow for a good decorrelation.

Thus, at present time, one starts to observe shortcomings to the rigid pseudoatom approximation.

The study of coupling between electrons and phonons is essential to the analysis of molecules in terms of pseudoatoms.

Few theoretical studies have been recently undertaken using a generalized pseudoatom scheme (ref. 12). Note finally that this effect is more pronounced in inelastic scattering (ref. 13).

## B. ANALYSIS OF DEBYE-WALLER FACTOR

From (II. 32) we can write :

$$W_k = \exp[-\tfrac{1}{2}\langle|\underset{\sim}{S}\cdot\underset{\sim}{u}_k|^2\rangle] \quad (13)$$
$$= \exp[-\tfrac{1}{2}\,\underset{\sim}{S}^t\underset{\approx}{B}_k\underset{\sim}{S}\,]$$

where $\underset{\sim}{S}$ is a (3x1) column vector, $\underset{\approx}{B}_k$ a symmetric tensor :

$$\underset{\approx}{B}_k = \langle\underset{\sim}{u}_k\underset{\sim}{u}_k^{\dagger}\rangle \quad (14)$$

It follows that

(15)

$$\underset{\approx}{B}_k = \frac{1}{Nm_k} \sum_{j\underset{\sim}{q}} \left[\frac{E_j(\underset{\sim}{q})}{\omega_j^2(\underset{\sim}{q})} \underset{\sim}{e}_{kj}(\underset{\sim}{q})\, \underset{\sim}{e}^{\dagger}_{kj}(\underset{\sim}{q})\right] \qquad (15)$$

It is obvious that the leading terms in (15) come from low frequency phonons. For those, one can often apply the classical approximation, ie $E_j(\underset{\sim}{q}) = k_B T$. Thus,

$$\underset{\approx}{B}_k = \frac{k_B T}{Nm_k} \sum_{j\underset{\sim}{q}} \left[\frac{\underset{\sim}{e}_{kj}(\underset{\sim}{q})\underset{\sim}{e}^{\dagger}_{kj}(\underset{\sim}{q})}{\omega_j^2(\underset{\sim}{q})}\right] \qquad (16)$$

If we make the assumption that the acoustic, elastic part of the spectrum is the major contribution, we have :

$$\omega_j(\underset{\sim}{q}) = V_j\, q$$

$$m_k^{-1/2} \underset{\sim}{e}_{kj}(\underset{\sim}{q}) = m^{-1/2} \underset{\sim}{\varepsilon}_j(\underset{\sim}{q}) \qquad \underset{\sim}{\varepsilon}_j(\underset{\sim}{q}) \text{ : polarisation vector}$$

$$\underset{\approx}{B}_k = \frac{k_B T}{Nm} \sum_{j\underset{\sim}{q}} \frac{\underset{\sim}{\varepsilon}_j(\underset{\sim}{q})\underset{\sim}{\varepsilon}^{\dagger}_j(\underset{\sim}{q})}{V_j^2\, q^2} = B_{el} \qquad (17)$$

(17) does not depend any more on the type of atom involved and it corresponds to an overall temperature factor. It is known to be a poor approximation. Calling $\alpha_j(\underset{\sim}{q})$ the angle between $\underset{\sim}{S}$ and $\underset{\sim}{\varepsilon}_j(\underset{\sim}{q})$ :

$$W_{el} = \exp\left[- \frac{1}{2} \frac{k_B T}{Nm} S^2 \sum_{\underset{\sim}{q}} q^{-2} \sum_{j=1}^{3} \frac{\cos^2\alpha_j(\underset{\sim}{q})}{V_j^2}\right] \qquad (18)$$

One can develop the theory for molecular crystals and relate $\underset{\approx}{B}_k$ to the quantities that describe the rigid body motion of the molecule (ref. 14). Let $\underset{\sim}{r}_p$ be the position of the pth atom from the mass center of the molecule :

$$\underset{\sim}{u}_p = \underset{\sim}{t} + \underset{\sim}{\theta} \times \underset{\sim}{r}_p$$

$$= \underset{\sim}{t} + \underset{\approx}{R}_p \underset{\sim}{\theta} \qquad (19)$$

with

$$\underset{\approx}{R}_p = \begin{pmatrix} 0 & z_p & -y_p \\ -z_p & 0 & x_p \\ y_p & -x_p & 0 \end{pmatrix}$$

From chapter I, we know that it is possible to define 6 branches corresponding to rigid body motion. Let $\underset{\sim}{V}_j(\underset{\sim}{q})$ be the eigenvector of the dynamical matrix and $\underset{\sim}{U}_j(\underset{\sim}{q}) = (\underset{\approx}{M}^{mol})^{-1/2} \underset{\sim}{V}_j(\underset{\sim}{q})$. $\underset{\sim}{U}_j(\underset{\sim}{q})$ has three components corresponding to the translation component $\underset{\sim}{t}_j(\underset{\sim}{q})$ and three components corresponding to the rotation $\underset{\sim}{\theta}_j(\underset{\sim}{q})$. We may define three fundamental tensors :

$$\underset{\approx}{T} = \sum_{jq} <\underset{\sim}{t}_j(\underset{\sim}{q})\underset{\sim}{t}_j^{\dagger}(\underset{\sim}{q})> \qquad \text{translation}$$
$$\underset{\approx}{\theta} = \sum_{jq} <\underset{\sim}{\theta}_j(\underset{\sim}{q})\underset{\sim}{\theta}_j^{\dagger}(\underset{\sim}{q})> \qquad \text{rotation} \qquad (20)$$
$$\underset{\approx}{S} = \sum_{jq} <\underset{\sim}{\theta}_j(\underset{\sim}{q})\underset{\sim}{t}_j^{\dagger}(\underset{\sim}{q})> \qquad \text{coupling}$$

One gets :

$$\underset{\approx}{B}_p = \underset{\approx}{T} + \underset{\approx}{R}_p\underset{\approx}{\theta}\underset{\approx}{R}_p^{\dagger} + \underset{\approx}{S}\underset{\approx}{R}_p^{\dagger} + \underset{\approx}{R}_p\underset{\approx}{S}^{\dagger} \qquad (21)$$

Pawley has developed constrained refinements where only the independent components of $\underset{\approx}{T}, \underset{\approx}{\theta}, \underset{\approx}{S}$ are determined. This theory is only valid for librations of small amplitude and becomes more complicated for large librations.

An important question : Is the contribution from internal modes negligible ? Fig.1 shows as a function of temperature $<u^2>$ and its components for the hydrogen atoms of hexamethylenetetrammine. The shapes are easy to predict from the previous theoretical discussions.

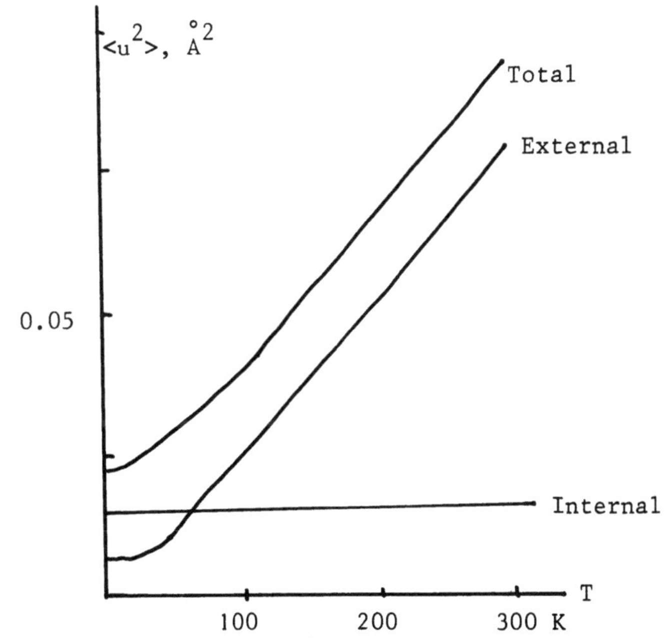


Figure 1
Mean square displacements of hydrogen atoms in HMT.

As predictable, the relative influence of internal modes increases when T is lowered. But if we neglect those, the absolute error is roughly constant, and not small. To correct for internal modes contribution, one only needs the I.R. spectroscopic characteristics of the molecule. Note that internal modes are non rigid modes and involve the mutual polarisability of component atoms.
The author (ref. 17) has shown, in the case of diatomic molecules, that the influence of internal vibration is important, anharmonic, but that the deformation density is not influenced by internal mode.

## C. ANHARMONICITY. COUPLING BETWEEN VIBRATIONAL AND ELECTRONIC DEFORMATIONS

If the classical approximation is applicable, and if V is the potential acting on the nuclei, we may write :

$$W_k = \frac{\int \exp(-V/k_BT)\, \exp(i\underset{\sim}{S}\cdot\underset{\sim}{u}_k)\, d\Omega}{\int \exp(-V/k_BT)\, d\Omega} \qquad (22)$$

where $d\Omega$ is the volume element in configurational space. Coupling between various nuclei is too complicated if V is expanded beyond the harmonic level. Willis, Cooper and Rouse (ref. 15) have assumed that V is approximated as :

$$V = \sum_{\text{atoms}} V_k \qquad (23)$$

This is equivalent to an Einstein model for the solid. At the harmonic level, the parameters are adjusted to give the correct Debye-Waller factors. (22) becomes :

$$W_k = \frac{\int \exp(-V_k/k_BT)\, \exp(i\underset{\sim}{S}\cdot\underset{\sim}{u}_k)\, d\underset{\sim}{u}_k}{\int \exp(-V_k/k_BT)\, d\underset{\sim}{u}_k} \qquad (24)$$

At the harmonic level, the expansion coefficients of $V_k$ are simply those of the p.d.f. of nucleus $\underset{\sim}{u}_k$. Therefore the decoupling of oscillators becomes only an approximation for the anharmonic part of the potential.
The application of the model is easier for simple structures with atoms on symmetry sites, since this restricts the number of parameters. But in the case of molecular crystals, one starts to consider also the influence of anharmonicity (see lectures by Coppens and Stewart).

The temperature dependence of V or $V_k$ is not an easy problem. It is related to thermal expansion. The simplest model is that of quasi-harmonic theory. One assumes that the relative frequency change is proportional to the expansion coefficient :

$$\Delta\omega/\omega = -\gamma\, \Delta v/v \qquad (25)$$

where $\gamma$ is a constant, the Grüneisen constant. If $\chi$ is the isothermal compressibility, and v the volume at 0°K

$$\Delta v/v = \chi T$$

Therefore :

$$\Delta\omega/\omega = -\gamma\chi T \tag{26}$$

In the Debye-Waller factor, $\omega_j^{-2}(\underset{\sim}{q})$ is involved, which becomes

$$\omega_j^{-2}(\underset{\sim}{q})\ \{1 + 2\gamma\chi T\} \tag{27}$$

Thus the quasiharmonic Debye-Waller factor is given by :

$$\underset{\approx}{B}_k = \underset{\approx}{B}_k^{\text{harm.}}\{1 + 2\gamma\chi T\} \tag{28}$$

This approximation has proved to be useful (see Fig. 2).
However the symmetry constraints are the same as in harmonic theory. There exists evidence that some reflections appear in neutron diffraction that should be forbidden in harmonic theory (ie (222) in Si). Dawson (ref. 16) has developed a simple and very general structure factor formalism, that has been at the origin of most of the work concerning deviations from isolated spherical atoms undergoing harmonic vibrations. This model clearly shows the coupling between the effect of anharmonic motion and electronic distortions, if one looks at the diffracted intensities. If the density of a pseudoatom is separated into a centrosymmetric and antisymmetric components $\rho_{kc}$ and $\rho_{ka}$, one gets : $f_k = f_{kc} + if_{ka}$.
If the p.d.f. of the nucleus is developed in the same manner, $W_k = W_{kc} + iW_{ka}$. Therefore $F(\underset{\sim}{S})$ is written :

$$F(\underset{\sim}{S}) = \sum_k (f_{kc}+if_{ka})(W_{kc}+iW_{ka})\exp(i\underset{\sim}{S}.\underset{\approx}{R}_k)$$

$$= A(\underset{\sim}{S}) + iB(\underset{\sim}{S}) \tag{29}$$

We obtain :

$$A(\underset{\sim}{S}) = \sum_k[(f_{kc}W_{kc}-f_{ka}W_{ka})\cos(\underset{\sim}{S}.\underset{\approx}{R}_k) - (f_{kc}W_{ka}+f_{ka}W_{kc})\sin(\underset{\sim}{S}.\underset{\approx}{R}_k)] \tag{30}$$

$$B(\underset{\sim}{S}) = \sum_k[(f_{kc}W_{kc}-f_{ka}W_{ka})\sin(\underset{\sim}{S}.\underset{\approx}{R}_k) - (f_{kc}W_{ka}+f_{ka}W_{kc})\cos(\underset{\sim}{S}.\underset{\approx}{R}_k)]$$

Equations (29) clearly show that X-Ray experiments alone do not allow for a good separation of the two effects. Many applications of this formalism have been done and will be specifically discussed during this school.

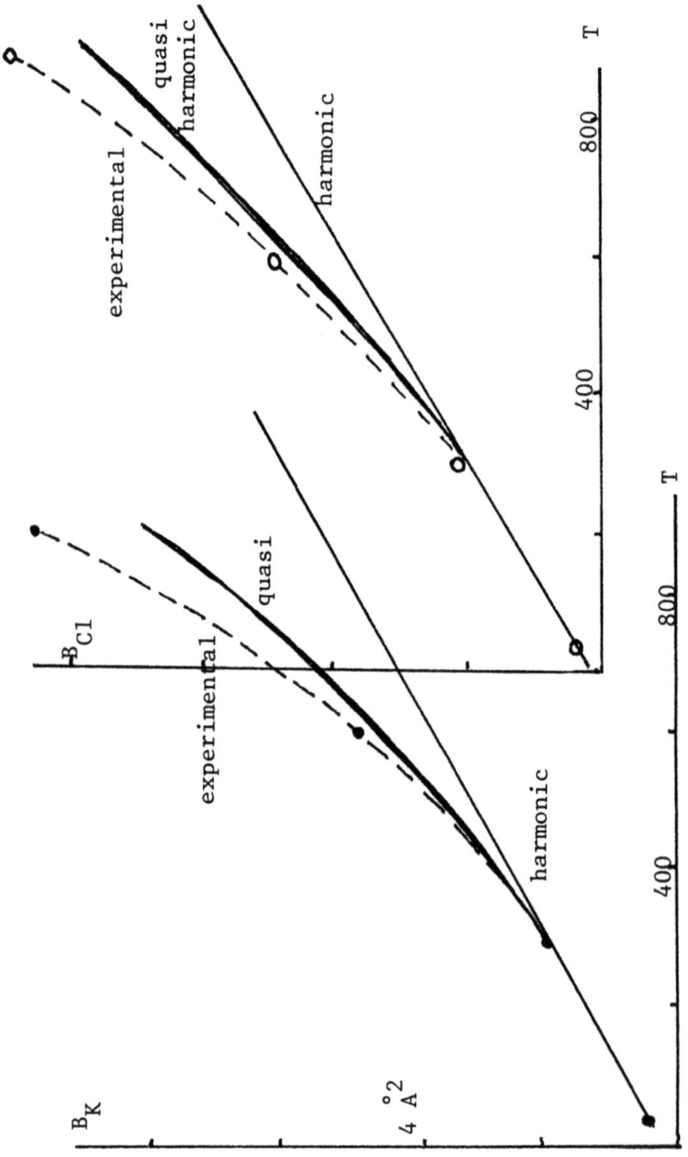


Figure 2 B factors for KCl versus temperature

References

1. Becker P.(1977). Physica Scripta, 15, 119

2. Coppens P., Rys J., Stevens E.D. (1977). Acta Cryst., A33, 333

3. Scheringer C. (1972). Acta Cryst., A28, 512

4. Becker P. (1975) Acta Cryst., A31, S227

5. Pryor, Willis, loc. cit.

6. March N.H. (1975). Orbital Theory for Molecules and Crystals. Academic Press.

7.-Anderson A.B., Parr R.G. (1971). J. Chem. Phys., 55, 5490
-Bartlett R.J., Parr R.G. (1978). J. Chem. Phys., 67, 5828

8.-Johnson F.A. (1969) Proc. Roy. Soc., A310, 79
-Johnson F.A. (1974) Proc. Roy. Soc., A339, 73

9. March N.H., Wilkins S.W. (1978). Acta Cryst., A34, 19

10. Born M. (1942). Proc. Roy. Soc., A180, 397

11. Bonnet M., Delapalme A., Becker P., Fuess H. (1977). Int. J. of Magnetism & Magnetic Materials, 7, 23

12. Luty T. (1977). J. Chem. Phys., 66, 1231

13. Ball M.A. (1975). J. Phys. C,8, 1

14.-Schomaker V., Trueblood K.N. (1968). Acta Cryst., B24, 63
-Johnson C.K. (1970) Thermal Neutron Diffraction. Oxford

15.-Willis B.T.M. (1970) Thermal Neutron Diffraction. Oxford
-Cooper M.J., Rouse K.D., Willis B.T.M. (1968) Acta Cryst., A24,484

16.-Dawson B. (1967) Proc. Roy. Soc. A298, 255, 264, 379
-Dawson B. (1976) Advances in Str. Res. by Diffract. Meth. 6, Oxford

17. Becker P., Stephens M.E. (1979) to be published

## IV. THERMAL DIFFUSE SCATTERING

### A. X-RAY CASE

We first recall the expressions for one phonon scattering from chapter II :

$$I_1(\underset{\sim}{S}) = \sum_{j\underset{\sim}{q}} I_1(j\underset{\sim}{q}) \qquad (1)$$

where each phonon contribution is :

$$I_1(j\underset{\sim}{q}) = S^2 \frac{k_B T}{N\omega_j^2(\underset{\sim}{q})} |G_{j\underset{\sim}{q}}|^2 \Delta^2(\underset{\sim}{q}+\underset{\sim}{S}) \qquad (2)$$

in the high temperature limit that we shall consider here. Scattering by phonon $(j\underset{\sim}{q})$ occurs at vectors :

$$\underset{\sim}{S} = \underset{\sim}{K} \mp \underset{\sim}{q} \qquad (3)$$

Since the wavelength of the photon is practically unchanged, $\underset{\sim}{S}$ ends on the Ewald sphere, which is the scattering surface (Fig.1). We have also for the one phonon structure factor :

$$G_{j\underset{\sim}{q}} = \frac{\underset{\sim}{S}}{S} \cdot \sum_k f_k W_k \frac{\underset{\sim}{e}_{kj}(\underset{\sim}{q})}{\sqrt{m_k}} \exp(i\underset{\sim}{K}.\underset{\sim}{R}_k) \qquad (4)$$

A great simplification occurs in the expression for $G_{j\underset{\sim}{q}}$ for phonons belonging to the elastic part of the spectrum.

$$\frac{\underset{\sim}{e}_{kj}(\underset{\sim}{q})}{\sqrt{m_k}} = \frac{\underset{\sim}{\varepsilon}_j(\underset{\sim}{q})}{\sqrt{m}}$$

Denoting the angle $\{\underset{\sim}{S},\underset{\sim}{\varepsilon}_j(\underset{\sim}{q})\}$ by $\alpha_j(\underset{\sim}{q})$, we obtain :

$$G_{j\underset{\sim}{q}} = F(\underset{\sim}{K}) \cos\alpha_j(\underset{\sim}{q})/\sqrt{m} \qquad (5)$$

We thus obtain :

$$I_1(j\underset{\sim}{q}) = F^2(\underset{\sim}{K})\Delta^2(\underset{\sim}{q}+\underset{\sim}{S}) \frac{k_B T S^2 \cos^2\alpha_j(\underset{\sim}{q})}{Nm\omega_j^2(\underset{\sim}{q})} \qquad (6)$$

where the term $\frac{k_B T S^2 \cos^2\alpha_j(\underset{\sim}{q})}{Nm\omega_j^2(\underset{\sim}{q})}$ is the $(j,\underset{\sim}{q})$ component of the elastic part of Debye Waller factor. For elastic scattering, we have seen that such a model is far from adequate. Here, we are only concerned by a small part of the BZ where the long wave approximation is the most adequate. Moreover, for branches of optic phonons, where $\omega_j(\underset{\sim}{q})$ does not depend on $\underset{\sim}{q}$ practically all the vectors $\underset{\sim}{q}$ in the vicinity of a reciprocal lattice point $\underset{\sim}{K}$ give an approximately constant contribution $I_1(j\underset{\sim}{q})$. Therefore, besides acoustic modes, thermal diffuse scattering is approximately included in the background.

Most of the present calculations are limited to the inclusion of the three acoustic branches.

In the case of molecular crystals, Cochran and Pawley (ref. 1) have introduced a model that only involves the rigid body lattice modes. We use the notations of a given molecule. $\underset{\sim}{R}_k$ is the position of the mass centre of the kth molecule, $\underset{\sim}{r}_p$ the position of atom p with respect to the mass centre. Each of the 6n branches has components $\underset{\sim}{t}_{kj\underset{\sim}{q}}$ for tranlation and $\underset{\sim}{\theta}_{kj\underset{\sim}{q}}$ for rotation. We obtain :

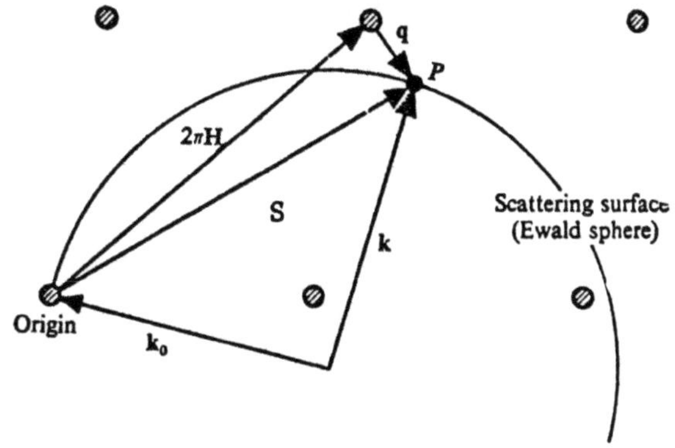


Figure 1
Scattering surface for one-phonon X-Ray scattering.

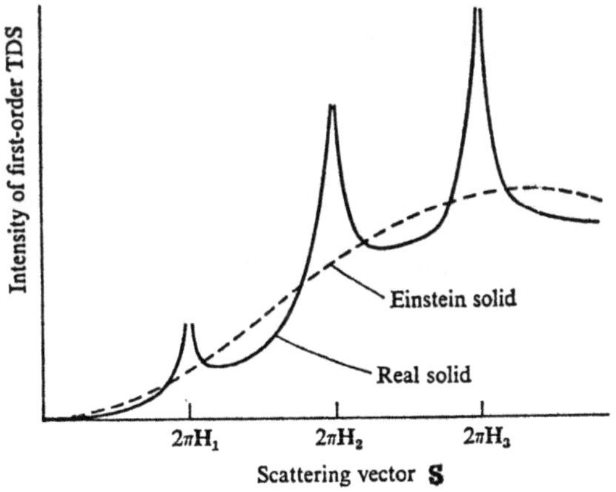


Figure 2
First order TDS intensity for an Einstein solid

$$G_{j\tilde{q}} = \frac{\tilde{S}}{S} \cdot \sum_k \sum_p f_p W_p \exp[i\tilde{K}.(\tilde{R}_k+\tilde{r}_p)][\tilde{t}_{kj\tilde{q}}+\tilde{\theta}_{kj\tilde{q}} x \tilde{r}_p]$$

$$= \frac{\tilde{S}}{S} \cdot \left\{ \sum_k F_k(\tilde{S}) \tilde{t}_{kj\tilde{q}} + \tilde{\theta}_{kj\tilde{q}} x \tilde{E}_k(\tilde{S}) \right\} \exp(i\tilde{K}.\tilde{R}_k)$$

$$F_k(\tilde{S}) = \sum_p f_p W_p \exp(i\tilde{S}.\tilde{r}_p) \qquad (7)$$

$$E_k(\tilde{S}) = \sum_p f_p W_p \tilde{r}_p \exp(i\tilde{S}.\tilde{r}_p)$$

It was found that the major contribution around $\tilde{K}$ comes from the 3 acoustic modes, that mainly contribute to translation.

If anharmonicity is important for Debye-Waller factors, it should be also true for the $(j\tilde{q})$ contributions, and thus for TDS. If we use the classical theory outlined in chapter III, we get :

$$\langle \exp[i\tilde{S}.(\tilde{u}_{k\ell}-\tilde{u}_{k'\ell'})] \rangle = W_k W_{k'} \qquad (8)$$

Thus such a theory does not allow for anharmonic contribution to TDS. Owing to the difficulty of applying full theory, one might at least use the quasi-harmonic theory, replacing $I_1(j\tilde{q})$ by

$$I_1(j\tilde{q})\ [\ 1 + 2\gamma\chi T\ ]\ .$$

Restricting ourselves to elastic modes, we obtain for $B(\tilde{q})$ (II 47) the simple expression :

$$B(\tilde{q}) = k_B T \sum_{j=1}^{3} \frac{S^2 \cos^2\alpha_j(\tilde{q})}{\rho\ \omega_j^2(\tilde{q})} \qquad (9)$$

where $\rho$ is the density of the sample.

<u>Variation of TDS with S and temperature</u>. Let us consider one term $I_1(jq)$. If we take the case of an optic branch, the major variation is :

$$I_1(j\tilde{q}) \sim S^2 T \exp(-1/2\ S^2 \langle u^2 \rangle) \qquad (10)$$

$$\langle u^2 \rangle = \eta T$$

The background thus varies with $\tilde{S}$ as shown in the dotted line of Fig. 2. Around a reciprocal vector, the acoustic modes add a $q^{-2}$ dependence, depicted by the full line in Fig. 2. But the function (10) is still multiplying the contribution of each mode so that the average inelastic scattering varies as (10). The maximum is obtained for $S = \langle u^2 \rangle^{-1/2}$. If $\langle u^2 \rangle \sim 0.01 Å^2$, as for Al at room temperature

$$S_{max} = 10\ Å^{-1} \qquad (\sin\theta/\lambda)_{max} = 0.75\ Å^{-1}$$

Thus one may have simultaneously extinction and TDS.

Looking for the temperature dependence, for each S, there exists a temperature $T_m$ of maximum diffuse scattering :

$$T_m = \frac{1}{S^2} \qquad \text{For Al : } T_m = 188/(\sin\theta/\lambda)^2$$

B. NEUTRON CASE

This case has been discussed by Willis (ref. 2). The energy and momentum conservation laws are :

$$\frac{\hbar^2}{2m}(k^2 - k_o^2) = -\varepsilon\hbar\omega_j(\tilde{q}) \tag{11}$$

$$\tilde{k} - \tilde{k}_o = \tilde{K} + \tilde{q}$$

$\varepsilon = +1$ corresponds to phonon creation
$\varepsilon = -1$ corresponds to phonon anihilation.
Scattering geometry is sketched in Fig. 3. Suppose that we fix the position of lattice vector $\tilde{K}$ (one position during scanning). We choose a length for q. From the lattice point $\tilde{K}$ we draw a sphere of radius q. From the dispersion curve, we get ω, thus k and then we draw a sphere of radius k from the centre of the Ewald sphere. Either there is no intersection or there is a circle. The locus of allowed vectors $\tilde{q}$ is a surface of revolution around the vector CK.

Let us assume the usual convention : isotropic elastic waves, that is to say a threefold degeneracy of the acoustic modes :

$$\omega_j(\tilde{q}) = V_s q$$

Since we only consider points near the Bragg point, $k \approx k_o$ . (11) becomes :

$$k - k_o = -\varepsilon\beta q \tag{12}$$

$$\beta = V_s/V_n$$

and we can replace the Ewald sphere by its tangential plane perpendicular to $\tilde{k}$ direction. (12) is thus the equation of a conic surface :

. an hyperboloïd of focus $\tilde{K}$ if $\beta < 1$ (faster than sound neutron) Each branch corresponds to $\varepsilon=+1$ or $\varepsilon=-1$. For any direction of $\tilde{k}$, there are two points of intersection, and two vectors $\tilde{q}$ and $\tilde{q}'$, one for each value of ε. See Fig. 4.

. an ellipsoid of focus $\tilde{K}$ if $\beta > 1$ (slower than sound neutron). There will be only either phonon creation or phonon anihilation, and that for some directions. The situation is much more complicated. The closer to the Bragg position, the smaller the scattering surface. If scanning involves a small angular range, one may neglect the effect of TDS. The greater β, the more accurate this approximation is.

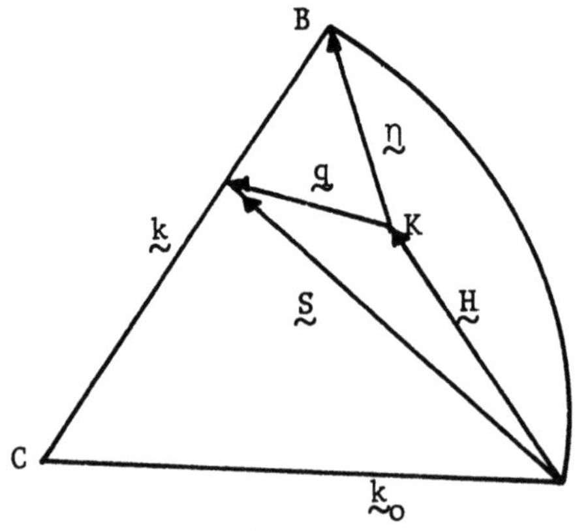


Figure 3
Vectors in reciprocal space, for neutron diffraction

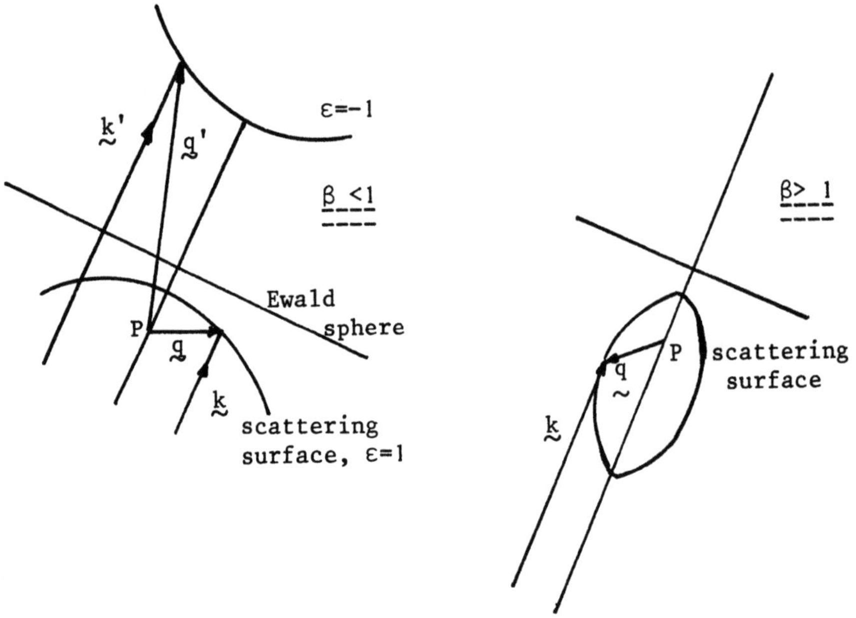


Figure 4
Scattering surfaces in neutron diffraction

Coming back to the case where $\beta < 1$, we note on Fig.3 that associated with $\underset{\sim}{q}$ is a vector $\underset{\sim}{\eta}$ which would be the phonon wave vector in the case of X-Rays. From (II.42) we have :

$$I_1(j\underset{\sim}{q}) = \frac{k}{k_o} \Delta^2(\underset{\sim}{q}+\underset{\sim}{S})S^2|G_j(\underset{\sim}{q})|^2 k_B T \frac{1}{2} \left[\frac{1}{\omega_j^2(\underset{\sim}{q})J_+} + \frac{1}{\omega_j^2(\underset{\sim}{q}')J_-}\right] \tag{13}$$

where $\underset{\sim}{q}$ and $\underset{\sim}{q}'$ are the two allowed vectors of Fig. 4.

$$J = 1 + \varepsilon\beta\frac{q_{\parallel}}{q}$$

where $q_{\parallel}$ is the component of $\underset{\sim}{q}$ along $\underset{\sim}{k}$. It can be shown that :

$$\frac{1}{2}\left[\frac{1}{J_+q^2} + \frac{1}{J_-q'^2}\right] = \frac{1}{\eta^2} \tag{14}$$

Since $k \sim k_o$, we conclude that (13) is equivalent to the expression of the $(j\underset{\sim}{q})$ component of one phonon scattering for X Rays. Therefore, for faster than sound neutrons, the TDS correction is the same as for X-Rays. For slower than sound neutrons, it is negligible in many cases or very difficult to calculate. This shows that if X-Rays or fast neutrons see an instantaneous distorted nuclear configuration, slow neutrons see only an averaged structure. This point was discussed by the author (ref. 3) and will be reconsidered when considering extinction.

## C. TDS CORRECTIONS

We shall only briefly consider this subject, which will be discussed practically by Dr Lehmann. Some very good papers exist on this subject (refs 4 to 8). The first three papers deal with analytical approximations, the last two with numerical computations. The basis of the calculation has been previously outlined (II.48).

The first problem is to define the part of the BZ that is spanned during the measurement. Fig. 5 shows the geometry in $\omega$ scan $\omega$-$2\theta$ scan and also in model scans (spherical and cylindrical). In the third direction, one should also consider the slit height.

1. Analytical Evaluation of $\alpha$. The most common assumption is that $B(\underset{\sim}{q})$ is considered isotropic. Considering (9), this requires the elastic isotropy. In cubic crystals, this can be sometimes the case but it still involves two velocities, one for longitudinal, one for transversal waves. Here we represent the two velocities by an average one $V_s$. In cubic crystals, isotropy is encountered when :

$$2\,C_{44} = C_{11} - C_{22}$$

$$V_l = (C_{11}/\rho)^{1/2} \qquad\qquad V_t = (C_{44}/\rho)^{1/2}$$

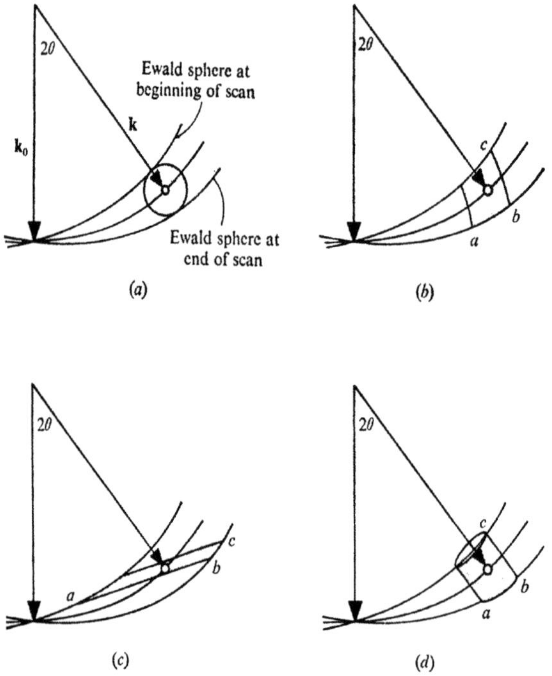


Figure 5

Diagram illustrating different types of diffractometer scan across a Bragg reflection.

a. spherical scan

b. ω-scan

c. ω/2θ scan

d. cylindrical scan

Shaded portions represent regions which are scanned. The dimension ab is determined by the aperture of the detector and bc by the rockong angle of the crystal.

Here we define a mean elastic constant C by :

$$V_s = (C/\rho)^{1/2} \tag{15}$$

This leads to the expression :

$$B(q) = \frac{S^2 k_B T}{Cq^2} \tag{16}$$

$\alpha$ is therefore given by :

$$\alpha = \frac{S^2 k_B T}{(2\pi)^3 C} \left[ \int_P \frac{d\underset{\sim}{q}}{q^2} - \int_B \frac{d\underset{\sim}{q}}{q^2} \right] \tag{17}$$

There are still great difficulties in evaluating the integrals in (17).

We shall take the case of an hypothetical spherical scan. Let $q_m$ be the maximum of q in the peak integration

$$\int_P \frac{d\underset{\sim}{q}}{q^2} = 4\pi q_m \tag{18}$$

We finally obtain :

$$\alpha = \frac{S^2 k_B T q_m}{3\pi^2 C} \tag{20}$$

In order to use this approximation we may estimate $q_m$ by :

$$4/3\pi q_m^3 = \Omega \tag{21}$$

where $\Omega$ is the volume spanned in a practical scan.

Let $\psi_1$ and $\psi_2$ be the horizontal and vertical angles subtended by the detector and $\omega$ be the angle of rotation of the crystal. For $\omega$-$2\theta$ scan we have :

$$\Omega_1 = (2\pi/\lambda)^3 2\sin^2\theta_B \ \psi_1\psi_2$$

For $\omega$ scan

$$\Omega_2 = (2\pi/\lambda)^3 \sin 2\theta_B \ \psi_1\psi_2 \tag{22}$$

Note that $\Omega_1 = \Omega_2$ for $\theta_B = \pi/4$, leading to :

$$\alpha = \frac{S^2 k_B T}{\pi \lambda C} \left(\frac{2}{9\pi}\omega\psi_1\psi_2\right)^{1/3} \tag{23}$$

Not correcting for TDS results in an apparent lowering of the Debye Waller factors, $\Delta B$ such that :

$$\alpha = 2\Delta B \sin^2\theta/\lambda^2 \qquad \Delta B = \frac{8\ k_B T}{\lambda C}\left(\frac{2}{9\pi}\omega\psi_1\psi_2\right)^{1/3} \tag{24}$$

We see that when $\lambda$ is decreased, $\alpha$ increases.

Other analytical expressions have been derived, which will not be reproduced here. Some results are shown for KCl and $BaF_2$, taken from ref. 4, where we compare the value of $\alpha$ correctly calculated, with those from spherical approximation and 'pseudo Debye-Waller' term. We see that the spherical approximation works well. Table 1.

2. Numerical Calculation of $\alpha$. Using the elasticity theory discussed in part I, one can show that :

$$B(\underset{\sim}{q}) = \frac{k_B T K^2}{q^2} \sum_{ij=1}^{3} h_i h_j (A^{-1})_{ij} \tag{25}$$

where $h_i$ is a direction cosine of K. The integration is done numerically, using the orientation matrices that monitor the scanning.

Stevens also included the contribution from two phonon process. It can be shown that the wave vector dependence is only $q^{-1}$ for this process but that for high scattering angles where, in order to integrate over $K\alpha_1$-$K\alpha_2$ doublet, one needs a fairly large scanning amplitude, the 2-phonon process may be nearly as important as one phonon.

Stevens has also shown that one might easily include some dispersion, replacing $\omega$=Vq by :

$$\omega = Vq\ D(q) \tag{26}$$

where

$$D(q) = \sin(\pi/2\ q/q_m)\ /\ (\pi/2\ q/q_m) \tag{27}$$

as deduced in the case of a simple cubic lattice. $q_m$ is the value of q at the limit of the BZ in the direction $\underset{\sim}{q}$. Dispersion is only important for high angles.

Coming back to the expression (II.30), we see that one phonon scattering is equivalent to a pseudo Debye-Waller anisotropic factor. Harada and Sakata have derived correct expressions for this pseudo-Debye-Waller factor.

## References


1. Cochran W., Pawley G.S. (1964) Proc. Roy. Soc. A280, 1
2. Willis B.T.M. (1970). Acta Cryst., A26, 396
3. Becker P. (1977) Physica Scripta 15, 119
4. Cooper M.J.,Rouse K.D. (1968) Acta Cryst., A24, 405
5. Rouse K.D., Cooper M.J. (1969) Acta Cryst. A25, 615
6. Cochran W. (1969) Acta Cryst., A25, 95
7. Stevens E.D. (1974) Acta Cryst., A30, 184
8. Harada J., Sakata J. (1974) Acta Cryst., A30, 77

Table 1

| h k l | θ | α | $\alpha_{sph.}$ | $\alpha_{\Delta B}$ |
|---|---|---|---|---|
| 4 0 0 | 13 | .033 | .036 | .033 |
| 6 0 0 | 19.7 | .079 | .081 | .076 |
| 4 4 4 | 22.9 | .109 | .108 | .102 |
| 8 0 0 | 26.7 | .147 | .146 | .139 |
| 10 0 0 | 34.2 | .227 | .226 | .224 |
| 6 6 6 | 35.7 | .244 | .244 | .244 |

TDS corrections in KCl at 20°C

| h k l | θ | α | $\alpha_{sph.}$ | $\alpha_{\Delta B}$ |
|---|---|---|---|---|
| 5 1 1 | 25.5 | .056 | .069 | .065 |
| 7 1 1 | 36.3 | .123 | .131 | .127 |
| 7 3 3 | 42.8 | .170 | .172 | .171 |
| 9 1 1 | 49.1 | .216 | .214 | .215 |
| 9 3 3 | 55.6 | .257 | .255 | .262 |
| 3 7 7 | 59.1 | .274 | .275 | .286 |

TDS corrections in $BaF_2$ at 400°C

# DIFFRACTION BY REAL CRYSTALS :

# II. EXTINCTION

Pierre Becker

Centre de Mécanique Ondulatoire Appliquée, CNRS and

University Paris VI, 23 rue du Maroc, 75019 Paris, France

The subject of extinction, that is to say the breakdown of kinematical theory, which results in a lowering of the diffracted intensity, is a fundamental one in order to understand the relationship between integrated reflectivities and structure factors. It has been the object of several recent studies, especially a review by the author (Becker, 1977a). I would like here to present a few fundamental aspects of the problem, rather than a comprehensive study. In particular, computational aspects will not be considered, since they have been extensively discussed elsewhere (Becker and Coppens, 1974a,b, 1975).

There are two main theories of diffraction of X-rays or neutrons :

a/ The kinematical theory assumes the Born approximation to be valid for the whole sample under study. Practically speaking, one assumes the incident beam to be unperturbed inside the sample. In fact, the periodic nature of a crystal imposes constraints to the waves that can travel inside it. Kinematical theory, which has been studied in the preceding lecture, is only valid for very thin crystals.

b/ The dynamical theory is the study of the waves that can propagate in a periodic system. The theory can be formulated for X-rays in terms of electromagnetic theory (James, 1963; Batterman and Cole, 1964; Authier, 1970; Kato, 1974). One of the great difficulties comes from the boundary conditions : no general solution can be found. In the case of neutrons, instead of starting from Maxwell's equations, one considers the Schrödinger one particle equation for the neutron : the results are simpler since one has to consider scalar waves (no polarisation factor) but are essentially equivalent.

The limit of dynamical theory for thin crystals is the kinematical theory. The frontier can be estimated in different ways and is called the 'extinction length' $\Lambda$, which is of the order of :

$$\Lambda = \frac{V_c}{aC}\,\frac{1}{|F_H|\lambda} = \left(\frac{\lambda}{Q\sin 2\Theta}\right)^{1/2} \tag{1}$$

$V_c$ is the volume of the unit cell, $F_H$ the structure factor, $\lambda$ the wavelength. For X-Rays, a is the classical radius of the electron (.28 $10^{-12}$cm) and C is the polarisation factor. For neutrons, a is $10^{-12}$cm, C is 1. Q is the kinematical integrated reflectivity per unit length :

$$Q = \left|\frac{aC}{V_c}\,F_H\right|^2 \frac{\lambda^3}{\sin 2\Theta} \tag{2}$$

If t is the thickness of the perfect part of the crystal :

if $t \ll \Lambda$ kinematical theory should be applied

if $t \gg \Lambda$ dynamical theory is to be applied.

$\Lambda$ is a coherence length, the necessary distance for dynamical effects to take place. It is obvious that $\Lambda$ varies with H and is the smallest for strong reflections or large wavelengths. It varies from a few to a few tens of microns.

Most experiments are done on small crystals (1/10 to 1 mm) the defect structure of which is unknown and very difficult to estimate. We need to understand the diffraction process in such real crystals and to find the fundamental equations governing the wave propagation. We shall here develop in some detail the derivation of these equations in the case of neutrons, the generalisation to X-Rays being briefly outlined. Besides the simplification due to scalar neutron waves, it provides a more direct comparison with band theory. Later on, we shall consider the methods of solution of these equations and some more difficult aspects : the case of anomalous absorption and the coupling between extinction and thermal vibrations.

## NEUTRON DYNAMICAL DIFFRACTION : PERFECT AND REAL CRYSTALS

Consider an assembly of nuclei at positions $\underset{\sim}{R}_i$. These nuclei exert on a neutron a pseudo potential that is written as (Marshall and Lovesey, 1971) :

$$V(\underset{\sim}{r}) = \frac{2\pi\hbar^2}{m} \sum_i b_i\, \delta(\underset{\sim}{r}-\underset{\sim}{R}_i) \tag{3}$$

where $b_i$ is the scattering length of the ith nucleus ( $\sim 10^{-12}$cm) For thermal neutrons of energy E, V/E is of the order of $10^{-5}$. The neutron propagation is governed by the one particle Schrödinger equation :

$$\{ - \frac{\hbar^2}{2m} \underset{\sim}{\nabla}^2 + V(\underset{\sim}{r}) - E \} \Psi(\underset{\sim}{r}) = 0 \tag{4}$$

A. PERFECT CRYSTAL

The Fourier representation of V is :

$$V(\underset{\sim}{r}) = \sum_{\underset{\sim}{H}} V_H \exp(-i\underset{\sim}{H}.\underset{\sim}{r}) \tag{5}$$

$$V_H = \frac{2\pi\hbar^2}{mV_c} F_H \tag{6}$$

where $F_H$ is the structure factor. In the case of magnetic materials, eqn (5) is still valid, but the expression for $F_H$ contains an extra contribution (see lecture notes by P.J. Brown).

Using Bloch's theorem, $\Psi(\underset{\sim}{r})$ can be expanded as :

$$\Psi(\underset{\sim}{r}) = \sum_{\underset{\sim}{H}} \psi_H \exp(-i\underset{\sim}{K}_H.\underset{\sim}{r}) \tag{7}$$

$$\underset{\sim}{K}_H = \underset{\sim}{K}_o + \underset{\sim}{H}$$

where $\underset{\sim}{K}_o$ is the wave vector associated with the incident wave in the crystal.

(4) becomes :

$$\{ \frac{\hbar^2}{2m} K_H^{\ 2} - E \} \psi_H = - \sum_{\underset{\sim}{H}'} V_{H-H'} \psi_{H'} \tag{8}$$

In the case of elastic scattering, we can write :

$$E = \frac{\hbar^2 k^2}{2m} \tag{9}$$

where k is the wavevector $(2\pi/\lambda)$ in free space. If we write :

$$\Gamma = \frac{\lambda^2}{\pi V_c} \qquad K^2 = k^2( 1 - \Gamma F_o) \qquad K = k( 1 - \frac{1}{2} \Gamma F_o) \tag{10}$$

then :

$$(K^2 - K_H^{\ 2}) \psi_H = k^2\Gamma \sum_{\underset{\sim}{H}' \neq \underset{\sim}{H}} \psi_{H'} F_{H-H'} \tag{11}$$

If one wave only is excited in the development (7), we have $K_o = K$. K is therefore the wavevector for a general wave propagating in the crystal : there is a refraction, but with an index of refraction less than one.

We are primarily interested to the case of two waves, $\psi_o$ and $\psi_H$.

Equations (11) become in this case :

$$\begin{pmatrix} K^2-K_o^2 & -k^2\Gamma F_{\bar{H}} \\ -k^2\Gamma F_H & K^2-K_H^2 \end{pmatrix} \begin{pmatrix} \psi_o \\ \psi_H \end{pmatrix} = 0 \tag{12}$$

Let us define :

$$\xi_o = K_o - K$$

$$\xi_H = K_H - K$$

Since $K_o \simeq K_H \simeq K$,

$$\begin{pmatrix} \xi_o & -\frac{1}{2} k\Gamma F_{\bar{H}} \\ -\frac{1}{2} k\Gamma F_H & \xi_H \end{pmatrix} \begin{pmatrix} \psi_o \\ \psi_H \end{pmatrix} = 0 \tag{13}$$

Consider the geometrical description of Figure 1. Let L be the point (tie point) from which we draw the vectors $\underset{\sim}{k}_o$ and $\underset{\sim}{k}_H$ corresponding the the Bragg condition in free space :

LO = LH = k

We draw the spheres centred at O and H, of radius K. Q is the corresponding tie point. $\xi_o$ and $\xi_H$ are small enough for the spheres around Q to be replaced by their tangent planes $(\Pi_o)$ and $(\Pi_H)$. We look for points M from which we will draw the vectors :

$\underset{\sim}{MO} = \underset{\sim}{K}_o$ $\underset{\sim}{MH} = \underset{\sim}{K}_H$, subject to (13).

$\xi_o$ and $\xi_H$ are approximately the distances from M to the planes $(\Pi_o)$ and $(\Pi_H)$. The condition (13) for non zero solution is :

$$\xi_o \xi_H = \frac{1}{4} k^2 \Gamma^2 F_H F_{\bar{H}} \tag{14}$$

The locus of M is an hyperboloid having $(\Pi_o)$ and $(\Pi_H)$ as asymptotic planes. It is the dispersion surface.

The discussion of the consequences of this result is the subject of dynamical theory (see for example the proceedings of the International Summer School on Dynamical Theory, Limoges, 1975).

We see from eqn (13) and (14) that there are two solutions, thus two wave fields, which have different properties, in particular concerning absorption. The physical meaning of these two solutions will be clear in the comparison with band theory.

How can one find the two wave-fields ? We must apply the condition of tangancial continuity of the wavevectors at the crystal surface. An incident wave is represented by $PO = \underset{\sim}{k}_o$, where P can be located on the plane $(\Pi_L)$ representing locally the sphere of center O and radius k. LP is the deviation from Bragg's condition. The normal $\underset{\sim}{n}$ to the entrance surface, drawn from P, cuts the dispersion surface at points $M_1$ and $M_2$. The amplitudes can be calculated from (13). The figure is for the Laue case, and can be drawn for

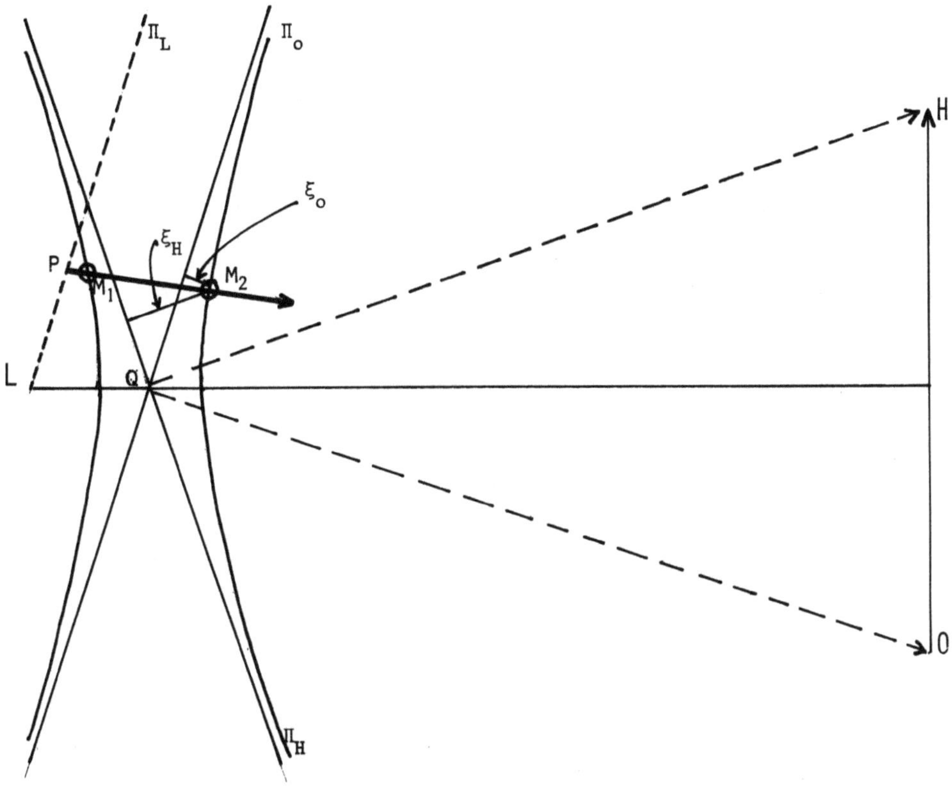


Figure 1

Bragg case too.

It is obvious, when the crystal shape is complex, and when the crystal is distorted, that the direct solution of the dynamical equations is extremely difficult. For statistically imperfect crystals (so called mosaic crystals) one shoud try to find more suitable equations. This is the purpose of the next paragraph.

DYNAMICAL EQUATIONS IN THE X-RAY CASE

In the X-Ray case, one gets from Maxwell's equations similar results. The value of $\Gamma$ is :

$$\Gamma = a\lambda^2/\pi V_c$$

The structure factor may contain a component due to anomalous dispersion, which is connected to absorption effects.

The major complication is related to the vectorial nature of the wave amplitudes. The plane $\underset{\sim}{K}_o$, $\underset{\sim}{K}_H$, defines the so called $\Pi$ plane. The displacement vector D can be decomposed in a $\pi$ component and a $\sigma$ component perpendicular to this $\Pi$ plane. The two components are nearly independent and, defining the polarisation constant C

$$\begin{aligned} C &= 1 \text{ for the } \sigma \text{ component} \\ C &= \cos 2\Theta \text{ for the } \pi \text{ component,} \end{aligned} \tag{15}$$

equations (13) and (14) become simply :

$$\begin{pmatrix} \xi_o & -\frac{1}{2} k^2 \Gamma C F_{\bar{H}} \\ -\frac{1}{2} k^2 \Gamma C F_H & \xi_H \end{pmatrix} \begin{pmatrix} \psi_o \\ \psi_H \end{pmatrix} = 0 \tag{16}$$

$$\xi_o \xi_H = \frac{1}{4} k^2 \Gamma^2 C^2 F_H F_{\bar{H}} \tag{17}$$

There are two dispersion surfaces corresponding to the two polarisation components. In the case of non polarised incident beam, one must take the arithmetical average of the intensities due to each component. Besides these small changes, the results are identical to the neutron case.

## B. PROPAGATION EQUATIONS IN IMPERFECT OR PERFECT CRYSTALS. TAKAGI'S EQUATIONS

Consider a real crystal, with a potential $V(\underset{\sim}{r})$. If $\underset{\sim}{u}(\underset{\sim}{r})$ is the local distorion from perfect periodicity, we write :

$$V(\underset{\sim}{r}) = V_p[\underset{\sim}{r} - \underset{\sim}{u}(\underset{\sim}{r})] \tag{18}$$

where $V_p$ is the perfect crystal potential. $V(\underset{\sim}{r})$ can be written :

$$V(\underset{\sim}{r}) = \sum_{\underset{\sim}{H}} V_H(\underset{\sim}{r}) \exp(-i\underset{\sim}{H}.\underset{\sim}{r}) \tag{19}$$

with

$$V_H(\underset{\sim}{r}) = V_H \exp[iG(\underset{\sim}{r})]$$

$$G(\underset{\sim}{r}) = \underset{\sim}{H}.\underset{\sim}{u}(\underset{\sim}{r}) \tag{20}$$

We look for a solution of the form :

$$\Psi(\underset{\sim}{r}) = \psi_o(\underset{\sim}{r}) \exp(-i\underset{\sim}{K}_o.\underset{\sim}{r}) + \psi_H(\underset{\sim}{r}) \exp(-i\underset{\sim}{K}_H.\underset{\sim}{r}) \tag{21}$$

$V_H(\underset{\sim}{r})$, $\psi_o(\underset{\sim}{r})$, $\psi_H(\underset{\sim}{r})$ are slowly varying functions of position.

Separating the two component modulated waves in the Schrödinger equation, we can write :

$$\begin{aligned} -\frac{\hbar^2}{2m} \underset{\sim}{\nabla}^2[\psi_o(\underset{\sim}{r})\exp(-i\underset{\sim}{K}_o\underset{\sim}{r})] + (V_o-E)\psi_o(\underset{\sim}{r})\exp(-i\underset{\sim}{K}_o.\underset{\sim}{r}) \\ + V_{\bar{H}}(\underset{\sim}{r})\psi_H(\underset{\sim}{r})\exp(-i\underset{\sim}{K}_o.\underset{\sim}{r}) = 0 \\ -\frac{\hbar^2}{2m} \underset{\sim}{\nabla}^2[\psi_H(\underset{\sim}{r})\exp(-i\underset{\sim}{K}_H.\underset{\sim}{r})] + (V_o-E)\psi_H(\underset{\sim}{r})\exp(-i\underset{\sim}{K}_H.\underset{\sim}{r}) \\ + V_H(\underset{\sim}{r})\psi_o(\underset{\sim}{r})\exp(-i\underset{\sim}{K}_H.\underset{\sim}{r}) = 0 \end{aligned} \tag{22}$$

Let us denote the Fourier transforms of $\psi_o$, $\psi_H$, $V_o$, $V_H$, $V_{\bar{H}}$, by $\phi_o$, $\phi_H$, $U_o$, $U_H$, $U_{\bar{H}}$. For example,

$$U_H(\underset{\sim}{p}) = \int V_H(\underset{\sim}{r}) \exp(-i\underset{\sim}{p}.\underset{\sim}{r})\, d\underset{\sim}{r}$$

(22) become :

$$\begin{aligned} \frac{\hbar^2}{2m}[(\underset{\sim}{K}_o+\underset{\sim}{p})^2-K^2]\phi_o(\underset{\sim}{p}) + \int U_{\bar{H}}(\underset{\sim}{p}-\underset{\sim}{q})\phi_H(\underset{\sim}{q})\, d\underset{\sim}{q} = 0 \\ \frac{\hbar^2}{2m}[(\underset{\sim}{K}_H+\underset{\sim}{p})^2-K^2]\phi_H(\underset{\sim}{p}) + \int U_H(\underset{\sim}{p}-\underset{\sim}{q})\phi_o(\underset{\sim}{q})\, d\underset{\sim}{q} = 0 \end{aligned} \tag{23}$$

If we replace the Ewald spheres around Q in Figure 1 by their tangent plane, we may write :

$$\begin{aligned} (\underset{\sim}{K}_o+\underset{\sim}{p})^2-K^2 = 2\underset{\sim}{K}_o.\underset{\sim}{p} \\ (\underset{\sim}{K}_H+\underset{\sim}{p})^2-K^2 = 2\underset{\sim}{K}_H.\underset{\sim}{p} \end{aligned} \tag{24}$$

We recall that the Fourier transform of $\underset{\sim}{\nabla}f(\underset{\sim}{r})$ is $-i\underset{\sim}{p}F(\underset{\sim}{p})$. Thus :

$\underset{\sim}{K}_o.\underset{\sim}{p}\phi_o(\underset{\sim}{p})$ corresponds to $iK(\partial\psi_o/\partial s_o)$ where $s_o$ is the coordinate along $\underset{\sim}{K}_o$.

Let us define the quantity :

$$\kappa_H = \frac{\lambda}{V_c} F_H \tag{25}$$

We obtain the so-called Takagi's equations (Takagi, 1969 ; Kato, 1973)

$$\begin{aligned} \frac{\partial \psi_o}{\partial s_o} &= i\, \kappa_{\bar{H}} \exp\{-iG(\underset{\sim}{r})\}\, \psi_H \\ \frac{\partial \psi_H}{\partial s_H} &= i\, \kappa_H \exp\{+iG(\underset{\sim}{r})\}\, \psi_o \end{aligned} \tag{26}$$

These equations are of course also valid for perfect crystals, where $G(\underset{\sim}{r})$ is constant. The only approximations are the representation of the crystal potential by (19) and the assumption that the Ewald sphere can be locally replaced by its tangent plane near the Bragg tie point.

If we consider the two wave-fields, each obey (26). Thus we can consider (26) to represent the propagation of waves for the total wave, assuming that the geometrical coordinates $s_o$ and $s_H$ are not perturbed by the small change in the wave vectors due to dynamical effects.

In the case of X-Rays, eqns (26) remain valid for each polarisation state, provided (25) is replaced by :

$$\kappa_H = \frac{aC\lambda}{V_c} F_H \tag{27}$$

It can be shown (Balibar, 1975) that eqns (26) are valid even for strong distortions from perfect periodicity. They are for us the basis for calculating the diffracted power.

## COMPARISON BETWEEN DYNAMICAL DIFFRACTION AND BAND THEORY

As previously stated, the comparison is more direct with neutron diffraction. Eqn (4) is valid for neutrons, where V is the pseudopotential (3). In elastic scattering, E is fixed, and we look for solutions of (4) that are compatible with this constraint.

In band theory, the motion of the electrons is governed by eqn (4) again, where $V(\underset{\sim}{r})$ is now the electronic potential. As we shall recall below, assuming a plane wave expansion of the crystal orbital, we look, for a given wave vector $\underset{\sim}{K}_o$, for the energy and wave amplitude satisfying (4).

In a sense, band theory is the problem of finding free modes of the electrons. Neutron diffraction is the problem of finding forced modes, when one imposes the energy. They are complementary facets of the same fundamental problem.

We write the electronic wavefunction $\Psi(\underset{\sim}{r})$ in terms of expansion (7), and expand $V(\underset{\sim}{r})$ as (5).

1. Assume $V(\underset{\sim}{r})$ to be constant : $V_o$. The solution of (4) is any plane wave :

$$\exp(-i\underset{\sim}{k}_o.\underset{\sim}{r}) \quad \text{with energy : } V_o + \hbar^2 k_o^2/(2m) \tag{28}$$

2. We now suppose that $V(r)$ is slowly varying throughout the crystal : in such a case, the plane wave (28) remains the major term in the expansion (7). We therefore choose any wavevector $\underset{\sim}{k}_o$ in the BZ and look for the perturbation of the associated plane wave. If we define $E_H$ to be $\hbar^2 K_H^2/(2m)$ :

$$(E-E_H)\ \psi_H = \sum_{H'} V_{H-H'}\ \psi_{H'} \tag{29}$$

3. $\psi_o$ is the dominant term and we may, as a first order solution, write :

$$\psi_H = \frac{V_H \psi_o}{E-E_H}$$

$$E - E_o = \sum_H \frac{|V_H|^2}{E - E_H} \tag{30}$$

Since E is of the order of $E_o$,

$$E = E_o - \sum_H \frac{|V_H|^2}{E_H - E_o} \tag{31}$$

4. This perturbation treatment remains valid as long as $E_H \neq E_o$. The condition $E_o = E_H$ is equivalent to Bragg's law. It occurs when $\underset{\sim}{k}_o$ ends at the boundary of one BZ. For $\underset{\sim}{k}_o$ at the vicinity of the BZ boundary, two terms in the expansion (7) are important $\psi_o$ and $\psi_H$. (29) becomes :

$$(E - E'_o)\ \psi_o = V_{\bar{H}}\ \psi_H$$

$$(E - E'_H)\ \psi_H = V_H\ \psi_o \tag{32}$$

with

$$E'_H = E_H + V_o$$

E must be solution of the equation :

$$(E - E'_o)(E - E'_H) = V_{\bar{H}} V_H \tag{33}$$

For the exact Bragg condition :

$E = E'_o \pm V_H$ . The two wave functions are simply :

$$\exp(-i\underset{\sim}{K_o}.\underset{\sim}{r}) + \exp(-i\underset{\sim}{K_H}.\underset{\sim}{r})$$
$$\exp(-i\underset{\sim}{K_o}.\underset{\sim}{r}) - \exp(-i\underset{\sim}{K_H}.\underset{\sim}{r}) \qquad (34)$$

At the Bragg condition, the crystal does not propagate any more plane waves, but, due to reflection on atomic planes, there are two standing waves. If the first has maxima on atomic positions, the second has minima at the atoms. Therefore, absorption properties are different, and this explains the anomalous absorption of X-Rays.

5. If we write :

$$E - V_o = \frac{\hbar^2 K^{*2}}{2m}$$

(33) can be transformed to :

$$(K^* - K_o)(K^* - K_H) = |V(H)|^2/K_o^2 \qquad (35)$$

which is the dispersion relation discussed in the previous paragraphs. We have two solutions, corresponding to approximately standing waves, having phase opposition. This completes the analogy between dynamical theory and band theory. For distorted crystals, it is possible to derive equations similar to Takagi's equations.

As stated at the beginning of the paragraph, the difference between (17) and (35) is the same as the difference between forced and free oscillations of a system.

## INTENSITY COUPLING EQUATIONS TO DESCRIBE EXTINCTION

Starting from Takagi's equations, Kato (1976, 1979) introduced a statistical description of the defects inside the crystal. He defines a correlation function f(z) :

$$f(z) = \langle \exp[i(G(r)-G(r+z)]\rangle \qquad (36)$$

and a coherence length $\tau$, which is a measure of the width of f(z). If $\tau$ is less than $\Lambda$, the extinction distance, Kato proves that for the statistical crystal, the energy propagation is governed by intensity coupling equations :

$$\frac{\partial\langle I_o\rangle}{\partial s_o} = -[2\tau Re(\kappa_H\kappa_{\bar{H}}) + \mu]\langle I_o\rangle + 2\tau|\kappa_{\bar{H}}|^2\langle I_H\rangle$$
$$\frac{\partial\langle I_H\rangle}{\partial s_H} = -[2\tau Re(\kappa_H\kappa_{\bar{H}}) + \mu]\langle I_H\rangle + 2\tau|\kappa_H|^2\langle I_o\rangle \qquad (37)$$

These eqns are discussed in some detail in Pr Kato's lecture notes. His important contribution fills the gap between kinematical and dynamical viewpoints for diffraction. Introducing high order correlation lengths, he can extend the validity of (37) (modified by a damping factor) to the case $\tau \sim \Lambda$. Kato's derivation is valid for spherical waves.

Independently, one can derive intensity coupling eqns in a phenomenological way. Assuming at a given point the diffracting power to be a function $\bar{\sigma}(\varepsilon)$, where $\varepsilon$ is the rocking angle, as defined in the article on kinematical theory, we can write :

$$\frac{\partial I_o}{\partial s_o} = -(\bar{\sigma} + \mu)I_o + \bar{\sigma}I_H$$
$$\frac{\partial I_H}{\partial s_H} = -(\bar{\sigma} + \mu)I_H + \bar{\sigma}I_o \qquad (38)$$

These eqns served as the basis of diffraction theories for a long time (Hamilton, 1957- Zachariasen, 1968- Becker, Coppens, 1974).

If presently we disregard the small difference between $|\kappa_H|^2$ and $\mathrm{Re}(\kappa_H\kappa_{\bar{H}})$, (37) and (38) are similar in form, except that (38) have an angular dependence. (38) are specific of a plane wave theory, though Kato's equations assume already an integration over all angles. (37) are valid for an homogeneous beam made of a superposition of incoherent waves : each wave excited at a given entrance point in the crystal is incoherent with waves excited at another point. In the averaging procedure of Kato, only those waves that originate from the same entrance point are coupled. If, on the contrary, one assumes coherence between different source points, a distance z apart, the coupling should be modulated by f(z). Going back to plane waves, by Fourier transformation, f(z) becomes a function $\phi(\varepsilon)$ of the rocking angle, and (37) take the same form as (38). Reality is certainly intermediate between the two viewpoints, and this important problem is open to discussion. Application to practical cases shows a preference for (38).

Werner(1974) pointed out some possible generalisations of (38) : inhomogeneities, by making $\bar{\sigma}$ a function of position, and the possibility of differenciating $\bar{\sigma}(\underset{\sim}{k}_o \rightarrow \underset{\sim}{k}_H)$ from $\bar{\sigma}(\underset{\sim}{k}_H \rightarrow \underset{\sim}{k}_o)$.

$\bar{\sigma}$ is a general function, submitted to :

$$\int \bar{\sigma}(\varepsilon)\, d\varepsilon = Q \qquad (39)$$

In the mosaic model, it can be related to an effect due to the size and shape of the perfect regions ($\sigma$), convoluted with the orientation distribution function of the blocks one to the others (W) :

$$\bar{\sigma} = \sigma * W \qquad (40)$$

and W are generally assumed to be Gaussians or Lorentzians (with a practical preference for Lorentzian shape). Kato's theory of the coherence length leads to nearly similar results, without assuming a discontinuous model. If W is narrow compared to $\sigma$, we have type II extinction (small coherent domains, or stacking faults). If $\sigma$ is narrow compared to W, we have type I extinction ( dislocations. It can be the case when the coherence length is large, which means that primary extinction is present). In practice, type I is the most often encountered case.

If primary extinction is present, it can be approximately taken into account by a factor $y_p$, which multiplies $\bar{\sigma}$ in (38). Becker (1977a, 1977b) has shown that $y_p$ can be rigorously calculated from Takagi's equations.

The exact domain of validity of intensity coupling equations is not known. The application to practical examples is very successful, even when $\tau$ is larger than $\Lambda$. Becker and Coppens have undertaken a direct numerical integration of Takagi's equations, simulating defects of various kinds, distributed through a random number generator program. It is thus possible to see when a statistical model becomes applicable, to see the effect of various sorts of defects. Moreover, the application to very small crystals (case of synchrotron radiation) is very important.

In the case of magnetic neutron scattering, equations (38) are replaced by 4 equations, since we must consider separately the two neutron spin states. Until now, one considers that the possibility of spin reversal during scattering is negligible. In such an event, one has to solve two independent sets of eqns (38), one for each polarisation state (Bonnet, Delapalme, Becker, Fuess, 1976). It is possible to include spin reversal process in a perturbation scheme and this has been performed by the author (to be published). For a parallel plate, the exact solution is even possible. It would be too lengthy to develop such a formalism here.

In a review paper (Becker, 1977a), I developed some possible criteria to test the validity of the extinction correction. They are of great practical importance and should be applied in all cases. It seems that the present theories are more valid for neutrons than for X-Rays, when extinction is severe : this is related to the fact that the extinction distance is larger in the neutron case. Due to larger crystals for neutrons, extinction is generally more important than in the X-Ray case, so that the present situation for X-N work is fairly favorable.

I think progress remains to be done to extend theory to more complex cases : the efforts of Pr Kato are remarkable. Another effort should be done in the description of statistically distorted crystals, in relation with growth thermodynamics : see the tentative work by Becker and Coppens. In a word, the physical situation is still not clear

## SOLUTION OF PROPAGATION EQUATIONS

Once we have appropriate equations to describe the propagation of incident and diffracted waves, the solution is still not easy, due to boundary conditions at the entrance and exit surfaces of the crystal (Becker, 1977b- Kato's lecture notes).

Using the Green function technique, it is possible to first calculate the solution corresponding to a unique source point on the entrance surface, and then to integrate over all possible source points. In the case of wave-coupling equations, this is related to the Fourier transformation between plane and spherical wave descriptions. For intensity-coupling equations, it is related to the description of the incident beam as a superposition of incoherent small beams. On Figure 2, the section of the crystal by a plane containing the incident and diffracted directions is shown. We define a point m in the augmented volume V' (AECBD). From m, drawing the lines that are parallel to the beams, we can define a unique entrance point S and a unique exit point M. Let I(m) be the intensity diffracted at M from a unit incident beam originating at S. We see that points m that are outside the crystal volume V correspond to situations where M cannot be reached from S through a single scattering (no kinematic analog). The results are the following :
For intensity-coupling equations :

$$P = \int_{V'} I(m)\, dv \qquad (41)$$

where P may depend on rocking angle, and thus has to be integrated to get the total integrated diffracted power.
For wave-coupling equations :

$$P_{\text{plane wave}} = (\lambda/\sin 2\Theta) \int_{V'} I_{\text{sph.wave}}(m)\, dv \qquad (42)$$

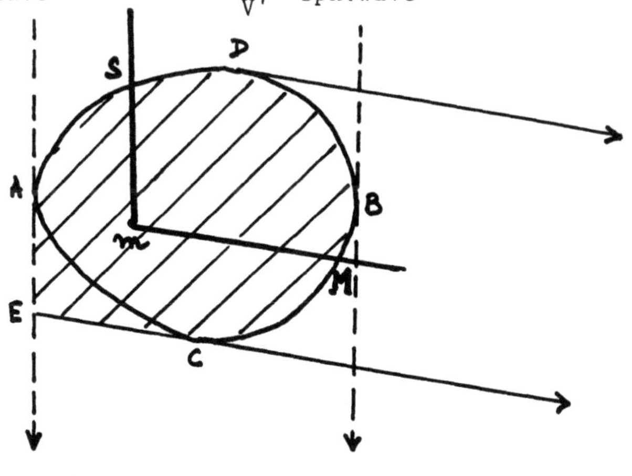


Figure 2

Using an operator technique, with the help of a linear operator that goes from (2n-1) to (2n+1) scattering, it is possible to show, when V' = V (Lauë geometry, corresponding to small Bragg angles), that :
For mosaic crystal :

$$P = \int \bar{\sigma} d\varepsilon \int_V \exp[-(\bar{\sigma}+\mu)(T_o+T_H')] I_o[2\bar{\sigma}(T_o T_H')^{1/2}]\, dv \qquad (43)$$

For a perfect crystal (primary extinction) :

$$y_p = V^{-1} \int_V |J_o[2(\kappa_H \kappa_{\bar{H}} T_o T_H')^{1/2}]|^2\, dv \qquad (44)$$

where $T_o$ = Sm, $T_H'$ = mM. $J_o$ and $I_o$ are zero order Bessel and modified Bessel functions.
In the case of more complex boundary conditions, one has to separate the crystal in volumes where I(m) is given by one unique expression. Expressions for I(m) are calculable in terms of higher order Bessel functions (Becker, to be published- see Werner, 1974). Presently under study is the estimation of the error made by using (43) for any geometry (which is the present assumption made in the routine least squares refinement program LINEX or MOLLY).
It should be noted that all kinds of propagation equations are of the same hyperbolic nature. I consider that the problem of getting a solution to these equations is now solved (Becker, Dunstetter, to be published).

## BORRMANN EFFECT IN REAL CRYSTALS

I would now like to comment on the anomalous absorption problem in imperfect crystals. In previous sections, we have seen that the two wave-fields for a perfect crystal give rise to standing waves, one having maxima at atomic positions (high absorption), the other minima (low absorption). This is the basis for the Borrmann effect. This effect is typically a perfect crystal effect. For each polarisation of the incident X-Ray beam, and for a given rocking angle ε, the effective absorption coefficient (see for example Zachariasen, 1968) is :

$$\mu_a = \mu_o \pm \mu_H \Phi(\varepsilon) \qquad (45)$$

corresponding to each wave-field.

$$\mu_H = \frac{C}{V_c} \sum_j \mu_j W_j \exp(i\underset{\sim}{H}.\underset{\sim}{R}_j) \qquad (46)$$

$\mu_j$ is related to the dispersive part of the form factor by :

$$\mu_j = 2a\lambda f_j'' \qquad (47)$$

$\Phi(\varepsilon)$ is given by :

$$\Phi(\varepsilon) = (1 + \alpha^2\varepsilon^2)^{-1/2} \tag{48}$$

$\alpha$ is the perfect crystal reflection width :

$$\alpha = \sin 2\Theta/(C\kappa_H)$$

Zachariasen assumed that $\mu_a$ applies to mosaic crystals :

$$\partial I_o/\partial s_o = -(\mu_a + \bar{\sigma}) I_o + \bar{\sigma} I_H \; .....$$

He supposed an average equidistribution of energy between the two wave-fields, which is reasonable, and thus took for the diffracted intensity :

$$I_H = \frac{1}{2}( I_+ + I_-) \tag{49}$$

This assumption is unreasonable : if the size of perfect regions becomes very small, there should not be any anomalous absorption!

We do propose another model for mosaic crystals :
Let us consider the propagation of energy in a given direction, on a distance T. The beam travels through a set of N perfect regions, each of mean size t, which are mutually misoriented. The effect in region i is :

$$\exp(-\mu_o t) \, \frac{1}{2} \, \{\exp[\mu_H \Phi(\varepsilon+\Delta_i)t] + \exp[-\mu_H \Phi(\varepsilon+\Delta_i)t]\} \tag{50}$$

$$= \exp(-\mu_o t) \, \cosh(\mu_H \Phi_i t)$$

where $\Phi_i$ stands for $\Phi(\varepsilon+\Delta_i)$. $\Delta_i$ is the misorientation of ith region.

Using T = Nt, the reduction factor is :

$$A = \exp(-\mu_o T) \prod_{i=1}^{N} \cosh(\mu_H t \Phi_i)$$

$$B = A \exp(\mu_o T) = \prod_{i=1}^{N} \{1 + \frac{1}{2} \mu_H{}^2 t^2 \Phi_i{}^2 + \ldots\}$$

$$\sim 1 + \frac{1}{2} \mu_H{}^2 t^2 \sum_i \Phi_i{}^2 \tag{51}$$

If W is the mosaic distribution function,

$$\sum_i \Phi_i{}^2 = N \, \Phi^2 * W \tag{52}$$

$\Phi^2$ is a Lorentzian. If W is a Lorentzian too, (52) still has the same shape. Energy transfer equations are valid if the width of W is larger than that of $\Phi$. Therefore $\Psi=\Phi^2*W$ is often similar to W.

$$B = 1 + \frac{1}{2}\mu_H{}^2 t\Psi T \sim \exp\{\frac{1}{2}\mu_H{}^2 t\Psi T\}$$

Thus, the apparent absorption coefficient to be used in the coupling equations is :

$$\mu = \mu_o - \frac{1}{2}\mu_H^2 \, t \, \Psi(\varepsilon) \tag{53}$$

The result (53) is now physically correct. In order to have a significant effect, $t/\Lambda$ must be of the order of 1, which means that primary extinction should be present, and that secondary extinction should be of type I. (53) has been applied with success in the case of Yttrium Iron Garnet (Bonnet, thesis, 1977) and for GaAs (Rigoult, Becker, to be published).

Finally let us consider the equations (37) of Kato. Specifically we consider the difference between $\mathrm{Re}(\kappa_H\kappa_{\bar{H}})$ and $|\kappa_H|^2$. We write :

$$F_H = F'_H + iF''_H \tag{54}$$

$$F_{\bar{H}} = F'^*_H + iF''^*_H$$

$$\mathrm{Re}(\kappa_H\kappa_{\bar{H}}) = (\lambda aC/V_c)^2\{|F'_H|^2 - |F''_H|^2\} \tag{55}$$

If we assume a centrosymmetric crystal ( which was supposed in the previous derivation, since $\mu_H$ was considered real)

$$|\kappa_H|^2 = (\lambda aC/V_c)^2\{|F'_H|^2 + |F''_H|^2\} \tag{56}$$

We thus get as an absorption coefficient

$$-4\tau\ (\lambda aC/V_c)^2|F''_H|^2 = -\tau\mu_H{}^2 \tag{57}$$

which is equivalent to (53).
Note that the expression (44) for primary extinction contains the Borrmann effect.

## THERMAL DIFFUSE SCATTERING AND EXTINCTION

We conclude this paper by some comments on the coupling between TDS and extinction. It is commonly assumed that the integrated reflectivity is given by :

$$F^2(1+\alpha)\ y(F^2) \tag{58}$$

where $\alpha$ is the TDS correction, y the extinction factor. Extinction theory is always derived on the basis of diffraction by a thermally averaged structure.

Assume that each perfect region is small enough for primary extinction to be negligible but large enough for the phonons to be identical to infinite crystal case. If we consider the integration procedure of the diffracting function $\sigma$ in each block, we may assume that in the parameter $x_s$ for secondary extinction, proportional to $F^2$, Q has to be replaced by $Q(1+\alpha)$ so that (58) becomes :

$$F^2 \ (1+\alpha) \ y\{F^2(1+\alpha)\} \tag{59}$$

For low angle reflections which govern extinction, $\alpha$ is small and this will not change the extinction parameter significantly, but $y_{calc}$ will be smaller for medium range reflections for which F will systematically be estimated higher.
In fact, the integration of $\Delta^2(\underset{\sim}{q}+\underset{\sim}{S})$ is done only over $\varepsilon_2$ and $\varepsilon_3$, and thus the thermal average should be done over those two angles. The last angular integration should be done simultaneously on TDS and extinction, which may be fairly difficult.

Take now a perfect crystal. We have seen that the integrated one-phonon intensity is the sum over all modes $(j,\underset{\sim}{q})$ of components that differ essentially by the apparent structure factor. If we take a Lauë geometry, P is given by :

$$P = |\kappa_H|^2(\lambda/\sin 2\Theta) \int_V dv \ |J_o\{2(\kappa_H\kappa_{\bar{H}}T_oT'_H)^{1/2}\}|^2$$

Assume a centric case :

$$\kappa_H = bF_H = \kappa_{\bar{H}} \qquad\qquad b = a\lambda C/V_c$$

For mode $(j,\underset{\sim}{q})$, the apparent structure factor is :

$$\xi_j(\underset{\sim}{q}) \sum_k \frac{\underset{\sim}{e}_{kj}(\underset{\sim}{q})}{\sqrt{Nm_k}} \cdot \underset{\sim}{H} \ f_k(\underset{\sim}{H}) \ \exp(i\underset{\sim}{H}.\underset{\sim}{R}_k) \ W_k(\underset{\sim}{H}) \tag{60}$$

which simply corresponds to scattering by the static periodic density

$$\xi_j(\underset{\sim}{q}) \sum_k \frac{\underset{\sim}{e}_{kj}(\underset{\sim}{q})}{\sqrt{Nm_k}} \cdot \underset{\sim}{\nabla}_k \ \bar{\rho}_k \tag{61}$$

where $\rho_k$ is the kth pseudoatom and $\bar{\rho}_k$ the averaged pseudoatom.
For an elastic wave, the apparent structure factor is

$$\Psi_j(\underset{\sim}{q}, \underset{\sim}{H}) = \xi_j(\underset{\sim}{q}) \frac{\underset{\sim}{\varepsilon}_j(\underset{\sim}{q})}{\sqrt{Nm}} \cdot \underset{\sim}{H} \ F(\underset{\sim}{H}) \tag{62}$$

The one-phonon scattering should be proportional to :

$$\sum_{j\underset{\sim}{q}} |\Psi_j(\underset{\sim}{q},\underset{\sim}{H})|^2 \ (b^2\lambda/\sin 2\Theta) \int dv |J_o(2b\Psi_j(\underset{\sim}{q},\underset{\sim}{H})\sqrt{T_o T_H'}\ )^2| \qquad (63)$$

where the summation is limited to the modes spanned during the reflection recording. Due to the relatively small value of the argument, the Bessel function should be limited to the first terms.

## References


Authier A. (1970) Adv. Struct. Res. Diff. Meth. 3, 1

Balibar F. (1975) Limoges Summer School on Dynamical Theory

Batterman B.W, Cole H. (1964) Rev. Mod. Phys. 36, 681

Becker P., Coppens P. (1974a) Acta Cryst., A30, 129

(1974b) Acta Cryst., A30, 148

(1975) Acta Cryst., A31, 417

Becker P. (1977a) Acta Cryst., A33, 243

(1977b) Acta Cryst., A33, 667

Bonnet M., Delapalme A., Becker P., Fuess H. (1976) Acta Cryst., A32, 945

Hamilton W.C. (1957) Acta Cryst., 10, 657

Kato N. (1973) Z. Naturforsch., 28a, 604

(1974) X-Ray Diffraction. Ed. by Azoroff. Mc Graw Hill

(1976) Acta Cryst., A32, 453, 458

(1979) Acta Cryst., under the press

Marshall W., Lovesey S.W. (1971) Theory of Thermal Neutron Scattering Oxford

Takagi S. (1969) J. Phys. Soc. Japon, 26, 1239

Werner S.A. (1974) J. Appl. Phys., 45, 3246

Zachariasen W.H. (1967) Acta Cryst., 23, 558

(1968) Acta Cryst. A24, 421

EXTINCTION AND FERROELECTRIC TRANSITION IN $RbH_2PO_4$

P. BECKER, P. BASTIE, J. LAJZEROWICZ

In the paraelectric phase, the system is quadratic : a = 7.586 Å, c = 7.250 Å. The transition temperature is $T_c$ = 144.92K Ray diffraction with = .03Å has been used to investigate the behaviour of the integrated reflectivity R with temperature. The reflections (200) and (600) have been measured, with a symmetric Laue geometry. The results are given in table 1. The results are reversible as a function of temperature variation.

1. Why can one use the approximation of an infinite parallel plate in the symmetric Laue geometry ?

2. Construct the two curves representing R(T).

3. By integration of energy transfer equations, one can show that the secondary extinction correction y is given by :

$$y = \sum_{n=0}^{\infty} \frac{(-1)^n}{n!} \overline{T^{(n)}} Q^n \int_{-\infty}^{\infty} \sigma^{n+1}(\varepsilon)\, d\varepsilon$$

$$\overline{T^{(n)}} = v^{-1} \sum_{j=0}^{n} \binom{n}{j}^2 \int_v t_1^j t_2^{n-j}\, dv$$

where $t_1$ and $t_2$ are given in Fig. 1. is the effective diffraction function, supposed here to be normalized, and moreover to be Lorentzian :

$$\sigma(\varepsilon) = \frac{2g}{(1 + 4\pi^2\varepsilon^2 g^2)}$$

Note that the kinematic reflectivity $R_k$ is : Qa.
Show that :

$$y = \exp(-u)[\, I_o(u) + I_1(u)]$$

$$u = 2R_k g$$

4. Show that extinction comes from two separate effects, one due to static defects, one due to dynamic effects near the transition.

If one calls the two parameters $g_s$ and $g_d$, and the overall parameter g, give the expression for g.

Note : for mosaic dominated extinction, g = constant,
for particle size dominated case, $g = t \sin 2\theta/\lambda$, where t is the the particle size.

5. We assume a critical dependence of $g_d$ :

$$g_d = \gamma (T-T_c)^p$$

For each of the two reflections, verify if the assumption is satisfied. Deduce a value for p. Examine the compatibility between the two experiments. Check the type of extinction, both for static and dynamic part.

The values of u are written in table 1, so as 1/u.

6. Theory of thermodynamic fluctuations leads to a shear angle $\eta$ that satisfies :

$$\eta = \alpha T_c/(T-T_c)$$

The internal strain S leads to a shear angle such that :

$$\eta = S\,C^{-1}/(T-T_c)$$

C : elastic constant

Conclude.

7. How can one interpret the strange behaviour of R about $T_c$ ? Let add that shoulders occur when one reaches the transition, on each side of the Bragg peak, being the most pronounced for (600). What can be their origin ? Can one propose another way of estimating the elastic component of the ferroelectric transition ?

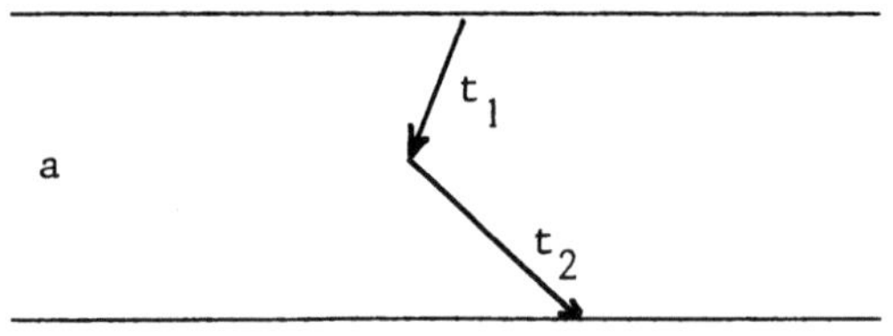


Figure 1

(200) $R_k = 2.65\ 10^{-5}$

| T | R(x10$^5$) | u | 1/u |
|---|---|---|---|
| 198.171 | .248 | 70 | .014 |
| 165.93 | .277 | 60 | .0166 |
| 153.82 | .311 | 45 | .022 |
| 153.13 | .329 | 38 | .026 |
| 147.403 | .474 | 20 | .05 |
| 146.334 | .596 | 12 | .083 |
| 145.751 | .777 | 7 | .142 |
| 145.462 | 1.022 | 4.1 | .244 |
| 145.320 | 1.231 | 2.6 | .384 |
| 145.227 | 1.380 | 2. | .5 |
| 145.135 | 1.545 | 1.5 | .666 |
| 145.069 | 1.722 | 1.1 | .91 |
| 145.023 | 1.862 | .85 | 1.176 |
| 144.984 | 1.970 | .7 | 1.428 |
| 144.928 | 2.267 | .35 | 2.857 |

(600) $R_k = 1.29\ 10^{-6}$

| T | R(x10$^6$) | u | 1/u |
|---|---|---|---|
| 197.698 | .716 | 1.7 | .59 |
| 165.93 | .778 | 1.35 | .74 |
| 153.13 | .849 | 1.04 | .96 |
| 146.40 | 1.056 | .44 | 2.27 |
| 145.72 | 1.167 | .20 | 5. |
| 145.40 | 1.258 | .04 | 25 |
| 145.30 | 1.276 | .02 | 50 |
| 145.086 | 1.281 | .01 | 100 |
| 144.999 | 1.256 | ... | ... |
| 144.962 | 1.218 | ... | ... |

Hint.

From Bonnet et al (1976, Acta Cryst., A32, 945, Appendix B), the expression for y can be calculated in a straightforward way.

$$1/g = 1/g_s + 1/g_d \qquad g_d = \gamma(T-T_c)^p$$

$$1/u = 1/(2R_k)\ \{1/g_s + 1/g_d\}$$

Let $u_\infty$ correspond to the effect of $g_s$ (T 200°K)

$$1/u - 1/u_\infty = \frac{1}{2R_k\gamma}\ (T-T_c)^{-p}$$

$$\ln(1/u - 1/u_\infty) = -\ln(2R_k\gamma) - p\ \ln(T-T_c)$$

For type I, $g_d$ constant with hkl. For type II, not true. Type I here.

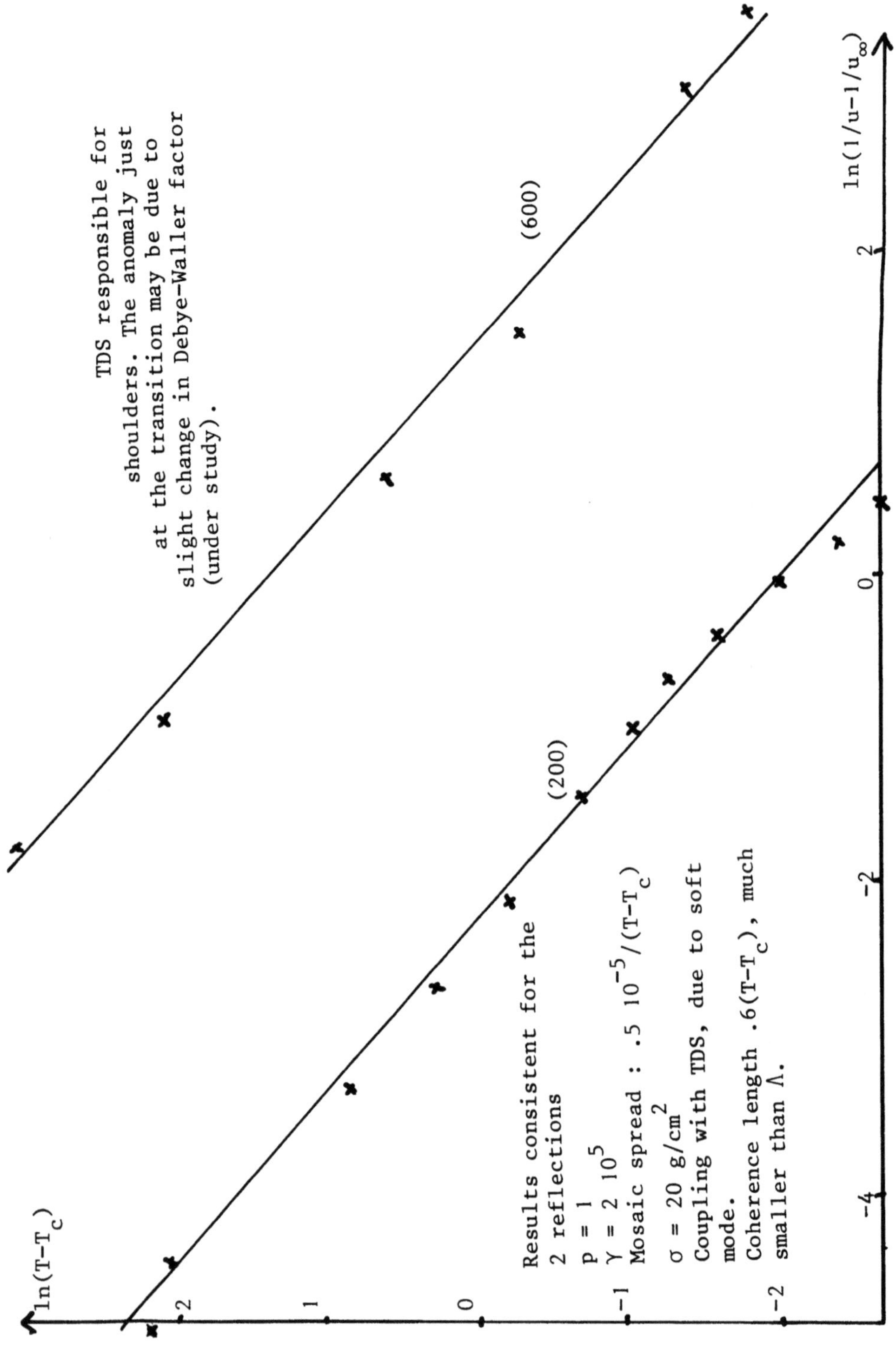

ln(T-T$_c$)
ln(1/u-1/u$_\infty$)
2
1
0
-1
-2
-4
-2
0
2
(600)
(200)
TDS responsible for shoulders. The anomaly just at the transition may be due to slight change in Debye-Waller factor (under study).
Results consistent for the 2 reflections
p = 1
γ = 2 10$^5$
Mosaic spread : .5 10$^{-5}$/(T-T$_c$)
σ = 20 g/cm$^2$
Coupling with TDS, due to soft mode.
Coherence length .6(T-T$_c$), much smaller than Λ.

# EXTINCTION TREATMENT IN POLARIZED NEUTRON EXPERIMENTS[1,2]

A. DELAPALME

Laboratoire Léon Brillouin

BP n°2, 91190 Gif-sur-Yvette, France

I The purpose of this training is to show that polarized neutron experiments, which do not depend on any scale factor, are very dependent on extinction and provide us with original tests for extinction treatment. Neutrons are two-spin state particles and polarized neutron beam is composed of neutrons in only one spin state. Collecting the intensities $P^+$ and $P^-$ for up and down spin states yields the "flipping ratio" $R = P^+/P^-$ which does not depend on any scale factor and very little on thermal parameters at low Bragg angles. These conditions are very favourable to disclose the extinction features of real centro-symmetric ferro-or ferrimagnetic crystals. The reading of reference [1] is necessary to get signification of notations and meaning of this exercise. We shall consider the primary extinction correction $y_p$ (Eqs.18 and 19) as a constant given in the data. Its formulation in Becker-Coppens treatment is only phenomenological, but has the advantage that the integrated intensity P becomes proportional to the structure factor when the extinction parameter $x_p \rightarrow \infty$ . Secondary extinction for a Lorentzian shape of the diffracting power $\bar{\sigma}$ will be taken from equation (14), but using $\alpha_G$ expression (equation 17) instead of $\alpha_L$ written in it. We shall use too the simplified expression (equation 6) for $\overline{T^{(n)}}$ , where n is given in data table I and determines the number of terms in $y_s$ expansion to get adequate convergence for this expression.

II <u>Extinction Correction in YIG[1](III)</u> : We give in data table I, a (unit cell), $F_N$ , $F_M$ , $\lambda$ , R, $T^{(1)}$ for the $2\bar{2}0$ and $4\bar{4}0$ reflexions and the corresponding calculated extinction coefficient $|e|$ (equation 22). 1) Knowing the correct refined value of $\eta$ and t, we can calculate the two components of $\alpha_G$ : $\frac{\lambda}{t\sin 2\theta} = \frac{d}{t\cos\theta} = 0.338$ and 0.171 for $2\bar{2}0$ and $4\bar{4}0$ reflexions respectively and $\eta\sqrt{2\pi} = 0.204$. We deduce that $\alpha_G$ varies with d (interplanar spacing for hkℓ) and this YIG crystal is type II at low Bragg angle and type I at higher angle. This label of "crystal type" is then irrelevant. 2) You calculate yourself (equation 22) the $|e|$ values given in table I (calculate $y^{\pm}_{hk\ell}$ according to equation 14 and 17 with n terms in expansion ; $y^{\pm}_{i} = y^{\pm}_{i_p}\ y^{\pm}_{i_s}$ ). You plot the result on diagram to compare with the theoretical line (Eq.27). Calculate the change in $|e|$ if you double the value of the mosaic spread $\eta$ . 3) One interesting feature of this y extinction correction is (Eq.26) to show that for $x_p$ and $x_s \to \infty$ , the integrated intensity could become proportional to $\sqrt{F}$ . Flipping Ratio $R_{2\bar{2}0} \sim R_D^{0.529}$ has been experimentally observed with large YIG single crystal ($\overline{T^{(1)}}$ ≠ 12 mm).


References

(1) M. BONNET, A. DELAPALME, P. BECKER and H. FUESS ; Acta Cryst. (1976) A 32, 945-953.

(2) N. KATO ; Acta Cryst. (1976) 10, 629-634.


Y.I.G : TABLE I : DATA

a = 12.376 Å ; $\eta$ = 0.028 min.of arcs ; t = 13μm; $g = \frac{1}{2\eta\sqrt{\pi}}$

($2\bar{2}0$) $F_N = -4.76$ ; $F_M = 79.7\mu_B * 0.27$ ; $F^{\pm} = F_N \pm F_M$

| λ(Å) | R | σ(R) | e | σ(e) | n | $y_p^+$ | $y_p^-$ | $\overline{T^{(1)}}$ (cm) |
|---|---|---|---|---|---|---|---|---|
| 0.81 | 0.550 | 0.003 | -0.352 | 0.039 | 11 | 0.994 | 0.986 | 0.0973 |
| 0.74 | 0.460 | 0.003 | -0.121 | 0.013 | 5 | 0.995 | 0.988 | 0.0275 |
| 0.50 | 0.430 | 0.003 | -0.058 | 0.006 | 3 | 0.998 | 0.995 | 0.0275 |
| ($4\bar{4}0$) | | | $F_N = 9.93$ ; $F_M = 58.2\mu_B * 0.27$ | | | | | |
| 0.81 | 17.0 | 0.40 | 0.156 | 0.017 | 4 | 0.987 | 1 | 0.0278 |
| 0.81 | 12.8 | 0.15 | 0.363 | 0.040 | 9 | 0.987 | 1 | 0.0975 |

# ON EXTINCTION

N. Kato

Department of Crystalline Materials Science, Faculty of Engineering, Nagoya University
Chikusa-ku, Nagoya 464, Japan

The extinction phenomena in crystal diffraction are not easy subjects to tackle. It is, however, very important for accurately determining crystal structure factors, in which most of crystallographers are interested. The situation is similar to that of studying "cold", from which most people are suffering but its pathological understanding is still poor. Nowadays, people will not die directly due to cold. Similarly, crystallographers can determine the atomic positions in unit cell, if they pay reasonable precautions to extinction. Nevertheless, once we intend to obtain the accurate charge distribution in crystalline states, we need to get rid of suffering from "extinction cold".

In this lecture note, the author intends to review three topics in the theory of extinction. After a brief preliminary description on the integrated intensity (§1), we are concerned with the approximate solution of Hamilton-Zachariasen's equations, which are regarded as the fundamental equation for secondary extinction (§2). Next, the wave-optical justification of the H-Z equations is described with an emphasis on the physical meaning (§3). Finally, we shall compare the results on the integrated intensities given by kinematical, dynamical and secondary extinction theories in the case of non-absorbing parallel-sided crystals (§4).

## §1. PRELIMINARY

### a. Integrated Diffracted Intensity

We shall start from a rather elementary consideration of integrated intensity. We usually use a small single crystal of about 0.1 mm size as specimen. The crystal is bathed in a wide incident beam. Then, the crystal is rotated over a sufficiently wide angular range. The aperture of the counter is also assumed to be sufficiently wide so that all diffracted beams due to a reflection plane can be collected in the counter. The diffuse scattering (for example TDS), however, is assumed to be eliminated by some technical considerations.

Under these circumstances (see Fig.1), the integrated intensity must be

$$R_g = \int_{-\infty}^{+\infty} d\theta \int_{C}^{D} I_g(\theta, X_g)\, dX_g \tag{1}$$

where $\theta$ is the glancing angle of the incident beam with respect to the net plane and $X_g$ is the coordinate perpendicular to the diffracted beam. Notice that the integration is two-fold; angular ($\theta$) and spatial ($X_g$).

### b. Kinematical Theory

If one assumes a single scattering, i.e. kinematical theory, one can write down $R_g$ per unit intensity immediately as

$$R_g^K = Q \cdot \bar{A} \cdot V \tag{2}$$

where the notations have the following meanings.

$Q$: The integrated intensity per unit volume.
$\bar{A}$: The mean absorption factor.
$V$: The crystal volume.

Their explicit formulae are given as follows.

$$Q = (r_c)^2 |CF|^2 \lambda^3 / v^2 \sin 2\theta_B \tag{3}$$

$$\bar{A} = V^{-1} \int_V \exp -\mu_o(s_o + s_g)\, dV \tag{4}$$

Since they are well-known and the notations used are standard, no further explanations are mentioned.

### c. Extinction Correction

If the product QV increases, no more the approximation of single scattering can be justified and all sorts of complications start. In any case, the integrated intensity of G-beam must be

$$R_g = \eta \cdot R_g^K \tag{5}$$

Here, $\eta$ is called extinction factor, which is less than unity, because the energy of the kinematical beams is transferred to O-beam in the direction of the incident beam. What we need to do is to evaluate the factor $\eta$ or more directly $R_g$ under the specified experimental conditions, taking multiple scatterings into account.

## §2. SECONDARY EXTINCTION THEORY

### a. Hamilton-Zachariasen's Equations

The conventional approach to treat multiple scatterings is so-called secondary extinction theory, which was initiated by Darwin (1922) and developed by Hamilton (1957) and Zachariasen (1967a,b). The intensities of O and G beams, $I_o$ and $I_g$ respectively, are assumed to satisfy the energy transfer equations

$$\frac{\partial I_o}{\partial s_o} = -(\mu_o + \sigma) I_o + \sigma I_g \tag{6a}$$

$$\frac{\partial I_g}{\partial s_g} = -(\mu_o + \sigma) I_g + \sigma I_o \tag{6b}$$

where $\mu_o$ is the normal absorption coefficient and $\sigma$ is the diffraction coefficient per unit length. As the result, $I_o$ and $I_g$ are the functions of the oblique coordinates $(s_o, s_g)$. Usually, they are also assumed to be dependent of the angle $\theta$ through the angular dependence of $\sigma$. This point will be discussed later on.

By the use of the transformations

$$I_o = J_o \exp - (\mu_o + \sigma)(s_o + s_g) \tag{7a}$$

$$I_g = J_g \exp - (\mu_o + \sigma)(s_o + s_g) \tag{7b}$$

one can rewrite eqs. (6a and b) as

$$\frac{\partial J_o}{\partial s_o} = \sigma J_g \quad , \quad \frac{\partial J_g}{\partial s_g} = \sigma J_o \qquad (8a,b)$$

which are also equivalent to

$$\frac{\partial^2 J_o}{\partial s_o \partial s_g} = \sigma^2 J_o \quad , \quad \frac{\partial^2 J_g}{\partial s_o \partial s_g} = \sigma^2 J_g \qquad (9a,b)$$

The differential equations of these types are called hyperbolic.

The boundary conditions for solving eqs.(6) must be

$I_o = 1$ on ACB (The entrance surface of O-beam) (10a)

$I_g = 0$ on CBD ( " of G-beam) (10b)

For parallel slab crystals, Werner et al (1965, 1966) gave the solutions both for the Laue and Bragg cases. A formal solution was discussed also by Zigan (1970) for convex bound crystals. Nevertheless, in spite of simplicity of the form, the complete solution of H-Z equations for crystals of arbitrary shape is still not yet available.

### b. The Fundamental Solution of H-Z Equations

We shall take up a special case that the crystal happens to be a parallelepiped with the edges along $s_o$ and $s_g$ directions, and an infinitesimally narrow beam impinges at the corner of the crystal (Fig.2). This solution is worthwhile to be mentioned because it plays the role of Green function in the hyperbolic differential equations of the type (6). The approach described here is arithmetic and useful to understand the optical meanings of multiple scatterings.

One can visualise the optical implication of H-Z equations by drawing zig-zag paths along $s_o$ and $s_g$ directions. For every infinitesimal interval "a", the O-beam attenuates by the factor $\{1-(\mu_o+\sigma)a\}$ due to the first term in the right of eq.(6a). Since $N=(s_o/a)$ intervals (cross points) are available for the events along a path of length $s_o$, the attenuation along $s_o$-direction must be

$$\{1-(\mu_o+\sigma)a\}^N = \sum_{n=0}^{N} \frac{N!}{n!(N-n)!} \{-(\mu_o+\sigma)a\}^n$$

$$\to \sum_{n=0}^{\infty} \frac{1}{n!} \{-(\mu_o + \sigma) s_o\}^n \qquad (a \to 0)$$

$$= \exp -(\mu_o + \sigma) s_o \tag{11}$$

The same argument can be applied to the path along $s_g$-direction. Therefore, the total attenuation factor must be

$$A = \exp -(\mu_o + \sigma)(s_o + s_g) \tag{12}$$

The similar argument can be applied also to the diffraction enhancement due to the second term of eqs.(6a and b). If the route possesses n white kink points, the intensity will increase by the factor

$$\sum_n \frac{1}{n!} (\sigma s_g)^n.$$

Similarly, the intensity increases due to (m+1) black kinks by the factor

$$\sigma \sum_m \frac{1}{m!} (\sigma s_o)^m .$$

Here, the last kink point is fixed after fixing other m points, it is treated separately. However, obviously, these two factors are not independent and $n = m$ must be satisfied, so that the total enhancement factor must be

$$E = \sigma \sum_n \frac{1}{n!} \frac{1}{n!} (\sigma^2 s_o s_g)$$

$$= \sigma I_o(2\sigma \sqrt{s_o s_g}) \tag{13}$$

where $I_o$ is the modified Bessel function of the zeroth-order. Therefore, one obtains the total diffracted beam for the present special case

$$i_g = A \cdot E$$

$$= \sigma \exp -(\mu_o + \sigma)(s_o + s_g)\, I_o\, (2\sigma \sqrt{s_o s_g}) \tag{14a}$$

Similarly, one can obtain the direct beam

$$i_o = \sigma \exp -(\mu_o + \sigma)(s_o + s_g) \sqrt{\frac{s_o}{s_g}}\, I_1\, (2\sigma \sqrt{s_o s_g}) \tag{14b}$$

where $I_1$ is the modified Bessel function of the first order. In fact, $i_o$ and $i_g$ satisfy the H-Z equations.

### c. Becker-Coppen's Formulae of the Integrated Intensity

Now, we shall consider a more realistic problem, namely a cylindrical crystal. If the incident point E and the observation point P locate as illustrated in Fig.3a, one need not bother with the crystal form at all. Therefore, the reasonable formula for the integrated intensity is

$$R_g^S = \int_B^A \int_D^C i_g(s_o, s_g)\, dX_o dX_g \qquad (15)$$

Here, meanwhile, we omit the integration over the angle θ. The coordinates $(s_o, s_g)$ must be regarded as the functions of $(X_o, X_g)$. Bearing in mind the relation

$$dX_o \cdot dX_g = \sin 2\theta_B\, dV \qquad (16)$$

eq.(15) can be written in the form

$$R_g^S = \sigma \sin 2\theta_B \int_V \exp -(\mu_o + \sigma)(s_o + s_g) \cdot I_o(2\sigma\sqrt{s_o s_g})\, dV \qquad (17)$$

This procedure has been suggested originally by Zachariazen, but with an erronious choice of $s_g$. Later on, Becker and Coppens revised this point and obtained the formula (17) (1974).

If one approximates the Bessel function $I_o$ by the first term ($I_o = 1$; $2\sigma\sqrt{s_o s_g} << 1$), one can see that eq.(17) is identical to the result of the kinematical theory(cf. eq.(2)), provided that $\sigma = Q / \sin 2\theta_B$ and $\mu_o >> \sigma$ can be assumed.

The solution (17), however, includes overestimation. If either one of the points E and P locates closely to the singular points A, B, C and D, the pertinent parallelepiped is cut off by the crystal (Fig.3b). This trouble may not be very serious, if the Bragg angle is sufficiently small. The extinction is significant only for low order reflection where the Bragg angle is small. Nevertheless, as mentioned above, we have not obtained complete analytical solution of H-Z equations for crystals of arbitrary shape.

## §3. THE FOUNDATION OF SECONDARY EXTINCTION THEORY

### a. The Wave Equation of Takagi-Taupin Type

Any optical theory must find its foundation on the wave equation. The H-Z equations are energy transfer equation so that they must be derived from a suitable wave equation. In the present problem, the equation of Takagi-Taupin type in the following form (Kato, 1973) is most convenient as the starting point.

$$\frac{\partial d_o}{\partial s_o} = i\kappa_{-g} \exp iG \cdot d_g \tag{18a}$$

$$\frac{\partial d_g}{\partial s_g} = i\kappa_g \exp -iG \cdot d_o \tag{18b}$$

Here, $\kappa_g$ is the diffraction amplitude per unit length; namely

$$\kappa_g = r_c \lambda C F_g / v \tag{19}$$

and G is the lattice phase given by $G = 2\pi(\underset{\sim}{g} \cdot \underset{\sim}{u})$ where $\underset{\sim}{g}$ is the reflection vector and $\underset{\sim}{u}$ is the displacement vector referred to a perfect lattice. If the lattice is perfect, G is constant throughout the crystal. For distorted crystals, however, it is a function of position through $\underset{\sim}{u}(\underset{\sim}{r})$.

T-T equations (18) are similar in mathematical structure to eqs.(8) which are equivalent to H-Z equations. Therefore, the solution must be the sum of the wavelets associated with the zig-zag paths in Fig. 2. Now, the beam intensity has to be read as the wave function. The black and white points imply the multiplication of $i\kappa_g \exp -iG$ and $i\kappa_{-g} \exp iG$, respectively. For perfect crystals, however, the phase factors of a pair of black and white points are cancelled out. The explicit form of the solution, therefore, for a sufficiently narrow incident wave is given by

$$d_g = [\, i\kappa_g \exp -iG \,]\, J_o\left(2(\kappa_g \kappa_{-g})^{1/2} \sqrt{s_o s_g}\,\right) \tag{20}$$

by the argument used in deriving eq.(13). Here, $J_o$ is the Bessel function of the zeroth order $[\, J_o(x) = I_o(ix) \,]$. Since no attenuation is expected due to diffraction, it is enough to consider only the photo-electric absorption, so that the absorption factor in intensity is

$$A = \exp -\mu_o(s_o + s_g) \tag{21}$$

and the diffracted intensity is given by

$$i_g = A|d_g|^2 \tag{22}$$

The result is exactly the same as that of spherical wave theory (Kato, 1961, 1968).

b. The Statistical Considerations in the Dynamical Theory

If the crystal is distorted, the role of the lattice phase is crucial. In any case, the amplitude and the intensity of the Bragg-reflected wave can be written in the following forms, respectively

$$D_g = \sum_R B_R \exp iQ_R \tag{23a}$$

$$|D_g|^2 = \sum_R |B_R|^2 + \sum_R \sum_{R'}{}' B_R B_{R'}^* \exp i(Q_R - Q_{R'}) \tag{23b}$$

where $B_R$ and $Q_R$ are the amplitude and the phase of each wavelet associated with a zig-zag route R. In addition, $\sum\sum'$ means the double sum omitting $R = R'$ terms and * indicates conjugate complex. The similar expression holds also for the direct wave.

The route R of the Bragg-reflected wave must have k+1 black kinks and k white kinks ($k = 0,1,\ldots.$)(Fig. 4). Therefore, one can see

$$B_R = (A/a)(i\kappa_g)^{k+1}(i\kappa_{-g})^k(a)^{2k+1} \tag{24}$$

where $(A/a)$ is the amplitude of the incident wave with width "a" which is equivalent to $A\delta(X_o)$. The phase factor also can be factorised as

$$\Phi_R \equiv \exp iQ_R = \phi_1 \cdot \phi_2^* \cdot \phi_3 \cdots\cdots\cdots \phi_{2k}^* \cdot \phi_{2k+1} \tag{25}$$

where $\phi_j$ is the individual lattice phase factor [ $\exp -iG$ ] at the j-th kink point.

When a lattice distortion is definitely given, the total phase $Q_R$ can be uniquely calculated. In this case, the calculation of $D_g$ with eq. (23a) is the final target in diffraction theory. The theory belongs to the category of <u>primary extinction theory</u>.

If, however, only the statistical nature of the lattice distortion is given, no way to calculate $D_g$ is conceivable. Then, one has to admit the statistical posturation that the observed intensity is an ensemble average of $|D_g|^2$ of (23b), i.e.,

$$I_{ob} = \sum_R |B_R|^2 + \sum_R \sum_{R'}{}' B_R B_{R'}^* \langle \exp i(Q_R - Q_{R'}) \rangle \tag{26}$$

The theory with this viewpoint is better to be called <u>secondary extinction theory</u>. With this view the distinction between primary and secondary extinctions is nothing to do with the severness of lattice distortion but the degree of information about lattice distortion. By this way, one can treat the intensities based on the wave functions and also get rid of the artificial model of "mosaic crystals" in extinction theories.

### c. The Correlation of the Lattice Phase Factors

In one extreme case (ideally imperfect crystals) the phase factors $\{\phi\}$ are independently random. Then, the double sum of eq.(26) can be omitted. In addition, as reducing the scale length "a" to zero in eq.(24), only the term specified by $k = 0$ remains as finite. The intensity $I_{ob}$, therefore, must be what is predicted by the kinematical theory. In another extreme case (ideally perfect crystals) where all $\{\phi\}$ are constant throughout the crystal, $I_{ob}$ is simply given by $(\Sigma B_R)(\Sigma B_{R'}^*)$. The sum $(\Sigma B_R)$ is already calculated in the form of eq.(20).

The real problem is to consider the intermediate cases. For simplicity, meanwhile, only the second order correlation of $\{\phi\}$ is taken into account. In general, it is written in the form

$$\langle \phi_p \cdot \phi_q^* \rangle = \langle \phi_p \rangle \langle \phi_q^* \rangle + \langle \partial\phi_p \cdot \partial\phi_q^* \rangle \tag{27}$$

where $\partial\phi = \phi - \langle\phi\rangle$, and p and q specify the kink positions. In most cases of lattice distortion, it is safe to assume $\langle\phi\rangle = 0$. Exceptional is the case of thermally vibrating perfect crystals, in which $\langle\phi\rangle = \exp -M$ is Debye-Waller factor. Omitting such special cases, we shall introduce the correlation function as

$$f(z) = \langle \phi_p \cdot \phi_q^* \rangle \tag{28}$$

where z is the distance between the positions p and q. By the definitions, one can see the following properties;

$$f(0) = 1, \qquad f(z \to \infty) = 0. \tag{29a,b}$$

Also, introduced are the various correlation lengths by

$$\tau_n = \int_0^\infty \{f(z)\}^n \, dz \qquad (30)^+$$

As mentioned above, our target is to obtain $I_{ob}$ (eq.26) in terms of the correlation length.

### d. The Fundamental Solution of $I_{ob}$

Here, we shall outline how to obtain the intensity $I_{ob}$ for a sufficiently narrow incident wave with a unit amplitude.

First of all, all kinks of the route R(R') are grouped into sequences of kink pairs (KP) and isolated kinks (IK). A sequence of KP bounded by IK forms a segment along $s_o$ or $s_g$ direction on a macroscopic scale. Since $\langle\phi\rangle = 0$ is assumed, every IK of the route R corresponds one by one to that of the route R'. The distance between the paired kinks in KP and IK must be less than $\tau$ in the order of magnitude. Otherwise, the total correlation $\langle \Phi_R \cdot \Phi_{R'}^* \rangle$ contains the factors $\langle \phi \rangle$ so that the combination of the wavelets specified by such routes will give no contribution to $I_{ob}$.

Thus, as illustrated in Fig.4, the significant beams are composed of the parallel segments of the routes R and R' and have a zig-zag form on macroscopic scale. This situation is the essence to convert the wave picture to the beam (intensity) picture.

Along a segment, the amplitude is multiplied by $(1 - \kappa_g\kappa_{-g}\tau a)$ per unit interval "a" owing to KP correlation. Similarly, the amplitude of the conjugate wave is multiplied by $(1 - \kappa_g^*\kappa_{-g}^*\tau a)$. By the similar argument used in deriving eq.(12), the attenuation factor for the total segments is given by

$$A = \exp -2\tau \mathrm{Re}(\kappa_g\kappa_{-g})(s_o + s_g) \qquad (31)$$

At the every IK of black type in the beam route, the intensity must be multiplied by a factor $2|\kappa_g|^2\tau a$. Similarly, at every IK of white type, the factor $2|\kappa_{-g}|^2\tau a$ is multiplied. According to the arguments in deriving eq.(13), therefore, the enhancement factor is given by

---

+ In the most parts of this article, however, we shall not explicitly specify n.

$$E = |\kappa_g|^2 \tau\, I_o(4|\kappa_g \kappa_{-g}| \tau \sqrt{s_o s_g}) \tag{32}$$

The similar procedures can be applied also to calculating $I_{ob}$ for the direct beam. Thus, finally, one obtains the fundamental solution of the observable intensity as

$$\langle i_o \rangle = \exp -2\tau_2 \mathrm{Re}(\kappa_g \kappa_{-g})(s_o + s_g)$$

$$\times\ |\kappa_g \kappa_{-g}| \sqrt{s_o/s_g}\ I_1\,(4\tau_2 |\kappa_g \kappa_{-g}| \sqrt{s_o s_g}\,) \tag{33a}$$

$$\langle i_g \rangle = \exp -2\tau_2 \mathrm{Re}(\kappa_g \kappa_{-g})(s_o + s_g)$$

$$\times\ |\kappa_g|^2\, I_o\,(4\tau_2 |\kappa_g \kappa_{-g}| \sqrt{s_o s_g}\,) \tag{33b}$$

Here, $\tau$ is replaced by $\tau_2$. This is required by taking into account properly the geometrical relations on the distances between the kinks of KP's and the pairs of IK's in a single segment. In connection with this, it must be mentioned that the factorisation

$$f(\Sigma z_i) = \Pi\, f(z_i) \tag{34}$$

is assumed in the above calculation. This approximation can be justified provided that the function f(z) decreases rapidly as increasing z.

## e. Remarks

First of all, the fundamental solution (33) satisfies the coupled differential equations

$$\frac{\partial \langle i_o \rangle}{\partial s_o} = -2\tau_2 \mathrm{Re}(\kappa_g \kappa_{-g}) \langle i_o \rangle + 2\tau_2 |\kappa_{-g}|^2 \langle i_g \rangle \tag{35a}$$

$$\frac{\partial \langle i_g \rangle}{\partial s_g} = -2\tau_2 \mathrm{Re}(\kappa_g \kappa_{-g}) \langle i_g \rangle + 2\tau_2 |\kappa_g|^2 \langle i_o \rangle \tag{35b}$$

These equations are identical to H-Z equations (6) in the mathematical structure. Thus, the energy transfer equations could be justified in the wave-optical sense. For arriving at this result, essential are the assumptions of (i) a sufficiently narrow incident wave and (ii) neglecting the higher order correlations of $\{\phi\}$ than the second order.

The assumption (i) is intimately related to the $\Theta$-independence of the coupling constants in eqs.(35). Since the narrow wave in real space implies a wide spread of the wave vector in Fourier space, the $\theta$-independence is a natural consequence of the assumption (i). Customarily, however, the assumption has been made for the incident beam to be monodirectional and for the coupling constant $\sigma$ of H-Z equations to be $\theta$-dependent. From the author's view, the monodirectional beam and the energy transfer equations satisfying a conservation rule [ $\frac{\partial I_o}{\partial s_o} + \frac{\partial I_g}{\partial s_g} = 0$ ] are conceptually contradictory. The narrower is the angular divergence of the incident beam, the more dissipates the beam energy from the original angular channel, so that the energy balance of O and G beams will be broken. The simple relation of the type (35) holds only when a sufficiently wide angular range is assumed at the beginning. For this reason, in fact, $\langle i_o \rangle$ and $\langle i_g \rangle$ of eqs.(35) should be interpreted as the angularly integrated intensity.

Of course, it is desirable to develop the theory in a way to describe the observable angular dependence of G-beams (rocking curve). For this purpose, one needs to start the incident wave which is angularly limited. Then, probably, the spatial correlation of the phase of the incident wave is crucial to the final result.

The present theory is a spherical wave theory. The plane wave theory is certainly worthy of consideration. As far as the integrated intensity is concerned, however, one can show that the both theories give the same result. The spherical wave theory is practical in the problem of extinction because the simple energy transfer equation can be used.

Regarding to the assumption (ii), one has to mention about the range of applicability of eq.(35). The assumption requires the density of KP and IK is less than $(1/\tau)$. On the other hand, the density must be larger than $(\tau|\kappa_g\kappa_{-g}|)$, for the contribution of the routes with the higher density to be neglected (see the power series of $I_o$ and $I_1$). For satisfying two conditions, required is the relation

$$\tau \ll |\kappa_g| = \Lambda \qquad \text{(extinction distance)} \tag{36}$$

This condition can be relaxed to the extent $\tau \simeq \Lambda$ by taking the higher order correlations of $\{\phi\}$ into account. Then, one can show that the coupling constants appearing in eqs.(35) must be multiplied by

$$R = 1 + (\tau_2)^{-1} \sum_{j=1}^{\infty} V_2^{(j+1)} (-\kappa_g \kappa_{-g})^j \tag{37}$$

where $V_2^{(j)}$, correlation volume, is defined for the 2j-th order correlation function. It is an extended concept of $\tau_2$ in a sense that $V_2^{(j)} = \tau_2$. The reduction factor R can be roughly interpreted by the so-called primary extinction within crystallites in the model of mosaic crystals. It is worth to note, however, that the present theory is free from such an artificial model and R must depend upon the degree of crystal perfection.

## §4. THE INTEGRATED INTENSITY $R_g$

To be exact and transparent, in this section, only the symmetrical Laue case of non-absorbing parallel-sided crystals are treated. Since the fundamental solution (33b) is independent upon the entrance point, the general expression (15) of $R_g^S$ can be reduced to an integral over the exit surface by setting $\int dX_o = 1$. Then, the integral is given by

$$R_g^S = K(L/\Lambda)\,\frac{\sinh 2(\tau/\Lambda)(T/\Lambda)}{2(\tau/\Lambda)(T/\Lambda)}\,\exp -2(\tau/\Lambda)(T/\Lambda) \tag{38}$$

where $K = \frac{\lambda}{\Lambda}(\sin 2\theta_B)^{-1}$ and $L = T/\cos\theta_B$. For comparison, the well-known expressions of the kinematical and dynamical theories are also presented with the same notations;

$$R_g^K = K(L/\Lambda) \sim \lambda^3|CF_g|^2T \tag{39}$$

$$R_g^D = \frac{1}{2}K\int_0^{2L/\Lambda} J_o(\rho)\,d\rho \tag{40a}$$

$$\rightarrow R_g^K \qquad (\text{for } (L/\Lambda) \rightarrow 0\ ) \tag{40b}$$

$$\rightarrow \frac{1}{2}K \qquad (\text{for } (L/\Lambda) \rightarrow \infty\ ) \tag{40c}$$

Fig.5 shows $R_g/K\cos\theta_B$ of eqs.(38), (39) and (40) versus two parameters $(\tau/\Lambda)$ and $(T/\Lambda)$. The kinematical theory holds in the range

$$(\tau/\Lambda)(T/\Lambda) \ll 1 \tag{41}$$

For a fixed value of $\tau/\Lambda$, $R_g^S$ increases monotonically up to the saturation value $\frac{1}{4}(K/\cos\theta_B)(\tau/\Lambda)$. On the other hand $R_g^D$ which is exact for $\tau \gtrsim T$ increases up to the value $\frac{1}{2}K$ with Pendellösung oscillation. Since $R_g^S < R_g^D$ is physically unacceptable, again one can conjecture that the applicability of eq.(38) is limited within the range (36) mentioned above. Nevertheless, if the reduction factor R which depends upon the crystal perfection ($\sim \tau/\Lambda$) is introduced, the applicability will be extended to the range $\tau \sim \Lambda$. It is interesting to note that the averaged curve of $R_g^D$ over the Pendellösung oscillation is very similar to $R_g^S$ for $\tau/\Lambda = \frac{1}{2}(\cos\theta_B)$ (see Fig.6).

In this lecture note, the fundamental problems of extinction were reviewed. The author hopes that such basic studies of "extinction cold" in pathological laboratories will connect with any reliable prescription of home doctors to crystallographers, with which they will enjoy their better life in crystallography.

## Literature

Becker, J.P. & Coppens, P. (1974). Acta Cryst. A30, 129-147.
Darwin, C.G. (1922). Phil. Mag. 43, 800-824.
Hamilton, W.C. (1957). Acta Cryst. 10, 629-634.
Kato, N. (1961). Acta Cryst. 14, 627-636.
Kato, N. (1968). J. Appl. Phys. 39, 2225-2230.
Kato, N. (1973). Z. Naturforsch. 28a, 604-609.
Werner, S.A. & Arrott, A. (1965). Phys. Rev. 140A, 675-686.
Werner, S.A., Arrott, A., King, J.S. & Kendrick, H. (1966). J. Appl. Phys. 37, 2343-2350.
Zachariasen, W.H. (1967a). Phys. Rev. Lett. 18, 195-196.
Zachariasen, W.H. (1967b). Acta Cryst. 23, 558-564.
Zigan, F. (1970). N. Jb. Miner. Monatsh., 374-384.

As to the details of the arguments presented in this lecture note, the following three papers are suggested to be referred to.

Kato, N. (1976a). Acta Cryst. A32, 453-457.
Kato, N. (1976b). Acta Cryst. A32, 458-466.
Kato, N. (1979). Acta Cryst. A35, in printing.

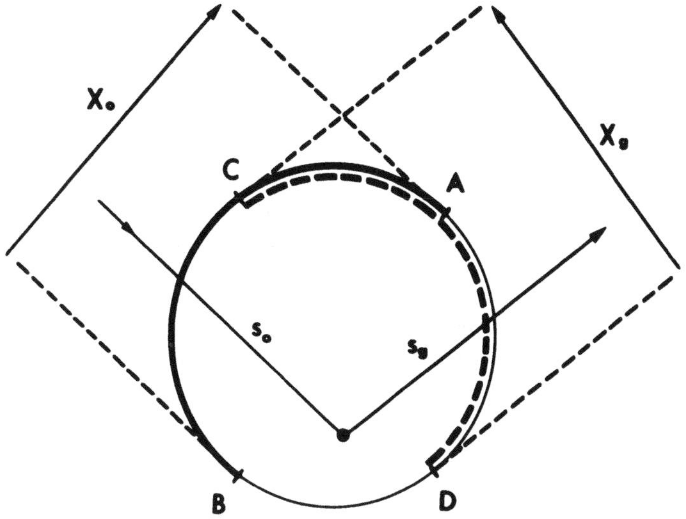


Fig.1 The diffraction from a finite crystal; arc ACB is the entrance surface of the incident beam and arc CAD is the exit surface of the diffracted beam.

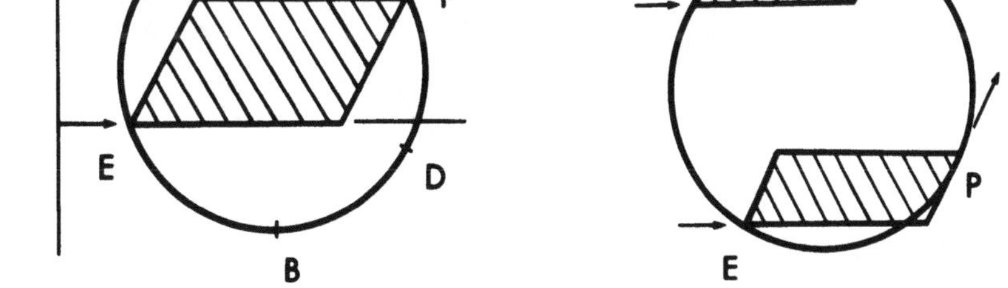

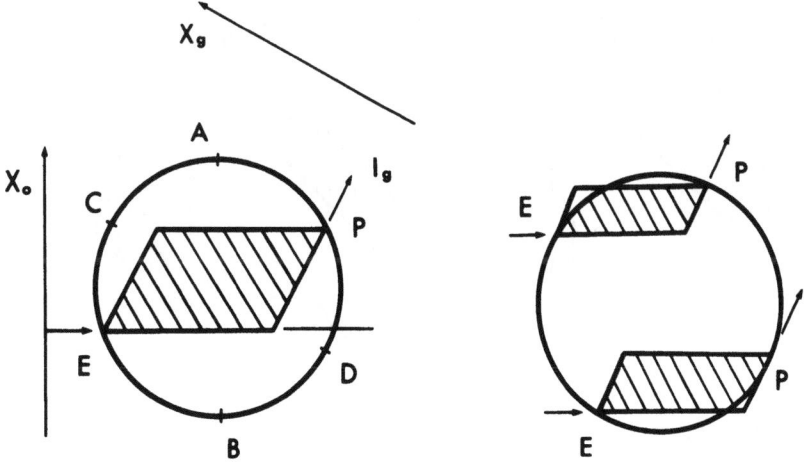

Fig.2 The optical meanings of H-Z equations.

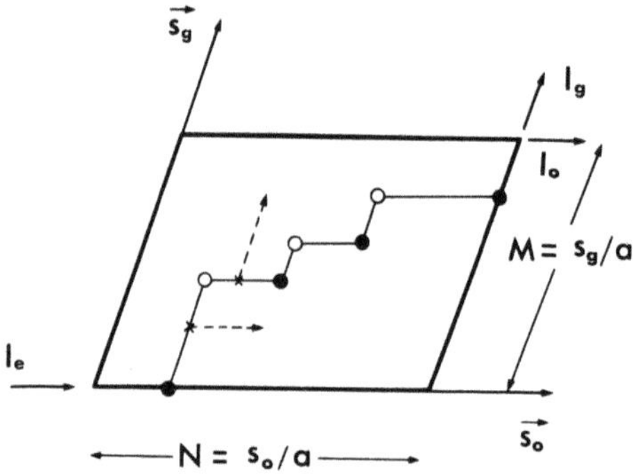


Fig. 3 The geometrical meaning of the approximation involved in Becker-Coppen's formula.

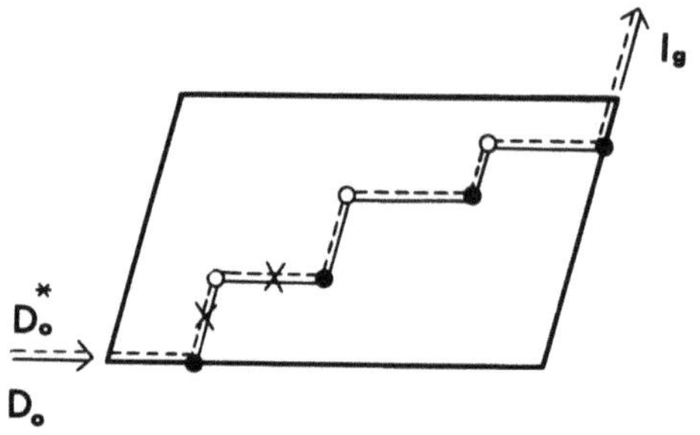


Fig. 4 The intensity (beam) picture based on the wave picture.

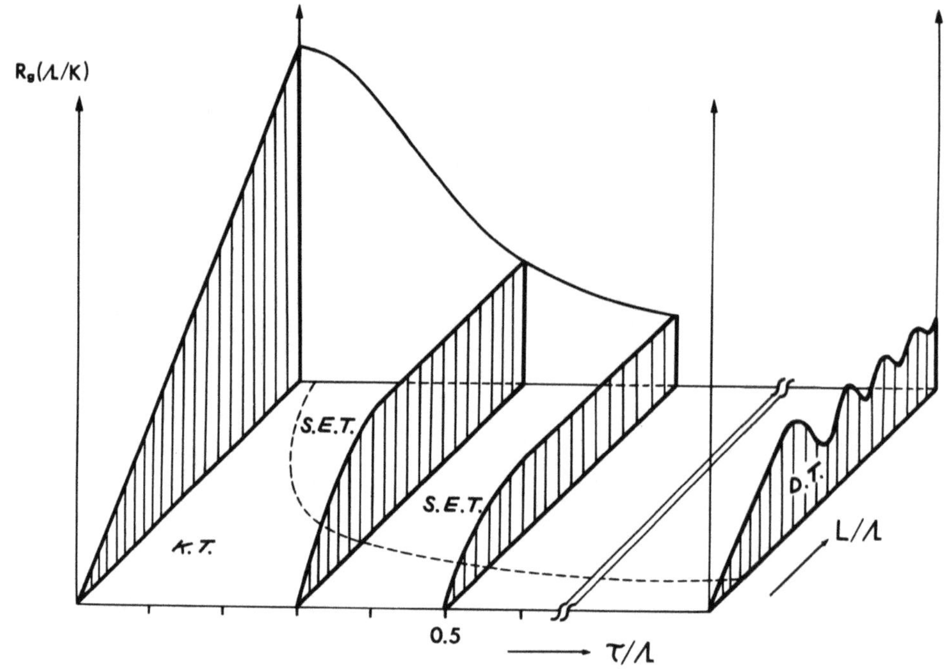


Fig. 5 Integrated intensity ($R_g \Lambda / K$) versus the crystal perfection ($\tau/\Lambda$) and crystal size ($T/\Lambda$) according to the kinematical theory (eq. 39), dynamical theory (eq. 40) and the secondary extinction theory (eq. 38).

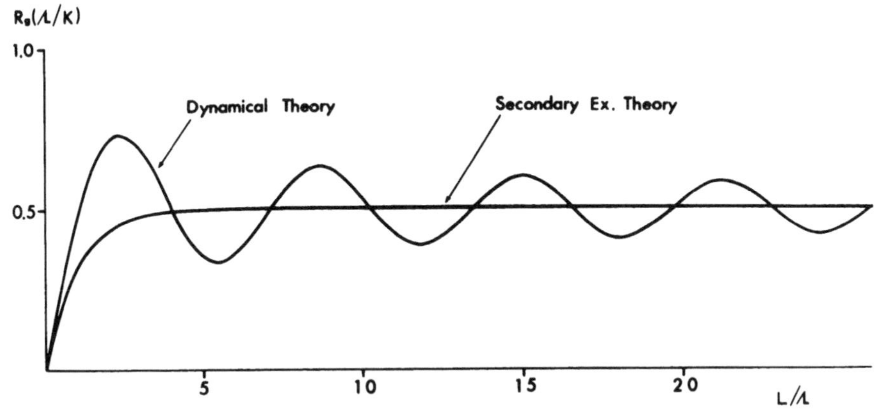


Fig.6 Comparison with the dynamical theory and the secondary extinction theory for $\tau/\Lambda = \frac{1}{2}(\cos\theta_B)$.

Problem : Obtain eq.(14b) in the same way as eq.(14a).

Solution: In the case of O-beam, the numbers of the black and white kinks must be identical. However, the last white point must be fixed and treated merely as a factor $\sigma$, when other kink points are fixed. Therefore, the enhancement factor is given by

$$E = \sigma\sum_{n=1}^{\infty} \frac{(\sigma s_o)^n}{n!} \frac{(\sigma s_g)^{n-1}}{(n-1)!}$$

$$= \sigma^2 s_o \sum_{n=0}^{\infty} \frac{(\sigma^2 s_o s_g)^n}{(n+1)!n!}$$

$$= \sigma\sqrt{\frac{s_o}{s_g}}\; I_1(2\sigma\sqrt{s_o s_g}\,)$$

As to the attenuation factor A, the same argument used in deriving eq.(13) can be applied. Thus, one can obtain eq.(14b).

# MAGNETIC NEUTRON SCATTERING

P.J. BROWN

Institut Laue-Langevin

Grenoble, France

## 1. BASIC THEORY

### a) The Born Approximation

In setting up a theory of scattering within the framework of quantum mechanics there are three basic ingredients. They are the incident wave, the scattered wave and the scatterer. Really only the incident wave and the scatterer are prerequisite. The scattered wave results from the interaction of these two and it may be better to consider scattering simply as the response of a radiation field to a non-uniform potential. In the case of interest in this lecture the radiation field is a beam of neutrons and the non-uniform potential is provided by the non-uniform magnetic field inside a magnetic material.

A popular technique for calculating scattering from a localised object such as an atom is the method of partial waves. In this technique the radiation field is expressed as the sum of the unperturbed field plus a development of the scattered wave in terms of functions based on the scattering centre with increasing values of angular momentum relative to that centre. If the scattering process is so weak that terms with non-zero angular momentum can be neglected, then the scattered wave has spherical symmetry with respect to the scattering centre. This is the s-wave or Born approximation. The essential requirement for the approximation to hold is that the perturbation of the incident wave by the scattered wave should be negligible at the level of accuracy required. Within the Born approximation the general equation for the differential scattering cross-section is :

$$\frac{\partial^2\sigma}{\partial\Omega\partial\epsilon} = \frac{k'}{k}\left(\frac{m_0}{2\pi\hbar}\right)^2 \left|\langle q'\sigma'|\int V(\underline{R})e^{i\underline{K}\cdot\underline{R}}dR^3|q\sigma\rangle\right|^2 \delta\left(\frac{\hbar^2}{2m_0}(k'^2-k^2)+E_{q'}-E_q\right) \quad (1)$$

The differential scattering cross-section is the scattering intensity into unit solid angle in unit energy range. The symbols k q $\sigma$ represent the incident neutron wave-vector and the initial states of the scattering system and the neutron, primed symbols represent the final states. V(R) is the potential of the neutron in the field of the scatterer ; K = k'-k the scattering vector and the final delta function ensures conservation of energy. To obtain the total cross-section one must calculate a cross-section for each initial state, summing over all accessible final states since normally the final state of the system is not determined. Finally the cross-section must be averaged over the distribution of initial states.

b) The Neutron's Interaction with Electrons

The variable part of the potential V(R) in a crystal is due to the interaction of the neutron with the atoms ; there are two major contributions to this interaction one of nuclear and the other of magnetic origin. In this lecture I consider mainly the magnetic part of the interaction.

The potential of a stationary neutron at the origin due to a single electron may be written :

$$4\gamma\mu_B\mu_N\left\{\frac{\underline{R}\times\underline{p}}{|R|^3} - \frac{\underline{s}}{|R|^3} + \frac{3(\underline{s}\cdot\underline{R})\underline{R}}{|R|^5} + \frac{8\pi\underline{s}}{3}\delta(\underline{R})\right\}\cdot S_N$$

where p and s are the momentum and spin operators for the electron and $S_N$ is the neutron spin operator. $\mu_B$ and $\mu_N$ are the Bohr magneton and nuclear magneton ; $\gamma$ is the magnetic moment of the neutron (-1.91) in nuclear magnetons.

The first term in equation 2 is due to the electron's orbital momentum, the second and third terms give the dipole-dipole interaction and the last term is the Fermi contact potential. This last term is given by the Dirac equations and ensures that the potential does not become infinite as R tends to zero. For a many electron system ; so long as the neutron velocity is small compared to that of the electrons ; we can replace $\underline{R}$ by $(\underline{R}-\underline{r}_i)$ p and s by $p_i$ and $s_i$ and sum over all electrons. $p_i$ $s_i$ and $r_i$ now represent the momentum, spin and position operators for the ith electron. On taking the Fourier transform of the potential in equation 2 the third and fourth terms cancel and so give no contribution to the scattering. It is rather easier to proceed from this point by recognising that the first two terms have somewhat the same form which can be recognised as potentials due to magnetic moments associated with the electron. In the first term the moment $\mu_B\sum_i(\underline{R}-\underline{r}_i)\times\underline{p}_i$ is due to the orbital motion of the electrons and in the second the moment which is simply $\mu_B \sum s_i$ is just due to their spin. If we introduce the concept of a magnetisation density which is the sum of moments due to spin and orbital motion then we can continue to deduce the scattering cross-section within the framework of a common formalism for spin and orbital moment. If we write the magnetisation density as $\underline{M}(\underline{r})$ then the vector potential due to such magnetisation is

$$\underline{A} = \int \frac{\underline{M}(r) \times (\underline{R} - \underline{r})}{|\underline{R} - \underline{r}|^3} \, dr^3$$

and the potential of the neutron is then

$$\gamma \mu_N S_N \cdot \text{Curl} A$$

Now one can express $\frac{\underline{R}}{|\underline{R}|^3} = -\nabla\left(\frac{1}{|R|}\right)$ and writing $\frac{1}{|\underline{R}|}$ as a Fourier integral $\frac{1}{|R|} = \int \frac{1}{2\pi^2 q^2} e^{iq \cdot \underline{R}} \, dq^3$

Hence the neutron potential becomes

$$V(\underline{R}) = 2\gamma \mu_N \underline{S}_N \cdot \int \text{Curl}\left(\frac{\underline{M}(\underline{r}) \times (\underline{R} - \underline{r})}{|\underline{R} - \underline{r}|^3}\right) dr^3$$

$$= 2\gamma \mu_N \underline{S}_N \cdot \int \nabla \times \underline{M}(r) \times \nabla\left(\int \frac{1}{2\pi^2 q^2} e^{i\underline{q} \cdot (\underline{R} - \underline{r})} dq^3\right) dr^3$$

$$= \frac{\gamma \mu_N \underline{S}_N}{\pi^2 q^2} \cdot \iint (\underline{q} \times M(\underline{r}) \times \underline{q}) e^{i\underline{q} \cdot (\underline{R} - \underline{r})} \, dq^3 \, dr^3$$

Now putting this potential into the Born approximation -equation 1 the matrix element involved becomes

$$\langle q'S' | \int V(R) e^{i\underline{K} \cdot \underline{R}} dR^3 | qS \rangle$$

$$= \langle q'S' | \frac{\gamma \mu_N \underline{S}_N}{\pi^2 q^2} \iiint (\underline{q} \times \underline{M}(\underline{r}) \times \underline{q}) e^{i\underline{q} \cdot (\underline{R} - \underline{r})} e^{iK \cdot R} dq^3 \, dR^3 \, dr^3 | qS \rangle$$

Now in this equation the integration over R will yield zero unless K = -q, so that after integration it reduces to

$$\langle q'S' | \gamma \mu_N \underline{S}_N \cdot \int (\hat{K} \times \underline{M}(\underline{r}) \times \hat{K}) e^{iK \cdot r} dr^3 | qS \rangle$$

In this equation $\hat{K}$ is a unit vector parallel to the scattering vector, and the result shows that the matrix element for neutron scattering is proportional to the projection onto the scattering plane (the plane perpendicular to K) of the Fourier transform of the magnetisation density. This result, that only the perpendicular components of magnetisation density can be measured in a scattering experiment is a direct consequence of Gauss' theorem, coupled with the fact that radiation is only scattered by variations in the scattering density. The scattering plane, which is defined by

$\underline{K}.\underline{r}$ = constant can be considered as the surface of a sphere of large radius to which Gauss' theorem must apply. We therefore can conclude that the total normal flux of magnetisation through this surface is zero and hence the resultant field projected along the scattering vector is zero and hence does not scatter into the reflection with wave-vector K.

### c) The Magnetic Structure Factor

If we now write the equation for the Born approximation for elastic scattering only this is

$$\frac{\partial\sigma}{\partial\Omega} = \left(\frac{m}{2\pi\hbar^2}\right)\left(\frac{\gamma e^2}{mc^2}\right)^2 \left|\langle q'S'|\underline{S}_N\cdot\int(\hat{K}\times\underline{M}(\underline{r})\times\hat{K})e^{i\underline{K}\cdot\underline{r}}dr^3|qS\rangle\right|^2$$

here the initial and final states of the system, represented by q are the same, but the spin state of the neutron, represented by S S' may change.

We may now usefully define a very general quantity which is closely related to the scattering cross-section and this is the Fourier transform of the magnetisation density for a particular quantum state of the system

$$\underline{M}(\underline{K}) = \langle q|\int M(r)e^{i\underline{K}\cdot\underline{r}}dr^3|q\rangle$$

This quantity is analogous to the chemical structure factor F(K) measured in an X-ray scattering experiment and can therefore be called the magnetic structure factor. One must note the added complexity introduced by the vector character of the magnetic interaction, $\underline{M}(\underline{K})$ is a vector each component of which may be complex.

### d) The Generalised Magnetic Interaction Vector

For the reasons just given the magnetic structure factor itself is not necessarily measurable in a scattering experiment. The scattering cross-section is proportional to

$$\left|\langle S'|S_N\cdot(\hat{K}\times\underline{M}(\underline{K})\times\hat{K})|S\rangle\right|^2$$

so it is useful to define the quantity

$$\underline{Q}(\underline{K}) = (\hat{K}\times\underline{M}(\underline{K})\times\hat{K})$$

which we call the magnetic interaction vector.

The only part of the interaction which remains to be evaluated is now that with the neutron spin. Taking the initial neutron spin direction parallel to a unit vector $\hat{\lambda}$, which we take as the z quantisation direction for the neutron, the cross-section for scattering without change of spin direction is given by $|\hat{\lambda}\cdot \underline{Q}(\underline{K})|^2$ and that for scattering with a change of spin direction by $|\hat{\lambda}\times \underline{Q}(\underline{K})|^2$ thus the sum over all final polarisations is just

$$\propto |Q(K)|^2$$

## 2. SIMPLE APPLICATIONS OF THE THEORY

I now want to use the concepts which we have just developed to calculate the scattering cross-sections for unpolarised neutrons of some simple magnetic crystals. We shall use the Heisenberg model for the magnetic system and make the following assumptions :

1 Atoms are non-overlapping
2 The ground state of the atom is an orbital singlet
3 There is no other orbital state close in energy
4 Within an atom the electron spins are all parallel to a single direction
5 Thermal motion can be accounted for by a Debye Waller factor

If assumptions 2 and 3 are valid then the magnetisation density arises from electron spin only and is given simply by the spin density

$$\underline{M}(\underline{r}) = \sum_i \sum_j \int \psi_i^*(r_1 s_1 \cdots r_j s_j \cdots)\, \tilde{S}\, \psi_i(r_1 s_1 \cdots r_j s_j \cdots)\, d\tau_j$$

where the integration is over all spin coordinates and over all space coordinates except these of the jth electron, the sum over i is over non-overlapping atoms.

If we define a scalar magnetisation density $m_i(r)$ associated with each atom, and define the magnitude and direction of the spin on that atom by a vector $S_i$ the total magnetisation density can be written :

$$\underline{M}(\underline{r}) = \sum_i \underline{S}_i\, m_i(r - r_i)$$

and the magnetic structure factor is

$$\underline{M}(K) = \int \sum_i \underline{S}_i\, m(\underline{r} - \underline{r}_i)\, e^{i\underline{K}\cdot\underline{r}}\, dr^3$$

which can be written $\sum_i m_i(\underline{K})\, \underline{S}_i\, e^{i\underline{K}\cdot\underline{r}_i}$

where $m_i(K) = \int m_i(r) e^{iK\cdot r} dr^3$ and is defined as the magnetic form factor.

a) Simple Ferromagnet

For a simple ferromagnet the spin direction is the same for all atoms and we can write the magnetic structure factor as

$$\underline{M}(\underline{K}) = \sum_{\text{Cells } \nu} \sum_{\substack{\text{Atoms in cell} \\ i}} m_i(K) \underline{S}_i e^{iK\cdot(\underline{R}_\nu + \underline{r}_i)}$$

where $R_\nu$ is now a vector from the origin to the origin of the $\nu$th unit cell and $r_i$ is the vector from the unit cell origin to the centre of the $i_{th}$ atom. Since $R_\nu$ is a lattice vector it can be written $n_1\underline{a} + n_2\underline{b} + n_3\underline{c} = R_\nu$ where $n_1$ $n_2$ $n_3$ are integers and $\underline{a}$ $\underline{b}$ $\underline{c}$ define the lattice. Because of this periodic property the sum over $\nu$ is zero unless $K = \tau$ where $\tau$ is a reciprocal lattice vector, and we have the well known result that for periodic structures the cross-section is only significantly non-zero when the scattering vector is a reciprocal lattice vector. We then have that for a simple ferromagnet

$$M(K) = N\delta(\underline{K} - \underline{\tau}) \sum_i m_i(K) S_i e^{i\underline{K}\cdot\underline{r}_i}$$

The result of the summation is the unit cell magnetic structure factor. Thus the scattering cross-section for a ferromagnet is proportional to

$$N^2 \delta(\bar{K} - \bar{\tau}) (\hat{K} \times \hat{\eta} \times \hat{K})^2 \left| \sum_i m_i(\underline{K}) \underline{S}_i e^{i\underline{K}\cdot\underline{r}_i} \right|^2$$

where now $\hat{\eta}$ is a unit vector parallel to the magnetisation direction and $S_i$ is the magnitude of the spin on the ith atom. The vector $(\hat{K} \times \hat{\eta} \times \hat{K})$ is sometimes called q and is the projection of the magnetisation direction onto the scattering plane.

If the magnetisation direction is parallel to the scattering vector $q = 0$ and no magnetic scattering occurs. If $\eta$ is perpendicular to $K, q = 1$ and $\underline{M}(\underline{K}) \equiv \underline{Q}(\underline{K})$. In ferromagnets in which the magnetisation direction can be changed by an applied field, a change in q from 1 to zero can be made by switching the field and may enable nuclear and magnetic scattering to be distinguished.

b) Antiferromagnets

There has been a variety of schemes proposed for classifying and describing antiferromagnetic structures. One method which has been quite widely adopted, and is convenient for the calculation of magnetic structure factors uses the propagation vectors of the spin ordering. A simple collinear antiferromagnetic structure can be defined by a single such propagation vector. This vector $\underline{\lambda}$ defines the spin directions of the magnetic atoms such that

$$\underline{S_i} = \underline{S_o}\, e^{i\underline{\lambda}\cdot(\underline{r_i}-\underline{r_o})}$$

if this is to have a simple interpretation $\underline{\lambda}\cdot(\underline{r_i}-\underline{r_o}) = 2\pi$ so that $S_i = \pm S_o$ for all $r_i$. The magnetisation density can then be written:

$$\underline{M}(\underline{r}) = \sum_{r_i} m_i(\underline{r}-\underline{r_i})\, \underline{S_o}\, e^{i\underline{\lambda}(\underline{r_i}-\underline{r_o})}$$

$$\underline{M}(\underline{K}) = \sum_{r_i} m_i(K)\, \underline{S_o}\, e^{i\underline{\lambda}(\underline{r_i}-\underline{r_o})}\, e^{i\underline{K}\cdot\underline{r_i}}$$

Now if we have a structure with magnetic atoms at distances $r_j$ from the unit cell origin then $\underline{r_i} = \underline{R_\nu} + \underline{r_j}$

$$\underline{M}(\underline{K}) = \sum_{\nu}\sum_{J} m_j(K)\, S_o\, e^{i\underline{\lambda}\cdot(\underline{R_\nu}+\underline{r_j}-\underline{r_o})}\, e^{iK\cdot(R_\nu + r_j)}$$

As usual the sum over $\nu$ gives zero unless the multiplier of $R_\nu$ is zero i.e. $\underline{\lambda}+\underline{K} = \tau$ a reciprocal lattice vector

$$\underline{M}(\underline{K}) = N\,\delta(\underline{K}+\underline{\lambda}-\tau)\, \underline{S_o} \sum_{J} m_j(K)\, e^{i(\underline{K}+\underline{\lambda})\cdot r_j}\left(e^{i\lambda\cdot r_o}\right)$$

and we can note that this is exactly the same as M(K) for a ferromagnet if $\lambda = 0$. It can be seen from the above that the propagation vector defines the positions in reciprocal space at which magnetic Bragg reflections can occur.

If we take the example of NiO the magnetic atoms are at the lattice points of a face-centred cubic cell in the magnetic structure magnetic atoms of opposite spin are separated by vector translations of (0 a/2 a/2) (a/2 0 a/2) (a/2 a/2 0). The propagation vector for this structure is $\underline{\lambda} = 2\pi/a(\tfrac{1}{2}, \tfrac{1}{2}, \tfrac{1}{2})$ this means that we expect magnetic reflections at positions separated by $2\pi/a(\tfrac{1}{2}, \tfrac{1}{2}\, \tfrac{1}{2})$ from the positions of the nuclear reflections. For the f.cc lattice these reflections are given by $2\pi/a(h, k, \ell)$ where hkl are integer and either all odd or all even.The magnetic reflections will correspond to indices on a doubled cell of h'k'l' where h'k'l' are either all of the form (4n+1) or all of the form (4n−1). The propagation vector lies along one of the triad axes of the originally cubic cell and selects this as a unique axis. The magnetic structure has therefore only rhombohedral symmetry. There are however four equivalent [111] axes in the nuclear structure and anyone of these may become the propagation axis of the magnetic structure ; four different configurational domains are therefore possible, each of which gives rise to a different set of magnetic reflections. For instance if the propagation vector is $2\pi/a(\frac{1}{2},-\frac{1}{2},\frac{1}{2})$ then magnetic reflections indexed on a doubled cell will have indices in which h and l are both 3n+1 with k 3n-1 or h and l are both 3n-1 with k 3n+1. Thus we see that knowledge of the propagation vector is sufficient to predict which magnetic vectors can occur. The intensity of these magnetic reflections will be strongly dependent on the orientation of the spin direction relative to the scattering vector because the cross-section is proportional to $|\hat{k} \times \underline{M}(\underline{K}) \times \hat{k}|^2$. If for the domain with [111] propagation vector the spin direction is parallel to [111] the (111) reflection will have zero intensity whereas if the spin lies in the (111) plane the 111 intensity will be large.

c) Incommensurable Structures

There is an increasingly large number of materials known in which the repeat distance of the magnetic structure is not an integral multiple of the chemical structure. The scattering from such incommensurable structures can be calculated using a simple extension of the concept of the propagation vector introduced above. I take as an example a case in which the spin direction is modulated as a function of its position in the lattice. We define a vector function $\underline{\eta}(\underline{r})$ which gives the magnitude and direction of the spin at each magnetic site in the crystal. Then the spin on the ith atom at distance $\underline{r}_i$ from the origin is given by $\underline{\eta}(\underline{r}_i)$. For an ordered structure $\underline{\eta}(r)$ must be periodic in r and we define the propagation vector $\underline{\lambda}$ such that $\underline{R}.\underline{\lambda} = 2\pi$ where $\underline{R}$ is the repeat distance of the magnetic structure. Because of its periodicity $\underline{\eta}(\underline{r})$ can be expressed as a Fourier sum $\underline{\eta}(\underline{r}) = \sum_n G(n) e^{in\underline{\lambda}\cdot\underline{r}}$

The magnetic structure factor -omitting the Debye-Waller factor is

$$\underline{M}(\underline{K}) = \sum_i m_i(K) \underline{S}_i e^{i\underline{K}\cdot\underline{r}_i} = \sum_i \sum_n m_i(K)\, G(n)\, e^{i(n\underline{\lambda}+\underline{K})\cdot\underline{r}}$$

Then as before we put in the periodicity of the nuclear structure

$$\underline{M}(\underline{K}) = \sum_n \sum_\nu \sum_j m_j(\underline{K})\, G(n)\, e^{i(n\underline{\lambda}+\underline{K})\cdot(R_\nu + r_j)}$$

again the sum over $\nu$ gives zero unless $(n\lambda + K) = 0$ so the result is similar to the previous one except that there is now a factor n multiplying $\lambda$ and a factor G(n) in the structure factor equation. This means that instead of a single magnetic reflection at distance $\lambda$ from each nuclear reflection there are now n "satellite" reflections at successive distances of $\lambda$. The number of such reflections is determined by the number of Fourier components needed to describe the function $\eta(r)$. We take the case of a simple conical structure in which the spin direction moves uniformly round the surface of a cone from atom to atom along the direction λ. The function η(r) in this case can be written

$$\underline{\eta}(\underline{r}) = \sin\phi\{\hat{q}\cos(\lambda\cdot r) + \hat{p}\sin(\lambda\cdot r)\} + \hat{t}\cos\phi$$

where $\underline{\hat{q}}$ $\underline{\hat{p}}$ $\underline{\hat{t}}$ are three orthogonal unit vectors not necessarily in the directions of the crystallographic axes. In this case $\underline{\hat{q}}$ and $\underline{\hat{p}}$ define the base plane of the cone and $\underline{\hat{t}}$ its axis ; φ is the cone semi-angle. The Fourier transform of this function is given by :

$$G(n) = \int \eta(r) e^{-\lambda\cdot r} dr^3$$
$$= \int \frac{\sin\phi}{2}\left\{(\hat{q} - i\hat{p}) e^{i\lambda(n-1)\cdot r} + (\hat{q} + i\hat{p}) e^{i\lambda(n+1)\cdot r}\right\} + \hat{t}\cos\phi\, e^{\lambda\cdot r}\, dr^3$$

the integral of the first term will be zero unless (n−1) = 0, the second unless (n+1) = 0 and the third is zero unless n = 0. Thus G(n) for the conical spiral is zero for all n except 0 ± 1. This means that in the diffraction pattern the magnetic scattering appears at the Bragg peaks (n = 0) and in pairs of satellites displaced by ± λ on each side of them. It should be noted that the scattering in the nuclear Bragg peaks is proportional to the axial component of the spiral which gives a ferromagnetic component in the magnetisation.

### d) Domains and Powders

It has been seen already how one kind of domain can occur in an antiferromagnetic structure. There are several other possibilities which have consequences for the calculation of the intensity of magnetic scattering. I take again the example of the nickel oxide type structure. If the spins do not lie parallel to the 111 propagation direction then there are a least three possible directions which will be equivalent at the Néel temperature. One will expect the antiferromagnetic crystal to contain a mixture of domains and the magnetic cross-sections measured will be the sum of contributions from different domains. It can be shown that because of this some information is lost about the absolute orientation of the moment. The same kind of effect is present when intensity data are collected from powdered samples. All reflections from a single form in the nuclear structure contribute to the same powder ring. If the spin direction of the magnetic structure is not a unique symmetry direction the magnetic interaction vector, and hence the intensity contributed, by different reflections to the ring will be different. To illustrate this point consider a structure in which the spin direction $\underline{\eta}$ is parallel to the zone axis uvw. The magnetic cross-section is proportional to :

$$|\,|M(\kappa)|\,\hat{\kappa}\times\hat{\eta}\times\hat{\kappa}\,|^2 = |M(\kappa)|^2\,(1-(\hat{\eta}\cdot\hat{\kappa})^2$$

$$= |M(hkl)|^2\left(1-\left(\frac{hu+kv+lw}{td}\right)^2\right)$$

for the hkl reflection. Here d is the length of the reciprocal lattice vector hkl and t is the length of the lattice vector uvw. The spin orientation dependence of the cross-section is contained in the term $1-\left(\frac{hu+kv+lw}{td}\right)^2$ which one may write as $\sin^2\alpha$ where $\cos\alpha = \frac{hu+kv+lw}{td}$ . It is this term which must be averaged over equivalent domains in a multidomain crystal, or equivalent reflections in a powdered sample. If the symmetry is orthorhombic the equivalent reflections are ± (hkl) ± ($h\bar{k}l$) ± ($\bar{h}kl$) so that the mean value of $\cos^2\alpha$ becomes $(h^2u^2+k^2v^2+l^2w^2)/t^2d^2$ and in principal the three components uvw of the magnetisation direction can be determined by measuring reflections with different hkl values. If the symmetry is tetragonal then in addition to the equivalent reflections given above there are ± (khl) etc. and the mean value of $\cos^2\alpha$ becomes: $\left(\frac{1}{2}(h^2+k^2)(u^2+v^2)+l^2w^2\right)/t^2d^2$

in this case u and v cannot be determined independently. Only the inclination of the magnetisation direction to the tetrad axis, given by $\tan^{-1}\sqrt{u^2+v^2/w^2}$ can be obtained. Extension of this argument shows that for all symmetries in which there is a unique axis, only the inclination of the spin direction to this axis can be determined. In cubic structures the mean value of $\cos^2\alpha$ is 1/3 independent of hkl so no information about the spin direction can be obtained from multidomain crystals and from powdered samples.

## 3. POLARISED NEUTRON SCATTERING

### a) The Cross-Section for Polarised Neutrons

The major interest in polarised neutron scattering comes from the interference between the nuclear and the magnetic scattering. To see how this arises we must add the neutron nuclear scattering to our expression for the cross-section. This can then be written :

$$\langle s'|\,N(\underline{\kappa}) + \tilde{S}\cdot\underline{Q}(\underline{\kappa})\,|s\rangle\langle s|\,N^*(\underline{\kappa})\cdot\tilde{S}^{\dagger}\cdot Q^*(\kappa)\,|s'\rangle$$

here as before S and S' represent the initial and final states of the neutron $\tilde{S}$ is the neutron spin operator and $\tilde{S}^{\dagger}$ its adjoint.

$N(\underline{K})$ represents the neutron nuclear structure factor which we presume for the moment is scalar and independent of the neutron spin direction. The expression above allows us to calculate the probability of scattering from one neutron state to another. For an arbitrary initial neutron spin direction $\hat{\lambda}$ the scattering without change of spin direction is given by :

$$|N(\underline{K})|^2 + |\hat{\lambda}\cdot\underline{Q}(\underline{K})|^2 + \hat{\lambda}\cdot N(\underline{K})\underline{Q}^*(\underline{K}) + N^*(K)\underline{Q}(\underline{K})\cdot\lambda$$

and the spin flip scattering by :

$$(\hat{\underline{\lambda}}\times\underline{Q}(\underline{K}))\cdot(\hat{\underline{\lambda}}\times\underline{Q}^*(K)) + \hat{\underline{\lambda}}\cdot(\underline{Q}(K)\times\underline{Q}^*(K))$$

the terms involving simple scalar products with $\hat{\lambda}$ are the polarisation dependent ones. Those in the non-spin-flip scattering are zero unless N(K) and Q(K) are simultaneously non-zero and also not in quadrature i.e. $N(\underline{K})\underline{Q}^*(\underline{K}) + N^*(\underline{K})\underline{Q}(\underline{K}) \neq 0$ . The term in the spin flip scattering is only finite when Q(K) is not parallel to $Q^*(K)$.

b) The Classical Polarised Neutron Technique

The first of these polarisation dependent terms is very well known as the nuclear magnetic interference scattering. It is used for the production of polarised neutron beams, and has been extensively exploited in the accurate determination of magnetic structure factors.

In the classical polarised neutron technique a magnetisable sample is magnetised in a direction parallel to the polarisation direction of a neutron beam and usually perpendicular to the plane of the incident and scattered beams. In such a case the magnetic interaction vector is parallel to the polarisation direction and the scattering cross-section is proportional to

$$|N(\underline{K})|^2 \pm 2N(\underline{K})\,Q(\underline{K}) + |Q(K)|^2$$

if we assume that both N(K) and Q(K) are real in this instance. The plus or minus corresponds to neutrons polarised parallel or antiparallel to $\underline{Q}(K)$. The quantity which is measured is the ratio between these two cross-sections. This "polarisation ratio" is given by

$$R = \frac{N(K)^2 + 2N(K)\,Q(K) + Q(K)^2}{N(K)^2 - 2N(K)\,Q(K) + Q(K)^2}$$

$$= \frac{1 + 2\gamma + \gamma^2}{1 - 2\gamma + \gamma^2} \quad \text{where} \quad \gamma = \frac{Q(K)}{N(K)}$$

solution for $\gamma$ gives

$$\gamma = \frac{(R+1) \pm \sqrt{4R}}{(R-1)}$$

In a real situation account must be taken of lack of complete polarisation of the up and down beams. It should be noted that the polarised neutron method allows one to determine the relative signs of magnetic and neutron scattering the uncertainty involved in the $\pm\sqrt{4R}$ corresponds to an uncertainty in whether $\gamma$ is greater or less than unity. In other words to an uncertainty in whether $\gamma = \frac{Q(\kappa)}{N(\kappa)}$ or $\frac{N(\kappa)}{Q(\kappa)}$. Note should also be taken of the special case where $\gamma << 1$ meaning that the magnetic scattering is very small. In this case $\gamma \simeq 1/4(R-1)$ a relationship which allows $\gamma$ to be measured with precision in cases where the change in the integrated intensity due to magnetic scattering would be negligible.

### c) Scattering with a Change of Polarisation

In reflections from antiferromagnetic structures to which nuclear scattering does not contribute the cross-section will in general not be polarisation dependent. Information about the orientation of the magnetic interaction vector cannot therefore be obtained from simple cross-section measurements. Such information can in principle be obtained by measuring the change of neutron polarisation in the scattering process. The simple interpretation of the two cross-sections given earlier is that the component of polarisation parallel to $\underline{Q}$ is not changed by the scattering process whereas that perpendicular to Q is reversed. A polarised neutron beam will only remain completely polarised after scattering by an antiferromagnet if the polarisation direction is parallel to Q. The term in $\hat{\lambda}\cdot\underline{Q}(\underline{\kappa})\times\underline{Q}^{*}(\kappa)$ which gives a polarisation dependent term in the antiferromagnetic cross-section is zero for most simple spin structures. It is non-zero for spiral and predicts that the intensity in the satellite reflections is polarised in the direction of the propagation vector.

## 4. EXTENSIONS OF THE SIMPLE THEORY

### a) Non-Localised Moments

In general the simple model of a magnetisation density confined to a single atom is not valid. The magnetic atoms are bound into a solid and the magnetisation will be transferred to some extent onto ligand atoms or perhaps into a conduction band. In such cases the concept of a magnetic form factor must be used with caution. Study of the delocalisation of magnetisation density is one of the most interesting aspects of magnetic scattering studies and will be covered more fully in future lectures.

b) Orbital Scattering

If the ground state of the magnetic ion is not an orbital singlet then there will be some contribution to the magnetisation from the orbital motion of the electrons. The orbital magnetisation density is given by the operator

$$\frac{1}{r}\int_r^\infty \hat{r} \times \underline{J_L}(\hat{r}x)\,dx$$

where $J_L$ is the operator for the orbital current density. This equation expresses the fact that the orbital magnetisation density at a point is not simply related to the electron density at that point, but has contributions from all electrons which complete orbits around that point. The orbital current density is given by :

$$J_L(\underset{\sim}{r}) = \frac{e}{2mc}\sum_i \int (\psi^* p_i \psi + \psi p_i \psi^*)\,d\tau_{-i}$$

$p_i$ is the electron momentum operator and the integration is over all space and spin coordinates except those of the ith electron.

From this it can be seen that the orbital contribution to the magnetisation density may be written :

$$M_L(r) = \frac{e\hbar}{2mc}\sum_i \int \psi^*\left\{\frac{1}{r}\int_0^\infty \tilde{L}_i \psi(\hat{r}_i x_i)\,dx_i\right\} + \psi\left\{\frac{1}{r}\int_r^\infty \tilde{L}_i \psi^*(\hat{r}_i x_i)\,dx_i\right\} d\tau_{-i}$$

For isolated magnetic ions the wave function can be written as a linear combination of products of n one-electron functions where n is the number of electrons (or holes) in the incomplete shell which leads to the magnetism. Each of the one electron functions has the form :

$$U(r_i)\,Y_m^\ell(\hat{r}_i)\,S_i$$

U(r) is the radial wave function and is to a first approximation the same for all electrons in the same shell ; $Y_m^1(\hat{r})$ is a spherical harmonic of order 1 (the electron angular momentum = 2 for d electrons) which describes the angular part of the wave function. The non zero terms in the orbital magnetisation density take the form

$$\frac{1}{2}\Big\{c^*(m_i s_i \cdots m_n s_n \cdots)\,c(m_j s_j \cdots m_n s_n \cdots)\,\delta(s_i s_j)\,\frac{1}{r}\int_0^\infty U^2(x)\,dx\,Y_{m_i}^{*\ell}(\hat{r})\,\tilde{L}\,Y_{m_j}^\ell(\hat{r})$$
$$+\,c(m_i s_i \cdots m_n s_n \cdots)\,c^*(m_j s_j \cdots m_n s_n \cdots)\,\delta(s_i s_j)\,\frac{1}{r}\int_0^\infty U^2(x)\,dx\,Y_{m_i}^\ell(\hat{r})\,\tilde{L}^+\,Y_{m_j}^{*\ell}(\hat{r})\Big\}$$

for comparison the spin part has the form :

$$2c^*(m_i s_i \cdots m_n s_n \cdots)\,c(m_j s_i \cdots m_n s_n \cdots)\,U^2(r)\,\langle s_i|\tilde{S}|s_j\rangle\,Y_{m_i}^{*\ell}(\hat{r})\,Y_{m_j}^\ell(\hat{r})$$

Thus each term in the magnetisation density consists of the product of two spherical harmonics of order 1 multiplying a radial function which is different for the spin and orbital contributions. It is usually convenient to transform the products of spherical harmonics into sums over single spherical harmonics in which case each component of the magnetisation density due to a single ion is given by a sum of the form :

$$M_i(r) = \sum_{i=x,y,z} \; \sum_{L=2\ell:2:0} \; \sum_{M=-L}^{L} \left\{ U^2(r)\, S_i(LM) + \frac{1}{r}\int_r^\infty U^2(x)\,dx \; L_i(LM) \right\} Y_M^L(\hat{r})$$

where the $S_i(LM)$ and $L_i(LM)$ are the sums of all contributions to the ith component of the magnetisation densities involving the spherical harmonic $Y_M^L(\hat{r})$. The magnetic structure factor is obtained from the Fourier transform of the magnetisation density by expanding $e^{ik.r}$ in spherical harmonics. After simplification the magnetic structure factor then becomes

$$M_i(k) = \sum_{i=xyz} \; \sum_{L=2\ell:2:0} \; \sum_{M=-L}^{L} i^L \left\{ S_i'(LM)\langle j_L(k)\rangle + L_i'(LM)\langle g_L(k)\rangle \right\} Y_M^{*L}(\hat{k})$$

where $\langle j_L(k)\rangle = \int_0^\infty j_L(kr)\, U^2(r)\, r^2 dr$ and $\langle g_L(k)\rangle = \int_0^\infty j_L(kr) \int_r^\infty U^2(x)\,dx \; r\,dr$

It has been shown that the integrals $\langle g_L(k)\rangle$ can be expressed as sums of the integrals $\langle j_L(k)\rangle$ and in particular that

$$\langle g_0(k)\rangle = \langle j_0(k)\rangle + \langle j_2(k)\rangle$$

which gives the dipole approximation to the orbital form factor. The whole problem of spin and orbital scattering from ions may be treated in a different way (see Marshall and Lovesey) using Racah algebra, but the result is the same. The magnetic interaction vector may be written as a sum of spherical harmonics functions multiplying radial integrals, the coefficients in the series must be calculated from the ionic wave function. Thus :

$$Q_i(k) = \sum_{i=xyz} \; \sum_{L=2\ell:2:0} \; \sum_{M=-L}^{L} \left\{ S_i'(LM)\langle j_L(k)\rangle + \sum_{Q=L:2:0} L'(LMQ)\langle j_Q(k)\rangle \right\} \cdot Y_M^{*L}(\hat{k})$$

c) Small Polarisation Dependent Effects

The spin and orbital scattering just considered give the main contributions the polarisation dependent cross-sections. However when measuring very small magnetisations using the polarised neutron technique there are a number of small effects which may become significant.

i) Nuclear polarisation. The alignment of nuclear spins may not be negligible if measurements are made with high applied fields at low temperatures. Nuclear polarisation will give additional polarisation dependent terms in the cross-section.

ii) Diamagnetism. Diamagnetic moment can contribute to the magnetisation density in much the same way as orbital magnetisation. When subject to high magnetic fields the diamagnetic moments associated with the atom cores may make a significant contribution to the magnetisation density. It can be shown (Stassis 1970) that this diamagnetic scattering has the form

$$m_d(K) \propto -\frac{Ze^2}{2mc^2}\,\frac{e}{\hbar c}\,\frac{1}{K}\,\frac{\partial}{\partial K}(f(K))\;\hat{\underset{\sim}{K}}\times\underset{\sim}{H}\times\hat{\underset{\sim}{K}}$$

where Z is the atomic number and f(K) is the X-ray form factor.

iii) Schwinger effect. This effect which is due to a neutron spin-neutron orbit interaction has been demonstrated in the scattering of slow neutrons by Vanadium. In this spin-orbit interaction the moving neutron magnetic moment senses the nuclear charge and this produces a small asymmetric scattering potential across the scattering atom. The scattering due to this interaction gives an imaginary scattering amplitude because of the asymmetry and this combines with the imaginary component of the nuclear scattering amplitude to give a polarisation dependent effect. The strength of the interaction is measured by $\frac{1}{2}(e^2/mc^2)Z(1-f(K))b'\mu_N$ where again f(K) is the X-ray form factor and b' is the imaginary part of the nuclear scattering length.

## 4. CONCLUSION

In a single lecture it is not possible to give more than the barest outline of the theory of magnetic neutron scattering. I hope that I have managed to cover enough here to give some idea of the possibilities and the pit-falls associated with magnetic neutron scattering. I hope also that these notes will provide a sufficient introduction to topics which will be covered more fully by other contributors to this course.

I have not given any references in the text since what I have given there is a synthesis of many different approaches to the description of magnetic scattering phenomena. I append a short reading list from which those interested may obtain more detailed information on subjects I have covered briefly here.

## 5. READING LIST

### a) General Theory

Halpern O. and Johnson M.H. (1939) Phys. Rev. 55 898-923
Marshall W. and Lovesey S.W. "Theory of Thermal neutron scattering" Oxford (1971)

b) Scattering by Ordered Structures

Izyunov and Ozerov, "Magnetic Neutron Diffraction" Plenum NY (1970)
Wilkinson C. (1975), Acta Cryst. A31 856-7
Shirane G. (1959), Acta Cryst. 12 282-5

c) Polarised Neutrons

Nathans R., Shull C.G., Shirane G. and Andresen A. (1959) J. Phys. Chem. Solids 10 138-146
Blume M. (1963), Phys. Rev. 130 1670-6
Brown P.J. and Forsyth J.B. (1964), Brit. J. Appl. Phys. 15, 1529-1533

d) Orbital Scattering

Trammel G.T. (1953), Phys. Rev. 92 1387-93
Johnston D.F. (1966), Proc. Phys. Soc. 88 37-52
Brown P.J., Welford P.J. and Forsyth J.B. (1973), J. Phys. C 6 1405-1421

e) Small effects

Shull C.G. (1963), Phys. Rev. Letters 10 297-8
Stassis C. (1970), Phys. Rev. Letters 24 1415-6

## 6. EXAMPLES

### Exercise 1

a) Four different kinds of magnetic ordering of a face centred cubic lattice have been described. The positions of oppositely oriented moments for one configuration domain are represented by black and white circles in the figure below.

For each structure

i) Identify the propagation vector.

ii) Find the configurational symmetry.

iii) Determine how many different configuration domains can exist and give the propogation vector for each.

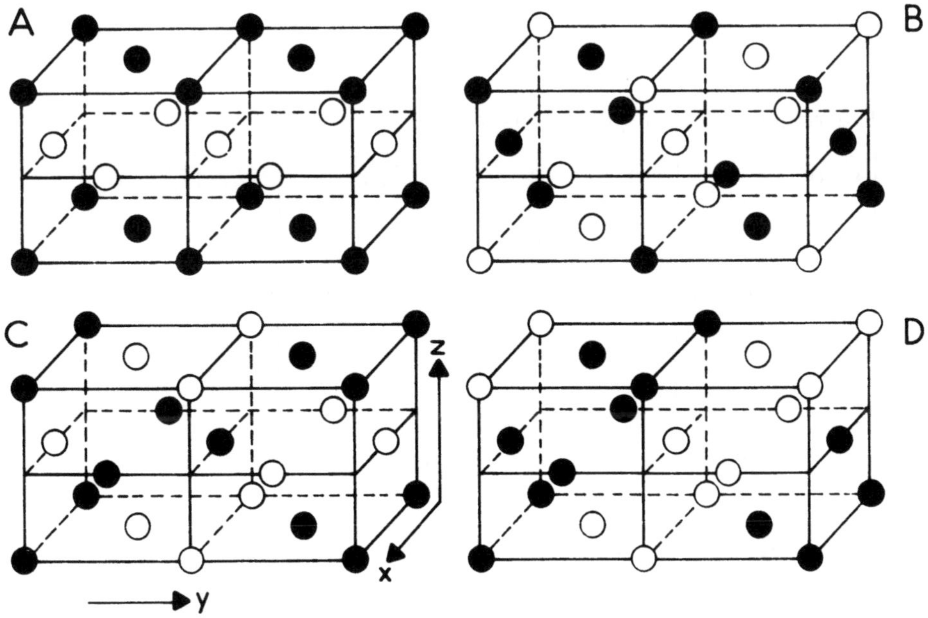


Four kinds of Magnetic Ordering found in face-centered cubic lattices

iv) Find the conditions on hkℓ for magnetic reflections from each of the domains.

v) What information about the orientation of moments in the structures could be obtained from powder diffraction patterns ?

b) In many orthorhombic structures, magnetically ordered ions occupy the positions 000, $^1/_2$ $^1/_2$ 0, 00$^1/_2$, $^1/_2$ $^1/_2$ $^1/_2$. If the magnetic cell is the same size as the chemical cell, determine the different ordering modes that are possible and identify the associated propagation vectors. What are the conditions on hkℓ for magnetic reflections to occur.

If the space group for a particular structure is Cmcm, which of the possible modes will give reflections with a polarisation dependent cross-section ? If for a particular mode a polarisation dependent cross-section occurs, identify the two types of 180° domain which will give opposite effects.

## Exercise 1 : Solution

A) Ordering of the first kind

From the diagram, planes on which the spin is the same are seen to be parallel to (001) $\therefore$ k = 001.

The magnetic atom positions are

| | | | | |
|---|---|---|---|---|
| + | 0 0 0 | 0 1 0 | $^1/_2$ $^1/_2$ 0 | $^1/_2$ $1^1/_2$ 0 |
| - | $^1/_2$ 0 $^1/_2$ | 0 $^1/_2$ $^1/_2$ | $^1/_2$ 1 $^1/_2$ | 0 $1^1/_2$ $^1/_2$ |

Check by calculating 2πk.r for each magnetic atom

| | | | | |
|---|---|---|---|---|
| + | 0 | 0 | 0 | 0 |
| - | π | π | π | π |

So Cos (2πk.r) = 1 for + spin and - 1 for - spin.

The k-vector is parallel to a tetrad axis of the cubic lattice so the configurational symmetry is tetragonal.

There are 3 equivalent $\{100\}$ directions in the cubic lattice hence 3 different configuration domains. Their k-vectors are (100), (010), (001).

Magnetic reflections will occur at $\pm$ k from nuclear reflections and the nuclear reflections only occur when h, k, ℓ, are all even or all odd. Therefore the conditions for magnetic reflections are

| | | | | |
|---|---|---|---|---|
| (100) | domain | k and ℓ | even | h odd |
| | | k and ℓ | odd | h even |
| (010) | domain | h and ℓ | even | k odd |
| | | h and ℓ | odd | k even |
| (001) | domain | h and k | even | ℓ odd |
| | | h and k | odd | ℓ even |

B) Ordering of the second kind

The planes on which the spin is the same are parallel to $(1\ 1\ \bar{1})$ so k is also parallel to $(1\ 1\ \bar{1})$.

The magnetic atom positions are

| | | | |
|---|---|---|---|
| + 0 0 0 | 0 $^1/_2$ $^1/_2$ | $^1/_2$ 0 $^1/_2$ | $^1/_2$ $1^1/_2$ 0 |
| - 0 1 0 | $^1/_2$ $^1/_2$ 0 | $^1/_2$ 1 $^1/_2$ | 0 $1^1/_2$ $^1/_2$ |

Calculate $2\pi\underline{k}.\underline{r}$ for each atom with $\underline{k} = (1\ 1\ \bar{1})$

| | | | |
|---|---|---|---|
| 0 | 0 | 0 | 0 |
| $2\pi$ | $2\pi$ | $2\pi$ | $2\pi$ |

to get $2\pi\underline{k}.\underline{r} = \pi$ for the planes of negative spin we must divide by 2 so $\underline{k} = (^1/_2,\ ^1/_2,\ -\ ^1/_2)$.

The k-vector is parallel to a triad axis of the cubic lattice so the configurational symmetry is trigonal. There are 4 equivalent $\{1\ 1\ 1\}$ directions in the cubic lattice and therefore 4 different configuration domains. Their k-vectors are : $(^1/_2\ ^1/_2\ ^1/_2)$, $(^1/_2\ ^1/_2\ -^1/_2)$, $(^1/_2\ -^1/_2\ ^1/_2)$, $(-^1/_2\ ^1/_2\ ^1/_2)$.

The conditions for magnetic reflections are

| | | | |
|---|---|---|---|
| $(^1/_2\ \ ^1/_2\ \ ^1/_2)$ | domain | $(h + ^1/_2,\ k + ^1/_2,\ \ell + ^1/_2)$ | |
| | | $(h - ^1/_2,\ k - ^1/_2,\ \ell - ^1/_2)$ | |
| $(^1/_2\ \ ^1/_2\ -^1/_2)$ | " | $(h + ^1/_2,\ k + ^1/_2,\ \ell - ^1/_2)$ | |
| | | $(h - ^1/_2,\ k - ^1/_2,\ \ell + ^1/_2)$ | with hkℓ all |
| $(^1/_2\ -^1/_2\ \ ^1/_2)$ | " | $(h + ^1/_2,\ k - ^1/_2,\ \ell + ^1/_2)$ | even or all odd |
| | | $(h - ^1/_2,\ k + ^1/_2,\ \ell - ^1/_2)$ | |
| $(-^1/_2\ \ ^1/_2\ \ ^1/_2)$ | " | $(h - ^1/_2,\ k + ^1/_2,\ \ell + ^1/_2)$ | |
| | | $(h + ^1/_2,\ k - ^1/_2,\ \ell - ^1/_2)$ | |

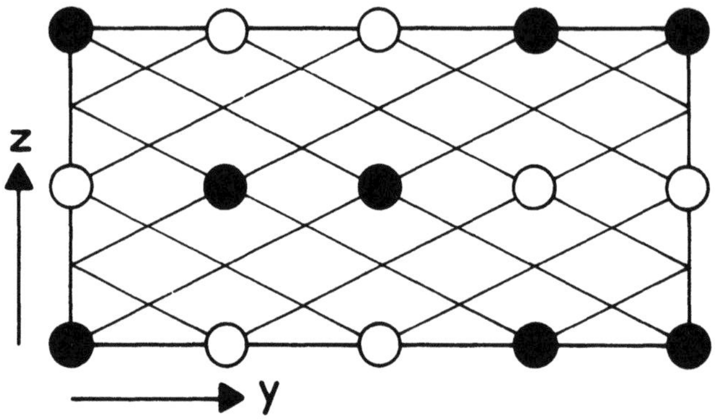


Projection of the structure C on (100) showing planes of constant spin

C) Ordering of the third kind

Projection of the structure onto (100).

The planes on which the spin is the same are parallel to (012) and $(0\bar{1}2)$.

This corresponds to two simultaneous k-vectors parallel to (012) and $(0\bar{1}2)$.

The magnetic atom positions are

| | | | |
|---|---|---|---|
| + 0 0 0 | 0 $^1/_2$ $^1/_2$ | $^1/_2$ 1 $^1/_2$ | $^1/_2$ $1^1/_2$ 0 |
| - 0 1 0 | $^1/_2$ $^1/_2$ 0 | $^1/_2$ 0 $^1/_2$ | 0 $1^1/_2$ $^1/_2$ |

Calculate $2\pi k.r$ for each atom

with k = (012)

| | | | |
|---|---|---|---|
| + 0 | $3\pi$ | $4\pi$ | $3\pi$ |
| - $2\pi$ | $\pi$ | $2\pi$ | $5\pi$ |

with k = $(0\bar{1}2)$

| | | | |
|---|---|---|---|
| + 0 | $\pi$ | 0 | $-3\pi$ |
| - $2\pi$ | $-\pi$ | $2\pi$ | $-\pi$ |

The phases and signs corresponding to atoms on successive planes are

| | | | | | |
|---|---|---|---|---|---|
| Sign | + | - | - | + | + |
| Phase for k = (0 1 2) | 0 | $\pi$ | $2\pi$ | $3\pi$ | $4\pi$ |
| k = (0 $^1/_2$ 1) | 0 | $\pi/2$ | $\pi$ | $3\pi/2$ | $2\pi$ |
| Sign | + | - | - | + | + |
| Phase for k = 0 $\bar{1}$ 2 | $-3\pi$ | $-2\pi$ | $-\pi$ | 0 | $\pi$ |
| k = (0-$^1/_2$ 1) | $-3\pi/2$ | $-\pi$ | $-\pi/2$ | 0 | $\pi/2$ |

We write $\underline{S} = \underline{So}\ \mathrm{Cos}2\pi(k.r + \phi)$ where $\phi$ is a phase shift then it $\phi = \pi/_4$  $S = So/2$ at all atoms with positive spin and $S = - So/2$ at all atoms with negative spin.

The two simultaneous k-vectors are related by the mirror plane parallel to (010) and are parallel to the (100) mirror plane. These two planes are therefore retained in the configurational symmetry which is thus orthorhombic. There are 24 equivalent vectors of the form $\{0\ h\ 2h\}$, two are needed to describe each domain and opposed vectors are not distinct so there are 6 distinct configuration domains. The pairs of k-vectors for the domains are

| | | |
|---|---|---|
| (1) | (0 $^1/_2$ 1) | (0 $-^1/_2$ 1) |
| (2) | (0 1 $^1/_2$) | (0 1 $-^1/_2$) |
| (3) | ($^1/_2$ 0 1) | ($-^1/_2$ 0 1) |
| (4) | (1 0 $^1/_2$) | (1 0 $-^1/_2$) |
| (5) | ($^1/_2$ 1 0) | ($-^1/_2$ 1 0) |
| (6) | (1 $^1/_2$ 0) | (1 $-^1/_2$ 0) |

Conditions for magnetic reflections

| | | |
|---|---|---|
| Domain (1) | $(h, k+{}^1/_2, \ell)$ | with h and k odd $\ell$ even |
| | $(h, k-{}^1/_2, \ell)$ | or h and k even $\ell$ odd |
| (2) | $(h, k, \ell+{}^1/_2)$ | with h and $\ell$ odd k even |
| | $(h, k, \ell-{}^1/_2)$ | or h and $\ell$ even $\ell$ odd |
| (3) | $(h+{}^1/_2, k, \ell)$ | with h and k even $\ell$ odd |
| | $(h-{}^1/_2, k, \ell)$ | or h and k odd $\ell$ even |
| (4) | $(h, k, \ell+{}^1/_2)$ | with k and $\ell$ even h odd |
| | $(h, k, \ell-{}^1/_2)$ | or k and $\ell$ odd h even |
| (5) | $(h+{}^1/_2, k, \ell)$ | with h and $\ell$ odd k even |
| | $(h-{}^1/_2, k, \ell)$ | or h and $\ell$ even k odd |
| (6) | $(h, k+{}^1/_2, \ell)$ | with k and $\ell$ even h odd |
| | $(h, k-{}^1/_2, \ell)$ | or h and $\ell$ odd h even |

D) Ordering of the 4th kind

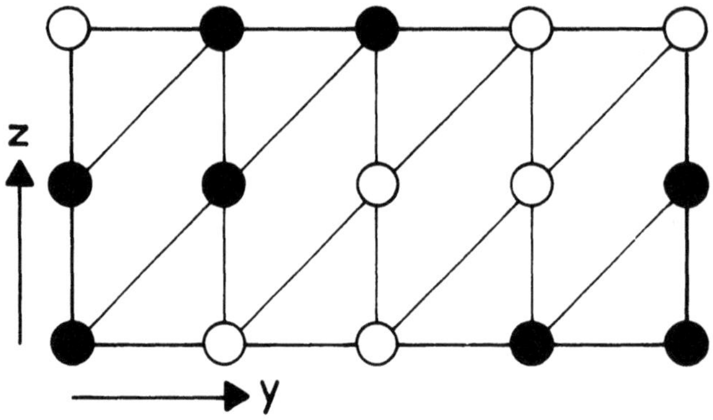


Projection of the structure D onto (100) showing planes of constant spin

The planes on which the spin is constant are parallel to $(01\bar{1})$ hence the k-vector is parallel to $(01\bar{1})$.

The positions of magnetic atoms are

| | | | |
|---|---|---|---|
| + 0 0 0 | $\frac{1}{2}$ 0 $\frac{1}{2}$ | 0 $\frac{1}{2}$ $\frac{1}{2}$ | $\frac{1}{2}$ $1\frac{1}{2}$ 0 |
| - 0 0 1 | $\frac{1}{2}$ $\frac{1}{2}$ 0 | $\frac{1}{2}$ 1 $\frac{1}{2}$ | 0 $1\frac{1}{2}$ $\frac{1}{2}$ |

Calculate $2\pi k.r$ for each atom

with $k = (01\bar{1})$

| | | | |
|---|---|---|---|
| + 0 | $-\pi$ | 0 | $3\pi$ |
| - $2\pi$ | $\pi$ | $\pi$ | $2\pi$ |

but a phase shift of $2\pi$ should give an atom of equivalent sign

therefore $k = (0\ \frac{1}{2}\ -\frac{1}{2})$ which gives for the phases

| | | | |
|---|---|---|---|
| + 0 | $-\pi/2$ | 0 | $3\pi/2$ |
| - $\pi$ | $\pi/2$ | $\pi/2$ | $\pi$ |

The phases and signs corresponding to successive planes are now

| Sign | + | + | - | - | + |
|---|---|---|---|---|---|
| | $-\pi/2$ | 0 | $\pi/2$ | $\pi$ | $3\pi/2$ |

The equation for the spin direction is therefore as in the previous example and the phase shift $\phi = \pi/4$.

$(01\bar{1})$ is a diad axis of the cubic lattice and also lies in a mirror plane. The configurational symmetry is therefore orthorhombic.

There are 12 equivalent $\{011\}$ vectors and therefore 6 distinct configuration domains. The k-vectors and magnetic reflections are

| | | | | | |
|---|---|---|---|---|---|
| (1) k-vector | $^1/_2$ | $^1/_2$ | 0 | $(h+^1/_2,\ k+^1/_2,\ \ell\ )$<br>$(h-^1/_2,\ k-^1/_2,\ \ell\ )$ | |
| (2) k-vector | $-^1/_2$ | $^1/_2$ | 0 | $(h-^1/_2,\ k+^1/_2,\ \ell\ )$<br>$(h+^1/_2,\ k-^1/_2,\ \ell\ )$ | |
| (3) k-vector | $^1/_2$ | 0 | $^1/_2$ | $(h+^1/_2,\ k\ ,\ \ell+^1/_2)$<br>$(h-^1/_2,\ k\ ,\ \ell-^1/_2)$ | with $hk\ell$ all even or all odd |
| (4) k-vector | $^1/_2$ | 0 | $-^1/_2$ | $(h+^1/_2,\ k\ ,\ \ell-^1/_2)$<br>$(h-^1/_2,\ k\ ,\ \ell+^1/_2)$ | |
| (5) k-vector | 0 | $^1/_2$ | $^1/_2$ | $(h\ ,\ k+^1/_2,\ \ell+^1/_2)$<br>$(h\ ,\ k-^1/_2,\ \ell-^1/_2)$ | |
| (6) k-vector | 0 | $^1/_2$ | $-^1/_2$ | $(h\ ,\ k+^1/_2,\ \ell-^1/_2)$<br>$(h\ ,\ k-^1/_2,\ \ell+^1/_2)$ | |

Information obtainable from powder measurements

A) Ordering of the first kind : the configurational symmetry is tetragonal.

The magnitude of the moment and its inclination to the k-vector can be determined.

B) Ordering of the second kind : is again uniaxial so only the magnitude and inclination of the moment to the k-vector can be determined.

C) Ordering of the third kind : the configurational symmetry is orthorhombic so all three components of the moment can be determined.

D) Ordering of the fourth kind : as for ordering of the third kind.

b)

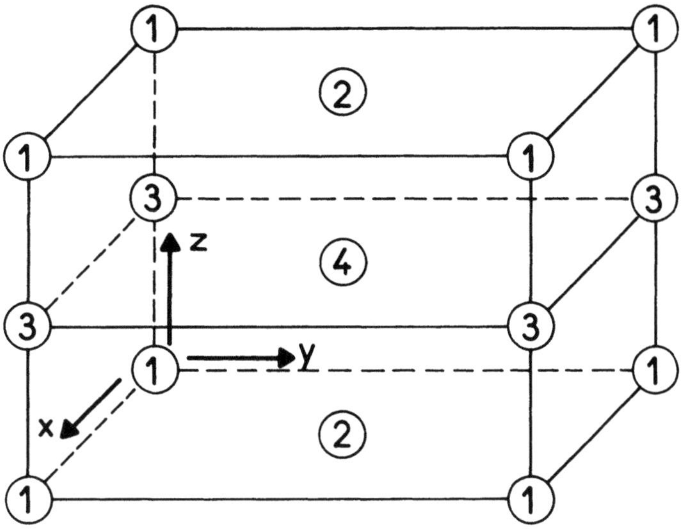


Positions of magnetic atoms in the orthorhombic structure of Exercise 1 b)

Let $S_1$ $S_2$ $S_3$ $S_4$ represent the moments on the atoms 1, 2, 3, 4 respectively in the figure. Then the possible ordering schemes may be described by the table given below.

| Atom<br>r | 1<br>(0,0,0) | 2<br>1/2,1/2,0 | 3<br>(0,0,1/2) | 4<br>(1/2,1/2,1/2) | |
|---|---|---|---|---|---|
| | $S_1$ | $S_2$ | $S_3$ | $S_4$ | k vector |
| A | + | + | − | − | 001 |
| B | + | − | + | − | 010 |
| C | + | − | − | + | 011 and $0\bar{1}1$ |
| F | + | + | + | + | 000 |

Schemes A, B, C are antiferromagnetic and scheme F is ferromagnetic.

Nuclear reflections from the magnetic lattice are restricted to reflections with h + k even and ℓ even. Since magnetic reflections are displaced by $\pm$ k from these the conditions for magnetic reflections are

| | | |
|---|---|---|
| A | k = 001 | h + k even ℓ odd |
| B | k = 010 | h + k odd ℓ even |
| C | k = (011) and $(0\bar{1}1)$ | h + k odd ℓ odd |
| F | k = 000 | h and k even ℓ even |

If the space group is Cmcm then the conditions on reflections from the nuclear structure are for general hkℓ h + k even
for h0ℓ reflections h and ℓ even

Mixed magnetic and nuclear reflections are hkℓ reflections in structure A with ℓ odd. The domains giving opposite effects correspond to the moment arrangements

| | $S_1$ | $S_2$ | $S_3$ | $S_4$ |
|---|---|---|---|---|
| | + | + | − | − |
| and | − | − | + | + |

Exercise 2

Polarisation effects in $MnF_2$

$MnF_2$ has the rutile structure, space group $P4_2/mnm$. The manganese atoms are at positions 000, $^1/_2$ $^1/_2$ $^1/_2$, and the fluorine atoms at positions $\pm$ (x x o) $(^1/_2+x, ^1/_2-x, ^1/_2)$

a = 4.873 Å c = 3.010 Å x = 0.310

In the antiferromagnetic phase stable below 73K the $Mn^{2+}$ ions have their spins aligned parallel to [001], those at 000 being oppositely orientated to those at $^1/_2$ $^1/_2$ $^1/_2$.

In a polarised neutron experiment carried out at 4.2K it was observed that the polarization ratios for the 210 and $1\bar{2}0$ reflections were respectively 7.04 and 0.1345. The polarization of the incident beam was known to be 0.96. Use these data to determine the fraction of the crystal which is of the predominant 180° domain type and find the neutron flipping efficiency of the diffractometer.

In the same experiment the values of polarization ratio given in the table below were measured for reflections of the form hh0. Use these results, together with the information deduced previously, to calculate a one-dimensional projection on [110] of the magnetisation density giving rise to these 'forbidden' reflections.

What can you deduce from those results about the spin transfer associated with the manganese fluorine interaction in $MnF_2$.

| h k ℓ | R |
|---|---|
| 1 1 0 | 0.8650 ± .0069 |
| 2 2 0 | 1.0134 ± .0040 |
| 3 3 0 | 0.9780 ± .0050 |
| 4 4 0 | 0.9896 ± .0060 |
| 5 5 0 | 1.0070 ± .0060 |

Exercise 2 : Solution

For $MnF_2$ the nuclear structure factor can be written

$$FN = b_{Mn}(1 + \mathrm{Cos}2\pi((h + k + \ell)/2) + 2b_F(\mathrm{Cos}2\pi(h + k)x + 2\mathrm{Cos}2\pi((h + k)/2 + (h - k)x)$$

and $b_{Mn} = -.373$ $\qquad x = 0.310$

$b_F = .566$

In the approximation of spherical symmetry the magnetic structure factor is

$$F_M = \pm\mu_{Mn} f_{Mn} (1 - \mathrm{Cos}2\pi((h + k + \ell)/2)$$

Where the + sign corresponds to the domain with positive moment on the atom at (0, 0, 0) and the - sign to the reverse domain.

If we define a domain ratio $\alpha$ such that $\alpha = \dfrac{f^+ - f^-}{f^+ + f^-}$ where $f^+$ and $f^-$ are the volume fractions of the crystal containing the positive and negative domains respectively.

The scattered intensity in the 210 reflection from the positive domain for spin up neutrons is proportional to

$$I^+f^+ = f^+(F_N + F_M)^2$$

and that from the negative domain to $I^-f^- = f^-(F_N - F_M)^2$.

The total number of spin up neutrons scattered is proportional to

$$(1 + P)\left[(1 + \alpha)I^{+} + (1 - \alpha)I^{-}\right] + (1 - P)\left[(1 - \alpha)I^{+} + (1 + \alpha)I^{-}\right]$$

and the number scattered when the polarisation is reversed to

$$(1+P)(1+\varepsilon)\left[(1+\alpha)I^{-} + (1-\alpha)I^{+}\right] + (1-P)(1-\varepsilon)\left[(1-\alpha)I^{-} + (1+\alpha)I^{+}\right]$$

Here the polarisation P and flipping efficiency $\varepsilon$ are defined in the usual way.

The polarisation ratio is thus given by

$$R = \frac{(I^{+} + I^{-}) + P\alpha(I^{+} - I^{-})}{(I^{+} + I^{-}) - P\alpha\varepsilon(I^{+} - I^{-})}$$

Now $I^{+} = F_N{}^2 + F_M{}^2 + 2F_MF_N$ and $I^{-} = F_N{}^2 + F_M{}^2 - 2F_MF_N$

so that $R = \dfrac{F_N{}^2 + F_M{}^2 + 2P\alpha F_NF_M}{F_N{}^2 + F_M{}^2 - 2P\alpha\varepsilon F_NF_M} = \dfrac{1 + \gamma^2 + 2P\alpha\gamma}{1 + \gamma^2 - 2P\alpha\varepsilon\gamma}$ were $\gamma = \dfrac{F_M}{F_N}$

For the 210 reflection $F_N = 2b_F(\mathrm{Cos}2\pi(3x) + \mathrm{Cos}2\pi(^3/_2+x))$

$= 2b_F(\mathrm{Cos}6\pi x - \mathrm{Cos}2\pi x) = 1.4410 \times 10^{-12}$ cms

$F_M = 2\mu_{Mn}f_{Mn}$

$\mu_{Mn} = 5\mu B = 5x.2715 \times 10^{-12}$ cms $\qquad f_{Mn} = 0.582$ at $\mathrm{Sin}\theta/\lambda = .2294$

Hence $F_M = 1.579 \times 10^{-12}$ cms and $\gamma = 1.096$

For the $(1\ \overline{2}\ 0)$ reflection $N = 2b_F(\mathrm{Cos}2\pi x - \mathrm{Cos}2\pi(-^1/_2+3x)) = -1.4410$

Hence $\gamma(1\ \overline{2}\ 0) = -\gamma(210)$

$$R(210) = \frac{1 + \gamma^2 + 2P\alpha\gamma}{1 + \gamma^2 - 2P\alpha\varepsilon\gamma} = 7.04 = R^{+} \qquad (1)$$

$$R(1\overline{2}0) = \frac{1 + \gamma^2 - 2P\alpha\gamma}{1 + \gamma^2 + 2P\alpha\varepsilon\gamma} = 0.1345 = R^{-} \qquad (2)$$

Equations (1) and (2) can be solved simultaneously for $\varepsilon$

and give $\varepsilon = \dfrac{(R^+ + R^- - 2)}{R^+ + R^- - 2R^+R^-} = 0.9878$

also $2P\alpha\gamma = \dfrac{(1+\gamma^2)(R^+ - 1)}{(R^+\varepsilon + 1)}$

from which using the value of $P = 0.96$ $\alpha = 0.794$

For reflections of the form (hh0) $F_N = 2b_{Mn} + 2b_F(1 + \text{Cos}4\pi(hx))$

Now $R = \dfrac{1 + \gamma^2 + 2P\alpha\gamma}{1 + \gamma^2 - 2P\alpha\varepsilon\gamma} \simeq 1 + 2P\alpha\gamma(1+\varepsilon)$ for $\gamma \ll 1$

in which case $\gamma \simeq \dfrac{R - 1}{2P\alpha(1+\varepsilon)}$ which in the present case $\dfrac{(R-1)}{3.03}$

Using the values of R given we therefore get

| hkℓ | $F_N$ | $\gamma$ | $F_M$ |
|---|---|---|---|
| 110 | - .439 | - 0.0445 | 0.0196 |
| 220 | .457 | 0.0044 | 0.0020 |
| 330 | 1.108 | - 0.0073 | - 0.0080 |
| 440 | - .737 | - 0.0034 | 0.0025 |
| 550 | 1.302 | 0.0023 | 0.0030 |

Fourier sum $\rho(r) = \frac{1}{V}\sum_k F(k)e^{ik.r}$

Projection of the density along R $\rho(R) = \int_{\perp R} dS \frac{1}{V} \sum_k e^{ik.r}$

Where the integral is taken over the planes perpendicular to R.

These integrals are zero if they contain an oscillating component, ie unless k is parallel to R.

In this case the value of the integral is simply the area of the cell perpendicular to R.

Hence $\rho(R) = \frac{1}{\ell} \sum_{k.R = 0} F(k)e^{ik.r}$ where $\ell$ is the cell length parallel to R.

In the present case R is parallel to [110] hence.

$$\rho(x)_{110} = \frac{2}{a\sqrt{2}} \sum_h F(hh0)\mathrm{Cos}4\pi hx$$

Only the cosine terms are needed since the projection is centro-symmetric.

From the equation above we see that $\rho(x)$ repeats with period of 0.5 and that $\rho(x) = \rho(0.5 - x)$ ; hence we need only calculate $\rho$ from x = 0 to x = 0.25.

| x | 0 | .025 | .050 | .075 | .100 | .125 | .150 | .175 | .200 | .225 | .250 |
|---|---|---|---|---|---|---|---|---|---|---|---|
| hh0 | $F(hh0)\mathrm{Cos}4\pi hx \times 10^4$ | | | | | | | | | | |
| 110 | 196 | 186 | 156 | 115 | 61 | 0 | - 61 | -115 | -156 | -186 | -196 |
| 220 | 20 | 16 | 6 | - 6 | - 16 | - 20 | - 16 | - 6 | 6 | 16 | 20 |
| 330 | - 80 | - 47 | 25 | 76 | 65 | 0 | - 65 | - 76 | - 25 | - 47 | 80 |
| 440 | 25 | 8 | - 20 | 20 | 8 | - 25 | - 8 | 20 | - 20 | 8 | 25 |
| 550 | 30 | 0 | - 30 | 0 | 30 | 0 | - 30 | 0 | 30 | 0 | - 30 |
| Σ | 191 | 163 | 137 | 205 | 132 | - 45 | -180 | -177 | -165 | -115 | -101 |

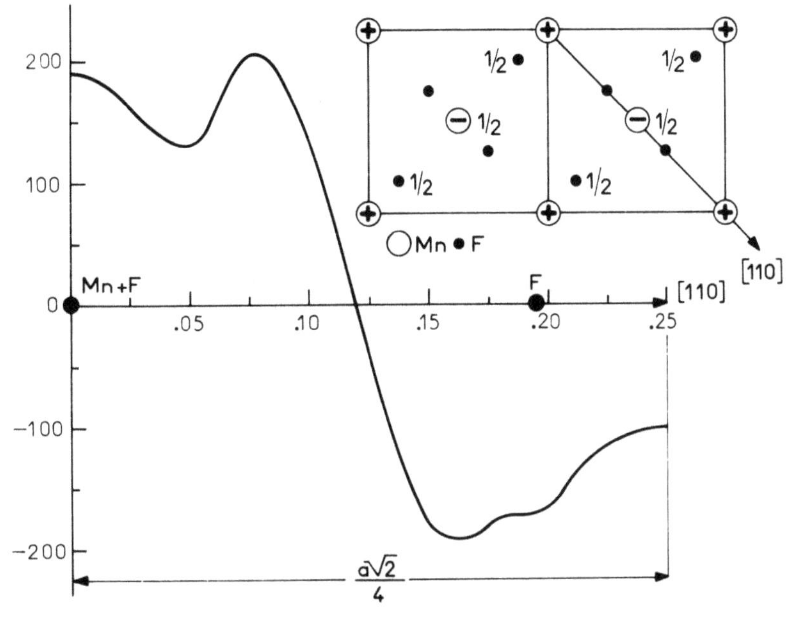


If we compare the projection of the density with the inset of the magnetic structure we see that the peak at the origin corresponds to the superposition of manganese ions of opposite spin and fluoride ions. The fluoride ions which project to the origin are those with two positive and one negative manganese neighbour. The other fluoride ions, those with one positive and two negative manganese neighbours project at x = .190 where the density is negative. Thus we may say that the density projection demonstrates the transfer of spin from manganese to fluorine by covalency.

# OPTIMAL EXPERIMENTAL CONDITIONS FOR CHARGE DENSITY DETERMINATION

M.S. LEHMANN

Institut Laue-Langevin, Grenoble, France

## INTRODUCTION - OPTIMAL CONDITIONS FOR AN EXPERIMENT

This chapter deals with the specific and well defined task : determining structure factor amplitudes from measurements of Bragg reflections on single crystals using X-ray or neutron radiation. It may however be worthwhile first to discuss in more general terms the optimal conditions for measuring any piece of information.

We can distinguish between two types of observations. On one side we have the experiment done once or a few times aimed at obtaining some quantity such as the velocity of light, the charge of the electron, the mass of the earth. On the other side we have nearly identical type measurements done over and over again on series of substances to determine their densities, to observe the phonon spectra arising from intermolecular interactions, to determine the molecular structure, etc. If we want to distinguish verbally the two types, we could characterize the first type as an experiment, whereas the second type is better described as a measurement indicating its nature of repetition and routine. Approximately the same rules for optimal conditions apply in the two cases, but we will concentrate on the second case.

In general we can assume that we have a standard set of equipment available, often of commercial origin, which we use frequently for the same type of measurements. Our first requirement should then be :

a) The experimental setup is stable, i.e. results obtained using the equipment should be insensitive to disturbances from the environment such as mechanical shocks and temperature variations.

The meaning of stable is two-fold. First we will require that the parts of the equipment that are not supposed to move in space or time do not move. It is for example of importance for an X-ray diffractometer that the X-ray beam tube holder is solid so that one does not risk a movement of the beam during the measurement. Secondly, however, one should aim at stabilising the measurements by making them insensitive to at least small changes in the equipment. If we again take the case of the beam we can by ensuring a sufficiently large, homogeneous beam guarantee that small variations in crystal centrering does not affect the measured reflected intensity (incidentally one should be able to misalign all parts of a measuring equipment and still obtain the same result). One additional problem that is always present is the long term stability. In most cases we can do nothing except monitor our results and correct for any drift. The only efficient way of reducing this problem is to reduce the measurement time when possible.

For the sample under study we must require :

b) The sample used for the measurement must be typical of the samples that are available or can be produced, and the results obtained using one sample should be indistinguishable from results obtained on any other sample from the same group.

This condition is applicable when we study intramolecular quantities (such as molecular electron density), as well as intermolecular quantities (such as stacking faults). In both cases we must require that other identical samples are available or can be produced in order for us if necessary to prove to ourself and others by repetition of the measurements that our observations are reproduceable and thus trustworthy.

The requirement of reproducibility leads directly to the following condition :

c) The measurement conditions should be well defined and constant, and they should be described completely in terms that are understood at least by persons working in the same field of research.

This condition should guarantee not only that one can repeat a series of measurements and obtain identical results to earlier observations, but the same must be possible for anybody else who wants to repeat the measurements. The task of describing well the experimental details for a measurement is formidable, and one should keep in mind - writing the experimental part of for example a diffraction paper - that the number of details given mainly reflects the care that was taken, but is no guarantee that the result is correct. There is for example normally not enough information to enable the reader to check all steps from the raw data to the final structure amplitudes.

Finally we must ensure that we know what we are observing, so a last condition is :
d) Theories must be available that describe what we are measuring.

In general we are not able to directly measure the quantity that we need, but we measure something that includes various side effects, which we must either correct for or minimize experimentally. It is therefore important to describe all that we measure with either parameters from the experiment or parameters obtained elsewhere.

When doing an experiment we have normally some limitations to the time we can spend on the measurements. There is as well a limit to how complex and complicated a series of operations we are willing to go through to obtain the results. To conclude the list of conditions it is therefore tempting to add that the experiment should be made so that it is short, or at least not too time consuming for us, that it is easy to carry out (this is often a purely technical problem), and that the number of corrections to be done afterwards is limited.

Because of limitations in space not every single point necessary for an optimal experiment can be discussed in detail, and the following considerations are aimed at giving a general idea of how to carry out a successful set of measurements. Recently other reviews of optimal experimental conditions have been given (Rees, 1977 ; Coppens, 1978) and the reader is referred to these papers for further references to work in this field.

## THE BEAM, ITS DISTRIBUTION IN REAL AND RECIPROCAL SPACE

The conventional diffraction geometry as we will discuss it here is described in figure 1. The scattering plane formed by the incoming and diffracted wavevectors $\vec{k}_0$ and $\vec{k}_1$ is normally coincident with the horizontal plane, and to bring the lattice point $\vec{\tau}$ through the scattering condition $\vec{k}_0 + \vec{\tau} = \vec{k}_1$ the crystal is rotated around an axis vertical to the scattering plane. During this process we assume that all parts of the crystal are within the beam and are illuminated by identical beams all with the same distribution of wavevectors (see figure 2), and by summing over the scattering for all orientations of the crystal during the scan we then obtain the integrated reflected intensity. An important underlying assumption for a reproduceable measurement is therefore that all parts of the beam have the same intensity and wavevector distribution.

We may also use a technique where we measure the intensity from plane parallel plates of crystals in transmission geometry with a beam smaller than the crystal plate size (see figure 2). This technique is often hard to use because of the difficulty in producing the crystal plates. Likewise when continuous X-ray radiation becomes generally available it might be more useful to obtain the integrated intensity by an integration over wavelength with fixed orientation of the crystal rather than as presently from a rotation of the crystal with fixed wavelength.

In any case the conditions of beam homogeneity remains the same. For neutron diffraction the condition is generally easily fulfilled with homogeneous beams of large cross sections. For X-rays it is more difficult to obtain an acceptable beam. In crystal-monochromatized beams the beam extension in the scattering plane of the monochromator is quite limited. This means that great care is required when crystals of irregular shape are used. Rees (1977) quotes that for a spherical crystal of 0.16 mm diameter a change of 2% was observed in the intensity when the crystal was shifted 0.09 mm. This is quite a large effect, and can be avoided by using filtered radiation or by placing a glass rod in the beam (Helmholdt and Vos, 1977 B). In the first case one is then however left with a worse peak-to-noise ratio as well as possible problems in data reduction. Whatever the choice is, and it could well be dictated by the accessible diffractometer, one must ensure that the beam is homogeneous, and this is quite easily checked by measurements of intensities of some reflections for a crystal displaced from the center of the diffractometer.

The incoming beam distribution is often produced in such a way that there is mirror symmetry with respect to the scattering plane. In this case we can reduce the discussion of the three-dimensional distribution of incoming wavevectors to a consideration of the distribution in the scattering plane, and then just keep in mind that the beam distribution has a component orthogonal to this plane, which will produce a vertical spot size of the reflected beam, but which will neither effect the scan width necessary to obtain the integrated intensity nor the relative movement of the diffracted beam during the scan. Further discussion will therefore be restricted to the scattering plane.

We can visualize the beam distribution in two ways, in an Ewald sphere construction or by considering the distribution of incoming wavevectors. As we assume the beam to be homogeneous our picture represents any part of the beam. The two ways of vizualization of the beam are shown in figure 3 for the special case of a beam obtained by crystal monochromatization of white radiation, as in the case of neutrons. Several authors have considered this case mainly because of the importance of understanding the various experimental aspects of neutron scattering (see for example Dachs, 1961 and Cooper and Nathans, 1968) and here we use expressions given by Nielsen and Møller (1969) for the lengths and

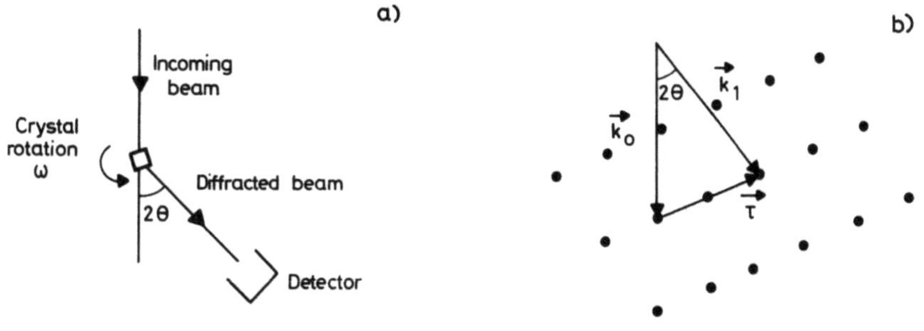


Figure 1a) - Beam diagram in a diffraction experiment. b) Scattering condition, $\vec{k}_o+\vec{\tau}=\vec{k}_1$. $\tau$ is a reciprocal lattice vector.

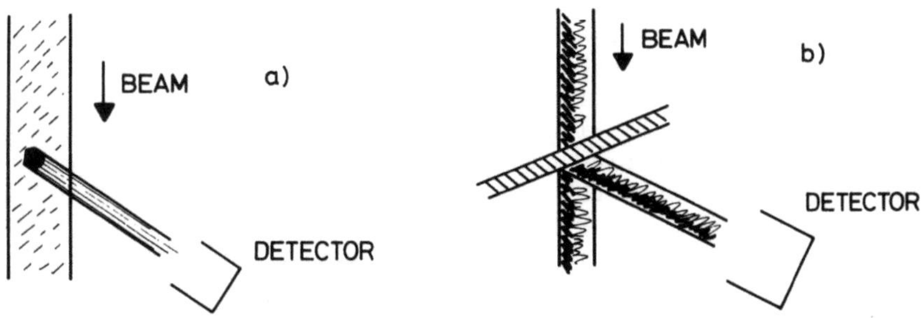


Figure 2a) - Beam larger than the crystal. Beam is assumed homogeneous. b) Crystal larger than beam in transmission geometry. Intensity distribution need not be homogeneous if the crystal is.

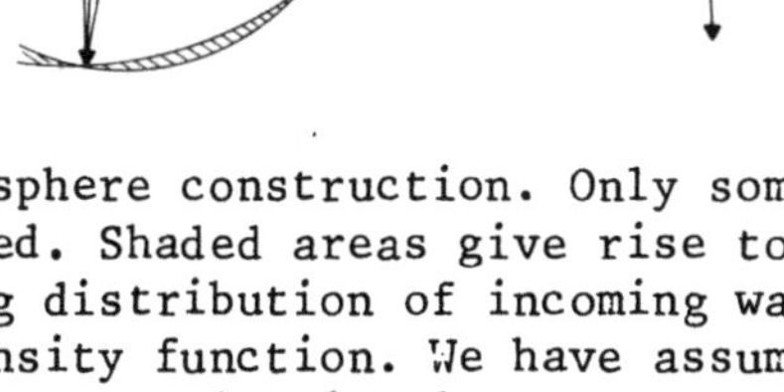

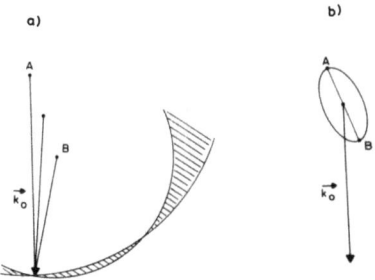

Figure 3a) - Ewald sphere construction. Only some of the limiting spheres are indicated. Shaded areas give rise to Bragg scattering. b) The corresponding distribution of incoming wavevectors given by a contour of the density function. We have assumed that all our functions determining the distribution are Gaussians, which then give rise to ellipsoidal contours. The size of the ellipse is much larger than found in a normal experiment.

orientations of the axis of the ellipse in figure 3b as a function of the monochromator mosaic and the various beam collimations. We note especially that the angle between the line AB and $\vec{k}_o$ is nearly equal to the monochromator angle $\theta_M$. For the X-ray case the situation is quite simple although the contours in this case might not be ellipses, as the wavevector distribution is not Gaussian. For crystal monochromatized radiation one main axis would be nearly along $\vec{k}_o$ with its length given by the wavelength spread $\Delta\lambda/\lambda$, and the other axis orthogonal to $\vec{k}_o$ would describe the angular divergence of the beam. The situation is similar for filtered radiation, and in this case we would always have two distinct distribution functions, one for the $\alpha_1$ radiation and one for the $\alpha_2$ radiation. The two distributions would be parallel but displaced with respect to each other along $\vec{k}_o$.

We will now follow the behaviour of the diffracted beam as we rotate the crystal through the diffraction condition, and we will first follow only the scattering from one small crystallite of the whole crystal (or similarly assume that the mosaic spread of the crystal is negligible). Figure 4 shows a situation during the scan. The condition for elastic scattering is that the incoming and diffracted wavevectors have the same size, so only wavevectors found on the line 1, which is orthogonal and bisecting to $\vec{\tau}$, can scatter. The part that gives rise to elastic scattering is therefore the intersection of the line 1 with the wavevector distribution. We can then calculate the relative scattering from this point, $P(\omega)$, which is proportional to the area of intersection. We can determine the direction of the diffracted beam which is a line from the midpoint of AB to C, and we can determine the angular width at half intensity of the diffracted beam which is $\angle$ ACB. Finally by rotating $\vec{\tau}$ around an axis orthogonal to the scattering plane we can determine the angle $\Delta\omega$, which brings the lattice point through reflection position (i.e. sweeps the line 1 through the ellipse), and this is indicated on figure 5.

Lebech and Nielsen (1975) have given an expression for $\Delta\omega$. They find, assuming all distributions involved to be Gaussian, and the mosaic to be negligible that

$$\Delta\omega = b \sin \beta / k_o \cos\theta$$

where the parameters are defined in figure 6. Intuitively we can see that $\Delta\omega$ will have a minimum when 1 is parallel to the longest axis of the ellipse. In the case shown in figure 3 this axis is at an angle $\theta_M$ to the incoming beam and $\Delta\omega$ will thus have a minimum for $\theta=\theta_M$.

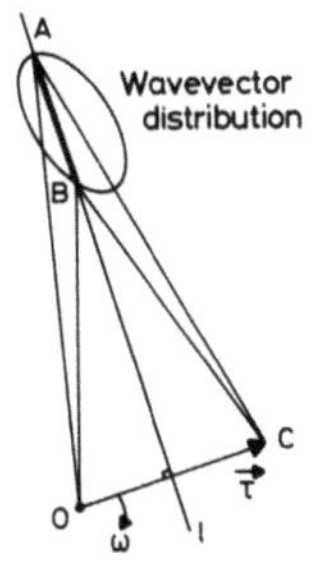


Figure 4 - Instantaneous picture of lattice point ($\vec{\tau}$) moving through reflection position. The contour encloses wavevector probability density which is higher than 50% of maximum. Only wavevectors originating from the line 1 will give rise to Bragg scattering as they are the only wavevectors for which $|\vec{k}_o|=|\vec{k}_1|$ (examples are AO=AC, BO=BC). The dark part of 1 will give rise to the major contribution to the intensity. Again it must be kept in mind that the ellipse is out of proportion.

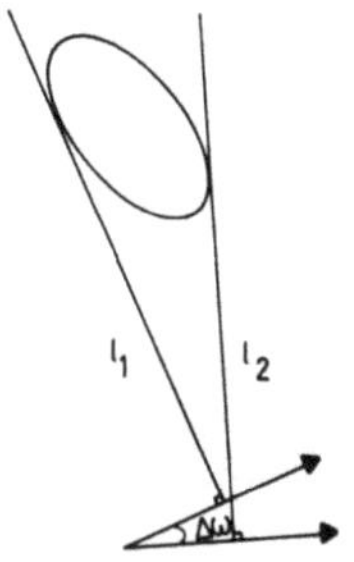


Figure 5 - The two positions ($l_1$ and $l_2$) where the intensity scattered is half of the maximum intensity. The angle between the two positions, Δω, corresponds to full width at half maximum of the Bragg profile.

From Lebech and Nielsen (1975) we can also find the factor between the angular velocity of the diffracted beam, d2θ/dt, and the angular velocity of the rotation of the crystal, dω/dt. We get

$$d2\theta/d\omega = 1+\sin^2\theta - \frac{\sin\theta\cos\theta}{tg\beta}$$

A discussion of the same problem has been given by Werner (1971,1972) and Sequeira (1975).

Clearly $d2\theta/d\omega$ can take nearly any value, but we find in general values between 0 and 1 for small $2\theta$ increasing to 2 for larger $2\theta$.

The above considerations do not include the crystal mosaicity. This will manifest itself as a distribution of $\vec{\tau}$ vectors (all with the same length of $2\sin\theta/\lambda$) around a mean value. We describe this distribution (in the horizontal plane) as $\mathbf{W}(\eta)$ where $\eta$ is the angular deviation of a $\vec{\tau}$ from the mean, and we obtain the reflection profile as a convolution of $\mathbf{W}(\eta)$ with the intensity distribution $P(\omega)$

$$I(\omega) = \int P(\omega-\eta)\mathbf{W}(\eta)\,d\eta$$

We are only interested in the integrated value of $I(\omega)$, and as this is equal to the product of the integrals of P and **W** we are less concerned with the shapes of these functions as long as we are sure that the integrations are reasonably complete. If however both functions are approximately Gaussian we can easily include the mosaic spread, $\sigma$, in the expression for $\Delta\omega$ and we get

$$\Delta\omega' = \sqrt{\Delta\omega^2+\sigma^2}$$

We can therefore either estimate $\sigma$ from reflection profiles knowing $\Delta\omega$ or we can make a priori estimates of the necessary scan width.

The important point is however the completeness of the integration and the assurance that we observe all the elastic intensity. This can be checked by a series of experiments which are based on the observations made above.

First we have to decide which type of scans to do, i.e. what type of coupling $X = d2\theta/d\omega$ do we want between the detector motion and the sample rotation. A preliminary experiment is naturally to measure X for a series of reflections, and this is done simply by determination of $X = (2\theta_2-2\theta_1)/(\omega_2-\omega_1)$ where $\omega_i$, $2\theta_i$ is a set of observed positions of $\omega$ and $2\theta$ during the scan. An example of measurements on NaCl is given in figure 7. In this case an $\omega$-$1.5\theta$ scan would obviously be most adequate. We tend however to use $\omega$-$2\theta$ scan, one main reason being that if we have white radiation background originating as streaks from the origin of reciprocal space then by using $\omega$-$2\theta$ scan we move along the streak and ensure a good background subtraction (Alexander and Smith, 1962). We must not forget, though, that by using $\omega$-$2\theta$ scan technique the diffracted beam will wander across the detector surface. We must therefore be sure that the detector has a constant response over the surface. Likewise we must ensure that the detector aperture is sufficiently large to accept the movement of the beam. One tends mainly to consider the beam size at high Bragg angle knowing from the above considerations that the spot size increases with angle. It should however be kept in mind that using $\omega$-$2\theta$ scan at low angle when it is better to use $\omega$ scan, care must be taken that the diffracted beam stays inside the detector aperture for the whole scan.

Recently Denne (1977) has suggested $\omega$-$\theta$ scan using unfiltered radiation with restricted detector size as a monochromation technique. The scan technique is described in figure 8. In figure 8a we consider the scan for one fixed wavelength, and as the triangle consisting of $\vec{k}_0$, $\vec{k}_1$ and $\vec{\tau}$ is of constant size the angular rotation of the crystal is the same as the angular shift in the beam, and we effectively carry out an $\omega$-$\theta$ scan. To accept a certain wavelength band we will then have to fix our detector size as shown in figure 8b, and to vary this as a function of the Bragg angle to keep the band-pass constant. The advantage of this technique, if we succeed in controlling the detector aperture, seems to be that it combines beam monochromatization with a homogeneous beam and preserves high flux.

Finally we should ensure that our scans are sufficiently large to give complete integration of the elastic scattering (Helmholdt & Vos, 1976), but if we use step scan measurement with recording of the individual points then an inspection of the recorded profile will tell us whether this condition is fulfilled.

## INTEGRATION OF REFLECTION PROFILES

Let us consider a reflection profile. It consists of a series of values $I_i$ measured at points $\theta_i$, which can either be observations for fixed $\theta_i$ or integrations over a small range, $\Delta\theta$, around $\theta_i$. The distinction is not important as long as sufficient points are available in ranges where $I(\theta)$ changes rapidly. A schematic drawing of such a profile is given in figure 9. To obtain an integrated intensity we assign some outer part of the profile as background, and we get an intensity

$$I = \text{peak-background (scaled)} = (B+P) - \frac{\Delta\theta_P}{2\Delta\theta_B}\, 2A = P$$

So obviously the intensity is independent of our background choice as long as this is sufficiently far from the peak. The error of our observation is however critically dependent on the exact choice of background position. Keeping in mind that for N counts $\mathrm{var}(N) \simeq N$ (following Poisson statistics) we get for the variance

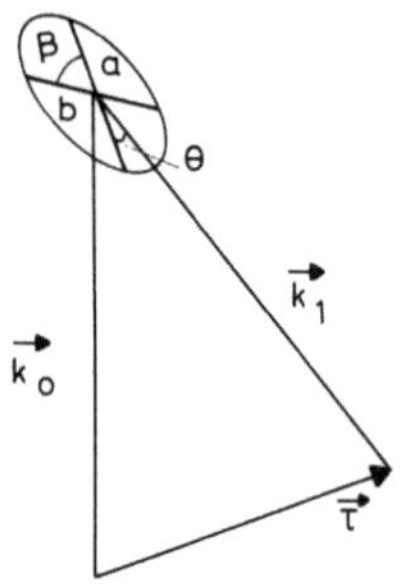


Figure 6 - Definition of various parameters defining the wavevector distribution. a and b are conjugate axis with a bisecting the angle between $\vec{k}_o$ and $\vec{k}_1$.

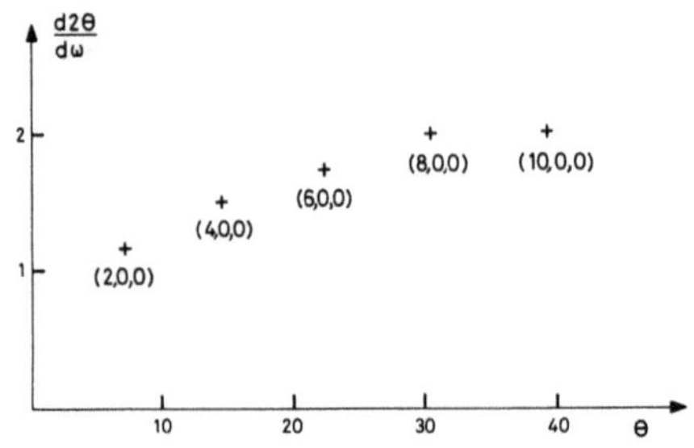


Figure 7 - The factor X=d2θ/dω for NaCl (mosaic width ∿ 1') using β-filtered Mo-radiation (courtesy of M. Thomas, Institut Laue-Langevin, Grenoble).

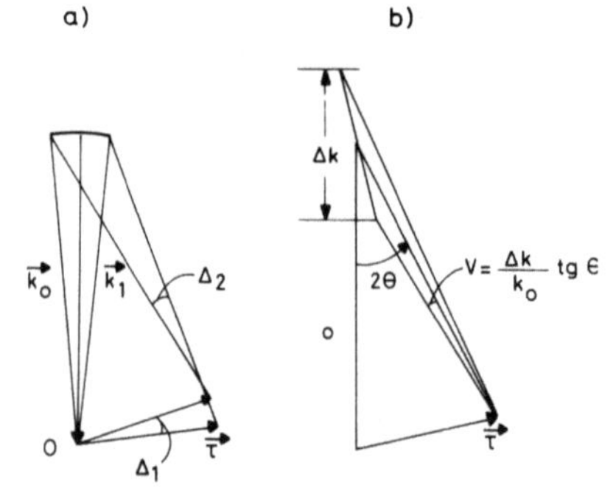


Figure 8a) - Using a constant wavelength we will perform an ω-θ scan ($\Delta_1 = \Delta_2$), as the triangle is constant during the scan.
b) To obtain a fixed wavevector spread, Δk, we must fix a certain detector size, v, which will be a function of the Bragg angle.

of the intensity

$$\mathrm{var}(I) = B+P+\left(\frac{\Delta\theta_p}{2\Delta\theta_B}\right)^2 2A$$

$$\simeq P+B+\frac{\Delta\theta_p}{2\Delta\theta_B} B$$

So in estimating var(I) the background is counted twice and var(I) increases with $\Delta\theta_p$. To ensure an optimal recording we must therefore require good estimate of the peak position, and this is only possible in all cases if the full recording of the profile is available. This is one of the main arguments for intensity determination from step-scan recordings rather than from an on-line integration during the scan. There are other arguments. If we have the full scan we can by inspection check that the peaks are well centrered, and we can easily detect any flaws in the reflection profiles which arise from errors in detection equipment and low quality of the crystal.

In many cases, especially for early models of diffractometers, it is not possible in routine measurements to get the full profile from step scanning (and it is normally not possible either to carry out any type of scan one would want to do). We will then make sure before the measurements start that the on-line integration of the peak resembles as much as possible the integration we would want to do on a step-scan recorded profile. This is of course best done by carrying out a limited set of measurements using "normally" recorded step-scan profiles ; it will in general give a good idea of the correct scan-range, and it will at the same time enable us to compare step-scan measured intensities with on-line recordings.

If step-scan recordings are available then our main problem is to locate the part of the profile that contains the Bragg scattering. Various techniques have been proposed. One can use information about the profile shape and by a least-squares fitting procedure position the profile. The intensity is then either directly obtained from the least-squares fit with the risk of a bias coming from errors in the a priori estimate of the profile, or one can, knowing the limits of the peak, return to a usual integration as described below. Such a method has been described by Diamond (1969) for use in protein crystallographic measurement and the same technique would clearly be appropriate in our case. Likewise Norrestam (1972) has suggested a method where the profile is assumed to be Gaussian. Other methods consists of determining directly the separation between peak and background by locating regions at the side of the scan which behave as background accordings to certain criteria (see for example Grant and Gabe,1978). Finally a technique has been proposed where the background is placed in such a way that the relative error $\sigma(I)/I$ of the observed intensity is minimized (Lehmann and Larsen, 1974). This last method tends to underestimate the width of weak peaks, but the bias can be described in terms of known quantities, and it is therefore easy to enlarge the peak width before integration is carried out. When treating step-scan data measured using radiation which has been filtered to remove the $K_\beta$ component a special problem occurs, namely a strong discontinuity in the spectra at the location of the $\beta$ filter. This problem has been treated by Blessing et al. (1974), who remove the points affected by the filter, and by Nelmes (1975) who uses an analytical description of the profile shape near the edge for background treatment.

As there is not any absolute criterion for determination of the peak position it is always important to follow carefully the working of any method by visual inspection of the result, as the human eye estimate seems in general to be superior to any other method.

After location of the peak it is relatively easy to carry out the integration once we have decided on an adequate type of background subtraction. In general we will assume that the background $B(\theta)$ is constant or varies linearly with the scanning angle, and we get

$$B(\theta) = s \times \theta + r$$

The constants s and k can be determined by least squares fit to the $n_b$ background points using the same weight, w, for all points and adjusting this so that the goodness-of-fit $(w \sum_i (B_i^{obs} - B_i^{calc})^2/(n_b-2))^{1/2}$ is unity.

For the integration we will use a trapezoidal sum, and we get

$$I = \Delta\Theta \sum_i (I_i - B_i)$$

where the sum is over $n_p$ points in the peak, $B_i = s x \theta_i + k$ and $\Delta\theta$ is the step-length, which we have assumed constant. This might not be true if for example we use longer steps in the part of the profile where $I(\theta)$ varies slowly, but the expression can then easily be changed.

The variance from counting statistics is given as a sum over the peak (Lehmann, 1975) :

$$\mathrm{var}(I) = \Delta\theta^2 \left[ \sum_i I_i + (\sum_i \theta_i)^2 \mathrm{var}(s) + n_p^2 \mathrm{var}(k) + 2 \sum_i \theta_i n_p \mathrm{cov}(s,k) \right]$$

where we have used the error estimates of s and k obtained from the least squares fit.

The correction for dead-time loss is normally done at this level. To a first approximation the corrected count-rate is given as $N_{corr}=N/(1-Nt_d)$ where $t_d$ is the counting chain dead-time and N is the observed countrate (see Chipman, 1969, for determination of $t_d$). This correction is easily carried out when the whole profile is available, but recourse to approximate expressions for the correction is necessary when only an on-line integration is available, and in addition in this case care must be taken that the countrates are not too large (Rees, 1977).

Finally to convert our intensities to squared structure amplitudes on a relative scale we must apply Lorentz and polarisation correction. The Lorentz factor for the normal equatorial recording geometry shown in figure 1 is $1/\sin 2\theta$, and for neutron diffraction the polarisation factor is 1. For X-rays the polarisation factor is different for filtered and crystal monochromatized radiation, and in the last case the factor depends on the perfection of the monochromator. An experimental technique for resolving this problem has been proposed by Coppens (unpublished) and is described in detail by Rees (1977).

## THE CHOICE OF CRYSTAL

Selecting a sample for the diffraction measurements is at least a three stage process.

In a first stage there is a selection of the topic to study. Clearly this choice is for the individual researcher, and it is difficult to give guidance.

In a second stage considerations of feasibility come in, and there our main concern is whether we are able to observe the effects that we hope to see. In the study of the charge density our main interest is to see the asphericity of the distributions, and we know from 50 years of diffraction studies that these effects are small ; indeed generally they are neglected. In choosing our sample for study we therefore have to estimate the possible outcome of the measurements from comparisons with previous successful studies of compounds of the same type. These types of comparisons can be standardized. Stevens and Coppens (1976) have suggested as a first estimate to use the quantity $S=V/\Sigma n^2_{core}$, where V is the unit cell volume, $n_{core}$ is the number of core electrons in an atom and the sum is over all atoms in the cell. They have shown that S is inversely proportional to the variance of the observed electron density, which is part of the error in the maps showing the asphericity. In general we expect the core electrons to be unchanged by chemical bonding so the scattering from these electrons is of no interest. The quantity S is inversely proportional to the amount of core scattering per unit volume and should therefore be as large as possible. For crystals containing first row atoms S is 3 to 5 falling to 0.1 to 0.3 for metals and alloys. Successful observation of electron distributions have been done on compounds until values of S of 0.05 (Staudenman et al., 1976), but of course S should only serve as a guide. The position of the electrons of interest are of importance as well. It will for example always be difficult to see changes near large concentrations of core electrons.

If the study is done by combination of X-ray and neutron diffraction measurements the scattering cross-sections for neutrons should be checked before the experiment is decided upon. Although some atoms have large scattering lengths for neutron scattering compared to the X-ray values (notably hydrogen), the inverse is true as well, and for example potassium has the same scattering length as hydrogen. Under these conditions we must accept that the thermal motions of the heavy atoms seem better determined from high order X-ray data than from neutron data.

The third step in choosing the sample consists in finding the crystal for the measurement. The size is determined on one hand by the amount of absorption and extinction that we can accept and on the other by the counting statistics required, but we can always request that the crystal is isometric and is bounded by surfaces that are easily described. We can thereby make sure that the variations in the corrections, that are a function of the crystal shape vary only little with crystal orientation. Naturally it often happens that the crystal cannot be obtained in an isometric form and we must then mount it so that the variations in the corrections are minimized. As example the correct mounting of a needle-shaped crystal on a four-circle diffractometer, using bisecting geometry, is with the needle axis near the $\phi$ axis. We must also require the crystal to be homogeneous in density and mosaicity, and preferably with a Gaussian mosaic distribution (Helmholtz & Vos, 1976).

## CORRECTION FACTORS

We have now come to the point of considering the various known errors that are associated with a diffraction experiment. We start by writing the general form of the observed integrated intensity at a lattice point $\vec{\tau}$ with indices $\vec{H}=(h,k,l)$ as

$$I(\vec{H})_{obs} = I(\vec{H})Ay(1+\alpha)+\Sigma_{\vec{H}'}\gamma_{\vec{H}'}I(\vec{H}')$$

where A is the absorption factor, y is the extinction factor, $\alpha$ is the contribution from thermal diffuse scattering and the second term is a sum over contributions from other Bragg reflections simultaneously in diffraction position with $\gamma'_{\vec{H}}$ being (partially unknown) factors depending on the diffraction geometry and the crystal. The corrected integrated intensity $I(\vec{H})$ is given as

$$I(H) = I_o N^2 \lambda^3 l^2 L_p V |F(\vec{H})|^2$$

where $I_o$ is the beam flux per unit area, N the number of unit cells per unit volume, $\lambda$ the wavelength, l the scattering length unit given as $10^{-12}$cm for neutrons and $e^2/mc^2 = 0.2818 \times 10^{-12}$cm for X-rays, V the crystal volume, $L_p$ the Lorentz-polarisation factors and $|F(\vec{H})|$ the structure amplitude. Apart from $L_p$ and $|F(\vec{H})|$ the parameters in the expression are of little interest if we do not aim at absolute measurements, and they can be kept constant during the measurements. We should note though that $\lambda^3 L_p$ is approximately proportional to $\lambda^2$ so that as we reduce the wavelength for the measurement the reflectivity $I(\vec{H})/I_o$ falls.

If we knew all physical parameters of the crystal and the beam (and could hope that they behaved according to known theories) then we could do an a priori correction for all the effects mentioned above. At present we can however only do this for the absorption correction and in a limited number of cases where the data are available for the thermal diffuse scattering. These corrections are therefore done first. We do not correct for multiple scattering yet, and we find it in general either impossible or too cumbersome to observe the parameters for a pre-calculated extinction correction, so instead we include the extinction as a variable parameter in the structure refinement.

## ABSORPTION

The absorption coefficient A is given as

$$A = \frac{1}{V} \int_V e^{-\mu T} dV$$

where $\mu$ is the linear absorption coefficient, and where the integration is over all beam paths T through the crystal. To do the correction well our only requirement is that the crystal size and shape is well known and that we know $\mu$. We can estimate the relative error in the intensity from this correction using an expression for the variance of the type

$$\mathrm{var}(f(X_1,X_2,\ldots X_n)) \approx \sum_i \left(\frac{\partial f}{\partial X_i}\right)^2 \mathrm{var}(X_i)$$

Approximating A by the simpler form $A = e^{-\mu\bar{T}}$ we get

$$\frac{\sigma(I)}{I} \approx (\bar{T}^2\sigma^2(\mu)+\mu^2\sigma^2(\bar{T}))^{1/2}$$

$$= \mu\bar{T}\left(\left(\frac{\sigma(\mu)}{\mu}\right)^2+\left(\frac{\sigma(\bar{T})}{\bar{T}}\right)^2\right)^{1/2}$$

where $\sigma(\mu)$ is the error in $\mu$ and $\sigma(\bar{T})$ is an average error for all possible path lengths T.

From this we see, not surprisingly, that we should keep $\mu\bar{T}$ low, preferably less than 1, and in fact this is often a criterion used for fixing the crystal size. We note too that the correlation coefficient between the correction on two different reflections coming from errors in $\mu$ is very high (Rees, 1977) as an error in $\mu$ has the same effect on all reflections, whereas this is only true for errors coming from errors in T when the calculation is for approximately identical orientations of the planes defining the crystal surface.

We know in general $\mu$ to an accuracy of better than 5% (a method of obtaining $\mu$ from transmission measurements has been given by Lawrence and Mathieson, 1976), and likewise the relative error on distances between crystal planes is probably of the same order of magnitude. So in absolute terms the error is quite large. If we only work on a relative scale the situation is much better as we are then concerned with the difference between the relative errors (Coppens, 1970), and if we ensure that the variation in absorption correction is sufficiently small, for example less than 10%, then we can obtain errors in the correction which are less than 1%. Again we should remember that this only applies to errors in $\mu$, and that errors in the description of the crystal surface are less likely to cancel in the same manner. A good test of the description of the crystal is then to compare observed and calculated intensities for several orientations of the crystal for one or more weak reflections not affected by extinction.

Various methods can be used for carrying out the integration. A method using numerical integration in a Gaussian grid has been described by Coppens et al. (1965) and by Coppens (1970), and an analytical method has been given by DeMeulenaer and Tompa (1965) and by Alcock (1970).

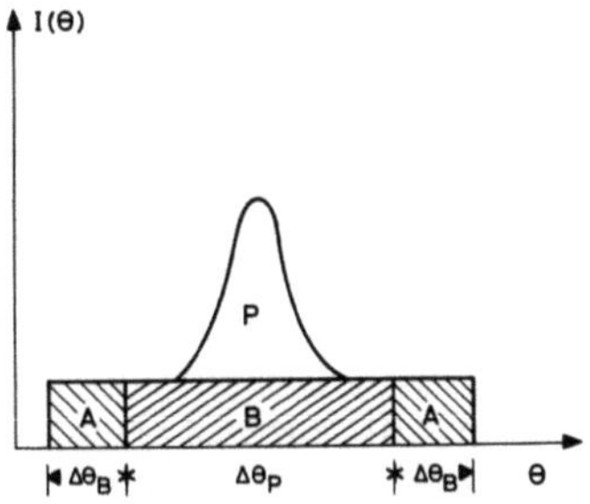


Figure 9 - Schematic drawing of reflection profile. A is background measurement. B+P is recording of peak. P is the elastic intensity.

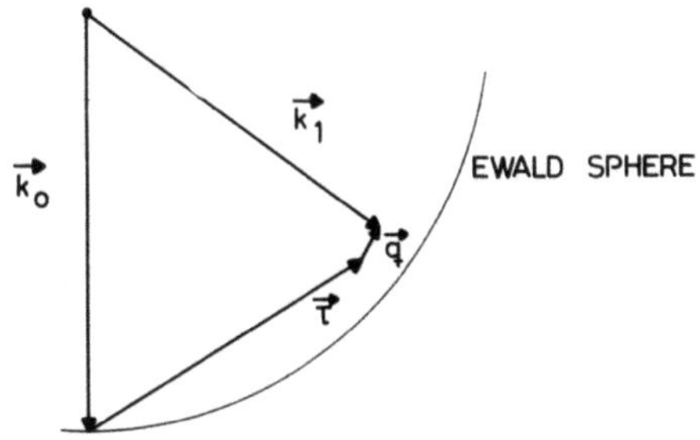


Figure 10 - Scattering diagram for phonon scattering. $\vec{\tau}$ is the lattice vector and $\vec{q}$ is the phonon wavevector. The magnitude of $\vec{q}$ is fixed by energy conservation and by the dispersion relation, which determines the frequency of the phonon. The figure is typical for neutron scattering : $|\vec{k}_0|$ is different from $|\vec{k}_1|$.

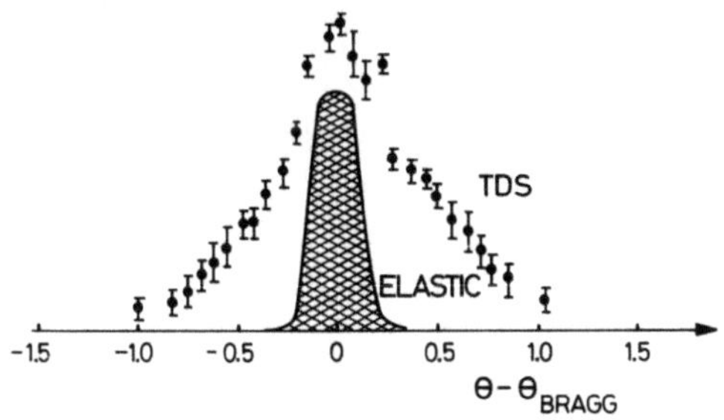


Figure 11 - TDS profile for perfect crystals of Si. The elastic peak is indicated crosshatched on arbitrary scale.

## THERMAL DIFFUSE SCATTERING

The thermal diffuse scattering contribution, $\alpha$, comes mainly from scattering processes involving low energy lattice vibrations ('acoustic modes' or 'acoustic phonons') of the crystal, and the main component is scattering interactions where one phonon is absorbed or created by the radiation. The scattering diagram is given in figure 10, and the condition for scattering is that $\vec{k}_1-\vec{k}_0=\vec{\tau}+\vec{q}$, where $\vec{q}$ is the phonon wavevector. The size of $\vec{q}$ is determined by energy conservation, $E(\vec{k}_0)-E(\vec{k}_1)=h\omega$. $E(\vec{k}_0)$ and $E(\vec{k}_1)$ are the energies of the incoming and scattered radiation, respectively, and the frequency $\omega$ of the phonon is fixed by the dispersion relation $\omega=\omega(\vec{q})$. This relation gives the frequency of the phonons with wavevector $\vec{q}$ propagating through the crystal. The size of $\alpha$ can be approximated by (Cochran, 1969)

$$\alpha = \frac{k_B T}{NV\rho} \sum_{\vec{q}} J(\vec{q})$$

where $k_B$ is the Boltzman constant, T the temperature, N the number of unit cells of volume V in the crystal, $\rho$ the density, and the sum is over all $\vec{q}$ that can contribute to the scattering during the scan. For $J(\vec{q})$ we have

$$J(\vec{q}) = \sum_j \frac{[(\vec{q}+\vec{\tau})\cdot\vec{e}_j(\vec{q})]^2}{\omega_j^2(\vec{q})}$$

where the sum is over the acoustic vibrations propagating along $\vec{q}$ (there are 3), $\omega_j(\vec{q})$ is the frequency of a vibration and $\vec{e}_j(\vec{q})$ is a unit vector in the direction of polarization of the acoustic mode j.

As mentioned above the main contribution to the scattering is from low energy phonons and for these we can assume a linear dispersion relation $\omega_j(\vec{q}) = V_j(\vec{q})\cdot q$ where $V_j(\vec{q})$ is the phonon velocity for the wavevector q. Approximating $V_j$ by a mean V (Nilsson, 1957) and assuming the solid to be isotropic we get

$$J(q) \approx \frac{|\vec{q}+\vec{\tau}|^2}{V^2 q^2} \sim \frac{\tau^2}{V^2 q^2}$$

as $q \ll \tau$. This shows that $J(\vec{q})$ peaks for q=0, i.e. at the lattice point, and the TDS contribution will therefore peak at the center of the reflection profile which makes it impossible, with standard measurement techniques, to remove it. In addition we note that J(q) is proportional to $\tau^2$ which is proportional to $(\sin\theta/\lambda)^2$, so we might expect that $\alpha$ is also proportional to $(\sin\theta/\lambda)^2$. This has indeed been observed (Nilsson, 1957 ; Cooper and Rouse, 1968), and we can write for the TDS factor

$$1+\alpha = 1+\beta(\sin\theta/\lambda)^2 \stackrel{\sim}{=} \exp(\beta(\sin\theta/\lambda)^2)$$

As this factor is applied to $|F(\vec{H})|^2 \approx \exp(-2B(\sin\theta/\lambda)^2)$, where B is the temperature factor, we observe

$$|F(\vec{H})|^2_{obs} = |F(\vec{H})|^2(1+\alpha) \approx \exp(-(2B-\beta)(\sin\theta/\lambda)^2)$$

so by not correcting for TDS we reduce our temperature factors by a quantity $\beta/2$.

A distinction must be made between thermal scattering using neutrons and X-rays. In the last case the relative change in energy is small because of the high energy of X-rays, and we have $|\vec{k}_o|=|\vec{k}_1|$. The scattering surface (i.e. the endpoints of the $\vec{q}$ vectors) fall on the Ewald sphere, the integration to obtain $\alpha$ is reasonably easy, and we can expect the TDS profile to peak under the Bragg peak as discussed above. For the neutron case the energies are comparable to the phonon energies, and we can have large differences between $|\vec{k}_o|$ and $|\vec{k}_1|$. The scattering surface is then a general surface in reciprocal space, the calculations are more complicated, and we can even expect the TDS profile to be flat so that there is no contribution to the integrated intensity (Cooper, 1971). This can be the case if the neutron velocity is lower than the velocity of the phonons. The actual form of the scattering surface will not be discussed here as it is the subject of one of the exercises.

To carry out the calculation of $\alpha$ we need to know the experimental geometry describing the part of reciprocal space over which to integrate (or sum), i.e. the part from which we can observe scattering for a given crystal orientation and detector size, and we must have the dispersion relationships, $\omega_j(\vec{q})$. Assuming as discussed above a linear form $\mathbf{V}_j(\vec{q}) = V_j \cdot q$ we only need to know the velocities $\underline{V}_j$ for the various modes (as well as the polarization vectors $\vec{e}_j(\vec{q})$), and these can be obtained from the elastic constants (see for example Kittel, 1971, p. 145). Computer programs are available (some of the more recent are : Stevens, 1974, Helmholdt and Vos, 1977A, Merisalo and Kurittu 1978. See also Walker and Chipman, 1970), and although the calculations can be rather lenghty the corrections can be obtained with sufficient precision. The main obstacle is to get the elastic constants, which are generally not available. Standard ultrasonic techniques for measurement of elastic constants require cm large crystals, and this rules out a large number of cases. The elastic constants can of course be obtained from measurements of phonon dispersion relationships for a number of directions in the reciprocal lattice, but again large crystals are required and the measurements are lengthy, and likewise observations of diffuse scattering by X-rays to obtain the information is not a technique that can be routinely carried out in a limited span of time.

Except for the cases where elastic constants can be obtained optimal experimental conditions are therefore to reduce TDS, and we note from the above expressions that $\alpha$ will be smaller at reduced temperatures. Measurements should therefore be carried out at as low a temperature as possible. We note too that the effect of TDS in the X-ray case (and most often in the neutron case) is to decrease the temperature factor by a constant, so that the effect is the same for high and low order data. When we determine the deformation density maps by combination of high and low order X-ray data we can therefore expect to a first approximation a cancellation of the bias in the temperature factors. In the case where we use temperature factors obtained from neutron diffraction data, we must make sure that the experimental parameters such as collimations are similar in the two sets of measurements in order to have the same bias.

Experimental techniques are available for X-rays for observing the TDS profile and thus correct for the error. The method consists in using Mössbauer resonance to detect radiation that has changed energy (Butt & O'Connor, 1967), but unfortunately the source of radiation is weak, which prevents its general application. Presently a neutron diffraction method using the spin-echo technique proposed by Mezei (1972) for neutron spectroscopy is being developed at the Institut Laue-Langevin (Hayter et al., 1978). This method combines high energy resolution with rather relaxed conditions on the beam divergence, so that high flux measurements should be possible, and it is expected that 95% of the inelastic scattering can be filtered out. Finally the TDS profile can sometimes be obtained by experimental "tricks", and an example of this is given in figure 11 (Lehmann et al., 1978). Perfect crystal plates of Si were used for these observations. The elastic intensity is independent of crystal thickness (dynamic behaviour) but TDS varies with crystal thickness (kinematical behaviour). By subtracting reflection profiles for the same reflection for two crystals with different thickness only the TDS scattering proportional to the difference between the two thicknesses is left over.

The profiles in figure 11 clearly show the difference in shape and width of the elastic and the TDS profile. As seen above $J(\vec{q})$ is inversely proportional to $q^2$ so that the TDS profile is of Lorentzian shape, but the observed TDS profile will be a convolution of this "theoretical" profile with the resolution function discussed above as in the case of the elastic scattering. The amount of TDS included in the integration of the reflection profile might therefore vary with Bragg angle if the TDS and the elastic profile do not vary in the same manner, and as the TDS profile seems to be quite large it is possible that more TDS is included at high angle where the elastic peak is wider. This would result in the temperature factors being smaller for the high order data, and this has been observed in the case of neutron diffraction, where the variation in peak widths is large (Mitschler et al., 1978, Lehmann, 1977). If a correction is done then resolution effects should be considered. A neglect can lead to overcorrection (Cochran, 1969 ; Scheringer, 1973).

A solution to these problems seems naturally to be real "profile analysis", where several profiles describing the TDS and the elastic scattering is fitted to the observed profile after background subtraction. This technique requires a good quality elastic profile and good knowledge of the TDS profile, and has therefore not yet been implemented for routine analysis.

At present the best we can do is to work at low temperatures, preferably on compounds where the elastic constants are known ; and if this is not the case, try to do the integration only over the elastic peak to reduce the reflect of TDS. But this might give temperature factors that are slightly dependent on the Bragg angle.

Finally it is worth noting that for many cases the relative error in the temperature factor, $\Delta B/B$, caused by TDS is virtually independent of temperature in the range generally accessible at present, i.e. down to liquid nitrogen temperatures. For an organic compound the thermal motion can be approximately described as a harmonic oscillator in the high temperature limit (see Coppens and Vos, 1971 ; Lehmann and Coppens, 1977), and B is then proportional to T. Likewise in the expression given above $\alpha$, and thereby $\Delta B$, is proportional to T, and thus $\Delta B/B$ is constant (these considerations do not take into account changes in elastic constants with temperature). One main reason why cooling is still of importance in these cases is then that it diminishes the thermal smearing of the density and it reduces thereby the effects from errors in B.

## EXTINCTION AND MULTIPLE REFLECTION

When reflection takes place from a Bragg plane $\vec{H}$ the primary beam is weakened and a new primary beam, the reflected beam, is created. Scattering processes involving this new beam will occur, and one obvious process is the Bragg scattering from the plane $-\vec{H}$ returning some of the intensity to the original primary beam.

This type of beam interactions will tend to reduce the intensity of the strong reflections and is called extinction. Other scattering processes will include other Bragg reflections which are in reflection position, and complicated exchange of energy will take place among these reflections. We call this multiple reflection, multiple Bragg-scattering or simulteneous reflection. The theory for extinction most used in practical cases is the theory developed by Becker and Coppens (1974), an extension of work by Zachariasen (1967) and based on the Darwin transfer equations, which describes the intensity coupling between the primary and reflected beams. In the theories (which are described elsewhere) a distinction is made between primary extinction, which comes from scattering processes within a perfect volume of the crystal and which includes both intensity and phase coupling of the beams involved, and secondary extinction which is the result of scattering processes from different parts of the crystal, where the phase information is lost. If only small perfect areas exist then primary extinction is negligible, and this is what we aim at obtaining. The critical parameter is the extinction length $\Lambda = (\lambda/Q\sin 2\theta)^{1/2}$, where Q is the scattering cross section (proportional to $F^2$), and the requirement for negligible primary extinction is that the size, d, of the perfect domains is much smaller than $\Lambda$ .

Unfortunately we have no experimental technique which will give us d.

In general we do not calculate the extinction correction, but we include parameters in our structural least squares refinement and adjust these simultaneously with the other parameters. The theory is based on a model which describes the crystal as consisting of perfect blocks of a certain size misaligned in a mosaic pattern, and the parameters used in the refinement describe the mosaic distribution and the crystallite size. In most cases one of the effects of mosaic or crystallite size is dominant, but our only means of distinction is from the agreement factors of the refinement. We find often that the mosaic distribution behaviour dominates the correction, and we are then able to compare with observations of the mosaic distribution. This is done by profile measurements using a highly collimated beam with little wavelength spread, for example γ-radiation (Schneider, 1974). Of course we have no evidence from our profile measurements that the crystal does really consist of mosaic blocks, as other types of crystal defects might result in a similar curve. A typical example of such "mosaic curves" are given in figure 12 (Lehmann and Schneider, 1977) for a Cu single crystal. The total volume of the crystal was 3x3x3 $mm^3$, and neutron data were measured till a $\sin\theta/\lambda$ of 1.52 $Å^{-1}$. There were 256 reflections and 8 parameters to determine. The final $R_F$ factor was 0.017 and the minimal extinction factor y was 0.70. The temperature factor was as expected, so it was in all ways a satisfactory experiment, despite the fact that the crystal defect structure does not follow any theory. We assign this to the fact that the extinction effects were not too severe. We do find better crystals, and figure 13 shows the full width at half height for the mosaic distribution of a Cu crystal where measurements were done the same way as described above. An anisotropic mosaic distribution was assumed and a formalism similar to the one given by Thornley and Nelmes (1974), which takes into account the beam divergence, was used in the refinement, that gave a final $R_F$ value of 0.009. Although the agreement between the mosaic spread obtained from the refinement and the one observed is far from perfect we see the same trends, and if we really had crystals following the theory we could undoubtedly calculate the extinction factors. It is interesting to note that acceptable values of the FWHH could be obtained from neutron profiles, so in principle we record all the necessary values for a correction while doing the experiment.

The conclusion that we reach, however, seems to be that we should try to reduce the effect of extinction, even considering that present theories and refinement techniques are very successful in dealing with the problem. The theories give clear guidelines for this. We should reduce the crystal size as much as possible, and if necessary measure on a small crystal at low angle and on a large crystal at high angle in order to reduce extinction in the regions where it is strongest. We should reduce the wavelength, and finally we should use all available means to increase the mosaic spread of the crystal. In many cases the beam divergence dominates the width of the profiles, and we would loose nothing by increasing the mosaic spread so that at least at low angle the mosaic was dominating. This would give us better estimates of the mosaic spread, and it would make the scan range vary less over the range of measurement.

The effects of multiple reflection follow much the same pattern as extinction being an extension of this to many beam interactions, and it will in most cases be more severely felt in neutron than in X-ray measurements because of the large crystals used in the neutron case. There has been no general theoretical solution worked out, but Moon and Shull (1964) have studied in detail the solution in the limit of thin crystal plates. They conclude that "the effects of simulteneous reflections may be present in all experiments where secondary extinction is not negligible, and the magnitudes of the two effects are frequently about the same". Recently Tanaka and Saito (1975) have proposed a quantitative approach for a correction to the intensities, and they find an improvement in 60% of the cases based on comparison of disagreement between symmetry related reflections. One must however underline the complexity of the problem. As we move one reflection through reflection

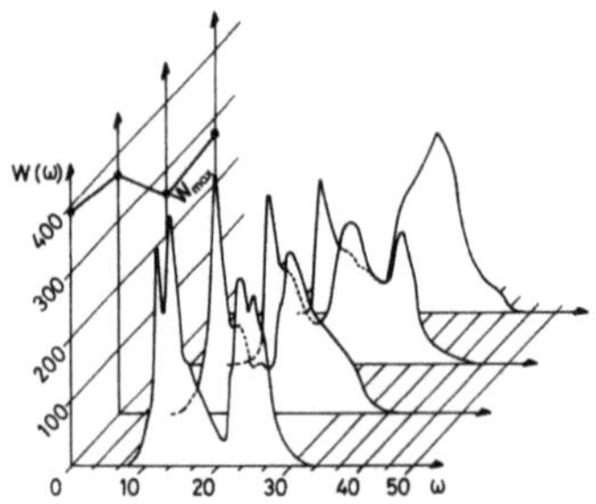


Figure 12 - Mosaic distribution W(ω) deduced from γ-ray rocking curves given for a series of sections of a Cu single crystal. Each section is 0.2 mm thick and the distance between sections is 0.5 mm.

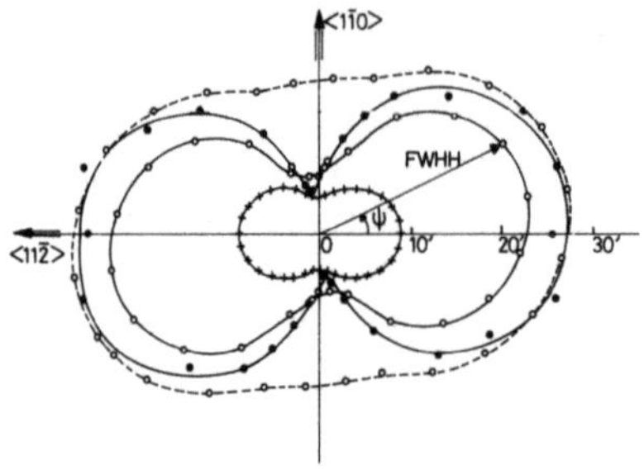


Figure 13 - Mosaic spread expressed as FWHH for a Cu crystal for the reflection 111 as a function of the rotation ψ around the scattering vector. The curves are :●—●—●: FWHH from γ-ray measurements, o--o--o--o-- : FWHH from neutron reflection profiles, o—o—o—o : the same after correction for experimental width,-x-x-x-x-: FWHH deduced from structure refinement of neutron data (Lehmann and Schneider, 1977).

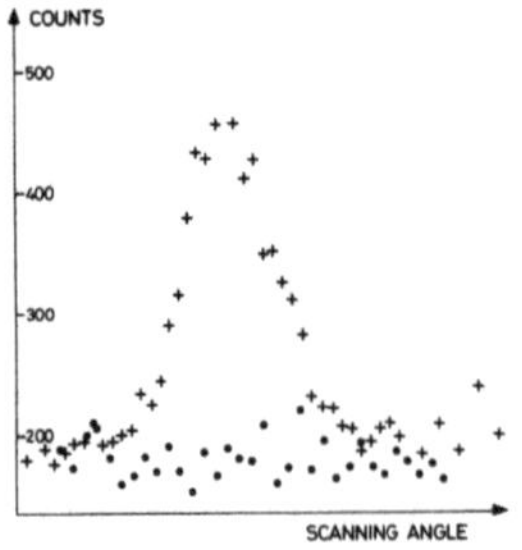


Figure 14 - Reflection profile for the extinct reflection 502 in oxalic acid, dihydrate measured at 100°K using neutrons. + : first observation (λ = 1.17 Å, step = 0.04°). • : later observation after change of experimental parameters (λ = 0.84 Å, step = 0.05°).

position other reflections will be partially excited and these excitations are related to the resolution function and the mosaic of the crystal.

Again the best approach is to avoid it, and the recommendations given above are also valid for the case of multiple scattering. In favourable cases we might even orient the crystal so that only one reflection is active, but in most cases we find that multiple reflection is unavoidable, especially at low measurement temperatures where a large number of reflections are very intense.

It is always advisable to measure all extinct reflections where multiple scattering can occur, and this gives an immediate estimate of the order of the effect.

If reflections occur then a second step is to measure the intensity for the reflection for a series of orientations of $\psi$, the rotation around the scattering vector. However, if several reflections are active all the time it is nearly impossible to eliminate the effect. A recent example is given by Thompson and Grimes (1977) who studied very weak "forbidden reflections" in spinel structures using neutron diffraction. They find that multiple scattering occurs for nearly all values of $\psi$. Likewise it is difficult to judge from a reflection profile whether it arises from multiple scattering. Figure 14 shows an extinct reflection of oxalic acid, dihydrate recorded using neutron radiation at 100°K (Feld, Brown and Lehmann, 1978). The shape corresponds well to normal, allowed reflections. In total about 50% of the extinct reflections were observed. By heating to room temperature the reflections disappeared slowly, so there was no indication of a sudden phase-transformation in the crystal. Measurements with another wavelength, another crystal orientation and other beam divergencies did not show any intensity in the extinct reflections, so it was finally concluded that the first observations came from multiple reflection.

In general multiple scattering will cause reduction in strong reflections and increase the weak reflections, so a good deal of what cannot be avoided may be accounted for in the extinction correction.

## CONCLUSION

After having discussed in detail some of the optimal conditions for a diffraction experiment we will briefly see whether we have followed the general guidelines set up in the introduction.

We have in general not been concerned with long term stability, as this will be discussed later. We have however tried to ensure that the beam is homogeneous and by requiring that step-scan technique is used, we should obtain stability and constancy in the data reduction. Likewise if we carry out a generalized scan we can keep the reflection at the same part of the detector surface and have both stability and reproducibility.

For the crystal we have required that it should have a reasonably large mosaic spread to reduce extinction. Our choice might therefore not be the typical crystal, but if we develop techniques for improving the mosaic at the same time we develop a method for standardizing the crystal.

We have not discussed in details how to describe the conditions of the experiment, but we have discussed test on various constants concerning the beam and measurement equipment, and the results of these should of course always be given.

Finally we do have theories describing most of the effects, and we have adapted a policy of trusting the corrections if the effect is small, and apart from this use the theory as a guide to a reduction of the effects. This will often be an approach that saves time and energy (if not money) and sometimes it allows us to neglect the correction.

## REFERENCES

Alcock, N.W. (1970). In Crystallographic Computing, 271-278, Munksgaard : Copenhagen.

Alexander, L.E. and Smith, G.S. (1962). Acta Cryst. 15, 983-1001.

Becker, P. and Coppens, P. (1974). Acta Cryst. A30, 129-147, 148-153.

Blessing, R.H., Coppens, P. and Becker, P. (1974), P. (1974). J. Appl. Cryst. 7, 488-492.

Butt, N.M. and O'Connor, D.A. (1967). Proc. Phys. Soc. 90, 247-252.

Chipman, D.R. (1969). Acta Cryst. A25, 209-214.

Cochran, W. (1969). Acta Cryst. A25, 95-101.

Cooper, M.J. (1971). Acta Cryst. A27 148-157.

Cooper, M.J. and Nathans, R. (1968). Acta Cryst. A24, 481-484, 619-624.

Cooper, M.J. and Rouse, K.D. (1968). Acta Cryst. A24, 405-410.

Coppens, P. (1970). In Crystallographic Computing, 255-270, Munksgaard : Copenhagen.

Coppens, P. (1978). Chapter 3 in "Neutron Diffraction" H. Dachs, Ed. Springer Verlag : Berlin.

Coppens, P., Leiserowitz, L. and Rabinovich, D. (1965). Acta Cryst. 18, 1035-1038.

Coppens, P. and Vos, A. (1971). Acta Cryst. B27, 146-158.
Dachs, H. (1961). Z. f. Krist. 115, 80-92.
De Meulenaer, J. and Tompa, H. (1965). Acta Cryst. 19, 1014-1018.
Denne, W.A. (1977). Acta Cryst. A33, 438-440, 987-992.
Diamond, R. (1969). Acta Cryst. A25, 43-55.
Feld, R., Brown, P.J. and Lehmann, M.S. (1978). Abstract, XI Congress of the I.U.Cr., Warsav.
Grant, D.F. and Gabe, E.J. (1978). J. Appl. Cryst. 11, 114-121.
Hayter, J.B., Lehmann, M.S., Mezei, F. and Zeyen, C. (1978). Acta Cryst., in press.
Helmholdt, R.B. and Vos, A. (1976). Acta Cryst. A32, 669.
Helmholdt, R.B. and Vos, A. (1977A). Acta Cryst. A33, 38-45.
Helmholdt, R.B. and Vos, A. (1977B). Acta Cryst. A33, 456-465.
Kittel, C. (1971). Introduction to Solid State Physics, Fourth Ed., John Wiley & Sons, New York.
Lawrence, J.L. and Mathieson, Mcl (1976). Acta Cryst. A32, 1002-1004.
Lebech, B. and Nielsen, M. (1975). RCN Report 234, Proceedings of the Neutron Diffraction Conference, Petten 1975, 466-488.
Lehmann, M.S. (1975). J. Appl. Cryst. 8, 619-622.
Lehmann, M.S. (1977). Unpublished results.
Lehmann, M.S. and Coppens, P. (1977). Acta Chem. Scand. A31, 530-534.
Lehmann, M.S. and Larsen, F.K. (1974). Acta Cryst. A30 580-584.
Lehmann, M.S. and Schneider, J.R. (1977). Acta Cryst. A33, 789-800.
Lehmann, M.S., Graf, H., Schneider, J.R. and Freund, A. (1978). Abstract, XI congress of the I.U.Cr., Warsav.
Mezei, F. (1972). Z. Physik 255, 146-160.
Mitschler, A., Rees, B. and Lehmann, M.S. (1978). J.A.C.S. 100, 3390-3397.
Merisalo, M. and Kurittu, J. (1978). J. Appl. Cryst. (1978) 11, 179-183.
Moon, R.M. and Shull, C.G. (1964). Acta Cryst. 17, 805-812.
Nelmes, R.J. (1975). Acta Cryst. A31, 273-279.
Nielsen, M. and Møller, H.B. (1969). Acta Cryst. A25, 547-550.
Nilsson, N. (1957). Ark. Fys. 12, 247-257.
Norrestam, R. (1972). Acta Chem. Scand. 26, 3226-3234.
Rees, B. (1977). Israel J. of Chem. 16, 154-159, 180-187.
Scheringer, C. (1973). Acta Cryst. A29, 283-290.
Schneider, J.R. (1974). J. Appl. Cryst. 7, 541-546, 547-554.
Sequeira, A. (1975). RCN Report 234, Proceedings of the Neutron Diffraction Conference, Petten 1975, 454-465.
Staudeman, J.-L., Coppens, P. and Muller, J. (1976). Solid State Comm. 19, 29-33.
Stevens, E.D. (1974). Acta Cryst. A30, 184-189.
Stevens, E.D. and Coppens, P. (1976). Acta Cryst. A32, 915-917.
Tanaka, K. and Saito, Y. (1975). Acta Cryst. A31, 841-845.
Thompson, P. and Grimes, N.W. (1977). Acta Cryst. 10 369-371.
Thornley, F.R. and Nelmes, R.J. (1974). Acta Cryst. A30 748-757.
Walker, C.B. and Chipman, D.R. (1970). Acta Cryst. A26 447-455.

Zachariasen, W.H. (1967). Acta Cryst. 23, 558-564.
Werner, S.A. (1971). Acta Cryst. A27, 665-669.

## EXERCISE ON "PROFILE ANALYSIS"

The integrated intensity of a Bragg reflection is given by

$$I = \Sigma(I(\theta)-B(\theta))\ d\theta$$

where $I(\theta)$ is a recording of the scattered intensity at the angle $\theta$, $B(\theta)$ is an estimate of the instrumental background, and the integral is over a range of angle covering all elastic scattering. Table 1 gives in column 2 and 3 $\theta$ and $I(\theta)$ for a series of measurement points.

a) Determine the background $B(\theta)$ and check whether it is constant.
b) Determine the integrated intensity and its standard deviation using a trapezoidal integration.
c) The diffractometer is worn down, and column 4 is the real position of the measurements. Estimate the change in the integrated intensity.
d) If one would study the setting errors of the diffractometer which part of these observations are most useful ?
e) Assume the profile to be a Gaussian truncated at $\pm$ 2 $\sigma$. How much error does one introduce by approximating the integral by a sum over 7 and 15 points, respectively ?

## EXERCISE ON A CRYOSTAT WITH A TEMPERATURE VARIATION

Let us consider a cryostat with a slow sinuisoidal variation in time of the temperature around a mean temperature $T_o$ and with a maximum deviation from $T_o$ of $\Delta T$. Using this cryostat we study a structure with structure factors $F(\sin\theta/\lambda) = f(\sin\theta/\lambda) \exp(-B_o x (\sin\theta/\lambda)^2)$. The temperature factor, $B_o$, varies with temperature, and we assume that $B_o$ is proportional to the temperature, which is generally not a bad approximation.

We will now study the effect of this temperature variation on the quality of our measurements.

a) Determine the temperature distribution for a time large compared to the variation time of the cryostat.

b) For a given reflection estimate the mean intensity for many measurements assuming $\Delta T$ to be much smaller than $T_o$, and compare it to the intensity if no temperature variation took place.

c) Determine the mean square deviation of the intensity coming from the temperature variation.

d) For $B_o$ = 1 Å$^2$ give curves that relate the relative error in the intensity $\sigma(I)/I$ to $\Delta T/T_o$ for $\sin\theta/\lambda$ = 0.3, 0.5, 07 and 0.9 Å$^{-1}$.

Table 1

A reflection profile

| 1 | 2: θ° | 3: I(θ) counts | 4: θ° |
|---|---|---|---|
| 1 | -3.0 | 66 | -3.01 |
| 2 | -2.8 | 44 | -2.80 |
| 3 | -2.6 | 56 | -2.59 |
| 4 | -2.4 | 42 | -2.41 |
| 5 | -2.2 | 44 | -2.20 |
| 6 | -2.0 | 46 | -2.01 |
| 7 | -1.8 | 50 | -1.82 |
| 8 | -1.6 | 52 | -1.61 |
| 9 | -1.4 | 59 | -1.40 |
| 10 | -1.2 | 103 | -1.18 |
| 11 | -1.0 | 162 | -1.00 |
| 12 | -0.8 | 298 | -0.77 |
| 13 | -0.6 | 431 | -0.61 |
| 14 | -0.4 | 589 | -0.42 |
| 15 | -0.2 | 807 | -0.20 |
| 16 | 0.0 | 808 | 0.00 |
| 17 | 0.2 | 763 | 0.21 |
| 18 | 0.4 | 640 | 0.41 |
| 19 | 0.6 | 381 | 0.63 |
| 20 | 0.8 | 247 | 0.81 |
| 21 | 1.0 | 180 | 0.99 |
| 22 | 1.2 | 89 | 1.21 |
| 23 | 1.4 | 76 | 1.40 |
| 24 | 1.6 | 56 | 1.60 |
| 25 | 1.8 | 44 | 1.83 |
| 26 | 2.0 | 48 | 2.00 |
| 27 | 2.2 | 52 | 2.20 |
| 28 | 2.4 | 54 | 2.41 |
| 29 | 2.6 | 42 | 2.61 |
| 30 | 2.8 | 50 | 2.79 |
| 31 | 3.0 | 62 | 3.00 |

## EXERCISE ON "THERMAL DIFFUSE SCATTERING"

When inelastic scattering (one phonon process) takes place three conditions must be fulfilled :

1) Momentum conservation

$$\vec{k}_1 - k_o = \vec{\tau} + \vec{q}$$

where $\vec{k}_0$ and $\vec{k}_1$ are wavevectors of the incoming and scattered X-rays or neutrons ($k = 2\pi/\lambda$) $\vec{\tau}$ is a reciprocal lattice vector and $\vec{q}$ is a phonon wavevector. The scattering situation is summarized in figure 1.

2) Energy conservation

$$E(\vec{k}_o) - E(\vec{k}_1) = h\omega(\vec{q})$$

where $E(k_o)$ and $E(k_1)$ are energies of the neutrons (X-rays) before and after the scattering process, respectively, and where $h\omega(q)$ is the phonon energy. The energy transfer $E(k_o)-E(k_1)$ is shown in figure 2 as a function of $k_1$ for two incoming neutron wavelengths. Values for conversion between the various units are given in table 1.

3) The dispersion relation

$$\omega = \omega(\vec{q})$$

i.e. the relation between the frequency (energy) of the phonon and the phonon wavevector. We assume this relationship to be linear, and figure 3 gives the dispersion relation for transverse acoustic phonons in Aluminium travelling along <100>.

Scattered wavevectors obeying the above conditions form a surface, the scattering surface, which is quite easy to construct using the following considerations :
We place the lattice point in question at a given distance from the Ewald sphere. We then choose (guess) a phonon with wavevector q, and obtain the frequency (energy) from figure 3. The corresponding $k_1$, using energy conservation, is then obtained from figure 2a,b. We now draw two circles, one with radius q and origin at the lattice point, and one with radius $k_1$ originating from the center of the Ewald sphere. Where the circles intersect we have momentum conservation, and the points of intersection are part of the scattering surface.

When starting the construction it is often worthwhile first to estimate the values of the q vectors which are parallel to the vector from the origin of the Ewald sphere to the lattice point. This is done by placing figure 3 on figure 2 with the origins displaced corresponding to the lattice point displacement from the Ewald sphere. Where the two curves intersect we have energy conservation, and the distance from the origin of figure 3 to the points of intersection is the length of the q vectors.

Construct the scattering surface for neutron scattering in the horizontal plane for transverse acoustic phonons around the lattice point 200 in Al (assuming the phonon behaviour to be isotropic) for the following cases :

a) $\lambda$ = 1.8 Å. The lattice point is 0.8 $Å^{-1}$ from the Ewald sphere, and on the inside of the sphere (let 1 $Å^{-1}$ be 5 cm).
b) $\lambda$ = 1.8 Å. The lattice point is 0.3 $Å^{-1}$ from the Ewald sphere, and the sphere.

c) $\lambda = 1.8$ Å. The lattice point is 0.4 $Å^{-1}$ from the sphere, and outside the sphere.

d) $\lambda = 1$ Å. The lattice point is 0.28 $Å^{-1}$ from the Ewald sphere inside this sphere (let 1 $Å^{-1}$ be 3 cm).

e) Indicate the cases of neutron energy loss and neutron energy gain.

f) What is the difference between the cases a) b) c) on one side and case d) on the other side ?

g) Indicate the scattering surface in case of X-ray scattering.

Most of the content of this exercise is taken from a paper by Dr. D. Hohlwein entitled : Observation of Phonon Scattering Surfaces by Neutron Film Method (to be published by IAEA, Vienna 1978). Dr. Hohlwein is hereby thanked for supplying the paper prior to publication.

Table 1

Conversion and some values

Wavevector and wavelength

$$k = 2\pi/\lambda \qquad \text{so 1 Å corresponds to } 2\pi\ Å^{-1}$$
$$\tau = 2\pi/d$$

Neutron wavelength, energy and velocity

$$\lambda = h/p \qquad p = hk$$
$$E = \frac{1}{2}mV^2 = \frac{1}{2m}p^2 = \frac{h^2}{2m}k^2 = \frac{h^2}{2m}\frac{1}{\lambda^2}$$

p and m are momentum and mass of neutron.

Velocity of 1 Å neutron : 3.956 km/sec

Energy of 1 Å neutrons : $1.311 \times 10^{-13}$ erg ; 81.8 meV ; 19.78 THz

Phonon velocity for small q in a special case

Transverse acoustic phonons travelling along <100>

$$\omega(q) = \sqrt{\frac{C_{44}}{\rho}}\, q$$

$$\sqrt{\frac{C_{44}}{\rho}} = V_S$$

where $C_{44}$ is an elastic constant, $\rho$ is the density and $V_S$ is the velocity of sound. For Al (300°K) : $C_{44} = 0.282 \times 10^{12}$ dyne/cm$^2$

$\rho = 2.70$ g/cm$^3$

$V_S = 3.230$ km/sec

Aluminium is cubic (fcc) with a = 4.05 Å

Conversion from km/sec to THz/Å$^{-1}$

$$1 \text{ km/sec} = 10/2\pi \quad \text{THz/Å}^{-1}$$

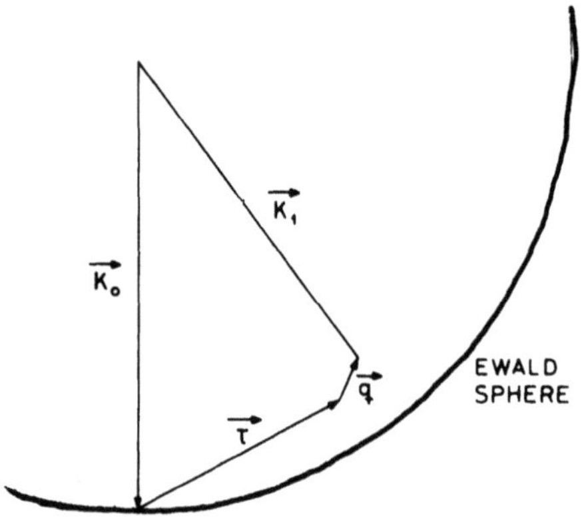


Figure 1 - Scattering diagram for one phonon process. $\vec{k}_o$ and $\vec{k}_1$ are wavevectors for incoming and diffracted radiation, $\vec{\tau}$ is a lattice vector and $\vec{q}$ is the phonon wavevector. The scattering diagram describes a case of neutron scattering.

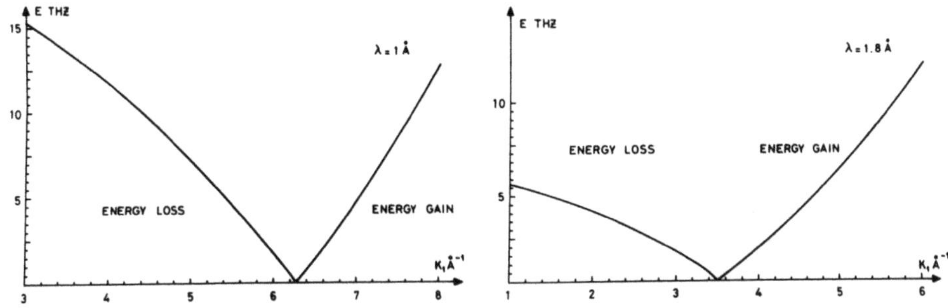


Figure 2 - Energy transfer $E = \frac{h}{2m}(k_o^2 - k_1^2)$ given as a function of $k_1$ for two neutron wavelength $\lambda = 1$ Å ($k_o = 2\pi$ Å$^{-1}$) and $\lambda = 1.8$ Å ($k_o = 2\pi/1.8 = 3.49$ Å$^{-1}$). Areas of neutron energy loss and gain are indicated.

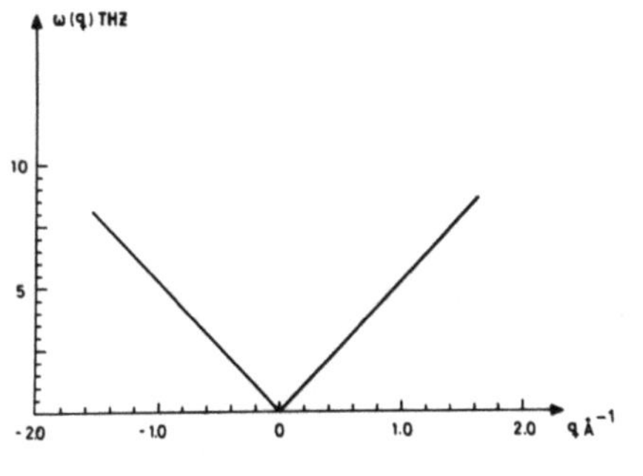


Figure 3 - Dispersion relationship for transverse acoustic phonons travelling along <100> in Al.

# SOLUTIONS TO EXERCISES

## Profile analysis

a) We choose points 1-8 and 24-31 as background, and get $\bar{B}$=50.5 and the variance var(B) $\cong$ B = 50.5. If B is constant we can assume the n observations to be taken from a normal distribution and

$$S = \sum_i \frac{(\bar{B}-B_i)^2}{\mathrm{var}(B)}$$

will then be distributed as $\chi^2_{n-1}$.

We find S=14.8, and as $\chi^2_{15,0.05}$ = 25.0 we cannot reject the hypothesis that B is constant on a 5% significance level

b)

$$I = \Delta\theta \left( \sum_9^{23} I_i - \frac{15}{16} \left( \sum_1^8 I_i + \sum_{24}^{31} I_i \right) \right) = 975$$

$$\sigma(I) = \Delta\theta \left( \sum_9^{23} I_i - \left(\frac{15}{16}\right)^2 \left( \sum_1^8 I_i + \sum_{24}^{31} I_i \right) \right)^{1/2} = 16$$

c)

$$I \cong \sum_8^{23} ((I_i + I_{i+1})/2 * (\theta_{i+1} - \theta_i) - \bar{B}\Delta\theta = 985$$

d) The intensity varies most rapidly with position at the sides of the peak. Assume a Gaussian peak

$$I(\theta) = I_0 \exp(-\theta^2/2\eta^2)$$

The change in intensity from setting errors is then

$$\frac{d(I(\theta))}{d\theta} = -I(\theta)\frac{\theta}{\eta^2}$$

The maximum change is found for

$$\frac{d}{d\theta}\left(\frac{d(I(\theta))}{d\theta}\right)_{\theta=\theta_0} = 0$$

$$-I(\theta_0)\frac{\theta_0^2}{\eta^4} + I(\theta_0)\frac{1}{\eta^2} = 0$$

$$\theta_0 = \pm\eta$$

So observations near half height of the peak are most interesting.

e) Assume normal distribution $G(\theta) = \frac{1}{\sqrt{2\pi}} \exp(-\theta^2/2)$

The integral $\int_{-2}^{2} G(\theta)\, d\theta$ is 0.9554

The values obtained by trapezoidal integration are for 7 points : 0.9465 and for 15 points : 0.9530.

Temperature variation of cryostat

a) $T = T_0 + \Delta T \sin t$
T is temperature and t is time. The distribution function is called $\phi(T')$, where T' is relative temperature, $T-T_0$.

We have $\phi(T')dt' = k\, dt$, where k is a constant. Using the above expression and normalizing we get

$$\phi(T') = \frac{1}{\pi} \frac{1}{\sqrt{\Delta T^2 - T'^2}}$$

b) The structure factor is

$$F = f \exp(-BQ^2)$$

with $Q = \sin\theta/\lambda$ and $B = B_0 + \Delta B T'$. The intensity is then

$$I = I_0 \exp(-2\Delta B T' Q^2) \cong I_0 (1 - 2\Delta B T' Q^2)$$

where $I = I_0$ for $T' = 0$.
The mean intensity is then

$$\bar{I} = \int_{-\Delta T}^{\Delta T} I(T') \phi(T') dT' = I_0 \int_{-\Delta T}^{\Delta T} (1 - 2\Delta B T' Q^2) \phi(T') dT' \approx I_0$$

c)

$$\mathrm{var}(I) = \int_{-\Delta T}^{\Delta T} (I_0 - I_0(1 - 2\Delta B T' Q^2))^2 \phi(T') dT'$$

$$= I_0^2 \int_{-\Delta T}^{\Delta T} k^2 T'^2 \phi(T') dT' = \frac{k^2 \Delta T^2}{2}$$

$$k = 2\Delta B Q^2$$

Thermal diffuse scattering

a)b)c) The scattering surfaces are ellipsoids that contract as the lattice point approaches the Ewald sphere. For a) and b) the ellisoid is inside the sphere, for c) outside

d) The scattering surface is two hyperboloids, one inside the Ewald sphere and one outside (examples of the surfaces are given by M.J. Cooper, Acta Cryst. (1971), A27, 151).

e) Points inside the Ewald sphere correspond to neutron energy loss, points outside the sphere correspond to neutron energy gain.

f) In case a) b) and c) neutron velocity is less than the phonon velocity. In case d) the inverse is the case.

g) For X-rays the scattering surface can be approximated by the Ewald sphere as the relative energy gain/loss is small for the radiation so that $\vec{K}_o = \vec{K}_1$.

THE MEASUREMENT OF MAGNETIC STRUCTURE FACTORS

J.B. FORSYTH

Neutron Beam Research Unit, SRC Rutherford Laboratory

Chilton, Oxon, OX11 0QX, England

## 1. EXPERIMENTAL METHODS

For an unpolarised neutron beam the diffracted intensities are proportional to the sum of the squares of the structure factors

$$I(\underline{k}) \propto F_N(\underline{k})^2 + F_M^{\perp}(\underline{k})^2$$

where $F_N$ denotes the nuclear structure factor and $F_M^{\perp}$ the component of the magnetic structure factor in the plane perpendicular to the scattering vector $\underline{k}$.

The form factor dependence of magnetic scattering gives rise to a sharp decrease in magnetic scattering intensity with increasing wavevector. Although both conventional powder and single crystal diffraction instruments are used to measure the magnetic scattering from ordered magnetic materials, they are not sufficiently sensitive to measure intensities which correspond to values of the form factor much less than some 0.3 μB. However, it is possible to produce beams of polarized neutrons; if the polarization of the beam if $\underline{P}_o$ and the probabilities of finding neutrons with spin parallel or antiparallel to the direction of $\underline{P}_o$ are $n^+$ and $n^-$ respectively, then the beam polarization is given by

$$P_o = \frac{n^+ - n^-}{n^+ + n^-}$$

The diffracted intensities are then given by

$$I(\underline{k}) \propto (F_N)^2 + 2\underline{P}_o\, F_N.F_M^{\perp} + (F_M^{\perp})^2$$

In comparison to the diffraction of unpolarized neutrons, there is an additional term in $F_N.F_M^{\perp}$ which depends on the polarization and which may give rise to significant scattered intensity even when $F_M$ is small. The measurement of weak magnetic structure factors is effectively confined to the use of the polarized beam technique and it should be emphasised that such small structure factors must be measured if an accurate magnetisation density distribution is to be determined. The use of a polarized incident neutron beam may also be combined with polarization analysis of the diffracted beam to give direct information on the directions of Fourier components of magnetisation within the unit cell or to separate magnetic from nuclear scattering which occurs in the same reflection. Although we shall be mainly concerned with the polarized neutron diffraction method without polarization analysis, techniques which use an unpolarized incident beam will be briefly mentioned, since they are also capable of yielding accurate magnetic structure factors in certain restricted cases.

1.1 Diffraction of an Unpolarized Neutron Beam by a Powdered Sample

The low angle magnetic lines are sometimes well resolved and can be measured with an accuracy which depends on their intensity and is comparable to that of the nuclear peaks. This is particularly true of substances in which the magnetic moment is high, such as in $BaTbO_3$ (S = 7/2) or MnO (S = 5/2), and where the structure is not too complicated. For more complicated structures such as $DyMn_2O_5$ (Dy S = 15/2, $Mn_I$ S = 2, $Mn_{II}$ S = 3/2) or for lower values of S such as NiO (S = 1) this is not possible, either due to the overlap of reflections in the first case (Figure 1) or because, for NiO, even the lowest angle magnetic reflections are weak compared with the nuclear intensities (Figure 2) and only the first few lines may be accurately measured.

The separation of magnetic from nuclear scattering in the same or an overlapped reflection can be made in one of several ways. Ferro- and ferri-magnetic materials which do not exhibit high magnetic anisotropy can have their moments aligned by the application of an external magnetic field. If this field is applied parallel to the scattering vector ($\underline{k}$), the magnetic interaction vector is zero and there is no magnetically scattered intensity. Conversely, if the specimen magnetisation is perpendicular to the plane of scattering, the magnetic scattering is at its maximum.

Measurements made above and well below the Néel transition temperature of an antiferromagnetic material allow the magnetic intensity to be determined by difference. However, this method of separation does not usually lead to very accurate results, since it assumes that the nuclear scattering remains the same at the two temperatures. Small structural changes in cell dimensions and atomic positional and thermal parameters limit the validity of this assumption.

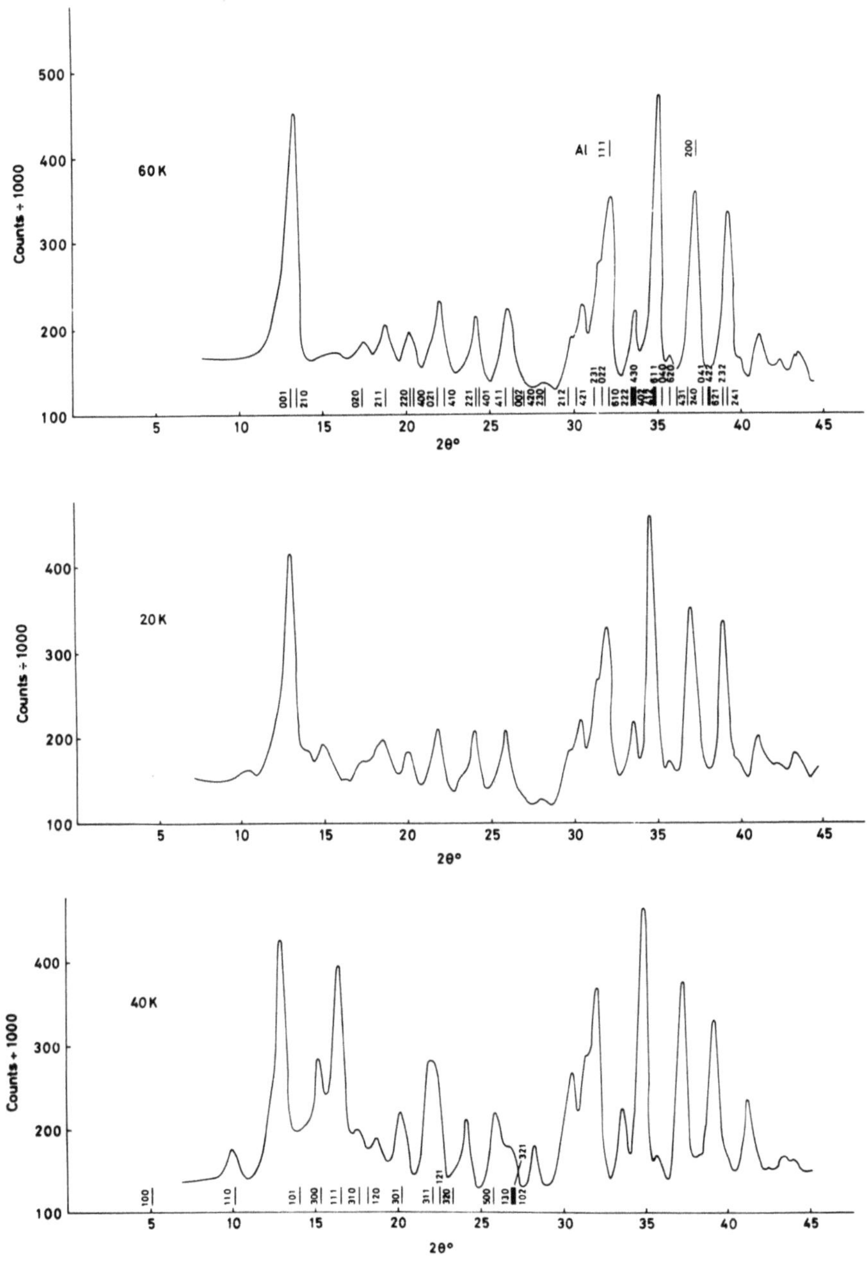


Figure 1 - The magnetic scattering in $DyMn_2O_5$ (Wilkinson et al, 1978). The chemical structure (upper pattern) and the two magnetic structures (lower patterns) are too complex for accurate magnetic structure factors to be derived from the powder diffraction data.

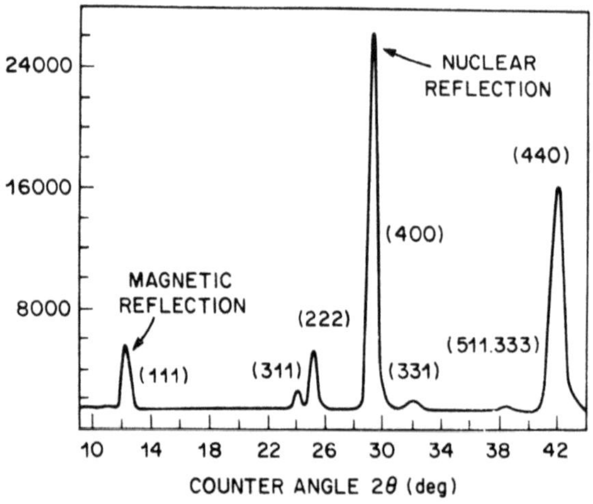


Figure 2 - Powder diffraction pattern of NiO showing that the magnetic peaks (odd indices) are weak even at low scattering angles. (Jacobsen, 1973).

## 1.2 Diffraction of an Unpolarized Neutron Beam by a Single Crystal

In single crystal experiments weaker reflections may be measured more easily and, indeed, the form factor of $Ni^{2+}$ in NiO was one of the first ionic form factors to be accurately determined (Alperin, 1962). The single crystal method normally provides sufficient resolution for adjacent reflections to be measured independently but significant difficulties remain in measuring weak magnetic reflections, in separating magnetic from nuclear intensity in the same reflection and in the estimation of extinction, which is a problem common to both nuclear and magnetic scattering.

## 1.3 The Polarized Neutron Diffraction Method

The principle of this method consists in bathing the sample in a beam of polarized neutrons in which the neutron spins are alternately plus with a polarization P close to 1 and then minus corresponding to a polarization P' close to -1. The diffracted intensities which correspond to the two spin states define the ratio.

$$R = \frac{I^+}{I^-}$$

R is normally termed the 'flipping' ratio. For a centro-symmetric structure it can be expressed in terms of the structure factors as

$$R(\underline{k}) = \frac{F_N(\underline{k})^2 + 2\underline{P}\, F_N(\underline{k})\, \underline{F}_M^{\perp}(\underline{k}) + F_M^{\perp}(\underline{k})^2}{F_N(\underline{k})^2 + 2\underline{P}\, F_N(\underline{k})\, \underline{F}_M^{\perp}(\underline{k}) + F_M^{\perp}(\underline{k})^2}$$

Figure 3 illustrates the relation between $\underline{F}_M^{\perp}(\underline{k})$ and $\underline{F}_M(\underline{k})$ where the angle between $\underline{F}_M(\underline{k})$ and the scattering vector $(\underline{k})$ is $\alpha$.

$$F_M^{\perp} = F_M \sin \alpha$$

$$\underline{P}.\underline{F}_M^{\perp} = p\, F_{M_z}^{\perp}$$

$$\underline{P}.\underline{F}_M^{\perp} = p\, F_M \sin^2 \alpha$$

If we replace $\sin^2 \alpha$ by q we may rewrite R as

$$R = \frac{F_N^{\,2} + 2pq^2\, F_N F_M + q^2\, F_M^{\,2}}{F_N^{\,2} + 2p'q^2\, F_N F_M + q^2\, F_M^{\,2}}$$

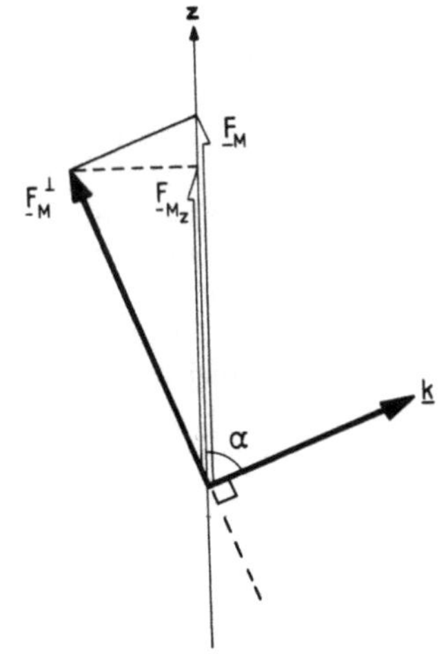


Figure 3 - The relationship between $\underline{F}_M^{\perp}(\underline{k})$ and $\underline{F}_M(\underline{k})$; $\underline{k}$ is the scattering vector which makes an angle $\alpha$ to the magnetisation direction $\underline{z}$.

In the simple case of ideal beam polarization and perfect polarization reversal, p = 1 and p' = -1. If the sample is also magnetised perpendicular to the plane of scattering then $\alpha = \pi/2$ and

$$R = \frac{F_N^2 + 2F_NF_M + F_M^2}{F_N^2 - 2F_NF_M + F_M^2}$$

which may be rewritten in terms of the ratio $\gamma = F_M/F_N$ as

$$R = \left(\frac{1 + \gamma}{1 - \gamma}\right)^2$$

This simple relationship immediately demonstrates the enhanced sensitivity to a magnetic structure factor which is weak compared to the corresponding nuclear one. For unpolarized neutrons one observes an intensity proportional to $F_N^2(1 + \gamma^2)$ whereas with a polarized beam the intensity is proportional to $F_N^2(1 + 2\gamma)$.

1.4 Restrictions on the Use of the Polarized Beam Method

It must be emphasised that the observation of a flipping ratio which is different from unity in the polarized beam method requires that neither the nuclear nor the magnetic structure factors for the reflection be zero or in phase quadrature to each other. This restriction means that the technique cannot be used to study antiferromagnetic structures if their magnetic cell dimensions are multiples of the chemical ones, since the magnetic reflections are then entirely separate from the nuclear ones. NiO is a simple example of this type of structure. Furthermore, the determination of an accurate magnetic structure factor from the observed flipping ratio is only possible if the nuclear structure is centrosymmetric; the situation in an acentric structure is illustrated in Figure 4, from which it can be seen that the flipping ratio does not define the ratio $F_M/F_N$.

However, there remain many materials which can be studied by polarized neutron diffraction since they are either ferromagnets, ferrimagnets or paramagnets in which the moments can be sufficiently aligned by the application of an external magnetic field. For those crystals the experimental determination of R is a simpler process than the measurement of an integrated intensity: it is not necessary to measure the complete diffraction profile, since it suffices to measure the intensity for the two spin states at the peak of the reflection and then correct these rates by a similar measurement in a background position after the crystal has been offset by a few

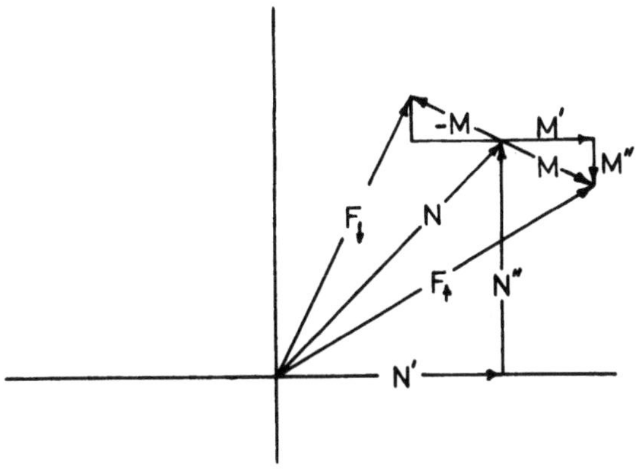


Figure 4 - The flipping ratio (F↑/F↓) does not determine $\underline{F}_M$ in an acentric structure even if $F_N$ is known exactly in amplitude and phase. The real and imaginary parts of $F_N$ and $F_M$ are N', N'', M', M'' respectively.

degrees. Similarly, the two intensities $I^+$ and $I^-$ are absorbed in the same fashion so no absorption correction need be applied to the experimentally observed value for R.

The two values of $\gamma$ which correspond to an experimental value of R are obtained as the roots of the quadratic equation (Figure 5). It is usually evident which root is to be chosen, particularly at high ($\underline{k}$), where $F_M$ is small and $\gamma$ is always less than unity. Since

$$F_M = \gamma F_N$$

it is obvious that accurate magnetic structure factor will only be obtained if $F_N$ is known to the appropriate precision. A determination of the nuclear positions and the thermal vibration parameters appropriate to the temperature at which the flipping ratios are measured is normally the subject of a separate experiment using an unpolarized neutron diffractometer. In crystals with sizeable moments, the unpolarized-beam, low ($\underline{k}$) integrated intensities will contain sizeable magnetic contributions and the nuclear structure factors are best extracted by using the experimental values of $\gamma$, together with the appropriate domain average, $\bar{D}$, to correct for the magnetic contribution

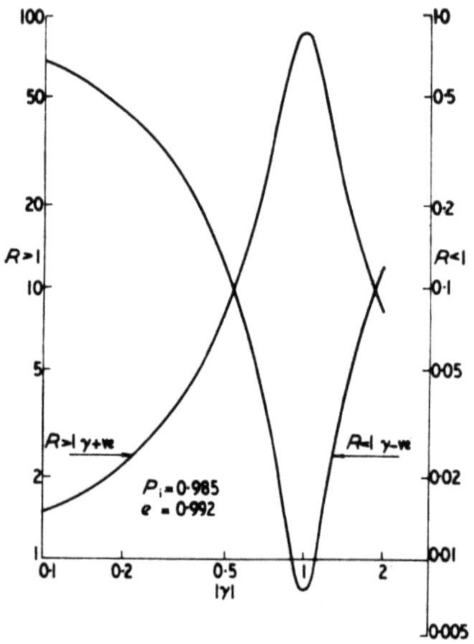


Figure 5 - The relationship between the flipping ratio R and $\gamma = F_M/F_N$ in a typical situation with incident beam polarization $P_i = 0.985$ and a flipping efficiency of 0.992.

$$I(\underline{k}) \propto F_N^2 + \bar{D}.F_M^{\perp 2}$$

$$I(\underline{k}) \propto F_N^2 (1 + \bar{D}\gamma^2)$$

Flipping ratios can only be exploited simply to derive the magnetic structure factors if the magnetic interaction vector $\hat{\underline{Q}}$ is parallel to the applied field. From the macroscopic point of view the magnetic moment must be aligned by the field $\underline{H}$, so materials which exhibit high anisotropy can only be studied with $\underline{H}$ parallel to an easy direction of magnetisation. At the atomic level it is necessary to verify that $\hat{\underline{Q}}(\underline{k})$ is collinear with the magnetic moment.

## 1.5 The Use of the Polarized Beam Technique with Powdered Samples

The observation of flipping ratios in powder diffraction patterns is limited by the considerations which apply to single crystals and which have been outlined in the previous section. However, the low intensity, poor peak/background ratio and the overlap of reflections associated with powder diffraction profiles make the application of the polarized beam technique to powders of very limited interest. Moreover, very high beam depolarization occurs on passing through the sample unless the material has a low magnetic anisotropy.

### 1.6 Polarization Analysis

The conventional polarized beam experiment is useful only for systems with polarization dependent cross sections. More information can often be obtained if the polarization of the diffracted neutrons can be measured. A comprehensive treatment of polarization analysis has been given by Marshall and Lovesey (1971). If the polarization analysis is carried out with respect to the direction of the incident beam polarization, the four partial cross sections corresponding to neutron spin transitions ++, --, +- and -+ can be separately determined.

Both nuclear coherent and nuclear incoherent scattering are always non spin-flip (++, --). Magnetic and nuclear spin scattering is also non spin-flip if the effective magnetisation components are along the neutron polarization direction, but are spin-flip (+-, -+) if the effective magnetisation components are perpendicular to the polarization direction. Further, because only atomic magnetic moment components perpendicular to $\underline{k}$ are effective in scattering neutrons, it follows that all magnetic scattering is spin flip if the neutron polarization direction is along $\underline{k}$. This difference from nuclear scattering can be used to distinguish magnetic from nuclear intensity in a powder diffraction pattern as was first demonstrated by Moon et al (1969), who were also able to measure the paramagnetic scattering from $MnF_2$ above its Néel point. (Figures 6 and 7). Such measurements are usually confined to instruments at high flux beam reactors such as Brookhaven, Oak Ridge and the Institut Laue-Langevin, Grenoble, since polarizers and analyzers for neutrons of wavelength 1 Å and less are inefficient and the final counting rates are low. Although polarization analysis does not help in the measurement of weak magnetic scattering, as does the flipping ratio method, it may be used to determine the absolute direction of $F_M^{\perp}(\underline{k})$ and is therefore a unique tool in the study of non-collinear magnetisation. An example of this type of study is reported by Brown and Forsyth (1977) who have searched for spin-orbit coupling effects in the magnetic neutron scattering by $FeCO_3$. The direction of the incident beam polarization $\hat{\underline{P}}$ is in the plane perpendicular to the scattering vector $\underline{k}$ of a reflection from the specimen crystal. The direction of $\underline{F}_M^{\perp}(\underline{k})$ is found by rotating the crystal about $\underline{k}$ and observing the change in the flipping ratio for the scattered beam after reflection by the analyser. The cross section for scattering without change of neutron spin is proportional to $|\underline{F}_M^{\perp}.\hat{\underline{P}}|^2$, so there is a minimum in $\sigma^{++}$ when $\underline{F}_M^{\perp}$ is perpendicular to the input polarization direction. The minimum in the $\sigma^{+-}$ cross section occurs when $\underline{F}_M^{\perp}$ is parallel to $\hat{\underline{P}}$.

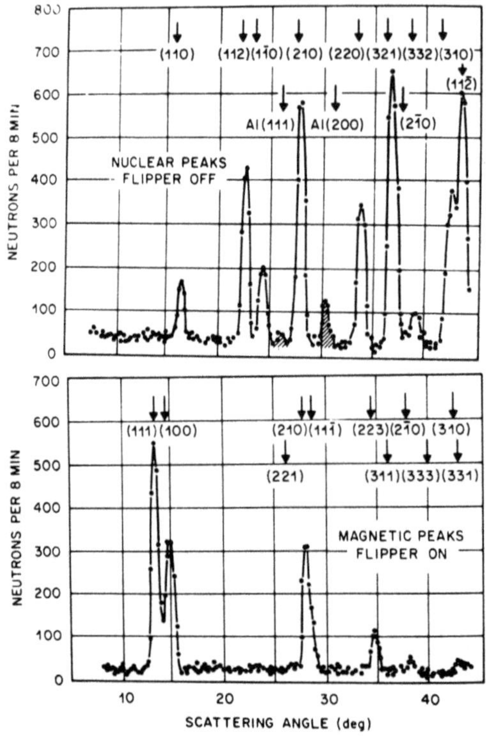


Figure 6 - Separation of magnetic from nuclear scattering in a powder diffraction pattern of $\alpha Fe_2O_3$ using polarization analysis (Moon et al., 1969). The incident neutron polarization is parallel to the scattering vector and all magnetic scattering give rise to spin flip.

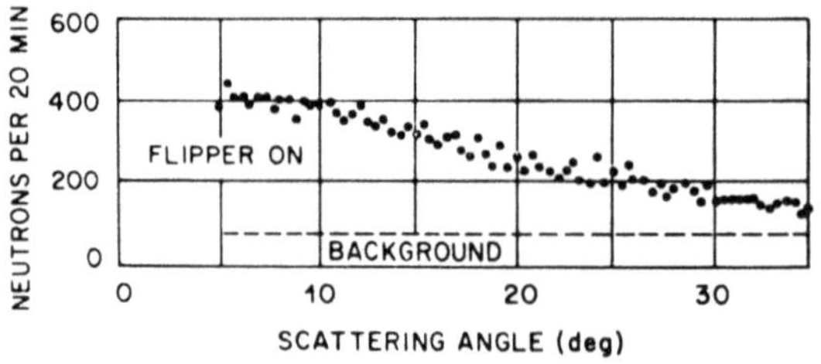


Figure 7 - Separation of paramagnetic scattering in $MnF_2$ by polarization analysis (Moon et al, 1969). The incident beam polarization is again parallel to the scattering vector and only magnetically scattered neutrons are detected with the flipper on.

## 2. THE POLARIZED BEAM DIFFRACTOMETER

Nathans et al (1959) were the first to describe the principles of the polarized beam diffractometer. Figure 8 is a schematic diagram of their apparatus and newer instruments have generally retained this form, save for some variety which has been introduced into the method of neutron spin reversal. Since data are normally required to values of sin $\theta/\lambda$ in excess of 0.8 $Å^{-1}$, the incident neutron beam has a wavelength of about 1 Å or less. As will be discussed in Section 3.4 extinction, that is the diminuition of the scattered intensity from that predicted by the kinematical theory, is the largest probable source of error in single crystal measurements both with polarized and unpolarized neutron beams. Extinction becomes smaller as the wavelength of the incident radiation is reduced and this is another strong reason for not using neutron wavelengths much in excess of 1 Å.

### 2.1 Polarizing the Incident Beam

The wavelength of the incident beam is selected by a magnetised monochromator crystal, which simultaneously introduces a high polarization along this magnetisation direction if its reflecting planes have $|F_N| \simeq |F_M|$. The (200) planes of fcc $Co_{0.92}$ $Fe_{0.08}$ meet this condition and the crystal is used in symmetrical transmission, since this geometry requires a smaller monochromator crystal at the normal take-off angles $\theta_M < 30^{\circ}$. The high absorption cross section of cobalt means that the most effective thickness of monochromator crystal is

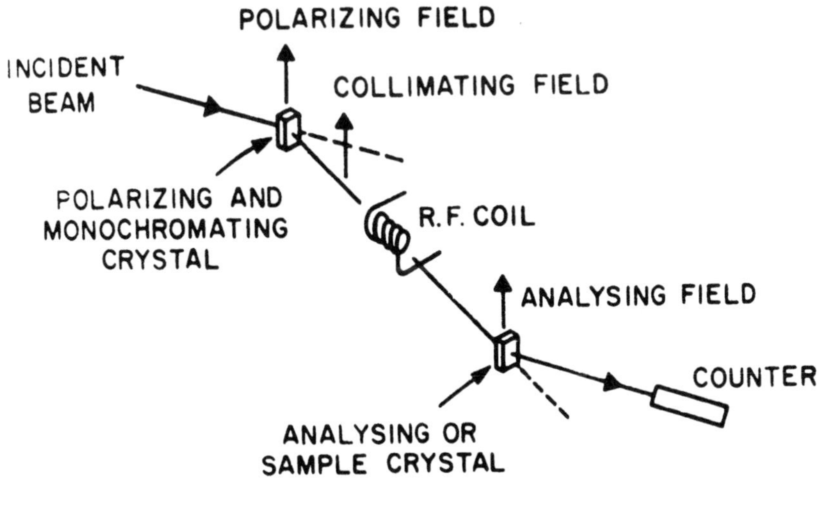


Figure 8 - Schematic diagram of a polarized beam apparatus.

only some 2-3 mm. Crystals have been heat treated and pressed to increase and homogenise their mosaic spread, but the reflectivity of this monochromator seldom exceeds 0.2 for neutrons of the correct spin state.

The Heusler alloy, $Cu_2MnAl$ and $Fe_3Si$ are two materials which have similar characteristics as polarizers and which can give higher reflectivities than $Co_{0.92}Fe_{0.08}$. Both form superlattices based on fcc structures and in each case the 111 reflection is matched. Matching of the nuclear and magnetic scattering depends critically on the state of order, since the nuclear amplitude is the difference between the scattering at two different atomic sites. In both materials the nuclear scattering amplitude of the 222 reflection is significantly higher than that of 111, so that half-wavelength contamination can be a problem (Delapalme et al, 1971). Table I compares the monochromator angle $\theta_M$ for a number of polarizing crystals. It can be seen that $Cu_2MnAl$ and $Fe_3Si$ have much lower focussing angles and will therefore be less suitable for use when high angle reflections from a second (specimen) crystal are to be measured.

| | $Co_{0.92}Fe_{0.08}$ | $Cu_2MnAl$ | $Fe_3Si$ | $^{57}Fe$:Fe | $HoFe_2$ | $(HoTb)Fe_2$ |
|---|---|---|---|---|---|---|
| Matched reflection | (200) | (111) | (111) | (110) | (620) | (444) |
| d-spacing (Å) | 1.76 | 3.43 | 3.27 | 2.03 | 1.16 | 1.06 |
| 2θ for 1 Å neutrons (degrees) | 33.1 | 16.7 | 17.6 | 28.6 | 50.9 | 56.2 |
| Maximum wavelength (Å) | 3.5 | 6.9 | 6.5 | 4.1 | 2.3 | 2.1 |

Table I - The properties of a number of different polarizing monochromators.

Iron is an alternative to $Co_{0.92}Fe_{0.08}$, since it has a similar low d-spacing and high focussing angle but lower absorption. Natural iron has too large a nuclear scattering amplitude ($0.951 \times 10^{-12}$ cm) but Koehler (1975) has made a single crystal of Fe mixed with $Fe^{57}$ ($b = 0.23 \times 10^{-12}$ cm) and 3% Si which gives excellent polarization in the (110) reflection. Clearly such a monochromator is very costly and beyond the reach of most establishments. A cheaper alternative for a high focussing angle polarizer may well result from current research into rare earth-transition metal alloys. Rare earth magnetic moments can be large, up to $\sim 10\ \mu B$, and their scattering factor curves fall more slowly with increasing $\sin\theta/\lambda$ than do those of the 3d electrons. It should therefore be possible to find a material with a matched reflection having a low d-spacing yet still giving good reflectivity. Rare earth alloys commonly exhibit high magnetic anisotropy which would limit the choice of magnetisation direction and might require a high field for saturation. Again, some rare earth elements are unsuitable because of their high absorption factors for neutrons. Schweizer et al (1977) are concentrating their efforts on the cubic Laves-phase $HoFe_2$ where the (620) and (444) reflections promise to give good results. The easy direction of magnetisation in $HoFe_2$ is [001], so the (620) reflection is to be preferred for this reason; unfortunately its reflectivity is some 50% that for (444). The latter reflection can, however, be used if some Ho is replaced by Tb, which produces an easy direction of magnetisation along [110] and only reduces the expected beam polarization to 0.97. It can be seen from Table I that the monochromator's take-off angle is significantly larger than for $Co_{0.92}Fe_{0.08}$. All monochromator housings should be equipped with a means of varying $2\theta_M$, not necessarily continuously, so that the wavelength of the monochromatic beam can be changed. Thermal reactor sources will normally give reasonable intensity down to 0.7 Å, but special hot sources, such as that available at the ILL, can produce high fluxes down to 0.5 Å or less.

## 2.2 The Control of Neutron Polarization

Neutron polarization can be maintained by providing a continuous magnetic guide field along the path of the neutron beam. Initially, both this magnetic field and the direction of neutron polarization will be common to the field direction in the polarizer. Subsequently, the direction of polarization can be changed by providing a slow rotation of the magnetic guide field, such that the rate of change of the field vector $\underline{H}$ acting on the neutron in this adiabatic guide region is slow compared with the Larmor frequency, $\omega_L$, of the neutron

$$\frac{d\theta_H}{dz} \cdot \frac{dz}{dt} << \omega_L = \gamma_n H$$

where $\theta_H$ is the angle between the magnetic field $\underline{H}$ and the neutron beam direction z and $\gamma_n$ ($-1.83 \times 10^4$ rad sec$^{-1}$ Oe$^{-1}$) is the neutron gyromagnetic ratio.

Spin flippers operate on one of two principles:

(a) systems in which the neutron spin undergoes a number of Larmor precessions whilst travelling a well-defined distance in a fixed magnetic field.

(b) systems in which the neutron spin experiences a fast magnetic field reversal, which it cannot follow adiabatically, so the neutron polarization is reversed with reference to the direction of the guide field.

The radio frequency spin flipper used by Nathans et al (1959) was first described by Stanford et al (1954) and it has since been extensively used in polarized neutron diffractometry. The radio frequency field, $H_1$, is applied perpendicular to the neutron polarization direction and is matched to $\omega_L$ in a region of highly uniform magnetic field, $H_o$. The rf current oscillations are adjusted in amplitude to achieve complete spin reversal in a well-defined coil length or neutron transit time. In practice $H_o$ is usually some 0.01 - 0.02T and $H_1 \sim 0.02\ H_o$. The wavelength dependence of the rf flipper is such that its use should be restricted to cases where $\Delta\lambda/\lambda_o < 0.1$. In particular any $\lambda/2$ component in an otherwise monochromatic beam is effectively depolarized on passing through the flipper, since its polarization is rotated by $\pi/4$. (See for example Jones and Williams, 1977).

The direct current spin flipper first suggested by Mezei (1972) is illustrated in Figure 9(a). It consists of two rectangular coils of thickness d mm, one placed immediately behind the other with their axes perpendicular to both the neutron beam direction and a magnetic guide field $H_o$. The coils produce magnetic fields $H_1$ along their axes which are equal and opposite in direction and, in the usual arrangement, these fields are also equal to the guide field $H_o$ ie, $H_1 = H_o = H$. If the coil width is such that

$$d = \pi v/\gamma_n\ H\sqrt{2}$$

where v is the neutron velocity in mm $sec^{-1}$, then all these neutrons undergo exactly half a Larmor precession around a $\pi/2$ cone in each coil and therefore undergo complete spin reversal on traversing both coils, Figure 9(b). Badurek et al (1973) have investigated the sensitivity of this 2 x $\pi/2$ flipper to magnetic field variations and van Laar et al (1975) have shown that the simple relationship given above is in error, since it neglects the presence of the return fields outside the two coils.

Non-adiabatic spin flippers reverse the magnetic guide field as seen by the neutron in a time which is much shorter than its Larmor period. A spin flip transition ($d\theta_H/dz = \pi$) is most easily achieved for fast neutrons, velocity v, in low magnetic fields where

$$\pi v >> \gamma_n H$$

In the flipping process the neutron spin vector can be considered as spatially fixed as it transverses the magnetic discontinuity, and it is therefore reversed with respect to its guide field after passing through the discontinuity. In addition to their usefulness with polychromatic beams, non-adiabatic spin flippers have the advantages that they have no special magnetic field homogeneity and

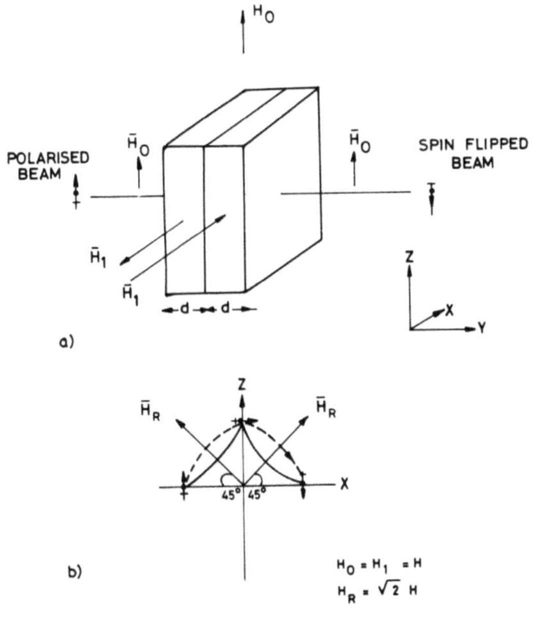


Figure 9(a) - Mezei flipper coils. (b) The rotation of the neutron spins in the two coils.

stability requirements and they can easily be made efficient over large beam areas ($\sim$ 10 cm$^2$). Recently, Tasset (1976) has produced a non-adiabatic spin flipper which uses a superconducting sheet of niobium to isolate the fields from two adjacent sections of guide field. Figure 10 shows the schematic arrangement of permanent magnet guides A and E which provide the initial and final vertical direction of polarization. Electromagnetic windings provide two short guide field regions, B and D, on either side of the superconducting sheet, C: the latter is cooled by conduction from a liquid He reservoir in a large capacity cryostat. Switching the current direction in the first of these coils introduces an adiabatic rotation of $\pi$ in the neutron spin direction in the gap between the regions A and B and thus reversed polarization is transmitted by the foil. This system is in routine use on the D3 diffractometer at the ILL.

## 2.3 The Diffractometer Geometry

The normal beam zero- and higher-layer diffraction geometry is the most convenient for polarized neutron diffractometry. The polarization direction of the incident beam is parallel or anti-

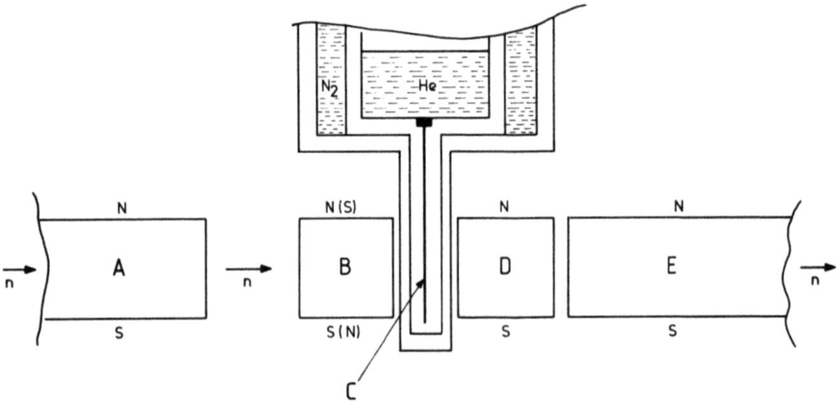


Figure 10 - Non-adiabatic spin flipper system which uses an isolating sheet of superconducting niobium (Tasset, 1976). (See text.)

parallel to the specimen rotation axis, $\omega$, which is also the direction of its magnetising field. The detector moves about two axes, $\gamma$ parallel to $\omega$ and $\nu$. Figure 11 shows the D3 diffractometer which has its $\omega$ axis vertical. The $\nu$ axis allows the detector to be raised out of the horizontal plane so that non-equatorial reflections can be measured. The polarization dependent cross term in the scattering is reduced for these reflections by a factor of sin $\chi$, where $\chi$ is the angle between the scattering vector $\underline{k}$ and the neutron polarization direction, so the angle $\nu$ is normally kept below some 25 - 30$^{\circ}$.

## 2.4 Specimen Environment - Field and Temperature

The simultaneous requirement for both a low specimen temperature and the presence of a high magnetic field is usually met, in conventional electromagnets, by boring out their iron pole pieces to accommodate the vertical tail of a cryostat. Figure 12 shows that a significant simplification in the construction of the cryostat can be obtained by increasing its tail diameter from the practical lower limit of about 18 mm to some 50 mm and incorporating further iron, which is still at room temperature, within the tail (Brown and Forsyth, 1967). The maximum field available from conventional electromagnets is usually limited to around 2T.

Superconducting Helmholtz coils usually have no pole pieces and a variable sample temperature is easier to obtain. The two coils are usually made asymmetric and coupled in series if field nodes in the neutron beam line must be avoided so that beam polarization is retained (Gilbert et al, 1973). The D3 polarized beam diffractometer

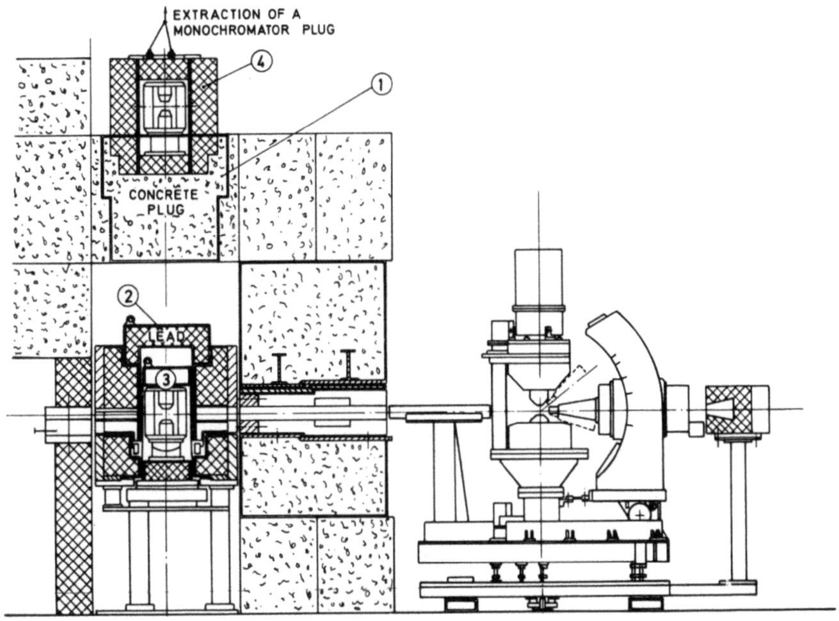


Figure 11 - The D3 polarized neutron diffractometer at the Institut Laue-Langevin, Grenoble. The monochromator crystal is magnetised by a permanent magnet (3). If $Co_{92}Fe_8$ is used it becomes very radioactive and must be well shielded by lead (4) when it is removed.

at the Institut Laue-Langevin, Grenoble operates with such a Nb-Ti superconducting magnet which gives a persistent field of 4.8T†. Niobium-tin tape has been used by Felcher at the Argonne National Laboratory, Chicago, to achieve fields of 10T, but the magnet will not work in the persistent mode and helium consumption is high. However, the advent of filamentary Nb-Sn wire should quickly lead to the production of a 10T magnet capable of persistent-mode operation.

The mechanical forces between pairs of superconducting coils become very large at the higher fields and the separators between the coils have to be massive. Systems of concentric rings introduce a large amount of unwanted scattering, so a number of discrete wedges are to be preferred in cases where all the angles in the plane of scattering need not be simultaneously available. A schematic plan of the superconducting magnet in use on the D3 and D5 diffractometers is given in Figure 13.

---

†Manufactured by the Oxford Instrument Co, Osney Mead, Oxford, UK.

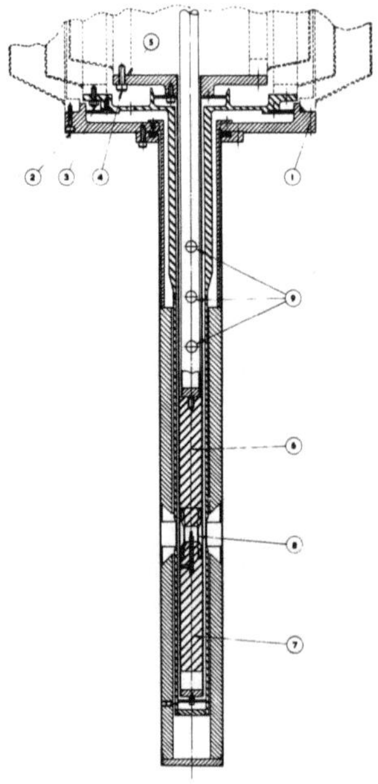

Figure 12 - Helium cryostat for insertion into an electromagnet (Brown and Forsyth, 1967). The outer iron pole pieces remain at ambient temperature.

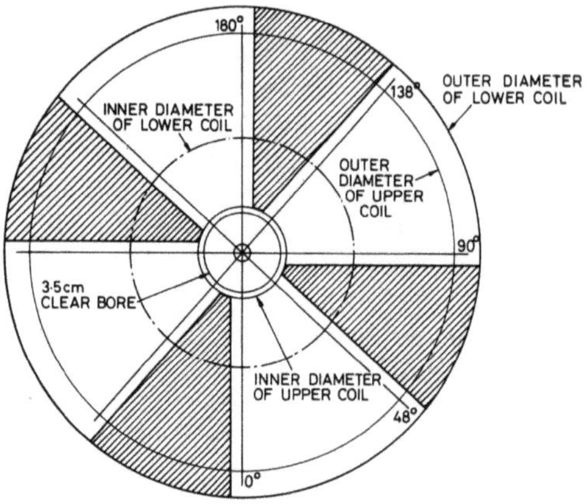


Figure 13 - Plan view of the D3 superconducting 4.8T magnet showing the arrangement of the coil separators. The obscured scattering angles become available on rotating the magnet by 45°.

## 2.5 Experiment Control

The time required to measure the flipping ratio of a reflection obviously depends on the accuracy to which the ratio is required, on the intensity of the peak of the reflection and on the intensity in the background. The ratio R is calculated from

$$R = \frac{PUCR - BUCR}{PDCR - BDCR}$$

where PUCR is the peak count rate for neutrons with spin up
PDCR " " " " " " " " " down
BUCR is the background count rate for neutrons with spin up
BDCR " " " " " " " " " down

It is usual to arrange for the times spent measuring the rates to be controlled separately, so that the statistical accuracy in R can be maximised in any given situation. In early polarized beam diffractometers, such as those at AERE Harwell, the diffractometer circles were not under computer control and the count times ratios for (peak up/peak down) and (peak up or peak down/background up and background down) were set manually for each reflection in the ranges 7/1 to 1/7 and 7/1 to 1/1 respectively. Measurement times varied between 15 minutes and several days per reflection.

The D3 and D5 diffractometers at the ILL are computer controlled and the high incident fluxes lead to relatively short measuring times for many reflections. The diffractometer shafts are set to the angles defined by the diffraction geometry, the incident neutron wavelength and the crystal orientation. The latter information is most conveniently described in terms of its orientation matrix, UB, or its transpose (Busing and Levy, 1967). A peak search routine may then be entered to ensure that the final settings correspond to the peak of the diffracted intensity. This is important in the measurement of weak magnetic structure factors when a high-field superconducting magnet is being used to align a paramagnetic crystal. Moon et al (1975) have demonstrated that magnetic forces which acts on a neutron as it enters a high-field magnet can influence the measured value of R. The force on the neutron is proportional to the gradient of the field modulus and the direction of the force is opposite for the two neutron spin states. This results in different velocity and position distributions for the two spin states at the sample position and affects the observed intensity ratio. Moon et al (1975) conclude that the most important of these effects is probably the change in velocity (or wavelength), resulting in a different Bragg angle (by about $10^{-1}$ degree) for the two spin states. The error in the flipping ratio is given by

$$\Delta R = -\frac{2|\mu_N|}{m_N v_o^2} \tan \theta H \frac{1}{S(\theta)} \frac{dS(\theta)}{d\theta}$$

where $\mu_N$ is the neutron moment, $v_o$ is the neutron velocity, H is the central field and D(θ) defines the shape of the crystal rocking curve. It is thus very important to position the crystal so that dS/dθ is as close to zero as possible. Figure 14 shows the calculated error in R as a function of wavelength for Gaussian peak shapes of various widths when the crystal is off the peak position by $0.02^o$. The task of setting and maintaining the crystal position to better than $0.02^o$ over the course of an experiment is non-trivial, especially when the crystal is mounted in the centre of a superconducting magnet. Sample support and control mechanisms are light weight to avoid excessive heat flow. Coolant levels vary slowly with time producing changing thermal gradients along the sample support. The ideal crystal for these measurements should therefore have a broad, flat-topped rocking curve.

Moon et al (1976) have also pointed out another field effect not discussed in their earlier paper. This is peculiar to split coil magnets which operate with unequal currents in the two coils or with

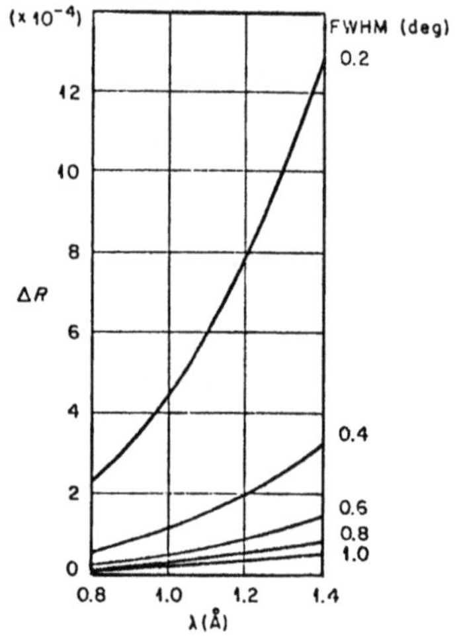


Figure 14 - Calculated error in flipping ratio for a crystal misset from the peak position by $0.02^o$, assuming Gaussian peak shapes of various widths H = 5.72T, sin θ/λ = 0.26 $Å^{-1}$ (Moon et al, 1976).

similar currents but unequal numbers of windings so as to avoid zero-field regions along the neutron path and the consequent depolarization of the beam. This cure for the depolarization problem has the undesirable side-effect that there is a vertical field gradient all along the neutron flight path. This produces a vertical separation of the two neutron spin states, as in the Stern-Gerlach experiment. The vertical angular deviation between two neutrons with opposite spins, each initially travelling on the same horizontal track with equal velocities, is given by

$$\delta(\uparrow) - \delta(\downarrow) = - \frac{2|\mu_N|}{m_N v_o^2} \int \frac{\delta H}{\delta z}\, d\rho$$

where the integration is along the neutron path, both into and out of the magnet. For their magnet operated at 6T this angular deviation is about $6 \times 10^{-4}$ deg, resulting in a vertical separation at the counter position of about $10^{-3}$ cm. Ordinarily this is not a problem because the counter diameter is large compared to the sample and the slits in front of the counter are wide enough to accept the entire reflected beam. However, if the crystal is slightly misaligned so that the reciprocal lattice vector is not exactly in the horizontal plane, the observed intensity can be sensitive to the vertical divergence with ` resulting error in the flipping ratio. This can produce an apparent front-back effect, that is, different results when the crystal is rotated $180^o$ about a vertical axis. If in one case the reflected beam is limited from above, it will be limited from below in the other case, and the error due to the vertical field gradient will be opposite in direction for the two cases. A confirmatory experiment on the (002) reflection from Ti, in which part of the detector was deliberately masked first from above and then from below, produced differences in the apparent residual flipping ratio of some 20 parts in $10^{-4}$.

Once the diffractometer circles are correctly set, the measurement of R begins. The incident beam is flipped for a fraction of the total period for a spin-up, spin-down cycle of some 1-2 seconds. Diffracted neutrons corresponding to the two incident spin states are collected in separate counters. After the peak measurement is completed, similar measurements are made at two background positions on either side of the peak crystal angle by moving the $\omega$ axis through $1^o - 4^o$. The sequence of peak and background measurements is then repeated after the control computer has estimated the four count rates PUCR etc and predicted the fraction of the total measuring time which should be assigned to their determination. The principle of optimization is that each rate should be measured to an accuracy which makes an equal contribution to the estimated error of the flipping ratio R. By making the measurement of R for each reflection over some five or ten peak-background cycles, the time wasted in the initial cycles through the lack of prior knowledge of R can be completely compensated. The up/down time ratio is set within the spin up - spin down cycle of 1 - 2 seconds and the time to measure the four different rates is controlled by counting the appropriate number of time or monitor pulses.

## 2.6 Multiple Scattering

In the Ewald construction, a diffraction peak occurs when a reciprocal lattice point intersects the reflecting sphere. Multiple scattering (the Renninger effect) occurs if a second reciprocal lattice point is simultaneously in contact with the reflecting sphere and such scattering can influence the measurement of R or the integrated intensity of a reflection. Clearly, the possibility of multiple reflection increases with shorter incident wavelengths, however, their effect is less severe since the reflectivity is proportional to $\lambda^3/\sin 2\theta$. In general, the crystal should not be oriented such that multiple reflection must occur. An example of such a situation would be an alignment with a principal zone axis exactly parallel to the $\omega$ axis of the diffractometer if non-zero layer reflections are to be measured and the reciprocal lattice has a translation vector parallel to this zone axis.

A rotation about the scattering vector keeps the principal reciprocal lattice point on the reflecting sphere but permits a secondary point to move away from contact. The diffractometer D5 at the ILL usually operates in the four-circle geometry with the detector in the horizontal plane. The three crystal rotations $\omega$, $\chi$ and $\phi$ can then be used to investigate the dependence of R on the angle of rotation about the scattering vector. In this way anomalous results due to multiple scattering can be eliminated from the data.

## 2.7 Data Processing

The first step in data processing is normally the correction for counts lost due to the finite dead-time of the detecting chain. After this, a number of other corrections have to be made and these are discussed in the next Section. All these computations are most easily carried out in a main-frame computer, so the raw data from the diffractometer specifying the reflection indices, the diffractometer angles and the observed rates PUCR etc (Section 2.5) are normally written on to magnetic tape by the experimental control computer so that they may be transferred easily.

## 3. CORRECTIONS AND ACCURACY

The formula for the flipping ratio given in Section 1.3, namely

$$R = \frac{F_N^2 + 2q^2 F_N F_M + q^2 F_M^2}{F_N^2 - 2q^2 F_N F_M + q^2 F_M^2}$$

corresponds to an ideal case. In the experimental determination of magnetic structure factors through the measurement of flipping ratios it is necessary to introduce two types of corrections which modify this formula. The first of these is concerned with the imperfections of the actual apparatus used in the measurements, whereas the second type are due to the specimen crystal itself. The most important correction of this later type is to take account of any extinction. Both types of correction will be described in the following paragraphs and the final formula for the flipping ratio is given in Appendix 1 to this account in a form due to Boucherle (1977).

### 3.1 Half Wavelength Contamination in the Polarized Incident Beam

A monochromator set to diffract neutrons of wavelength $\lambda$ from a set of Bragg planes ($hk\ell$) will also be correctly oriented to diffract neutrons of wavelength $\lambda/2$ from the planes ($2h$, $2k$, $2\ell$). If the structure factor for this reflection is not zero, there will be a $\lambda/2$ contamination in the monochromatic beam. The $\lambda/2$ component is very weak for a $Co_{0.92}Fe_{0.8}$ monochromator working at $\lambda < 1$ Å on a thermal reactor hole. However, in the case of a hot source instrument, such as D5 at the Institut Laue-Langevin, the $\lambda/2$ contribution may need to be attenuated and corrected for. The maximum intensity in the neutron spectrum from the hot source occurs between 0.4 - 0.5 Å and the $\lambda/2$ contribution becomes important for wavelengths greater than 0.7 Å. Fortunately, there are a number of elements which exhibit strong resonance absorptions in the wavelength range 0.24 - 0.53 Å and which are suitable filters for $\lambda/2$ in 0.48 - 1.08 Å monochromatic beam. Table II gives values for their total scattering cross section at the resonance and at a wavelength twice that of the resonance: the use of such a filter clearly imposes certain restrictions on the choice of monochromatic wavelength.

By measuring the integrated intensity of the same reflection at both $\lambda$ and $\lambda/2$ one obtains the fluxes $J(\lambda)$ and $J(\lambda/2)$ which allow corrections to be made in deriving a true value of $\gamma$ from the observed flipping ratio via a constant

$$C = J(\lambda/2)/8J(\lambda).$$

| Element | Resonance eV | $\sigma_T$(resonance) barns | $\lambda$ Å | $\sigma_T(\lambda)$ barns | $\frac{\sigma(\lambda/2)}{\sigma(\lambda)}$ |
|---|---|---|---|---|---|
| $_{68}Er$ | 0.46 | 2300 | 0.84 | 125 | 18.4 |
| | 0.58 | 1500 | 0.75 | 127 | 11.8 |
| $_{77}Ir$ | 0.66 | 4950 | 0.70 | 183 | 27.0 |
| $_{91}Pa^{231}$ | 0.39 | 4900 | 0.92 | 116 | 42.2 |
| $_{94}Pu^{239}$ | 0.29 | 3800 | 1.06 | 500 | 7.6 |
| $_{94}Pu^{240}$ | 1.08 | 115000 | 0.55 | 145 | 14.50 |
| $_{90}Th^{229}$ | 0.61 | 6200 | 0.73 | <100 | >62.0 |
| $_{45}Rh$ | 1.27 | 4500 | 0.51 | 76 | 59.2 |
| $_{72}Hf$ | 1.10 | 5000 | 0.55 | 58 | 86.2 |
| $_{63}Eu$ | 0.46 | 10100 | 0.84 | 1050 | 9.6 |
| $_{49}In$ | 1.45 | 30000 | 0.48 | 94 | 319 |

Table II - Resonance absorption data for some thermal neutron filters.

### 3.2 Incident Beam Polarization and Flipping Efficiency

If the initial beam polarization given by the monochromator is P and the flipping efficiency is e, the polarization of the flipped beam is -Pe. The sign of P depends on the choice of monochromator crystal and is positive for $Co_{0.92}Fe_{0.08}$ and negative for $Cu_2MnAl$. The flipping ratio given by

$$R = \frac{F_N^2 + 2Pq^2 F_N F_M + q^2 F_M^2}{F_N^2 - 2Peq^2 F_N F_M + q^2 F_M^2}$$

is appropriate to the use of a radio frequency flipper where the efficiency for $\lambda/2$ is zero (Section 2.2).

Brown and Forsyth (1964) have shown that the values of P and e can be derived from measurements of the flipping ratios for two special analyser crystals namely the (200) reflection from $Co_{0.92}Fe_{0.08}$ and the (111) reflection from the Heusler alloy $Cu_2MnAl$, which have $\gamma = +1$ and $\gamma = -1$ respectively. Boucherle (1977) has included the correction for significant $\lambda/2$ contamination in the polarized beam. The pair of simultaneous equations which must be solved to find P and e depend on the particular choice of monochromator:

For $Co_{0.92}Fe_{0.08}$, the (400) reflection has $\gamma = 0.286$ and gives a polarization P of 0.457: the equations are then

$$P = 1 - (1 - e)\, R_H + 1.97C$$

$$e = 1 - \frac{(1 + P)}{R_{Co}} + 0.54C$$

where $R_{Co}$ and $R_H$ are the flipping ratios for the two analyser crystals.

For the monochromator $Cu_2MnAl$, (222) has $\gamma = 4.50$ and gives a polarization of -0.380: the simultaneous equations are then

$$P = -\left[1 - (1 + e)R_{Co} - 0.42C\right]$$

$$e = -\left[1 - \frac{(1 - P)}{R_H} + 2.43C\right]$$

### 3.3 Depolarization in the Specimen

The incident beam may suffer depolarization in traversing the specimen. Depolarization is reduced if the specimen is both elongated and magnetised along an easy direction of magnetisation. Strongly-magnetic, metallic crystals should have any misoriented surface material removed by careful polishing. The depolarization can be measured by measuring the polarization, P', of the beam transmitted by a sample of thickness, t, in the manner described in previous section. Brown and Forsyth (1964) have shown that correction for volume depolarization takes the form of replacing P by PD in the equations for R and $\gamma$. D is different for each reflection and is found by calculating

$$D = \int \exp\,(-\beta d)\, dv$$

over the whole crystal volume, where d is the path length of the incident beam and $\beta$ is given by

$$\beta = \frac{1}{t}\,\ell n\,\frac{P}{P'}$$

The calculation is similar to the calculation of x-ray or neutron absorption coefficients but the integration is only carried out over the incident beam path.

### 3.4 Extinction

If the sample crystal exhibits extinction, the intensities $I^+$ and $I^-$ will be affected to different extents if their ratio, R, is not unity. However, it is possible to make a satisfactory correction for small extinction without a prior knowledge of the crystal structure and its moment distribution, since in this case the extinction is proportional to $\lambda^3/\sin 2\theta$. A comparison of measurements made at several different wavelengths is therefore a good means of determining the extinction parameters. In the ordinary measurement of integrated intensities, this comparison can only be made after the intensity measurement at different wavelengths have been normalised to each other. Since such measurements are on a relative scale, extinction can only be treated by a global examination of all the diffraction data.

In the polarized beam technique, however, the measurement of R provides an absolute value which can be compared with other measurements made at different wavelengths, $\lambda_j$, for the same reflection, $h_i$. The values of $R(h_i, \lambda_j)$ depend only on the magnetic structure factor $F_M(h_i)$, which is considered to be unknown, and the extinction parameters which describe the crystal. Darwin (1922) defines an imperfect crystal as the juxtaposition of small perfect blocks which are slightly misoriented, each to the others. Two parameters are sufficient to describe this mosaic model: t the dimension of the blocks and g which determines the distribution $W(\Delta)$ of the orientations of the small blocks according to a Gaussian distribution

$$W(\Delta) = \sqrt{2}\, g \exp(-2\pi g^2 \Delta^2) \qquad \text{with } g = 1/2\sqrt{\pi}\eta$$

where $\eta$ represents the crystal mosaic.

If one therefore measures $R(h_i, \lambda_j)$ for the same $h_i$ and at two or more wavelengths, it is possible to extract the parameters t and g and the true value for $F_M(h_i)$ by a refinement process (Brown, 1970; Boucherle, 1977). This method is not suitable for experiments in which all the magnetic structure factors are very weak compared to the nuclear ones, and the ratios R lie in the range, say, 0.99-1.01. The relative errors of the magnetic structure factors measured in this sort of experiment are much larger than in experiments to measure the distribution of moment corresponding to several Bohr magnetons and long measurement times would be necessary to achieve sufficient accuracy for a meaningful refinement of $R(h_i, \lambda_j)$ to be made. In these cases, integrated intensity measurements depend little on the magnetic distribution, which is unknown, and they can be used to determine the extinction parameters from a refinement of nuclear structure in the classical manner.

The determination of accurate magnetic structure factors is difficult if the specimen crystal exhibits strong extinction. Indeed, the flipping ratios from crystals of yttrium iron garnet of various thicknesses have been measured by Bonnet et al (1976), who have shown that the results form a sensitive test for the suitability of the extinction model adopted. However, every step should be taken to minimise or eliminate extinction effects unless they are themselves the object of study. The advent of pulsed neutron sources with high fluxes of shorter wavelength neutrons, coupled to data collection systems which facilitate the determination of flipping ratio as a function of wavelength, should improve the experimental situation.

## REFERENCES

Alperin H A (1962), J Phys Soc Japan Suppl BIII, 17, 12.
Badurek G, Westphal G P and Zeigler P (1973), Nucl Instr and Methods 120, 351.
Bonnet M, Delapalme A, Becker P and Fuess H (1976), Acta Cryst A32, 945.
Boucherle J-X (1977), Thèse pour docteur es-sciences physiques Université de Grenoble.
Brown P J (1970), in Thermal Neutron Diffraction (Ed B T M Willis), London Oxford Univ Press p179.
Brown P J and Forsyth J B (1964), Brit J Appl Phys 15, 1529.
Brown P J and Forsyth J B (1967), Proc Phys Soc 92, 125.
Brown P J and Forsyth J B (1977), J Phys Soc C 10, 3157.
Busing W R and Levy H A (1967), Acta Cryst 22, 457.
Darwin C G (1922), Phil Mag 43, 800.
Delapalme A, Schweizer J, Couderchon G and Perrier de la Bathie R (1971), Nucl Instr and Methods 95, 589.
Gilbert E, Hanley P, Hayter J B and White J W (1973), J Phys Soc E 6, 714.
Jacobson A J (1973), in Chemical Applications of Neutron Scattering (Ed B T M Willis) London Oxford Univ Press p270.
Jones T J L and Williams W G (1977), Rutherford Laboratory Report RL 77-079A.
Koehler W C (1975), private communication.
van Laar B, Maniawski F and Mijnarends P E (1976), Nucl Instr and Methods 133, 241.
Marshall W and Lovesey S W (1971), Theory of Thermal Neutron Scattering, London Oxford Univ Press.
Mezei F (1972), Z Physik 255, 146.
Moon R M, Koehler W C and Shull C G (1975), Nucl Instr and Methods 129, 515.
Moon R M, Koehler W C and Cable J W (1976), Proceedings of the Conference on Neutron Scattering, Gatlinburg, USA.
Moon R W, Riste T and Koehler W C (1969), Phys Rev B5, 997.

Nathans R, Shull C G, Shirane G and Andresen A (1959), J Phys Chem Solids 10, 138.
Schweizer J, Gregory A and Givord D (1977), private communication.
Stanford C P, Stephenson T E, Cochran L W and Bernstein S (1954), Phys Rev 94, 374.
Tasset F (1976), private communication.
Wilkinson C, Sinclair F, Forsyth J B and Wanklyn B M Y (1978), submitted for publication in J Phys Soc C.

## APPENDIX

## GENERAL FORMULA FOR THE CORRECTION OF OBSERVED FLIPPING RATIOS

This appendix follows the work of Boucherle (1977), who has based his treatment on methods developed by Delapalme (1967) and Tasset (1975).

### A. Complete Formula for the Flipping Ratio R

In the case where it is possible to replace $F^{\perp}_{Mz}$ by $q^2F_M$ and $F^{\perp 2}_M$ by $q^2F_M{}^2$, one obtains the global formula for R by introducing the different experimental corrections for extinction; the flipping ratio has been defined previously as

$$R = \frac{PUCR - BUCR}{PDCR - BDCR}$$

where PUCR is the count rate in the peak measured with the flipper off, PDCR is the rate when the flipper is on. BUCR and BDCR are the corresponding background rates. In these conditions the calculated flipping ratio is given by

$$R = \frac{F_N{}^2 + q^2\ F_M{}^2 + 2pq^2\ F_NF_M + ESP + ESFP + APR}{F_N{}^2 + q^2\ F_M{}^2 - 2peq^2\ F_NF_M + ESM + ESFM + BPR}$$

where

- $F_N$ is the nuclear structure factor
- $F_M$ is the magnetic structure factor
- $q^2 = \sin^2\alpha$ ($\alpha$ is the angle between the magnetisation and the scattering vector $\underline{k}$)
- p is the polarization in the specimen
- e is the flipping efficiency
- ESP, ESFP, ESM and ESFM represent the extinction corrections

and APR and BRP correspond to the $\lambda/2$ contributions.

### B. Extinction

ESFP and ESFM may be non-zero only if $q^2 \neq 1$ and they take account of the modifications introduced into the extinction correction by diffraction which involves spin flip. Writing

$$f^{+2} = F_N{}^2 + q^2\ F_M{}^2 + 2q^2\ F_NF_M$$

$$f^{-2} = F_N{}^2 + q^2\ F_M{}^2 - 2q^2\ F_NF_M$$

one obtains the following expressions for the extinction corrections:

$$ESP = -\frac{1}{2}\left[f^{+2}\ ES^{+}\ (1 + p) + f^{-2}\ ES^{-}\ (1 - p)\right]$$

$$ESM = -\frac{1}{2}\left[f^{+2}\ ES^{+}\ (1 - pe) + f^{-2}\ ES^{-}\ (1 + pe)\right]$$

where

$$ES^{+} = ES\ f^{+2}$$

$$ES^{-} = ES\ f^{-2}$$

and ES is expressed to first order as

$$ES = \frac{\lambda^{3}}{V^{2}\sin2\theta}\left[\frac{2}{3}\ \frac{t^{2}\sin2\theta}{\lambda} + \frac{1}{\sqrt{2}}\ \frac{T}{\sqrt{\left(\frac{\lambda}{t\sin2\theta}\right)^{2} + \left(\frac{1}{\sqrt{2}\ g}\right)^{2}}}\right]$$

In cases where $q^2 \neq 1$, the terms ESFP and ESFM are given by

$$ESFP = 4pq^{4}\ (1 - q)\ F_N F_M^{\ 3}\ ESF$$

$$ESFM = -4\ peq^{4}\ (1 - q^{2})\ F_N F_M^{\ 3}\ ESF$$

where

$$ESF = \frac{\lambda^{3}}{V^{2}\sin2\theta}\left[\frac{1}{\sqrt{2}}\ \frac{T'}{\sqrt{\left(\frac{\lambda}{t\sin2\theta}\right)^{2} + \left(\frac{1}{\sqrt{2}\ g}\right)^{2}}}\right]$$

and T' is the effective path length in the crystal after scattering. The extinction parameters t and g in the above expressions may be obtained directly from a set of observations of each flipping ratio as a function of wavelength providing the extinction is weak.

### C. Correction for $\lambda/2$ Contamination in the Beam

The $\lambda/2$ correction coresponds to the contribution of the reflection (2h, 2k, 2ℓ) for neutrons of wavelength $\lambda/2$; because of the $\lambda^3$ dependence of the reflectivity, this contribution is characterised by

$$C = \frac{1}{8}\ \frac{J(\lambda/2)}{J(\lambda)}$$

where $J(\lambda)$ is the flux for the wavelength $\lambda$.

By defining $ff^{\pm 2}$ for the reflection (2h, 2k, 2ℓ) in the same manner as one has defined $f^{\pm 2}$ for the reflection (hkℓ), one obtains

$$APR = C \left[ ff^{+2} \frac{1+p'}{2} (1-ESS^{+}) + ff^{-2} \frac{1-p'}{2} (1-ESS^{-}) \right]$$

$$BPR = C \left[ ff^{+2} \, 1/2 \, (1-ESS^{+}) + ff^{-2} \, 1/2 \, (1-ESS^{-}) \right]$$

here $ESS^{+}$ and $ESS^{-}$ are defined in an analogous manner to $ES^{+}$ and $ES^{-}$ by replacing $f^{\pm 2}$ by $ff^{\pm 2}$ and $\lambda$ by $\lambda/2$. p' is the polarization of the $\lambda/2$ neutrons. It should be noted that these formula are appropriate to an rf flipper which has zero efficiency, e', for $\lambda/2$ radiation.

D. Calculation of $\gamma$

Starting from the global formula, it is possible to derive $\gamma = F_M/F_N$ as the solutions of a quadratic equation

$$\gamma = U \pm \left[ U^2 - 1/q^2 - \frac{R}{R-1} \frac{ESM+ESFM+BPR}{q^2 F_N^2} + \frac{1}{R-1} \frac{ESP+ESFP+APR}{q^2 F_N^2} \right]^{\frac{1}{2}}$$

where

$$U = p \left( \frac{Re+1}{R-1} \right)$$

The choice between the two roots can usually be made on physical grounds, save for some instances where $\gamma \sim 1$.

REFERENCES

Boucherle J-X (1977), These pour docteur es-sciences physiques Universite de Grenoble.
Delapalme A (1967), These Universite de Grenoble.
Tasset F (1975), These Universite de Grenoble (AO 10916).

# ERROR ANALYSIS IN EXPERIMENTAL DENSITY DETERMINATION

M.S. LEHMANN

Institut Laue-Langevin, Grenoble, France

## INTRODUCTION

Whenever we use experimental data to obtain some quantity we will during this process be presented with the problem of assessing the relative and absolute reliability of our data. In general we will assign an error to each observation and use this error in the mathematical analysis. For an observed squared structure amplitude $F^2_{obs}$ we estimate an error $\sigma(F^2_{obs})$.

In the study of electron densities we employ the two types of analysis common in diffraction work, namely Fourier transformation of the structure amplitudes to get an electron density, and adjustment by least squares techniques of parameters describing the electron density distribution. In the first case observations are included with equal weight in the summations giving the density, and the errors of the observations are then used to estimate the error in the density. In the second case we obtain estimates of the parameters from a set of equations, and we assign to each equation a weight $w(F^2_{obs}) = 1/\sigma^2(F^2_{obs})$. Indeed we can reject observations ($\sigma=\infty$) if we suspect that they will seriously bias the outcome. This seems at first glance an advantage and would influence the outlay of the experiment and the error analysis. Reflections affected by extinction could for example be excluded from the least squares treatment. By rejecting data we risk however to reject vital information. Inclusion of all data in both types of analysis is therefore desirable, and we will use the same experimental approach and methods for estimating the errors in the two cases. We will not make any distinction between X-ray and neutron data.

The object is therefore for us together with the observation to supply an error. This is done by adding together the various errors $\sigma_i$ that can be thought of

$$\sigma = (\sum_i \sigma_i^2)^{1/2}$$

assuming that each $\sigma_i$ is the standard deviation of a probability distribution function for some experimental variable. Using the central limit theorem we can state that our total error distribution will approach a normal distribution with standard deviation $\sigma$.

The experimental errors can be divided into statistical errors and systematical errors, but the line of division between the two types is floating. Statistical errors which follow the above assumption are most often encountered in the early parts of the experiment. The most prominent is the counting statistical error. Other errors of this type where we can assume the experimental variable to vary randomly are short term fluctuations in measurement equipment and temperature as well as errors in the setting of the crystal during a scan. Characteristic of these errors are that the standard deviation can be reduced by prolonged or repeated measurements. Systematic errors are associated with basic errors in the measurement equipment such as inhomogeneities in the beam, long term drift of the flux and recording equipment, and the magnitude of these errors can often be revealed by repetition of the measurement after the necessary improvements or adjustments of the equipment. In addition there is a series of errors associated with the last part of the experiment where the corrections for absorption, thermal diffuse scattering, multiple scattering and extinction are done, and errors of this nature are generally only discovered when the measurement is repeated by somebody else somewhere else.

Many of the above errors will, as discussed later, lead to standard deviations which are proportional to the integrated intensity. It is therefore customary in diffraction studies to assign to the intensity I a standard deviation $\sigma(I)$ (Busing and Levy, 1957)

$$\sigma(I) = (\sigma^2_{count}(I)+(k\ I)^2)^{1/2}$$

where the first term describes the counting statistical contribution, and k is a parameter fixed for the data set. We then avoid a tedious and probably rather arbitrary book-keeping of the various error contributions. The factor k is often fixed so that the agreement among reflections measured many times (standard reflections or eventually symmetry related reflections) is comparable to the standard deviation (see McCandlish et al., 1975). This is done by ensuring that the sample variance defined as

$$s^2 = \frac{1}{n-1} \sum_j (\bar{I}-I_j)^2$$

where $\bar{I}$ the mean of n measurements $I_j$ is equal to the mean of the standard deviations defined above.

The quantities and relations used in this discussion can be found in many textbooks (for example Hamilton, 1964), and a short summary of the definitions and relations is given in the Appendix together with a remark on the relation between $\sigma(F_{obs})$ and $\sigma(F^2_{obs})$.

## ERRORS OCCURING DURING THE MEASUREMENT

The step-scan measurement has already been treated earlier. We record the intensity $I_i$ at positions $\theta_i$ and the integrated intensity is then given as

$$I = \sum_{p1}^{p2} \frac{1}{2}(\theta_{i+1}-\theta_i)(I_{i+1}-B_{i+1}+I_i-B_i)$$

where the sum is over the points in the peak. $B_i$ is estimated from the background points in form of a least squares line, $B_i = s\,\theta_i+k$. Using the above expression we can estimate the counting statistical error of I (given in the previous talk on optimal experimental conditions), which has two terms. One term is a sum of the counts in the peak. The other reflects the errors in B, and in case we use a least squares line description it will be related to the quality of the fit of the line to the points in the background.

An additional term occurs if there are setting errors in the scanning arc. We then rewrite the expression for I so that $\theta_i$ becomes the variable (Lehmann, 1975), and we find as expected that the error in I is most sensitive to the setting position where the profile changes rapidly. Assuming the profile to be a Gaussian we can show that the standard deviation is proportional to $\sigma(\theta)\times I$, where $\sigma(\theta)$ is the setting error. Numerical estimates give that using approximately 20 points to measure a peak with a full width at half height of $0.40^\circ$ the relative error is less than $\sigma(\theta)$ (in degrees). For more steps or larger peaks the error is smaller. In the analysis it is assumed that the setting angle $\theta$ acts as a random variable when repeatedly positioned at the same point, and it is generally found that $\sigma(\theta)$ is well below $0.01^\circ$, which renders the error negligible. Dietrich (1976) has proposed a technique for estimating the magnitude of the error as well as errors coming from the recording equipment. A profile is measured many times, and the counting statistical variance $\sigma^2(I_i) \simeq I_i$ is compared with the sampling variance $s^2$ (described in the previous section) for a point in the profile. If there is a variation in time of the position then $s^2 > \sigma^2(I_i)$, and one finds in general that the effect is largest at points on the slopes of the profile. In addition one can follow the drift in the position with time. A much more serious error is systematic missettings at specific positions of $\theta$ for example because of mechanical errors. The effect is difficult to discover, and can often only be revealed by careful study of the reflection profile.

The standard technique for recording of a profile is by constant step and constant time per step. From the expressions for the counting statistical variance it is clear that not all points are of equal importance and Mackenzie and Williams (1973) have pointed out that reduction in variance can be obtained if measurement time is redistributed. They show that a minimum is obtained when the counting time is proportional to the product of the square root of the intensity and the steplength (for trapezoidal integration). For normal peaks with a peak to background ratio of less than 25 a 40% reduction in variance can be achieved. For larger peak to background ratios further improvements are obtained. Similar gain can be obtained by varying the steps. The simplest is here to measure with large steps in the background and short steps in the peak. One could envisage further advances where data are measured according to their intensity following some criterium, but this is not frequently done, probably because there is no general agreement on the criterium to use. Normally all reflections are measured the same time, and one will generally aim at a precision as judged from counting statistics of 1 to 2%.

An important part of the second term in the often used expression for the variance given in the introduction can be attributed to the uncertainty in bringing the observations to a common scale (McCandlish and al., 1975). During the measurements a set of standard reflections is recorded with regular intervals, and by a scaling procedure the reflections are brought to scale. This has a result that $\sigma^2(I)$ must now be written

$$\sigma^2(I)=S^2(K)I^2+(kI)^2+\sigma^2_{count}(I)$$

where S(K) is a standard deviation that reflects the scatter of the scale factors, $k_i$, for the various standard reflections around the mean scaling factor K, that is applied to the data. S(K) will be a function of time. During the experiment the $k_i$ might develop differently, so S(K) will tend to increase, with a degradation of the $\sigma(I)$ as a result. The changes of the standards are caused either by instrumental changes whereby all $k_i$ will stay identical or by changes in the crystal, notably changes in the degree of perfectness, whereby the $k_i$'s might diverge. McCandlish et al. suggest for this reason that the most important subsets should be collected first. Considering that mean values of for example extinction correction parameters will be used afterwards it might seem more appropriate to refer the scaling and the important measurements to the middle of the experiment. The above considerations apply especially to X-ray diffraction. In the case of neutron diffraction the beam flux impinging on the crystal is monitored and the crystal will generally not degrade under radiation. An expression where S(K) is constant, and where we add the two terms $S^2(K)I^2+(kI)^2$ is therefore more adequate, and we revert to the technique proposed in the introduction. The same might be the case in the X-ray measurement where crystals are stable, and good agreement with the above considerations were found in a study of pyrazine (DeWith and Feil, 1976).

Temperature fluctuations is another source of error. Although agreement of temperature is very important when combining X-ray and neutron data this subject is hardly discussed at all in the literature. If we assume that the size of the temperature factor is proportional to the temperature (the harmonic oscillator in the high temperature approximation) then clearly we must require stability of better than 1° at 100°K. This is difficult to obtain in any open system of the cold gas flow type. In addition we must require a knowledge of the sample temperature with the same accuracy if we want to avoid systematic errors. It is here possible to calibrate the temperature recorder using a known phase transition (for example the ferroelectric transition in $KH_2PO_4$ at 122.5°K, which gives rise to an intensity overshoot especially in the h00 reflections (Zeyen, 1976)). For many soft compounds it is necessary to measure at reduced temperatures, and although there is little direct evidence it would seem reasonable to assume that in most cases the temperature factors for this reason are not determined with an accuracy of better than 1%.

## ERRORS ARISING FROM THE CORRECTIONS.

The optimal conditions for an experiment is when the effects of the various corrections are minimized. This has been discussed in the previous chapter. The corrections done after the measurements include in the ideal case the following effects : absorption, thermal diffuse scattering (TDS), multiple Bragg scattering and extinction.

The correction for absorption should be straight forward. The crystal surface can be described, and the absorption coefficient, μ, is either known or can be measured. The relative variations of the calculated transmission factors are small if the crystal is isometric. This allows rather relaxed conditions on the error in μ as long as we are only interested in data on a relative scale. The conditions are of course more stringent when attempts are made to observe data on absolute scale (Stevens and Coppens, 1975). Rees (1977) has discussed the error in the correction coming from errors in $\mu$, and he points out that the covariance for two reflections, $cov_{\mu}(I_i, I_k)$, is unity. This means that if we have to take the error into account then we would have to include correlation terms, which complicate any subsequent analysis.

In order to do the correction for TDS knowledge of the elastic constants of the crystal is required. In most cases they are not available, and it is often beyond the means of the experimenter to obtain them. We must then estimate the effect that a neglect of a correction has on the data. The contribution from TDS to the elastic intensity is approximately proportional to $(\sin\theta/\lambda)^2$, so the neglect of the correction results in an increase in intensity at high angle, and a corresponding reduction of the temperature parameters, $\Delta B$, is the result. The effect has been estimated for several compounds. Göttlicher (1968) gives for NaCl $\Delta B$=0.203 Å$^2$,0.138 Å$^2$ for two different detector apertures and Cooper (1969) gives 0.16 Å$^2$ for KCl and 0.2 Å$^2$ for hexamethyltetramine (room temp.). Helmholdt and Vos (1977A) give for ammonium hydrogen oxalate values around 0.1 Å$^2$ at room temperature falling to around 0.006 Å$^2$ at 15° K. The effect on the deformation densities is most pronounced near the atomic positions, and in NaCl (Göttlicher, 1968) no effects were observed beyond 0.5 Å from the atomic position. For organic materials (ammonium hydrogen oxalate) at 110°K Helmholdt and Vos (1977A) find effects in carbon-carbon bonds of 0 to 0.03 e/Å$^3$, and only 0.05 to 0.08 e/Å$^3$ at atomic positions when high order X-ray data are used to fix the temperature parameters. A much larger effect was found by De With et al. (1976) for measurements on pyrazine at 184°K. By correcting for TDS the density in the bonds and lone pairs were reduced by up to 0.15 e/Å$^3$, but again the really large effects were found within 0.5 Å of the atomic positions. The effect of incorrect thermal factors is however not important for comparison with theoretical dynamic densities when these are based on temperature parameters adjusted as much as possible to the experimental values as pointed out by Helmholdt and Vos (1977B). Finally one should note that the relative error falls with reduced temperatures, and that often the best correction is simply to do the measurement at low temperatures.

One can expect that the neglect of extinction correction might influence the temperature parameters, at least for simple compounds. Strong reflections for which extinction is most serious are mainly found at low scattering angle. If they are underestimated because of a neglect of the correction, and if there is no experimental determination of the scale factor, then the temperature factors will be observed too small. Low order reflections are an important contributor to details in deformation density, as they are affected by non spherical components of the density. If we use refinement techniques to correct for extinction effects then it is exactly the strong, low order reflections that determines the extinction parameters, and we risk correlation effects between electron density and extinction. From neutron diffraction, where extinction is often a more serious problem than with X-rays, we know that present formalisms account well for the observed reduction in intensity ; the main effect in correcting for extinction is mostly found to be an improvement in the agreement factor R. The temperature factors do generally not change. We can therefore hope that the same is true in the X-ray case, and that the form of the two effects in question are sufficiently disparate to obtain low correlation. De With et al. (1976) report for example that correction for extinction in pyrazine has only little effect on the deformation density. Thomas (1977) reports that in dimethylammonium hydrogen oxalate (room temp.) removal of the reflections with large extinction corrections reduces the density, but does not change relative features.

Multiple Bragg scattering, which is much more complex to describe, follows a similar pattern to extinction. At present the best correction is to avoid it by experimental techniques, and one can only hope that one day it is possible as well to do a calculated correction. An important problem, that is yet partially unsolved is the disagreement that is often found between temperature parameters from high order X-ray data and from neutron data. This has been discussed by Coppens (1978). In some cases the disagreement in temperature is the reason for differences, but if the difference in temperature is small, then one would expect that one could scale the temperature factors, and that the scale factors for the three sets of $U_{ii}$, i=1,2,3 would be the same. In several cases this is found not to be so. For $Cr(CO)_6$ (Rees and Mitschler, 1976) the scale factors were 0.864, 0.936 and 0.936, and in sulfamic acid (Bats et al., 1977) the scale factors were 1.01, 1.14 and 1.03. We can only guess the reason for these discrepancies, and presently the solution is to scale the $U_{ii}$. This is generally a successful solution. It does however give us an idea of the accuracy of the temperature parameters, which is obviously not better than a few percent.

## ESTIMATION OF ERRORS USING COMPARISONS

Repeated measurements of reflections, observations of a series of symmetry related reflections or the collection of several complete, independent sets of data on the same compound are methods for learning about various components of the error in the observation.

The measurements of one or several reflections at intervals all through a data collection is a standard technique in diffraction measurements. This serves as a check of the stability of crystal setting and quality as well as of recording equipment, and has been discussed above. It will not tell anything about errors such as beam inhomogeneities or the size of the thermal diffuse scattering, but it might well signal a change in crystal quality during the measurements as reflected by a change in the extinction effect.

In general at the stage of refinement or electron density calculations only symmetry averaged reflections are used. During the process of averaging information is obtained about the general internal quality as it is assumed that the data has been corrected for all effects which are crystal orientation dependent. Corrections are made for absorption, thermal diffuse scattering, and the agreement of symmetry related data reflects the success of these corrections in describing the anisotropy of the crystal as well as the quality of experimental parameters such as beam homogeneity. The internal variance

$$\sigma^2_{int} = \frac{\Sigma(F^2_{obs}-\bar{F}^2_{obs})^2}{n-1}$$

where $\bar{F}^2_{obs}$ is the mean of a group of n symmetry related reflections, can be compared with the counting statistical variance, $\sigma^2_{count}$, and we can either add the two or use $\sigma^2_{int}$ as a guide in estimating the part proportional to $F^2_{obs}$ that we add to $\sigma^2_{count}$. If we have measured a large number of symmetry related reflections we can use $\sigma^2_{int}$ in the further analysis as it will represent a large fraction of the random errors occuring (even long term instability if the symmetry related reflections are measured at long intervals). Errors not represented in $\sigma^2_{int}$ are mainly of systematic nature, and it is questionable whether they should be included in the error estimate and how this is done. In the case for example of TDS the error in a deformation density from a neglect of the correction is not the same for combination of X-ray and neutron data as for the use of X-ray data only.

In most cases the low crystal symmetry prevents measurements of sufficient symmetry related reflections to make $\sigma_{int}$ reliable. If we believe the data only to suffer from errors which can be related to parameters such as intensity, $\sin\theta/\lambda$ or time when the measurement took place, then with the use of only two symmetry related reflections we can still construct an estimate of $\sigma$ based on the internal consistency. This is done using analysis-of-variance technique (see Hamilton, 1964, or for a short summary Abrahams et al., 1967). An example of this is given in an analysis of neutron data on DL-serine (Frey et al., 1973, spacegroup $P2_1/a$). First $\sigma^2_{int}$ was calculated using the above expressions (in most cases n=2), and it was then attributed to two experimental quantities, $F^2_{obs}$ and $\sin\theta/\lambda$. So

$$\sigma_{int} = \sigma_i(F^2_{obs})+\sigma_j(\sin\theta/\lambda)$$

where $\sigma_i(F^2_{obs})$ is a table function in $F^2_{obs}$, and $F^2_{obs}$ is categorized as

$$F^2_i \leq F^2_{obs} < F^2_{i+1} \qquad i=1,12$$

The $\sigma_j(\sin\theta/\lambda)$ was described in a similar manner in a table with 22 entries. A least squares calculation (setting $\sigma_j$=0 for j=1) was then carried out fixing the values in the tables from 1248 $\sigma^2_{int}$ corresponding to the symmetry independent reflections. The weight used in the structure refinement was then given as

$$w = 1/\ (\sigma_i(F^2_{obs})+\sigma_j(\sin\theta/\lambda))^2$$

The refined parameters using this technique showed no significant deviations from results obtained using conventional estimates of $\sigma^2(=\sigma^2_{count}+(0.02\ F^2_{obs})^2)$.

In most cases of averaging the internal consistence factors $R_{int}$ and $wR_{int}$ are obtained. $wR_{int}$ can be defined

$$wR_{int} = \Sigma\Sigma_{hj} w_{hj}|F^2_{hj}-\bar{F}^2_h|/\Sigma\Sigma w_{hj}F^2_{hj}$$

where $\bar{F}^2_h$ (or $\bar{F}_h$) is the mean of the symmetry related reflections $F^2_{hj}$, j=1,n. The weight is $w_{hj}$, and $R_{int}$ is defined in a similar manner with $w_{hj}$=1. These R factors give a general indication of the quality of the data, but they are only one measure of consistency and take no account of the many types of systematical errors.

By doing several measurements of full sets of data on the same crystal using different measurement techniques and samples one can get an overall estimate of the obtainable absolute accuracy. Two projects have been conducted for single crystal measurements : The American Crystallographic Association undertook a study on $CaF_2$ using the same crystal for seven sets of measurements in different laboratories (Abrahams et al., 1967) and the I.U.Cr later carried out a similar project on D(+)-tartaric acid (Abrahams et al., 1970, Hamilton and Abrahams, 1970). These projects were started shortly after the advent of automatic diffractometers with the aim of assessing "the accuracy of the integrated intensities and resultant structure factors measured by current diffractometer methods, with data to be collected by the participant's normal routine procedure" (Abrahams et al., 1970). The data were measured 1965. Since then measurement techniques have advanced, and the type of diffractometers used then were often of different geometry than presently, so we might expect that better accuracy would be obtained now. Many of the conclusions reached are probably however still valid.

The result of studies on $CaF_2$ were that routine data collection in most cases produced values which were within 5% of the true values, but that there was no evidence for any $F^2_{obs}$ to have been measured to better than 2%. In general the largest systematic errors were associated with scattering angle. These conclusions were further elaborated by Mackenzie and Maslen (1968) who studied subsets of the data after having removed the data with highest deviations from the mean. They found that especially the low order was affected by errors, but that they could not find any evidence for any systematic errors apart from dependence on angle and intensity. Similar conclusions were reached in the study of D(+)-tartaric acid. In this case internal and external agreement factors were calculated, and the results for the external agreement factors between experiments were that probable differences were 6% with outer limits of 3% and 10%. Subgroups showed better agreement. Among these was a group mainly consisting of 4-circle diffractometers which gave an average external agreement of 5.2%. It must be kept in mind, though, that some systematic errors might well be masked when comparisons are made among a specific type of instruments. Again the general observation was that the errors were dependent on intensity and angle. The internal agreement factors were in all cases much smaller than the agreement with the mean of all the experiments. The factor between the two was in the range from 2 to 10. This must be kept in mind when using the internal agreement factor cited before as a measure of the quality of the data.

A very informative way of studying discrepancies between measurements on identical systems has been proposed by Abrahams and Keve (1971, Hamilton and Abrahams, 1972). The technique is to use normal probability plots. From the two sets of observations $(O_{1i},\sigma(O_{1i}))$, $(O_{2i}, \sigma(O_{2i}))$, $i=1,n$ is calculated the quantities $\delta m_i=(O_{1i}-O_{2i})/(\sigma^2(O_{1i})+\sigma^2(O_{2i}))^{1/2}$. If the $O_{ij}$ are normally distributed then $\delta m_i$ will be values from a normal distribution with zero mean and unit variance. We can now rank the list of $\delta m_i$ according to size and compare the values with a similar set of theoretical values. This is done by plotting one against the other. If the above assumptions are correct then the curve will be linear with unit slope and go through zero. If the curve is linear, but with slope different from 1 then this can be taken as an indication that the $\sigma$ values are misestimated.

This type of plot can conveniently be used to compare refined parameters and thus give an external estimate of the standard deviations. An example is given from comparison of two neutron and one X-ray study of ℓ-asparagine, monohydrate (Ramanadham et al., 1972, Verbist et al., 1972 ; Kartha, 1971). Comparing atomic position parameters and thermal parameters gave that the standard deviations were underestimated by a factor 1.3 to 1.6. A comparison between distances in the molecule of serine (Frey et al. (1973)) in two crystal modifications gave a similar result. The factor was 1.4. These measurements were probably of reasonably good quality, so when comparing parameters from a least squares refinement this type of factors should probably be taken into account.

## ESTIMATES OF ERRORS IN THE OBSERVED AND CALCULATED ELECTRON DENSITIES.

In the study of electron density using the deformation density (either from Fourier transformation of reflection data or based on parameters from a refinement of density parameters) two levels of questions are asked in much the same manner as when we are concerned with molecular geometry. In a first instance we ask where maxima in the deformation density are located like we would ask for atomic positions. We determine the lone pair positions and find out whether they point towards some metal atom or a hydrogen in a hydrogen bond. There is little doubt that the density is in a given position, and that the position is rather insensitive to improvements in the data similar to what is found for atomic positions. Estimates of the error in the peak height is only of little interest. In a second stage interactions between molecular units become interesting (similar to thermal motion in molecules), and for a quantitative discussion the knowledge of standard deviations is vital.

The deformation (or valence) density is given as

$$\Delta\rho=\rho_{obs}-\rho_{calc}=\rho'_{obs}/k-\rho_{calc}$$

where $\rho_{obs}$ is the observed density on absolute scale, $\rho'_{obs}$ is the unscaled density, k is a scalefactor ($F_{obs}$=k $F_{calc}$) and $\rho_{calc}$ is the density based on some model, usually a spherical atom model.

The variance of $\Delta\rho$ is

$$var(\Delta\rho)=var(\rho_{obs})+var(\rho_{calc})-2cov(\rho_{obs},\rho_{calc})$$

The densities $\rho_{obs}$ and $\rho_{calc}$ are correlated via the scalefactor k and we get

$$cov(\rho'_{obs}/k,\rho_{calc})=\frac{d(\rho'_{obs}/k)}{dk}cov(k,\rho_{calc})=\frac{-\rho_{obs}}{k}cov(k,\rho_{calc})$$

so

$$var(\Delta\rho)=var(\rho_{obs})+var(\rho_{calc})+2\rho_{obs}cov(\rho_{calc},k)/k$$

Rees (1976, 1977, 1978) has discussed this expression in detail for various conditions. The first term can be written

$$var(\rho_{obs})=\frac{1}{k^2}(var(\rho'_{obs})+\rho^2_{obs}var(k)-2\rho_{obs}cov(\rho'_{obs},k))$$

Normally k is determined from a least squares procedure involving observed and calculated structure factors $F_{obs}$ and $F_{calc}$ as well as $\sigma(F_{obs})$, and $var(\rho_{obs})$ therefore becomes a function of both the observed and calculated density as well as the derivatives of the structure factors with respect to the parameters. The same applies for the two last terms in the expression for $var(\Delta\rho)$. In addition a term has to be incorporated for description of the model errors in the determination of the scale factor.

At first sight the situation seems rather complex, but a determination of the errors in a real case (Rees, 1977, estimates on $Cr(CO)_6$) clarifies the picture. It is found that the scalefactor error is predominant in the heavy atom region. The errors due to the calculated density are largest near the nuclei, but in the heavy atom region there is a reduction due to the scaling procedure. The error from observations is nearly constant and is dominating till approximately 0.5 Å from the nuclei. We have already seen that errors arising from neglect of TDS corrections can be serious within 0.5 Å from the atomic centers, so if we restrain ourself to study areas far from the nuclei then the situation is much simplified.

In the centrosymmetric case ($P\bar{1}$) the error in $\rho'_{obs}(x,y,z)$ can be written

$$var(\rho'_{obs})=\frac{4}{V^2}\sum_{1/2}\sigma^2(F_{obs})\cos^2 2\pi(hx+ky+\ell z)$$

summing over a hemisphere of reciprocal lattice. Writing $\cos^2 x = \frac{1}{2} + \frac{1}{2}\cos 2x$ we get

$$\text{var}\,(\rho'_{obs}) = \frac{2}{V^2} \sum_{1/2} \sigma^2(F_{obs})\,(1+\cos 2\pi(h2x+k2y+\ell 2z))$$

The second term is a Patterson-type function, and is similar in nature to the implication diagram (Buerger, 1959, p. 145). It will have peaks at the atomic positions, but will in addition have a widespread distribution of peaks. For a sufficiently complicated structure of low symmetry the peak distribution will be random, and $\sigma(\rho'_{obs})$ will effectively be constant (Coppens and Hamilton, 1968, Rees 1977), except where symmetry elements create high density of peaks for example at a center of symmetry or on Harker sections (Buerger, 1959, p. 132). The expression for $\sigma^2(\rho'_{obs})$ is then

$$\text{var}\,(\rho'_{obs}) = \frac{2}{V^2} \sum_{1/2} \sigma^2(F_{obs})$$

as given originally by Cruickshank (1949). To obtain the total variance of $\rho_{obs}$ we neglect the correlation between $\rho_{obs}$ and k and we add a term proportional to $\rho^2_{obs}$. The proportionality factor $\text{var}(k)/k^2$ is generally so small that this term can be neglected except at the center of the atom.

Unfortunately we will often seek information in the environment of 0.5 Å from the nucleus, and we must then further study the calculated density. General expression are given (Stevens and Coppens, 1976 ; Rees, 1977), and here we will only try by an approximation to estimate a priori the error in the map coming from errors in the atomic parameters. We will write the electron density for one electron as a Gaussian

$$\rho(x) = \frac{1}{\sqrt{2\pi U}} \exp(-(x-x_o)^2/2U)$$

where $x_o$ is the origin of the atom and U includes both the static electron density distribution and the thermal smearing ($B=8\pi^2 U$). Describing the electron density distribution as Gaussian is of course a rough approximation, but if we adjust the Gaussian scattering factor to agree with the usual scattering factor for $\sin\theta/\lambda$ around 1 Å$^{-1}$ then $B_{electron}$ for light atoms will be between 2 and 3 Å$^2$. At low temperatures $B_{thermal}$ will be 1, so the influence from $B_{thermal}$ is only around 1/4 (assuming the two distributions to be convoluted : $B=B_{electron}+B_{thermal}$). The errors in $\rho(x)$ from errors in $x_o$ and B are then

$$\frac{\sigma_{x_o}(\rho(x))}{\rho(x)} = \frac{8\pi^2(x-x_o)^2}{B} \frac{\sigma(x_o)}{x-x_o}$$

and

$$\frac{\sigma_B(\rho(x))}{\rho(x)} = \frac{1}{2}\left(\frac{8\pi^2(x-x_o)^2}{B} - 1\right)\frac{\sigma(B)}{B} \approx \frac{8\pi^2(x-x_o)^2}{B}\frac{\sigma(B)}{2B}$$

For $x-x_o=0.5$ Å and typical values $\sigma(x_o)=0.005$ Å, $\sigma(B_{thermal})=0.03$ Å$^2$ and $B=4$ Å$^2$ we have $\sigma(x_o)/(x-x_o) = 0.01$ and $\sigma(B)/2B = 0.004$, so the error in the density coming from errors in the parameters is of least importance at this distance from the nucleus. We can now estimate the error per electron at $x$. It is 0.006 e/Å$^3$ for B = 3 Å$^2$ and 0.008 e/Å$^3$ for B = 4 Å$^2$. For oxygen this corresponds to errors in the range 0.03 to 0.05 e/Å$^3$ comparable to $\sigma(\rho_{obs})$ values given by Stevens and Coppens (1976) for a series of compounds. At this distance the error is still manageable. A priori estimates using more realistic electron densities can easily be done and has been discussed by Stevens and Coppens (1976).

From the expression for $\sigma(\rho_{obs})$ it is clear that as more terms are included in the Fourier summation the $\sigma(\rho_{obs})$ increases. Simultaneously the resolution in the map is improved leading to sharper features and increased peak values (Coppens and Lehmann, 1976 ; Scheringer, 1977 ; Lehmann and Coppens, 1977). Beyond a certain limit in $\sin\theta/\lambda$ where the resolution is sharper than the features the noise added is larger than the information, and it is then reasonable not to include more terms in the Fourier summation. At this level we can recognize density features with a certain resolution, and we then understand the error estimate as an estimate for a region corresponding to this resolution. This has been expressed by Rees (1977), who takes into account the covariance of neighbouring points, $\text{cov}(\rho_a,\rho_b)$. This quantity contains a correlation function which is determined by the maximum $\sin\theta/\lambda$ involved, and for low resolution the correlation is high. Another approach is to "smooth" the density by considering the mean density within a volume element (Brown, 1977). The density is convoluted with a step function which is unity within a defined volume (a box or a sphere) and zero elsewhere. This corresponds to multiplying the structure factors with a zero or first order Bessel function (Coppens and Hamilton, 1968 and references therein). The error is easily obtained, and is now the error of the mean electron density within the volume element.

## FINAL REMARKS

Intra and interexperimental comparisons can supply us with estimates of the precision and accuracy of the observed intensity. Presently it seems that $F^2_{obs}$ can be obtained with a 2% standard deviation or better, and that estimated standard deviations from structural refinements must be multiplied by a factor larger than 1 to be reliable.

In electron deformation densities the estimated error is normally of reasonable size outside a radius of 0.5 Å from the nuclei. Inside this the error can be larger partially because of errors in the scale factor and partially because of neglect of corrections (for example for thermal diffuse scattering).

When refinement techniques are used a variance-covariance matrix for the parameters is obtained nearly automatically. This allows immediate comparisons. We must however keep in mind that the general magnitude of the values in the matrixes might be underestimated. In addition the errors become model dependent, which complicates any statistical analysis. Further discussions of these effects are desirable along with the general development of this type of techniques for the study of electron density distributions.

## APPENDIX

Some definitions are given. For further discussion see Hamilton (1964). Let us consider a random variable x with a probability density function $\phi(x)$. The probability of finding a value x between $x^o$ and $x^o+dx$ is then

$$P(x^o \leq x \leq x^o+dx) = \phi(x^o)dx$$

The mean of x is

$$\bar{x} \equiv \int_{-\infty}^{\infty} x\,\phi(x)\,dx$$

The variance for the distribution is

$$\sigma^2(x) \equiv \mathrm{var}(x) \equiv \int_{-\infty}^{\infty} (x-\bar{x})^2 \phi(x)\,dx$$

The standard deviation is $\sigma(x)$, and we frequently call this the error in x. If we have a multi variate probability density function $\phi(x_1, x_2, \ldots, x_n)$ then

$$\bar{x}_i \equiv \int_{-\infty}^{\infty} x_i \phi(x_1, \ldots x_n)\,dx_1 \ldots dx_n$$

is the mean of $X_i$ and the variance and covariances are

$$\mathrm{var}(x_i) \equiv \mathrm{cov}\,(x_i, x_i) \equiv \int_{-\infty}^{\infty} (x_i-\bar{x}_i)\,\phi dx_1 \ldots dx_n$$

and

$$\mathrm{cov}(x_i,x_j) \equiv \int_{-\infty}^{\infty} (x_i-\bar{x}_i)(x_j-\bar{x}_j)\phi dx_1\ldots dx_n$$

If f is a function of the random variables $x_1$, $x_2$, $x_3 \ldots x_n$

$$f = f(x_1,\ldots,x_n)$$

then we can estimate a standard deviation for f as

$$\mathrm{var}(f(x_1,\ldots,x_n)) \approx \sum_i \sum_j \frac{\partial f}{\partial x_i} \frac{\partial f}{\partial x_j} \mathrm{cov}(x_i,x_j)$$

In cases where $x_i$ and $x_j$ are non correlated then

$$\mathrm{var}(f(x_1,\ldots,x_n)) \approx \sum_i \left(\frac{\partial f}{\partial x_i}\right)^2 \mathrm{var}(x_i) \equiv \sum_i \left(\frac{\partial f}{\partial x_i}\right)^2 \sigma^2(x_i)$$

A limiting case is

$$f = \Sigma a_i x_i$$

which gives

$$\sigma^2(f) = \Sigma a_i^2 \sigma^2(x_i)$$

Most commonly we use for a distribution function a Gaussian (or normal) distribution

$$G(x) = \frac{1}{2\pi\sigma} \exp(-(x-\bar{x})^2/2\sigma^2)$$

where $\bar{x}$ is the mean and $\sigma$ is the standard deviation.

In counting statistics we employ the Poisson distribution function

$$p(k) = \frac{e^{-\lambda}\lambda^k}{k}$$

where p(k) is the probability of k counts occuring in a unit of time. The mean value and the variance are both equal to the distribution parameter $\lambda$. For reasonable count rates the difference between an observation k and $\lambda$ is small so

$$\sigma(k) = \sqrt{\lambda} = \sqrt{k}$$

Let us finally consider n observations of x : $x_i$, i=1, n. We then commonly define

The mean : $\bar{x} = \frac{1}{n}\sum_{i=1}^{n} x_i$

The sample variance : $s^2 = \frac{1}{n-1}\sum_{i=1}^{n} (x_i-\bar{x})^2$

In diffraction measurement it is normally $F^2_{obs}$ that is observed with standard deviation $\sigma(F^2_{obs})$. In some cases we need $\sigma(F_{obs})$. From the above consideration we have

$$\sigma(F^2_{obs}) \approx 2F_{obs}\sigma(F_{obs})$$

or

$$\sigma(F_{obs}) \approx \frac{\sigma(F^2_{obs})}{2F_{obs}}$$

This relation holds well for $\sigma(F_{obs})/F_{obs}$ small, i.e. for strong reflections. For weak reflections another expression has been suggested by Rees (1976)

$$\sigma(F_o) = \sqrt{\sigma(F_o^2)/2}$$

REFERENCES


Abrahams, C.C., Alexander, L.E., Furnas, T.C., Hamilton, W.C., Ladell, J., Okaya, Y., Young, R.A. and Zalkin, A. (1967). Acta Cryst. 22, 1-6.

Abrahams, S.C., Hamilton, W.C., and Mathieson, A. Mcl. (1970). Acta Cryst. A26, 1-17.

Abrahams, S.C. and Keve, E.T. (1971). Acta Cryst. A27, 157-165.

Bats, J.W., Coppens, P. and Koetzle, T.F. (1977). Acta Cryst. B33, 37-45.

Brown, P.J. (1977). Private communication.

Buerger, M.J. (1959). Vector Space, John Wiley and Sons : New York

Busing, W.A. and Levy, M.A. (1957). J. Chem. Phys. 26, 563-568.

Cooper, M.J. (1970). Thermal Neutron Scattering, 51-67, Ed. B.T.M. Willis, Oxford University Press.

Coppens, P. and Hamilton, W.C. (1968). Acta Cryst. B24, 925-929.

Coppens, P. and Lehmann, M.S. (1976). Acta Cryst. B32, 1777-1784.
Coppens, P. (1978). Chapter 3 in "Neutron Diffraction", H. Dachs, Ed., Springer Verlag : Berlin.
Cruickshank, D.W.J. (1949). Acta Cryst. 2, 65-82.
DeWith, G. and Feil, D. (1976). Acta Cryst. B32, 1011-1012.
DeWith, G., Harkema, S. and Feil, D. (1976). Acta Cryst. B32, 3178-3184.
Dietrich, H. (1976). J. Appl. Cryst. 9, 205-208.
Frey, M.N., Lehmann, M.S., Koetzle, T.F. and Hamilton, W.C. (1973). Acta Cryst. B29, 876-884.
Göttlicher, S. (1968). Acta Cryst. B24, 122-129.
Hamilton, W.C. (1964). Statistics in Physical Science ; The Ronald Press Company. New York.
Hamilton, W.C. and Abrahams, S.C. (1970). Acta Cryst. A26, 18-24.
Hamilton, W.C. and Abrahams, S.C. (1972). Acta Cryst. A28, 215-218.
Helmholdt, R.B. and Vos, A. (1977A). Acta Cryst. A33, 38-45.
Helmholdt, R.B. and Vos, A. (1977B). Acta Cryst. A33, 456-465.
Kartha, G. (1971). Private communications.
Lehmann, M.S. (1975). J. Appl. Cryst. 8, 619-622.
Lehmann, M.S. and Coppens, P. (1977). Acta Chem. Scand. A31, 530-534.
Mackenzie, J.K. and Maslen, V.W. (1968). Acta Cryst. A24, 628-639.
Mackenzie, J.K. and Williams, E.J. (1973). Acta Cryst. A29, 201-204.
McCandlish, L.E., Stout, G.H. and Andrews, L.C. (1975). Acta Cryst. A31, 245-249.
Ramanadham, M., Sikka, S.K. and Chidambaram, R. (1972). Acta Cryst. B28, 3000-3005.
Rees, B. (1976). Acta Cryst. A32, 483-488.
Rees, B. (1977). Israel J. of Chem. 16, 180-187.
Rees, B. (1978). Acta Cryst. A34, 254-256.
Rees, B. and Mitschler, A. (1976). J.A.C.S. 98, 7918-7924.
Scheringer, C. (1977). Acta Cryst. A33, 588-592.
Stevens, E.D. and Coppens, P. (1975). Acta Cryst. A31, 612-619.
Stevens, E.P. and Coppens, P. (1976). Acta Cryst. A32, 915-917.
Thomas, J. (1977). Acta Cryst. B33, 2867-2876.
Verbist, J.J., Lehmann, M.S., Koetzle, T.F. and Hamilton, W.C. (1972). Acta Cryst. B28, 3006-3013.
Zeyen, C. (1976). Private communication.

ACKNOWLEDGEMENTS

Dr. J.C. Speakman and D.J. Lehmann are thanked for careful reading of the manuscripts, and professors P. Coppens and B. Rees are thanked for supplying the author with manuscripts prior to publication. Mme Pollak is thanked for typing of the manuscript.

Part III

# Analysis of Experimental Densities

ANALYSIS AND PARTITIONING OF ELECTRONIC DENSITIES

Pierre Becker

Centre de Mécanique Ondulatoire Appliquée

CNRS and University Paris VI, 23 rue du Maroc

75019 Paris, France

## INTRODUCTION

A. The importance of the electronic density is related to :

1. Hellmann-Feynman theorem (ref.1)

$$\frac{\partial E}{\partial \lambda} = \langle \Psi | \frac{\partial H}{\partial \lambda} | \Psi \rangle \qquad (1)$$

where $\lambda$ is a parameter of the Hamiltonian. In particular, the force acting on a given nucleus A is :

$$\underset{\sim}{F}_A = \underset{\sim}{F}_{A,\text{nuclei}} + Z_A \int \frac{\rho(\underset{\sim}{r},\underset{\sim}{R})\ \underset{\sim}{r}_A}{r_A^3}\, d\underset{\sim}{r} \qquad (2)$$

$\rho(\underset{\sim}{r},\underset{\sim}{R})$, where $\underset{\sim}{R}$ represents the nuclear configuration, is the fundamental quantity.

Most wave functions do not satisfy the theorem (ref.2). Therefore the first order density matrix $\rho\ (\underset{\sim}{r},\underset{\sim}{r}')$, including the non diagonal terms, and the two particle density matrix $\rho_2$ may play an important role.

2. Hohenberg-Kohn theorem (ref.3) : $\rho(\underset{\sim}{r},\underset{\sim}{R})$ characterizes the ground state.

This theorem is not valid for approximate charge densities.

B. One should also insist on the statistical nature of quantum mechanics :

$\langle A\rangle = \langle\Psi|A|\Psi\rangle$ is the only measurable quantity associated with an observable A. We may define the dispersion $\Lambda(A)$ as :

$$\Lambda^2(A) = \langle A^2\rangle - \langle A\rangle^2 = \|A\Psi\| - \|\rho_N A\Psi\| \qquad (3)$$

where $\rho_N$ is the projector on $\Psi$ (Nth order density operator).

$\Lambda = 0$ if and only if $\Psi$ is an eigenstate of A.

C. When partitioning a system into subunits :

1. The partitioning should be associated with an hermitian operator A, so that $\langle A\rangle$ is an invariant.
2. $\Lambda(A)$ should be estimated : the physical meaning of the partitioning depends on the value of $\Lambda(A)$.

-These requirements are often underestimated in the literature.

D. I would like to make few comments on the problem of interaction between two systems A and B (ref.4)

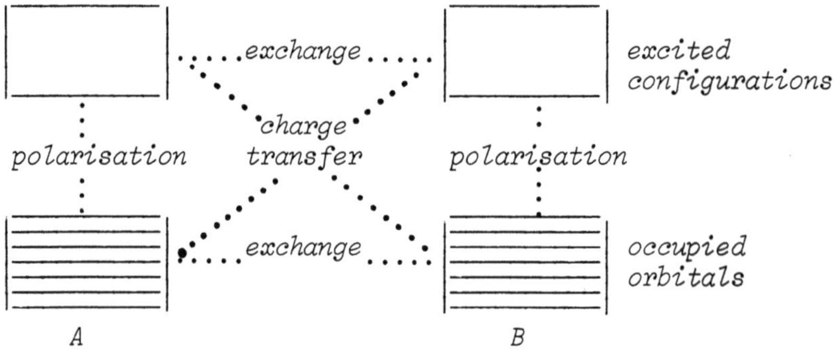


A and B may be :

| | |
|---|---|
| two molecules in a crystal<br>two molecules under reaction<br>two atoms in a molecule | } the major difference is the magnitude of the overlap S. |

- In X-Ray analysis, one uses as the simplest model to take care of bonding effects variable populations of the valence shell of atoms. Such a charge transfer implies also a variation of the shape of each valence shell, due to polarisation. In fact the simplest reasonable model should be to represent the valence shell as

$P_v \kappa^2 \rho_v(\kappa r)$, where $\kappa$ is an adjustable parameter. This model has been developed by Coppens and the author (ref. 5) and leads to quite interesting results (similarity with Slater's rules of screening), and later by Stewart and Pople (ref. 6).

One key question is : What minimal information does one need to correctly describe a system or its interaction with other species ? As we shall see, the answer depends on the specific problem one is trying to solve.

E. Let us consider an example that shows that charge is not uniquely defined in chemistry and depends on the problem one has in mind. This example deals with the use of charge density in chemical reactivity problems (ref. 7, 8).

The hamiltonian H can be written :

$$H = \sum_i h(i) + \frac{1}{2} \sum_{ij}{}' g(i,j) \tag{4}$$

where h(i) is a one electron operator and $g(i,j) = r_{ij}^{-1}$.

1. Consider the action of an approaching reactant. It's main effect is a change in h(i) :

$$\delta h(i) = \delta V(\underset{\sim}{r}_i) \tag{5}$$

To first order, the energy change is :

$$\delta E = \int \rho(\underset{\sim}{r})\delta V(\underset{\sim}{r})d\underset{\sim}{r} \tag{6}$$

where $\rho(\underset{\sim}{r})$ is supposed to be exact. If one develops on an atomic basis :

$$\delta E = \sum_{\mu\nu} P_{\mu\nu}\delta h_{\mu\nu} \tag{7}$$

For conjugated systems, if the site $\mu$ is attacked,

$\delta h_{\mu\mu} = \delta\alpha_\mu$ all other terms being zero.

The charge $q_\mu^\pi$ is defined as :

$$q_\mu^\pi = \frac{\partial E}{\partial \alpha_\mu} \quad \text{(static index of reactivity)} \tag{8}$$

This shows that the charge depends on the type of reaction.

2. We can now express the problem in a different way :

$$E = \mathrm{Tr}[\rho_1 h] + \frac{1}{2}\mathrm{Tr}[\rho_2 g] \tag{9}$$

$$\frac{\partial E}{\partial \alpha_\mu} = \mathrm{Tr}[\rho_1 \frac{\partial h}{\partial \alpha_\mu}] + \mathrm{Tr}[\frac{1}{2}\frac{\partial \rho_2}{\partial \alpha_\mu} g + \frac{\partial \rho_1}{\partial \alpha_\mu} h] \tag{10}$$

If Hellmann-Feynman's theorem is satisfied, the second term in (10) is zero, and we retrieve :

$$\frac{\partial E}{\partial \alpha_\mu} = P_{\mu\mu}$$

But this is not true in other cases, where $\rho_2$ plays a major role.

3. We consider the protonation of N-containing conjugated molecules (ref. 7). Let B be such a molecule :

$$B + H^+ \rightleftarrows A$$

is an acid-base reaction. The goal is to compare the pK of ground state and first excited state (singlet and triplet)

$$pK_{S,T} - pK_G = \frac{1}{2.3} \frac{\Delta E^* - \Delta E}{RT} \tag{11}$$

$$\Delta E = E_A - E_B$$

$$pK_{S,T} - pK_G = \frac{1}{2.3} \frac{(E_A^* - E_A) - (E_B^* - E_B)}{RT} \tag{12}$$

Experiment shows that :

$$pK_G < pK_T < pK_S$$

Let $\alpha$ be the Coulombic integral for nitrogen during the reaction process. For most molecules the results are as in Fig. 1.

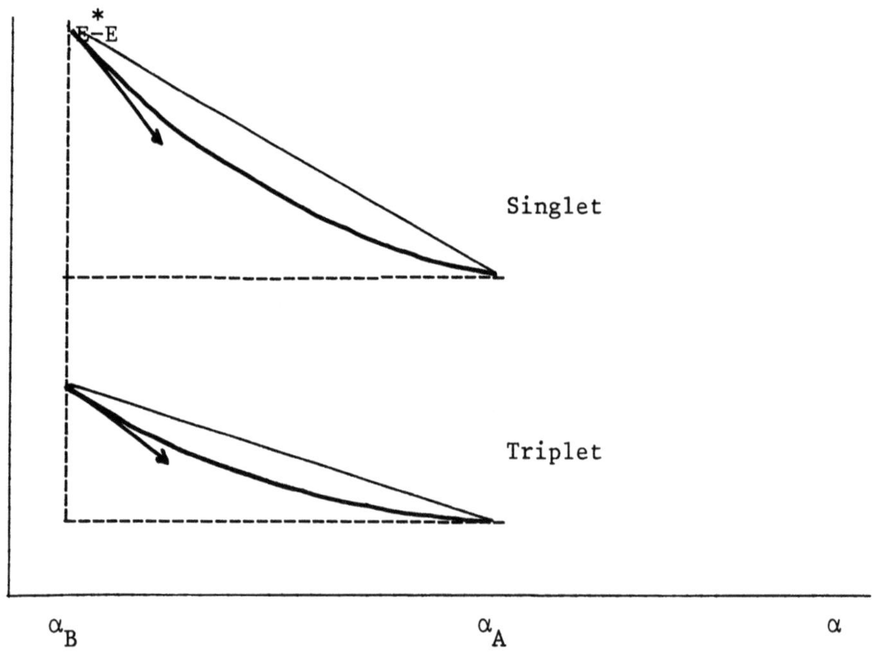


Figure 1

- $\partial E/\partial\alpha$ is the slope at the origin of the curve. The correct reactivity index should be $\Delta E/\Delta\alpha$ . In Fig. 1, the qualitative result is correctly predicted by the static index of reactivity.

- The most striking effect comes from the fact that, at the SCF level :

$$\rho_S(\underset{\sim}{r}) \equiv \rho_T(\underset{\sim}{r})$$

Therefore, the difference can only come from the two particle densities that are different.

---

Exercise 1. Show, for the $\pi$ electron wave function of ethylene, that the one particle densities are identical for singlet and triplet states.

---

- In the case of indole, the reaction path is given in Fig. 2. It is obviously a case where the static index of reactivity gives a wrong answer.

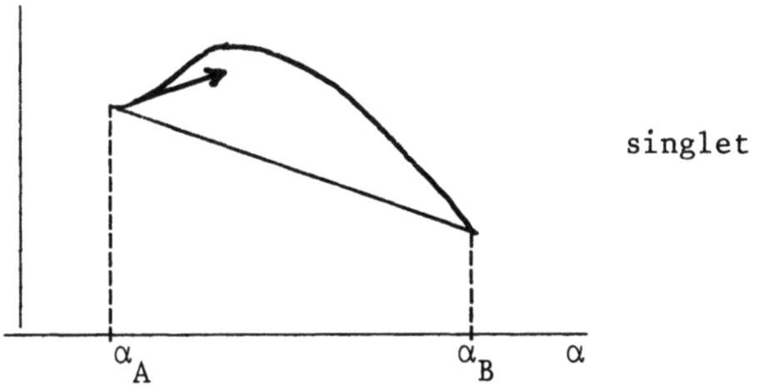


Figure 2

- Therefore, charge as defined by chemists not only is not unique, but involves the two particle density and may give wrong indications of reactivity.

F. Outline of partitioning procedures.

1. Exact partitioning

$$\rho(\underset{\sim}{r}) = \sum_A \rho_A(\underset{\sim}{r}) \tag{13}$$

where A is a subunit (atomic region for example)

2. Mean squares methods

$$\int |\rho(\underset{\sim}{r}) - \rho_m(\underset{\sim}{r})|^2 d\underset{\sim}{r} \qquad \text{minimum} \tag{14}$$

where $\rho_m(\underset{\sim}{r})$ is a model density.

$$\iint |\rho(\underset{\sim}{rr'}) - \rho_m(\underset{\sim}{rr'})|^2 d\underset{\sim}{r}d\underset{\sim}{r}' \qquad \text{minimum} \tag{15}$$

where $\rho(\underset{\sim}{rr'})$ is the one particle density matrix.

- Methods 2 are projection methods

- In methods 1, one further expands $\rho_A(\underset{\sim}{r})$ in multipolar components, ie 1 → 2
- 2 → 1 if one uses a complete basis of expansion.

## I. MULLIKEN TYPE SCHEMES

$$\rho(\underset{\sim}{rr'}) = \sum_{\mu\nu} \sum P_{\mu\nu}\phi_\mu(\underset{\sim}{r})\phi^*_\nu(\underset{\sim}{r'}) \quad (16)$$

$$\rho(\underset{\sim}{r}) = \sum_\mu \sum_\nu P_{\mu\nu}\phi_\mu(\underset{\sim}{r})\phi^*_\nu(\underset{\sim}{r})$$

1. Cross terms are approximated with one center terms (ref. 9)

$$\phi_\mu\phi_\nu \simeq \frac{1}{2} S_{\mu\nu}(\phi_\mu{}^2 + \phi_\nu{}^2) \quad (17)$$

Pseudoatom A is formed only with $\phi_\mu{}^2$, the population of which is changed from the isolated atom. We have seen that this cannot be correct in the introduction.

$$\rho_A(\underset{\sim}{r}) \simeq \sum_{\mu \in A} (P_{\mu\mu} + \sum_\nu{}' P_{\mu\nu}S_{\mu\nu})\phi_\mu{}^2 \quad (18)$$

$$n_A = \sum_{\mu\in A} (PS)_{\mu\mu} = Tr_A(PS)$$

is the electronic population of atom A.

2. An exact partitioning can be defined (ref. 10) :

$$\rho(\underset{\sim}{r}) = \sum_A P_A(\underset{\sim}{r}) + \sum_{AB} P_{AB}(\underset{\sim}{r}) \quad (19)$$

where $P_A(\underset{\sim}{r})$ contains one centre terms, $P_{AB}(\underset{\sim}{r})$ two centre terms.

$$\rho(\underset{\sim}{r}) = \sum_A \rho_A(\underset{\sim}{r})$$

$$\rho_A(\underset{\sim}{r}) = P_A(\underset{\sim}{r}) + \frac{1}{2}\sum_{AB} P_{AB}(\underset{\sim}{r}) \quad (20)$$

$$n_A = \int\rho_A(\underset{\sim}{r})d\underset{\sim}{r} = Tr_A(PS)$$

- 2 gives exact multipolar moments, but is a very delocalized scheme.
- No operator is associated with this type of partitioning. $n_A$ is neither invariant, nor bounded.
- The method is based on P matrix, and not only on $\rho(\underset{\sim}{r})$. $\sigma$-$\pi$ separability is preserved

- This partitioning scheme is only valid when comparing similar molecules, from wavefunctions of equivalent quality.
- As an example, we take FH. From two wavefunctions of RHF quality (ref. 11) :

$\varepsilon_1 = -100.0571$ au $\qquad$ $n_H = 0.25$

$\varepsilon_2 = -100.0575$ au $\qquad$ $n_H = 0.48$

This example shows the weakness of the method.

Modified Schemes

$$1.\ \phi_\mu\phi_\nu \simeq \frac{1}{2}(\xi_{\mu\nu}\phi_\mu^{\ 2} + \xi_{\nu\mu}\phi_\nu^{\ 2}) \tag{21}$$

$\xi_{\mu\nu} + \xi_{\nu\mu} = 2S_{\mu\nu}$ $\qquad$ ( is generally not symmetric)

$$n_A = \mathrm{Tr}_A(P\xi) \tag{22}$$

is defined as to give right dipole moments from point charges (ref. 12).

$$\xi_{\mu\nu}\underset{\sim}{r}_\mu + \xi_{\nu\mu}\underset{\sim}{r}_\nu = 2S_{\mu\nu}\underset{\sim}{r}_{\mu\nu} \tag{22'}$$

where $\underset{\sim}{r}_\mu$, $\underset{\sim}{r}_\nu$, $\underset{\sim}{r}_{\mu\nu}$ are centroids of $\phi_\mu^2$, $\phi_\nu^2$, $\phi_\mu\phi_\nu$.
(22') can only be satisfied in one direction, the bond $\mu$-$\nu$ axis.

2. Rivail and coworkers (ref. 13) have given a generalization. The dipole moment $\mu$ can be decomposed as :

$$\underset{\sim}{\mu} = \sum_A \underset{\sim}{\mu}_A - \sum_A q_A \underset{\sim}{R}_A + \underset{\sim}{\mu}_\dagger \tag{23}$$

$\underset{\sim}{\mu}_A$ is the hybrid part of the dipole moment. $q_A\underset{\sim}{R}_A$ corresponds to charge transfer and is calculated from Lowdin's scheme (ref. 12). $\underset{\sim}{\mu}_\dagger$ is the component coming from the fact that (22') is not satisfied in directions perpendicular to the bonds.
The method outlined is valid only if $\underset{\sim}{\mu}_\dagger$ is small.

3. Jug (ref. 14) introduces a peculiar method, based on the commutation between two operators $\hat{t}$ and $\hat{x}$.

$$[\hat{t},\hat{x}] = 0 \tag{24}$$

In a complete basis $\{\phi_\lambda\}$ we get :

$$\sum_{\lambda\lambda'} (t_{\mu\lambda}S^{-1}_{\lambda\lambda'}x_{\lambda'\nu} - x_{\mu\lambda}S^{-1}_{\lambda\lambda'}t_{\lambda'\nu}) = 0 \tag{25}$$

For a finite basis, (25) is still valid for :

$\hat{x} = 1$ $\quad \hat{t}\ \forall$ , ie charge conservation

$\hat{x} = \underset{\sim}{r}$ $\quad \hat{t} = \underset{\sim}{r}$ , ie dipole conservation.

- Using only one function, one retrieves Mulliken's scheme.
- Using two functions :

$$x_{\mu\nu} = \frac{1}{t_{\mu\mu} - t_{\nu\nu}} [(t_{\mu\mu} - St_{\nu\nu})x_{\mu\mu} + (St_{\mu\mu} - t_{\nu\nu})x_{\nu\nu}]$$

which is the Lowdin's scheme.

- With more functions, polarisation is included and one gets very good dipole moments, based on gross-charges.
- An example is given in table 1, that shows a comparison between various methods. Lowdin's scheme gives intuitively bad results for the series $H_2O$, $NH_3$, $CH_4$. This shows again the importance of polarisation.

This example proves the relative arbitrariness of the definition of the charge of an atom in a molecule.

Table 1

| Molecule | Atom | Mulliken | Lowdin | Jug |
|---|---|---|---|---|
| LiH | Li | .35 | .62 | .45 |
| | H | -.35 | -.62 | -.45 |
| FH | F | -.23 | -.06 | -.20 |
| | H | .23 | .06 | .20 |
| $H_2O$ | O | -.40 | -.01 | -.30 |
| | H | .20 | .005 | .15 |
| $NH_3$ | N | -.49 | .15 | -.36 |
| | H | .16 | -.05 | .12 |
| $CH_4$ | C | -.08 | .79 | .06 |
| | H | .02 | -.20 | -.015 |

Atomic charges in some molecules, from various partitioning schemes.

Important Remark

The one electron density operator is :

$$\hat{\rho} = |\Phi>P<\Phi| \qquad (26)$$

where

$$|\Phi>= |\phi_1\phi_2\ldots\ldots\phi_\mu\ldots.>$$

In space representation, it becomes $\rho(\underset{\sim}{r}\underset{\sim}{r}')$

Writing :

$$S = \langle\Phi|\Phi\rangle = (1 + \Delta)$$

Lowdin has proposed a symmetric atomic orbital orthogonalisation (ref. 15) :

$$|\Phi'\rangle = |\Phi\rangle\, S^{-1/2} \qquad (27)$$

Obviously :

$$\langle\Phi'|\Phi'\rangle = 1$$

$\{\Phi'\}$ can be shown to be the orthonormal set the most similar to $S^{-1/2}$ is obtained by a series expansion of $(1 + \Delta)^{-1/2}$.
We get :

$$\hat{\rho} = |\Phi'\rangle P'\langle\Phi'|$$

$$P' = S^{1/2} P S^{1/2} \qquad (28)$$

Mulliken populations become :

$$n'_A = \sum_{\mu\in A} P'_{\mu\mu}$$

One can easily prove :

$$n_A = n'_A - \frac{1}{4} \sum_{\mu\in A} (\Delta P'\Delta)_{\mu\mu} + \ldots \qquad (29)$$

- In semi-empirical techniques, one generally uses the zero differential overlap approximation (ZDO), implicitely assuming to employ orthogonal basis functions. Population analysis thus refers to Lowdin's orthogonalized orbitals and not to real atomic orbitals. These $\phi'_\mu$ are obviously more delocalized than $\phi_\mu$ and before interpretation, one should deorthogonalize the basis.

## II. SPACIAL PARTITIONING

1. Politzer (ref. 16) introduced a method which unfortunately is only valid for linear molecules. Let z be the molecular axis. $\rho(\underset{\sim}{r})$ is the density and $\rho°(\underset{\sim}{r})$ the promolecule density (superposition of non interacting spherical atoms at the experimental geometry). The atomic region of A is limited by two planes perpendicular to z such that :

$$\int_{V_A} \rho°(\underset{\sim}{r})d\underset{\sim}{r} = Z_A \qquad (30)$$

The first plane corresponds to terminal atom, the other limit being at infinity.

$$\int_{V_A} \rho(\underset{\sim}{r})d\underset{\sim}{r} = Q_A \tag{31}$$

defines the charge of atom A in the molecule. Examples are given in Fig. 3, showing the function :

$$G(z) = \int_{\infty}^{z} \rho(\underset{\sim}{r})d\underset{\sim}{r}$$

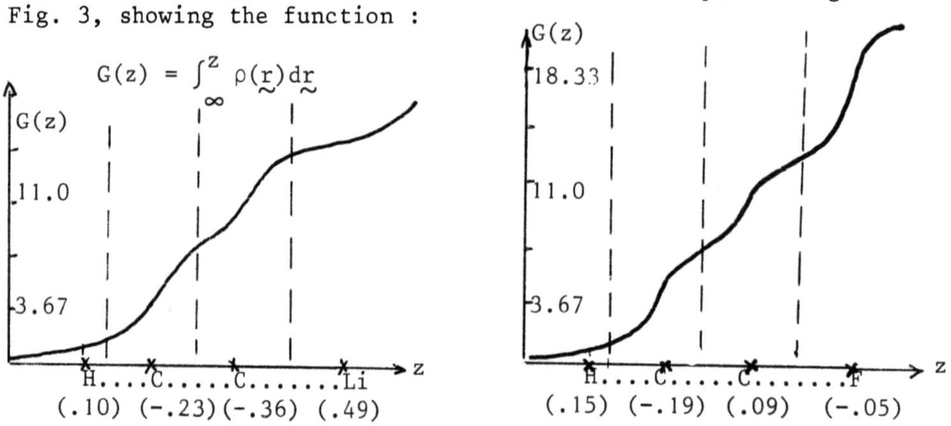


Integrated charge G(z) for HCCLi and HCCF. The numbers into parentheses are the net charges.

Net charges are :

$$q_A = -\ (\ Q_A - Z_A\ )$$

- The boundary occurs often in regions where dG/dz is minimum.
- The model is based on the promolecule and does not need a wavefunction, but is limited to linear molecules.
- In fluoroacetylene, one finds F nearly neutral, in agreement with inner shell energy measurements.
- Coming back to the previous example of FH molecule, the two wavefunctions discussed lead now to net charges of 0.26 and 0.27 for hydrogen.

2. Coppens (ref. 17) uses also a very general procedure for crystals. A current point M is joined to the various atoms $A_i$. If $R_i$ is a characteristic radius of atom $A_i$ (van der Waals radius for example) M is affected to the atom $A_k$ such that

$$r_k/R_k = \text{Min}\ (r_i/R_i)\ \ i=1,2....$$

-The method is very general, applicable to any $\rho(\underset{\sim}{r})$

- Partitioning fills the space.
- It satisfies crystal symmetry.
- It occurs that dQ/dv is often minimum at the boundary, if one slightly varies its limit.

3. Bader (ref. 18) has introduced a method that generalizes the previous one, and is based on strong quantum mechanical grounds. One difficulty in partitioning a molecule is that kinetic energy is not well defined for a subspace. If $\Omega$ is a subspace :

$$\int_{\Omega} \Psi^*(\underset{\sim}{p}^2\Psi)d\underset{\sim}{r} \neq \int_{\Omega} [\underset{\sim}{p}\Psi]^2 d\underset{\sim}{r} \qquad (31)$$

unless $\Omega$ is the total space or is such that at the boundary,

$$\underset{\sim}{\nabla}\rho.\underset{\sim}{n} = 0 \qquad (32)$$

if $\underset{\sim}{n}$ is the normal to the surface.

---

Exercise 2. Prove eqns 31 and 32.

---

If one assumes that between two bonded atoms there exists one and only one saddle point in $\rho(\underset{\sim}{r})$, one can uniquely define 'atomic fragments' by these conditions.
It can be proved that the virial theorem applies to each fragment.

$$T(\Omega) = - E(\Omega) = - V(\Omega) / 2$$

- It is the only partitioning method that is based on such quantum mechanical conditions. It has been applied to many small systems.
- It is an a posteriori method but can be applied to an experimental density.
- In a crystal

$$\rho(\underset{\sim}{r}) = V_c^{-1} \sum_{\underset{\sim}{H}} F(\underset{\sim}{H}) \exp(-i\underset{\sim}{H}.\underset{\sim}{r})$$
$$\underset{\sim}{\nabla}\rho(\underset{\sim}{r}) = -iV_c^{-1} \sum_{\underset{\sim}{H}} \underset{\sim}{H}\, F(\underset{\sim}{H}) \exp(-i\underset{\sim}{H}.\underset{\sim}{r}) \qquad (33)$$

Therefore, $\nabla\rho$ gives more weight to high order data, that are of lower accuracy. Moreover, series limitation effects are amplified. Our experience is that it seems presently necessary to first model the experimental density, then apply the virial partitioning to the model density, and finally calculate properties on experimental density (ref. 19).
Or one may define the shape function of the fragments, or its Fourier components, to a given precision : eqns (31) and (32) have to be Fourier transformed.

4. Daudel has introduced the loge partitioning (ref. 20). Schematically, one decomposes space in p subspaces $V_1, \ldots, V_p$ :

$$R^3 = V_1 \cup V_2 \cup \ldots \cup V_p \qquad V_i \cap V_j = \qquad (34)$$

An event is defined such that $n_1$ electrons are in $V_1$, ..., $n_p$ in $V_p$. We call $p_\lambda$ the probability of such an event.

$$n_1 + \ldots + n_p = N$$

Taking all the possible events $\lambda$, one defines the missing information function as :

$$I = \sum_\lambda p_\lambda \ln(p_\lambda^{-1}) \qquad (35)$$

I depends on the boundaries of $V_i$. The best partitioning is that which minimizes I.
The application is extremely difficult for large systems. It can be compared with Bader's method (see below and Daudel's lectures). Table 2 shows the results for a few molecules.

Table 2

| | BeH | BH | $BeH_2$ |
|---|---|---|---|
| $N_H$ | 1.87 | 1.95 | 1.86 |
| $\Lambda(N_H)$ | 0.22 | 0.34 | 0.18 |
| Virial partitioning | | | |
| $N_H$ | 1.99 | 1.99 | 1.96 |
| $\Lambda(N_H)$ | 0.18 | 0.26 | 0.14 |
| Loge partitioning | | | |

Loge partitioning, by construction, minimizes the fluctuations of electron content of each loge. Virial partitioning gives fairly similar results, though it cannot be rigorously proved. Fig. 4 shows that the function I and the fluctuation $\Lambda$ have parallel shapes, at least for the partitioning of atoms.

5. Hirshfeld (ref. 21) has also introduced a very general method of partitioning of the density, based on the promolecule. It should be compared with the method of Politzer. It in fact gives fairly similar results. Details can be found also in the lectures of Hirshfeld.

Figure 4

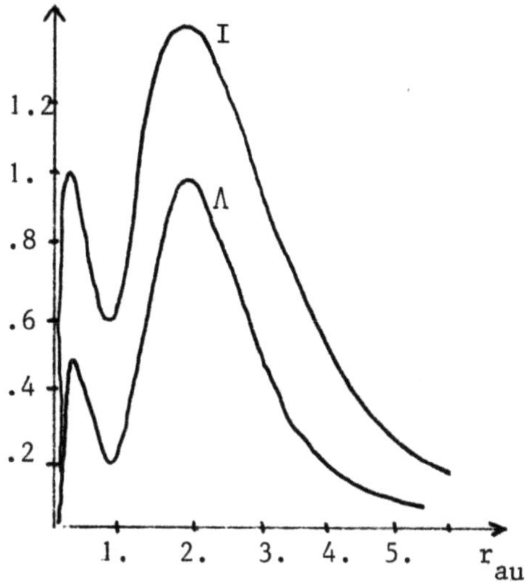


The variation of I and Λ for a spherical loge of radius R centred on B in the molecule BH

Pseudoatom $\rho_A$ is defined as :

$$\rho_A(\underset{\sim}{r}) = \frac{\rho_A^\circ(\underset{\sim}{r})}{\rho^\circ(\underset{\sim}{r})}\,\rho(\underset{\sim}{r}) \tag{36}$$

It is obvious that in (36) the core density is slightly perturbed.

One can propose the following modification, based on the assumption, common in crystallography, of an unperturbed core.

$$\begin{aligned} \rho^\circ(\underset{\sim}{r}) &= \rho_C(\underset{\sim}{r}) + \rho_V^\circ(\underset{\sim}{r}) \\ \rho(\underset{\sim}{r}) &= \rho_C(\underset{\sim}{r}) + \rho_V(\underset{\sim}{r}) \end{aligned} \tag{37}$$

The pseudoatom would therefore be defined by $\rho'_A$ :

$$\rho'_A(\underset{\sim}{r}) = \rho_C(\underset{\sim}{r}) + \rho_V(\underset{\sim}{r})\,\frac{\rho_{V,A}^\circ(\underset{\sim}{r})}{\rho_V^\circ(\underset{\sim}{r})} \tag{38}$$

For a first row homonuclear molecule, one gets for $\Delta = \rho'_A - \rho_A$, the features of Fig. 5.

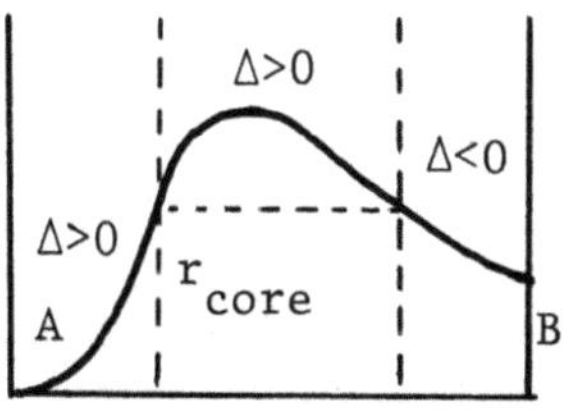


Figure 5

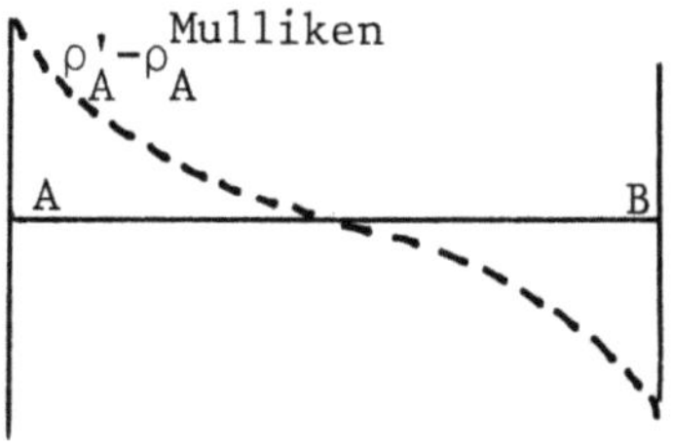


Figure 6

Definition (38) is more contracted in the valence region of the atom.
Comparison with Mulliken's scheme (eqn. 20) shows also a much higher localisation (Fig. 6).

## III. MEAN SQUARES PARTITIONING

A. We start with the minimization of

$$\varepsilon = \int [\rho(\underset{\sim}{r}) - \sum_i P_i f_i(\underset{\sim}{r})]^2 \, d\underset{\sim}{r} \quad (39)$$

If we define the vector $\vec{V}$ by its components :

$$v_i = \int \rho f_i d\underset{\sim}{r} \quad (40)$$

and

$$S_{ij} = \int f_i f_j d\underset{\sim}{r} \quad (41)$$

the populations $P_i$, defining a vector $\vec{P}$, are given by :

$$\vec{P} = S^{-1}\vec{V} \quad (42)$$

- $P_i$ are the expectation values of an operator only if $S = 1$.
  . either $f_i$ are localized functions
  . either $f_i$ are orthogonal functions.

$$\varepsilon = <\rho - \rho_m>$$

1. Kurki-Suonio (ref. 22) developed an expansion about an arbitrary center :

$$\rho_m = \sum_{\ell,m} R_\ell(r) Y_{\ell m}(\Omega) \quad (43)$$

$P_i$ are in this case the functions $R_\ell(r)$. S is unit matrix. Hence :

$$R_\ell(r) = \int \rho \, Y_{\ell m}(\Omega) \, d\Omega \quad (44)$$

In a crystal :

$$R_\ell(r) = 4\pi V_c^{-1} (-1)^\ell \sum_{\underset{\sim}{H}} F(\underset{\sim}{H}) \, j_\ell(Hr) \, Y_{\ell m}(\Omega_H) \quad (45)$$

- (45) is a rapidly convergent series.
- $\varepsilon \to 0$ when the basis increases.
- The method can be used to give information about the radial shape of the density near the nuclei (important for spin density

where there is no core density to start with in a refinement).

- Spacial partitioning may be added, to define ions...

2. Stewart (ref. 23 and his lectures) generalized the method in developing around each centre :

$$\rho_{mod.} = \sum_p \rho_p(\underset{\sim}{r}-\underset{\sim}{R}_p) \qquad F_{mod.} = \sum_p f_p(\underset{\sim}{K})\exp(i\underset{\sim}{K}.\underset{\sim}{R}_p) \tag{46}$$

$$\rho_p = \sum_{\ell,m} R_{p\ell}(r_p)Y_{\ell m}(\Omega_p)$$
$$\rightarrow \quad f_p(\underset{\sim}{K}) = \sum_{\ell,m} f_{p\ell}(K)Y_{\ell m}(\Omega_K) \tag{47}$$

$\varepsilon$ can be written as :

$$\varepsilon = \int |F(\underset{\sim}{K}) - F_{mod.}(\underset{\sim}{K})|^2 d\underset{\sim}{K} = \int_0^\infty \varepsilon(K)K^2 dK \tag{48}$$

One makes $\varepsilon(K)$ minimum for each K. In this case S is given by :

$$S_{p\ell m,qkn} = \int Y_{\ell m}Y^*_{kn} \exp[i\underset{\sim}{K}.(\underset{\sim}{R}_p-\underset{\sim}{R}_q)]d\Omega_K \neq 1 \text{ or } 0 \tag{49}$$

And :

$$v_{p\ell m} = \int F(\underset{\sim}{K})Y_{\ell m}(\Omega_K) \exp(-i\underset{\sim}{K}.\underset{\sim}{R}_p)d\Omega_K \tag{50}$$

is the projection of all terms having a given symmetry around $\underset{\sim}{R}_p$.

- $S\neq 1$, so the results are not invariant.

- $\int \rho(\underset{\sim}{r})g(r_p)Y_{\ell m}(\Omega_p)d\underset{\sim}{r}$ are exactly reproduced if $\ell$ and m are less than the limit of the expansion.

- In the case of crystals, we must assume radial functions, which contain some arbitrariness. The problem is similar to the choice of a suitable orbital basis set in quantum chemistry.

- Application to experimental data is in general very satisfactory. However some of the multipoles are very diffuse.

B.

1. Davidson (ref.24) proposes to minimize :

$$\varepsilon = \iint |\rho(\underset{\sim}{r}\underset{\sim}{r}') - \sum_{i\,j} \gamma_{ij} f_i(\underset{\sim}{r})f^*_j(\underset{\sim}{r})|^2 d\underset{\sim}{r}d\underset{\sim}{r}' \tag{51}$$

If we define :

$$n_{ij} = \langle f_i|\hat{\rho}|f_j\rangle = \iint f^*_i(\underset{\sim}{r})\rho(\underset{\sim}{r}\underset{\sim}{r}')f_j(\underset{\sim}{r}')d\underset{\sim}{r}d\underset{\sim}{r}'$$
$$S_{ij} = \int f^*_i f_j d\underset{\sim}{r}$$

We obtain :

$$\underset{\approx}{\gamma} = S^{-1}\,\underset{\approx}{n}\,S^{-1} \qquad \text{or} \qquad \underset{\approx}{n} = S\,\underset{\approx}{\gamma}\,S \tag{52}$$

2. Roby (ref. 25) works directly on the fisrt order density operator,

$$\hat{\rho} = \sum_i |\lambda_i>\lambda_i<\lambda_i| \tag{53}$$

in a natural expansion, where $0 \leqslant \lambda_i \leqslant 1$ .
The meanvalue of any one electron operator $\Omega$ is :

$$<\Omega> = Tr(\Omega\hat{\rho}) \tag{54}$$

Let us now define a projector P on the subspace subtended by a basis

$$|\chi> = |\chi_1\chi_2\cdots\cdots\chi_n>$$

$$P = |\chi>S^{-1}<\chi| \tag{55}$$

---

Exercise 3 : Prove (54) and (55)

---

For example, we may use the projector on an atomic subspace :

$$P_A = \sum_{\mu\in A} |\mu_A><\mu_A|$$

or on a two-atom subspace :

$$P_{AB} = \sum_{\mu,\nu\in A,B} |\mu>S^{-1}_{\mu\nu}<\nu|$$

In the subspace $E_P$, $\hat{\rho}$ is represented by :

$$\begin{aligned}\hat{\rho}_P &= P\hat{\rho}P \\ &= |\chi>S^{-1}<\chi|\hat{\rho}|\chi>S^{-1}<\chi|\end{aligned} \tag{56}$$

We define the probability of occupancy of $E_P$ by $N_P$:

$$N_P = Tr(\hat{\rho}_P) = Tr(\hat{\rho}P) \tag{57}$$

---

Exercise 4 : Prove (56) and (57), and the following results

---

- Take one function

$$n_\mu = \sum_i \lambda_i|<\mu|\lambda_i>|^2 = (SPS)_{\mu\mu} \tag{58}$$

We get the boundary $0 \leqslant n_\mu \leqslant 1$

- For one atom :

$$n_A = \sum_{\mu \in A} n_\mu = Tr_A(SPS) \tag{59}$$

(59) can be compared to the Mulliken population.

- For two atoms :

$$n_{AB} = Tr(\hat{\rho} P_{AB}) \geqslant n_A, n_B \tag{60}$$

In general, one observes that :

$$n_{AB} \leqslant n_A + n_B$$

but this cannot be rigorously proved. The shared density is :

$$s_{AB} = n_A + n_B - n_{AB} \leqslant n_{AB} \tag{61}$$

- If we take the total atomic basis, P is the projector :

$n = Tr(\hat{\rho}P)$ is in general $\leqslant N$

$\varepsilon = N - n$ is an indication of the inadequacy of the basis.

One can define the net charge of an atom :

$$q_A = Z_A - Q_A$$

where

$$Q_A = n_A - \frac{1}{2!} \sum_B s_{AB} + \frac{1}{3!} \sum_{BC} s_{ABC} + \dots$$

All these numbers are invariants, being expectation values of an hermitian operator. But we see that one needs $\hat{\rho}$ and not only $\rho(\underset{\sim}{r})$. In order to define such invariant methods, one needs $\rho(\underset{\sim}{r}\underset{\sim}{r}')$ and thus Compton scattering as well as Bragg scattering.

---

Exercise 4 : Consider $H_2$ molecule, with a minimal basis set. For the LCAO wavefunction, show that :

$n_A = 1 + S$ $(S \sim 0.7)$

For the valence bond wavefunction, show that :

$n_A = (1 + 3S^2)/(1 + S^2)$

Compare the shared densities, $Q_A$, $q_A$.

---

From exercise 5, we see that $n_A = 1.7$, $s_{AB} = 1.4$.

Such a scheme therefore is close to the principle of shell completion of an atom during molecule formation.

- One should realise that if the basis around one atom is too large, $n_A \rightarrow N_A$, and no chemistry can be retrieved (ref. 26). Taking

$Li_2$ and $Be_2$ with a (1s, 2s, $2p_\sigma$) basis :

$$s_{LiLi} = 1.74 \qquad s_{BeBe} = 1.69$$

which is absurd. Alrichs and Heinzman have thus proposed to restrict the basis to be minimal on each atom, and to modify it until $Tr(\hat{\rho}P)$ is maximal. For $Be_2$, this leads to

$$s_{BeBe} = -.006. \qquad \varepsilon \text{ goes from } 0.2 \text{ to } 0.005$$

3. There are other interesting features.

$\|Pf\|$ is a measure of the localisation of a function f in $E_P$.

$\|P_A \phi_\nu\| = \{\sum_{\mu\in A} S_{\mu\nu}^2\}^{1/2}$ measures the hybridization of $\phi_\nu$ around A.

$\|P_A \Psi_\iota\|$ where $\Psi_\iota$ is a M.O. measures the localisation of $\Psi_\iota$ around A.

- If we take $Q = P_A P_B P_A$ and look for the eigenvector of maximum eigenvalue of Q, it gives the most localized functions of A around B.

$P_A P_{mol} P_A$ gives localisation of A in the molecular subspace.

$P_{mol} P_A P_{mol}$ tells about localisation of the M.O.s around A.

## IV. ANALYSIS OF MOLECULES OR CRYSTALS

After having reviewed some partitioning methods of $\rho(\underset{\sim}{r})$ and $\rho(\underset{\sim}{r}\underset{\sim}{r}')$, I would like to discuss various applications.

If we want to define an atom in a molecule, or a molecule in a crystal, we want the subunit to be as independent as possible from the rest of the system. The fluctuation of properties is thus an essential quantity.

Take a one particle operator $\Omega$ :

$$\langle\Omega\rangle = Tr(\hat{\rho}\Omega) \tag{62}$$

$$\Lambda^2(\Omega) = Tr(\Omega^2\hat{\rho}) + Tr[\hat{\rho}_2\Omega(1) \otimes \Omega(2)] - [Tr(\Omega\hat{\rho})]^2 \tag{63}$$

---

Exercise 6 : Prove (62) and (63).

---

1. Space Partitioning

Following McWeeny (ref. 27) the pair distribution function $P_2(\underset{\sim}{r},\underset{\sim}{r}')$ is written as :

$$P_2(\underset{\sim}{r},\underset{\sim}{r}') = \rho(\underset{\sim}{r})\rho(\underset{\sim}{r}') [1 + f(\underset{\sim}{r},\underset{\sim}{r}')] \tag{64}$$

where f is the correlation function.
If $\Omega$ is the volume of a fragment,

$$\Lambda^2 = \int_\Omega \rho(\underset{\sim}{r})d\underset{\sim}{r} [1 + \int_\Omega \rho(\underset{\sim}{r}')f(\underset{\sim}{r},\underset{\sim}{r}')d\underset{\sim}{r}'] \tag{65}$$

$\Lambda$ is zero if

$$\int_\Omega \rho(\underset{\sim}{r}')f(\underset{\sim}{r},\underset{\sim}{r}')d\underset{\sim}{r}' \sim -1 \text{ for any } \underset{\sim}{r}.$$

- This is true for $\Omega = R^3$, and corresponds to the definition of the Fermi hole, the radius of which is :

$$\left[\frac{3}{4\pi\rho(\underset{\sim}{r})}\right]^{1/3}$$

- For a finite volume, $\Lambda$ is small if exchange-correlation is maximized inside $\Omega$ and minimized outside

- The fact that $\Lambda$ is often small in virial partitioning method is a strong point in favor of this method.

- We may remark that when $\Omega$ is very small,

$$\Lambda^2 \sim <N(\Omega)>$$

$$\Lambda(\Omega)/ N(\Omega) \gg 1 \tag{66}$$

showing the little significance of $\rho(\underset{\sim}{r})$ at a given pint in space. In the interaction between two systems, $\rho(\underset{\sim}{r})$ will be the leading information only if each system does not feel the fluctuations of density in the other, as for example in the approach of a proton. In building a bond, the answer is less obvious (ref. 28).

2. Take any operator and a SCF function :

$$\Lambda^2(\Omega) = <\Omega> - \sum_i \sum_j |\Omega_{ij}|^2 \tag{67}$$

where i and j denote spinorbitals.

---

Exercise 7 : Take a two electron case.

$$|\Psi_1> = a_1|\Phi_1> + a_2|\Phi_2> \} \begin{matrix}|\alpha> \\ |\beta>\end{matrix}$$

and

$$\Omega_1 = |\Phi_1> <\Phi_1|$$

a. Show that :

$$<\Omega_1>= 2\ [1 - a_2^2(1 - S^2)]$$

$$\Lambda^2(\Omega_1) = <\Omega_1>(1 - \frac{1}{2} <\Omega_1>)$$

Discuss.

b. Apply to $H_2$, in the MO theory. Discuss the asymptotic behaviour when $R \to \infty$.

c. Consider now the Valence Bond method. Show that

$\Lambda^2 = \dfrac{2S^2(1 - S^2)}{(1 + S^2)^2}$. Compare the asymptotic behaviour with the MO case. Conclusions.

- One should remark that if we take 2 systems with a small overlap, the VB function is equivalent to the allowance of exchange between the two systems. In that case, exchange does not increase the fluctuations much.

---

3. If we want to compare mean square partitioning on $\rho(\underset{\sim}{r})$ and on $\rho(\underset{\sim}{rr'})$, we can do it in the case where :

$$\hat{\rho}_m = \sum_i P_i |f_i><f_i| \qquad \text{and} \quad <f_i|f_j> = \delta_{ij}.$$

The two methods are the most alike when $f_i$ are the most localized in the von Niessen's sense (ref. 29)

$$\sum_{i \neq j} \int f_i^{\,2} f_j^{\,2} d\underset{\sim}{r} \qquad \text{minimum}$$

or

$$\sum_i \int f_i^{\,4} d\underset{\sim}{r} \qquad \text{maximum} \tag{68}$$

4. Let us consider the coupling with nuclear vibrations. If we do a mean squares expansion, for each geometry $\underset{\sim}{R}$,

$$\varepsilon(\underset{\sim}{R}) = \int |\rho(\underset{\sim}{r},\underset{\sim}{R}) - \rho_m(\underset{\sim}{r},\underset{\sim}{R})|^2\ d\underset{\sim}{r} \tag{69}$$

where the optimized parameters will depend on $\underset{\sim}{R}$.

Most methods assume rigid pseudoatoms, working on the average density :

$$\rho(\underset{\sim}{r}) = \int \rho(\underset{\sim}{r},\underset{\sim}{R})\ P(\underset{\sim}{R})\ d\underset{\sim}{R}.$$

The results obtained by using $\rho(\underset{\sim}{r})$ are not the average values of those obtained from geometry dependent density. Assume :

$$\rho_m(\underset{\sim}{r},\underset{\sim}{R}) = \sum_n P_n(\underset{\sim}{R})\ f_n(\underset{\sim}{r}-\underset{\sim}{R}_n)$$

we get :

$$\sum_n P_n(\underset{\sim}{R})\int \exp[i\underset{\sim}{K}.(\underset{\sim}{R}_m-\underset{\sim}{R}_n)]f_m(\underset{\sim}{K})f_n(\underset{\sim}{K})d\Omega$$

$$= \int\rho(\underset{\sim}{r},\underset{\sim}{R})f_m(\underset{\sim}{K})\exp(-i\underset{\sim}{K}.\underset{\sim}{R}_m)d\Omega \qquad (70)$$

From $\rho(\underset{\sim}{r})$, we should get coefficients $P'_n$.

$$\langle P'_n\rangle = \langle P_n\rangle \quad \text{only if}$$

$$\langle P_n \exp[i\underset{\sim}{K}.(\underset{\sim}{R}_m-\underset{\sim}{R}_n)]\rangle = \langle P_n\rangle\langle\exp[i\underset{\sim}{K}.(\underset{\sim}{R}_m-\underset{\sim}{R}_n)]\rangle$$

which implies that the pseudoatom has a low polarisability.

- It should be possible to employ the following method :

$$\varepsilon = \int \varepsilon(\underset{\sim}{R})P(\underset{\sim}{R})d\underset{\sim}{R} \qquad (71)$$

is minimized, using rigid pseudoatoms : that may serve as one definition of undeformable cores. Then $\rho - \rho_{m,rigid}$ is analyzed using deformable functions. This would give information about the non rigid following part of the density.

- Use of isotopic replacement, when possible, should be useful in order to elucidate that problem.

- One should recommend the development of multitemperature experiments, as proposed by Coppens.

## V. CHEMICAL REACTIVITY

A very spectacular example of the role of 'charges' in chemical reactivity is a study of saturated hydrocarbons by Fliszar (ref. 30). Charge in an atom is defined using the principle of charge alternation :

$$Q_i = \sum_{\text{neighbouring } j} (\alpha_j Q_j + \beta_j) \qquad \alpha_j < 0$$

$$\sum_i Q_i = 0 \qquad \text{Define} \quad 1 + \alpha_H = -n\alpha_C \qquad (72)$$

Reactivity of saturated hydrocarbons is related to inductive effect, that is usually defined in terms of Taft's equation :

$$Q_H = a\sigma^*(R) + b \qquad \text{in R-H}$$

$$Q_{CH_3} = -\, a'\sigma^*(R) \qquad \text{in } R\text{-}CH_3$$

Using eqn (72) on various isomers, it can be shown that the problem is overdetermined. Assuming :

$$\sigma^*(CH_3) = 0 \qquad \sigma^*(C_2H_5) = -0.1$$

as in experimental scales, one finds :

$\sigma^*$(isopropyl) = - 0.2 (experimental value : - 0.19)

$\sigma^*$(tertiobutyl) = - 0.3 (experimental value : - 0.30)

Assuming $\alpha_C$ to be the same from primary to quaternary carbons, one gets :

$$a' = - a = \frac{10}{3n}\beta_C$$

If $n > 0$, one should have $\overset{-}{C}\overset{+}{H}$ in methane
If $n < 0$, one should have $\overset{+}{C}\overset{-}{H}$ in methane.

Most of chemistry of alkanes is related to this inductive parameter. If n is small, inductive effect plays a strong role, if n is large, it plays little role.

- The use of various quantum chemical methods gives population charges that correlate well with Taft's constants. But the slope of the linear correlation varies much :

| Method | Del Re | EHMO | PCILO | INDO | STO3G | STO431G |
|---|---|---|---|---|---|---|
| n | 34 | 9.5 | 0.5 | -2 | 1 | 4.2 |

In particular, depending on the method, the charge of C in methane varies from -0.8 to +0.8 !
Does a preferential value of n exist ?

- Take molecules with carbons of various multiplicity. We should expect their charges to be as similar as possible :

$$\sum_{i \neq j} (q_i - q_j)^2 \quad \text{minimum}$$

where the sum is over many molecules (even cyclic).
This leads to $n = -4.4$ and thus to $C^+$ in methane.
From that point, Fliszar proposes to modify Mulliken's scheme in order to give a good correlation with $n = -4.4$.

- Now take any property P depending linearly on $q_C$

$$P_{calc} = Aq_C + B$$

Minimizing

$\Sigma\,(P_i - Aq_C^{\,i} - B)^2$ with respect to n, A, B leads to an optimal n.

. From chemical shifts of $C^{13}$ $n = -4.4$
. From ionisation potentials $n = -4.41$

This value of -4.4 seems a very reasonable one for n.

- It is also possible to correlate enthalpies of formation and vibrational energies with charges. In particular, rotational motions in cyclic molecules can be studied.

This simply shows the importance of charges in chemistry. But how is such a charge related to that obtained from a partition of $\rho(r)$ ? Using Bader's method, a correlation between the so defined charges and should give an answer. Preliminary results kindly given by Fliszar show that the value of n depends on the quality of the wavefunction, and n goes in the direction of -4.4 when the quality increases.

## VI. INTERMOLECULAR EFFECTS

1. The electrostatic approximation is based on the assumption, if $\Psi_A{}^{\circ}$ and $\Psi_B{}^{\circ}$ are isolated wavefunctions, that :

$$\Psi \sim \Psi_A{}^{\circ} \cdot \Psi_B{}^{\circ} \tag{73}$$

This approximation has been used by several authors.

- Scrocco (ref. 31) calculates electrostatic potentials to study protonation or hydration of molecules.
- A. Pullmann (ref. 32) is able to predict reasonably well the first hydration shell of molecules like uracil or cytosine, with a good estimate of energies of hydration, compared with the supermolecule method.
- Rein (ref. 33) studied sucessfully the guanine-cytosine pair.
- See also Hirshfeld's lectures for the study of crystalline cohesion (also studied by Berkovitch-Yellin and Leizerowitz).

In order to simulate the potential energy surface, one should try to speed up the calculation. It occurs that the multipolar expansion of the density in the Stewart's sense leads to a fast convergence if one stops the expansion at the quadrupole-quadrupole interaction.

Fig. 7 shows the electrostatic versus total component of the energy, as a function of the distance 0....0, for $H_2O...H_2O$ (ref. 34).

When R increases, the electrostatic component becomes predominant. But there is a limit under which one cannot use this approximation. However, adding an exchange term in the $\rho^{1/3}$ scheme, one should retrieve a minimum. This should not be difficult to do. If we consider the rotational degrees of freedom, one sees (Fig. 8 and 9) that the electrostatic approximation gives a very good answer.

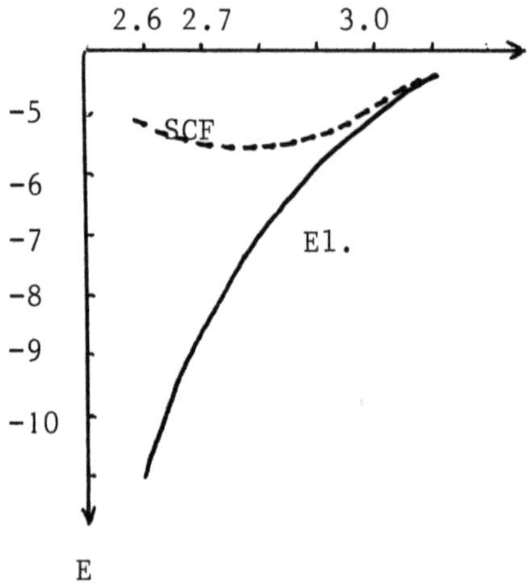


Figure 7

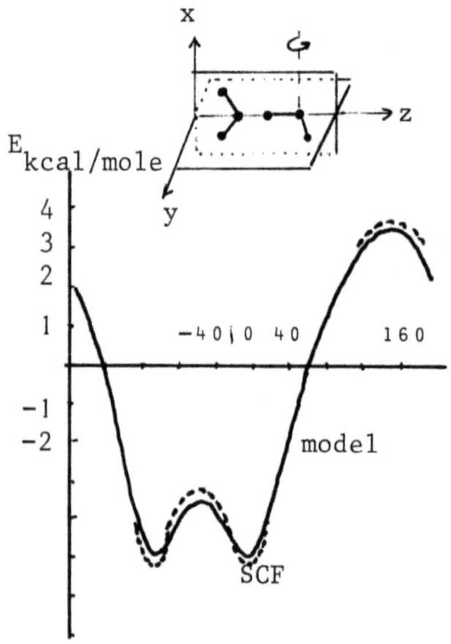


Figure 8
SCF(dotted line) and model interactions for rotation of molecule B around A , about $X_A$, $R_{O...O}$=2.76 Å

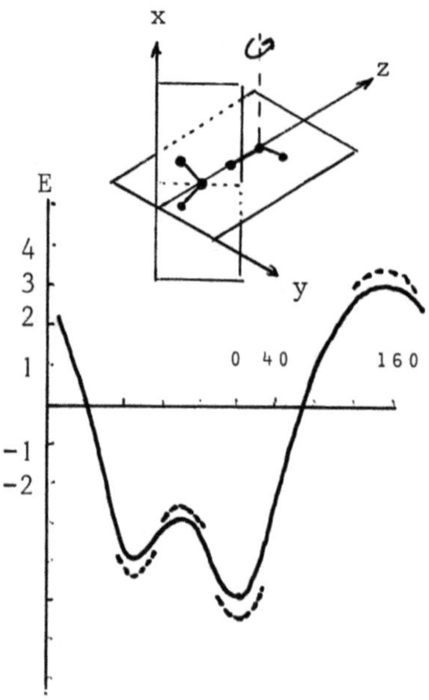


Figure 9
SCF(dotted line) and model interactions for rotation of molecule B around A, about $X_A$. $R_{O...O}$=2.76 Å. Angle of -50°.

The hydration shells of uracil and cytosine are given in Figs 10 and 11. The lability of water can be studied.

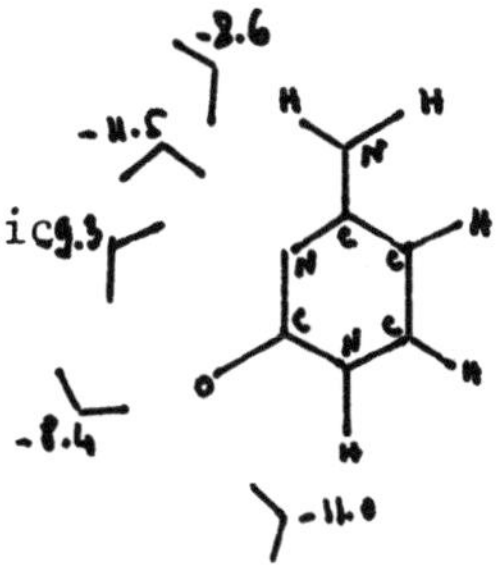

Electrostatic Model

SCF Calculation Supermolecule

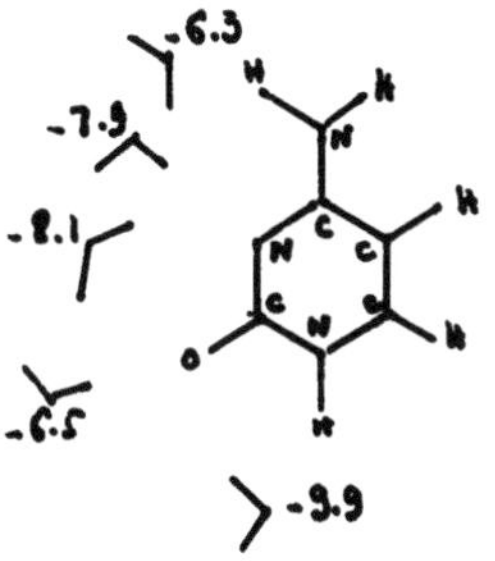

CYTOSINE

Figure 10
O...Cytosine = 2.85 Å

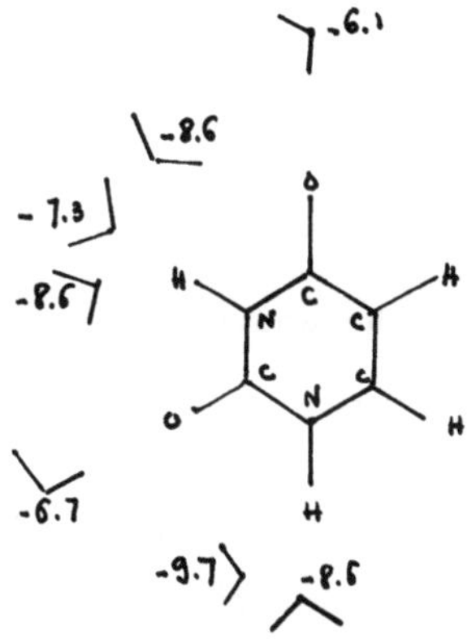

Electrostatic Model

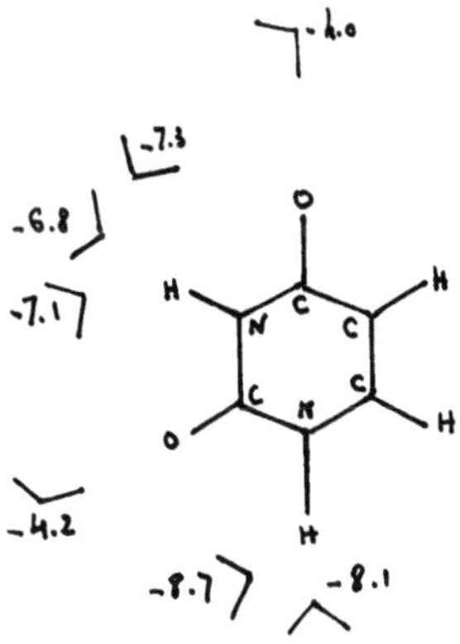

URACIL

SCF Calculation Supermolecule

Figure 11
O...Uracil = 3.0 Å

Obviously, a great deal might be learned by doing that kind of work systematically from experimental charge densities, in the field of chemical reactivity, intermolecular forces, crystalline cohesion.

2. Murrell and Morokuma propose a partitioning of the components of interaction (ref. 35 and 4).

$\Psi_1 = \Psi_A{}^{\circ}\Psi_B{}^{\circ}$ $\qquad \rho_{el} = 0$

$\Psi_2 = \mathcal{A}[\Psi_A{}^{\circ}\Psi_B{}^{\circ}]$ allows for exchange $\qquad \rho_{ex} = \rho_2 - \rho_1$

$\Psi_3 = \Psi_A\Psi_B$ $\qquad$ where $\Psi_A$ is calculated in the field of B...
This allows for polarisation

$$\rho_{pol} = \rho_3 - \rho_1$$

$\Psi_4 = \mathcal{A}[\Psi_A\Psi_B]$ $\qquad$ allows for charge transfer. (74)

The deformation density corresponding to molecular interaction is :

$$\delta\rho = \rho_4 - \rho_1$$

$$\rho_{ct} = \rho_4 + \rho_1 - \rho_2 - \rho_3$$

For water dimer, results are shown in Fig. 12. One sees the effect of exchange, that can be shown to be generally in the same direction. We also see that the overall effect is :

$$\rho_{pol} \sim \delta\rho$$

Looking at the energy components, we get :

| $E_{el}$ | $E_{ex}$ | $E_{pol}$ | $E_{ct}$ | $E_{total}$ |
|---|---|---|---|---|
| -9 | 4.2 | -0.5 | -2.5 | -7.8 |

The energy of polarisation is very small in this case : $\rho_{pol}$ stabilizes the supersystem but destabilizes each component.
We see that the energy does not follow the same trends as charge density

3. If we summarize the intermolecular components of energy we obtain :

$$E_{el} = -\sum_b Z_b \int \rho_{oo}^A(\underset{\sim}{1}) r_{b1}^{-1} d\underset{\sim}{r}_1 - \sum_a Z_a \int \rho_{oo}^B(\underset{\sim}{1}) r_{a1}^{-1} d\underset{\sim}{r}_1$$

$$+ \iint \rho_{oo}^A(\underset{\sim}{1}) \rho_{oo}^B(\underset{\sim}{2}) r_{12}^{-1} d\underset{\sim}{r}_1 d\underset{\sim}{r}_2 + \sum_a \sum_b Z_a Z_b r_{ab}^{-1} \qquad (76)$$

where a refers to A, b to B. $\rho_{oo}^A$ is the independent molecule density.

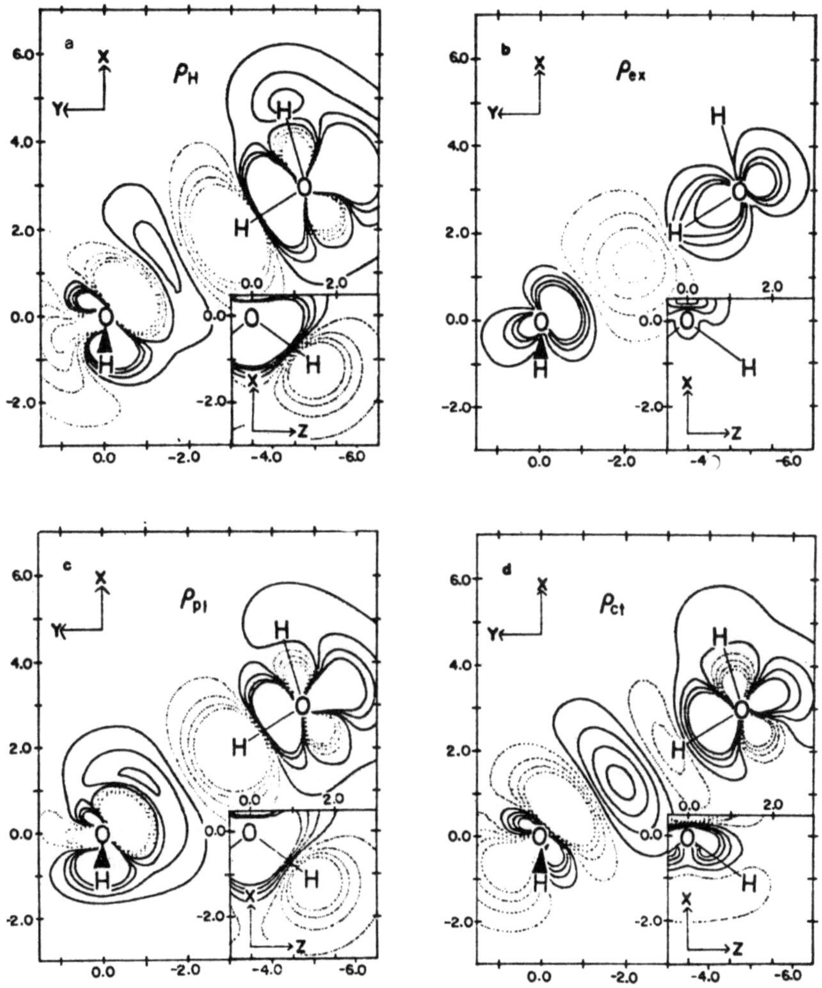

Figure 12
The electronic density change and its components in the water dimer. Full lines correspond to density increase. Values of the contours are ±.2, ±.6, ±1.0, ±1.4 $10^{-3}au^{-3}$.

- $E_{pol}$ is approximated if $E_{A,i}-E_{A,o} \sim \Delta E_A$ constant, where i refers to an excited state. It involves the ground state, the polarisabilities and the dipole moments :

$$\alpha_{ij}^{A} = 2(\Delta E_A)^{-1} \langle A_o|\mu_i \mu_j|A_o\rangle \qquad (76)$$

where $A_o = |a_1 \ldots\ldots a_m|$

$B_o = |b_1 \ldots\ldots\ldots b_n|$

$\mu_i$ and $\mu_j$ are dipole moment components.

$$E_{ex} = \sum_{i=1}^{m} \sum_{j=1}^{n} [\iint \rho_{ij}^{A}(\underset{\sim}{1}) r_{12}^{-1} \rho_{ij}^{B}(\underset{\sim}{2}) d\underset{\sim}{r}_1 d\underset{\sim}{r}_2 \qquad (77)$$

$$+ S_{ij}\int \rho_{ij}^{A}(\underset{\sim}{1}) V_B(\underset{\sim}{1}) d\underset{\sim}{r}_1 + S_{ij}\int \rho_{ij}^{B}(\underset{\sim}{1}) V_A(\underset{\sim}{1}) d\underset{\sim}{r}_1 ]$$

where :

$S_{ij} = \langle a_i|b_j\rangle$

${}_{ij}^{A} = a_i(b_j - P_A b_j)$ $\qquad P_A$: projector on A

$$V_B(\underset{\sim}{1}) = -\sum_b Z_b r_{b1}^{-1} + \int \rho_{oo}^{B}(\underset{\sim}{2}) r_{12}^{-1} d\underset{\sim}{r}_2$$

It is related to Roby's analysis in each molecule. It may also be estimated through statistical exchange.

- $E_{ct}$ involves $S^2$ terms and is often small. It might be perhaps estimated from the knowledge of $\rho - \rho°$, that we get experimentally.

4. Solid state studies by Ellis, Zunger and Freeman (ref. 36) using statistical exchange-correlation are becoming a very powerful tool. See the lectures of Freeman and Ellis.

It would be interesting to introduce as input experimental density and calculate a wavefunction and properties compatible with this charge density approach.

## CONCLUSION

Good partitioning schemes involve at least $\rho(\underset{\sim}{r}\underset{\sim}{r}')$ and fluctuations should be considered as important informations.

Nevertheless, $\rho(\underset{\sim}{r})$ is the fundamental information in studying intermolecular effects, and even chemical reactivity. It is also fundamental in the study of chemical bonding and crystal cohesion. The 'charge' is not uniquely defined and the 'chemical charge' as used by chemists is not simply related to that obtained from a partitioning of $\rho(\underset{\sim}{r})$.

References

1. Feynman R.P. (1939) Phys. Rev., 56, 340

2. Slater J.C. (1963) Quant. Theory of Molecules and Solids, Vol 1 Mc Graw Hill

3. Hohenberg P.C, Kohn W. (1964) Phys. Rev., B136, 864

4. Morokuma K. (1977) Acc. Chem. Res., 10, 294
. Kitaura K., Morokuma K. (1976) Int. J. Quant. Chem., 10, 325

5. Coppens P., Gururow T.N., Leung P., Stevens E.D, Becker P., Yang Y.W. (1979) Acta Cryst, under the press

6. Yanez M., Stewart R.F., Pople J.A. (1978) Acta Cryst., A34, 641, 649

7. McWeeny R. (1976) in 'Orbitals in Molecules and Crystals', edited by March N.H., Oxford

8. Constanciel R. (1972) Theo. Chim. Acta, 26, 249
. Chalvet O., Constanciel R., Rayez J.C. (1974) Jerusalem Symposium on Quant. Chem. and Biochem. VI, 77

9. Mulliken R. (1955) J. Chem. Phys., 23, 1833, 1841, 2338, 2343

10. Steiner (1976) Det. and Interpret. of Molec. Wavefunctions Cambridge Un. Press

11. Nesbet R.K. (1962) J. Chem. Phys., 36, 1518
. Clementi E. (1962) J. Chem. Phys., 36, 33

12. Lowdin P.O. (1952) J. Chem. Phys., 20, 374

13. Rinaldi D., Rivail J.L., Barriol (1971) Theor. Chim. Acta, 22, 298

14. Jug K. (1971) Theor. Chim. Acta, 23, 183
(1972) 26, 231
(1973) 29, 9
(1973) 31, 63

15. Lowdin P.O. (1950) J. Chem. Phys., 18, 365

16. Politzer P., Harris R.R. (1970) J. A. C. S., 92, 6451

17. Coppens P. (1975) Phys. Rev. Let., 35, 98

18. Bader R.F.W. (1975), Local. and Deloc. in Quantum Chemistry, edited by Chalvet, Daudel, Diner, Malrieu Reidel, Vol 1, 15

19. Becker P., Stephens M.E., to be published

20. Daudel R. (1979) see lecture notes and ref. therein

21. Hirshfeld F.L. (1977) Theor. Chim. Acta, 44, 129

22. Kurki Suonio K. (1977) Isreal J. Chem., 16, 115, 132

23. Stewart R.F. (1977) Israel J. Chem., 16, 124, and lecture notes

24. Davidson R. (1967) J. Chem. Phys., 46, 3320

25. Roby K.R. (1974) Mol. Phys., 27, 81 ; 28, 1441

26. Heinzmann R., Alrichs R. (1976) Theor. Chim. Acta, 42, 33

27. McWeeny R. (1960) Rev. Mod. Phys., 32, 335

28. Claverie P., Diner S. (1976) Localisation and Delocalisation in Quantum Chemistry, vol II, Reidel, p 395

29. Von Niessen W. (1972) J. Chem. Phys., 56, 4290

30. Fliszar S., and coworkers (1972) J. A. C. S., 94, 1068
(1972) 94, 7386
(1974) 96, 4353
(1977) 99, 5889

31. Bonnaccorsi R., Cimiraglia R., Scrocco E., Tomasi J. (1974) Theor. Chim. Acta, 33, 97

32. Pullmann A., Perhaia D. (1978) Theor. Chim. Acta, 48, 29
. Pullmann A., Berthod H. (1978) Theor. Chim. Acta, 48, 269

33. Rein R. (1973) Adv. in Quant. Chem., 7, 325

34. Bonnaccorsi R., Petrongolo C., Scrocco E., Tomasi J. (1971) Theor. Chim. Acta, 20, 331

35. Murrel W. (1976) Orbitals in Molecules and Crystals, edited by March N.H., Oxford

36. Zunger A., Freeman A.J. (1976) Int. J. Quant. Chem., S10, 383

# MULTIPOLAR EXPANSION OF ONE-ELECTRON DENSITIES

Robert F. Stewart

Department of Chemistry, Carnegie-Mellon University

4400 Fifth Ave., Pittsburgh, Pa. 15213 USA

## ABSTRACT

One and two-center atomic orbital products can be expanded with a finite number of multipoles on the several centers to high accuracy. The problem is formally solved for the charge density of a diatomic molecule by the method of least squares. In this case radial scattering factors for the two centers are directly determined from functional equations. Both the molecular form factor and the elastic x-ray scattering intensity are accurately given for a small set of multipoles on each center. The generalized x-ray scattering factors are not a property unique to the molecule, but a large number of static charge molecular properties are correctly given. Several applications of generalized x-ray scattering factors, with restricted radial functions, to x-ray diffraction data are briefly discussed.

## INTRODUCTION

Many-centered finite multipole expansions of x-ray structure factors or molecular form factors can provide useful insight into bonding effects, as revealed in reciprocal space, and also can provide a firm basis for studying vibrational or phonon effects on the structure factors. There is some merit in a formal study of the expansion problem, where molecular formfactors are accurately determined from good molecular wavefunctions. Applications to real x-ray diffraction data can follow with an understanding of the model limitations. For this lecture, I think it may prove useful to follow my own thoughts on the problem in a chronological order. In this context the lecture is by no means a review of the subject, but rather, is the evolution of an idea.

## A STARTING POINT

About ten years ago I proposed that the Fourier transform of atomic orbital products be explored as a basis for quantitative analysis of charge density information from x-ray diffraction data by the method of least squares.[1] These orbital products are the same if not similar basis functions which span the first order denity matrix from ab initio quantum chemical calculations. This approach to x-ray diffraction analysis has several difficulties, as is abundantly clear from published applications.[2,3,4]

The valence charge density model introduced represents the one-electron density function for a molecule or crystal as,

$$\rho(\vec{r}) = \sum_j \rho_j^c(r_j) + \sum P_{\mu\nu}\phi_\mu\phi_\nu \tag{1}$$

$r_j = |\vec{r} - \vec{R}_j|$ , ($\vec{R}_j$ is the nuclear position of atom j), where $\rho_j^c(r_j)$ is a spherically symmetrical core charge assigned to atom j and $\phi_\mu$

and $\varphi_\nu$ are valence type atomic orbitals such as 2s,2p,3s, 3p or 3d. For a second row atom (i.e. Li to F) $\rho_j^c(r_j)$ is derived from a spin-restricted, canonical Hartree-Fock, core atomic orbital. (For a density localized atomic orbital, see my exercise #1.) The $P_{\mu\nu}$ coefficients in (1) are adjustable coefficients (population parameters) in a least squares treatment of x-ray diffraction data. The actual basis functions in x-ray analysis are the generalized x-ray scattering factors

$$\Phi_{\mu\nu}(\vec{S};\vec{R}) = \int \varphi_\mu \varphi_\nu e^{i\vec{S}\cdot\vec{r}} d^3\vec{r} \tag{2}$$

where the scattering vector $|\vec{S}| = 2\pi|\vec{H}| = 4\pi \frac{\sin\theta}{\lambda}$ and $\vec{R}$ is an internuclear vector. The incorporation of (2) into a structure factor equation presents several new problems. First we need to know if the several $\Phi_{\mu\nu}$ are linearly independent or nearly so. Secondly we must choose $\{\Phi_{\mu\nu}\}$ which render (1) rotationally invariant and satisfy the site symmetry of the atom. On this latter point Barrie Dawson had emphasized this important invariant principle for charge density analysis of x-ray data.[5] [Sadly, there are examples in the recent literature on charge density analysis where the site symmetry of an "atom" has not been respected. In some cases the neglect of this important symmetry principle can lead to erroneous conclusions comparable to structure refinement errors because a higher symmetry for an ambiguous space group was chosen.]

A key physical assumption in the derivation of the x-ray crystallographic structure factor is that the charge density assigned to an atomic site (presumably the nucleus) is a perfectly following function of the nuclear motion. Thus for a two-center case in (2) $(|\vec{R}| \neq 0)$ such an assumption may have to be drastically modified. But first we can explore facets of (1) for x-ray diffraction analysis with a nuclear fixed system and defer vibrational averaging to a later effort.

For a systematic study of the applicability of the $\Phi_{\mu\chi}$ in spanning a least squares function, it is convenient to represent the atomic orbital products in terms of surface harmonics (for the one-center case) or in terms of sin $m\phi$ and cos $m\phi$ for the two-center case. In the former case the functions $P_\ell^m(\cos\theta)\cos m\phi = y_{\ell m}{}^+$ and $P_\ell^m(\cos\theta)\sin m\phi = y_{\ell m}{}^-$ form a basis for the irreducible representation of the full rotation group and in the latter case the functions span the cylindrical point groups. It is important that we remember that the charge density function must transform as the totally symmetric representation of the point group or of the space group for a crystal.

If we consider the orbital products among 2s and 2p orbitals on the same atom, ten generalized x-ray scattering factors are generated, but only nine are needed to span the full rotation group. (We assume here that the site symmetry of the atom is 1.) So at this level a possible linear dependency is evident; only the radial behavior of these density basis functions can make the spherically symmetrical component from $2p_x2p_x$, $2p_y2p_y$ and $2p_z2p_z$ distinguishable from 2s2s. In the case of Slater-type atomic orbitals, $R_{2p}(r) = R_{2s}(r)$, so that radial scattering factors are identically the same. For canonical, atomic Hartree-Fock radial orbitals, it also turns out that the radial x-ray scattering factors are virtually the same.[1] With the inclusion of 3d type orbitals for the same center, one introduces further linear dependencies.[6] For two-center orbital products, a similar but more complicated case of linear dependencies arises. It is convenient to represent the 2p orbitals as linear combinations of $2p\pi$, $2p\bar{\pi}$ and $2p\sigma$. Among the 2s and 2p orbital products between center a and center b fifteen unique two-center functions can be written in terms of sin $m\phi$ and cos $m\phi$ and functions in the confocal elliptical coordinates $\xi$ and $\eta$. For the case outlined here m is 0, 1 or 2 so that only five rotation type basis functions span the cylindrical point groups. For m = 0, there are five different functions in $\xi$ and $\eta$. In a typical case of a CN fragment with standard molecular Slater type orbitals at a bond length of 1.37Å the five functions are very nearly "parallel" to each other in spanning the space of $\xi$ (1 to $\infty$) and $\eta$ (-1 to +1). A measure of the linear dependency among these density functions is the projection coefficient,

$$P_{jk} = \int \rho_j \rho_k \, d\tau \Big/ \left\{ \int \rho_j^2 \, d\tau \int \rho_k^2 \, d\tau \right\}^{1/2} \tag{3}$$

where the integration is over all space and j and k label the two-center orbital product. For the case mentioned above, the $2p\sigma_a 2p\sigma_b$ and $(2p\pi_a 2p\pi_b + 2p\bar{\pi}_a 2p\bar{\pi}_b)$ are the least parallel with a projection coefficient of .604. One could choose these two functions (the $\sigma$ and $\pi$ type) as a possible bond scattering basis function for the totally symmetric case (m = 0). A projection analysis can be used for a similar cautious choice of sin $\phi$ and cos $\phi$ type scattering factors.

Further ambiguity in the selection of generalized x-ray scattering factors for the least squares problem arises by the partial representation of two-center charge distributions with one-center multipole functions. Ruedenberg[7] has shown that for multipoles about the N nuclei of a molecule,

$$\phi_{\mu a} \phi_{\nu b} = (1/N) \sum_{\ell}^{\infty} \sum_{q=1}^{N} C_{\ell q} \rho_{\ell q}(\vec{r}_q) \tag{4}$$

where $\rho_{\ell q}(\vec{r}_q)$ comprises $2\ell+1$ functions for the $\ell^{th}$ multipole. Note that (4) is N-fold redundant. By restricting q to centers a and b and truncating at $\ell = 3$ (an octopole term) Newton[8] has carried out a detailed projection analysis of the expansion (4). Relative root mean square fits varied from .05 to .6 for a variety of diatomic fragments. In these studies radial functions were restricted to Slater-type orbitals.

It should be clear, therefore, that the expansion (1) above in terms of atomic orbital products is virtually overcomplete and introduces unwanted linear dependencies. One approach to resolve the present dilemma is to form canonical density basis functions by the method of Lowdin.[9,10] However, this procedure will result in highly delocalized density functions which will considerably complicate the modeling of vibrational motion. An approach based on orthogonalized atomic orbitals was reported by Roby.[11] It would appear that with a complete set of orbitals on each center, one in principle could represent the charge density of a molecule with only one-center density functions. King, Newton and Stanton, however, point out that the Roby partitioning does not insure zero for off diagonal type Coulomb repulsion integrals.[12] This problem, however, is not directly related to efficient expansions of the charge density or of the x-ray structure factor.

## MULTIPOLES

It became apparent that an efficient model for generalized x-ray scattering factors need only incorporate multipole functions centered on the several nuclei of the molecule. For any molecule, one may represent the total charge density with an infinite multipole expansion about a given nucleus. The series would be very slowly convergent. Moreover the multipoles would provide an entire description of the molecular charge density and would not be restricted to the local charge near the nucleus. Such a scheme is contrary to Kurki-Suonoi's principle of locality[13] for the description of an "atom" in a molecule or crystal. Eventually, we must deal with the nuclear motion and it would be useful to realistically assign the local charge cloud the same amplitude of vibration as the nucleus to which it is associated. On the other hand, a second center of infinite multipoles will comprise an overcomplete set of density basis functions, so that the radial functions would be indeterminant. A small, finite set of multipoles about the several nuclear centers of a molecule can be determined, however.

We start with the one-electron density function of a diatomic molecule and seek to decompose the total charge density into one-center atomic contributions as a function of multipole expansion length. If the molecule is in a $\Pi$ or $\Delta$ state, we will take the cylindrical average. Thus,

$$\rho_{mol}^{c}(\vec{r}) = (4\pi)^{-1}\left[\sum_{j=0}^{J} \rho_j(r_a)\, P_j(\cos\theta_a) + \sum_{k=0}^{K} \rho_k(r_b)\, P_k(\cos\theta_b)\right] \tag{5}$$

We recast (5) by Fourier transforms to,

$$F_{mol}^{c}(\vec{r}) = \sum_{j=0}^{J} i^{j} f_{a,j}(S)\, P_j(\eta')\, e^{-ic\eta'} + \sum_{k=0}^{K} i^{k} f_{b,k}(S)\, P_k(\eta')\, e^{ic\eta'} \tag{6}$$

where $\eta'$ is the direction cosine between the scattering vector $\vec{S}$ (parallel to the Bragg vector) and the internuclear vector $\vec{R}$, $c = \frac{1}{2}|\vec{S}||\vec{R}|$ and $|\vec{S}| = 2\pi|\vec{H}| = 4\pi \sin\theta/\lambda$. The origin for $F_{mol}$ is taken at the bond midpoint, the $P_k(\eta')$ are Legendre polynomials and $f_{b,k}(S)$ are the Fourier-Bessel transforms of $\rho_k(r_b)$ :

$$f_{b,k}(S) = \int_0^\infty \rho_k(r_b)\, j_k(S r_b)\, r_b^2\, dr_b \tag{7}$$

Since the $P_j(\eta')$ span the totally symmetrical representation of the cylindrical point groups, we need only determine the radial functions $f_{a,j}(S)$ and $f_{b,k}(S)$. If we use (6) as a mean square fit to the true $F_{mol}(\vec{S};\vec{R})$, then the $\{f_a\}$ and $\{f_b\}$ can be determined by minimizing the residual,

$$\varepsilon = \int |F_{mol}(\vec{S};\vec{R}) - F_{mol}^{c}(\vec{S};\vec{R})|^2\, d^3\vec{S} = \int_0^\infty \varepsilon(S)\, S^2\, dS \tag{8}$$

where

$$\varepsilon(S) = \int |F_{mol} - F_{mol}^{c}|^2\, d\Omega_S \tag{9}$$

in (8) will be a minimum if for each $S$, $\varepsilon(S)$ in (9) is a minimum. Since the $\{f_a\}$ and $\{f_b\}$ are functions only of $S$ and not the vector $\vec{S}$, we solve directly for the f's at every $S$ by minimizing (9):

$$\frac{\partial \mathcal{E}(\underline{S})}{\partial f_{a,\ell}} = 0, \ \ell = 0, 1, \cdots J \quad \text{and} \quad \frac{\partial \mathcal{E}(S)}{\partial f_{b,m}} = 0, \ m = 0, 1, \cdots K \tag{10}$$

(10) results in a series of linear equations for $\{f_a\}$ and $\{f_b\}$:

$$I_{a,\ell}(\underline{S};\vec{R}) = (2\ell+1)^{-1} f_{a,\ell}(\underline{S}) + \sum_{k=0}^{K} \Big[\sum_{n=|\ell-k|}^{\ell+k} (-1)^{(n+k-\ell)/2} b_n(k,\ell) j_n(\underline{S}\,\underline{R})\Big] f_{b,k}(\underline{S})$$

$$\ell = 0, 1, \cdots J$$

$$I_{b,m}(\underline{S};\vec{R}) = \sum_{j=0}^{J} \Big[\sum_{n=|j-m|}^{j+m} (-1)^{(n+m-j)/2} b_n(j,m) j_n(\underline{S}\,\underline{R})\Big] f_{a,j}(\underline{S}) + (2m+1)^{-1} f_{b,m}(\underline{S})$$

$$m = 0, 1, \cdots K \tag{11}$$

In (11) the $b_n(j,m)$ are (n + 1/2) times Gaunt coefficients,[14] the $j_n(\underline{S}\,\underline{R})$ are spherical Bessel functions and the inhomogeneous terms,

$$I_{a,\ell} = (4\pi)^{-1} \int F_{mol}(\vec{S};\vec{R})(-i)^{\ell} P_\ell(\eta')e^{ic\eta'}\, d\Omega_S$$

$$I_{b,m} = (4\pi)^{-1} \int F_{mol}(\vec{S};\vec{R})(-i)^{m} P_m(\eta')e^{-ic\eta'}\, d\Omega_S \tag{12}$$

If the determinant of the $\underline{S}$ dependent coefficients for f's in (11) is non-zero then the solution (11) for a given [J/K] expansion is unique. Except for S = 0, this is apparently the case so long as J and K are finite. At $\underline{S}$ = 0 the $I_{a,\ell}$ and $I_{b,m}$ ($\ell > 0$ and $m > 0$) also vanish so that special methods are demanded for a solution. I will return to this problem shortly. It is at $\underline{S}$ near zero where we have Fourier components that give us the gross features of the pseudoatom such as the charge.

Equations such as (11) and (12) can appear to be forbidding and downright intimidating (at least my graduate students tell me that). Let's consider a [0/0] expansion; a monopole at center $\underline{a}$ and at center $\underline{b}$. Then (11) is simply

$$I_{a,o} = f_{a,o} + j_o(SR) f_{b,o}$$

$$I_{b,o} = j_o(SR) f_{a,o} + f_{b,o} \tag{13}$$

and the inhomogeneous terms, i.e. (12) are proportional to the electron-nuclear interference terms in gas-phase, high energy electron diffraction,

$$I_{a,o} = \int_0^\infty \rho_{mol}(\vec{r}) j_o(\underline{S}\, \underline{r}_a)\, d\tau$$

$$I_{b,o} = \int_0^\infty \rho_{mol}(\vec{r}) j_o(\underline{S}\, \underline{r}_b)\, d\tau \tag{14}$$

We first note that the $f_{a,o}$ and $f_{b,o}$ satisfy the electron-nuclear interference term. Secondly, note that the determinant of the least squares matrix in (13) is simply $1 - (j_o(SR))^2$, so that it is non-zero for all $\underline{S} > 0$. [Recall that $j_o(SR) = \sin(SR)/(SR)$.] The solutions for (13) are,

$$f_{a,o} = (I_{a,o} - j_o(SR) I_{b,o}) / (1 - (j_o(SR))^2)$$

$$f_{b,o} = (I_{b,o} - j_o(SR) I_{a,o}) / (1 - (j_o(SR))^2) \tag{15}$$

For a homonuclear diatomic, $I_{a,o} = I_{b,o}$, so that

$$f_{a,o} = I_{a,o}(1 - j_o(SR)) / (1 - (j_o(SR))^2) \tag{16}$$

As an example, let's take $\rho_{mol}(\vec{r})$ for the $H_2$ molecule at $R_e$. A rather accurate wavefunction was reported by Kolos and Roothaan[15] in 1960 and the first natural-spin orbitals were published by Davidson and Jones[16] two years later. The Davidson - Jones natural orbitals was converted to $I_{a,o}(S;R)$ (see (14)) and used as a basis to determine a hydrogen "atom" in the hydrogen molecule.[17] The actual hydrogen atom form factor is the solution to (16), but $\underline{R} = g\underline{R}_e$ was chosen where (8) is the minimum and this effectively led to $g = .81$. ($I_{a,o}$, however, was fixed for $\underline{R} = \underline{R}_e$.) What we learned in that endeavor is that $f_H$ in $H_2$ is considerably more extended in $\underline{S}$ space compared to the isolated H atom and that this extension could account for the anomalously low Debye - Waller factors for hydrogen atoms in organic molecules as reported by Jensen and Sundaralingam.[18] Clearly here was a bonding effect to be reckoned with.

We noted that the best monopoles from (13) satisfy the electron-nuclear-interference terms. We may wonder what other properties of $\rho_{mol}(\vec{r})$ are recovered at the [0/0] level. The mean square equation (8) can be stated equivalently in direct space,

$$\varepsilon = (2\pi)^3 \int (\rho_{mol} - \rho^c_{mol})^2 \, d\tau = min \tag{17}$$

The minimum condition for (17) includes condition (10) for each $\underline{S}$ which shows us that $\underline{a}$ and $\underline{b}$ centered multipoles of $\rho_{mol}$ are built into the properties of the pseudoatom radial multipole functions of $\rho_j(\underline{r}_a)$ and $\rho_k(\underline{r}_b)$. The $\underline{a}$ centered multipole expansion for $\rho_{mol}$ is defined,

$$\rho_{mol} = (4\pi)^{-1} \sum_{\ell=0}^{\infty} \rho_{mol,\ell}(\underline{r}_a) P_\ell(\cos\theta_a) \tag{18}$$

For $\ell \leq J$ the functional derivative condition $\partial\varepsilon/\partial\rho_\ell(\underline{r}_a) = 0$ is

$$\rho_{mol,\ell}(\underline{r}_a) = \rho_\ell(r_a) + (4\pi)^{-1}(2\ell+1)\sum_{k=0}^{K} \int \rho_k(\underline{r}_b) P_k(\cos\theta_b) P_\ell(\cos\theta_a) \, d\Omega_a \tag{19}$$

The derivative condition with respect to the J radial functions $\rho_\ell(\underline{r}_a)$ and $f_{a,\ell}(\underline{S})$ in $\underline{r}_a$ and $\underline{S}$, respectively, are equivalent. Thus (19) with its counterpart in $\underline{r}_b$ is equivalent to eq. (11). In fact the expressions in (12) are really the Fourier transforms of the $\ell^{th}$ multipole of $\rho_{mol}$ at center $\underline{a}$ and the $m^{th}$ multipole of $\rho_{mol}$ at center $\underline{b}$. It follows that averages of the form $\int \rho_{mol}(\vec{r}) g(\underline{r}_a) P_\ell(\cos\theta_a) \, d\tau$ are reproduced correctly for $\ell \leq J$ and for arbitrary functions $g(\underline{r}_a)$ for which the integral exists.

From this very general argument we see that a [0/0] expansion will satisfy $\langle r_a^{-1} \rangle$ and $\langle r_b^{-1} \rangle$ as well as the molecular charge density at nucleus $\underline{a}$ and at nucleus $\underline{b}$. (The notation $\langle h(\vec{r}) \rangle$

denotes the average $\int \rho_{mol} h(\vec{r})\, d\tau$ .) Since $\langle r_a^2 \rangle$ and $\langle r_b^2 \rangle$ are satisfied, the molecular dipole movement is also correctly given.

We now ask what is the charge of a pseudoatom. From (7) we see that it is $f_{b,o}(\underline{S})$ at $\underline{S} = 0$. But it is precisely this value of $\underline{S}$ where (11) appears to be indeterminant. Let's consider [0/0] only and study (13) for small $\underline{S}$. For $\underline{S}$ near zero we can write,

$$I_{a,o} = \sum_{p=0}^{\infty} \frac{(-1)^p S^{2p} \langle r_a^{2p} \rangle}{(2p+1)!} \tag{20}$$

$$j_o(SR) = \sum_{q=0}^{\infty} \frac{(-1)^q S^{2q} R^{2q}}{(2q+1)!} \tag{21}$$

and

$$f_{a,o} = \sum_{t=0}^{\infty} \frac{(-1)^t S^{2t} (r_a^{2t})_o}{(2t+1)!} \tag{22}$$

where

$$(r_a^k)_\ell = \int_0^\infty \rho_\ell(\underline{r}_a)\, r_a^{k+\ell+2}\, d\tau \tag{23}$$

With (20), (21), (22) and corresponding expressions for $I_{b,o}$ inserted into (13) we have,

$$\sum_{p=0}^{\infty} \frac{(-1)^p \langle r_a^{2p} \rangle S^{2p}}{(2p+1)!} = \sum_{t=0}^{\infty} \frac{(-1)^t (r_a^{2t})_o}{(2t+1)!} S^{2t} + \sum_{q=0}^{\infty} \frac{(-1)^q}{(2q+2)!} \left( \sum_{k=0}^{q} \binom{2q+2}{2k+1} (r_b^{2k})_o R^{2(q-k)} \right) S^{2q}$$

$$\sum_{p=0}^{\infty} \frac{(-1)^p \langle r_b^{2p} \rangle S^{2p}}{(2p+1)!} = \sum_{q=0}^{\infty} \frac{(-1)^q}{(2q+2)!} \left( \sum_{k=0}^{q} \binom{2q+2}{2k+1} (r_a^{2k})_o R^{2(q-k)} \right) S^{2q} + \sum_{t=0}^{\infty} \frac{(-1)^t (r_b^{2t})_o S^{2t}}{(2t+1)!} \tag{24}$$

where (24) is valid as $S \rightarrow 0$. From (24) we can equate coefficients of like powers of S to get the equations,

$$\langle 1 \rangle = (1_a)_o + (1_b)_o$$

$$\langle r_a^2 \rangle = (r_a^2)_o + R^2 (1_b)_o + (r_b^2)_o$$

$$\langle r_b^2 \rangle = R^2 (1_a)_0 + (r_a^2)_0 + (r_b^2)_0$$

$$\langle r_a^4 \rangle = (r_a^4)_0 + R^4 (1_b)_0 + \frac{10}{3} R^2 (r_b^2)_0 + (r_b^4)_0$$

$$\bullet \quad \bullet \quad \bullet \qquad (25)$$

The equations above are sufficient to solve for the charges of pseudoatoms a and b.

$$(1_b)_0 = \tfrac{1}{2}\left[\langle 1 \rangle + R^{-2}(\langle r_a^2 \rangle - \langle r_b^2 \rangle)\right]$$

and

$$(1_a)_0 = \tfrac{1}{2}\left[\langle 1 \rangle - R^{-2}(\langle r_a^2 \rangle - \langle r_b^2 \rangle)\right]$$

We see that pseudoatom charges give the total molecular charge and dipole moment. Notice that for a homonuclear the pseudoatom charge is just one-half the molecular charge. To find the mean square radii of the pseudoatoms, $(r_a^2)_0$ and $(r_b^2)_0$, one needs the $\langle r_a^4 \rangle$ and $\langle r_b^4 \rangle$ conditions from (25).

The multipole expansion in (6) is overcomplete for $J \rightarrow \infty$ and $K \rightarrow \infty$. As J and K are increased, the approach to overcompleteness is manifest by the near zero in the determinant of the S dependent coefficients of $\{f_a\}$ and $\{f_b\}$ . Even for small expansions, such as [2/2], the inverse least squares matrix is difficult to determine numerically for small S values. Nevertheless, for a fixed [J/K] expansion it is possible to fully characterize a pseudoatom in terms of its charge, centroid, eccentricity, mean square radius and so forth. With expansions such as (24), it is possible to determine all the moments of the pseudoatom. I leave it as an exercise for the student to show that the lowest moment of the quadrupole of pseudoatom b at [2/2] expansion is,

$$
\begin{aligned}
(l_b)_2 = & -\frac{2655}{256}R^2\langle 1\rangle - \frac{110535395}{1558656}R\left[\langle r_a P_1(\cos\theta_a)\rangle + \langle r_b P_1(\cos\theta_b)\rangle\right] \\
& + \frac{1575}{128}(\langle r_a^2\rangle + \langle r_b^2\rangle) - \frac{5268115}{97416}\langle r_a^2 P_2(\cos\theta_a)\rangle + \frac{15954995}{194832}\langle r_b^2 P_2(\cos\theta_b)\rangle \\
& + \frac{197435}{1152}R^{-1}\left[\langle r_a^3 P_1(\cos\theta_a)\rangle + \langle r_b^3 P_1(\cos\theta_b)\rangle\right] - \frac{2205}{256}R^{-2}\left[\langle r_a^4\rangle + \langle r_b^4\rangle\right] \\
& + \frac{1150}{9}R^{-2}\langle r_a^4 P_2(\cos\theta_a)\rangle - \frac{11225}{72}R^{-2}\langle r_b^4 P_2(\cos\theta_b)\rangle \\
& - \frac{10725}{64}R^{-3}\left[\langle r_a^5 P_1(\cos\theta_a)\rangle + \langle r_b^5 P_1(\cos\theta_b)\rangle\right] + \frac{525}{128}R^{-4}\left[\langle r_a^6\rangle + \langle r_b^6\rangle\right] \\
& - \frac{2275}{24}R^{-4}\langle r_a^6 P_2(\cos\theta_a)\rangle + \frac{5075}{48}R^{-4}\langle r_b^6 P_2(\cos\theta_b)\rangle + \frac{175}{2}R^{-5}\langle r_a^7 P_1(\cos\theta_a)\rangle \\
& + \frac{2975}{32}R^{-5}\langle r_b^7 P_1(\cos\theta_b)\rangle + \frac{525}{16}R^{-6}\left[\langle r_a^8 P_2(\cos\theta_a)\rangle - \langle r_b^8 P_2(\cos\theta_b)\rangle\right] \\
& - \frac{4725}{128}R^{-7}\left[\langle r_a^9 P_1(\cos\theta_a)\rangle + \langle r_b^9 P_1(\cos\theta_b)\rangle\right] + \frac{945}{128}R^{-8}\left[\langle r_a^{10}\rangle - \langle r_b^{10}\rangle\right]
\end{aligned}
$$

With the introduction of a multipole component for each pseudoatom, another molecular multipole is projected out so that we can expect the pseudoatom itself to change in gross features, but at the same time we expect a better fit to the molecular form factor and to the elastic x-ray scattering intensity, $I_{el}^{xR}(\underline{S}) = (4\pi)^{-1}\int |F_{mol}|^2 \, d\Omega_S$, for the gas-phase molecule. Our experience for several diatomic molecules, such as the diatomic hydrides, BH to FH, and $CO(^1\Sigma^+)$, $CO(^3\Pi)$ and $BF(^1\Sigma^+)$ is that convergence to the x-ray form factor is accurately achieved with rather small multipole expansions.[19] With six functions, i.e. [2/2] expansion, the relative root mean square difference is .2% and the maximum relative error in the elastic intensity is less than or equal to $4 \times 10^{-5}$. For a [4/4] expansion of $O_2(^3\Sigma_g^-)$ (i.e. five multipole functions), the relative root mean square error is .025% and the maximum relative error is $2 \times 10^{-7}$ for the elastic x-ray scattering intensity.[20]

A summary of convergence properties of pseudoatoms in NH for several expansions is given below in Table 1. $R_m$ is the maximum value for $1 - I_{el}^{a}/I_{el}$, where $I_{el} = (4\pi)^{-1}\int |F|^2 d\Omega_S$ and because $I_{el}^{c}$ is based on least squares fits, $I_{el} \geq I_{el}^{c}$. $R_w = \left[\int_0^{S_m} \varepsilon(S) S^2 dS / \int_0^{S_m} I_{el} S^2 dS\right]^{1/2}$ where $S_m$ corresponds to $(\sin\theta/\lambda)_{max} = 3.067$ Å$^{-1}$. $\mu$ and Q are the molecular dipole and N-centered quadrupole moments, respectively. Notice how the total charge of the N pseudoatom varies from 7.18 to 6.77 for the several [J/K] expansions. Atomic charges of the pseudoatoms are not a property unique to a molecule; they are unique only for given [J/K] expansions. The diffuse radial components of the pseudoatom on one center are sensitive to the total number of multipoles assigned to the other pseudoatom. We see that rapid convergence in the mean square fit of the molecular form factor is achieved with rather small expansions, but the pseudoatom description, particularly for low $\underline{S}$ values, is strongly dependent on [J/K]. For illustrations in the variations of the generalized x-ray scattering factors see references 19 and 20. On the other hand, the pseudoatoms in superposition represent a vast number of molecular properties. For example, the recovery of the electric field gradient for $N_2$ at [2/2] expansion is illustrated in reference 21. Extension of equations (11) and (12) to polyatomic systems has been reported.[22] Solutions to these equations are difficult and tedious to obtain.

From a formal study of the least squares model for multipole expansions, we see that molecular or crystal properties are projected into the pseudoatoms. With reasonable balance of the density basis functions, convergence to a small relative root mean

Table 1. Radial moments in NH as a function of [J/K] expansions. N is center a and H is center b.

| J/K | 0/0 | 1/0 | 2/0 | 1/1 | 2/1 | 2/2 |
|---|---|---|---|---|---|---|
| $(1_N)_0$ | 7.3263 | 7.1783 | 7.2994 | 6.9360 | 6.9158 | 6.7721 |
| $(r_N^2)_0$ | 13.77 | 12.74 | 13.29 | 12.12 | 11.62 | 10.92 |
| $(1_N)_1$ | – | -.8712 | -.1583 | -1.5840 | -1.6632 | -2.2329 |
| $(1_N)_2$ | – | – | 2.3302 | – | -.1294 | -.7502 |
| $(1_H)_0$ | .6737 | .8217 | .7006 | 1.0640 | 1.0842 | 1.2778 |
| $(r_H^2)_0$ | .855 | 1.325 | 1.224 | 1.948 | 2.422 | 3.126 |
| $(1_H)_1$ | – | – | – | -.7128 | -.7524 | -1.3222 |
| $(1_H)_2$ | – | – | – | – | – | .6211 |
| $R_m x10^3$ | 2.580 | 2.312 | .699 | .666 | .124 | .031 |
| $R_w$ | .01708 | .01515 | .00841 | .00829 | .00371 | .00209 |
| $\mu$ | .64005 | .64005 | .64005 | .64006 | .64006 | .64005 |
| Q | 1.2553 | .68594 | .68594 | .68594 | .68594 | .68594 |

square is accomplished. On the other hand, properties of the pseudoatoms are not unique and depend on the choice of the several centered multipoles. In actual applications to real diffraction data, this suggests that we look for static charge properties of the crystal system, but we need not dwell excessively on the pseudoatoms themselves.

## APPLICATIONS

I will briefly discuss three examples of multipole analysis of x-ray diffraction. The multipole model has been reviewed and applied by several crystallographers.[23,24,25,26] In all cases one must use restricted radial functions, in contrast to the radial functions determined by equations (11) and (12). Monopole functions are usually based upon spin-restricted Hartree-Fock atomic wavefunctions or some partitioned part, such as the core, and then supplemented with Gaussian or single exponential type radial functions. Electron population parameters are then determined by the method of least squares. One can then display the results as an electron density map and use them to make static charge property estimates. Most models assume that the pseudoatom (as represented by multipoles) is rigid with respect to nuclear vibrations. In this context the convolution approximation is used.

The rigid pseudoatom model was applied to Brown's diffraction data of crystalline beryllium[27] for an electron population analysis.[28] In this study it was found that $y_{53-}$, $y_{60+}$ and $y_{73-}$ and a valence monopole contributed to significant reduction in the least squares residuals. Final figures of merit, $R_W$ and R were .0081 and .0066, respectively, as compared to .0370 and .0323 for a free atom ($^1S$) Be scattering factor. A charge density map, constructed from the least squares determined multipoles, revealed a small build-up of charge near the tetrahedral vacancy with a smaller local charge in the octahedral "hole." This mildly surprising bonding density effect can be thought of crudely as a five-center bonding scheme. It is interesting that the multipole analysis of the measured structure factors yielded an electron density map, sans Be core density, in near quantitative agreement with a pseudopotential calculation reported by Inoue and Yamashita.[29] At this stage, however, the Be charge density analysis is only indicative of the probable bonding scheme. It remains to be shown whether the model, within the convolution approximation, correctly accounts for the known phonon dispersion curves and the elastic constants.

Monopole analyses have been applied to x-ray diffraction data of uracil.[30,31] In one case the valence monopole was constructed from an L-shell, standard molecular Slater-type function.[32] In the second case the pseudoatom charges were based on a set of "standard" spherical atoms.[33] The corresponding net charges found are displayed in Figure 1. Not only do the standard atom functions lead to a more polar picture of the constituent atoms, but the net charges on O and N are reversed in magnitude (not sign) from the standard molecular L-shell values. We see then at the monopole level how the charge of the pseudoatom can depend on the nature of restricted radial function employed. On the other hand, note that the molecular dipole moment estimates from these two different sets of pseudoatom charges are virtually the same. (See Fig. 1) The estimated standard deviation of the dipole moment is 1.3D, which is a relative error of 32%. The largest contributor is relatively large standard deviation of the H atom time-average position. A recent publication on ESCA spectra for pyrimidine and purine bases[34] reports an atomic charge analysis of the spectra. The results for uracil would favor the L-shell STO scattering factors as a basis for atomic charges that correlate with ESCA chemical shifts. The main point here is that pseudoatom charges are not a unique property of the molecule but the molecular dipole moment is. In the case I give here,

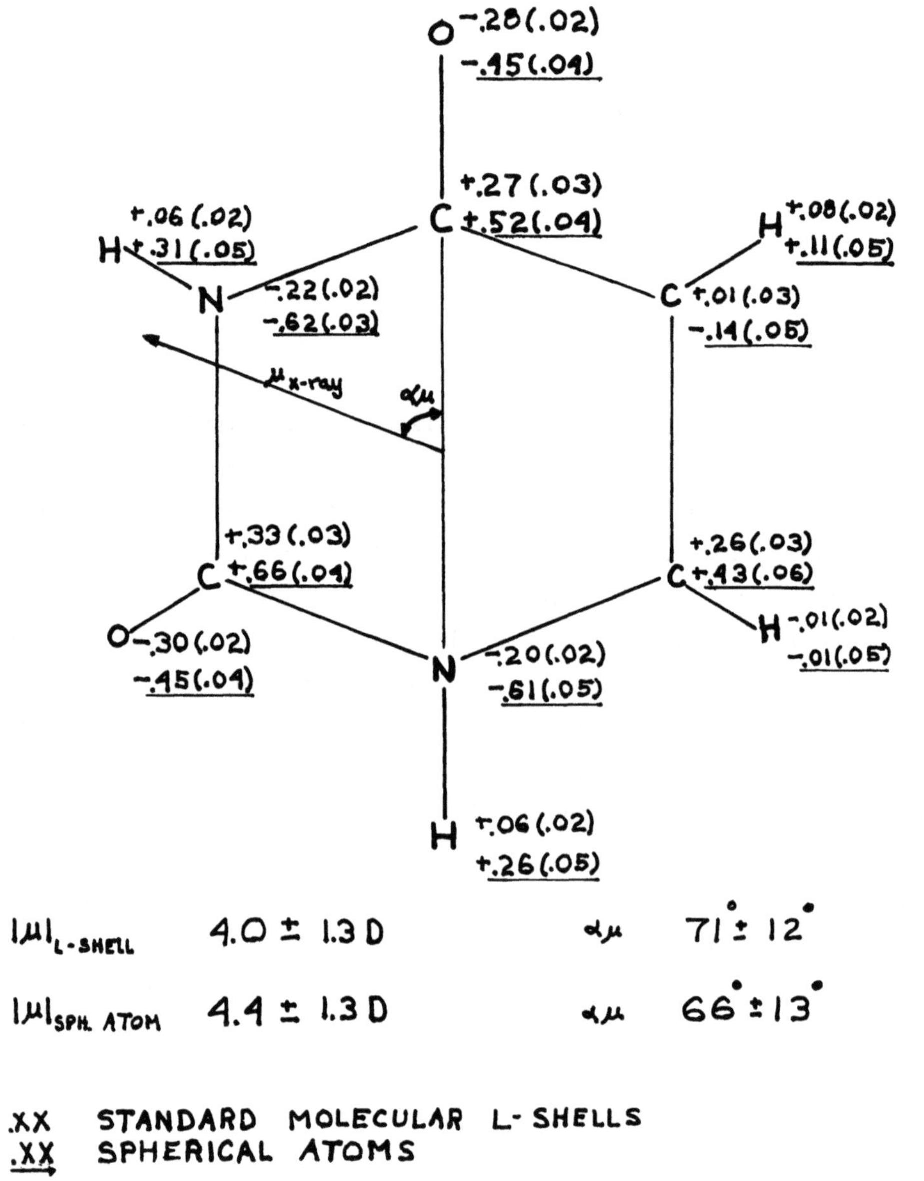


Figure 1. Pseudoatom Charges and the Molecular Dipole Moment from Monopole Analyses of X-ray Diffraction Data of Uracil.

apparently, both monopole functions project out the molecular dipole moment for the uracil molecule in its own crystal.

The last example I choose to give is a multipole analysis of crystal diffraction data for 1,1-azobiscarbamide and melamine.[35] In that work we found that pseudoatom octopoles played a dominant role in lowering the least squares residuals for these two organic molecular crystals. In this work the multipoles are plotted with stereo views (see reference 35). It was gratifying to see that the multipoles populated such that charge density was built up in the expected covalent bonding directions. Electric field gradient estimates for all the atoms were determined from the refinement models. The amide N in azobiscarbamide apparently has a principal field gradient component, $\lambda zz$, perpendicular to the amide plane. The lone pair charge density of the amide N(2) was built up by a superposition of dipoles and octopoles. These bases make no contribution to the field gradient of N(2). Strong inferences about the static charge densities in these crystals, however, are hampered by the large amplitudes of motion of the nuclei. The extent of deconvolution is not known so that thermal motion can be confused with electric field gradients. A reduced temperature study is clearly desirable and necessary before quantitative studies of electric field gradients from x-ray data can be seriously undertaken.

## Acknowledgement

This research was supported by NSF Grant CHE-77-09649.

## References

1. R.F. Stewart, J. Chem. Phys. 51, 4569-4577 (1969).
2. P. Coppens, T.V. Willoughby and L.N. Csonka, Acta Cryst. A27, 248-256 (1971).
3. D.S. Jones, D. Pautler and P. Coppens, Acta Cryst. A28, 635-645 (1972).

4. D.A. Matthews, G.D. Stucky and P. Coppens, J. Am. Chem. Soc. 94, 8001-8008 (1972).

5. B. Dawson, Proc. Roy. Soc. A298, 264-288 (1967).

6. R.F. Stewart, J. Chem. Phys. 58, 1668-1676 (1973).

7. K. Ruedenberg, J. Chem. Phys. 19, 1433-1434 (1951).

8. M.D. Newton, J. Chem. Phys. 51, 3917-3926 (1969).

9. P.O. Lowdin, Intern. J. Quantum Chem. 1S, 811-827 (1967).

10. P.O. Lowdin, Advan. Phys. 5, 1-172 (1956); Advan. Quantum Chem. 5, 185-199 (1970).

11. K.R. Roby, Chem. Phys. Lett. 12, 579-582 (1972).

12. H.F. King, M.D. Newton and R.E. Stanton, Chem. Phys. Lett. 31, 66-69 (1975).

13. K. Kurki-Suonio, "Charge Densities in Crystals as a Superposition of Non-Spherical Atomic Densities," Invited Lecture at Stony Brook, Report Series in Physics, 3/69, University of Helsinki (1969).

14. J.A. Gaunt, Philos. Trans. R. Soc. London Ser. A228, 151-196 (1929). (For appendix of interest, see pp. 192-196.)

15. W. Kolos and C.C.J. Roothaan, Rev. Mod. Phys. 32, 219-232 (1960).

16. E.R. Davidson and L.L. Jones, J. Chem. Phys. 37, 2966-2971 (1962).

17. R.F. Stewart, E.R. Davidson and W.T. Simpson, J. Chem. Phys. 42, 3175-3187 (1965).

18. L.H. Jensen and M. Sundaralingam, Science 145, 1185-1187 (1964).

19. J. Bentley and R.F. Stewart, J. Chem. Phys. 63, 3794-3803 (1975).

20. J. Epstein, J. Bentley and R.F. Stewart, J. Chem Phys. 66, 5564-5567 (1977).

21. R.F. Stewart, Chem Phys. Lett. 49, 281-284 (1977).

22. R.F. Stewart, Israel J. Chem. 16, 111-114 (1977).

23. M. Harel and F.L. Hirshfeld, Acta Cryst. B31, 162-172 (1975).

24. P. Coppens and N.K. Hansen, Israel J. Chem. 16, 163-167 (1977).

25. R.F. Stewart, Acta Cryst. A32, 565-574 (1976).

26. P.F. Price, E.N. Maslen and W.T. Delaney, Acta Cryst. A34, 194-203 (1978).

27. P.J. Brown, Phil. Mag. 26, 1377-1394 (1972).

28. R.F. Stewart, Acta Cryst. A33, 33-38 (1977).

29. S.T. Inoue and J. Yamashita, J. Phys. Soc. Japan 35, 677-683 (1973).

30. R.F. Stewart, J. Chem. Phys. 53,205-213 (1970).

31. M. Yanez and R.F. Stewart, Acta Cryst. A34, (1978).

32. W.J. Hehre, R.F. Stewart and J.A. Pople, J. Chem. Phys. 51, 2657-2664 (1969).

33. M. Yanez, R.F. Stewart and J.A. Pople, Acta Cryst. A34, (1978).

34. J. Peeling, F.E. Hruska and N.S. McIntyre, Can. J. Chem. 56, 1555-1561 (1978).

35. D.T. Cromer, A.C. Larson and R.F. Stewart, J. Chem. Phys. 65, 336-349 (1976).

# PARTITIONING OF HARTREE-FOCK ATOMIC FORM FACTORS INTO CORE AND VALENCE SHELLS

R.F. STEWART

For the solution to the Hartree-Fock equations of motion for electrons in the field of fixed nuclei, one may choose the matrix of Lagrange multipliers to be diagonal, leading to canaonical orbitals[1]. These orbitals are reasonable approximations to excitation and ionization processes[2]. But a single determinant can be represented by an infinite number of unitary transformations of its component orbitals[3].

For an atom, one might expect the orbitals with the smallest eigenvalue to be localized near the nucleus, compared with the others. Some years ago, I proposed that the valence electron density of molecular crystals be defined by subtracting out the core density of the first row atoms. The core density was defined as the product of the canonical spin restricted HF orbital for the ground state atom[4]. In this sense, the residual is the valence density. On the other hand, when the HF atom is partitioned in this manner, a valence shell orbital has a residual non zero amplitude on the nucleus. The $(\chi_{2s})^2$ of N ($^4$S), as an example, has a charge density of 32 eÅ$^{-3}$ on the nucleus as compared to .663 for $(\chi_{1s})^2$. For a plot of these densities, see ref. 5. We may address ourselves to other possible orbitals within the HF scheme and ask how such orbital products express themselves as components of the HF atomic scattering factor.

The literature on localized atomic and molecular orbitals is very rich and finds its genesis in the Cambridge group of quantum chemistry in the early 1950's[6-11]. One may choose those orbitals which are energy localized[12] (maximize the sum of the $\langle r_{ii}^{-1}\rangle$ ) or those orbitals which are density localized[13] (maximize the sum of $\int\rho_\mu d\tau$). Von Niessen has emphasized that the two methods give rather similar results.

In charge density work it is natural to choose a localization method based on densities. For any closed shell atom described by HF atomic orbitals one may seek the unitary transf. which minimizes the sum of overlap integrals of the different orbital products (or maximizes the sum of the charge densities of like orbital products).

For like orbital products, there is a corresponding x-ray form factor

$$f = \int \rho \exp(i\underset{\sim}{S}.\underset{\sim}{r})dv, \quad \text{where } \rho = \chi_\mu^{\,2}. \text{ Since}$$

$$\int \chi_\mu^{\,2} \chi_\nu^{\,2}\, dv = (2\pi)^{-3} \int f_\mu^* f_\nu d\underset{\sim}{S},$$

a minimum of

$$\sum_{\mu<\nu} \int \mu^2\, \nu^2 dv \quad \text{is a minimum of} \quad \sum_{\mu<\nu} \int f_\mu^* f_\nu d\underset{\sim}{S}.$$

Correlation among form factors will be minimum for density localized orbitals.

For open shells, it is essential that one have spin unrestricted HF orbitals so that $\alpha$ spin orbitals may be transformed independent of $\beta$ spin orbitals. In the present exercice, spin restricted HF atomic orbitals will be localized as if they were spin unrestricted. Moreover the orbitals of different L(L+1) (different angular momenta) will be localized separately.

-In Li ($^2$S)

$$\begin{pmatrix} is \\ os \end{pmatrix} = U \begin{pmatrix} 1s \\ 2s \end{pmatrix}$$ , such that $\int (is)^2 (os)^2$ is a minimum. The spherically averaged density is :

$$\rho_{Li} = (is)^2 + (1s)^2 + (os)^2$$

$$\rho_{core} = (is)^2 + (1s)^2 \qquad \rho_{val} = (os)^2$$

-For N ($^4$S)

$$\begin{pmatrix} is \\ os \end{pmatrix} = U \begin{pmatrix} 1s \\ 2s \end{pmatrix}$$ , such that $\int (is)^2 (os)^2$ is a minimum. The density

$\rho_N = 2(is)^2 + 2(os)^2 + 3(2\bar{p})^2$, where $2\bar{p}$ is the radial HF 2p.

$$\rho_{core} = 2(is)^2 \qquad \rho_{val} = 2(os)^2 + 3(2\bar{p})^2.$$

-For Na ($^2$S)

$$\begin{pmatrix} is \\ ms \\ os \end{pmatrix} = U \begin{pmatrix} 1s \\ 2s \\ 3s \end{pmatrix} \quad \text{with} \quad \int [(is)^2 (ms)^2 + (is)^2 (os)^2 + (ms)^2 (os)^2] \text{min}.$$

Also,

$$\begin{pmatrix} i's \\ m's \end{pmatrix} = U' \begin{pmatrix} 1s \\ 2s \end{pmatrix} \quad \text{with} \quad \int (i's)^2 (m's)^2 \quad \text{minimum. Then :}$$

$$\rho_K = (is)^2 + (i's)^2$$

$$\rho_L = (ms)^2 + (m's)^2 + 6(2\bar{p})^2$$

$$\rho_M = (os)^2$$

-For P ($^4$S),

$$\begin{pmatrix} is \\ ms \\ os \end{pmatrix} = U_s \begin{pmatrix} 1s \\ 2s \\ 3s \end{pmatrix} \qquad \begin{pmatrix} ip \\ op \end{pmatrix} = U_p \begin{pmatrix} 2p \\ 3p \end{pmatrix}$$

$$\rho_K = 2(is)^2 \; ; \; \rho_L = 2(ms)^2 + 3(ip)^2 + 3(2p)^2 \; ; \; \rho_M = 2(os)^2 + 3(op)^2$$

The spherical component of each orbital product is only considered :

$$f_{shell} = \int_0 \rho_{shell}(r)\, j_o(Sr)\, r^2 dr \quad (1)$$

$$\rho_{shell} = \int \rho_{shell}(\underset{\sim}{r})\, d\Omega \quad (2)$$

A useful ordering parameter for the various like-orbital products is the compactness

$$C_\mu = \int \rho_\mu^2 dv \quad (3)$$

Since $\int \rho_\mu dv = 1$, $C_\mu$ is a density in units of e $au^{-3}$. We may think of $C_\mu^{-1}$ as the quantum volume of the orbital product $\mu^2$, and define the atom volume as :

$$V = \sum_\mu N_\mu \,/[\sum_{\mu\nu} N_\mu N_\nu \int \rho_\mu \rho_\nu dv] \quad (4)$$

where $N_\mu$ is the orbital occupancy. $C_\mu$ is proportional to the area under a $S^2 f_\mu f_\nu$ curve versus $(\sin 2\theta/\lambda)$.

For a 2x2 rotation

$$\begin{pmatrix} \mu' \\ \nu' \end{pmatrix} = \begin{pmatrix} \cos\alpha & -\sin\alpha \\ \sin\alpha & \cos\alpha \end{pmatrix} \quad (5)$$

it is easy to verify that the angle for minimization of $\int \mu'^2 \nu'^2$ is :

$$\tan(4\alpha_{min}) = \frac{4\int[\mu'\nu'^3 - \mu'^3\nu']dv}{\int[\mu'^4 + \nu'^4 - 6\,\mu'^2\nu'^2]dv} \quad (6)$$

This condition can be used in an algorithm to minimize

$$D = \sum_{\mu'<\nu'} \int \mu'^2 \nu'^2\, dv \quad (7)$$

where more than 2 orbitals are included. The algorithm chosen here is:

1. Form a n component vector, $V_o$, of the canonical HF orbitals. Initialize the nxn matrix $U_o$ to be identity.

2. Determine all $\alpha_{min}$ for every pair of unlike orbitals from (6). Select the largest $\alpha_{min}$ to construct

$U_n = U'U_{n-1}$, where U' contains $u_{ab} = \delta_{ab}$ except for the ls pair, where $u_{ll} = u_{ss} = \cos\alpha$ , $-u_{ls} = u_{sl} = \sin\alpha$ .

3. Form the vector $V_n = U_n V_{n-1}$ of rotated orbitals. Cycle back to 2 and stop when all $\alpha_{min}$ are less than $10^{-7}$ radians.

Then one calculates the form factors of $\mu'^2$, starting from HF functions of Clementi[14]. During the school, outputs corresponding to Li,N,Na,P, Fe were provided. Compactness parameters are tabulated in table 1.

Except for Fe, compactness parameters differ by orders of magnitude so that one can easily sort the various orbital products into shells.

Table 1. Compactness for density localized and Canonical HF atomic orbitals Only orbitals with the same $L^2$ are localized. Power of 10 in ( )

| Atom | Li | N | Atom | Na | P | Atom | Fe |
|---|---|---|---|---|---|---|---|
| $1s^2$ | 7.6822(−1) | 1.1707(1) | $1s^2$ | 4.7464(1) | 1.2312(2) | $1s^2$ | 6.6137(2) |
| $2s^2$ | 1.9599(−3) | 5.9323(−2) | $2s^2$ | 3.1108(−1) | 1.0986 | $2s^2$ | 8.1469 |
| $2p^2$ | | 4.9532(−2) | $3s^2$ | 1.3928(−3) | 1.7051(−2) | $3s^2$ | 2.6149(−1) |
| $is^2$ | 7.9920(−1) | 1.2611(1) | $is^2$ | 5.2073(1) | 1.3851(2) | $4s^2$ | 3.1301(−3) |
| $op^2$ | 1.7109(−2) | 4.4899(−2) | $ms^2$ | 2.3002(−1) | 8.2071(−1) | $is^2$ | 7.7063(2) |
| | | | $os^2$ | 1.3311(−3) | 1.4343(−2) | $mis^2$ | 6.1489 |
| | | | $i's^2$ | 5.1934(1) | | $mos^2$ | 1.9875(−1) |
| | | | $m's^2$ | 2.2300(−1) | | $os^2$ | 2.9873(−3) |
| | | | $2p^2$ | 2.2624(−1) | 9.3865(−1) | $2p^2$ | 7.7140 |
| | | | $3p^2$ | | 9.0533(−3) | $3p^2$ | 1.9276(−1) |
| | | | $ip^2$ | | 1.0095 | $ip^2$ | 8.9318 |
| | | | $op^2$ | | 8.2318(−3) | $op^2$ | 1.6003(−1) |
| | | | | | | $3d^2$ | 9.4377(−2) |

Valence shell scattering factors, when localized, no longer have sizable amplitudes of scattering higher than .7 $Å^{-1}$. Core scattering factors are 5% larger above this limit.

Valence shell scattering factors have a structure similar to that of a monopole, single exponential radial function.

For first row atoms, the localized core scattering factor will necessarily increase the least square, mean square amplitudes of vibration compared to fits with canonical core scattering factors and single exponential valence type scattering factors. For first row atoms and data that extends to 1 $Å^{-1}$, the increase is of the order of +.001 $Å^2$ in $U^2$.

References


1. C. Edminston and K. Ruedenberg, J. Chem. Phys., 43, S97 (1965)
2. T. Koopmans, Physica, 1, 104 (1933)
3. V. Fock, Ann. der Physik 16, 126 (1930)
4. R.F. Stewart, J. Chem. Phys.,48, 4882 (1968)
5. R.F. Stewart, Acta Cryst., A24, 497 (1968)
6. J.E. Lennard Jones, Proc. Roy. Soc., A198, 14 (1949)
7. G.G. Hall, J.E. Lennard Jones, Proc. Roy. Soc., A202, 155, (1950)
8. J.E. Lennard Jones, J.A. Pople, Proc. Roy. Soc., A210, 166, (1950)
9. J.E. Lennard Jones, J.A. Pople, Proc. Roy. Soc., A220, 446, (1953)
10. A.C. Hurley, J.E. Lennard Jones, J.A. Pople, Proc. Roy. Soc., A220, 446 (1953)
11. G.G. Hall, Rep. Prog. Phys., 23, 1 (1959)
12. Edminston, K. Ruedenberg, Rev. Mod. Phys., 35, 457, (1963)
13. W. von Niessen, J. Chem. Phys., 56, 4290 (1972)
14. E. Clementi, IBM J. Res. Dev., 9, 2, supp., (1965)
15. R.F. Stewart, J. Chem. Phys., 53, 205, (1970)

RESTRICTED RADIAL FUNCTIONS FOR ANALYSIS OF MOLECULAR FORM FACTORS

C. Ceccarelli and R.F. Stewart

Part I

The simplest function for the analysis of charge density centred on a source point is

$r^n y_{\ell m\pm}(\theta,\phi)$. We know that $n \geq \ell$ to satisfy the Poisson equation.

$$y_{\ell m+}(\theta,\phi) = P_\ell^m(\cos\theta)\cos m\phi \qquad y_{\ell m-}(\theta,\phi) = P_\ell^m(\cos\theta)\sin m\phi$$

At large r, in the absence of other potentials, the Schrödinger equation predicts a charge density of $\exp(-\alpha r)$. A radial function

$r^n \exp(-\alpha r)$ will be explored here.

We start with some results from the previous exercise In this case we deal with spherically symmetrical functions or monopoles. In particular the focus will be on N ($^4$S) and P ($^4$S) atomic form factors. How well can the density localized valence scattering factors be represented with a single exponential function ? We seek to answer this question with a mean square fit to the shell scattering factor over a finite range of $\sin\Theta/\lambda$. Let

$$\varepsilon = \sum_x x^2(f_{shell} - f_n(\alpha,x))^2 \tag{1}$$

where $x = \sin\Theta/\lambda$, $f_{shell}$ is the shell form factor from a HF atomic wave function and:

$$f_n(\alpha,x) = \frac{\alpha^{n+3}}{(n+2)!} \int_0^\infty r^{n+2}\exp(-\alpha r)j_o(Sr)dr \tag{2}$$

$S = 4\pi a_o x$, $a_o$= .529177 Å. In (2) note that the radial function is normalized. For a given n, we solve for α that minimizes (1)

$$\sum_x x^2 f_{shell}\frac{\partial f_n}{\partial \alpha} = \sum_x x^2 f_n \frac{\partial f_n}{\partial \alpha} \tag{3}$$

The procedure to solve (3) is the following :

1. Generate $f_{shell}$ at intervals of .05 up to x = 2.00 $\mathring{A}^{-1}$
2. Compute $f_n(\alpha,x)$, $\partial f_n/\partial\alpha$ and $\partial^2 f_n/\partial\alpha^2$ with a trial $\alpha^o$.
3. Calculate :

$$\Delta\alpha = \alpha - \alpha^o = \frac{\sum_x x^2 \Delta f \frac{\partial f_n}{\partial\alpha}}{\sum_x x^2 \{ (\frac{\partial f_n}{\partial\alpha})^2 - \Delta f \frac{\partial^2 f_n}{\partial\alpha^2} \}} \qquad (4)$$

4. Add $\Delta\alpha$ to $\alpha^o$ and return to (2). Terminate the process when $\Delta\alpha/\alpha < 10^{-5}$.

The analytical evaluation of $f_n(\alpha,x)$ is outlined in the next exercise.

Let

$$I_{n,o}(\alpha,x) = \int_0^\infty r^{n+2} \exp(-\alpha r) j_o(4\pi a_o xr) dr$$

and

$$N = \alpha^{n+3}/(n+2)!$$

$$f_n(\alpha,x) = N\, I_{n,o}(\alpha,x) \qquad (4')$$

$$\frac{\partial f_n}{\partial\alpha} = \{(n+3)/\alpha\} f_n - N I_{n+1,o}(\alpha,x) \qquad (5)$$

$$\frac{\partial^2 f_n}{\partial\alpha^2} = (n+3)\{ 2 \frac{\partial f_n}{\partial\alpha} - (n+4) \frac{f_n}{\alpha} \}\alpha + N I_{n+2,o}(\alpha,x) \qquad (6)$$

Once $I_{n,o}$ can be determined (next problem), then (4) (5) and (6) are easily coded as algorithms. The results for fits to valence shells of N and P are summarized in table 1. The best fits from the scans on n are included as copies of computer output.

In all cases single exponential fits to localized valence shell scattering factors are superior to corresponding fits to canonical valence shell scattering factors. The best fits, however, are not exceptional with $R_w$ factors of .065, .034, .077. Even with localized core scattering factors, one can expect only modest success in recovering a HF atomic scattering factor with single exponential function for the valene shell. In the example given here, the maximum x of 2.0 is rather large. One could obviously improve the fits by restricting x to a smaller domain.

Part II

For the second part, we will use least squares to determine single exponential function fits to the molecule PN ($^1\Sigma^+$). In the first case, we seek the best mean square fit of the two monopoles, one on P, the other on N, to the molecular form factor of PN. The model x-ray form factor is simply :

$$F_c(\underset{\sim}{S}) = f_P(S)\exp(-ic\eta') + f_N(S)\exp(ic\eta') \qquad (7)$$

Table 1

Least Squares Fits of Single Exponential Functions
to HartreeOFock Shell Scattering Factors for $N(^4S)$ and $P(^4S)$,
up to $\sin\theta/\lambda = 2\text{Å}^{-1}$. $\epsilon = \Sigma x^2[f_{shell} - f_n(\alpha,x)]^2$. $R_W = (\bar{\epsilon}/\Sigma x^2 f^2_{shell})$

| $N(^4S)$ | Canonical L-Shell Fits | | | | Localized L-Shell Fits | | |
|---|---|---|---|---|---|---|---|
| n | $\alpha$ | $\epsilon$ | $R_W$ | n | $\alpha$ | $\epsilon$ | $R_W$ |
| 1 | 2.8483 | 2.08(-3) | .1941 | 1 | 2.8840 | 5.99(-4) | .1034 |
| 2 | 3.7711 | 2.48(-3) | .2118 | 2 | 3.8538 | 2.34(-4) | .0645 |
| 3 | 4.6749 | 3.86(-3) | .2641 | 3 | 4.7925 | 1.64(-3) | .1710 |
| 4 | 5.5637 | 5.74(-3) | .3223 | 4 | 5.7048 | 3.88(-3) | .2629 |

| $P(^4S)$ | Canonical L-Shell Fits | | | | Localized L-Shell Fits | | |
|---|---|---|---|---|---|---|---|
| n | $\alpha$ | $\epsilon$ | $R_W$ | n | $\alpha$ | $\epsilon$ | $R_W$ |
| 1 | 7.8942 | 3.82(-3) | .0578 | 1 | 8.0237 | 9.21(-3) | .0870 |
| 2 | 10.516 | 4.77(-3) | .0647 | 2 | 10.697 | 1.38(-3) | .0336 |
| 3 | 13.118 | 3.09(-2) | .1648 | 3 | 13.360 | 2.08(-2) | .1305 |
| 4 | 15.993 | 7.06(-2) | .2488 | 4 | 16.004 | 5.50(-2) | .2125 |
| 5 | 18.237 | 1.17(-1) | .3204 | 5 | 18.631 | 9.72(-2) | .2824 |

| | Canonical M-Shell Fits | | | | Localized M-Shell Fits | | |
|---|---|---|---|---|---|---|---|
| n | $\alpha$ | $\epsilon$ | $R_W$ | n | $\alpha$ | $\epsilon$ | $R_W$ |
| 1 | 1.6423 | 3.27(-3) | .4714 | 1 | 1.6764 | 1.42(-3) | .3201 |
| 2 | 2.1726 | 2.92(-3) | .4455 | 2 | 2.2666 | 5.37(-4) | .1967 |
| 3 | 2.7084 | 2.68(-3) | .4271 | 3 | 2.8487 | 1.58(-4) | .1068 |
| 4 | 3.2476 | 2.57(-3) | .4181 | 4 | 3.4235 | 8.24(-5) | .0771 |
| 5 | 3.7878 | 2.57(-3) | .4182 | 5 | 3.9912 | 1.98(-4) | .1195 |

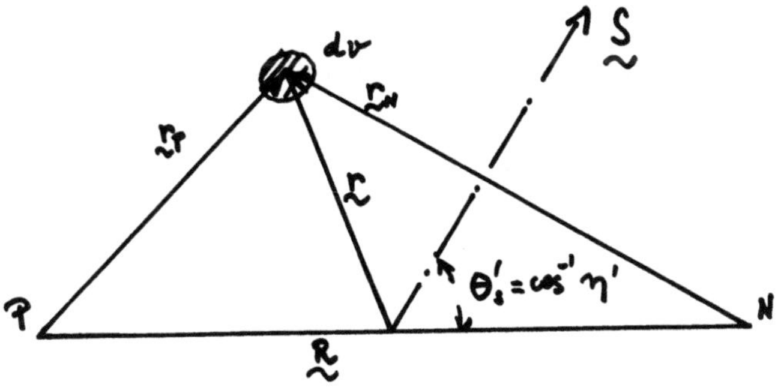

where the origin is chosen at the midpoint of the PN vector. Both $f_P(S)$ and $f_N(S)$ are spherically symmetric, $\eta'$ is the direction cos between S and R, and c = 1/2SR. Define the mean square error function

$$\varepsilon = (4\pi)^{-1} \int |F_{mol}(\underset{\sim}{S}) - F_c(\underset{\sim}{S})|^2 \, d\underset{\sim}{S} = \int_0^{\infty} \varepsilon(S) S^2 dS \qquad (8)$$

In (8) $\varepsilon$ is a minimum if $\varepsilon(S)$ is a minimum for each S. Since $f_P(S)$ and $f_N(S)$ are only functions of S, one can minimize $\varepsilon(S)$ to solve for the monopole scattering factors. The solution is :

$$f_P(S) = \{I_{P,o} - j_o(SR) I_{N,o}\} / \{1 - j_o^2(SR)\}$$
$$f_N(S) = \{j_o(SR) I_{P,o} - I_{N,o}\} / \{1 - j_o^2(SR)\} \qquad (10)$$

with

$$I_{P,o} = (4\pi)^{-1} \int F_{mol}(\underset{\sim}{S}) \exp(ic\eta') d\Omega_s$$
$$I_{N,o} = (4\pi)^{-1} \int F_{mol}(\underset{\sim}{S}) \exp(-ic\eta') d\Omega_s \qquad (11)$$

The limiting solution for S=0 is :

$$f_{P,o}(0) = \tfrac{1}{2}\{\int \rho_{mol} dv + R^{-2}[\int \rho_{mol} r_P^2 dv - \int \rho_{mol} r_N^2 dv]\}$$
$$f_{N,o}(0) = \tfrac{1}{2}\{\int \rho_{mol} dv - R^{-2}[\int \rho_{mol} r_P^2 dv - \int \rho_{mol} r_N^2 dv]\} \qquad (12)$$

Thus the total charge as well as the molecular dipole of PN are correctly given by these best monopoles. In general, it can be shown that (10) will correctly give all possible expectations of $\rho_{mol}(\underset{\sim}{r})$ that are of spherical symmetry around P or N. For example, the density at the nuclei is exact and the average potential $\langle r_N^{-1} \rangle$ or $\langle r_P^{-1} \rangle$ are reproduced.

For single exponential studies we represent the atomic form factors as

$$f_P(S) = f_{K,P}(S) + f_{L,P}(S) + f_{M,P}(S)$$
$$f_N(S) = f_{K,N}(S) + f_{L,N}(S) \qquad (13)$$

where K-shell form factors are taken from the density localized shells

$$f_{L,P} = C_{L,P} \int_0^{\infty} r^{n_{LP}} \exp(-\alpha_{LP} r) \, j_o(Sr) r^2 dr \qquad (14)$$

.........

From part I, we will choose $n_{LP} = n_{LN} = 2$ ; $n_{MP} = 4$. We seek the values of the C's and $\alpha$'s which minimize

$$\varepsilon = (4\pi)^{-1} \sum_x \int |F_{mol}(\underset{\sim}{S}) - F_c(\underset{\sim}{S}, P)|^2 d\Omega_s \, x^2 \Delta x \qquad (15)$$

where $P$ is the vector of the six parameters C and $\alpha$. (15) involves (11)

$$\sum_x x^2 \{I_{P,o} - f_P - f_N \, j_o(SR)\} \, \partial f_P / \partial P_P \, \Delta x = 0$$

$$\sum_{x} x^2 \{I_{N,o} - f_P j_o(SR) - f_N\} \partial f_N/\partial P_N \Delta X = 0 \quad (16)$$

Eqns (16) can be solved by Newton- Raphston techniques where initial $\alpha$ are taken from part I and the C start with values corresponding tp eight and five electrons in the L and M shells of P, and five in the N L shell.

Term (11) is computed from the HF wavefunction. The method is iterated until a convergence of parameters to $10^{-5}$. The phosphorous M shell changes the most with a contraction from $\alpha$= 3.424 in the free atom to $\alpha$= 3.797 in PN

The P fit is .0069, while that for N is .0125. Overall the monopole form factors with restricted radial functions mimic the exact best monopole reasonably well. On the other hand, note that the total charge of the valence shells is 17.309, which with the fixed K shell charges of 4, gives 21.309 electrons for the molecule. The exact $f_{P,o}$ and $f_{N,o}$ give 22 on the other hand (from 12). The molecular dipole moment from the restricted monopoles is -.980 au while the exact value from $f_{Po}$ and $f_{No}$ is -1.271. A second task is to minimize (15) subject to the constraint :

$$C_{LP}(n_{LP}+2)!/\alpha_{LP}^{n_{LP}+3} + C_{MP}(n_{MP}+2)!/\alpha_{MP}^{n_{MP}+3} + C_{LN}(n_{LN}+2)!/\alpha_{LN}^{n_{LN}+3} = 18 \quad (17)$$

Applying (17) the total molecular charge of 22 is exactly satisfied by the restricted monopoles. The charge of the P monopole is 14.751 and that of N is 7.249. The exact 'best' monopoles have values $f_{Po}(0)$ = 14.5488 and $f_{No}(0)$ = 7.4512. The restricted monopoles give a dipole of -.702 au, less accurate than the unconstrained one. The $R_W$ values are now .0091 for P and .0117 for N. Constrained $f_N$ is slightly closer to exact monopole, but $f_P$ is worse.

Summary of results.

a. The density localized valence shells of N and P can be roughly represented by single exponentials, better than for the canonical valence shells. Error of several %.

b. For the molecule PN, with fixed K shells, the single exponential approximation for the valence monopoles is rather close to the exact solution. The total charge and dipole moment differ by 3% and 23% from exact values. Electron nuclear interference terms are reproduced to 1%.

c. With a least squares solution constrained to satisfy the total charge, the estimated dipole differs by 45%. Other propertis are less accurately represented.

## ALGORITHMS FOR FOURIER TRANSFORMS OF ANALYTICAL DENSITY FUNCTIONS

R.F. Stewart

The charge density about a nucleus may be represented with multipole basis functions of the type :

$$\rho_{\ell,m\pm}(\underset{\sim}{r}_a) = \rho_{\ell,m\pm}(r_a)\ y_{\ell,m\pm}(\theta_a,\phi_a)\ /4\pi \qquad (1)$$

The x-ray form factor is :

$$f_{\ell,m\pm}(S) = (-1)^{\ell/2}\{\int_0^\infty r_a{}^2\rho_{\ell m\pm}(r_a)j_\ell(Sr_a)dr_a\}\ y_{\ell m\pm}(\theta_s,\phi_s) \qquad (2)$$

The task is to evaluate the radial integral (2) in terms of $d^* = S/2\pi$. Let

$$f_{\ell m\pm}(S) = \int_0^\infty \rho_{\ell m\pm}(r_a)j_\ell(Sr_a)r_a{}^2dr_a \qquad (3)$$

At small S

$$\lim_{S\to0}\{f_{\ell m\pm}(S)\} = \lim_{S\to0}\{S^\ell \int_0^\infty \rho_{\ell m\pm}(r_a)r_a^{2+\ell}dr_a\ /(2\ell+1)!!\} \qquad (4)$$

This suggests that we normalize

$\int_0^\infty \rho_{\ell m\pm}(r)r^{2+\ell}dr$ to unity. Our choice of normalisation is :

$$\lim_{S\to0}\{(2\ell+1)!!f_{\ell m\pm}(S)/S^\ell\} = 1 \qquad (5)$$

We shall now evaluate (3) for various choices of radial functions.

a. Single exponential functions.

$$\rho_\ell(r) = r^n \exp(-\alpha r)\ /(4\pi) \qquad (6)$$

$n \geqslant \ell$ since $\lim_{r\to0}\rho_\ell(r)$ is $r^\ell$, in order to satisfy the coulomb potential $r^{-1}$. When $n \geqslant \ell$, the evaluation of (3) is simplified. The answer can be found in Watson's book on Bessel functions (Cambridge University, 1966). Let

$$I_{n\ell}(\alpha,S) = \int_0^\infty x^{n+2}\exp(-\alpha x)j_\ell(Sx)dx. \qquad (6')$$

$$I_{n\ell}(\alpha,S) = \frac{(S/\alpha)^{\ell}(n+\ell+2)!}{(2\ell+1)!!\alpha^{n+3}\{1+(S/\alpha)^2\}^{n+2}}\ {}_2F_1\{\frac{\ell-n-1}{2}, \frac{\ell-n}{2}, \ell+\frac{3}{2}; -(S/\alpha)^2\} \qquad (7)$$

The hypergeometric function in (7)

$${}_2F_1(a,b;c;x) = \sum_{k=o}^{\infty} \frac{(a)_k(b)_k}{(c)_k\ k!}\ x^k \qquad (8)$$

where $(a)_k = a(a+1)\ldots.(a+k-1)$ and $(a)_o = 1$, is a small finite polynomial due to the constraint $n \geqslant \ell$. Applying (5) :

$$\lim_{S\to 0}\{(2\ +1)!!I_{n\ell}(\alpha,S)/S^{\ell}\} = \frac{(n+\ell+2)!}{\alpha^{n+\ell+3}}$$

we see that we want $\dfrac{\alpha^{n+\ell+3}}{(n+\ell+2)!}\ I_{n\ell}(\alpha,S)$ for the radial scattering factor

$$f_{n\ell}(\alpha,S) = \frac{S^{\ell}}{(2\ell+1)!!\{1+(S/\alpha)^2\}^{n+2}}\ {}_2F_1\{\frac{\ell-n-1}{2}, \frac{\ell-n}{2}; \ell+\frac{3}{2}; -(S/\alpha)^2\} \qquad (9)$$

(9) can be coded very efficiently. The derivative of (9) with respect to :

$$\partial f_{n\ell}(\alpha,S)/\partial\alpha = (n+\ell+3)(f_{n\ell} - f_{n+1,\ell})/\alpha \qquad (10)$$

$$\partial^2 f_{n\ell}/\partial\alpha^2 = \{(n+\ell+2)f'_{n\ell} - (n+\ell+4)(n+\ell+3)(f_{n+1\ell}-f_{n+2\ell})/\alpha\}/\alpha \qquad (11)$$

$f'_{n\ell}$ being (10). A Fortran code is added at the end.

b. <u>Single gaussian functions</u> $r^n \exp(-\gamma r^2)$.

Let

$$G_{n\ell}(\gamma,S) = \int_0^{\infty} x^{n+2}\exp(-\gamma x^2)j_{\ell}(Sx)dx \qquad (12)$$

$$= \frac{\Gamma\{(\ell+n+3)/2\}S^{\ell}\exp(-S^2/4\gamma)}{2(\sqrt{\gamma})^{\ell+n+3}\ (2\ell+1)!!}\ {}_1F_1(\frac{\ell-n}{2}; \ell+\frac{3}{2}; \frac{S^2}{4\gamma}) \qquad (13)$$

$${}_1F_1(a;b;x) = \sum_{k=o}^{\infty} \frac{(a)_k}{(b)_k\ k!}\ x^k \qquad (14)$$

is a polynomial of same parity as $\ell$. With (5), the form factor is :

$$g_{n\ell}(\gamma,S) = \frac{S^{\ell}}{(2\ell+1)!!}\exp(-S^2/4\gamma)\ {}_1F_1(\frac{\ell-n}{2};\ell+\frac{3}{2}; \frac{S^2}{4\gamma}) \qquad (15)$$

$$\partial g_{n\ell}/\partial\gamma = \frac{\ell+n+3}{2}(g_{n,\ell} - g_{n+2,\ell}) \qquad (16)$$

$$\partial^2 g_{n\ell}/\partial\gamma^2 = \{(\ell+n+1)g'_{n\ell}-(\ell+n+3)(\ell+n+5)(g_{n+2,\ell}-g_{n+4,\ell})/(2\gamma)\}/(2\gamma) \qquad (17)$$

${}_1F_1$ is usually rapidly convergent.

c. <u>Laguerre functions</u>. $\rho_{n\ell}(r) = r^{\ell}\, L_n^{2\ell+2}(\gamma r)\, \exp(-\frac{\gamma r}{2}) \,/(4\pi)$ (18)

L is a Laguerre polynomial of order n and degree (2ℓ+2). The Fourier Bessel transform of (18) is :

$$H_{n\ell}(\gamma,S) = \int_0^{\infty} \rho_{n\ell}(r)\, r^2 j_{\ell}(Sr)\,dr \qquad (19)$$

$$H_{n\ell}(\gamma,S) = \frac{2^n (2/\gamma)^{\ell+3} (n+2\ell+2)!\,(2S/\gamma)^{\ell}}{\{2(\ell+n)+1\}!!\,\{1+(2S/\gamma)^2\}^{\ell+2}}\, P_n^{(\ell+3/2,\ell+1/2)}(t) \qquad (20)$$

$P_n^{(a,b)}$ is a Jacobi polynomial and note that $-1 \quad t \quad 1$.

$$t = \{(2S/\gamma)^2 - 1\}/\{(2S/\gamma)^2 + 1\}$$

Normalized radial function for (18) is, after some manipulation :

$$R_{n\ell} = \frac{(-1)^n n!\,(\gamma/2)^{2\ell+3}}{(n+2\ell+2)!}\, r^{\ell} L_n^{2\ell+2}(\gamma r)\, \exp(-\gamma r/2)\,/(4\pi) \qquad (21)$$

The Fourier Bessel transform is :

$$h_{n\ell}(\gamma,S) = \frac{(-1)^n n!\, 2^n S^{\ell}}{\{2(\ell+n)+1\}!!\,\{1+(2S/\gamma)^2\}^{\ell+2}}\, P_n^{(\ell+3/2,\ell+1/2)}(t) \qquad (22)$$

Working out recurrance formulas, one gets :

$$h_{n+1,\ell} = -\frac{1}{n+2\ell+3}\left[\left\{\frac{+1}{n+\ell+1} + 2(n+\ell+2)t\right\} h_{n,\ell} + \frac{n(n+\ell+2)}{n+\ell+1}\, h_{n-1,\ell}\right] \qquad (23)$$

$$h_{o,\ell} = \frac{S^{\ell}}{(2\ell+1)!!\,\{1+(2S/\gamma)^2\}^{\ell+2}}$$

$$h_{1,\ell} = -\, h_{o,\ell}\{(2\ell+4)t+1\}/(2\ell+3) \qquad (24)$$

Subroutine to calculate the F.B. transform of a single exponential

```
      FUNCTION FEXP(N,L,A,X)
C     THIS FUNCTION IS PROPORTIONAL TO THE FOURIER BESSEL TRANSFORM
C     OF A SINGLE EXPONENTIAL. INTEGRAL (T**(N+2)+EXP(-A*T)*JL*(X*T)
C     *DT FOR T FROM 0 TO INFINITY. THE FACTOR (N+2)FACTORIAL/(A**
C     (N+3)) IS NOT INCLUDED. L IS THE ORDER OF THE SPHERICAL BESSEL.
C     N MUST BE EQUAL TO OR GREATER THAN L. IF X IS SIN(THETA)/LAMBDA
C     IN A-1, THEN A IS ALPHA/(4*PI*AO), WHERE ALPHA IS AU, AND AO=
C     .529177 A. SEE J. CHEM. PHYS. 66, 4057 (1977).
      Q=X/A
      Q2=Q*Q
      Q3=1.+Q2
      K=N+2
      P=1.
      DO 1 I=1,K
    1 P=P/Q3
      IF(L)4,4,2
    2 DO 3 I=1,L
    3 P=X*P/FLOAT(2*I+1)
    4 P1=1.
      L1=L-N-1
      L2=L1+1
      IF(2 (L1/2)-L1)5,6,5
    5 LL=-L2/2
      K1=3
      GOTO 7
    6 LL=-L1/2
      K1=1
    7 IF(LL)10,10,8
    8 K2=2*(L+LL)+1
      K3=LL
      DO 9 I=1,LL
      P1=1.-P1*FLOAT(K1)*FLOAT(I)+Q2/(FLOAT(K2)*FLOAT(K3))
      K1=K1+2
      K2=K2-2
    9 K3=K3-1
   10 FEXP=P*P1
      RETURN
      END
```

EXERCISES ON FORM FACTOR MODELS

Professor J. A. Avery

Problem 1

a. Suppose that the charge density in an atom is given by :

$$\rho(\underset{\sim}{x}) = \sum_{\ell m} \rho_{\ell m}(r) Y_{\ell m}(\theta,\phi)$$

Use the expansion :

$$\exp(i\underset{\sim}{k}.\underset{\sim}{x}) = 4\pi \sum_{\ell=o}^{\infty} i^{\ell} j_{\ell}(kr) \sum_{m=-\ell}^{\ell} Y^{*}_{\ell m}(\theta,\phi) Y_{\ell m}(\theta_k,\phi_k)$$

to show that the Fourier transform of $\rho$ is given by :

$$\rho^{t}(\underset{\sim}{k}) = (2\pi)^{-3/2} \int d\underset{\sim}{x} \exp(i\underset{\sim}{k}.\underset{\sim}{x}) \rho(\underset{\sim}{x})$$

$$= \sum_{\ell m} a_{\ell m}(k) Y_{\ell m}(\theta_k,\phi_k)$$

$$a_{\ell m}(k) = i^{\ell}(2/\pi)^{1/2} \int_{o}^{\infty} dr\, r^2 j_{\ell}(kr)\, \rho_{\ell m}(r)$$

*No solution needed here.*

b. Suppose that

$$\rho_{\ell m}(r) = \sum_{j} C_j\, r^{n_j} \exp(-\zeta_j r)$$

Show that :

$$a_{\ell m}(k) = i^{\ell}(2/\pi)^{1/2} \sum_{j} C_j J_{n_j+2,\ell}(\zeta_j,k)$$

$$J_{\mu\nu}(\zeta,k) = \int_{o}^{\infty} dr\, r^{\mu} j_{\nu}(kr) \exp(-\zeta r)$$

c. Using the fact that $j_o(x) = \sin x/x$, show :

$$J_{1,o}(\zeta,k) = \frac{1}{\zeta^2 + k^2}$$

Calculate the Fourier transform of $r^{-1}$. *( $(2/\pi)^{1/2} k^{-2}$ )*

d. Use the recursion formulas :

$$J_{\nu+1,\nu} = \left(\frac{2\nu k}{k^2+\zeta^2}\right) J_{\nu,\nu-1}$$

$$J_{\nu+2,\nu} = (2\nu+2)\zeta(k^2+\zeta^2)^{-1} J_{\nu+1,\nu}$$

$$(k^2+\zeta^2)J_{\mu+1,\nu} + (\mu+\nu)(\mu-\nu-1)J_{\mu-1,\nu} = 2\mu\zeta J_{\mu,\nu}$$

$$kJ_{\mu,\nu-1} + (\mu-\nu-1)J_{\mu-1,\nu} = \zeta J_{\mu,\nu}$$

to generate the functions $J_{\mu,\nu}$ shown in the following table :

Table 1

$$J_{\mu\nu} = \int_0^\infty dr\, r^\mu j_\nu(kr)\, e^{-\zeta r}$$

| ↑ ν | | | | |
|---|---|---|---|---|
| 3 | | | | $\frac{48\,k^3}{(k^2+\zeta^2)^4}$ |
| 2 | | | $\frac{8k^2}{(k^2+\zeta^2)^3}$ | $\frac{48k^2\zeta}{(k^2+\zeta^2)^4}$ |
| 1 | | $\frac{2k}{(k^2+\zeta^2)^2}$ | $\frac{8k\zeta}{(k^2+\zeta^2)^3}$ | $\frac{8k(5\zeta^2-k^2)}{(k^2+\zeta^2)^4}$ |
| 0 | $\frac{1}{(k^2+\zeta^2)}$ | $\frac{2\zeta}{(k^2+\zeta^2)^2}$ | $\frac{2(3\zeta^2-k^2)}{(k^2+\zeta^2)^3}$ | $\frac{24\zeta(\zeta^2-k^2)}{(k^2+\zeta^2)^4}$ |
| | 1 | 2 | 3 | 4 μ → |

e. Use the recursion formula

$$j_{\ell+1}(\eta) = -\eta^\ell \frac{\partial}{\partial\eta}\{\eta^{-\ell} j_\ell(\eta)\}$$

to show that :

$$j_1(\eta) = \sin(\eta)/\eta^2 - \cos(\eta)/\eta$$

$$j_2(\eta) = (3/\eta^3 - 1/\eta)\sin(\eta) - 3/\eta^2\cos(\eta)$$

$$j_3(\eta) = (15/\eta^4 - 6/\eta^2)\sin(\eta) - (15/\eta^3 - 1/\eta)\cos(\eta)$$

f. The 1s orbital of hydrogen is : $\chi_{1s} = (\pi a_o^3)^{-1/2}\exp(-r/a_o)$

Calculate the FT of the charge density of hydrogen atom.

*The result is :* $(2/\pi^3)^{1/2} 4\,(a_o^2 k^2 + 4)^{-2}$.

g. Suppose that the following orbitals in a carbon atom are doubly filled :

$$\chi_{1s} = (\zeta_1^3/\pi)^{1/2}\exp(-\zeta_1 r)$$

$$\chi_{2s} = (\zeta_2^3/8\pi)^{1/2}\exp(-1/2\,\zeta_2 r)\,(1 - 1/2\,\zeta_2 r)$$

$$\chi_{2p_z} = (\zeta_3{}^5/32\pi)^{1/2}\exp(-1/2\ \zeta_3 r)\ r\cos\Theta$$

Calculate the FT of the charge density of the atom. Use :

$$Y_{2,0} = (5/16\pi)^{1/2}(3\cos^2\Theta - 1) \qquad Y_{0,0} = (4\pi)^{-1/2}$$

h. Use the Poisson equation $\nabla^2 V = -4\pi\rho$
to show that the FT of V is related to the FT of $\rho$ by :

$$V^t(\underset{\sim}{k}) = \frac{4}{k^2}\rho^t(\underset{\sim}{k}).$$

Application to carbon.

i. Use the relations

$$\int_o^r dr'\ r'^n\exp(-\zeta r') = \frac{n!}{\zeta^{n+1}}\{\ 1 - \exp(-\zeta r)\sum_o^n \frac{(\zeta r)^s}{s!}\ \}$$

$$\int_r^\infty dr'\ r'^n\exp(-\zeta r') = \frac{n!}{\zeta^{n+1}}\exp(-\zeta r)\sum_o^n \frac{(\zeta r)^s}{s!}$$

$$r_{12}^{-1} = \sum_{\ell=o}^{\infty}\frac{4\pi}{2\ell+1}\ \frac{r_<^\ell}{r_>^{\ell+1}}\ \sum_{-\ell}^{\ell} Y_{\ell m}^*(\theta_1,\phi_1)Y_{\ell m}(\theta_2,\phi_2)$$

$$r_<^\ell/r_>^{\ell+1} = r_1^\ell/r_2^{\ell+1} \text{ if } r_1 < r_2\ ;\ = r_2^\ell/r_1^{\ell+1} \text{ if } r_1 > r_2$$

to calculate the potential V produced by the carbon atom.

j. Show that if the spherically averaged charge density on a neutral atom is :

$$\rho(r) = \sum_j C_j\ r^{n_j}\exp(-\zeta_j r)$$

where $n_j$ are integers, then the total potential of the atom (electronic + nuclear) is :

$$V(r) = \frac{4\pi}{r}\sum_j C_j(n_j+1)!\exp(-\zeta_j r)(\zeta_j)^{-(n_j+3)}\sum_{s=0}^{n_j+1}(n_j-s+2)(\zeta_j r)^5(s!)^{-1}$$

k. Calculate the FT of the potential of the hydrogen atom.

Problem 2

a. Use the relation

$$\{x^j y^\ell z^m r^n\exp(-\zeta r)\}^t = (-i\frac{\partial}{\partial k_x})^j(-i\frac{\partial}{\partial k_y})^\ell(-i\frac{\partial}{\partial k_z})^m(-\frac{\partial}{\partial\zeta})^n(\frac{2}{\pi})^{1/2}[\frac{2\zeta}{(k^2+\zeta^2)^2}]$$

to calculate the FT of the following functions :

$\exp(-\zeta r)$ ; $r^n\exp(-\zeta r)$ with n up to 4

$\exp(-\zeta r)$ multiplied by $x$, $x^2$, $xy$, $xyz$, $x^2y$, $x^3$.

b. Establish a similar relation in the case of $\exp(-\alpha r^2)$ and apply to the FT of :
$\exp(-\alpha r^2)$ multiplied by 1, $r^2$, $r^4$, $x$, $x^2$, $xy$, $xyz$, $x^2y$, $x^3$.

## REAL VS. RECIPROCAL SPACE ANALYSIS OF ELECTRON DENSITIES

Philip Coppens & Edwin D. Stevens

Chemistry Department

State University of New York at Buffalo

Buffalo, New York 14214

### I. Introduction

X-ray scattering measurements give the squared scattering amplitudes, at discrete values of the scattering factor S, which are related to the electron density in real space $\rho(r)$ by a simple Fourier transform relationship. Thus, X-ray scattering may be analyzed in either real or reciprocal space, the choice often being made on the basis of personal preference. Much early work in electron density analysis, done on high symmetric crystals with small unit cells, employed a small number of low order structure factors. The method is still being used in comparison of theory and experiment on structures such as silicon[1] and vanadium.[2] Reciprocal space analysis is also extremely well suited to least-squares adjustment of a parametrized model because of the discrete nature of the Fourier transform of a periodic (though continuous) function.

But analysis of the real space electron density maps has the undisputable advantage that we may apply chemical intuition to interpret observations, and thus span the gap between the physical experiment and its pertinence to those of our colleagues who live in real space. More generally, we can not afford to discard one way of analysis in preference to another, but will have to use each in conjunction with the other.

We shall discuss a number of recent studies which illustrate the merits of both approaches and their use in obtaining properties derived from the one-electron density.

## 2. From metals to molecular crystals: three examples of charge density analysis in real and reciprocal space

### 2.1 Electron density in beryllium metal

As noted by Brown[3] beryllium is a particularly suitable material for charge density analysis because two of its four atomic electrons are valence electrons which participate in the bonding. Brown collected the intensities of 27 reflections on an absolute scale and compared the corresponding structure amplitudes with those based on both a free atom model and an orthogonalized plane wave model for the valence electrons. Small but significant deviations in the structure factors could be eliminated by occupation of 2p-like atomic orbitals, especially those directed along the c axis of the hexagonal closed packed structure.

A more extensive reciprocal space analysis of the same accurate data set was performed by Stewart,[4] using a model composed of the atom-centered spherical harmonics $y_{00}$, $y_{53-}$, $y_{60}$ and $y_{73-}$, selected through a systematic search of triplets of multipole density functions. Stewart's analysis leads to a lowering of the R factor from 0.034 for the best previous model to 0.0081 with the introduction of four charge density parameters beyond the spherical atom model. The model fit allows representation in real space by inverse Fourier transform of the valence charge-density functions.

It is of interest to compare the model valence density map with a valence charge density obtained by direct Fourier transform of the observed amplitudes after correction for core scattering. Both maps support Brown's conclusion of covalency in the z-axis direction and further reveal valence density peaks in the tetrahedral holes which are located directly above and below the beryllium atoms (Fig. 1). The preference for the tetrahedral hole is stronger in the directly calculated valence density then in the model map, the ratios of $\rho$(tetrahedral)/$\rho$(octahedral) being 1.5 and 1.13 respectively in the two maps, indicating a possible bias introduced by incompleteness of the model. On the other hand, the model maps contain less noise than the direct valence density. Both agree remarkably well with a valence density calculated by Inoue and Yamashita[6] from the results of an APW (augmented plane wave) calculation, the peak heights in the tetrahedral holes being almost quantitatively equal in the three maps (model: 0.33 $e\mathring{A}^{-3}$; direct: 0.37 $e\mathring{A}^{-3}$; theory: 0.32 $e\mathring{A}^{-3}$). The two experimental maps show electron-deficient areas at or near the atomic cores, an indication of possible core polarization or thermal anharmonicity.

We note that the results provide possible explanations for the deviation from the ideal c/a ratio in beryllium, for the brittle-

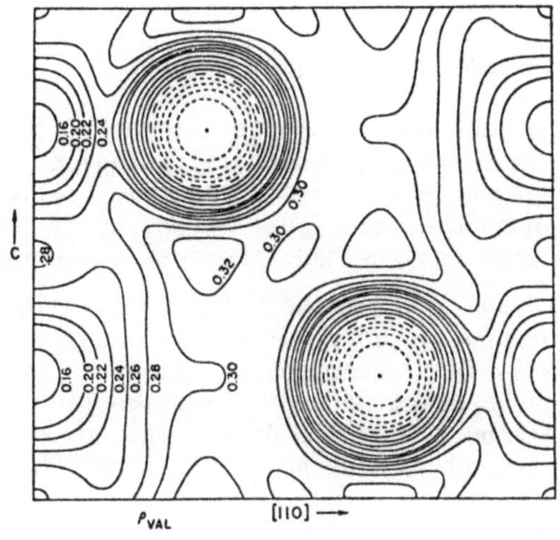


(a)

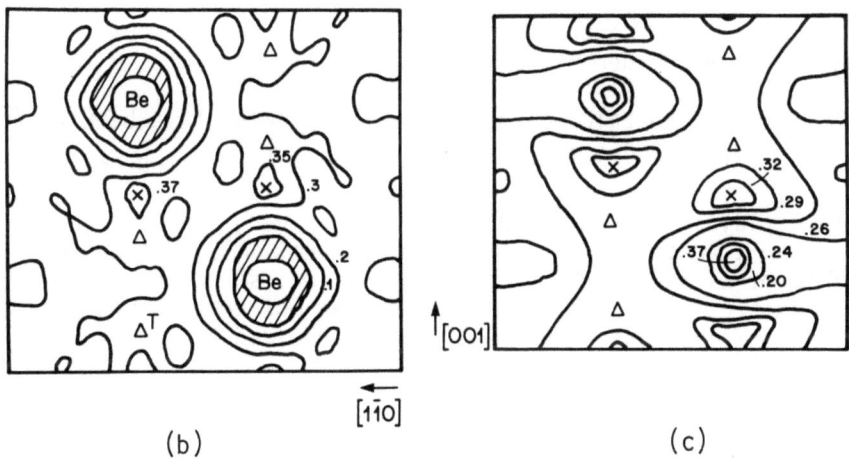

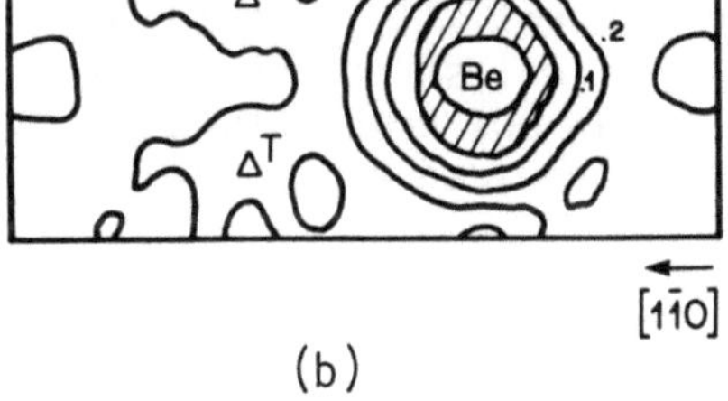


(b)

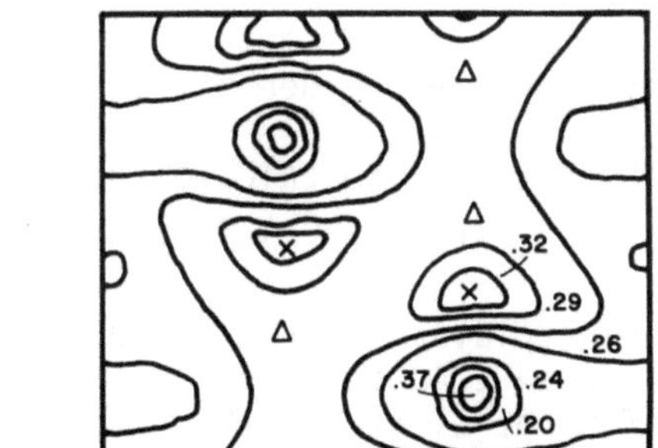


(c)

Fig. 1. Valence electron density in the (110) plane of Be metal
a) According to least squares model (Stewart, ref. 4)
b) Experimental valence density
c) Theoretical map based on APW calculation (Inoue and Yamashita, ref. 6).

ness of the metal, and also correlate nicely with calculations on the location of hydrogen atoms on beryllium surfaces by Schaefer and coworkers, which show the lowest energy for H atoms located above tetrahedral sites.[7]

## 2.2 Electron density in vanadium metal

Vanadium is a much heavier metal than beryllium and its valence electron density is therefore less easily studied. But measurements of the X-ray structure factors of the "paired" reflections 330-411 and 600-442 (which should have equal intensity if the thermally smeared vanadium atom density had spherical symmetry) by Weiss and Demarco[8a] and by Diana and Mazzone[8b] indicate a significant anisotropy. The anisotropy and Compton profile measurements have been interpreted by Weiss in a recent publication[2] as giving evidence of contraction of the vanadium 3d radial wavefunction in the direction of the nearest neighbors (the 111 direction) and expansion in the direction of the second nearest neighbors (the 100 direction).

A more complete picture is obtained by Fourier transform into real space of the accurate 80K powder diffraction data on vanadium measured by Korhonen, Rentavuori and Linkoaho.[9] In both the valence and deformation density maps (fig. 2) an accumulation of charge density is observed near 1/4, 1/4, 1/4, i.e. halfway between nearest neighbors. The maps seem to indicate a covalent contribution to the bonding rather than merely a change in the radial dependence of the atomic orbitals. A contraction as proposed by Weiss[2] would more likely increase the density closer the vanadium nucleus.

Band structure calculations on vanadium metal have been performed by Klein et al.[10a] and Wang and Callaway.[10b] The two calculations differ signicantly in detail: while the calculation by Klein and coworkers employs a $X\alpha$ APW method, the work by Wang and Callaway is based on an LCAO expansion as described in ref. 11. But a Fourier transform of either of the sets of theoretical structure factors, smeared isotropically with the experimental 80K temperature parameter (fig. 2) shows no signs of anisotropy as prediced by the experiment.

Obviously a large discrepancy remains between theory and experiment, the resolution of which awaits either additional experiments or new theoretical methods.

## 2.3 Electron density in the metal-metal bond in dichromium tetraacetate dihydrate

Very short metal-metal bonds are found in dinuclear molecular complexes of transition metals. In 1970 Cotton and coworkers redetermined the structure of the hydrate of chromous acetate dihydrate

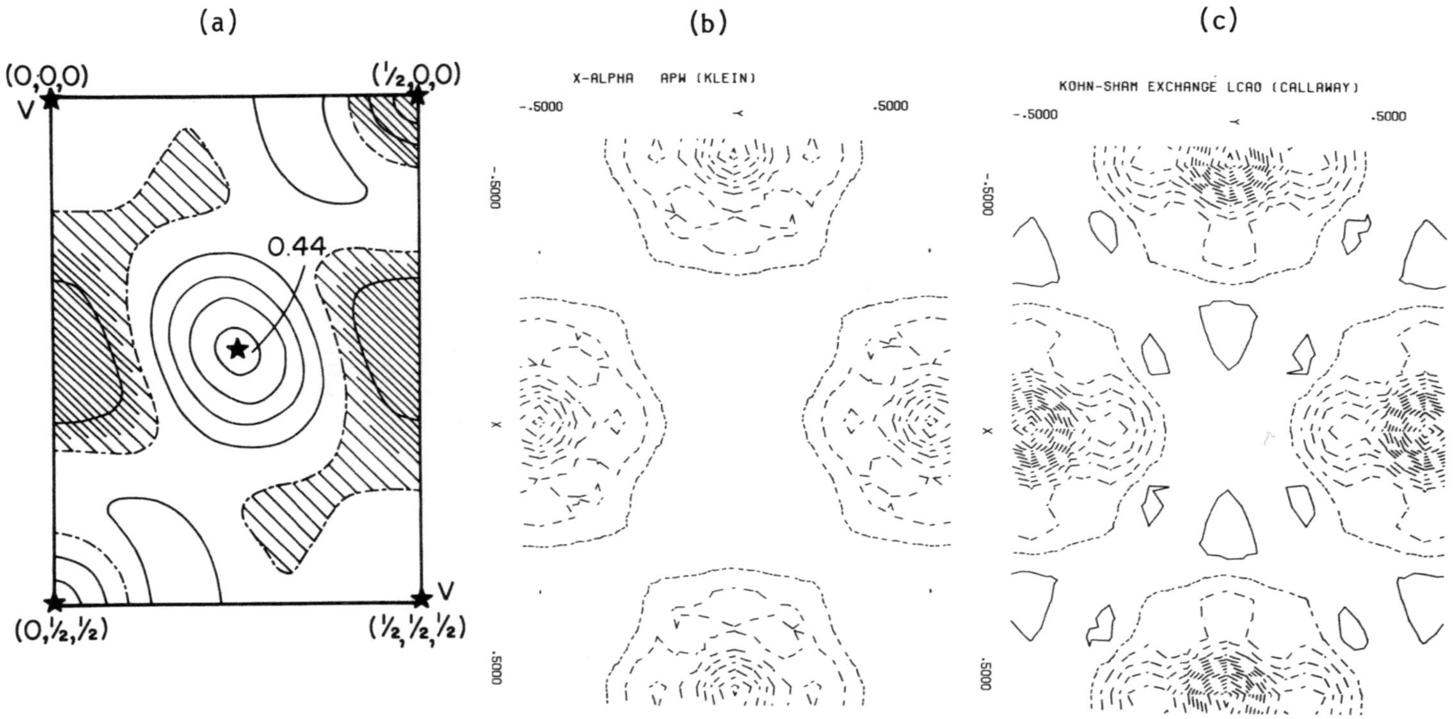


Fig. 2. Thermally smeared deformation electron density in V metal, $(\sin\theta/\lambda)_{max} = 0.85$ Å$^{-1}$
a) From experimental data by Korhonen et al (ref. 9a, b) b) Theoretical (ref. 10a, c); contours at 0.05 eÅ$^{-3}$ c) Theoretical (ref. 10b, c); contours as in b)

and assigned quadruple bond character to the Cr-Cr bond of 2.36 Å length;[12] the bond being composed of a $\sigma$ component, formed by the $d_{z^2}$ orbital, two $\pi$ ($d_{xz}$, $d_{yz}$) and one $\delta$ component ($d_{xy}$). The assignment, subsequently supported by SCF-X$\alpha$ and ab-initio calculations,[13] has been followed by the identification of a series of molecules containing Cr-Cr bonds from 2.541 to as low as 1.847 Å (as summarized in ref. 13).

Though the metal-metal bonds are quadruple, the designation strictly implies only that eight electrons are paired in four molecular orbitals with considerable bonding character in the metal-metal bond. But the contributions of each of the four orbitals to the strength of the bond will in general not be equal, nor is the bond necessarily of great strength, i.e. corresponds to a large bond energy.

The electron density analysis of the hydrate of chromous acetate was based on two data sets with respectively 5885 and 2885 independent reflections measured (at 90 and 120K) on a crystal supplied by D. F. Chodosh.[14] Clearly, examination of structure factors as applied in studies of simple metals becomes impractical at this point. Since the number of structure factors is also very large and least-squares modeling therefore expensive, Fourier methods were selected for the analysis. The molecule lies on a crystallographic inversion center but the molecular symmetry ($D_{4h}$) is much higher, so that chemically unique regions are determined several times. Examples are planes I and II, IV and V (Fig. 3) which each individually have left-right symmetry. The deformation maps show the expected peaks in the bond and lone pair regions of the acetate group. But in addition two large density peaks of approximately 0.6 $e\text{Å}^{-3}$ are observed in plane I on each side of the Cr atom. The most likely explanation of these peaks, which do not follow the $D_{4h}$ symmetry and are evident in both data sets, is anharmonicity the molecular motion of the complex which would be especially evident around the heavier chromium atom. Such anharmonicities are further discussed in the chapter on thermal smearing and chemical bonding. But regardless of what is the proper explanation for the peaks, they are effectively eliminated by symmetry averaging of chemically equivalent sections.

The results shown in Fig. 4, indicate a lack of deformation density in the $\sigma$-bond region but a small and probably significant accumulation in the regions associated with the $\pi$ and $\delta$ bonds ($\sigma$ in the <u>averaged</u> maps is estimated at 0.02 $e\text{Å}^{-3}$).

How are these results to be interpreted? Certainly, we must conclude that the bond is weak and that its $\sigma$-component does not

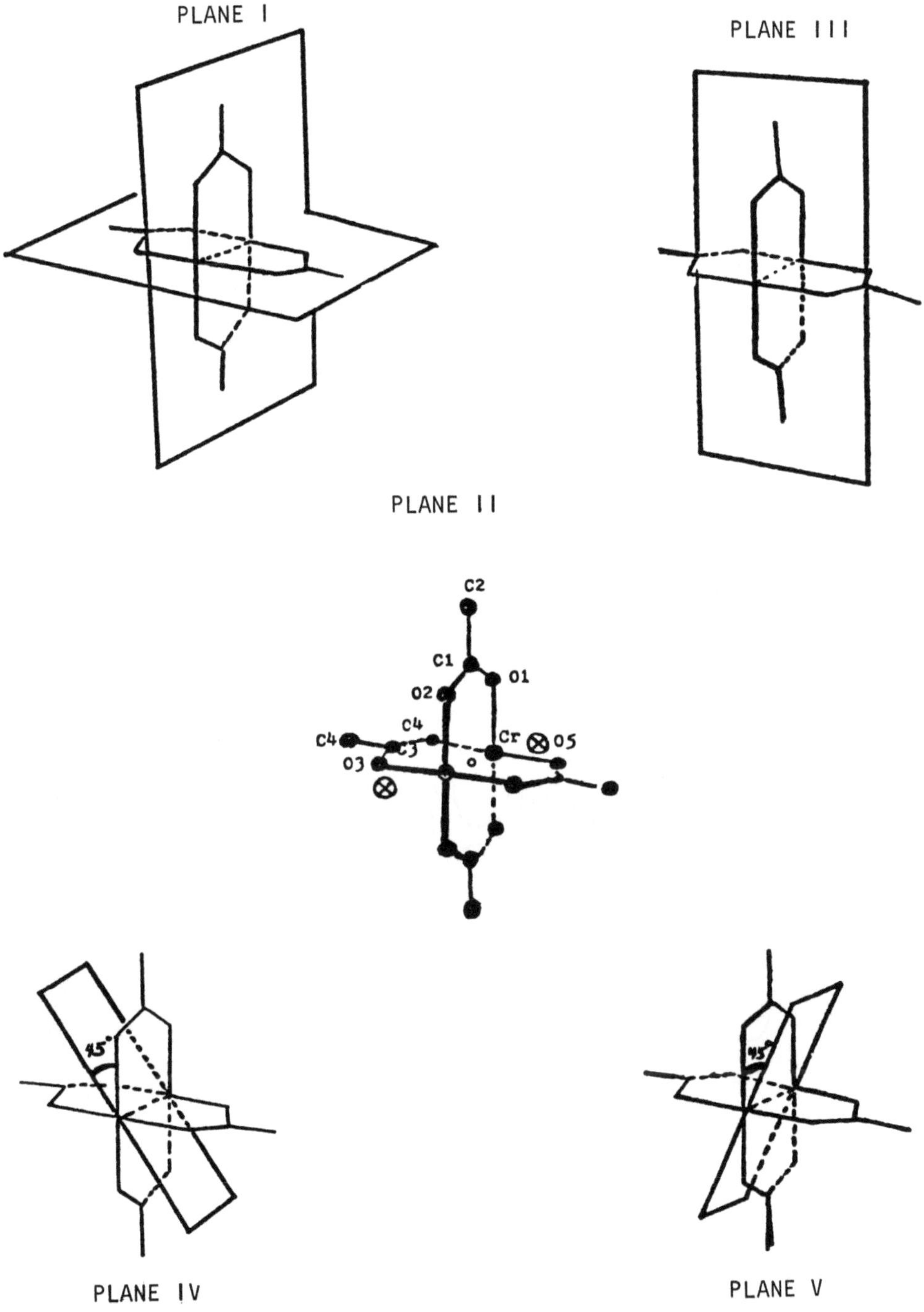


Fig. 3. Molecular structure of chromous acetate and definition of planes.

Fig. 4. Deformation density in the planes through the Cr atoms and the non-hydrogen atoms of the acetate group. Contours at 0.1 $e\mathring{A}^{-3}$ (ref. 14).

contribute perceptibly to its strength as judged by the overlap density. This is not in contradiction with the chemical theory as recently developed by Cotton et.al., who have specifically stated that "bond multiplicity is simply a measure of the number of electron-pair interactions and not a measure of bond strength".[15] The weakness (or essential absence) of the σ-contribution to bond strength can be attributed to the presence of an axial ligand which competes for the bonding character of the $d_{z^2}$ orbitals.[13]

It is of interest to note that a similar absence of density in the σ-bond region of binuclear complexes has been found in recent electron density studies of μ-octatetraenyl-bis (cyclopentadienylchromium) $[(\eta^5\text{-}C_5H_5Cr)_2C_8H_8]$ by Goddard and Kruger,[16] and in the carbonyl-bridged iron compound $[\eta^5\text{-}C_5H_5Fe(CO)_2]_2$ by Mitschler, Rees and Lehmann.[17]

To complement experimental information on multiple metal-metal bonds, further analyses are needed on the very short dinuclear chromium compounds,[18] and also on Mo-Mo complexes which show a much larger resistance to axial bonding and may therefore be expected to have stronger σ-components.

## 3. Choice of basis set in reciprocal space analysis

When large numbers of structure factors are available systematic modeling techniques become necessary. The model describes the electron density in terms of density functions of limited flexibility and variable populations, such that

$$\rho(\underset{\sim}{r}) = \sum_i P_i D_i\ (\underset{\sim}{r}-\underset{\sim}{r}_i,\ p_{1i},\ \cdots\ p_{ni}) \tag{1}$$

where $\underset{\sim}{r}_i$ defines the origin of the function $D_i$ with population $P_i$ and adjustable parameters p. The corresponding reciprocal space expression

$$F_{cal} = \sum_i (\int D_i\ (\underset{\sim}{r})\ \exp 2\pi i\ \underset{\sim}{S}.\underset{\sim}{r}\ d\underset{\sim}{r}) T_i \tag{2}$$

is the observational equation commonly used in minimization of the least squares error functions of the type

$$\varepsilon = \Sigma w\ (F_{obs}-F_{calc})^2 \tag{3}$$

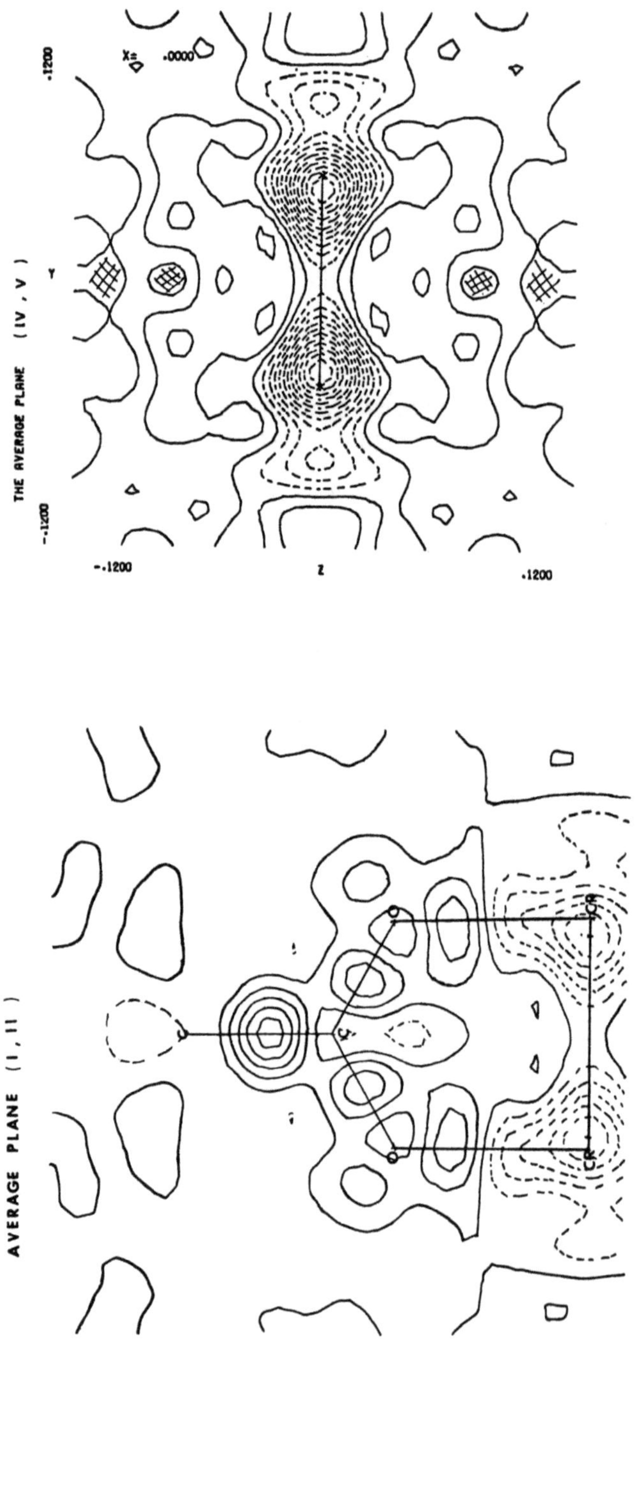


Fig. 5. Deformation density in averaged planes (I + II) and (IV + V) in chromous acetate. Contours at 0.1 and 0.05 eÅ$^{-3}$ respectively.

The choice of density function in (1) is obviously of crucial importance. We may demand that the functions be space filling and sufficiently flexible to provide a good fit to the observations.* We distinguish:

a) atom-centered expansions with angular functions which are equal or closely related to the spherical harmonics. Beyond the spherical atom approximation the minimal basis set of this type has a monopolar valence density function of adjustable population and radial dependence, in addition to a fixed spherical core density. Some applications of this treatment are discussed in section 4.

b) atom + bond centered expansions, such as the orbital product formalism in which the density is expressed as a sum over products of atomic orbitals, in analogy with the density representation obtained theoretically from LCAO (linear combination of atomic orbitals) molecular orbitals. A simpler formalism of this type, developed recently by Hellner and coworkers, uses rigid, spherical, atom-centered functions plus Gaussian shaped bond- and lone pair centered charge clouds.[20]

Which type of basis set is most suitable? Beyond the requirement of goodness of fit measuable by R factors and residual density maps (Fig. 6) the answer depends on the purpose of the study. Strong correlation between parameters which occur in the orbital product formalism are of little importance if only an analytical fit to the density is required. Such an analytical fit may be used in a real space plot, in which much of the experimental noise is filtered out, or for the calculation of derived molecular properties. But when individual density functions are to be interpreted in terms of concepts like atomic charge or shape or bond density, strong correlation between parameters are not permissable. It is for this reason that the minimal-basis monopole model is more suitable for the study of atomic charge and radial dependence. As shown in ref. 21, the much more complete expansion in spherical harmonics up to the $\ell = 4$ (hexadecapole level) contains density terms with large amplitudes on adjacent atoms. Thus the "pseudo-atoms" are too delocalized to be useful in the interpretation of the density in terms of component atoms.

### 3.1 An example: Valence density basis sets for transition metal atoms

The main asymmetry around transition metal atoms is frequently

*We note that the reciprocal and real space error functions are minimized at the same time (see ref. 19)

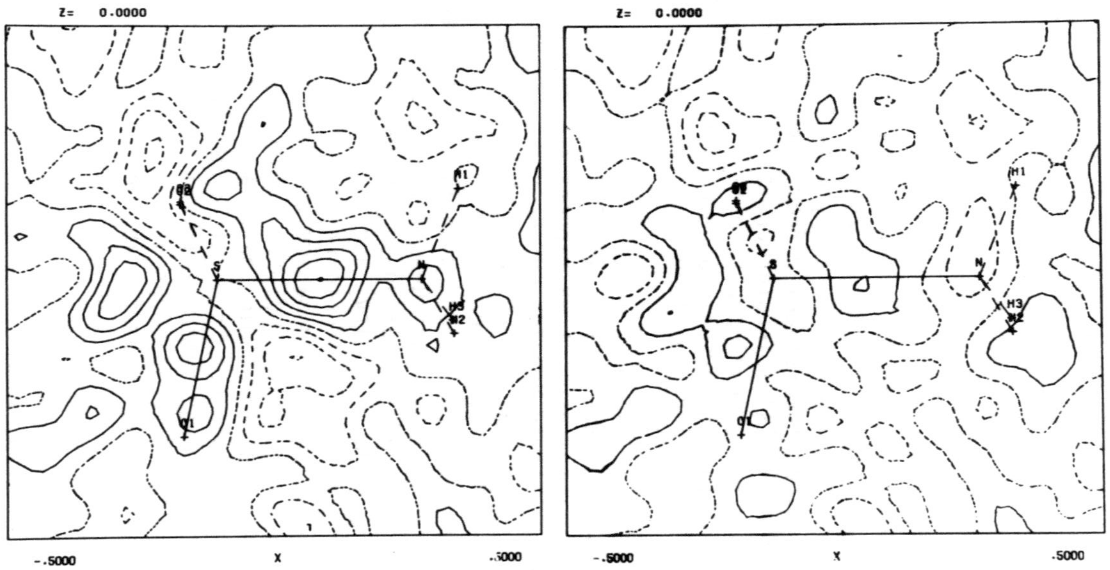


Fig. 6. Residual maps after multipole refinement on sulfamic acid
a) including density functions through the octapole level
b) including density functions through the hexadecapole level.
Contours at 0.05e$Å^{-3}$, ref. 33

a result of preferential filling of the d orbitals, whose degeneracy is lifted in a non-spherical crystal field. If the crystal field is strong and the energy separation between the no-longer-degenerate d orbitals overrides the electron spin-electron spin interactions, which demand maximum spin multiplicity, and a low-spin state results with a large electron asphericity. In the crystal field approximation of inorganic chemistry covalency in metal-atom to ligand bonds is neglected relative to the electrostatic field exerted by the ligands. In this approximation the orbital product formalism discussed above reverts to the one-center multipolar expansion. We will discuss two cases:

1. Octahedral symmetry: example $CrCO_6$

In an octahedral field the d orbitals are split into double degenerate $e_g$ and lower-lying triply degenereate $t_{2g}$ orbitals which relate as follows to the real spherical harmonics $y_{\ell m}$

$$e_g: \; y_{20} = 3z^2-r^2; \quad y_{22-} = (x^2-y^2)/2$$

$$t_{2g}: \; y_{21+} = xz; \quad y_{21-} = yz; \quad y_{22+} = xy$$

Since orbitals of different symmetry do not mix in the density representation and mixing of orbitals of the same degenerate representation does not change the density, the d orbital density is described as

$$\rho(d)=R(d)[1/2P(e_g)\{y_{20}^2+y_{22-}^2\}+1/3P(t_{2g})\{y_{21+}^2+y_{22+}^2+y_{21-}^2\} \tag{4}$$

where R(d) represents the radial dependence of the density.

The products of spherical harmonic functions are again spherical harmonics, for example[22]

$$y_{20}y_{20} = 0.2418\, y_{40} + 0.1802\, y_{20} + 0.2821\, y_{00} \tag{5}$$

In expression (5) all spherical harmonics are normalized to obey the condition $\int y^2 d\tau = 1$. However as discussed in reference 21 a more appropriate normalization for density functions is

$$\int |y_{\ell m}| d\tau = 2 \text{ for } \ell \neq 0 \text{ and } \int |y_{\ell m}| d\tau = 1 \text{ for } \ell = 0.$$

The normalization factors in both cases are given in ref. 21. If we label their ratio for each value of $\ell$ and m with the symbol $M_{\ell m}$, (5) becomes

$$y_{20}y_{20} = 0.2418\ M_{40}y_{40} + 0.1802\ M_{20}y_{20} + 0.2821\ M_{00}y_{00} \qquad (6)$$

where the $y_{\ell m}$ on the right hand side are now density functions and

$$M_{\ell m} = N_{\ell m}(\text{wave function})/N_{\ell m}(\text{density function}).$$

For the other products in (4) we get similarly:

$$y_{21\pm}y_{21\pm} = \pm\ 0.1802\ M_{42}\ y_{42+} \pm 0.1561\ M_{22}\ y_{22+} - 0.1612\ M_{40}y_{40} + $$
$$+ .0901\ M_{20}y_{20} + 0.2821\ M_{00}y_{00} \qquad (7)$$

and

$$y_{22\pm}y_{22\pm} = \pm\ 0.2384\ M_{44}\ y_{44+} + 0.0403\ M_{40}y_{40} - 0.1802\ M_{20}y_{20}$$
$$+ 0.2821\ M_{00}y_{00} \qquad (8)$$

Combining (6), (7) and (8) gives for the angular part in (4):

$$\rho(d) = \tfrac{1}{2}P\ (e_g)\ \{-0.2384\ M_{44}\ y_{44+} + 0.2821\ M_{40}y_{40} + 0.5642\ M_{00}y_{00}\}$$
$$+ \tfrac{1}{3}P\ (t_{2g})\ \{+0.2384\ M_{44}\ y_{44+} - 0.2821\ M_{40}y_{40} + 0.8463\ M_{00}y_{00}\}$$
$$\equiv P_4\ \{y_{44+} + \gamma y_{40}\} + P_0y_{00} \qquad (9)$$

with $\gamma = -0.2821\ M_{40}/0.2384\ M_{44} = 1.352$ or, substituting numerical values for $M_{44}$, $M_{40}$ and $M_{00}$ (which are 1.335, 1.5238 and 3.545 resp.):

$$\rho(d) = P(e_g)\ \{-0.1592\ y_{44+} + 0.2150\ \ y_{40} + y_{00}\}$$
$$+ P(t_{2g})\ \{+\ 0.1061\ y_{44+} - 0.1433\ \ y_{40} + y_{00}\} \qquad (9a)$$

Expression (9) implies the relation between the multipole population coefficients with $\ell = 4$ and $\ell = 0$, which are the only-ones allowed in the octahedral point group and the orbital populations $P(e_g)$ and $P(t_{2g})$. We note that if $P(e_g) = 2$ and $P(t_{2g}) = 3$ the non-spherical terms cancel and the undistorted spherical free-atom configuration is recovered.

Expression (f) is being applied in the refinement of $Cr(CO)_6$ which has a very close to octahedral configuration.

2. <u>The trigonal point group $\bar{3}$m:</u> example $FeS_2$ (Stevens & Coppens, ref. 23)

The coordination sphere of the iron atom in iron pyrite is a distorted octahedron with point symmetry $\bar{3}$. In this environment it is convenient to transform the spherical harmonic functions such that the 3-fold axis becomes the axis of quantization. The appropriate expressions have been given by Ballhausen[24] and are reproduced in ref. 23. When only nearest neighbor sulfur atoms are considered the symmetry of the coordination sphere is $\bar{3}$m. The symmetry species of the d orbitals in this point group are $a_0(t_{2g})$ directed along the main symmetry axis and $e_g(t_{2g})$ and $e'_g(e_g)$, where the label in parenthesis refers to the parent octahedral point group. As for the octahedral case the electron density can again be expressed in terms of the orbital populations with the additional feature that the $e_g$ and $e_{g'}$ orbitals are in principle allowed to mix. The resulting expression in analogy with (4) is:

$$\rho(d) = R(d)[P_1(a_0)^2 + \frac{1}{2} P_2 \{(e_{g-})^2 + (e_{g+})^2\} + \frac{1}{2} P_3 \{(e_{g-}')^2 + (e_{g+}')^2\} + P_4 \{e_g e_g'\} \quad (10)$$

where the product $e_g e_g'$ is normalized to contain one electron when $P_4 = 1$. The four orbital population parameters in (10) are related to the population parameters of the symmetry allowed multipoles $y_{00}$, $y_{20}$, $y_{40}$ and $y_{43+}$ through expression like 6-8 as described in reference 23. In addition the refinements included as variables the allowed multipole parameters on sulfur, positional, anisotropic and anharmonic (third and fourth cumulant) thermal parameters, an isotropic secondary extinction parameter and the radial dependence $\zeta$ of the multipole parameters.

Since the iron 4s orbital contributes very little to the X-ray scattering, its population cannot be reliably determined, and has been assumed to be zero in this study.

The results of several refinements are listed in Table 1. In refinement I, the multipoles were fixed corresponding to a spherical atom model for comparison, while in II, the total population of the iron 3d shell was kept constant at 6.0e. In all cases the populations are constrained to maintain electroneutrality in the crystal.

For a low-spin configuration of iron with pure octahedral coordination, the expected values of $P_1$, $P_2$, $P_3$ are 2.0, 4.0, and 0.0. In pyrite, the distorted octahedral field may be expected to favor the population of the $a_0$ orbital over the $e_g$ orbitals since the S-Fe-S angle of 94.3° is greater than 90° for S atoms related by the 3-fold axis, thus increasing the mean distance between the ligands and electrons in the $a_0$ orbital. This effect is indeed found to be significant. Both refinements II and III show considerable distortion from a spherical charge distribution toward the low-spin configuration with preference for population of the $a_0$ orbital. The extent of mixing of the $e_g$ and $e_g{}'$ orbitals is found to be significant.

The experimental deformation density about iron is plotted in Figure 7 along with a plot of the model density corresponding to the parameters obtained from refinement III. Both maps have been calculated including all X-ray measurements with $I > 3\sigma(I)$ and $0 < \sin\theta/\lambda \leq 1.45$ Å$^{-1}$ and include the smearing due to thermal motion of the atoms. The preference for the $a_0$ orbital pointing toward the larger of the S-Fe-S angles is evident in the model map but practically absent in the experimental density. Possible causes for this discrepancy include differences in the anisotropic and anharmonic temperature factor which is applied to the core density subtracted out from both maps. It appears therefore from the real space distribution that the conclusion regarding preferential occupancy of the $a_0$ orbital is subject to the correctness of the thermal motion model in the reciprocal space analysis.

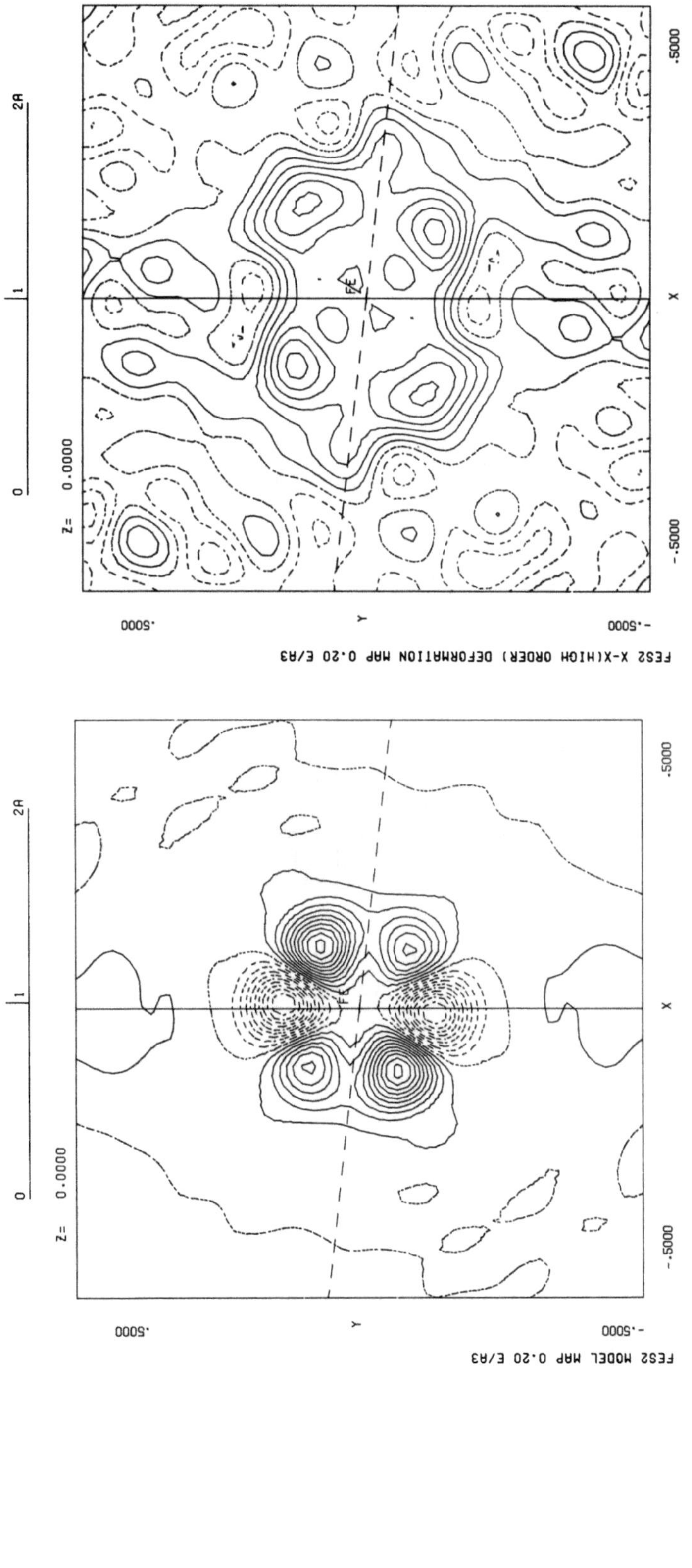


Fig. 7. Deformation density in $FeS_2$ (ref. 23)

a) according to least squares model
b) directly from observations. Contours at 0.20 $e\AA^{-3}$

Table 1. Orbital occupancies of Fe in pyrite from refinement of X-ray data.

| Refinement | I | II | III |
|---|---|---|---|
| $\zeta$(a.u.$^{-1}$) | 7.46(6) | 7.58(5) | 7.87(7) |
| $P_{3d}(=P_1 + P_2 + P_3)$ | 6.0 | 6.0 | 5.22(13) |
| $P(S_2)$ | 2.0 | 2.0 | 2.77(13) |
| $P_1$ | 1.2 | 1.98(12) | 1.82(13) |
| $P_2$ | 2.4 | 3.36(15) | 3.06(17) |
| $P_3$ | 2.4 | 0.66(15) | 0.34(17) |
| $P_4$ | 0.0 | -0.04(16) | 0.02(16) |
| R(F) | 2.35% | 1.81% | 1.79% |
| $R_w$(F) | 1.92% | 1.31% | 1.29% |

## 4. Application of the minimal basis set model: the radial refinement

In earlier refinements of atomic populations the atomic valence electrons were projected into a rigid spherical valence shell centered at the nucleus.[25] But rigidity of the valence shell in this model is at variance with chemical theory according to which electron-electron repulsions and exchange interactions increase and decrease with electron population. The minimal density basis set therefore must include a radial parameter. Such a parameter may be defined as a scale factor $\kappa$ which allows expansion or contraction of the spherical valence shell according to the expression:[26]

$$\rho'_{valence}(r) = P_{valence}\,\kappa^3\,\rho_{valence}(\kappa r) \tag{11a}$$

or in reciprocal space

$$f'_{valence}(S) = P_{valence}\,f\left(\frac{S}{\kappa}\right) \tag{11b}$$

Application of expression (10) to several data sets are discussed in ref. 26. For hydrogen atoms a strong correlation exists between $\kappa$ and the thermal parameter. Taking hydrogen temperature parameters from neutron diffraction experiments, a value of $\kappa_H$ = 1.40 was selected in cases where no neutron data were available. This is considerably higher than the best $\kappa$ value for the molecular hydrogen (1.16) which is the basis of the Stewart, Davidson and Simpson scattering factor.[27]

The change in $\kappa$ with variable electron population is very clear-cut for the nitrogen atom, the valence shell expanding with increasing negative charge as predicted by theory. The slope of the curve for nitrogen is almost exactly in agreement with the Slater rules for analytical atomic orbitals, a remarkable finding given the simplicity of the Slater recipe. Of course, the least squares parameters $\kappa$ and P are correlated as illustrated in Figure 8 in which the probability ellipsoids[29] related to the variance covariance matrix $M_X$ by the expression

$$\tilde{X}\, \tilde{M}_X^{-1}\, \tilde{X} = 1 \qquad (12)$$

are plotted in the $\kappa$-q plane (q = net charge) for the four carbon atoms of the glycylglycine molecule, based on liquid nitrogen data collected by Kvick, Koetzle and Stevens.[30] The probability ellipsoids indicate that a positive error in $\kappa$ is likely to be accompanied by positive error in the electron population, so that the slope of the $\kappa$/q line is less accurately known than otherwise would have been the case.

One of the merits of the "radial" refinement is that it provides at a relatively elementary level the opportunity to calculate derived properties such as net charges and molecular dipole moments. The comparison of these values with those derived from the more complete multipole expansion up to and including terms with $\ell = 4$, and with those from direct integration in real space are discussed in the following section.

## 5. Comparison of properties derived by reciprocal space refinement and by real space density integration

With the analytical description provided by least squares modelling, functions of the one-electron density, defined in general by the operator equation

$$\langle O \rangle = \hat{O}\rho(\tilde{r}) \qquad (13)$$

can be derived.

The dipole moment of a molecule or molecular fragment for example is obtained from the multipole density through the expression:[19]

$$\mu_{mol} = \sum_{\substack{\text{all} \\ \text{atoms}}} (\tilde{D}_j - 4.803\, q_j\, \tilde{r}_j) \text{ Debye} \qquad (14a)$$

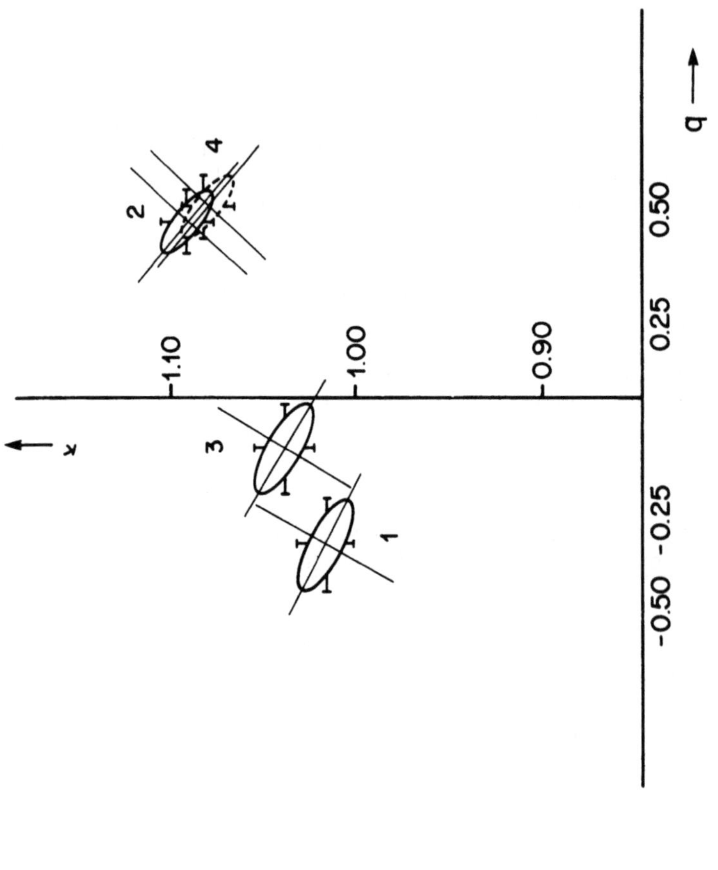


Fig. 8. Expansion-contraction parameters vs. net charge for the four carbon atoms in glycylglycine. Ellipses are defined as $\underset{\sim}{X}\,\underset{\sim}{M}^{-1}\,\underset{\sim}{X} = 1$, M = variance covariance matrix.

where $q_j$ is the net charge on atom j in electron units and

$$\mathbf{D}_j = \int r\, R(\kappa r) \sum_{m=-1}^{1} P_{1mj}\, Y_{1m}(\mathbf{r}/r)\, d\mathbf{r} \qquad (14b)$$

when the dipolar functions have been included in the refinement, and

$$\mathbf{D}_j = 0 \text{ (nonhydrogen atoms)}$$

$$\mathbf{D}_j = \mathbf{r}_X - \mathbf{r}_N \text{ (hydrogen atoms)} \qquad (14c)$$

when the calculations are based on the results of the monopole-only radial refinement.

The analogous expressions in real space, given in ref. 19, are based on the evaluation of

$$\mu_{\text{electronic}} = \int_V \mathbf{r}\, \rho(\mathbf{r})\, d\mathbf{r} \qquad (15)$$

where V is the molecular volume, which may be defined by discrete boundary planes[30] or by overlapping density functions.[31]

Expressions analogous to (14) and (15) apply to other moments of the charge density distribution such as net charges and quadrupole moments. To what extent are the results of (14) and (15) compatible given the different ways of space partitioning involved and the approximations implied on the effect of thermal smearing?

### 5.1 Comparison of net charges and dipole moments from monopole refinement and direct space integration

#### 5.1.1 Net charges

Net charges in the organic salt TTF-TCNQ (at 100K) and in the superconducting material $V_3Si$ (at room temperature) are compared in Table 2.

Table 2. Net charges (in positive electron units)

| compound | fragment | radial refinement | direct space integration |
|---|---|---|---|
| TTF-TCNQ | TTF | 0.47(15) | 0.47(15) |
| | TCNQ | -q(TTF) | |
| $V_3Si$ | Si | +1.19(24) | 1.5-2.5 |
| | V | -q(Si)/3 | |

Table 3 Comparison of dipole moments (D)

| | X-ray | | | Theoretical | Other experimental values |
|---|---|---|---|---|---|
| | radial refinement | multipole refinement | direct space integration | | |
| sulfamic acid (78K) | 9.6(6) | 9.9(6) | 9.1 | 9.33[a] | 10.2[1,b]<br>12.2[2,b]<br>13.3[3,b] |
| formamide (90K) | 4.4(5) | 4.8(5) | 5.3 | 4.3[c]<br>4.07[d] | 3.69[4,e] |

[a]J. Rys, 4-31G unpublished results

[b]Sears, Fortune & Blumenshine, J. Chem. Eng. Data, 11, 406 (1972)

[c]Snyder & Basch, Molecular Wave Functions and Properties, Wiley and Sons, New York, double zeta;

[d]Christensen, Kortzeborn, Bak & Led, J. Chem. Phys., 53, 3912 (1970)

[e]Kurland & Wilson, J. Chem. Phys., 27, 585 (1957)

1) in dimethylsulfoxide. 2) in N-methyl 2 pyrrolidine. 3) in N,N dimethyl acetamide. 4) gas phae

The direct space integration is based on a polyhedral volume the dimensions of which are determined by the ratio of the radii of neighboring atoms in adjacent fragments. For molecular crystals the exact location of the boundary planes is less important, as the electron density is low in the intermolecular region. Accordingly the results on TTF-TCNQ obtained by the two different methods are in satisfactory agreement with each other and with values from other physical measurements.[32] For $V_3Si$ in which the components are much less separated differences are larger though both methods agree regarding the direction of the transfer and its magnitude. In such a case the volume definition inherent in monopole formalism, which allows for overlap of fragments, is to be preferred.

### 5.1.2 Dipole moments

For both sulfamic acid ($NH_3SO_3$) and formamide ($HCNH_2O$) molecular dipole moments have been obtained by both computational methods, while some theoretical values are also available. Results, summarized in Table 3 indicate that the dipole moments from different techniques compare well within the fairly large experimental standard deviations. The radial refinement even though it does not include dipolar atomic functions gives a very good account of the molecular dipole moment. As it is a computationally much faster method than either the multipole model treatment or the direct space integration, it provides a routine method for the evaluation of molecular dipole moments from diffraction data.

A similar conclusion evolves when the directions of the dipole moments are compared. For sulfamic acid, which has close to three-fold symmetry, all dipole moments are within a few degrees of the symmetry axis. The magnitude and directions of the dipole moments are shown in Figure 9. The agreement between the X-ray experiment, the microwave measurement and theory is within the estimated standard deviations.

Finally a check of procedures is possible by analysis of the theoretical dynamical X-ray structure factors calculated from the extended basis set wave function.[34] The dipole moment for this wave function is 4.7 D. A radial refinement of the theoretical structure factors gives a value of 5.2 D, while the direct space integration result is 5.1 D. The differences are typical for the uncertainties in X-ray dipole moments that may be expected. Nevertheless, the values agree within the uncertainties predicted by the experimental error analysis.[26]

Though net charges and dipole moments from X-ray scattering are not exceedingly accurate quantities, at least at present, they are physically meaningful and a measure of the quality of the

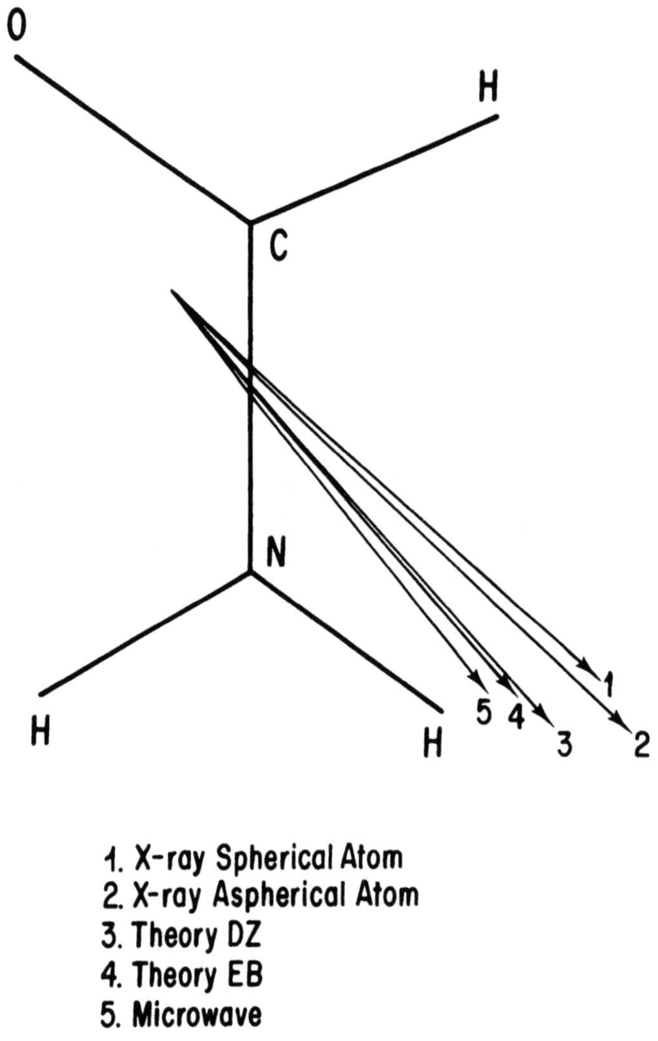


Fig. 9. Experimental and theoretical dipole moments in the plane of the formamide molecule.

experimental density function.

Acknowledgement: The authors would like to thank their collaborators and especially Drs. T. N. Guru Row, M. L. DeLucia and Mr. P. Leung who contributed greatly to the research described in this article. Support by the National Science Foundation (CHE 7613342A01) is gratefully acknowledged.

## References

1. P.J.E. Aldred & M. Hart, Proc. Roy. Soc. (London) A332, 239 (1973).

2. R. J. Weiss, Phil. Mag., In Press.

3. P. J. Brown, Phil. Mag., 26, 1377 (1972).

4. R. F. Stewart, Acta Cryst., A33, 33 (1977).

5. Y. W. Yang & P. Coppens, Acta Cryst., A34, 61 (1978).

6. S. T. Inoue and J. Yamashita, J. Phys. Soc. Japan, 35, 678 (1973).

7. C. W. Bauschlicher, Jr., C. F. Bender, H. F. Schaefer III & P. S. Bagus, Chem. Phys., 15, 227 (1976).

8. a) R. J. Weiss & J. J. DeMarco, Phys. Rev., 140, A1223 (1965).
   b) M. Diana & G. Mazzone, Phil. Mag., 32, 1227 (1975).

9. a) U. Korhonen, E. Rantavuori & M. V. Linkoaho, Ann. Acad. Sci. Fenn. Ser. A,. VI, 361 (1971).
   b) J. L. Staudenmann & P. Coppens, unpublished results.

10. a) B. M. Klein, D. A. Papaconstantopoulos & L. L. Boyer, Abstract APS Solid State Meeting, Atlanta (1976).
    b) C. S. Wang & J. Callaway, private communication.
    c) P. W. Leung, R. Trueman & P. Coppens, unpublished results.

11. C. S. Wang & J. Callaway, Phys. Rev. B., 15, 298 (1977).

12. F. A. Cotton, B. G. DeBoer, M. D. LaPrade, J. R. Pipal & D. A. Veko, J. Am. Chem. Soc., 92, 2926 (1970); Acta Cryst. B 27, 1664 (1971).

13. F. A. Cotton, Acc. of Chem. Res., 11, 225 (1978).

14. M. L. DeLucia, Thesis, State University of New York at Buffalo (1977).

15. F. A. Cotton, & G. G. Stanley, Inorg. Chem., 16, 2668 (1977).

16. R. Goddard & C. Krüger, Angew. Chemie, In Press.

17. A. Mitschler, B. Rees & M. S. Lehmann, J. Am. Chem. Soc., In Press.

18. A study of one of the very short Cr-Cr dinuclear complexes is being undertaken by B. Rees & A. Mitschler (private communication).

19. P. Coppens and N. K. Hansen, Isr. J. of Chem., 16, 163 (1977).

20. E. Hellner, Acta Cryst., B33, 3813 (1977).

21. N. K. Hansen & P. Coppens, Acta Cryst. A, In Press.

22. Proc. Summer School in Quantum Chemistry, Upsala, Sweden, 1968.

23. E. D. Stevens & P. Coppens, to be published.

24. C. J. Ballhausen, Introduction to Ligand Field Theory, McGraw-Hill, New York (1962).

25. a) R. F. Stewart, J. Chem. Phys., 53, 205 (1970).
    b) P. Coppens, D. Pautler & J. F. Griffin, J. Am. Chem. Soc., 93, 1051 (1971).

26. P. Coppens, T. N. Guru Row, P. Leung, E. D. Stevens, P. J. Becker & Y. W. Yang, Acta Cryst., In Press.

27. R. F. Stewart, E. R. Davidson & W. T. Simpson, J. Chem. Phys., 42, 3175 (1965).

28. C. A. Coulson, Valence, Oxford University Press, Oxford (1961).

29. A. Kvick, T. F. Koetzle & E. D. Stevens, to be published.

30. P. Coppens, Phys. Rev. Lett., 35, 98 (1975).

31. F. L. Hirshfeld, Isr. J. of Chem., 16, 198 (1977).

32. P. Coppens & T. N. Guru Row, Annals of the N. Y. Academy of Sciences, 313, 244 (1978).

33. P. Coppens & T. N. Guru Row, to be published.

34. E. D. Stevens, J. Rys & P. Coppens, J. Am. Chem. Soc., 100, 2324 (1978).

IMPOSING NON-CRYSTALLOGRAPHIC SYMMETRY CONSTRAINTS ON MULTIPOLE POPULATIONS AND DEFINING LOCAL ATOMIC COORDINATE SYSTEMS

Edwin D. Stevens

Chemistry Department
State University of New York at Buffalo
Buffalo, New York 14214

## I. Introduction

In molecular crystals which contain symmetry elements which are not incorporated into the crystallographic site symmetry, the number of parameters in a multipole refinement may be reduced by constraining the multipole parameters of chemically equivalent parts to be equal. For example, the crystal structure of $S_4N_4$ contains molecules at general positions, and it is reasonable to constrain the 4 sulfur and 4 nitrogen atoms to be equivalent. The allowed multipoles can be further reduced by (non-crystallographic) mirror planes passing through alternate sulfurs and 2-fold axes passing through alternate nitrogens.

In order to conveniently apply the constraints in the least-squares refinement, local atomic coordinate systems should be defined for each atom which make the multipole parameters refer to chemically equivalent deformations throughout the molecule. Included below are the input instructions for defining the local atomic coordinate systems and a table of the angular part (spherical harmonic functions in real form) of the multipole density functions used in the refinement program MOLLY (Hansen & Coppens, Acta Cryst., A34, 909-921 (1978)).

(Part 1.) What is the total number of population parameters (all terms from $\ell = 0$ to $\ell = 4$) for the $S_4N_4$ molecule without symmetry constraints?

(Part 2.) Prepare the input for MOLLY defining the local coordinate system for each atom. Assume the atoms are numbered as indicated in the figure and appear in the atom list in the order $S_1, S_2, S_3, S_4$, $N_1, N_2, N_3, N_4$.

(Part 3.) List the allowed multipole populations of sulfur and of nitrogen assuming all of the chemical symmetry has been incorporated.

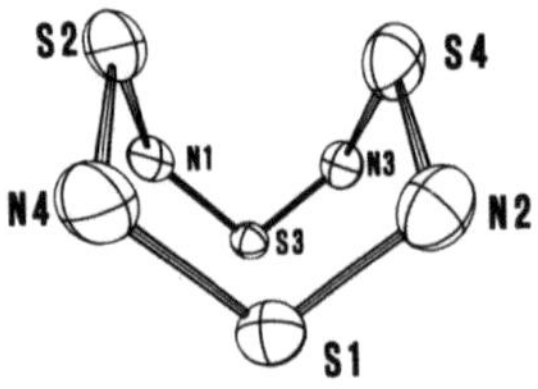


Input Instructions for MOLLY - Local Atomic Coordinate System.
For each atom specify NAY, NAX1, NAX2, ICO1, ICO2.

The local coordinate system is defined for each atom by 5 integers:

NAY, NAX1, & NAX2 define other atoms by their sequence number in the atom list; ICO1 and ICO2 label of the axis and may take the values 1, 2, or 3 where 1 = x, 2 = y and 3 = z.

Three vectors are defined as follows:

1. $V_1$ is a vector from the atom for which the coordinate system is being defined to atom NAY.

2. $V_2$ is a vector from atom NAX1 to atom NAX2.

3. $V_3 = V_1 \times V_2$

The coordinate system will be defined by:

1. Axis ICO1 along V1

2. Axis ICO2 along V3XV1

3. Axis ICO3 along V3

(The system is right handed unless NAY is made negative in which case V3 in inverted.)

Table 1 Real Spherical-harmonic functions (x,y,z are different-cosines)

| Order | Symbol | Angular function | Normalization<br>Density function | <br>Wave function |
|---|---|---|---|---|
| 0 | 00 | $1$ | $1/4\pi$ | $(1/4\pi)^{1/2}$ |
| 1 | 11+ | $x$ | | |
| | 11- | $y$ | $1/\pi$ | $(3/4\pi)^{1/2}$ |
| | 10 | $z$ | | |
| 2 | 20 | $2z^2-(x^2+y^2)$ | $\frac{3\sqrt{3}}{8\pi}$ | $(\frac{5}{16\pi})^{1/2}$ |
| | 21+ | $zx$ | | $(\frac{15}{4\pi})^{1/2}$ |
| | 21- | $zy$ | $3/4$ | |
| | 22+ | $xy$ | | |
| | 22- | $(x^2-y^2)/2$ | | |
| 3 | 30 | $2z^3-3z(x^2+y^2)$ | $\frac{10}{13\pi}$ | $(\frac{7}{16\pi})^{1/2}$ |
| | 31+ | $x(4z^2-(x^2+y^2))$ | $(ar+14/5-\pi/4)^{-1}$ | $(\frac{21}{32\pi})^{1/2}$ |
| | 31- | $y(4z^2-(x^2+y^2))$ | | |
| | 32+ | $z(x^2-y^2)$ | $1$ | $(\frac{105}{16\pi})^{1/2}$ |
| | 32- | $2xyz$ | | |
| | 33+ | $x^3-3xy^2$ | $4/3\pi$ | $(\frac{35}{32\pi})^{1/2}$ |
| | 33- | $y^3-3yx^2$ | | |
| 4 | 40 | $8z^4-24z^2(x^2+y^2)+3(x^2+y^2)^2$ | ! | $(\frac{9}{256\pi})^{1/2}$ |
| | 41+ | $x(4z^3-3z(x^2+y^2))$ | | |
| | 41- | $y(4z^3-3z(x^2+y^2))$ | $\frac{735}{512\sqrt{7}-196}$ | $(\frac{45}{32\pi})^{1/2}$ |
| | 42+ | $(x^2-y^2)\ (6z^2-(x^2+y^2)$ | | |
| | 42- | $2\ xy\ \ (6z^2-(x^2+y^2$ | $\frac{105\sqrt{7}}{4(136+28\sqrt{7}}$ | $(\frac{45}{64\pi})^{1/2}$ |
| | 43+ | $z(x^3-3xy^2)$ | | |
| | 43- | $z(y^3-3yx^2)$ | $5/4$ | $(\frac{315}{32\pi})^{1/2}$ |
| | 44+ | $x^4-6x^2y^2+y^4$ | | |
| | 44- | $4x^3y-4xy^3$ | $15/32$ | $(\frac{315}{256\pi})^{1/2}$ |

*ar - arctan(2)

! $N^{-1}_{ang.} = (14A_-^5-14A_+^5-20A_-^3+20A_+^3\,\_6A_- -6A_+)2\pi$

where: $A\pm = (\frac{30\pm\sqrt{480}}{70})^{1/2}$

## II. Solution

Number of Population Functions

| | |
|---|---|
| 1 | Monopole |
| 3 | Dipoles |
| 5 | Quadrapoles |
| 7 | Octapoles |
| 9 | Hexadecapoles |
| 25 | parameters/atom x 8 atoms = 200 population parameters |

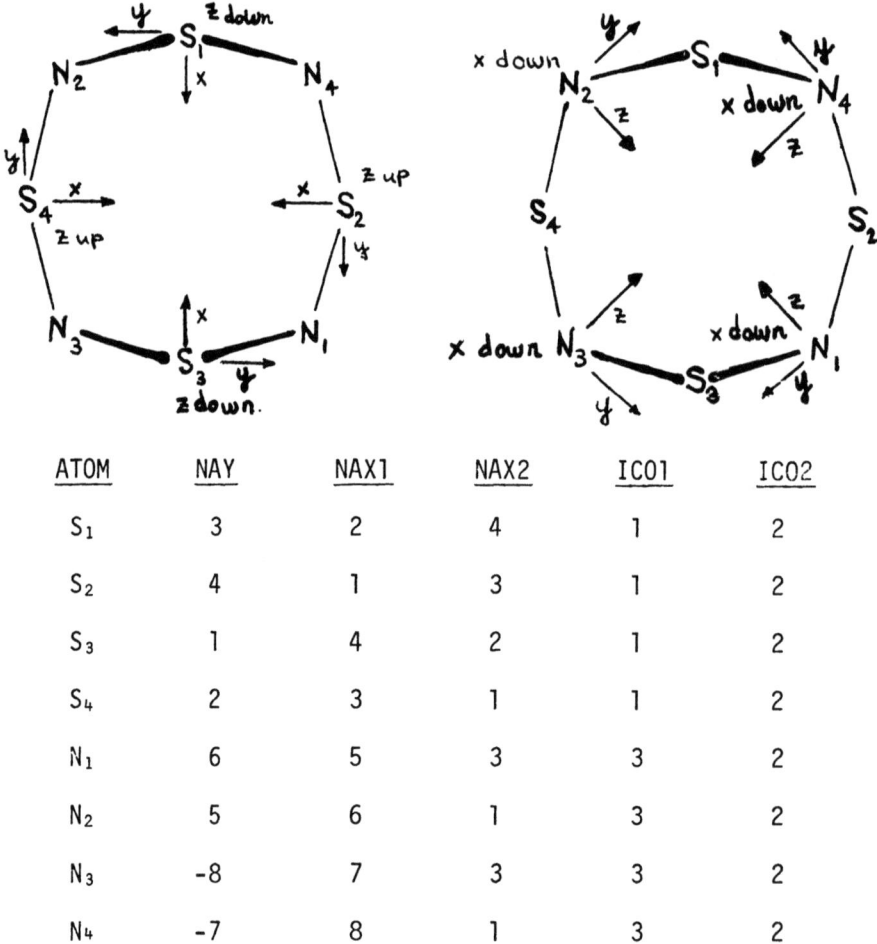

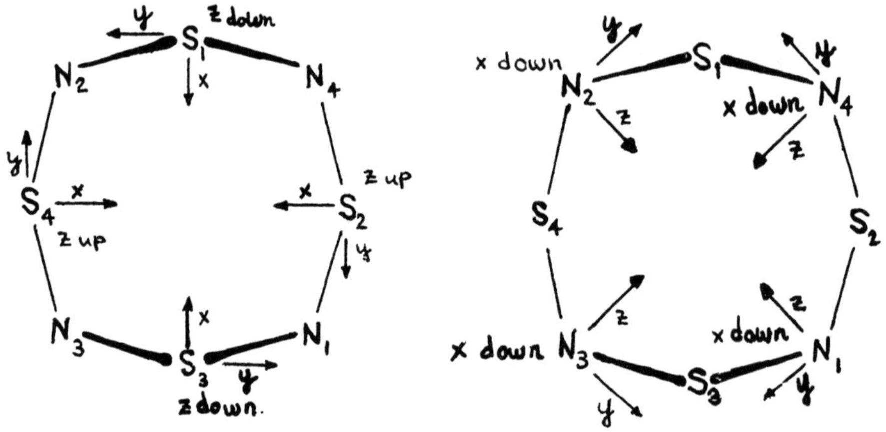


| ATOM | NAY | NAX1 | NAX2 | ICO1 | ICO2 |
|---|---|---|---|---|---|
| $S_1$ | 3 | 2 | 4 | 1 | 2 |
| $S_2$ | 4 | 1 | 3 | 1 | 2 |
| $S_3$ | 1 | 4 | 2 | 1 | 2 |
| $S_4$ | 2 | 3 | 1 | 1 | 2 |
| $N_1$ | 6 | 5 | 3 | 3 | 2 |
| $N_2$ | 5 | 6 | 1 | 3 | 2 |
| $N_3$ | -8 | 7 | 3 | 3 | 2 |
| $N_4$ | -7 | 8 | 1 | 3 | 2 |

ALLOWED MULTIPOLE POPULATIONS ON NITROGEN
(2-fold axis along z-axis)

| | | | | |
|---|---|---|---|---|
| $P_{00}$ | $P_{22+}$ | $P_{30}$ | $P_{40}$ | $P_{44+}$ |
| $P_{10}$ | $P_{22-}$ | $P_{32+}$ | $P_{42+}$ | $P_{44-}$ |
| $P_{20}$ | | $P_{32-}$ | $P_{42-}$ | |

ALLOWED MULTIPOLE POPULATIONS ON SULFUR
(mirror plane $\perp$ to y-axis)

| | | | | |
|---|---|---|---|---|
| $P_{00}$ | $P_{20}$ | $P_{30}$ | $P_{40}$ | $P_{43+}$ |
| $P_{11+}$ | $P_{21+}$ | $P_{31+}$ | $P_{41+}$ | $P_{44+}$ |
| $P_{10}$ | $P_{22-}$ | $P_{32+}$ | $P_{42+}$ | |
| | | $P_{33+}$ | | |

200 possible population parameters reduced to only 28 by imposing non-crystallographic symmetry constrains.

# INTERPRETATION OF THE SPIN DENSITIES IN METALS AND ALLOYS

J. Schweizer.
Institut Laue Langevin and
DN/RFG Centre d'Etudes Nucléaires
38042 Grenoble, France

As the magnetic structure factors are the Fourier components of the magnetization density, one can construct this density using the values measured for the structure factors

$$m(\vec{r}) = \frac{1}{V} \sum_{H} F(\vec{H}) e^{+i\vec{H}\vec{r}} \qquad (1)$$

$\Sigma$ over all the diffusion vectors $\vec{H}$ including (000) which represents the total magnetization per cell.

In many cases the magnetization density is mainly a spin density, the orbital moment being either quenched by the crystal field (transition elements) or zero due to the Hund's rule (gadolinium). This spin density represents in that case directly the presence probability of electrons, but here, unpaired electrons only. We shall see how to analyze such an experimental spin density when it is first localized close to the nucleus and then when it is far from it.

a) Spin density localized around the nuclei

A first approach is to compare the experimental spin distribution to that deduced from a one electron wave function $\psi(r)$ calculated for free ions by an Hartree Fock method.

1) Symmetry of the distributions - crystal field effects

The shape and asphericities of a spin distribution are striking features. Generally the comparison with the given wave function $\psi$ is performed in terms of structure factors

$$F(\vec{H}) = \mu \int \psi^{*}(r) e^{i\vec{H}\vec{r}} \psi(r) d^3r \qquad (2)$$

where the one electron wave function $\psi$ is the product of a radial part and an angular part:

$$\psi(\vec{r}) = R(r) \sum_m a_m Y_m^{\ell}(\Theta , \Phi) \qquad (3)$$

If there is no anisotropy of the distribution around the nucleus, the measured structure factor values lie on a smooth curve versus $|\vec{H}|$ or $\sin\theta/\lambda$. This is the case of cobalt measured by Moon(1) (figure 1 and 2).

If the spin density distribution presents anisotropies as in the case of nickel (fig. 3) measured by Mook(2), the points representing the structure factors do not lie anymore on a smooth curve versus $\sin\theta/\lambda$ (fig. 4). To evaluate the structure factors according to (2) and using (3) one may(3) expend the exponential:

$$e^{i\vec{H}\vec{r}} = 4\pi \sum_{KQ} i^K j_K(Hr) Y_Q^K(\Theta,\Phi) Y_Q^{K*}(\theta,\phi) \qquad (4)$$

where $j_K(Hr)$ are sperical Bessel functions, $\Theta$ and $\Phi$ correspond to the direction of $\vec{r}$ and $\theta$ and $\phi$ to the direction of $\vec{H}$.

Using the integral over the sphere of the product of 3 spherical harmonics(3)

$$\int_0^{2\pi} \int_0^{\pi} Y_{m_1}^{\ell_1}(\Theta,\Phi) Y_{m_2}^{\ell_2}(\Theta,\Phi) Y_{m_3}^{\ell_3}(\Theta,\Phi) \sin\Theta d\Theta d\Phi =$$

$$= \left[\frac{(2\ell_1+1)(2\ell_2+1)(2\ell_3+1)}{4\pi}\right]^{\frac{1}{2}} \begin{pmatrix} \ell_1 & \ell_2 & \ell_3 \\ 0 & 0 & 0 \end{pmatrix} \begin{pmatrix} \ell_1 & \ell_2 & \ell_3 \\ m_1 & m_2 & m_3 \end{pmatrix} \qquad (5)$$

where the 3j coefficients are connected to the Clebsch Gordon coefficients by

$$\langle \ell_1 m_1 \ell_2 m_2 | \ell_3 m_3 \rangle = (-1)^{\ell_1 - \ell_2 + m_3} (2\ell_3+1)^{\frac{1}{2}} \begin{pmatrix} \ell_1 & \ell_2 & \ell_3 \\ m_1 & m_2 & -m_3 \end{pmatrix} \qquad (6)$$

and defining the radial integrals $\langle j_K(H) \rangle$ by

$$\langle j_K(H) \rangle = \int_0^{\infty} r^2 j_K(Hr) |R(r)|^2 dr \qquad (7)$$

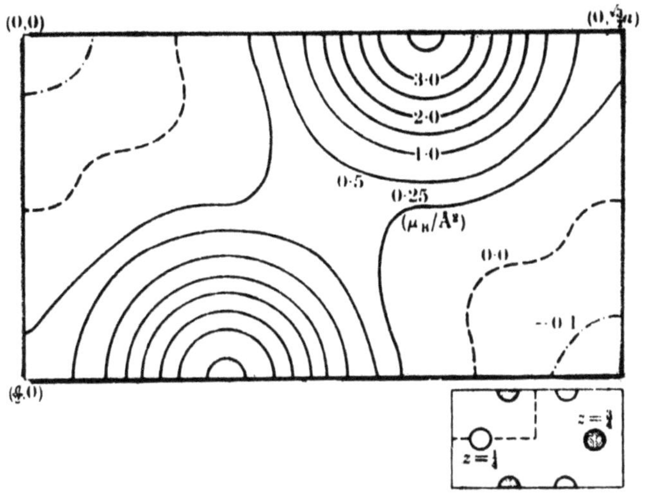


Fig. 1: Spin density of hexagonal cobalt projected onto the basal plane[1]

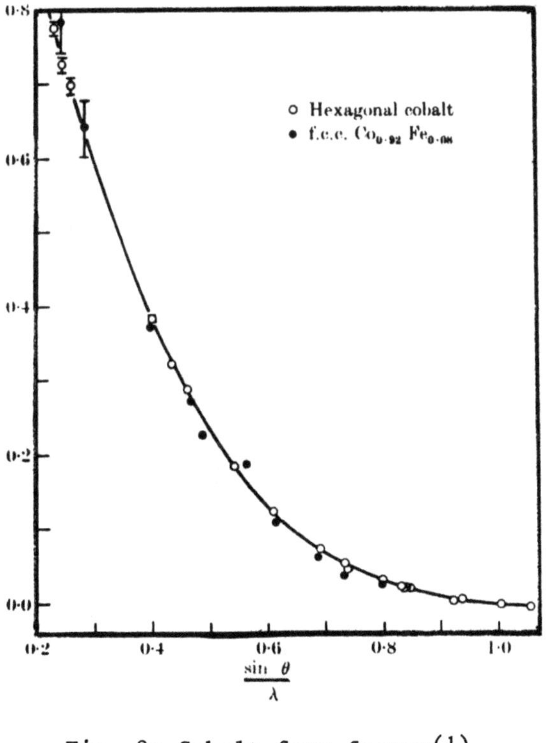


Fig. 2: Cobalt form factor[1]

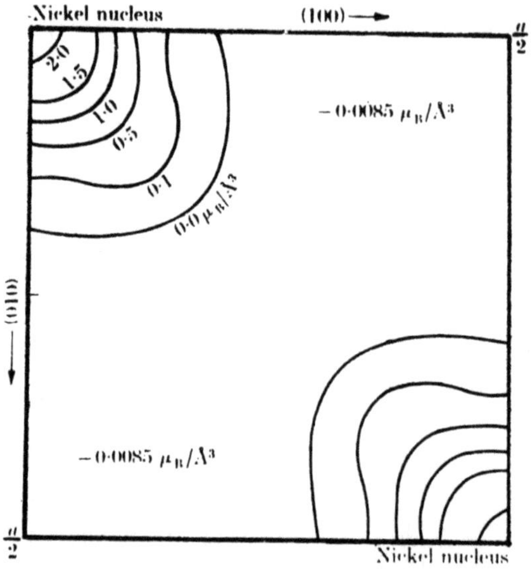


Fig. 3: Spin density in the [100] plane for nickel(2)

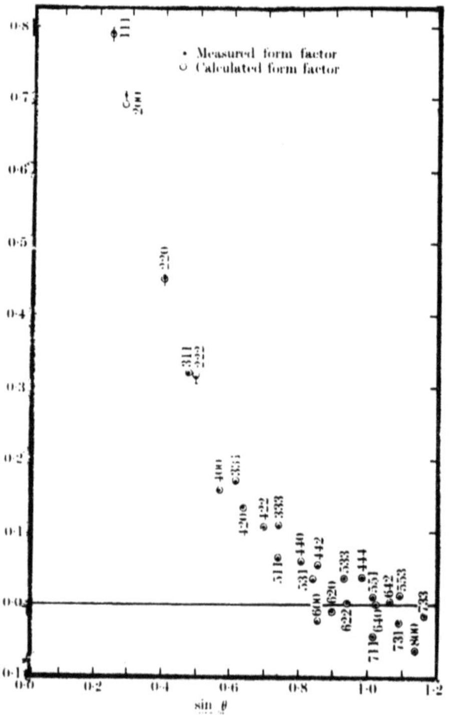


Fig. 4: Nickel form factor(2)

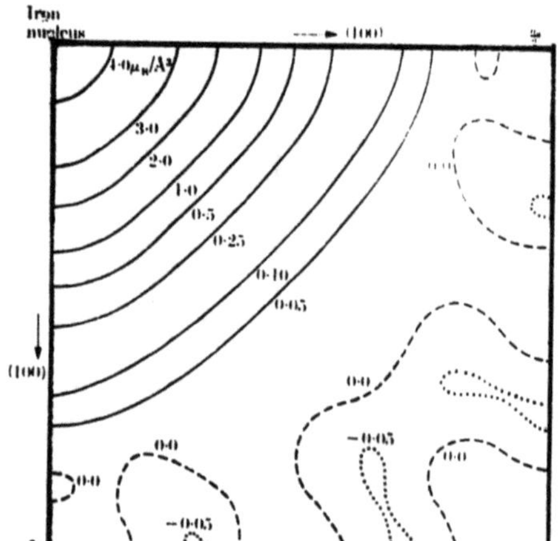


Fig. 5: Spin density in the [100] plane for iron(5)

one arrives to the expression of the structure factor.(32):

$$F(\vec{H}) = \mu \sum_K \langle j_K(H) \rangle [\sum_Q c_Q^K Y_Q^{K*} (\Theta, \phi)] \tag{8}$$

with

$$c_Q^K = i^K (2\ell+1)[4\pi(2K+1)]^{\frac{1}{2}} \sum_{mm'} (-1)^m a_m^* a_{m'} \begin{pmatrix} \ell & K & \ell \\ o & o & o \end{pmatrix} \begin{pmatrix} \ell & K & \ell \\ -m & Q & m' \end{pmatrix} \tag{9}$$

The 3j coefficients are tabulated in reference(4). The properties of $\begin{pmatrix} \ell & K & \ell \\ o & o & o \end{pmatrix}$ and $\begin{pmatrix} \ell & K & \ell \\ -m & Q & m' \end{pmatrix}$ imply that $C_Q^K$ is different from zero only for $K \leq 2\ell$, even, and for $-m+Q+m' = 0$.

d electrons in a cubic crystal field

The magnetic d electrons correspond to $\ell=2$ with a 5fold degenerate level. The presence of a cubic crystal field removes the degeneracy within an Eg doublet and a $T_{2g}$ triplet with the following wave functions:

$$\text{Eg} \begin{cases} d_{z^2} = Y_0^2(\Theta,\Phi) \\ d_{x^2-y^2} = \frac{1}{\sqrt{2}} [Y_2^2(\Theta,\Phi) + Y_{-2}^2(\Theta,\Phi)] \end{cases}$$

$$T_{2g} \begin{cases} d_{xz} = \frac{1}{\sqrt{2}} [Y_1^2(\Theta,\Phi) + Y_{-1}^2(\Theta,\Phi)] \\ d_{yz} = \frac{-i}{\sqrt{2}} [Y_1^2(\Theta,\Phi) - Y_{-1}^2(\Theta,\Phi)] \\ d_{xy} = \frac{-i}{\sqrt{2}} [Y_2^2(\Theta,\Phi) - Y_{-2}^2(\Theta,\Phi)] \end{cases} \tag{10}$$

Application of formulae (8) and (9) gives for the structure factors

$$F(Eg) = \langle j_o \rangle + \frac{3}{2} A(\Theta,\phi) \langle j_4 \rangle$$

$$F(T_{2g}) = \langle j_o \rangle - A(\Theta,\phi) \langle j_4 \rangle \tag{11}$$

with

$$A(\Theta,\phi) = \frac{1}{8} (35 \cos^4\theta - 30 \cos^2\theta + 3) + \frac{5}{8} \sin^4\theta \cos 4\phi \tag{12}$$

and, in term of Miller indices for a cubic cell:

$$A(\Theta,\phi) = \frac{h^4+k^4+\ell^4-3(h^2k^2+h^2\ell^2+k^2\ell^2)}{(h^2+k^2+\ell^2)^2} \qquad (13)$$

For a distribution of electrons with 40% Eg and 60% T2g which would correspond to a spherical distribution the <j4> term would disappear.

In nickel the structure factor has been analyzed in terms of 19% Eg and 81% T2g which explains the expansion of the spin density in the direction of the diagonals of the cube face.

The asymmetry of the spin density around the Ni nucleus is clearly demonstrated by the comparaison of reflections corresponding to the same $\sin\theta/\lambda$ but for which very different values of the structure factor have been observed:

| | |
|---|---|
| F(511) = .0056 $10^{-12}$ cm | F(333) = .0170 $10^{-12}$ cm |
| F(600) = −.0038 | F(442) = .0081 |
| F(551) = .0014 | F(711) = −.0073 |
| F(731) = −.0041 | F(553) = .0018 |

In the case of iron investigated by Shull and Yamada(5), the results are in marked contrast with those obtained for Nickel. The spin density map is also asymmetric (fig. 5) but expands along the edges of the cube which is characteristic of Eg orbitals. By matching the observed and calculated structure factors using equations (11) a value of 53% has been found for the occupancy of the Eg state and 47% for T2g.

<u>In the case of hexagonal cobalt</u> the crystal field splits the 3 d electrons into 3 substates: 2 doubly degenerate levels E2g and E1g and one single level A1g corresponding to the following wave function

$$A_{1g} \quad d_{z^2}$$

$$E_{1g} \begin{cases} d_{xz} \\ d_{yz} \end{cases} \qquad (14)$$

$$E_{2g} \begin{cases} d_{xz} \\ d_{x^2-y^2} \end{cases}$$

Formulae (8) and (9) give for the structure factors (15)

$$F(A_{1g}) = \langle j_o \rangle - \frac{10}{14}(3\cos^2\theta-1)\langle j_2 \rangle + \frac{18}{56}(35\cos^4\theta-30\cos^2\theta+3)\langle j_4 \rangle$$

$$F(E_{1g}) = \langle j_o \rangle - \frac{5}{14}(3\cos^2\theta-1)\langle j_2 \rangle - \frac{12}{56}(35\cos^4\theta-30\cos^2\theta+3)\langle j_4 \rangle$$

$$F(E_{2g}) = \langle j_o \rangle + \frac{10}{14}(3\cos^2\theta-1)\langle j_2 \rangle + \frac{3}{56}(35\cos^4\theta-30\cos^2\theta+3)\langle j_4 \rangle$$

in which angle $\phi$ is not present due to the axial symmetry of the site. Refinement of the data had led to the following occupancy:

| | |
|---|---|
| E1g | 0.416 |
| E2g | 0.394 |
| A1g | 0.190 |

which is very close to the spherical occupancy: 40% 40% 20%

The reasons for which the distribution of the spin density is so different for the 3 elements Fe Co Ni is not obvious. Their crystal structure is different but it is not enough as shown by the study of fcc alloys CoNi(6) and NiCu(7,8).

In fig. 6 is represented the aspherical part of the moment which is either of T2g or of Eg type. An attempt to understand this results by a rigid band model starting from pure Ni and filling with the electrons, keeping the total spin moment equal to the experimental value was not satisfactory. A coherent potential approximation(9-10) which calculates the spin and orbital dependent densities of states from which are deduced the asphericities of the spin distribution gives results in agreement with the data.

Formulae (8) and (9) may be applied in the cases of more complex crystal structures including different magnetic atoms. If there is a large number of asphericity parameters to determine, the number of observations has to be increased consequently.

2) The radial part of the wave function

The agreement between observed and calculated structure factors was surprisingly good for the examples previously given. In these

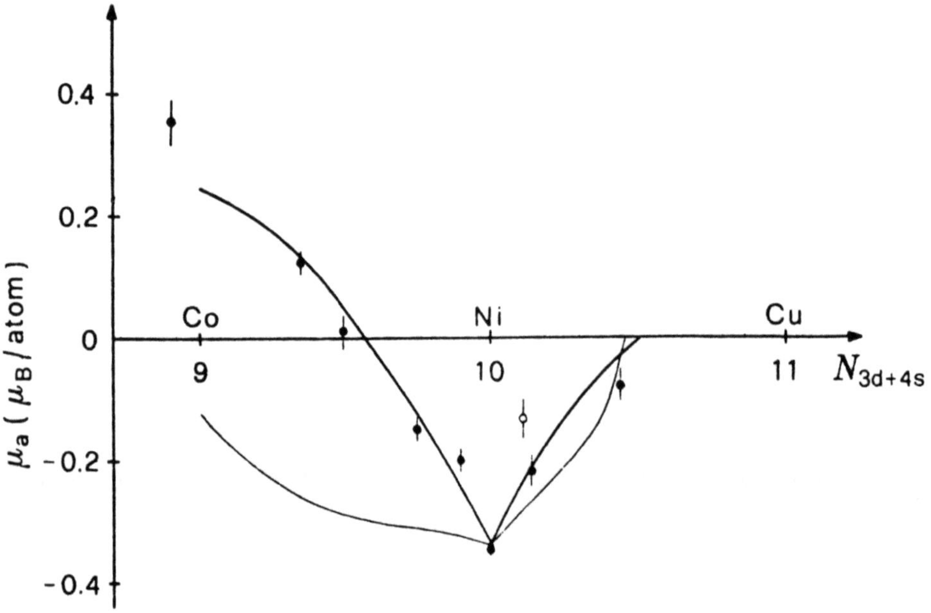


Fig. 6: Aspherical part of the moment for Co-Ni and Ni-Cu alloys(10)

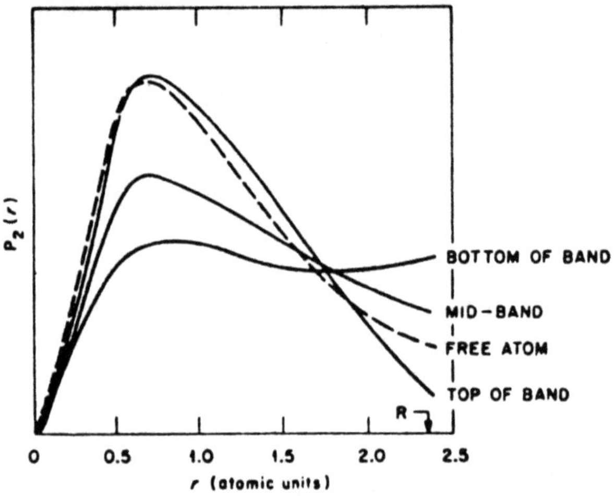


Fig. 7: Energy dependence of 3d radial wave functions in Fe(13)

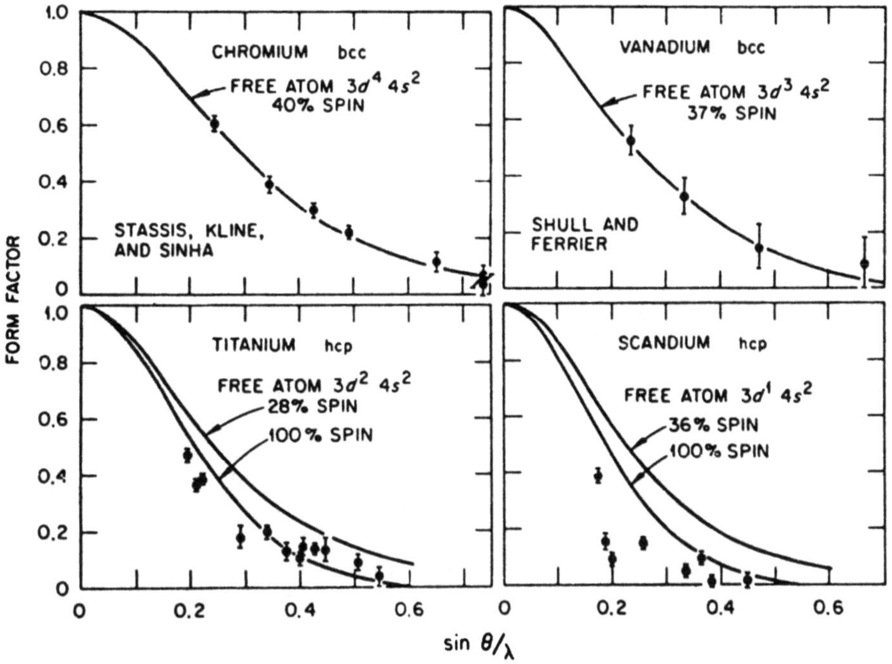


Fig. 8: Paramagnetic form factors for 3d elements(14,15,16)

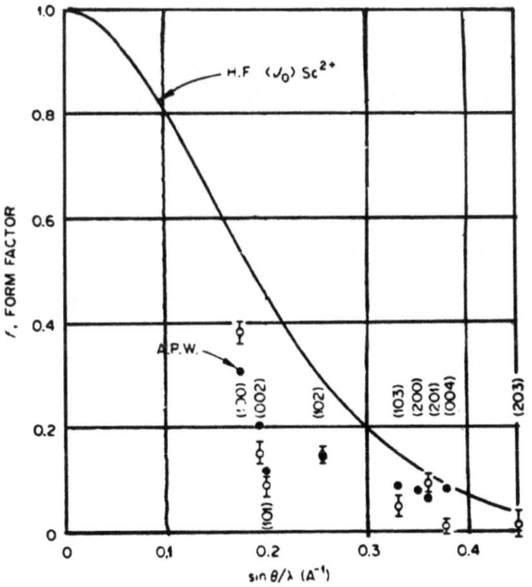


Fig. 9: Paramagnetic form factor for Sc: experimental(16) (open circles), APW(17) (closed circles) and free ion (solid line) calculations

cases, the radial part of the wave function was calculated by an Hartree Fock method for a free atom(11-12), and not for atoms arranged in a solid. This leads to the following question wether the radial part of the wave function is or not affected by the band structure of the solid. The answer has been first given by Wood(13) when calculating the band structure of iron: the radial wave functions corresponding to the top of the band are as much and even more localized than that of the free atom while those corresponding to the bottom of the band are less concentrated (fig. 7). This may be understood considering that the bottom of the band electrons are bonding electrons while the top of the band electrons are antibonding electrons.

A very good illustration of this effect is given by the form factor measurements of the paramagnetic transition metals. In these measurements, the ordered magnetic moments measured with polarized neutrons were aligned by a magnetic field produced by a superconducting Helmholtz split coil. The wave function expansion which is not visible for vanadium(14) and chromium(15) is clear for titanium and scandium(16) (fig. 8). These results show the limit of adequacy of atomic models to describe the electron density. Only calculations taking into account the solid state may give in that case satisfactory results. This is the case for the augmented plane wave (APW) calculation made for the scandium(17) the results of which are represented in figure 9.

For heavy atoms another problem arises when comparing experimental and calculated radial integrals. The measured structure factors did not agree with those calculated by the Hartree Fock method: they decrease faster than the calculated ones, indicating that the actual magnetization distribution is more expanded than the calculated one. An example of it is shown in figure 10 for metallic Gd(18). As in this case the band of gadolinium is just half filled, it cannot be a bottom of the band effect. The explanation has been given by Freeman and Desclaux(19) who have shown that it was an indirect relativistic effect: the s electrons which are relativistic are more concentrated than in the Hartree Fock calculations; it results that their screening effect is larger and that the 4 f and 5 f electron distribution is more expanded.

Therefore, relativistic calculation (Hartree Fock Dirac) are necessary to calculate the radial integrals $\langle j_K(H) \rangle$ for heavy atoms like rare earth and actinides. Figure 11 compares for neodymium the results of the two calculations: by Blume et al(20) for the non relativistic and by Stassis et al(21) for the relativistic one.

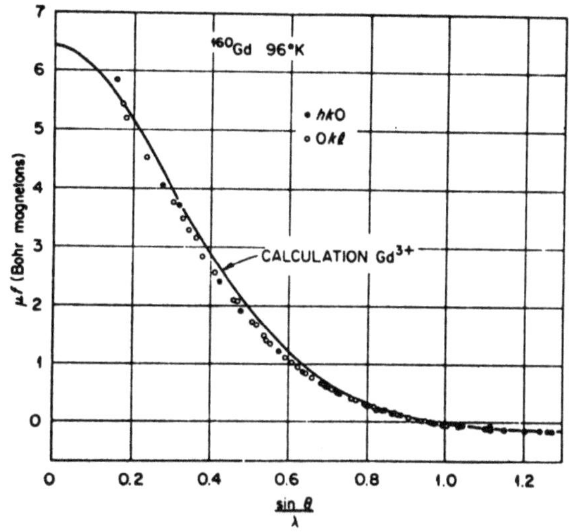


Fig. 10: Magnetic amplitudes for Gd(18)

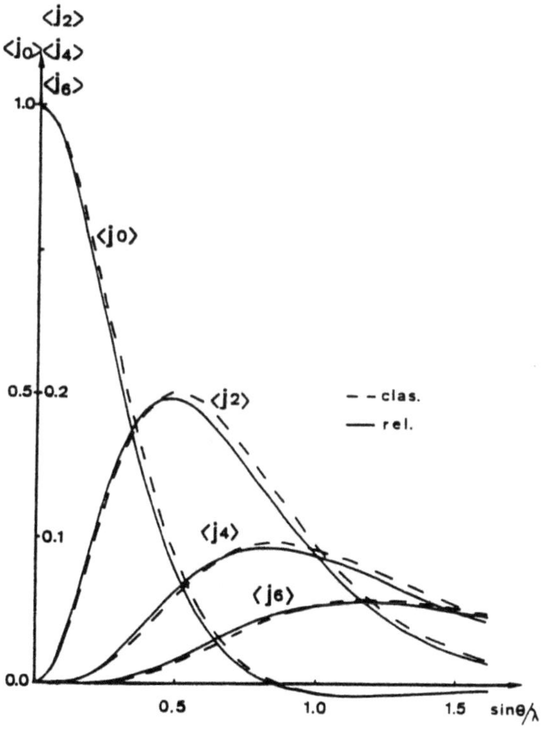


Fig. 11: Radial integrals for $Nd^{3+}$: non relativistic(20) (broken lines) and relativistic(21) (solid lines) calculations

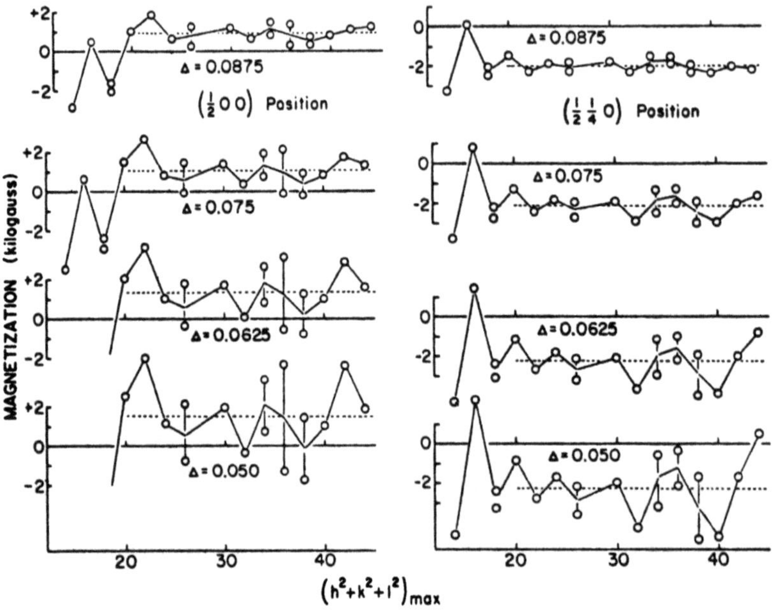


Fig. 12: Convergence of the spin density at positions($\frac{1}{2}$00) and ($\frac{1}{2}$ $\frac{1}{4}$ 0) in Fe, for different sizes of the averaging cube(23)

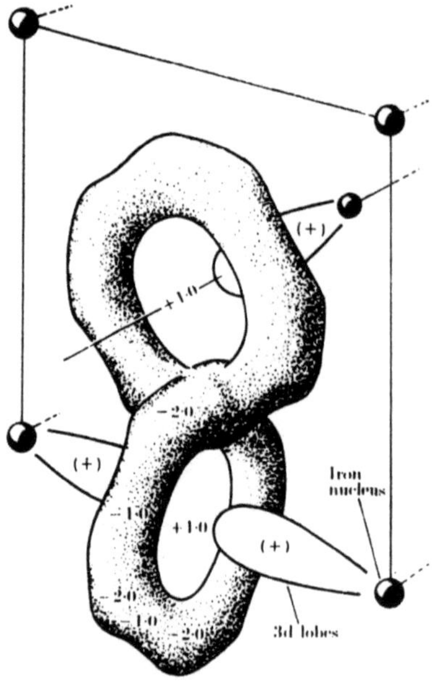


Fig. 13: Zone of negative spin density in Fe(23)

3) Small orbital effects

If the orbital moment is not completely quenched by the crystal field, the magnetization density produced by the orbital motion of the electrons give rise to a contribution to the scattering of the neutrons. Scattering by orbital magnetization densities will be treated later but we shall examine here, for sake of completeness the way of taking into account these effects when they are small, as it is the case for almost all transition elements.

It has been shown[22] that a first approximation for the orbital structure factor (dipole approximation) is:

$$F_L(H) = \mu_L\langle g_o(H)\rangle = \mu_L[\langle j_o(H)\rangle+\langle j_2(H)\rangle] \tag{16}$$

which has to be compared for the spin structure factor to

$$F_S(H) = \mu_S\langle j_o(H)\rangle \tag{17}$$

Due to the shape of functions $\langle j_K\rangle$, these formulae show that orbital form factors decrease more slowly than spin form factors as a function of $\sin\theta/\lambda$, which indicates for a same electronic distribution that the orbital magnetization density is more concentrated than the spin one.

When the orbital contribution is small, one uses to write, using for the spin part equation (8)

$$F(\vec{H}) = \mu_S\sum_K \langle j_K\rangle \sum_Q C_Q^K Y_Q^{K^*}(\theta,\phi) + \mu_L[\langle j_o\rangle + \langle j_2\rangle] \tag{18}$$

$\mu_S$ and $\mu_L$ representing the spin and the orbital parts of the magnetic moment. If S is a good quantum number one can write $\vec{L} = (g-2)\vec{S}$ and the g factor can be determined by gyromagnetic experiments

$$\frac{\mu_L}{\mu_S} = \frac{g-2}{2} \tag{19}$$

If this ratio is unknown, for instance in the case of several magnetic atoms, it can be determined by matching with the experimental structure factors.

b) Delocalized spin density far from the nuclei

Far from the nuclei the spin density is of great interest as connected to electrons taking part in the bonding process between atoms. But being generally very weak this density cannot be treated independently from the uncertainties connected with the Fourier transformation.

1) Series termination errors

The spin or magnetization density is given by Fourier series of type (1) only if the sum is extended to all the points $\vec{H}$ of the reciprocal lattice. Practically it is not the case and the experimental limitation of the observed Bragg reflections has for result a broadening of the density peaks and erratic oscillations.

If 2 points are very close one to the other in the real space and cannot be separated because of finite number of informations, one has to discuss the density in term of average on a volume or on a surface.

Let us evaluate the density $m^v(\vec{r})$ averaged on a volume v around the point $\vec{r}$. We write the density at the point $\vec{r} + \vec{r}'$ close to $\vec{r}$:

$$m(\vec{r}+\vec{r}') = \frac{1}{V}\sum_H F(\vec{H})e^{i\vec{H}(\vec{r}+\vec{r}')}$$

$$= \frac{1}{V}\sum_H F(\vec{H})e^{i\vec{H}\vec{r}}e^{i\vec{H}\vec{r}'} \tag{20}$$

The averaged density on the integration volume v is

$$m^v(\vec{r}) = \frac{1}{v}\int_v m(\vec{r}+\vec{r}')d^3r'$$

$$= \frac{1}{V}\sum_H \tilde{F}^v(\vec{H})e^{i\vec{H}\vec{r}} \tag{21}$$

with

$$\tilde{F}^v(\vec{H}) = F(\vec{H}).\frac{1}{v}\int_v e^{i\vec{H}\vec{r}'}d^3r' \tag{22}$$

The averaged density on an integration surface s is as well obtained with:

$$\tilde{F}^s(\vec{H}) = F(\vec{H}).\frac{1}{s}\int_s e^{i\vec{H}\vec{r}'}d^2r' \tag{23}$$

For a cubic cell average of side length $2\delta a$ (if a is the unit cell length)

$$\tilde{F}^v(\vec{H}) = F(\vec{H})\frac{1}{\delta^3}\frac{\sin 2\pi h\delta}{2\pi h}\frac{\sin 2\pi k\delta}{2\pi k}\frac{\sin 2\pi \ell\delta}{2\pi \ell} \tag{24}$$

For a spherical average of radius R

$$\tilde{F}^v(\vec{H}) = F(\vec{H}).\frac{3}{(HR)^3}(\sin HR - HR\cos HR) \tag{25}$$

For an average on a circle surface of radius R

$$\tilde{F}^{s}(\vec{H}) = F(\vec{H}).2\,\frac{J_1(HR)}{HR} \tag{26}$$

where $J_1$ is the normal Bessel function of first order.

For an average on the perimeter of a circle of radius R

$$\tilde{F}^{s}(\vec{H}) = F(\vec{H})\; J_0(HR) \tag{27}$$

where $J_0$ is the normal Bessel function of order zero.

In the averaged Fourier series, compared to the non averaged ones, each term is multiplied by a factor which is a decreasing function of $|\vec{H}|$. Therefore the relative importance of the neglected terms is lowered and, if the averaging surface or volume is large enough, the oscillations related to the cut-off disappear. Fig. 12 shows in the case of Fe[23] the improvement of convergence resulting for an averaging of the spin density on a small cube when its edge $2\delta$ is increasing from 0.10 to 0.175. Forcing the convergence by integrating on a volume, one has to keep in mind that one is loosing the ponctual aspect of the Fourier series: when the density varies rapidly the procedure introduces a systematic deformation and has to be used with care.

The results concerning the spin density far from the atoms in the case of transition metals differ from one metal to the other: it is either constant or fluctuating throughout the cell. In cobalt[1] ($-0.085\ \mu B/Å^2$ in projection on the basal plane) and in nickel[2] ($-0.0085\ \mu B/Å^3$) it is constant and negative. It fluctuates in iron[23]: positive, with exception of negative rings as shown in figure 13.

The case of nickel is puzzling: when alloyed with other elements (2% Ti[21] or 3% V[25]) the density far from the nuclei has been found not constant and changing sign. As a small error on the first Bragg reflections, as unperfect extinction correction, would strongly affect the value of this density, one must consider that this question for nickel and nickel alloys is not clear for the moment.

2) Separation between localized and delocalized density

It is often possible to go one step further and make a clear separation between the density which is localized around the nuclei and the delocalized density, and this everywhere in the cell. One considers the total experimental density $m^T(\vec{r})$ as the sum of 2 densities: $m^L(\vec{r})$ concentrated around the magnetic atoms and

$m^D(\vec{i})$ diffuse in the cell

$$m^T(\vec{r}) = m^L(\vec{r}) + m^D(\vec{r}) \qquad (28)$$

Each of these periodic densities is the sum of Fourier series

$$m^\alpha(\vec{r}) = \frac{1}{V}\sum_H F^\alpha(\vec{H})e^{i\vec{H}\vec{r}} \qquad (29)$$

and the observed total structure factors $F^T(\vec{H})$ can also be decomposed as a sum:

$$F^T(\vec{H}) = F^L(\vec{H}) + F^D(\vec{H}) \qquad (30)$$

A first method to separate localized-delocalized spin density is to use a model which represents correctly the localized density $m^L(\vec{r})$. If one considers that the high angle Bragg reflections have completely determined the wave function of the localized magnetic electrons, a Fourier difference map between the observed structure factors and those calculated by the model shows the density which exists besides the localized model.

An other method of separation is based on the fact that $m^D(\vec{r})$ varies slowly throughout the cell and therefore only the first corresponding structure factors $F^D(\vec{H})$ will be non zero, and one takes

$(\sin\theta/\lambda <$ limit $\quad F^T(H) = F^L(H) + F^D(H)$

$(\sin\theta/\lambda >$ limit $\quad F^T(H) = F^L(H)$

An example of this method is given for the total density projection of $NdAl_2$ (26) represented in figure 14. Having assumed that $F^D(H)$ is zero above $\sin\theta/\lambda = 0.39$ Å$^{-1}$, the 4 structure factors $F^L(000)$, $F^L(220)$, $F^L(400)$ and $F^L(440)$ are determined in such a way that when making a Fourier sum using these 4 unknown $F^L$ $(\sin\theta/\lambda < 0.39$ Å$^{-1})$ and the other $F^L(\sin\theta/\lambda > 0.39$ Å$^{-1})$ taken equal to the observed values, one gets a localized density which, in the interatomic zones, is as flat and as close to zero as possible.

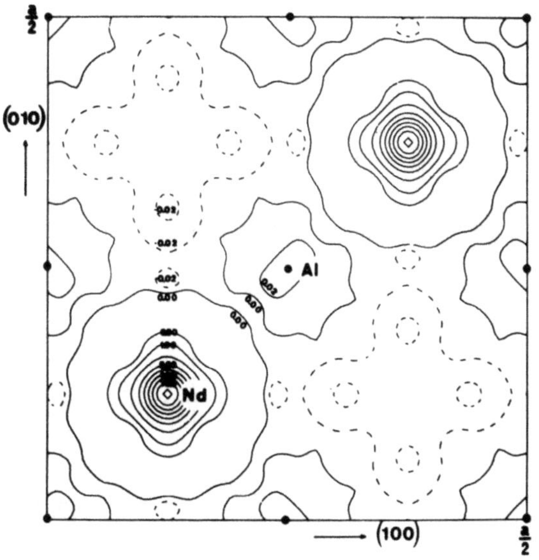


Fig. 14: Total magnetization density projected onto plane (001) of $NdAl_2$ (26)

(a) (b)

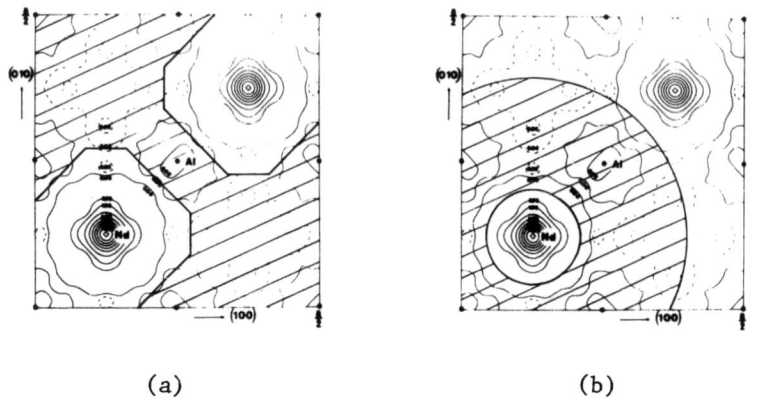

Fig. 15: $NdAl_2$: Interatomic zones in which the magnetization density is assumed to be zero[27]

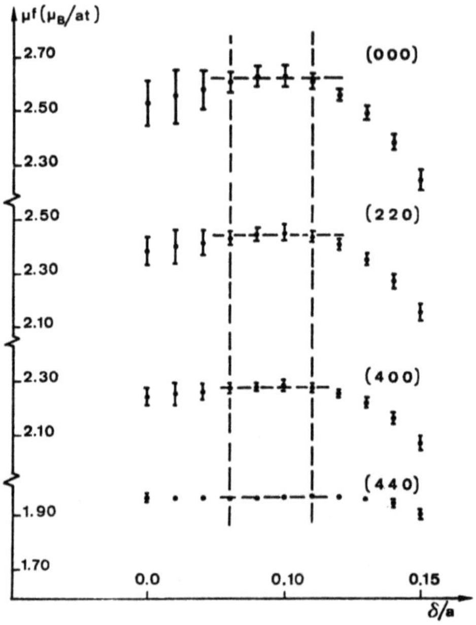


Fig. 16: NdAl : Results of the refinement of $^{2}F_{L}$ (hkl)[27]

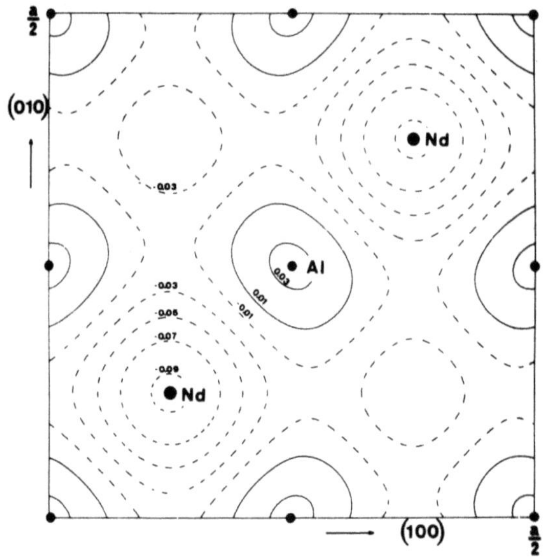


Fig. 17: $NdAl_2$: Difference between the total and the ized magnetization density[26]

The quantity which is minimized is $\Sigma|m^L(r_i)|^2$ on a certain number of points i shown on figure 15(a), which are far from the Nd atoms. The resulting structure factors are affected by the series termination if the averaging squares around each point is too small and then affected by overlapping with the magnetic atoms if these squares are too large (figure 16). An other possibility was to minimize the area included between two circles around the Nd atom (fig. 15b). The same results were found:

| | $F^T$ measured | $F^L$ points+ squares | $F^L$ between 2 circles |
|---|---|---|---|
| F(000) | 2.47(2)μB | 2.63(3)μB | 2.61(3)μB |
| F(220) | 2.40(2) | 2.44(2) | 2.43(3) |
| F(400) | 2.19(2) | 2.28(1) | 2.28(1) |
| F(440) | 2.00(3) | 1.97(1) | 1.97(2) |

These results show that for the (440) reflection there is almost no more difference between the total and the localized structure factors which justifies the method and the limit of 0.39 $Å^{-1}$. Once the separation performed, it is easy to build the Fourier transform of each of the densities : the localized and the delocalized one. This last one is represented on figure 17 which shows a negative spin density at the Nd atom location and a positive one on the Al atoms. For other rare earth compounds, the diffuse magnetization found on the rare earth site has been found:

| | Diffuse magnetization per rare earth atom | J | sign of Γ |
|---|---|---|---|
| $CeAl_2$ [27] (H=48oe//100) | + 0.01 μB | L-S | - |
| $NdAl_2$ [26,27] | - 0.14 μB | L-S | + |
| $SmAl_2$ [28] | - 0.21 μB | L-S | + |
| $HoAl_2$ [26] | + 1.60 μB | L+S | + |
| $GdAl_2$ [27] | + 0.20 μB | S | + |
| Gadolinium [18] | + 0.55 μB | S | + |

The origin of this delocalized density is a polarization of the conduction electrons (s and d electrons). Actually in the metallic compounds, particularly with rare earth where the 4 f density is very concentrated around the nucleus, the magnetic order is due to indirect exchange through the polarization of conduction electrons. The coupling between these electrons spin $\vec{s}$ and the ionic spin $\vec{S}$ may be described by a simple model

$$H = -2\Gamma\vec{s}.\vec{S}$$

As the spin of the rare earth atoms is antiparallel to the moment for the first half one finds a positive sign for the coupling constant Γ, with the exception of Cerium. In this case the negative value of Γ proposed some years before(29) and explained by a resonant diffusion of conduction electrons, allows to understand some of the unusual properties of cerium alloys as Kondo effect. The value of the polarization however is not understood for the moment. In the case of pure gadolinium, an APW calculation(30) found a spin density having the same general shape as the experimental one, but without quantitative agreement.

As a conclusion for the spin densities, one notes that in comparison with the very accurate experimental data, the theory is at least as precise for the localized part of the density. On the other hand for the diffuse part, theory gives only qualitative answers and one may expect that this situation will be improved in the future.

Bibliography


(1) R.M. Moon, Phys.Rev. 136 (1964) A 195

(2) H.A. Mook, Phys.Rev. 148 (1966) 495

(3) A.R. Edmonds, Angular momentum in quantum mechanics - Princeton University Press

(4) H. Appel, Numerical tables for 3 j symbols, Landolt-Börnstein, Vol. 3, Springer Verlag (1968)

(5) C.G. Shull, Y. Yamada, J. Phys. Soc. Japan 17 BIII (1962) 1

(6) B. Antonini, F. Menzinger, A. Paoletti, F. Sacchetti, Intern. J. Magnetism 1 (1971) 183

(7) F. Sacchetti, P. de Gasperis, F. Menzinger, Phys. Stat. Solidi b76 (1976) 309

(8) Y. Ito, J. Akimitsu, J. Phys. Soc. Japan 35 (1973) 1000

(9) F. Leoni, F. Sacchetti, Nuovo Cim. 21B (1974) 97

(10) F. Menzinger, F. Sacchetti, Proceedings of the Conference on Neutron Scattering, Gatlinburg 1976, p. 677

(11) R.E. Watson, A.J. Freeman, Acta Cryst. 14 (1961) 27

(12) A.J. Freeman, R.E. Watson, Acta Cryst. 14 (1961) 231

(13) J.H. Wood, Phys.Rev. 117 (1960) 714

(14) C.G. Shull, R.P. Ferrer, Phys. Rev. Letters 10 (1963) 295

(15) C. Stassis, G.R. Kline, S.K. Sinha, Phys. Rev. B11 (1975) 2171

(16) W.C. Koehler, R.M. Moon, Phys. Rev. Letters 36 (1976) 616

(17) R.P. Gupta, A.J. Freeman, Phys. Rev. Letters 36 (1976) 613

(18) R.M. Moon, W.C. Koehler, J.W. Cable, H.R. Child, Phys. Rev. B5 (1972) 997

(19) A.J. Freeman, J.P. Desclaux, Int. J. Magnetism 3 (1972) 311

(20) M. Blume, A.J. Freeman, R.E. Watson, J. Chem. Phys. 37 (1962) 1245

(21) C. Stassis, H.W. Deckmann, B.N. Harmon, J.P. Desclaux, A.J. Freeman, Phys. Rev. B 15 (1977) 369

(22) J. Brown, Magnetic Neutron Scattering, these proceedings

(23) C.G. Shull, H.A. Mook, Phys. Rev. Letters 16 (1966) 184

(24) F. Livet, P. Radhakrishna, Sol. Stat. Com. 18 (1976) 331

(25) F. Maniawski, L. Dobrzynski, D. Sikorska, Physica 51 (1971) 627

(26) J.X. Boucherle, J. Schweizer, Physica B 86 (1977) 174

(27) J.X. Boucherle, Thèse Université de Grenoble 1977

(28) J.X. Boucherle, D. Givord, J. Laforest, J. Schweizer, F. Tasset Proceeding of the Colloque: La Physique des Terres Rares à l'état metallique St. Pierre de Chartreuse 1978

(29) J.R. Schrieffer, P.A. Wolf, Phys. Rev. 149 (1966) 491

(30) B.N. Harmon, A.J. Freeman, Phys. Rev. B 10 (1974) 1979

(31) R.J. Weiss, A.J. Freeman, J. Phys. Chem. Solids 10 (1959) 147

(32) F. Tasset Thèse, Université de Grenoble (1975)

# INTERPRETATION OF MAGNETIZATION DENSITIES

J. Schweizer

Institut Laue Langevin and
DN/RFG Centre d'Etudes Nucléaires
38042 Grenoble, France

In the cross section expression for polarized neutrons:

$$N(\vec{H})^2 + 2N(\vec{H})\vec{P}.\vec{F}(\vec{H}) + \vec{F}(\vec{H})^2 \qquad (1)$$

$\vec{P}$ is the polarization of the neutrons, $N(\vec{H})$ the nuclear structure factor and $\vec{F}(\vec{H})$ the magnetic structure factor given by: (1)

$$\vec{F}(\vec{H}) = \frac{\gamma e^2}{mc^2} < \psi \, |\vec{D}^{\perp}| \psi > \qquad (2)$$

with operator $\vec{D}^{\perp}$ defined by:

$$\vec{D}^{\perp} = \sum_{\nu \text{electrons}} e^{i\vec{H}.\vec{r}_\nu} \left[ \frac{\vec{H}}{|H|} \wedge (\vec{s}_\nu \wedge \frac{\vec{H}}{|H|}) - \frac{i}{\hbar |H|} \frac{\vec{H}}{|H|} \wedge \vec{p}_\nu \right] \qquad (3)$$

where $\vec{p}_\nu$ is the linear momentum of the electron and $\frac{\gamma e^2}{mc^2} = -0.54 \; 10^{-12} \; cm^{-1}$. The first half of the expression of $\vec{D}^{\perp}$ corresponds to the scattering of neutrons by the spin part of the electronic moment and the second half to the orbital part of the moment. This second contribution to the scattering vanishes when the orbital moment of the magnetic electrons is zero. This is almost the case for the transition ions where the orbital moment of the d electrons is very often quenched by the crystal field, but not at all for the f electrons of rare earth and actinides where the scattering of neutrons by the orbital magnetization is as important as the scattering by the spin magnetization.

a) Structure factor for spin and orbit

The expression of the magnetic structure factors for spin and orbit have been calculated by Johnston, Lovesey and Rimmer using the tensor operators and the Racah algebra. In this method a vector operator is a tensor $\vec{S^1}$ of rank 1, with spherical components:

$$\begin{cases} S^1_o = S_z \\ S^1_1 = \frac{-1}{\sqrt{2}} (S_x+iS_y) \\ S^1_{-1} = \frac{1}{\sqrt{2}} (S_x-iS_y) \end{cases} \quad (4)$$

and when a tensor operator operates in different spaces (for instance electron and neutron), one may expand this operator into components which are themselves tensor operators, but acting only in one of the spaces, and connected by a Clebsch Gordan coefficient:

$$T^K_Q(e,n) = \sum_{\substack{K'K'' \\ Q'Q''}} V^{K'}_{Q'}(e)\, W^{K''}_{Q''}(n)\, \langle K'Q'K''Q''|KQ\rangle \quad (5)$$

with $\langle K'Q'K''Q''|KQ\rangle$ different from zero only for $K \geqslant |K'-K''|$, $K \leqslant K'+K''$ and for $Q = Q'+Q''$.

If the wave function of the magnetic ion can be written as a sum of states of same quantum numbers L,S,J

$$|\psi\rangle = \sum_M a_M |JM\rangle \quad (6)$$

then, the spherical components of the magnetic structure factor $\vec{F}(\vec{H})$, which is a vector, are:

$$F_q(\vec{H}) = \frac{\gamma e^2}{mc^2} \sqrt{4\pi} \sum_{K''Q''} Y^{K''}_{Q''}(\hat{H}) \sum_{K'Q'} \sum_{MM'} a_M a^*_{M'} \left[A(K''K')+B(K''K')\right]$$
$$\langle K'Q'JM'|JM\rangle\langle K''Q''K'Q'|1q\rangle \quad (7)$$

In this formula $Y^{K''}_{Q''}(\hat{H})$ depends on the 2 angles $\theta,\phi$ which characterize the orientation of the scattering vector $\vec{H}$.

A (K"K') and B(K"K') represent respectively the orbital and the spin contribution to the scattering, independently of the wave expansion of the ion given by formula (6).

A (K"K') is different from zero for

$$\left.\begin{array}{l} K'\text{odd} \leqslant 2\ell+1 \\ K''\text{even} \leqslant 2\ell \\ K'-1 < K'' < K'+1 \end{array}\right\} \quad (8)$$

| K' | K" | A(K",K') | B(K",K') |
|---|---|---|---|
| 1 | 0 | $A'(0,1)\left[\langle j_0\rangle + \langle j_2\rangle\right]$ | $C'(0,1)\,\frac{2}{3}\langle j_0\rangle - C'(2,1)\,\frac{\sqrt{2}}{3}\langle j_2\rangle$ |
| | 2 | $\sqrt{\frac{1}{2}}\,A'(0,1)\left[\langle j_0\rangle + \langle j_2\rangle\right]$ | $\sqrt{\frac{1}{2}}\left[C'(0,1)\,\frac{2}{3}\langle j_0\rangle - C'(2,1)\,\frac{\sqrt{2}}{3}\langle j_2\rangle\right]$ |
| 3 | 2 | $A'(2,3)\left[\langle j_2\rangle + \langle j_4\rangle\right]$ | $-C'(2,3)\,\frac{4}{\sqrt{21}}\langle j_2\rangle + C'(4,3)\,\frac{2}{\sqrt{7}}\langle j_4\rangle$ |
| | 4 | $\sqrt{\frac{3}{4}}\,A'(2,3)\left[\langle j_2\rangle + \langle j_4\rangle\right]$ | $\sqrt{\frac{3}{4}}\left[-C'(2,3)\,\frac{4}{\sqrt{21}}\langle j_2\rangle + C'(4,3)\,\frac{2}{\sqrt{7}}\langle j_4\rangle\right]$ |
| 5 | 4 | $A'(4,5)\left[\langle j_4\rangle + \langle j_6\rangle\right]$ | $C'(4,5)\,\frac{6}{\sqrt{33}}\langle j_4\rangle - C'(6,5)\sqrt{\frac{10}{11}}\langle j_6\rangle$ |
| | 6 | $\sqrt{\frac{5}{6}}\,A'(4,5)\left[\langle j_4\rangle + \langle j_6\rangle\right]$ | $\sqrt{\frac{5}{6}}\left[C'(4,5)\,\frac{6}{\sqrt{33}}\langle j_4\rangle - C'(6,5)\sqrt{\frac{10}{11}}\langle j_6\rangle\right]$ |
| 7 | 6 | | $-C'(6,7)\,\frac{7}{\sqrt{39}}\langle j_6\rangle$ |
| | 8 | | $\sqrt{\frac{7}{8}}\left[-C'(6,7)\,\frac{7}{\sqrt{39}}\langle j_6\rangle\right]$ |

Table I: Structure of A(K"K') and B(K"K') for rare earth ions

| | | A'(0,1) | A'(2,3) | A'(4,5) | C'(0,1) | C'(2,1) | C'(2,3) | C'(4,3) | C'(4,5) | C'(6,5) | C'(6,7) |
|---|---|---|---|---|---|---|---|---|---|---|---|
| Ce | $^2F_{5/2}$ | -1.1268 | -0.5216 | -0.2173 | 0.4226 | 0.4781 | -0.2391 | -0.6901 | 0.1040 | 1.1396 | 0 |
| Pr | $^3H_4$ | -1.7688 | 0 | 0.3077 | 0.8944 | 0.3654 | -0.2525 | 0.3975 | -0.1473 | -0.7035 | 0.1201 |
| Nd | $^4I_{9/2}$ | -2.1106 | 0.3476 | -0.1568 | 1.3568 | 0.1357 | -0.0996 | 0.2369 | -0.1001 | 0.8585 | -0.1948 |
| Pm | $^5I_4$ | -2.0870 | 0.3236 | -0.1281 | 1.7888 | -0.1342 | 0.0927 | -0.2205 | 0.0817 | -0.7010 | 0.1197 |
| Sm | $^6H_{5/2}$ | -1.6903 | 0 | 0.0855 | 2.1129 | -0.3453 | 0.1726 | -0.2718 | 0.0410 | 0.1957 | 0 |
| Tb | $^7F_6$ | -1.0802 | -0.4020 | -0.1164 | -3.2404 | 0.2546 | -0.4606 | -0.1450 | 0.3343 | 0.0470 | -0.1525 |
| Dy | $^6H_{15/2}$ | -1.7743 | 0 | 0.2763 | -2.6614 | 0.2509 | -0.4213 | 0.1769 | -0.3528 | -0.1858 | 0.4758 |
| Ho | $^5I_8$ | -2.1213* | 0.3596 | -0.1737 | -2.1213 | 0.1000 | -0.1648 | 0.1297 | -0.2496 | 0.3505 | -0.8481 |
| Er | $^4I_{15/2}$ | -2.1291 | 0.3678 | -0.1842 | -1.5969 | -0.1004 | 0.1685 | -0.1327 | 0.2646 | -0.3716 | 0.9517 |
| Tm | $^3H_6$ | -1.8002 | 0 | 0.3493 | -1.0801 | -0.2546 | 0.4606 | -0.1934 | 0.4458 | 0.2348 | -0.7627 |
| Yb | $^2F_{7/2}$ | -1.1339 | -0.5471 | -0.2770 | -0.5670 | -0.2673 | 0.6268 | 0.1974 | -0.7957 | -0.1118 | 1.2368 |

Table II: Coefficients A'(K"K) and C'(K"K') for the trivalent rare earth ions(2)

and one can write

$$A(K''K') = A'(K''K') \left[<j_{K'+1}(H)> + <j_{K'-1}(H)>\right] \qquad (9)$$

with the following relation between the coefficients A'

$$\frac{A'(K''=K'-1,\ K')}{A'(K''=K'+1,\ K')} = \left(\frac{K'+1}{K'}\right)^{1/2} \qquad (10)$$

B(K"K') is different from zero for

$$\begin{array}{l} K' \text{ odd} \leqslant 2\ell+1 \\ K'' \text{ even} \leqslant 2\ell+2 \\ K'-1 < K'' < K'+1 \end{array} \qquad (11)$$

and one can write

$$B(K''=K'-1,\ K') = C'(K'-1,\ K') < j_{K'-1}(H) > i^{K'-1} \frac{(K'+1)}{\left[3(2K'+1)\right]^{1/2}}$$

$$+C'(K'+1,\ K') < j_{K'+1}(H) > i^{K'+1} \left[\frac{K'(K'+1)}{3(2K'+1)}\right]^{1/2}$$

$$(12)$$

with the following relation

$$\frac{B(K''=K'-1,K')}{B(K''=K'+1,K')} = \left(\frac{K'+1}{K'}\right)^{1/2} \qquad (13)$$

The structure of quantities A(K"K') and B(K"K') is given in table I for the rare earth ions ($\ell=3$). For this case coefficients A' and C' have been calculated by Lander and Brun(2) and are reported in table II. For completeness the spherical harmonics up to K=8 are listed in table III.

Considering now the cross section of polarized neutrons (1) for the two possible directions of the incoming neutron polarization:

$$N(\vec{H})^2 \pm 2N(\vec{H}).F_o(\vec{H}) + \left[F_o(\vec{H})^2 + 2|F_1(\vec{H})|^2\right] \qquad (14)$$

it differs from the usual Halpern and Johnson form:

$$N(\vec{H})^2 \pm 2\ N(\vec{H})\ M(\vec{H}) + M(\vec{H})^2 \qquad (15)$$

where the magnetic structure factor was supposed to be scalar.

Table III: Spherical harmonics. (All the functions have to be multiplied by $\frac{1}{\sqrt{4\pi}}$).

**Ordre 0**

$$Y_{0\,0} = 1$$

**Ordre 1**

$$Y_{1\,0} = \sqrt{3}\cos\theta$$

$$Y_{1\pm1} = \mp\sqrt{\frac{3}{2}}\sin\theta \exp(\pm i\phi)$$

**Ordre 2**

$$Y_{2\,0} = \frac{\sqrt{5}}{2}(3\cos^2\theta - 1)$$

$$Y_{2\pm1} = \mp\sqrt{\frac{15}{2}}\sin\theta\cos\theta \exp(\pm i\phi)$$

$$Y_{2\pm2} = \sqrt{\frac{15}{8}}\sin^2\theta \exp(\pm 2i\phi)$$

**Ordre 3**

$$Y_{3\,0} = \frac{\sqrt{7}}{2}(5\cos^3\theta - 3\cos\theta)$$

$$Y_{3\pm1} = \mp\frac{\sqrt{21}}{4}\sin\theta(5\cos^2\theta - 1)\exp(\pm i\phi)$$

$$Y_{3\pm2} = \frac{\sqrt{210}}{4}\sin^2\theta\cos\theta \exp(\pm 2i\phi)$$

$$Y_{3\pm3} = \mp\frac{\sqrt{35}}{4}\sin^3\theta \exp(\pm 3i\phi)$$

**Ordre 4**

$$Y_{4\,0} = \frac{3}{8}(35\cos^4\theta - 30\cos^2\theta + 3)$$

$$Y_{4\pm1} = \mp\frac{3\sqrt{5}}{4}\sin\theta(7\cos^3\theta - 3\cos\theta)\exp(\pm i\phi)$$

$$Y_{4\pm2} = \frac{3\sqrt{10}}{8}\sin^2\theta(7\cos^2\theta - 1)\exp(\pm 2i\phi)$$

$$Y_{4\pm3} = \mp\frac{3\sqrt{35}}{4}\sin^3\theta\cos\theta \exp(\pm 3i\phi)$$

$$Y_{4\pm4} = \frac{3\sqrt{70}}{16}\sin^4\theta \exp(\pm 4i\phi)$$

**Ordre 6**

$$Y_{6\,0} = \frac{\sqrt{13}}{16}(231\cos^6\theta - 315\cos^4\theta + 105\cos^2\theta - 5)$$

$$Y_{6\pm1} = \mp\frac{\sqrt{546}}{16}\sin\theta(33\cos^5\theta - 30\cos^3\theta + 5\cos\theta)\exp(\pm i\phi)$$

$$Y_{6\pm2} = \frac{\sqrt{1365}}{32}\sin^2\theta(33\cos^4\theta - 18\cos^2\theta + 1)\exp(\pm 2i\phi)$$

$$Y_{6\pm3} = \mp\frac{\sqrt{1365}}{16}\sin^3\theta(11\cos^3\theta - 3\cos\theta)\exp(\pm 3i\phi)$$

$$Y_{6\pm4} = \frac{3\sqrt{182}}{32}\sin^4\theta(11\cos^2\theta - 1)\exp(\pm 4i\phi)$$

$$Y_{6\pm5} = \mp\frac{3\sqrt{1001}}{16}\sin^5\theta\cos\theta \exp(\pm 5i\phi)$$

$$Y_{6\pm6} = \frac{\sqrt{3003}}{32}\sin^6\theta \exp(\pm 6i\phi)$$

**Ordre 8**

$$Y_{8\,0} = \frac{\sqrt{17}}{128}\left[6435\cos^8\theta - 12012\cos^6\theta + 6930\cos^4\theta - 1260\cos^2\theta + 35\right]$$

$$Y_{8\pm1} = \mp\frac{3\sqrt{34}}{64}\left[715\cos^7\theta - 1001\cos^5\theta + 385\cos^3\theta - 35\cos\theta\right]\sin\theta \exp(\pm i\phi)$$

$$Y_{8\pm2} = \frac{3\sqrt{595}}{64}\left[143\cos^6\theta - 143\cos^4\theta + 33\cos^2\theta - 1\right]\sin^2\theta \exp(\pm 2i\phi)$$

$$Y_{8\pm3} = \mp\frac{\sqrt{39270}}{64}\left[39\cos^5\theta - 26\cos^3\theta + 3\cos\theta\right]\sin^3\theta \exp(\pm 3i\phi)$$

$$Y_{8\pm4} = \frac{3\sqrt{2618}}{128}\left[65\cos^4\theta - 26\cos^2\theta + 1\right]\sin^4\theta \exp(\pm 4i\phi)$$

$$Y_{8\pm5} = \mp\frac{3\sqrt{34034}}{64}\left[5\cos^3\theta - \cos\theta\right]\sin^5\theta \exp(\pm 5i\phi)$$

$$Y_{8\pm6} = \frac{\sqrt{7293}}{64}\left[15\cos^2\theta - 1\right]\sin^6\theta \exp(\pm 6i\phi)$$

$$Y_{8\pm7} = \mp\frac{3\sqrt{24310}}{64}\cos\theta\,\sin^7\theta \exp(\pm 7i\phi)$$

$$Y_{8\pm8} = \frac{3\sqrt{24310}}{256}\sin^8\theta \exp(\pm 8i\phi)$$

However, if it is possible to write the following equation:

$$F_o^2 = \sin^2\theta\ (F_o^2+2|F_1|^2) \tag{16}$$

one can define a $\gamma$ ratio as

$$\gamma = \frac{F_o(\vec{H})/\sin^2\theta}{N(\vec{H})} \tag{17}$$

and have the following relation for the polarization ratio

$$R = \frac{1 + 2\ \gamma\sin^2\theta + \gamma^2\sin^2\theta}{1 - 2\ \gamma\sin^2\theta + \gamma^2\sin^2\theta} \tag{18}$$

from which one can obtain

$$\gamma = \frac{R+1}{R-1} \pm \sqrt{\left(\frac{R+1}{R-1}\right)^2 - \frac{1}{\sin^2\theta}} \tag{19}$$

If equation (16) is not fulfilled one cannot define a $\gamma$ value and extract it from the classical polarization technique. This is connected to the fact that $\vec{F}(\vec{H})$ is a vector and one experiment cannot measure all its components. Assumption (16) is satisfied when the angle $(\vec{F},\vec{z})$ is equal to $\frac{\pi}{2}-\theta$ (fig. 1). In that case it is possible to measure $F_o$ with the classical polarization technique.

Most of the time, equation (16) is either exactly fulfilled, (particularly when $\theta=\frac{\pi}{2}$, which corresponds to diffraction in the basal plane, when symmetry reasons impose $F_1(H)=0$) or verified with enough accuracy to allow the use of equations (17), (18) and (19) without introducing a noticeable error.

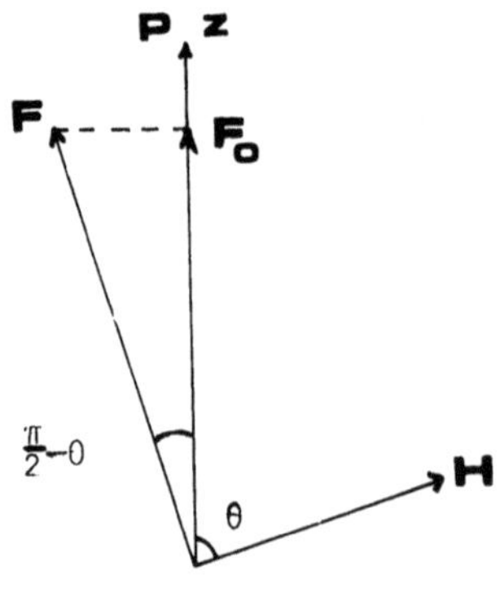


Fig. 1: Relative positions of the magnetic structure factor $\vec{F}$, the scattering vector $\vec{H}$ and the polarization $\vec{P}$ ($\vec{P}$ // to $o\vec{z}$) to fulfill condition (16).

An example of this last case is provided for $NdAl_2$. The Nd wave function has been first determined with the help of (hko) reflections. Then the quantity $F_o^2 \cos^2\theta - 2|F_1|^2 \sin^2\theta$ has been calculated for some ($hk\ell$) reflections with $\ell \neq 0$. It represents how far are these reflections from exact fulfillment of condition (16).

| h k ℓ | sinθ/λ (Å$^{-1}$) | θ | $F_o$ ($\mu_B$/at) | $\|F_1\|$ ($\mu_B$/at) | $F_o^2\cos^2\theta-2\|F_1\|^2\sin^2\theta$ x $10^6$ |
|---|---|---|---|---|---|
| 1 1 1 | 0.109 | 54.74 | 1.6848 | 0.8423 | -5 |
| 3 1 1 | 0.208 | 72.45 | 2.1478 | 0.4803 | -6 |
| 1 1 3 | 0.208 | 25.23 | 0.4049 | 0.6069 | -9 |
| 3 3 1 | 0.273 | 76.74 | 2.0952 | 0.3491 | -4 |
| 3 1 3 | 0.273 | 46.51 | 1.1012 | 0.7387 | -8 |
| 4 2 2 | 0.307 | 65.90 | 1.7309 | 0.5475 | -34 |
| 3 3 3 | 0.326 | 54.74 | 1.3096 | 0.6547 | -3 |
| 5 1 1 | 0.326 | 78.90 | 1.9968 | 0.2770 | -12 |
| 5 3 1 | 0.371 | 80.27 | 1.8774 | 0.2276 | -25 |
| 5 1 3 | 0.371 | 59.23 | 1.3642 | 0.5675 | -49 |
| 5 3 3 | 0.411 | 62.77 | 1.3569 | 0.4938 | -125 |

It is clear in such an example that the use of the usual procedure of polarized neutrons would not bias the results.

b) Spin and orbit distribution: application to the $f^5$ ions

As already shown for the dipolar approximation, spin magnetization and orbit magnetization have not the same spatial extension. This is more striking when their directions are opposed to each other as in the first half or rare earth or actinide series. Particularly, when the 2 moments almost cancel each other out, the different distributions result in regions of space with positive magnetization density and regions with negative magnetization density. Consequently, the Fourier transform $\mathcal{F}(\vec{H})$ presents a maximum for values of the scattering vector $\vec{H}$ which are not zero.

The dipoles approximation[1,3] which corresponds to limit expression (7) to K'=1 and K"=0,2, gives

$$F_o(\vec{H}) = \sin^2\theta < \psi \, |(<j_o> + <j_2>)L_z + 2 <j_o> S_z|\psi> \qquad (20)$$

where θ is the angle between the scattering vector $\vec{H}$ and the direction of quantization z. Applied to a free ion, with scattering on the basal plane ($\theta = \frac{\pi}{2}$), this formula becomes

$$F_o(\vec{H}) = \mu\left[<j_o(H)> + <j_2(H)> \frac{J(J+1)+L(L+1)-S(S+1)}{3J(J+1)+S(S+1)-L(L+1)}\right] \qquad (21)$$

approximation found by Trammel[4] in 1953. The $C_2$ coefficient of of $<j_2>$ is tabulated here for the different configurations $f^n$

| | L | S | J | $C_2=\frac{J(J+1)+L(L+1)-S(S+1)}{3J(J+1)+S(S+1)-L(L+1)} = \frac{2}{g_J} - 1$ |
|---|---|---|---|---|
| $f^1$ | 3 | 1/2 | 5/2 | 1.333 |
| $f^2$ | 5 | 1 | 4 | 1.500 |
| $f^3$ | 6 | 3/2 | 9/2 | 1.750 |
| $f^4$ | 6 | 2 | 4 | 2.333 |
| $f^5$ | 5 | 5/2 | 5/2 | 6.000 |
| $f^6$ | 3 | 3 | 0 | no moment |
| $f^7$ | 0 | 7/2 | 7/2 | 0 |
| $f^8$ | 3 | 3 | 6 | 0.333 |
| $f^9$ | 5 | 5/2 | 15/2 | 0.500 |
| $f^{10}$ | 6 | 2 | 8 | 0.600 |
| $f^{11}$ | 6 | 3/2 | 15/2 | 0.667 |
| $f^{12}$ | 5 | 1 | 6 | 0.714 |
| $f^{13}$ | 3 | 1/2 | 7/2 | 0.750 |

It varies greatly with the number of f electrons with a sharp maximum for the configuration $f^5$ where the spin moment and the orbit moment almost cancel each other out.

This property has been used to characterize the ionization state of atoms. In actinides where the question of ionization is a difficult problem, structure factor analysis provides a very good answer. Lander and Reddy(5) have shown that the best fit for Pu in PuP and $PuF_2$ corresponds to a $Pu^{3+}$ ion (figure 2).

The unusual behaviour of the structure factor may be enhanced by the addition in the ground state of other J values as shown in $SmCo_5$(6). It is well known that the unusual magnetic behaviour of the ion $Sm^{3+}$ is due to the relatively small separation between the $^6H_{5/2}$, $^6H_{7/2}$ and $^6H_{9/2}$ levels in the free ion and that applied magnetic field, exchange field or crystal field result in an admixture of the higher levels in the ground state. In $SmCo_5$ the ground state wave function of $Sm^{3+}$ is

$$\psi_G = 0.978|^5/_2, ^5/_2> -0.205|^7/_2, ^5/_2> +0.038|^9/_2, ^5/_2> \quad (22)$$

with a structure factor shown in figure 3a and the excited states are

$$\text{at 329 K} \quad \psi = 0.973|^5/_2,^3/_2> -0.232|^7/_2,^3/_2> +0.013|^9/_2,^3/_2 >$$
$$\text{at 541 K} \quad \psi = 0.942|^5/_2,^1/_2> -0.329|^7/_2,^1/_2> +0.068|^9/_2,^1/_2 > \quad (23)$$

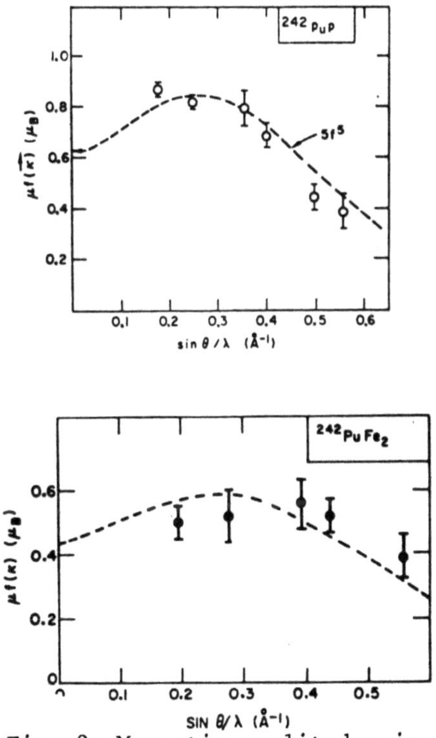


Fig. 2: Magnetic amplitudes in PuP and $PuFe_2$(5). The broken lines correspond to a $5f^5$ configuration

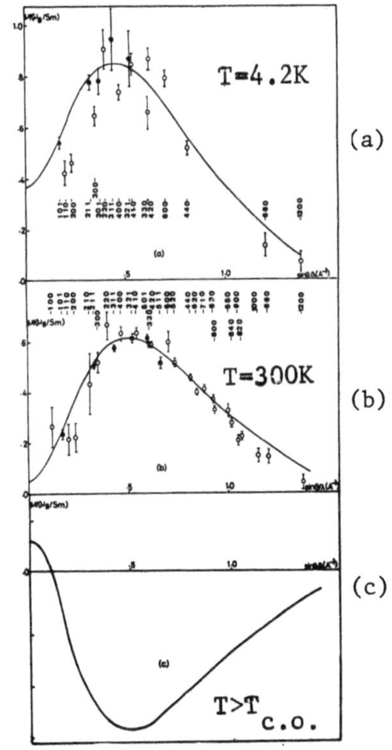


Fig. 3: Magnetic amplitudes of Sm in $SmCo_5$ (a) at 4.2 K, (b) at 300 K and (c) above the cross over temperature(6)

and other states at higher energies. When the temperature increases and the excited levels become more populated, and play a larger part in the form factor the total moment decreases as shown by the room temperature experiment (fig. 3b). It has been calculated that at 350 K a cross over takes place which means that the moment stops being more orbit than spin, cancels and then, at higher temperatures becomes more spin than orbit, with a form factor looking as in figure 3c.

One cannot leave that area of $f^5$ configuration without mentioning the experiment of Moon et al(7) on SmS. In this compound, at atmospheric pressure, lattice constant and other properties are characteristic of $Sm^{2+}$. At about 6 kbar it transforms to a mixed valence phase with a contraction of the lattice corresponding to $Sm^{2.8+}$. This state is due to the coexistence at the Fermi level of an ionic 4f level and a wide s-d conduction band and the electrons transfer between them with a fluctuation of the valence between integral values. For experiments having a long measuring time compared with the fluctuation time (lattice constant) only an intermediate quantity is measured. If the measuring time is short, quantities corresponding to both state are observed.

The structure factors, measured on an isotope powder, show no difference for the two states at low and high pressure (fig. 4), in contrast to what was expected for $Sm^{2+}$ and $Sm^{3+}$. The results have been interpreted, noting that the form factor of $Sm^{3+}$ is very sensitive to temperature and that at high temperature it tends towards that of $Sm^{2+}$. A model has been built in which the interconfiguration fluctuations are simulated by an artificial temperature: in all the thermal averages T has been replaced by $T+\Delta_2$ for $Sm^{2+}$ and $T+\Delta_3$ for $Sm^{3+}$. $\Delta_2$ and $\Delta_3$ have been determined by fitting the data: $\Delta_2$ = 170 K, $\Delta_3$ = 720 K.
The agreement for SmS in the high pressure state is shown in figure 5. Obviously the last word about this experiment has not yet been spoken.

c) Ground state determination

Knowing the magnetic structure factors (and the magnetization density) it is in general possible to go back to the source, i.e. the wave function of the ionic ground state.

In a first case, different wave functions are possible and the comparison between observed and calculated structure factors allows to choose the wave function. We shall illustrate such a case with US(8) which is a good conductor and where the ionization and the outer electron configuration were uncertain. Using for simplicity Hund's rules and Russel-Saunders coupling to find the

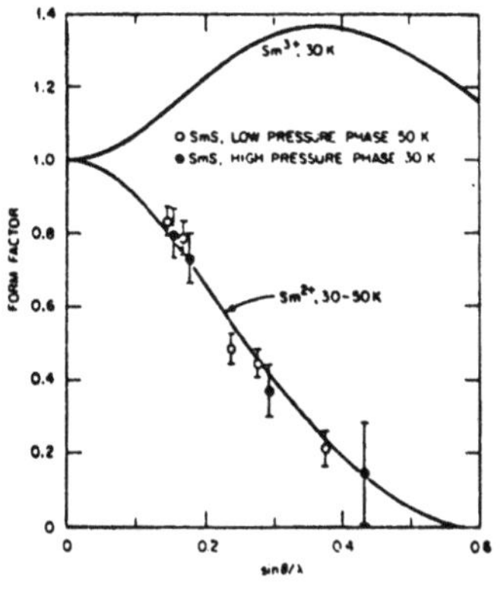


Fig. 4: Form factor of high pressure phase of SmS compared with the low pressure results(7)

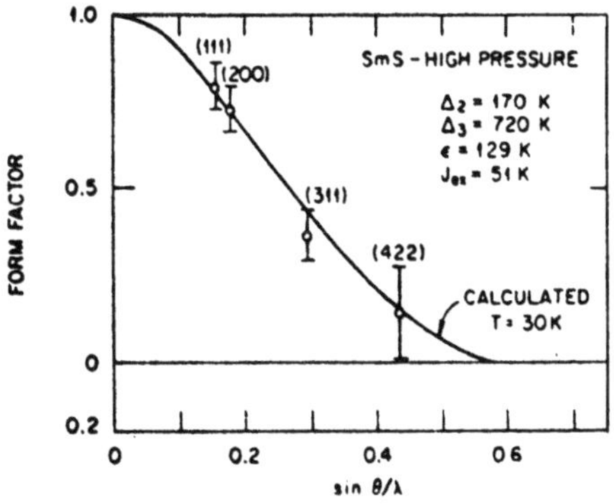


Fig. 5: Comparison of experimental results for the high pressure phase of SmS with the model calculation(7)

ground state and taking into account the crystal field (neglecting the small sixth order term), the possible wave functions were as follows:

$$
\begin{array}{ll}
5f^1\Gamma_7\text{+Ex} & 0.8819|^5/_2,^5/_2\rangle +0.4714|^5/_2,-^1/_2\rangle \\
5f^2\Gamma_7\text{+Ex} & 0.8971|4,3\rangle -0.4270|4,0\rangle -0.2566\ |4,-3\rangle \\
5f^3\Gamma_6 & 0.8459|^9/_2,^7/_2\rangle +0.4095|^9/_2,^1/_2\rangle -0.3417|^9/_2,-^5/_2\rangle \\
5f^3\Gamma_8^{(1)} & 0.6784|^9/_2,^9/_2\rangle +0.7347|^9/_2,^3/_2\rangle \\
5f^4\Gamma_5 & 0.6993|4,4\rangle -0.7005|4,1\rangle -0.1423|4,-2\rangle
\end{array}
\qquad (24)
$$

In the case of $5f^1$ and $5f^2$, an exchange potential was added to obtain a correct value of the moment (in $5f^2$ for instance the ground state was a singlet with zero moment). To compare observed and calculated structure factors (fig. 6a), Wedgwood has analyzed them into isotropic (fig. 6b), θ dependent (fig. 6c) and ϕ dependent (fig. 6d). The isotropic part is not very instructive, but from the anisotropic part it is clear that from the 5 trial wave function, only $5f^2\Gamma_1$+Ex can fit the data.

In another case, one knows exactly the configuration of the ion but not the coefficiants $a_M$ of equation (6). In that case the expression (7) for the structure factor may be written

$$F_q(\vec{H}) = \sum_{MM'} a_M a_M^* \, H_{MM'}(\vec{H}) \qquad (25)$$

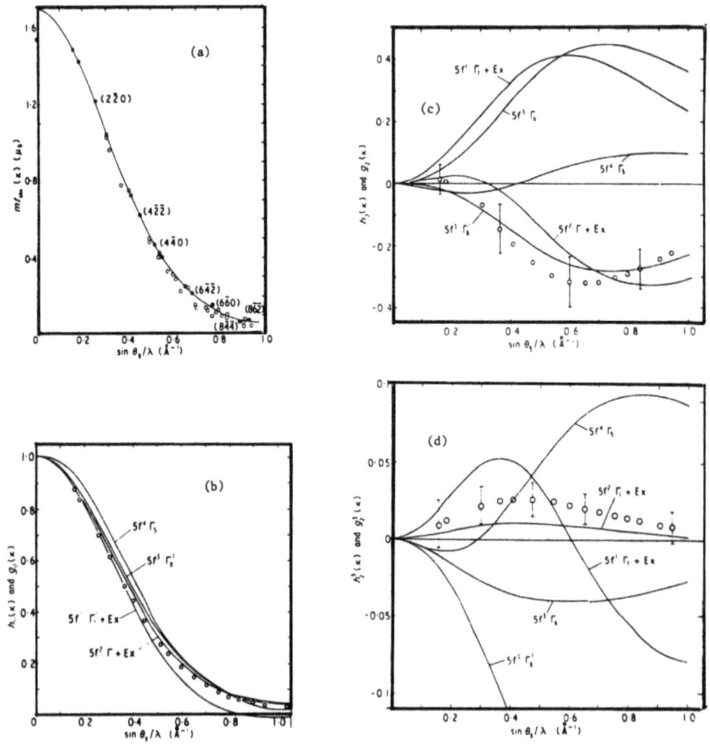


Fig. 6: Comparison of experimental results with model calculations for US(8) (a) total results, (b) isotropic part, (c) $\cos^2\theta$ part and (d) $\cos\theta\sin\theta\cos3\phi$ part

where $H_{MM'}$ includes the spherical harmonics, the A' and C' coefficients, the radial integrals $\langle j_K(H)\rangle$ and the Clebsch-Gordan coefficients. This formula shows that it is possible to refine the coefficients $a_M$ of the ground state, minimising the quantity $\sum_{H} W(\vec{H})\ [F(\vec{H})cal-F(\vec{H})obs]^2$ where $W(\vec{H})$ is the weight associated to the observation $F(\vec{H})$.

We illustrate this case with cubic $NdAl_2$ (9,12) in which we have seen that besides the 4f localized electrons there exists a polarization of the conduction electrons and this extra magnetization (here reversed) makes it impossible to deduce the ionic wave function from the magnetization measurements. Here the degeneracy of the multiplet $J={}^9/_2$ is raised by the crystal and the exchange field, and with the moments aligned parallel to the [001] axis of the cube, the ground state $\Gamma_6$ has the following structure

$$\Psi\ (\Gamma_6)= a_{9/2}|{}^9/_2,{}^9/_2\rangle + a_{1/2}|{}^9/_2,{}^1/_2\rangle + a_{7/2}|{}^9/_2,-{}^7/_2\rangle \quad (26)$$

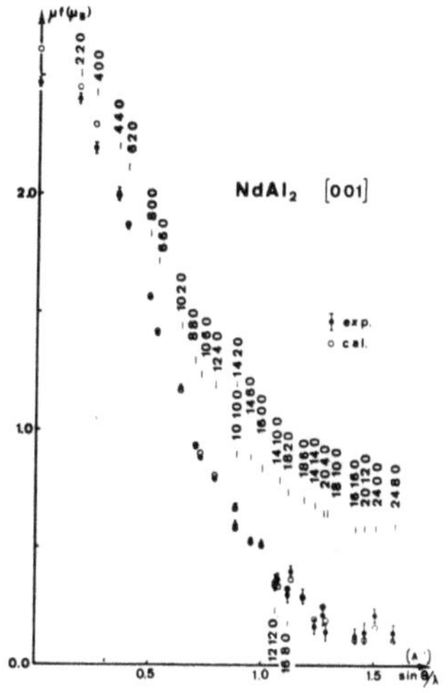

Fig. 7: Comparison of observed (full circles) and calculated (open circles) magnetic structure factors for $NdAl_2$ with field parallel to [001] (9).

Refinement of these 3 quantities gives the following wave function

$$\Psi\ (\Gamma_6) = 0.885|^9/_2,{}^9/_2\rangle + 0.451|^9/_2,{}^1/_2\rangle - 0.112|^9/_2 - {}^7/_2\rangle \qquad (27)$$

and a very good agreement for the structure factors (fig. 7) with the exception of the first reflections. This is not surprising: the ionic wave function represents only the localized structure factor $F^L$ and not the total structure factor $F^T$. Comparison with the localized structure factor, separated by Fourier techniques only (12) is completely satisfactory:

| h k ℓ | $F^T$observed | $F^L$separated | $F(\Gamma_6)$ |
|---|---|---|---|
| 0 0 0 | 2.47 (2) | 2.63 (3) | 2.61 |
| 2 2 0 | 2.40 (2) | 2.44 (2) | 2.45 |
| 4 0 0 | 2.19 (2) | 2.28 (2) | 2.29 |
| 4 4 0 | 2.00 (3) | 1.97 (1) | 1.99 |

If at the temperature of the experiment the excited states are populated, one has to take into account their influence on the structure factors. For low population, it is convenient to correct the observed structure factors first and to carry out the ground state refinement then.

d) Anisotropy of the magnetization distribution

Within an energy level of given J, the introduction of exchange, applied field or crystal field removes the (2J+1) degeneracy of the wave function of the magnetic ions. The ground state then corresponds to an expression of type (6), which, depending on the coefficients $a_M$, may correspond to a magnetization distribution either close to a spherical symmetry, or far from it. In turn the structure factor must reflect these anisotropies when they exist and show in these cases departures from a smooth curve. It is what we shall examine here in the case of a cubic system and for different orientations of the magnetic moments.

In formula (7) it is the $\theta$ and $\phi$ dependence of $Y^{K''}_{Q''}(\theta,\phi)$ which is responsible for the anisotropy and which corresponds to terms containing

$$Y^{K''}_{Q''}(\theta,\phi) \ \langle j_{K'\pm 1}\rangle\langle K'Q'JM'|JM\rangle\langle K''Q''K'Q'|1q\rangle \tag{28}$$

If one is interested in $F_o(\vec{H})$ only (scattering in the horizontal plane with $\theta=\frac{\pi}{2}$) these terms become

$$Y^{K''}_{Q''}(\tfrac{\pi}{2},\ \phi) \ \langle j_{K'\pm 1}\rangle\langle K'Q'JM'|JM\rangle\langle K''Q''K'Q'|10\rangle \tag{29}$$

Let us consider the moments (and the quantization axis) aligned parallel to the different directions of the cube. When parallel to the fourfold axis [001], the wave function (6) corresponds to a selection rule $\Delta M=4$:

$$\psi = \alpha|J,M_o\rangle + \beta|J,M_o+4\rangle + \gamma|J,M_o+8\rangle + \tag{30}$$

The first Clebsch Gordan coefficient in (28) $\langle K'Q'JM'|JM\rangle$ implies $Q'=0,\pm4\ldots$ which in turn requires for the second coefficient $\langle K''Q''K'Q'|10\rangle$ either $Q''=Q'=0$ or $Q''=-Q'=\pm4$ or... Let us look for the term containing the lowest order radial integral $\langle j_\ell(H)\rangle$ presenting a $\phi$ dependence. $Y^{K''}_{o}(\frac{\pi}{2},\phi)$ has no (see table III). $Y^4_4(\frac{\pi}{2}\phi)$ has such a dependence. In that case $K'' = Q'' = 4$ and $Q' = -4$. Then $K' \geqslant 4$ and the first term representing an anisotropy is (see table I) $Y^4_4\ (\frac{\pi}{2},\phi)\langle j_4(H)\rangle\langle 5\text{-}4\ JM'|JM\rangle\langle 445\text{-}4|10\rangle$.

If the moment is parallel to the two fold axis [110] the selection rule for the wave function is $\Delta M=2$. The first coefficient implies $Q' = 0,\ \pm2\ldots.$ and the second $Q'' = Q' = 0$ or $Q'' = -Q' = \pm2\ldots.$and the first term with a $\phi$ dependence is $Y^2_2(\frac{\pi}{2},\phi)\langle j_2(H)\rangle\langle 3\text{-}2\ JM'|JM\rangle\langle 223\text{-}2|10\rangle$.

If the moment is parallel to the three fold axis [111], $\Delta M=3$ and $Q'=0,\ \pm3\ldots.$ which causes $Q'' = Q' = 0$ or $Q'' = -Q' = \pm3\ldots$ As $Y^4_3\ (\theta,\phi)$ is zero for $\theta = \frac{\pi}{2}$, the first non zero term with a $\phi$ dependence is

$Y_6^6(\frac{\pi}{2},\phi)\langle j_6\rangle\langle 7\text{-}6\ JM'|JM\rangle\langle 667\text{-}6|10\rangle$

As the radial integrals $\langle j_\ell(H)\rangle$ diminish in magnitude when $\ell$ increases, the terms with $\langle j_4\rangle$ and $\langle j_6\rangle$ are more difficult to observe than terms with $\langle j_2\rangle$. The result is that, of the 3 principal directions, it is with moments parallel to [110] that most anisotropies are expected. But of course, it depends on coefficients $a_M$ in expression (6).

This feature is clearly demonstrated on the 2 following examples:

1) In TmSb(10), the ground state of Tm is a singlet and therefore non magnetic. However, by applying a magnetic field on the crystal, the ground state receives a contribution from the excited states and becomes magnetic (it is a perturbation effect at low temperature and not a population effect). Knowing the crystal field coefficients from inelastic neutron experiments and magnetization measurements one may calculate the wave function when a 12.5 kOe field is applied along [001] or [110]

$$\psi_{001} = 0.7526\ |6,4\rangle - 0.3493|6,0\rangle + 0.5581|6,-4\rangle \tag{31}$$

$$\psi_{110} = 0.4114\ |6,6\rangle - 0.4718|6,4\rangle + 0.2392|6,2\rangle$$
$$+0.5637|6,0\rangle + 0.2059|6,-2\rangle - 0.3494|6,-4\rangle + 0.2621|6,-6\rangle \tag{32}$$

and the corresponding structure factors, observed and calculated, are represented on figure 8. Besides the good agreement one notices that the points corresponding to the [001] direction lie on a smooth curve and those related to [110] do not.

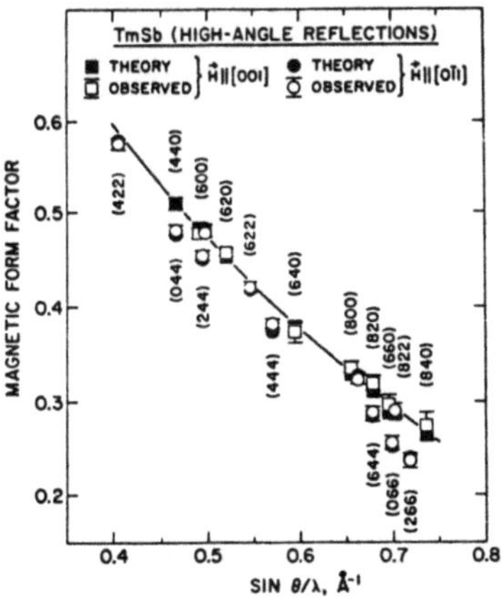


Fig. 8: Form factor for TmSb with both applied field directions(10)

2) $CeAl_2$ has a complex antiferromagnetic structure and it is also possible to induce a ferromagnetic moment by applying a magnetic field along any direction of the crystal[9,11]. The difference of anisotropy between the structure factors related to the 2 orientations of the field is tremendous (fig. 9) and, for equivalent reflections, one gets very different values of the form factor depending on the direction of the moments:

| H//[001] | | H//[01$\bar{1}$] | |
|---|---|---|---|
| 4 4 0 | 0.748 | 0 4 4 | 0.544 |
| 8 0 0 | 0.609 | 8 0 0 | 0.596 |
| 8 8 0 | 0.328 | 0 8 8 | 0.054 |
| 16 0 0 | 0.177 | 16 0 0 | 0.197 |
| 12 12 12 | 0.081 | 0 12 12 | -0.073 |

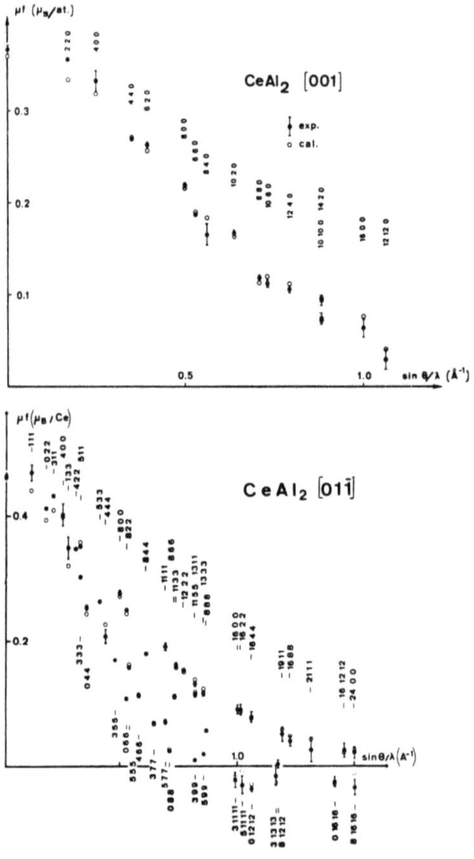


Fig. 9: Comparison of observed (full circles) and calculated (open circles) magnetic structure factors for $CeAl_2$ with field parallel to [001] and to [01$\bar{1}$][9]

The wave functions determined from the data are the following:

$$\psi_{001} = 0.945|5/2,\ 3/2> -0.328|5/2,-5/2> \quad (33)$$

$$\psi_{110} = 0.691|5/2,5/2>-0.627|5/2,1/2>-0.358|5/2,-3/2> \quad (34)$$

and,in the real space by Fourier transform of the data the anisotropy of the magnetization density projection is striking (fig. 10).

This feature represents a very important difference which exists between d electrons spin densities and f electrons magnetization densities. In the case of transition elements, the spin orbit coupling is small compared to the crystal field. The crystal field determines the shape of the density by constraining the wave function. The spin direction is practically uncoupled with the wave function and the spin density does not depend on the orientation of the spin as schematically presented in fig. 11a.

For rare earth and actinides the spin orbit is larger than the crystal field. In absence of a crystal field, as in the case of free ions, the ground state in a field (applied, exchange) is the state $|J,M=J>$ and the magnetization density is a ellipsoid of revolution with axis parallel to the field. A rotation of the field (of the moment) results in a rotation of the ellipsoid in the cell without deforming it (fig. 11b).

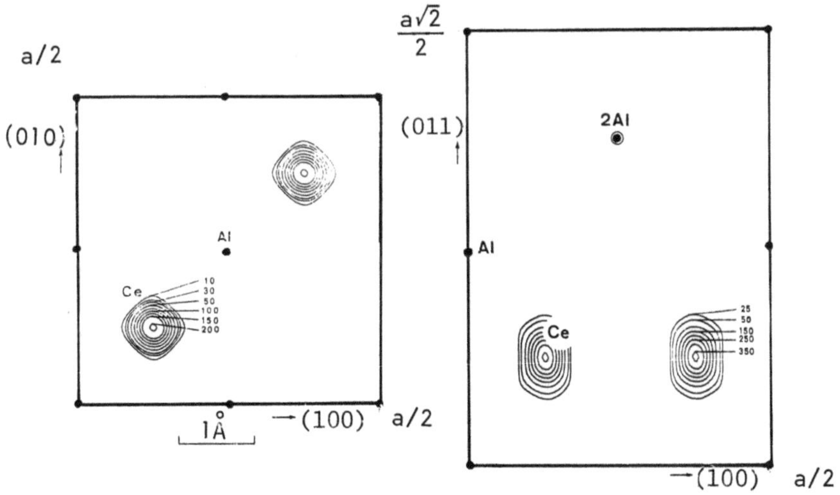


Fig. 10: Projected magnetization density with field parallel to [001] and [01$\bar{1}$] (11)

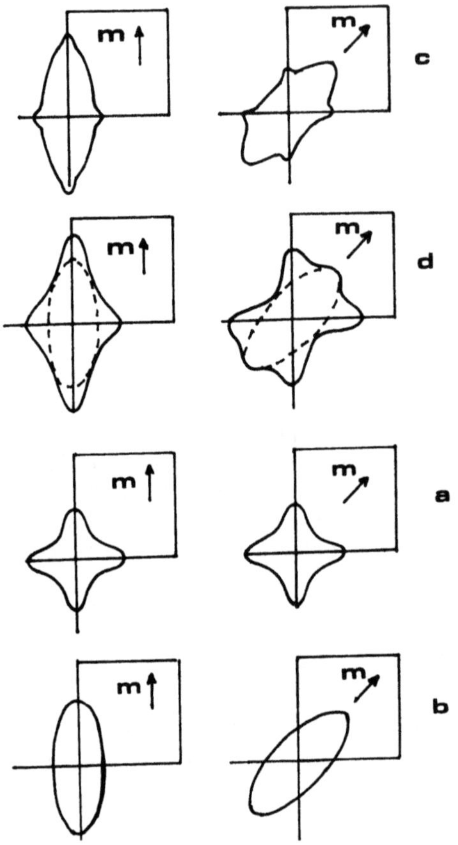


Fig. 11: Schematic representation of the magnetization density (a) spin only, (b) spin and orbit, without crystal field, (c) spin and orbit, crystal field <<exchange and applied field and (d) spin and orbit, crystal field ~ exchange and applied field.

For a crystal field small compared to the exchange coupling, which is the case in general for rare earth of the second series, the ellipsoid is slightly deformed by the crystal field. A rotation of the direction of the moments changes the direction of the ellipsoïd, changing as a consequence the way the crystal field distorts it. But the distortion remains small (fig. 11c). It is for instance the case in TmSb.

When the crystal field is as important as the exchange field as it is the case for $CeAl_2$ and often for the rare earth of the first serie, the crystal field twists the magnetization ellipsoid so much that the shape of the magnetization density is completely different for different orientations of the moment.

It is therefore possible to say that, magnetization densities including spin and orbit offer more fancy and more fun than spin densities.

Bibliography

(1) S.W. Lovesey, D.E. Rimmer, Rep. Prog. Phys. 32 (1969) 333

(2) G.H. Lander, T.O. Brun, J. Chem. Phys. 53 (1970) 1387

(3) W. Marshall, S.W. Lovesey, Theory of thermal neutron scattering, Clarendon press 1971, p. 152

(4) G.T. Trammel, Phys. Rev. 92 (1953) 1387

(5) G.H. Lander, J.F. Reddy, Proceedings of the Conference of Neutron Scattering, Gatlimburg 1976, p. 623

(6) J.X. Boucherle, D. Givord, J. Laforest, J. Schweizer, F. Tasset, Proceedings of the Colloque: La Physique des Terres Rares à l'Etat Métallique, St. Pierre de Chartreuse, 1978

(7) R.M. Moon, W.C. Koehler, D.B. McWhan, F. Holtzberg, J. Appl. Phys. 49 (1978) 2107

(8) F.A. Wedgwood, J. Phys.C, Solid State Phys. 5 (1972), 2427

(9) B. Barbara, J.X. Boucherle. J.P. Desclaux. M.F. Rossignol, J. Schweizer, Crystal Field Effects in Metals and Alloys 1977, p. 168; A. Furrer, Plenum Press, New York

(10) G.H. Lander, T.O. Brun, O. Vogt, Phys. Rev. B7 (1973) 1988

(11) J.X. Boucherle, Thèse, Université de Grenoble 1977

(12) J. Schweizer, Interpretation of the spin densities in metals and alloys; these proceedings.

THERMAL SMEARING AND CHEMICAL BONDING

Philip Coppens

Department of Chemistry, State University of New York

Buffalo, New York 14214

## 1. Introduction

As the experimentally accessible electron density is averaged over the thermal vibrations of the crystal, detailed knowledge and adequate modeling of the crystal dynamics is required if details of the experimental charge distribution are to be interpreted.

We are concerned here with the effect of thermal motion on the elastic scattering amplitudes $F(\underset{\sim}{S})$, which are the Fourier transform of the time-averaged density $\langle\rho(\underset{\sim}{r})\rangle$:

$$F(\underset{\sim}{S}) = \mathcal{F}\langle\rho(\underset{\sim}{r})\rangle \qquad (1)$$

where $\mathcal{F}$ is the Fourier Transform operator and

$$\langle\rho(\underset{\sim}{r})\rangle = \rho_{dynamic}(\underset{\sim}{r}) = \int\rho_{static}(\underset{\sim}{r}-\underset{\sim}{u})\, P(\underset{\sim}{u},\underset{\sim}{r}-\underset{\sim}{u})d\underset{\sim}{u} \qquad (2)$$

and $P(\underset{\sim}{u},\underset{\sim}{r})$ is the thermal probability distribution function at $\underset{\sim}{r}$.

When the crystal is assumed to consist of perfectly following density units such as atoms or rigid molecules (the convolution approximation) expression (2) simplifies to

$$\rho_{dynamic}(\underset{\sim}{r}) = \int\rho_{static}(\underset{\sim}{r}-\underset{\sim}{u})\, P(\underset{\sim}{u})\, d\underset{\sim}{u} \qquad (2a)$$

in the case of translations, and

$$\rho_{dynamic}(\underset{\sim}{r}) = \int \rho_{static}(\underset{\sim}{R}\,\underset{\sim}{r})\, P(\underset{\sim}{w})\, d\underset{\sim}{w} \tag{2b}$$

in the case of librational motion with angular displacement w and a rotation matrix $\underset{\sim}{R}$ defined by $\underset{\sim}{R}\,\underset{\sim}{r} = \underset{\sim}{r} + \underset{\sim}{u}_{lib}$.[1]

According to the Fourier convolution theorem

$$F(\underset{\sim}{S}) = F_{static}(\underset{\sim}{S})\, \mathcal{F}\,[P(\underset{\sim}{u})] = F_{static}(\underset{\sim}{S})\, T(\underset{\sim}{S}) \tag{3}$$

which is the well-known derivation of the X-ray temperature factor T(S).

In the aspherical atom scattering formalisms such as the multipolar expansion the static density is represented by density functions the scattering of which is multiplied by the temperature factor $T(\underset{\sim}{S})$. If $T(\underset{\sim}{S})$ is restricted to the harmonic approximation the population and nature of the density functions may easily be biased. A more general formalism should take into account higher non-harmonic terms in the potential V, expressed (in the one dimensional case) as a series in powers of the displacement coordinate Δ.

$$V = V_0 + \alpha\Delta + \beta\,\Delta^2 + \gamma\Delta^3 + \delta\Delta^4 + \ldots =$$
$$V_0 + V_0'\Delta + \frac{1}{2!}V_0''\,\Delta^2 + \frac{1}{3!}V_0'''\,\Delta^3 + \frac{1}{4!}V_0^{iv}\,\Delta^4 + \cdots \tag{3a}$$

where $V_0$ is the potential energy at the equilibrium position. At equilibrium the first derivative term is zero, unless rotation-vibration interactions are important which is generally not the case in crystals.

In most studies thermal motion has been treated in the harmonic approximation by truncating the series (1) after the term in $\Delta^2$. Inclusion of higher order terms is not unambiguous as the symmetry of the corresponding probability distribution functions in three dimensions is the same as that of the multipolar functions used in charge density analysis. Nevertheless there is hope that charge density asphericity and thermal anisotropy may be separated because: (a) the radial dependence of the thermal probability distribution function may differ from that of the bonding anisotropy, leading to different |S| dependence in scattering space, (b) the amplitudes of thermal vibration are temperature dependent while bonding effects are in first approximation temperature independent, (c) a different technique such as neutron diffraction may be used to provide independent information on thermal vibrations.

A prerequisite for any of these methods is the availability of generalized thermal motion formalisms. We will give some examples of anharmonicity and discuss generalized temperature factor formalisms which may be applied in such studies.

## 2. Anharmonicity of vibrational modes

### 2.1 High-frequency modes

The potential curves for high-frequency internal modes generally have pronounced anharmonicity. A Morse function fit for a typical molecule such as CO, shown in figure 1, is very steep towards shorter atomic separations where repulsions dominate, but shallow in the region of bond extension which is much more easily accomplished. Even though at temperatures of the diffraction experiment the excited vibrational levels are not populated ($\nu$ = 2169.8 $cm^{-1}$, vs. kT = 69.5 $cm^{-1}$ at 100 K and 218.5 $cm^{-1}$ at 300 K) the asymmetry of thermal motion is evident in density space.

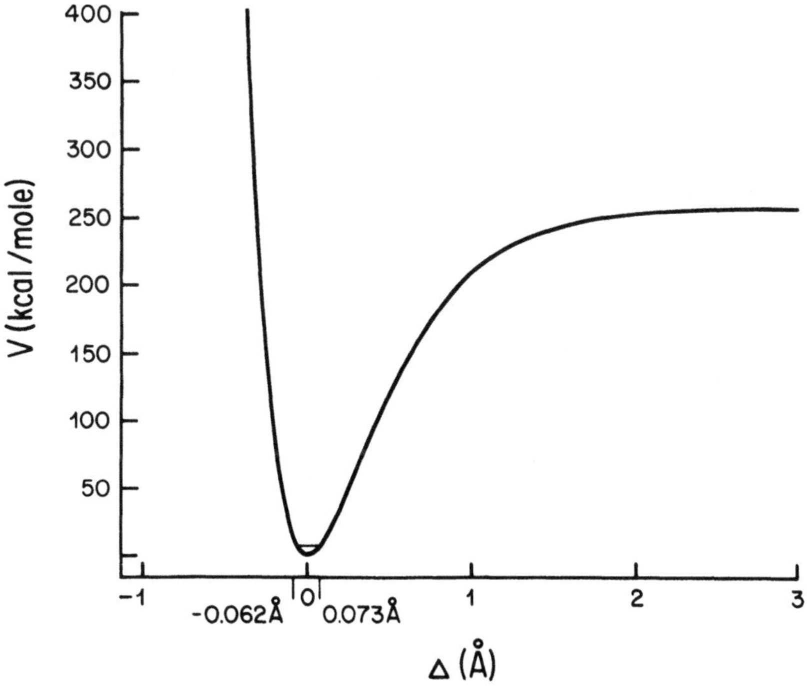


Fig. 1. Morse potential curve for the carbonmonoxide molecule. The zero level is indicated. Note the asymmetry of the curve even at the zero level.

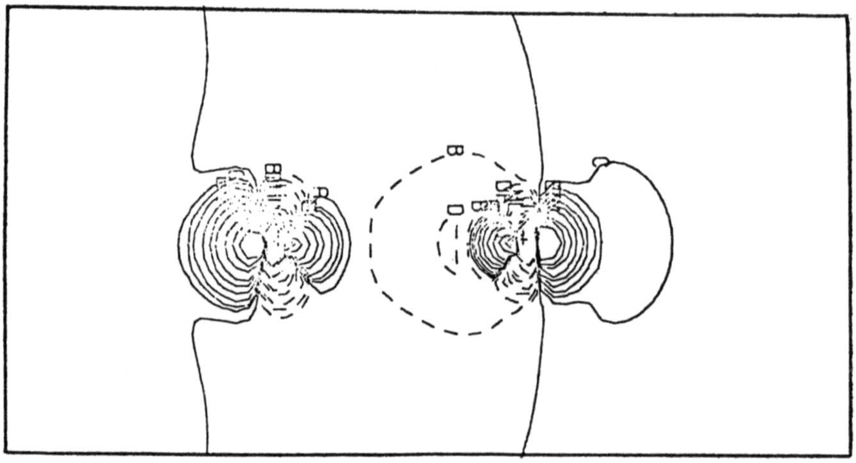


Fig. 2. Difference between thermal averaged anharmonic density and density at rest for the CO molecule according to Becker (ref. 3). Contours at $2 \times 10^{-3}$, $4 \times 10^{-3}$, $8 \times 10^{-3}$ $eau^{-3}$ (ACEG...for positive contours, BDFH...for negative contours).

Calculations by Becker[3] on BeH and CO (figure 2) show very clearly that the thermally averaged density increases more at the back of the atoms than in the bonding region.

It is a quite general phenomenon that the antisymmetric terms in the potential expression (3) tend to reduce the electron density asphericity in the thermally averaged deformation maps. A second rather dramatic example occurs for silicon. The intensity of the "forbidden" 442 reflection is due, at low temperature, to the density in the covalent Si-Si bonds. As the temperature increases away from the bonds, the intensity of the 442 reflection decreases to become zero at about 250°K where bonding effects and anharmonicity cancel. At even higher temperatures the intensity increases because anharmonicity dominates.[4] (figure 3)

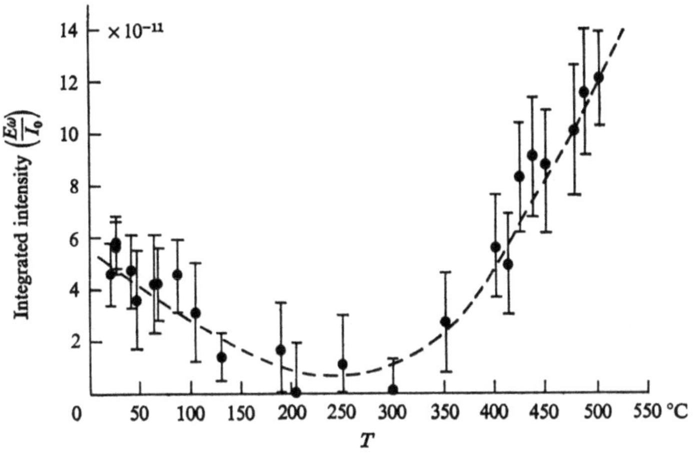


Fig.3. Si(442) integrated intensity vs. temperature according to Trucano and Batterman (ref. 4).

2.2 Low frequency modes.

The external or lattice modes of molecular crystals have much lower frequencies, as have internal modes such as ring puckering and torsional oscillations of molecular fragments. Some typical values are given in Table 1. Since these frequencies are comparable with kT the distribution over the energy levels and therefore the X-ray temperature parameters are temperature dependent, and anharmonic terms in the potential function may be as important as harmonic terms even at liquid nitrogen temperature. Two examples which we will use in model calculations are:

1) the angle bending potential in cyclobutane is of importance in the molecular ring puckering mode. It has been estimated to be approximately[6]

$$V = 15.5\ (\text{kcal/mole rad}^{-2})\ (\theta-\theta_0)^2 - 50\ (\text{kcal/mole rad}^{-3}) - (\theta-\theta_0)^3. + 40\ (\text{kcal/mole Å}^{-2})\ r_{ij}-r_{ij}{}^{\circ})^2 \qquad (4)$$

by Harel[6a]; where $r_{ij}$ is the distance between the non bonded atoms in the $C_1$-$C_2C_3$ group.

For the purpose of creating a simple model potential we convert this expression to a function of the $C_1$--$C_3$ distance only and obtain:

$$V = 59.9\ (\text{kcal/mol}\ \text{Å}^{-2})\Delta^2 + 72.6\ (\text{kcal/mole}\ \text{Å}^{-3})\ \Delta^3 \qquad (4a)$$

2) the torsional oscillations of the external phenyl rings in p-terphenyl for which a potential curve has been derived by Rietveld, Maslen and Clews (fig. 4)[7]. A reasonable two-term analytical fit to this potential curve given by

$$V = 0.7\ (\text{kcal/mol rad}^{-2})\ \Delta^2 + 41.0\ (\text{kcal/mole rad}^{-3})\ \Delta^3 \quad (5)$$

shows the preponderance of the quartic term over the quadratic harmonic contribution in this type of oscillatory motion. Application of an harmonic temperature factor in such a case would clearly be incorrect and may introduce artifacts in the multipole deformation density.

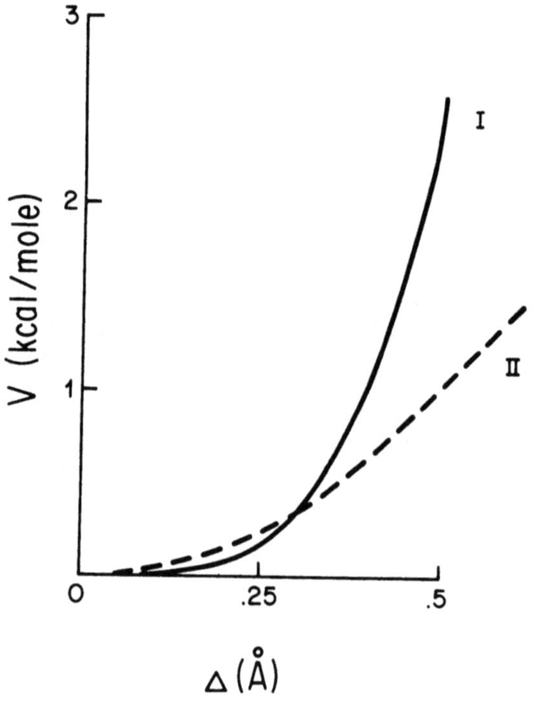


Fig. 4. Potential curve for the terminal ring in p-terphenyl(I) as in ref. 7; and harmonic potential with the same mean square displacement in high temperature limit at 100K(II). Displacements are for atom at 1Å from the libration axis.

Table 1 Some representative values of $\nu$ for molecular vibrations

| | $\nu$ ($cm^{-1}$) | ref. |
|---|---|---|
| nitrogroup torsional oscillations in p-nitroaniline | 72 | 5 |
| ring puckering in cyclobutane | ∿200 | 6[a,b] |
| rigid body librations and torsional oscillations of phenyl groups in p-terphenyl | 17,35,42,80,94 | 7 |
| rigid body librations in anthracene | 48,68,120 | 8 |
| in benzene | 35,63,69,105 | 8 |
| in imidazole | 44,63,88,104,118,176 | 9 |

## 3. Generalized X-ray and neutron temperature factors

### 3.1 The higher cumulant probability density function.[10]

The anisotropic temperature factor commonly used in crystallographic work is the Fourier transform of a trivariate normal distribution function. As we shall see in the next section this distribution function results from a harmonic potential when $kT \gg \beta$.* Without further reference to the nature of the potential function a general distribution may be expanded about the normal distribution using the Edgeworth series expansion. As described by Johnson[10] such an expansion can be expressed in Cartesian coordinate systems with multidimensional Hermite polynomials, the coefficients of which contain the cumulants $\kappa$, related to the moments $\mu$ of the probability density distribution. In the one dimensional case the relations are:

$$^1\kappa = {}^1\mu$$

$$^2\kappa = {}^2\mu-({}^1\mu)^2$$

$$^3\kappa = {}^3\mu- 3\,{}^1\mu\,{}^2\mu + 2\,({}^1\mu)^3$$

$$^4\kappa = {}^4\mu- 4\,{}^1\mu\,{}^3\mu + 6\,{}^2\mu({}^1\mu)^2-3({}^1\mu)^4-3[{}^2\mu-({}^1\mu)^2]^2 \qquad (6)$$

The second cumulant for example is the expectation value of $(x-{}^1\mu)^2$ or

---

*As the wave function for the zero level of the harmonic oscillator corresponds to a normal probability distribution the treatment is also exactly valid when only the zero level is populated.

$$\varepsilon\{(x-\mu)^2\} = \varepsilon\{x^2\} - \varepsilon\{x^1\mu\} = \varepsilon\{x^2\} - [\varepsilon\{x\}]^2$$

For a normal distribution with mean zero all but the second cumulant vanish and the model reduces to the zero-order harmonic approximation. For more general distributions expanded around their mean value expressions (6) reduce to

$$^2\kappa = {}^2\mu; \quad ^3\kappa = {}^3\mu; \quad ^4\kappa = {}^4\mu - 3(^2\mu)^2 \tag{6a}$$

The great advantage of the cumulant expansion is the simplicity of its Fourier transform, which, in tensor notation, may be written as

$$\Psi = \exp\,(i\,{}^1\kappa^j t_j + \frac{i^2}{2!}\,{}^2\kappa^{jk}\,t_j t_k + \frac{i^3}{3!}\,{}^3\kappa^{jk\ell}\,t_j t_k t_\ell + \frac{i^4}{4!}\,{}^4\kappa^{jk\ell m} t_j t_k t_\ell t_m) \tag{7}$$

where $t_j = 2\pi h_j$, and the $h_j$ are the covarient reciprocal lattice coordinates.

A disadvantage of (7) is that its terms are not easily related to physical properties such as the vibrational potential function and that the corresponding three dimensional probability distribution is extremely complicated. Even if (7) is truncated after the fourth cumulant the probability distribution function contains a term of hexacontatetrapole (64) symmetry. Thus there is no simple one-to-one relation between the terms in this distribution function and the multipoles defined by spherical harmonic functions. Its application in charge density studies using only X-ray data seems therefore of limited usefulness.

### 3.2 One particle potentials: Generalized Willis formalism.

Under the assumptions of classical statistics which are valid when $kT > (V-V_0)$ the probability distribution function for a potential function $V(\tilde{r})$describing a single oscillator is given by

$$P(\tilde{r}) = \exp\,(-V(\tilde{r})/kT)/\int\exp(-V(\tilde{r})/kT)d\tilde{r} \tag{8}$$

where the denominator represents the normalization factor. The corresponding temperature parameter is then given by the Fourier transform

$$T = \int P(\tilde{r})\,\exp\,2\pi i\;\tilde{h}.\tilde{r}\;d\tilde{r} = \frac{\int\exp-V(\tilde{r})/kT\;\exp2\pi i\;\tilde{h}.\tilde{r}\;d\tilde{r}}{\int\exp-V(\tilde{r})/kT\;d\tilde{r}} \tag{9}$$

If the crystal is treated as a system of <u>independent</u> anharmonic oscillators (an "Einstein solid"), expression (9) represents the appropriate classical temperature factor. Willis and coworkers in a number of pioneering studies have applied expression (9) to neutron diffraction data on crystals with the cubic fluorite structure,[11] while applications to hexagonal structures have been described more recently by Moss, Whiteley and Barnea.[12]

Using the same approximations a temperature factor valid in any crystal system may be derived as follows. In tensor notation the generalized three-dimensional potential function can be written as

$$V = V_0 + \alpha_j u^j + \beta_{jk} u^j u^k + \gamma_{jk\ell} u^j u^k u^\ell + \delta_{jk\ell m} u^j u^k u^\ell u^m + \cdots \quad (10)$$

where the equilibrium condition again leads to $\alpha_j = 0$.

Substituting in (9) leads to an expression which may be simplified by the assumption that the leading term is the harmonic component represented by $\beta_{jk}$, thus if

$$T \equiv N/D$$

the numerator N becomes, with $\exp-\Delta \sim (1-\Delta)$:

$$N = \int_{-\infty}^{\infty} (\exp-\beta'_{jk} u^j u^k)(1 - \alpha'_j u^j - \gamma'_{jk\ell} u^j u^k u^\ell - \delta'_{jk\ell m} u^j u^k u^\ell u^m \ldots)$$

$$\exp 2\pi i\, h_j u^j d\underset{\sim}{u} \quad (11)$$

where $\alpha' = \alpha/kT$, $\beta' = \beta/kT$ etc.

This integral is reduced to standard expressions if integration is performed in the coordinate system which diagonalizes $\beta_{jk}$ (see section 3.6 for a more general treatment). We will use the expressions as formulated recently by Chandler & Spackman.[13]

$$\int_{-\infty}^{\infty} x^n \exp(-px^2 + 2qx)\, dx = \exp\left(\frac{qx^2}{p}\right) \left(\frac{\pi}{p}\right)^{1/2} g_n(p,q) \quad (12)$$

with the following definitions of $g_n(p,q)$:

| n | $g_n(p,q)$ |
|---|---|
| 0 | 1 |
| 1 | $q/p$ |
| 2 | $1/2p + (q/p)^2$ |

$$3 \qquad 3q/2p^2 + (q/p)^3$$

$$4 \qquad 3/4p^2 + 3q^2/p^3 + (q/p)^4 \tag{12a}$$

which gives with $q_j = \pi i h_j$ and $p_j = \beta'_{jj}$

$$N = \frac{\pi^{3/2}}{|\beta'|^{1/2}} \exp-(\pi^2 h_j^2/\beta_{jj}')(1-\alpha'_j g_{1j} - [\underset{3}{\gamma'_{jjj} g_{3j}} + \underset{3}{\gamma'_{jjk} g_{2j} g_{1k}}$$
$$\underset{6}{\phantom{x}}$$

$$+ \underset{1}{\gamma'_{jk\ell} g_{1j} g_{1k} g_{1\ell}}] - [\underset{3}{\delta'_{jjjj} g_{4j}} + \underset{6}{\delta'_{jkkk} g_{1j} g_{3k}} + \underset{3}{\delta'_{jjkk} g_{2j} g_{2k}}$$

$$+ \underset{3}{\delta'_{jk\ell\ell} g_{1j} g_{1k} g_{2\ell}}] \tag{13a}$$

with the repeated summation of the tensor notation excluding the terms for which two or more indices are equal (such as j = k), and the number of terms in each summation indicated on the second line.

Similarly for the denominator (for which q in expression 12 equals zero)

$$D = \frac{\pi^{3/2}}{|\beta'|^{1/2}} [1 - \delta'_{jjjj} \frac{3}{4(\beta'_{jj})^2} - \delta'_{jjkk} \frac{1}{4\beta'_{jj}\beta'_{kk}}] \tag{13b}$$

Expressions (13) are a generalized anharmonic temperature parameter valid in any symmetry for a system of independent high-temperature oscillators. In the following sections we will address two issues: a) the relation between the Cartesian displacement coordinates in (11) and the multipole expansion, and b) the predicted temperature dependence.

### 3.4 Relation between the thermal probability density distribution and the spherical harmonic expansion.

The probability density distribution

$$P(u^j,u^k,u^\ell) = N_{norm} \exp(-\beta'_{jk}u^ju^k)(1-\alpha'_ju^j-\gamma'_{jk\ell}u^ju^ku^\ell- \delta'_{jk\ell m}u^ju^ku^\ell u^m.....) \quad (14)$$

which is the basis of expression (11) is closely related to the spherical harmonic expansion. A term such as

$$\gamma'_{jk\ell}u^ju^ku^\ell \exp(-\beta'_{jk}u^ju^k)$$

is closely related to thermally smeared density functions of the Hirshfeld type. As far as angular symmetry is concerned the expansion is redundant as illustrated by the following table:

Table 2

| Cartesian functions | # | type of spherical harmonics included |
|---|---|---|
| $x^j$ | 3 | 3(p) |
| $x^jx^k$ | 6 | s + 5 (d) |
| $x^jx^kx^\ell$ | 10 | 3(p) + 7(f) |
| $x^jx^kx^\ell x^m$ | 15 | s + 5(d) + 9(g) |

The functions of order 3 for example, span all 7 octopoles and three dipoles through the linear combinations

$$x^3 + xy^2 + xz^2 = xr^2$$
$$x^2y + y^3 + yz^2 = yr^2$$
$$x^2z + y^2z + z^3 = zr^2 \quad (15)$$

As noted previously $\alpha_j$ must be zero for an equilibrium state.

Consistency demands that the $3^d$ order functions are similarly constrained such that the combinations (15) are zero. This is not necessarily true for the s and d functions implicit in the fourth order Cartesians because terms in the potential like

$$x^3y + xy^3 + xyz^2 = xyr^2$$

have a radial dependence different from that of the second order Cartesians. But whenever these terms correlate strongly with functions already included additional constraints may be called for.

Fully constrained the p.d.f. (11) will become identical with the multipole expansion with the important reservation that it applies to all the atomic electrons rather than being restricted to the valence shell. The similarity implies however that core deformation is not detectable with X-ray data restricted to a single temperature.

### 3.5 Temperature dependence of the cubic and quartic terms in the high temperature limit

Expression (13a) directly predicts the temperature dependence of the various terms. Since $\beta'_{jj} = \beta_{jj}/kT$, $\gamma'_{jjj} = \gamma_{jjj}/kT$ etc we obtain, ignoring the first order term, for the quadratic-cubic potential:

$$T = \exp(-\pi h_j^2 kT/\beta_{jj})\{1 - i\gamma_{jjj}[kT(\frac{3\pi h_j}{2\beta_{jj}^2}) - (kT)^2(\frac{\pi h_j}{\beta_{jj}})^3] + i\gamma_{jjk}[kT\frac{\pi h_k}{2\beta_{jj}\beta_{kk}} - (kT)^2\frac{\pi^3 h_j^2 h_k}{\beta_{jj}^2\beta_{kk}}] - i\gamma_{jk\ell}(kT)^2\frac{\pi^3 h_j h_k h_\ell}{\beta_{jj}\beta_{kk}\beta_{\ell\ell}}\} \quad (16)$$

The extension to quartic terms is elaborate but straightforward. Expresion (16) present the possibility of testing the physical basis of the algorithms through multi-temperature experiments.

### 3.6 Hermite polynomial formulation of the generalized Willis formation

A general expression for the Fourier transform of the function

$$P(\underset{\sim}{u}) = \Phi(\underset{\sim}{u})\ [C_0 + C_j\ u^j + C_{jk}\ u^j u^k/2! + \ldots]$$

with $\Phi(u)$ being a trivariate normal distribution has been given by Johnson.[15] If $f(\underset{\sim}{t}) = \mathcal{F}[P(u)] = \exp(-\sigma^{jk} t_j t_k/2)$ and $\underset{\sim}{t} = 2\pi \underset{\sim}{h}$, the Fourier transform becomes

$$T = f(\underset{\sim}{t})\,[C_o + iC_j G^j(\underset{\sim}{t}) + i^2 C_{jk}\, G^{jk}(\underset{\sim}{t})/2! + \ldots] \qquad (18)$$

where the G's are three-dimensional Hermite polynomials as defined by Erdelyi.[16] The first three terms are

$$G^j = z^j$$
$$G^{jk} = z^j z^k - \sigma^{jk}$$
$$G^{jk\ell} = z^j z^k z^\ell - (z^h \sigma^{k\ell} + z^k \sigma^{j\ell} + z^\ell \sigma^{jk}) \qquad (19)$$

with $z^j = \sigma^{jk} t_k$. This equation reduces to expression (11) when $\sigma^{jk}$ represents a diagonal matrix, i.e. when the expansion is performed in the coordinate system which diagonalizes $\beta_{jk}$.

## 4. Extension to strongly anharmonic potentials

The example of the librating phenyl ring in p-terphenyl discussed in section 2 shows that the harmonic term in the potential expression may not always be the leading term. In such a case a different series expansion must be selected. If the quartic term is dominant as in (5) we get

$$P = N_{norm} \exp(-\delta'_{jk\ell m}\, u^j u^k u^\ell u^m)\,(1 - \beta'_{jk}\, u^j u^k - \gamma'_{jk\ell}\, u^j u^k u^\ell - \varepsilon'_{jk\ell mn}\, u^j u^k u^\ell u^m u^n \ldots.) \qquad (20)$$

The Fourier transform of this function is complicated and in general numerical methods may be required (see section 7.2).

## 5. X-ray scattering formalism

Following Dawson[17] atomic scattering and temperature factors can be separated into real and imaginary parts labelled with the subscripts c (centric) and a (acentric) respectively. The dynamic atomic scattering factor <f> is then equal to

$$\langle f\rangle = f_c T_c - f_a T_a + i(f_c T_a + f_a T_c) \qquad (21)$$

in which $f_c$ and $f_a$ include the scattering $F_{\ell m}$ of density functions

$Y_{\ell m}$ with $\ell$, even and odd respectively:

$$f_c = \sum_{\ell \text{ even}} P_{\ell m} F_{\ell m}; \quad f_a = \sum_{\ell \text{ odd}} P_{\ell m} F_{\ell m}$$

Expression (11) can be formulated as

$$T = T_2\,(1 - iT_3 - T_4), \text{ or}$$
$$T_c = T_2\,(1 - T_4) \text{ and}$$
$$T_a = T_2 T_3 \qquad (22)$$

Thus the cubic thermal parameter which is often the leading perturbation affects both the real and imaginary parts of the static scattering factor. As the functions $g_{nj}$ of (12) are similarly linear combinations of reciprocal space spherical harmonics, the dynamic atomic scattering is a product of static and thermal spherical harmonic functions. To a good approximation only the latter are temperature dependent.

## 6. Limitations of the classical model: Effect of quantization of the energy levels.

The classical statistics which we have employed up to this point are valid when $\Delta < kT$, i.e. for low frequency modes at moderately elevated temperatures (Figs. 5 and 6).

Using quantum-statistics the vibrational energy of a harmonic oscillator may be expressed as a function of $u = h\nu/kT$:[18]

$$E = E_0 + \frac{h\nu}{e^u - 1} = \frac{1}{2} h\nu + \frac{h\nu}{e^u - 1}$$

According to the virial theorem for a harmonic oscillator[19] $E = 2V$, $V$ being the potential energy which is also equal to $\frac{1}{2} k_e \Delta^2$. Thus,

$$\langle \Delta^2 \rangle = \frac{h\nu}{k_e}\left[\frac{1}{2} + \frac{1}{e^u - 1}\right] = \frac{h}{4\pi^2 m\nu}\left[\frac{1}{2} + \frac{1}{e^u - 1}\right] \qquad (23)$$

This function is plotted for several frequencies and a librator with the mass of the cyanuric acid molecule in figure 5.[20] A linear behavior is observed in this figure in the space where $kT \geq h\nu$,

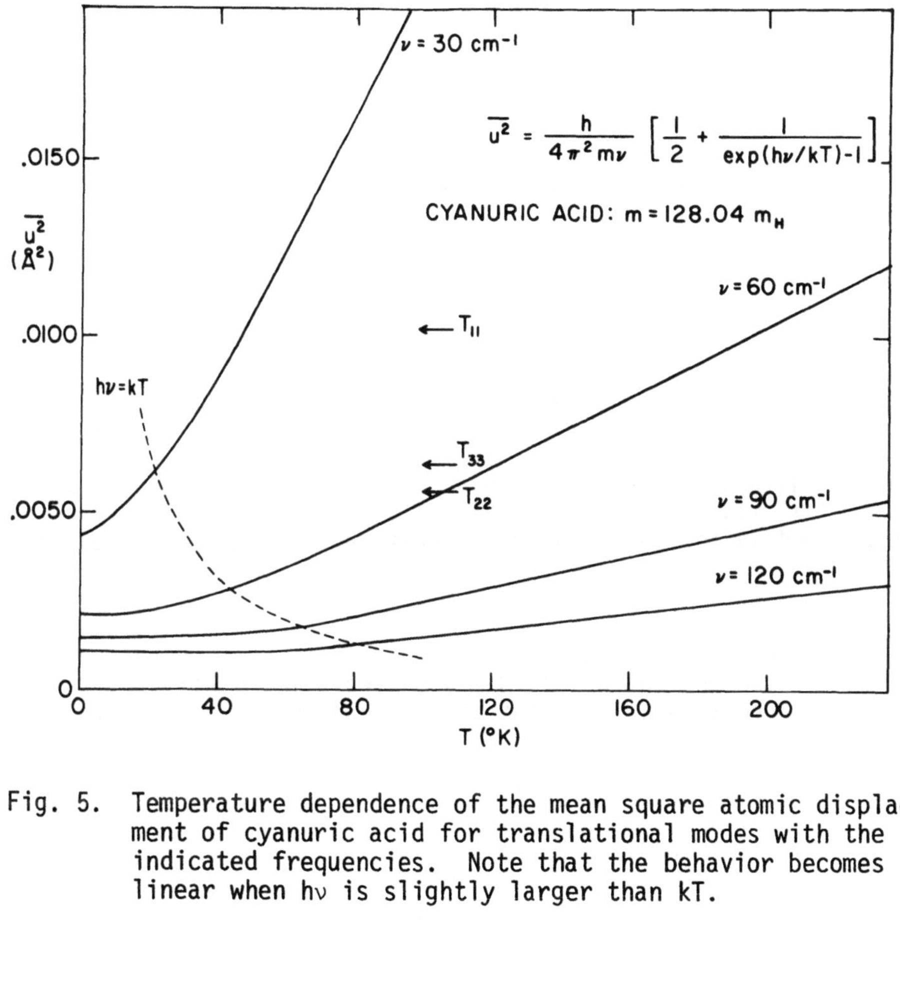


Fig. 5. Temperature dependence of the mean square atomic displacement of cyanuric acid for translational modes with the indicated frequencies. Note that the behavior becomes linear when hν is slightly larger than kT.

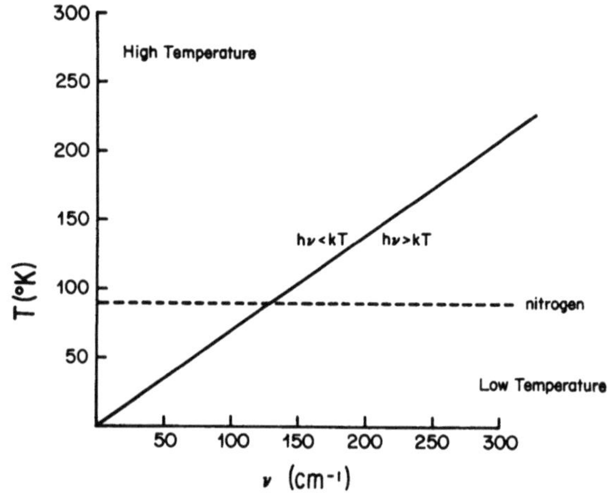


Fig. 6. High and low temperature regions in υ-T space.

while quantum effects are apparent only at low temperatures. Clearly for the low frequency librations in cyanuric acid (indicated by the arrows) the high temperature statistics are adequate at least down to liquid nitrogen temperature. This is of course not true for internal modes for which a temperature factor based on (23) must be used in the analysis of temperature dependence of thermal motion.

Can a similar treatment be applied to the anharmonic oscillator? For the anharmonic oscillator including the cubic but not the quartic term in the potential, Davidson (ref. 18, p. 119) gives the following correction term to the energy (excluding rotational terms):

$$E_{corr} = 2x \frac{h\nu_e\,(2ue^u - e^u - 1)}{(e^u - 1)^3} \tag{24}$$

where $x = \frac{(\nu_e x_e)}{\nu_e}$ is the ratio of the anharmonicity term $(\nu_e x_e)$ and the harmonic frequency in the spectroscopic expression for the energy levels of the anharmonic vibrator:

$$\varepsilon_v = h\nu_e\,(v + 1/2) + h\nu_e x_e\,(v + 1/2)^2 + \ldots \tag{25}$$

The virial theorem in the form E = 2V is not strictly valid for the anharmonic oscillator which is not homogeneous of degree 2 in the displacement coordinate $\Delta$. But if the total energy were proportional to $\Delta^3$ the relation between the total and the kinetic energy would be E = (5/2)V, not a very different result. Thus if the anharmonic term is just a perturbation, the harmonic relation E = 2V will be approximately valid. Thus

$$E_{corr} = 2\gamma\langle\Delta^3\rangle = 4x\,\frac{h\nu(2ue^u-e^u+1)}{(e^u-1)^3} = \frac{4(\nu_e x_e)\,h(2ue^u-e^u+1)}{(e^u-1)^3} \tag{26}$$

As the anharmonic term $(\nu_e x_e)$ and $\gamma$ are related, $\nu_e x_e$ being proportional to $\gamma^2$ (ref 2, p. 150), we obtain

$$\langle x^3\rangle \sim h\gamma\,\frac{2ue^u-e^u+1}{(e^u-1)^3} \tag{27}$$

A curve of the temperature dependent part of (27) is given in figure 7.

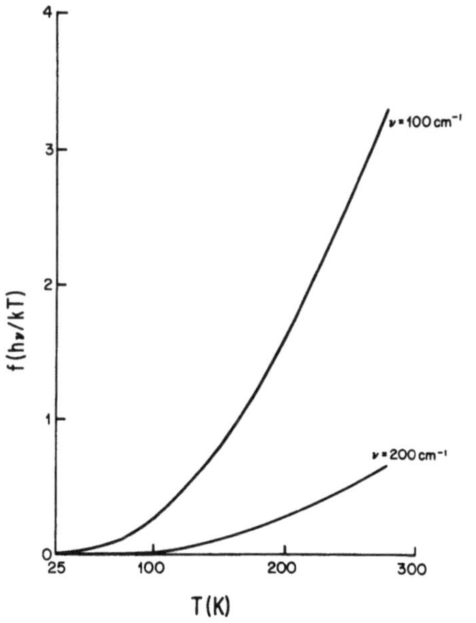


Fig. 7. Temperature dependence of $\langle x^3\rangle$ beyond its value at the zero point level as given by expression 27 for frequencies of 100 and 200 $cm^{-1}$.

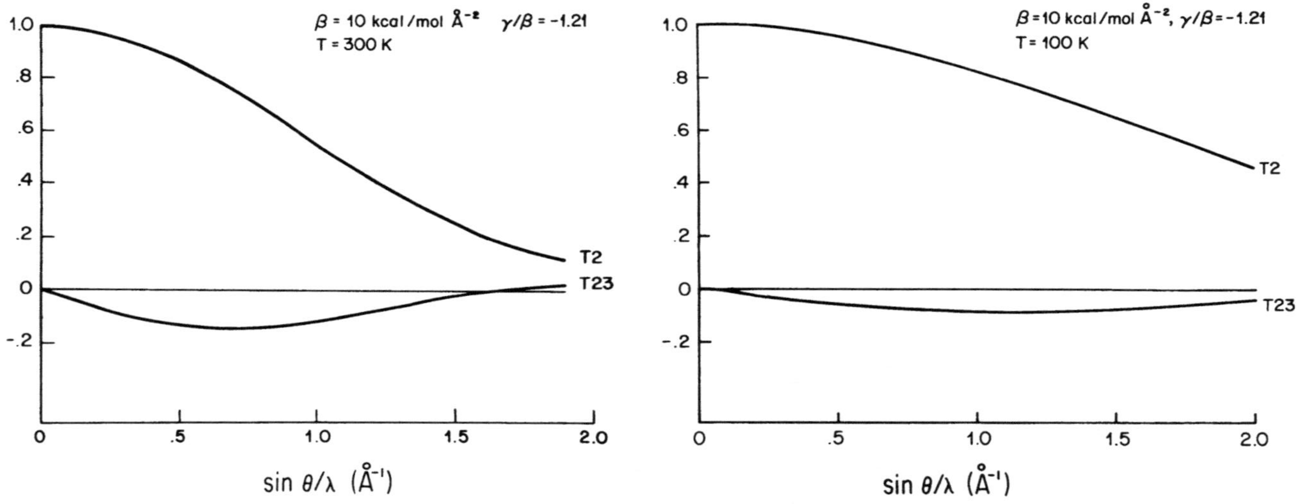


Fig. 8. Harmonic and anharmonic (cubic) components of the temperature factor for $\beta$ = 10 kcal/mol $Å^{-2}$ and $\gamma$ = 12.1 kcal/mole $Å^{-3}$ at 100 and 300°K.

## 7. Some model calculations

### 7.1 Harmonic vs Anharmonic temperature factor for bending-type vibrational mode

In order to obtain an estimate of the relative importance of the terms $T_2$ and $T_2T_3$ in expression (22) the temperature factor given by (13) was evaluated for a one-dimensional potential with the ratio of third and second order coefficients as calculated for the bending mode in cyclobutane (expression 4a).

For $\beta$ = 10 kcal/mol $\mathring{A}^{-2}$ and T = 100 and 300K, results are given in figure 8. As expected the anharmonicity is more pronounced at the higher temperature, though it is by no means negligible at 100K. The magnitude of the product $T_2T_3$ goes through a maximum which lies at about 0.7 $\mathring{A}^{-1}$ at 300K and shifts to larger reciprocal radii when $T_2$ increases with decreasing temperature.

The values of $T_2$ and $T_2T_3$ at $\sin\theta/\lambda$ = 0.8 $\mathring{A}^{-1}$ as a function of $\beta$ are plotted in figure 9 which for a given $\gamma/\beta$ ratio shows the decreasing importance of anharmonicity, with increasing depth of the potential well. The ring puckering mode in cyclobutane has a frequency of about 200 $cm^{-1}$. For an oscillator with the same mass fig. 9 corresponds to frequencies of 35 - 160 $cm^{-1}$.

### 7.2 Calculations for a quartic terphenyl-type potential

The potential function for the torsional oscillations of the external phenyl rings in p-terphenyl (fig. 4, expression 5) is almost purely quartic. The thermal probability density function at 300K for an oscillator at 1Å of the libration axis is shown in figure 10, together with the p.d.f. of an oscillator in a harmonic field with the same mean square displacement. The differences in the distributions are very clear, the quartic function falling off less rapidly near the equilibrium position and faster at larger distances. It is clear that the choice of an incorrectly restricted thermal parameter in such a case may lead to serious artifacts in the "deformation" density. The corresponding temperature factors (obtained by numerical Fourier transform in the quartic case) at 100 and at 300K are shown in fig. 11. At high scattering angles part of the quartic distribution scatters out of phase and the temperature factor becomes negative. In general differences become pronounced at high angles. We conclude that, as more high-angle data are included in electron density studies, anharmonic effects will increasingly have to be accounted for in the scattering formalism.

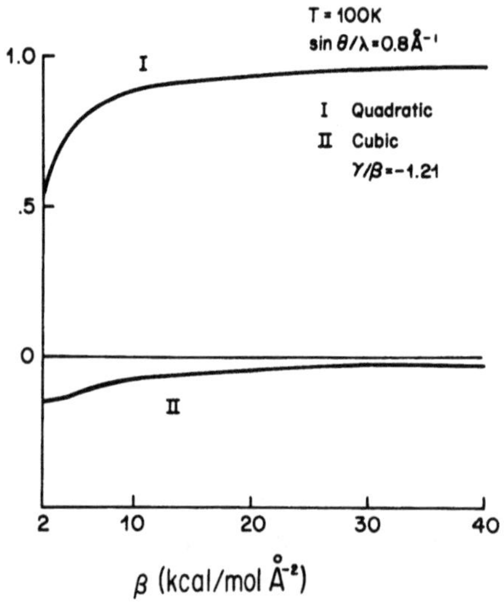


Fig. 9. Dependence of harmonic and anharmonic (cubic) components of the temperature factor as a function of $\beta$ with $\gamma/\beta$ = -1.21 at $\sin\theta/\lambda$ = 0.8 $Å^{-1}$ and T = 100 K.

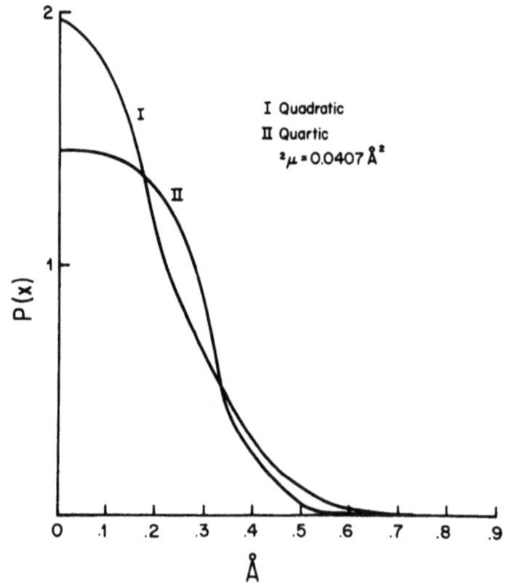


Fig. 10 Quadratic and quartic probability density distributions with $\langle\Delta\rangle$ = 0.0406 $Å^2$ (as in terphenyl at 300K for an atom at 1Å from the libration axis).

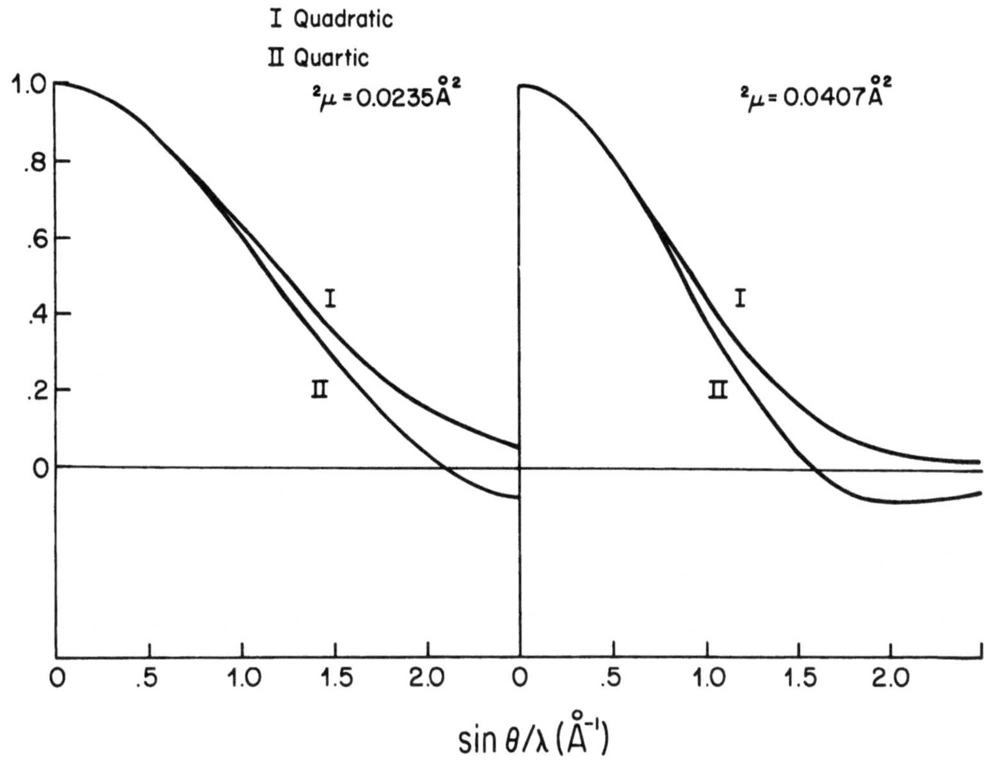


Fig. 11. Quartic and quadratic temperature factors corresponding to probability density distributions as in figure 10: a) $\langle\Delta^2\rangle = 0.0235\text{Å}^2$; b) $\langle\Delta^2\rangle = 0.0407\text{Å}^2$.

Acknowledgements

The author would like to thank Mr. Eldad Coppens for programming and performing the model calculations described in this article. Financial support by the National Science Foundation (CHE 7613342A01) is gratefully acknowledged.

References

1. E. D. Stevens, J. Rys and P. Coppens, Acta Cryst A33, 333 (1977).

2. I. N. Levine, Quantum Chemistry Volume II: Molecular Spectroscopy, Allyn & Bacon, Boston (1970).

3. P. Becker, Physica Scripta 15, 119 (1977).

4. P. Trucano & B. W. Batterman, Phys. Rev., 6, 3659 (1972).

5. K. N. Trueblood in Molecular Dynamics and Structure of Solids, J. J. Carter & J. J. Rush, Eds., National Bureau of Standards Special Publication 301, p. 371 (1969).

6. a) M. Harel, Thesis Weizmann Institute of Science, Israel, 1974.

   b) T. Ueda & T. J. Shimanouchi, J. Chem. Phys., 49, 470 (1968).

7. H. M. Rietveld, E. N. Maslen & C.J.B. Clews, Acta Cryst B26, 693 (1970).

8. A. Fruhling, Ann. Phys. Paris, 6, 401 (1951).

9. C. Scheringer, Paper presented at VIII IUCr Congress, Stony Brook, N.Y. (1969).

10. C. K. Johnson in Thermal Neutron Diffraction, B.T.M. Willis, Ed., Oxford University Press, (1970).

11. B.T.M. Willis, Acta Cryst, A25, 277 (1969); B.T.M. Willis in Thermal Neutron Diffraction, B.T.M. Willis, Ed., Oxford University Press; B. Dawson in Advances in Structure Research by Diffraction Methods, Vol. 6, W. Hoppe & R. Mason, Eds., Pergamon Press (1975).

12. a) G. Moss, Thesis, University of Melbourne (1977).

    b) B. Whiteley, G. Moss & Z. Barnea, Acta Cryst., A34, 130 (1978).

13. G. S. Chandler & M. A. Spackman, Acta Cryst. A34, 341 (1978).

14. F. L. Hirshfeld, Isr. J. of Chem., 16, 226 (1977).

15. C. K. Johnson, ACA Program and Abstracts, Series I, Tulane University, Winter Meeting, p. 60 (1970).

16. A. Erdelyi (Editor), Bateman Manuscript Project, Higher Transcendental Functions, Vol. 11, p. 264ff, McGraw-Hill, New York (1953).

17. B. Dawson, Proc. Roy. Soc. A298, 255 (1967).

18. N. Davidson, Statistical Mechanics, McGraw-Hill, New York (1962).

19. I. N. Levine, Quantum Chemistry, Allyn & Bacon, Inc., Boston, 2nd Ed. (1974).

20. P. Coppens & A. Vos, Acta Cryst B27, 146 (1971).

ELASTIC X-RAY SCATTERING FROM SOLIDS CONTAINING NON-RIGID PSEUDOATOMS

S.W. Wilkins

CSIRO, Division of Chemical Physics, PO Box 160, Clayton

Victoria, Australia 3168

Background : March, Wilkins, Acta Cryst. (1978), A34, 19-26, referred to as MW, and in particular page 21.

Problems and questions.

1. Briefly note what you think is meant by non-rigid pseudoatoms and what you think are some of the consequences for X-Ray scattering ?

2. Give a schematic illustration of dynamic deformation of a pseudoatom as $\{\underset{\sim}{u}_{\ell'\kappa'}\}$ changes from $\{0\}$ to a finite set of values $\{\underset{\sim}{u}^{o}_{\ell'\kappa'}\}$. Use a set of axes with fixed directions in space and located on the nucleus of the pseudoatom in question.

3. Illustrate the change in electron density

$$\Delta\sigma(\underset{\sim}{r}-\underset{\sim}{R}_{\ell\kappa},\{\underset{\sim}{u}^{o}_{\ell'\kappa'}\}) = \sigma(\underset{\sim}{r}-\underset{\sim}{R}_{\ell\kappa},\{\underset{\sim}{u}^{o}_{\ell'\kappa'}\}) - \sigma(\underset{\sim}{r}-\underset{\sim}{R}_{\ell\kappa},\{0\})$$

due to the deformation of the given pseudoatom. Use the same set of axes as in question 2. Note that the first argument in the σ's is constant since we are illustrating effects relative to a set of axes fixed on the nucleus.

4. For a function $f(\underset{\sim}{r})$, fill in the tableau for the odd/even and real/imaginary character of its Fourier transform $F(\underset{\sim}{k})$

| $f(\underset{\sim}{r})$ | even | odd |
|---|---|---|
| real | F real, even | |
| imaginary | | |

5. Using the expression for $\Delta\sigma$ in question 3 and the result of question 4, discuss the symmetry and complex nature of $\Delta\sigma(\mathbf{k})$, the Fourier transform of $\Delta\sigma(\mathbf{r}-\mathbf{R}_{\ell\kappa})$. What is the predominant character of the $\Delta\sigma(\mathbf{k})$ you have assumed ? Note that $\Delta\sigma(\mathbf{k})$ is closely related to the Born $\boldsymbol{\beta}$ parameter defined in eqn (5) of MW. Discuss the nature of $\boldsymbol{\beta}$ defined there.

6. By Fourier transforming eqn (3) of MW with respect to the first argument and using eqns (4) and (5), show that the X-Ray unit cell structure factor appropriate to coherent scattering is given by :

$$F(\mathbf{k},\{\mathbf{u}_{\ell'\kappa'}\}) = \sum_{\kappa} f^{o}_{\kappa}(\mathbf{k})\ [1 + \sum_{\ell'\kappa'} \mathbf{u}_{\ell'\kappa'}\cdot\boldsymbol{\beta}(\ell'\ell,\kappa'\kappa,\mathbf{k})] \times \exp(i\mathbf{k}\cdot\mathbf{u}_{\ell\kappa})\ \exp(i\mathbf{k}\cdot\mathbf{R}_{\ell\kappa}) \qquad (12)$$

Answers

In many cases there is no unique answer to a given question, however, suitable answers are given below.

1. See sections 1, 2, 5, 6 in MW.

2.

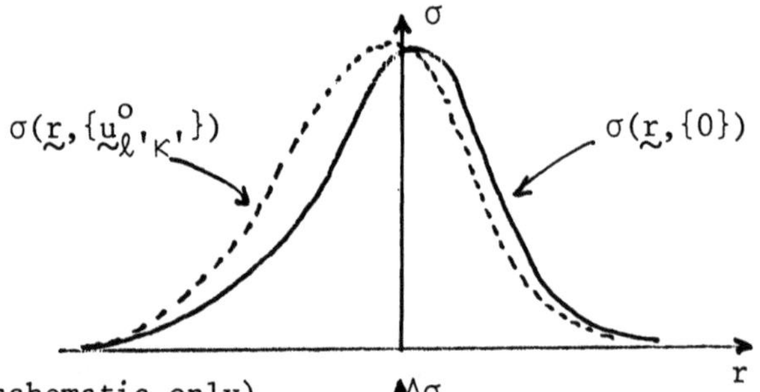


3. (schematic only)

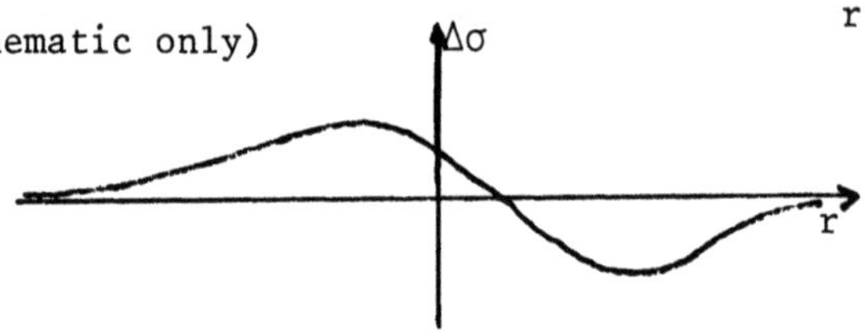


Note :$\Delta\sigma$ need not pass through origin since one can have breathing-type modes of deformation.

4.

| $f(\mathbf{r})$ | even | odd |
|---|---|---|
| real | real and even | imag. and odd |
| Imaginary | Imag. and even | Real and odd |

5. From the sketch in question 3, it is clear that for the drawn there $\Delta\sigma(\mathbf{k})$ will be essentially imaginary and odd but with a small real and even component. The quantity $\Delta\sigma(\mathbf{r})$ when divided by an appropriate small displacement $\Delta u$ is a finite-difference estimate of $\boldsymbol{\beta}$. The Born parameter $\boldsymbol{\beta}(\mathbf{k})$ gives the ratio of the rate of change of the scattering factor at $\mathbf{k}$ as an atom is moved, relative to $f^{o}(\mathbf{k})$ see section 3 in MW.

6. Answer is contained in the question. Point to note is that the Fourier transformation is with respect to the first argument $\underset{\sim}{r} - \underset{\sim}{R}_{\ell\kappa} - \underset{\sim}{u}_{\ell\kappa}$ only.

7.

$$F(\underset{\sim}{k}) = f^{o}(\underset{\sim}{k})[\ 1 + \langle \exp(i\underset{\sim}{k}.\underset{\sim}{u}_{\ell})\underset{\sim}{u}_{\ell'}\rangle.]\underset{\sim}{\beta}(\ell'\ell,\underset{\sim}{k})$$

for the case where there is only one atom per unit cell. Assuming no correlation between atomic displacements on different sites (ie Einstein model) and using the given identity (appendix of MW)

$$F(\underset{\sim}{k}) = f^{o}(\underset{\sim}{k})\langle \exp(i\underset{\sim}{k}.\underset{\sim}{u})\rangle[\ 1 + i\langle u^{2}\rangle(\underset{\sim}{k}.\underset{\sim}{\beta})]$$

so that the elastic scattering intensity is given by :

$$I = \langle F\rangle^{*}\langle F\rangle = (f^{o}\langle \exp(i\underset{\sim}{k}.\underset{\sim}{u})\rangle)^{*}\ (f^{o}\langle \exp(i\underset{\sim}{k}.\underset{\sim}{u})\rangle\ )$$
$$[1 - i\langle u^{2}\rangle(\underset{\sim}{k}.\underset{\sim}{\beta}^{*} - \underset{\sim}{k}.\underset{\sim}{\beta}) + O(\beta^{2})$$

and is to be evaluated at the Bragg reflections. From this result it can be seen that the intensity consists of the thermally smeared scattering factor multiplied by an extra term (in square brackets) due to dynamic deformation. One can immediately see that, within the given approximations, one only gets a dynamic deformation contribution to the intensity from the imaginary and odd (rigid shell deformation) component of $\underset{\sim}{\beta}$.

One can then see that to a rough approximation the deformation contribution will act like a modification to the usual Debye-Waller factor, since $\beta \sim iak$ to a first approximation, where a is constant.

# VIBRATIONAL STUDIES OF X-RAY MOLECULAR FORM FACTORS AND INTENSITIES

Joel Epstein and Robert F. Stewart
Dept. of Chemistry, Carnegie-Mellon University
4400 Fifth Ave., Pittsburgh, Pa. 15213 USA

## ABSTRACT

The vibrational average of $I_{el}^{XR}(\underset{\sim}{S}\ ;\ \underset{\sim}{Q})$ is an observable. It is not clear that a corresponding vibrational average of $F(\underset{\sim}{S}\ ;\ \underset{\sim}{Q})$ is an observable. Vibrationally averaged X-Ray and electron scattering intensities are determined theoretically for the diatomic molecules $N_2(^1\Sigma_g^+)$, CO $(^1\Sigma_g^+)$, BF $(^1\Sigma_g^+)$ and FH $(^1\Sigma_g^+)$. These are averages for ground vibrational states. The intensities are also modeled with generalized X-Ray scattering factors (pseudoatoms). It is found that anharmonic effects are about ten times larger than non-rigid effects. Generalized X-Ray scattering factors are extracted from $\langle F(\underset{\sim}{S}\ ;\ \underset{\sim}{R})\rangle$ and compared to the same functions from $F(\underset{\sim}{S}\ ;\ \underset{\sim}{R}_e)$. If anharmonic terms are included in the pseudoatom model, then deconvolution is found to be excellent. Vibrational force constants for the above diatomics are determined from [2|2] rigid pseudoatoms. In all cases $k_e$ is too large (by as much as 58%). The $\{\partial f_a/\partial R\}$ and $\{\partial f_b/\partial R\}$ functions necessarily give the correct $k_e$ when included in the pseudo atom model.

## INTRODUCTION

Although the Thomson cross section for the scattering of X-Rays by nuclei is much smaller than for electrons ($Z^4 m^{-2}$, where m is the mass of the nucleus in units of the electron rest mass), the motion of the nuclei can have a profound effect on the X-Ray scattering intensity. The nuclear motion influences the motion of the electrons which gives rise to a change in the electron charge density. The corresponding molecular or crystal form factor will change when the nuclei move.

In this lecture we will assume that the electron density for a particular nuclear configuration, $\underset{\sim}{Q}$, can be correctly given with an adiabatic wavefunction. In this case the many-electron, many-nuclear wavefunction

$$\Psi(\underset{\sim}{X},\underset{\sim}{Q}) = \sum_k \chi_k(\underset{\sim}{Q})\ \phi_k(\underset{\sim}{X},\underset{\sim}{Q}) \qquad (1)$$

where $\phi_k(\underset{\sim}{X},\underset{\sim}{Q})$ is the electronic wavefunction of state k that parametrically depends on $\underset{\sim}{Q}$. The coordinate $\underset{\sim}{X}$ refers to the space and spin of all the electrons. $\chi_k(\underset{\sim}{Q})$ is the vibrational wavefunction for the crystal; for a free molecule, $\chi_k(\underset{\sim}{Q})$ includes rotational states. $\phi_k(\underset{\sim}{X},\underset{\sim}{Q})$ for $\underset{\sim}{Q} = \underset{\sim}{Q}_e$ (the equilibrium configuration) and for $\underset{\sim}{Q}$'s near $\underset{\sim}{Q}_e$, continuously changes but is assumed to have negligible mixing of other electronic states. Fot small coupling of different electronic states, the $\chi_k(\underset{\sim}{Q})$ are solutions to the Schrödinger equation :

$$[\ T_N + U_N(\underset{\sim}{Q}) - E\ ]\chi_k(\underset{\sim}{Q}) = 0 \qquad (2)$$

where $T_N$ is the nuclear kinetic energy operator and $U_N(\underset{\sim}{Q})$ the potential function :

$$U_N(\underset{\sim}{Q}) = \Phi_k(\underset{\sim}{Q}) + \int \phi_k^* \, T_N \, \phi_k \, d^{4N}\underset{\sim}{X} \qquad (3)$$

In eq. (3), $\Phi_k(\underset{\sim}{Q})$ is the total electronic energy.

We restrict our attention to the ground electronic state k=0 with associated vibrational wavefunctions $\chi_{on}(\underset{\sim}{Q})$. (At this point we suppress the subscript 0 for $\chi$ and write $\chi_n(\underset{\sim}{Q})$.)

The one electron density function for the adiabatic wavefunction at some $\underset{\sim}{Q}$ near $\underset{\sim}{Q}_e$, is :

$$\rho(\underset{\sim}{r}\ ;\ \underset{\sim}{Q}) = N \int \phi_o^*(\underset{\sim}{X},\underset{\sim}{Q})\ \phi_o(\underset{\sim}{X},\underset{\sim}{Q})\ d^{4N-3}\underset{\sim}{X} \qquad (4)$$

where the integration is over all electrons but one and all spins. The corresponding X-Ray form factor is :

$$F(\underset{\sim}{S}\ ;\ \underset{\sim}{Q}) = \int \rho(\underset{\sim}{r}\ ;\ \underset{\sim}{Q})\ \exp(i\underset{\sim}{S}.\underset{\sim}{r})\ d\underset{\sim}{r} \qquad (5)$$

and the elastic X-Ray scattering intensity

$$I^{XR}_{el}(\tilde{S} ; \tilde{Q}) = F^*(\tilde{S} ; \tilde{Q}) \, F(\tilde{S} ; \tilde{Q}) \qquad (6)$$

In eq.6, the Thomson cross-section for classical scattering by an electron is incorporated into $I^{XR}_{el}$.

The observed X-Ray intensities for molecules or crystals are a canonical ensemble average of $I^{XR}_{el}$ over the states for $\tilde{Q}$ (ref.1) :

$$I^{XR}_{av}(\tilde{S}) = \sum_n W_n \int \chi^*_n(\tilde{Q}) I^{XR}_{el}(\tilde{S} ; \tilde{Q}) \chi_n(\tilde{Q}) \, d^{3P}\tilde{Q} \qquad (7)$$

for P nuclei. The weigth factor in (7) is :

$$W_n = \exp(-E_n/kT) \, / \, [\sum_n \exp(-E_n/kT)] \qquad (8)$$

Equation (7) represents what is fundamentally measured in most Bragg diffraction experiments for which kinematic diffraction conditions are applicable. $I^{XR}_{av}(\tilde{S})$ contains inelastic components that represent coupling of the phonons or vibrons with the scattered X-Ray by virtue of the response of $\rho(\tilde{r} ; \tilde{Q})$ to $\tilde{Q}$. Those inelastic components are very small energy differences compared to the incident X-Ray frequency. The inelastic scattering contribution is often called thermal diffuse scattering. It is also clear that (7) is not related to the vibrational average of the one-electron density function in any simple way. The task is to seek a way to unscramble (7) into an average of F, which is then squared as opposed to the average of $|F^2|$.

Born[1] published an explicit evaluation of (7) with an harmonic oscillator model for the $\chi_n(\tilde{Q})$ and a rigid pseudoatom model for $F(\tilde{S} ; \tilde{Q})$. The final result :

$$I_{av} = \sum_p \sum_{p'} F_p F_{p'} \prod_j \exp \{- \tfrac{1}{2} (\mu^j_{pp'})^2 \, \bar{\varepsilon}_j \, / \, \omega^2_j \} \qquad (9)$$

where

$$F_p = f_p(\tilde{S} ; \tilde{Q}^\circ) \exp(i\tilde{S}.\tilde{Q}^\circ_p) \qquad (10)$$

and

$$(\mu^j_{pp'})^2 = S^2[(e^j_{ps})^2 + (e^j_{p's})^2 - 2 \, (e^j_{ps} \, e^j_{p's})] \qquad (11)$$

In (10) $f_p(\tilde{S} ; \tilde{Q}^\circ)$ is a generalized scattering factor. For (11)

$$e^j_{ps} = \frac{\tilde{S}}{S} . \, \tilde{e}^j_p \, / \sqrt{m_p}$$

If the cross term in (11) can be neglected, then (9) factors into the usual structure factor expression. The tensor components for the mean square amplitude of motion for atom p,

$$U_{p,\alpha\beta} = m_p^{-1} \sum_j \bar{\varepsilon}_j \, e^j_{p\alpha} \, e^j_{p\beta} \, / \, \omega_j^2 \qquad (12)$$

$$\alpha,\beta = x,y,z$$

We then have,

$$I_{av} = \sum_p \sum_{p'} F_p F^*_{p'} \exp(-1/2 \; \underset{\sim}{S}^t U_p \underset{\sim}{S}) \exp(-1/2 \; \underset{\sim}{S}^t U_{p'} \underset{\sim}{S})$$

$$= |\sum_p F_p \exp(-1/2 \; \underset{\sim}{S}^t U_p \underset{\sim}{S})|^2 \qquad (13)$$

Born[1] defines a scattering matrix

$$S_{pp'} = \sum_j e^j_{p\alpha} \, (\bar{\varepsilon}_j/\omega_j^2) \, e^j_{p'\beta}$$

$$= \frac{\hbar}{2} \, D_{pp'}^{-1/2} \coth\left(\frac{\hbar \, D_{pp'}^{1/2}}{kT}\right) \qquad (14)$$

where $D_{pp'}$ is the dynamical matrix of force constants. In terms of the scattering matrix $S_{pp'}$, the cross term in (9) (and shown in (11)) can be written :

$$\ddot{M}_{pp'} = (m_p m_{p'})^{-1/2} \; \underset{\sim}{S}^t S_{pp'} \underset{\sim}{S} \qquad (15)$$

so that (9) can be written as :

$$I_{av} = \sum_p \sum_{p'} F_p F^*_{p'} \exp(-1/2 \; \underset{\sim}{S}^t U_p \underset{\sim}{S}) \exp(-1/2 \; \underset{\sim}{S}^t U_{p'} \underset{\sim}{S}) \exp(\ddot{M}_{pp'}) \qquad (16)$$

When $\exp(\ddot{M}_{pp'})$ is expanded in power series, we get the result,

$$I_{av} = \sum_k \sum_p \sum_{p'} F_p \, [\underset{\sim}{S}^t S_{pp'} \underset{\sim}{S} \, / (m_p m_{p'})^{1/2}]^k / k! \; F^*_{p'} \qquad (17)$$

where

$$F_p = f_p(\underset{\sim}{S} \; ; \; \underset{\sim}{Q}^\circ) \exp(i \underset{\sim}{S} . \underset{\sim}{Q}^\circ_p) \exp(-1/2 \; \underset{\sim}{S}^t U_p \underset{\sim}{S})$$

In (17) for k=0, one has pure Bragg scattering and the higher terms represent TDS. In going from a large molecule to a crystal lattice, $S_{pp'}$ becomes a periodic function of $\underset{\sim}{S}$, where atoms p and p' are equivalent, but in different unit cells.

Thus with harmonic theory and a rigid pseudoatom model one can use (17) in an attempt to reduce $I_{av}$ to a vibrational average of F and then squared. The k $\geq$ 1 terms in (17) are associated with 1-phonon, 2-phonon... scattering for k=1,2... If the inelastic, phonon induced scattering can be subtracted from the Bragg peak,

$$I_{av}(\text{corrected}) = I_{Bragg}(\underset{\sim}{S}) = |F_{Bragg}(\underset{\sim}{S})|^2 \tag{18}$$

where $\underset{\sim}{S}=2\pi\underset{\sim}{H}$ for a crystal lattice. If we work backwards from (18) or (17) with k=0, we discover that :

$$I_{Bragg}(\underset{\sim}{S}) = | \sum_n W_n \int \chi_n^*(\underset{\sim}{Q}) F(\underset{\sim}{S} ; \underset{\sim}{Q}) \chi_n(\underset{\sim}{Q})\, d^{3P}\underset{\sim}{Q}|^2 \tag{19}$$

where $W_n$ is given by (8), $\chi_n(\underset{\sim}{Q})$ are the harmonic oscillators and $F(\underset{\sim}{S} ; \underset{\sim}{Q})$ is for rigid pseudoatoms. Eq. (19) is a curious result. I don't know if it is applicable to full adiabatic theory where in general $\chi_n(\underset{\sim}{Q})$ are not harmonic oscillator functions and $F(\underset{\sim}{S} ; \underset{\sim}{Q})$ is given more exactly by (5). An explicit evaluation of (7) in full adiabatic theory is immensely difficult. Vibrational wavefunctions for anharmonic potentials do not lead themselves to normal coordinate analysis. If $F(\underset{\sim}{S} ; \underset{\sim}{Q})$ is given with generalized X-Ray scattering factors to high accuracy, then $f_p(\underset{\sim}{K} ; \underset{\sim}{Q})$ must be known as a function of $\underset{\sim}{Q}$.

In the zero temperature limit eq. (7) is :

$$I_{av} = \langle\chi_o|FF^*|\chi_o\rangle \qquad T \to 0 \tag{20}$$

We can rewrite (20), for a complete set of vibrational wavefunctions,

$$I_{av} = \sum_{n'} \langle\chi_o|F^*|\chi_{n'}\rangle \langle\chi_{n'}|F|\chi_o\rangle \tag{21}$$

For n' = 0

$$I_{Bragg} = |\langle\chi_o|F|\chi_o\rangle|^2 \tag{22}$$

and for n' $\neq$ 0

$$I_{TDS} = \sum_{n'\neq 0} |\langle\chi_o|F|\chi_{n'}\rangle|^2 \tag{23}$$

For the rest of this lecture we will evaluate (20) and (22) (and other properties of $\langle\chi_o|F|\chi_o\rangle$ ) for small diatomic molecules in order to learn something of anharmonic contributions and of the breakdown of the rigid pseudoatom approximation. In this way we isolate the problem to a single vibration, in zero pint motion, in diatomic molecules for which rather good $\rho(\underset{\sim}{r} ; \underset{\sim}{Q})$ can be obtained for $\underset{\sim}{Q}$ near $\underset{\sim}{Q}_{eq}$.

## THE PSEUDOATOM MODEL

For pseudoatoms we use a finite set of multipoles. Each set is centered on the nuclei. For diatomics there are only two nuclear centers. By methods described previously[2,3] a unique decomposition, for a given finite multipole expansion, of $\rho_{mol}(\underset{\sim}{r} ; \underset{\sim}{Q})$ can be determined. We designate the expansion [J|K] where J is the largest multipole for center $\underline{a}$ and K for center $\underline{b}$. The generalized X-Ray scattering factors, $f_{a,j}(\underset{\sim}{S} ; \underset{\sim}{Q})$ are determined directly by a solution to functional equations. These functions comprise a minimum mean square fit to the molecular form factor. In superposition the pseudoatoms obey all averages of the form

$$\int \rho_{mol}(\underset{\sim}{r} ; \underset{\sim}{Q})\, h(r_a)\, P_j(\cos\theta_a)\, d\underset{\sim}{r} \qquad j \leqslant J$$

$$\int \rho_{mol}(\underset{\sim}{r} ; \underset{\sim}{Q})\, g(r_b)\, P_k(\cos\theta_b)\, d\underset{\sim}{r} \qquad k \leqslant K$$

where $h(r_a)$ and $g(r_b)$ are arbitrary functions of $r_a$ and $r_b$ for which the integral exists. In previous work[3] we found that [2|2] expansions reproduce the gas phase, elastically scattered X-Ray intensity with a maximum relative error fit to the molecular form factor of 0.2%. For most vibrational studies [2|2] expansions (6 functions for a heteronuclear diatomic) are adequate.

The procedure is to determine $f_{a,j}(\underset{\sim}{S} ; R)$ and $f_{b,k}(\underset{\sim}{S} ; R)$† from several $F_{mol}(\underset{\sim}{S} ; R)$, which are determined from theoretical wavefunctions of the diatomic. Once the $\{f_a(\underset{\sim}{S} ; R)\}$ and $\{f_b(\underset{\sim}{S} ; R)\}$ are found, they can be expanded about the theoretical $R_e$ with a Taylor expansion.

$$f_{a,j}(\underset{\sim}{S} ; R) = f_{a,j}(\underset{\sim}{S} ; R_e) + \left(\frac{\partial f_{a,j}}{\partial R}\right)_{R_e}(R-R_e) + \frac{1}{2}\left(\frac{\partial^2 f_{a,j}}{\partial R^2}\right)_{R_e}(R-R_e)^2 + \ldots \qquad (24)$$

We take the leading term, $f_{a,j}(\underset{\sim}{S} ; R_e)$, as the rigid pseudoatom. The higher order terms are the deforming pseudoatom due to the vibration of the nuclei. In general we expect the derivative terms in (24) to be non zero since the $\{f_a\}$ and $\{f_b\}$ satisfy a large (formally infinite) number of static charge properties of $\rho_{mol}(\underset{\sim}{r} ; R)$ at each R. In pratice we found that first and second derivative terms were sufficient to reconstruct a number of vibrational averages to four or five figures accuracy.

† R, the internuclear distance, is the only relevant Q variable.

## THE ELECTRONIC WAVEFUNCTIONS

The diatomic molecules we have studied were $N_2(^1\Sigma_g^+)$, $CO(^1\Sigma_g^+)$, $BF(^1\Sigma_g^+)$ and $FH(^1\Sigma_g^+)$. The molecules in this series undergo a single internal vibration with increasing root mean square amplitude. The electronic wavefunctions are near Hartree-Fock quality spanned with extensive basis sets of STO. The $N_2$ wavefunction is that tabulated by Cade et al[4,5], for CO, Huo[6], for FB, McLean and Yoshimine[7] and for FH, Cade and Huo[8] (plus Bader bootlegging other Ψ's of FH at different R to us). The conversion of these wavefunctions to molecular form factors was done by procedures previously given[9]. The $\{f_a\}$ and $\{f_b\}$ at [2|2] expansion were determined by the methods given in ref.3.

## THE VIBRATIONAL WAVEFUNCTIONS

A vibrational potential function, U(R), was determined from the theoretical electronic energy of the electronic wavefunctions discussed above. In this case we chose,

$$U(R) = \frac{1}{2}(R-R_e)^2\left(\frac{\partial^2 U}{\partial R^2}\right)_{R_e} + \frac{1}{6}(R-R_e)^3\left(\frac{\partial^3 U}{\partial R^3}\right)_{R_e} \qquad (25)$$

For an harmonic potential (inclusion only of the first term) the vibrational wavefunction is simply

$$\phi_n = \left(\frac{\alpha}{\sqrt{\pi}\,2^n\,n!}\right)^{1/2}\frac{1}{R}H_n[\alpha(R-R_e)]\exp[-\alpha^2(R-R_e)^2/2] \qquad (26)$$

where : $\alpha^2 = \frac{1}{2\langle X_{12}^2\rangle}$ $\qquad \langle X_{12}^2\rangle$ : mean square amplitude of vibration

$H_n(y)$ : Hermite polynomial of order n

The cubic term in (25) is treated as a perturbation. The Rayleigh-Schrödinger perturbation theory was used to get the ground-state vibrational wavefunction :

$$\chi_o(R) = \phi_o(R) + C_1\,\phi_1(R) + C_3\,\phi_3(R) \qquad (27)$$

where the $\phi_n$ are given in (26). The coefficients $C_1$ and $C_3$ are :

$$C_1 = -\left(\frac{\partial^3 U}{\partial R^3}\right)_{R_e}\mu\,\langle X_{12}^2\rangle^{5/2}/\hbar^2$$

$$C_3 = \sqrt{2/27}\;C_1 \qquad (28)$$

In (28) μ is the reduced mass of the molecule and ħ is Planck's constant divided by 2 . The vibrational parameters for (27) are summarized in table 1. Eq. (27) or its truncation to the first term was used for all our vibrational averages. Note in Table 1 how small the $<X_{12}^2>^{1/2}$ are compared to the usual Debye-Waller factor in structure analysis.

Table 1. Vibrational Parameters

| Molecule | $<X_{12}^2>^{1/2}.10^2$(au) | $C_1.10^2$ | $C_2.10^2$ |
|---|---|---|---|
| $N_2$ | 5.6125 | 4.9393 | 1.3443 |
| CO | 6.0099 | 4.9185 | 1.3386 |
| BF | 7.5833 | 7.0979 | 1.9318 |
| FH | 11.864 | 11.533 | 3.1390 |

(to convert $<X_{12}^2>^{1/2}$ to a Debye Waller factor, $B = 22.1<X_{12}^2>$ )
For $N_2$, B=.070 Å$^2$, for FH, B=.311 Å$^2$.

## AVERAGES OF INTENSITIES

For a freely rotating molecule (appreciable populations of all rotational levels)

$$I_{el}^{XR}(S\ ;\ R) = (4\pi)^{-1} \int |F(\underset{\sim}{S}\ ;\ R)|^2\, d\Omega_S \tag{29}$$

The gas phase elastic X-Ray scattering intensities were computed for $N_2$, CO, BF and FH at several R about $R_e$. It is instructive to represent $I_{el}(S\ ;\ R)$ as a polynomial :

$$I_{el}^{XR}(S\ ;\ R) = \sum_{k=0} a_k(S)\ (R-R_e)^k \tag{30}$$

We then fit our results from (29) to a fifth order polynomial, (30), where $a_k(S)$ were determined by least squares. With (30) and (27) put into (20),

$$I_{av}(S) = I_{el}^{XR}(S\ ;\ R_e) + 2a_1(S)\ C_1\ <X_{12}^2>^{1/2} + a_2(S)\ <X_{12}^2>$$
$$+\ a_3(S)\ (6C_1 + 2\sqrt{6}\ C_3)\ <X_{12}^2>^{3/2} + 3a_4(S)\ <X_{12}^2>^2 \tag{31}$$

where terms of order $C^2$ are neglected. For $C_1=C_3=0$, the average for a harmonic potential,

$$I_{av,\ harm}(S) = I_{el}^{XR}(S\ ;\ R) + a_2(S)\ <X_{12}^2> + 3a_4(S)\ <X_{12}^2>^2 \tag{32}$$

Let us equate (32) to a known model result. For rigid spherical atoms with X-Ray form factors $f_a(S)$ and $f_b(S)$, (the IAM) Debye[10] showed that the harmonic average is :

$$I^{IAM}_{av,\, harm}(S) = f_a^2(S) + f_b^2(S) + 2f_a f_b \exp(-1/2\langle X_{12}^2\rangle S^2)\, \frac{\sin(SR_e-\beta)}{SR_e\cdot(1+\langle X_{12}^2\rangle^2)\frac{S^2}{R_e^2}}$$

$$\beta = \tan^{-1}(\langle X_{12}^2\rangle S/R_e) \tag{33}$$

When (33) is expanded in powers of $\langle X_{12}^2\rangle$,

$$I^{IAM}_{av,\, harm}(S) = I^{IAM}_{el}(S\ ;\ R_e) - f_a f_b\left[j_o(SR_e)S^2 - \frac{2\cos(SR_e)}{R_e^2}\right]\langle X_{12}^2\rangle$$

$$+ \frac{1}{2} f_a f_b\left[j_o(SR_e)S^4 + \frac{\cos\ (SR_e)-2j\ (SR_e)}{R_e}\ S^2\right]\langle X_{12}^2\rangle \tag{34}$$

If $I_{av,\, harm}(S)$ were correctly given by the independant atom model (IAM), then on comparing (34) and (32),

$$a_2(S) = -f_a f_b\left[\frac{j_o(SR_e)S^2 - 2\cos(SR_e)}{R_e^2}\right] \tag{35}$$

$$a_4(S) = \frac{1}{6} f_a f_b \left[j_o(SR_e)S^4 + \frac{\cos(SR_e) - 2j_o(SR_e)}{R_e^2}\ S^2\right]$$

and so on. The inclusion of anharmonic terms can be seen in (31). They occur as odd powers of $\langle X_{12}^2\rangle^{1/2}$. Equation (31) tells us that if we want to think of $I_{av}$ as close to $I_{el}(S\ ;\ R_e)$ in powers of $\langle X_{12}^2\rangle^{1/2}$, we must first consider the anharmonic term before the second contribution from the harmonic oscillator. One suspects that anharmonic effects will play an important role in $I_{av}$. It is clear from (33) that the small polynomial expansion in $(R-R_e)$ from (30) will fail at large values of S.

From previous work we know that $I^{XR}_{el}(S\ ;\ R_e)$ can be accurately represented with generalized X-Ray scattering factors at the [2|2] level. This compels us to use the pseudoatom model for $I_{el}(S\ ;\ R)$. For a [2|2] expansion

$$I^{XR}_{el}(S\ ;\ R) = \sum_{j=0}^{2} f_{a,j}(S,R) I_{a,j}(S,R) + \sum_{k=0}^{2} f_{b,k}(S,R) I_{b,k}(S,R) \tag{36}$$

In (36) the $\{f_a\}$ and $\{f_b\}$ are the generalized X-Ray scattering factors of the pseudoatoms and

$$I_{a,j}(S\ ;\ R) = (4\pi)^{-1} \int F_{mol}(\mathbf{S}\ ;\ R)(-i)^j P_j(\eta')\exp(ic\eta')d\Omega_S$$

$$I_{b,k}(S\ ;\ R) = (4\pi)^{-1} \int F_{mol}(\mathbf{S}\ ;\ R)(-i)^k P_k(\eta')\exp(-ic\eta')d\Omega_S \tag{37}$$

For (37) $\eta'$ is the direction cosine between $\mathbf{S}$ and $\mathbf{R}$ and $c = SR/2$ Eq. (36) is a compact way to write $I_{el}(S\ ;\ R)$ in terms of generalized scattering factors since, as shown by eq. (11) in the previous lecture on multipolar expansions of one-electron densities,

$$I_{a,\ell} = (2\ell+1)^{-1} f_{a,\ell}(S,R) + \sum_{k=0}^{K} \left[ \sum_{n=|\ell-k|}^{\ell+k} (-1)^{(n+k-\ell)/2} b_n(\ell,k) j_n(SR)\right] f_{b,k}(S,R)$$

$$I_{b,m} = \sum_{j=0}^{J} \left[ \sum_{n=|j-m|}^{j+m} (-1)^{(n+m-j)/2} b_n(j,m) j_n(SR)\right] f_{a,j}(S,R) + (2m+1)^{-1} f_{b,m}(S,R) \tag{38}$$

For rigid pseudoatoms, the R dependence of $I_{a,l}$ and $I_{b,m}$ is explicitely given by the spherical Bessel functions, $j_n(SR)$, in (38). Thus

$$I_{av}^{rigid}(S) = \sum_{j=0}^{2} f_{a,j}(S,R_e)\langle I_{a,j}^{rigid}\rangle + \sum_{k=0}^{2} f_{b,k}(S,R_e)\langle I_{b,k}^{rigid}\rangle \tag{39}$$

In (39) $I_{a,j}^{rigid}$ and $I_{b,k}^{rigid}$ are given by (38) with $f_{a,j}$ and $f_{b,k}$ replaced by $f_{a,j}(S\ ;\ R_e)$ and $f_{b,k}(S\ ;\ R_e)$. The brackets in (39) represent the vibrational average with (27) as a wavefunction. When anharmonic terms are neglected ($C_1=C_2=0$), we use the notation $\langle..\rangle_{harm}$. The dominant contributor to $I_{el}$ is due to the monopoles, $f_{a,0}(S\ ;\ R)$ and $f_{b,0}(S\ ;\ R)$, so that it is instructive to study $\langle j_o(SR)\rangle$ and $\langle j_o(SR)\rangle_{harm}$. Plots of $\langle j_o(SR)\rangle_{harm} - \langle j_o(SR_e)\rangle$ compared to $\langle j_o(SR)\rangle - \langle j_o(SR_e)\rangle$ for $N_2$ reveal substantial differences for $S < 12\mathring{A}^{-1}$ ($\sin\theta/\lambda < 1\mathring{A}^{-1}$). The anharmonic terms clearly play a dominant role for these vibrational averages over an appreciable range of the Bragg vector magnitude.

For deforming pseudoatoms,

$$I_{av} = \sum_{j=0}^{2}\langle f_{a,j}(S,R) I_{a,j}(S,R)\rangle + \sum_{k=0}^{2}\langle f_{b,k}(S,R) I_{b,k}(S,R)\rangle \tag{40}$$

The R dependence of $\{f_a\}$ and $\{f_b\}$ is given by (24). In practice we found that when $\{f_a\}$ and $\{f_b\}$ are truncated at second derivative terms, then (40) reproduces (31) except at large S, where we expect (31) to fail since it is based on the small polynomial expansion (30). Moreover at large S, the derivative terms for f as a function of R are small for monopoles.

The elastic scattering intensity by high energy electrons for a gas phase diatomic molecule,

$$S^4 <I_{el}^{ED}(S ; R)> = Z_a^2 + Z_b^2 + 2Z_aZ_b<j_o(SR)> - 2[Z_a<I_{a,o}(S ; R)> + Z_b<I_{b,o}(S ; R)>] + <I_{el}^{XR}(S ; R)> \qquad (41)$$

$Z_a$ and $Z_b$ are the nuclear charges of a and b, respectively. It is of some interest to study the vibrational averages of $I_{a,o}$ and $I_{b,o}$, the electron-nuclear interference terms. Since generalized X-Ray scattering factors necessarily satisfy $I_{a,o}$ and $I_{b,o}$, we can work directly with $\{f_a\}$ and $\{f_b\}$. At the [2|2] level,

$$<I_{a,o}^{rigid}> = f_{a,o}(S ; R_e) + \sum_{k=0}^{2} (-1)^k<j_k(SR)>f_{b,k}(S ; R_e)$$
$$<I_{b,o}^{rigid}> = f_{b,o}(S ; R_e) + \sum_{j=0}^{2} (-1)^j<j_j(SR)>f_{b,j}(S ; R_e) \qquad (42)$$

and for the deforming pseudoatoms,

$$<I_{a,o}> = <f_{a,o}(S ; R)> + \sum_{k=0}^{2} (-1)^k<j_k(SR)f_{b,k}(S ; R)>$$
$$<I_{b,o}> = <f_{b,o}(S ; R)> + \sum_{j=0}^{2} (-1)^j<j_j(SR)f_{a,j}(S ; R)> \qquad (43)$$

A number of various difference functions can be plotted against S. The most informative for our studies are :

$$\Delta_1(S) = <I_{el}^{XR}> - <I_{el}^{rigid}> \qquad (44)$$

$$\Delta_2(S) = <I_{el}^{XR}> - <I_{el}^{XR}>_{harm} \qquad (45)$$

and corresponding difference functions for the electron-nuclear interference terms $I_{a,o}$ and $I_{b,o}$. Plots of (44), (45) and related difference functions for $N_2$, CO, BF, FH consistently emphasize the dominant role of anharmonicity ($\Delta_2(S)$) over the deformation of pseudoatoms ($\Delta_1(S)$). The anharmonic difference is typically ~10 times larger than non rigid effects and of opposite sign for low values of S ($\sin\theta/\lambda < .4A^{-1}$). A summary of maximum absolute differences is given in tables 2 and 3.

Table 2. Maximum non-rigid corrections for $<I^{rigid}(S\ ;\ R)>$ and corresponding value of ($\sin\theta/\lambda$, $Å^{-1}$)

| Intensity | $N_2$ | | CO | | BF | | FH | |
|---|---|---|---|---|---|---|---|---|
| $<I^{XR}_{el}>$ | .04% | .21$Å^{-1}$ | .03% | .16$Å^{-1}$ | .01% | .15$Å^{-1}$ | .06% | .21$Å^{-1}$ |
| $<I_{a,o}>$ | .03% | .25 | .03% | .20 | .04% | .21 | .04% | .25 |
| $<I_{b,o}>$ | .03% | .25 | .03% | .20 | .01% | .33 | .84% | .49 |

Table 3. Maximum anharmonic corrections for $<I(S\ ;\ R)>$ and corresponding values of ($\sin\theta/\lambda$, $Å^{-1}$)

| Intensity | $N_2$ | | CO | | BF | | FH | |
|---|---|---|---|---|---|---|---|---|
| $<I^{XR}_{el}>$ | -.23% | .16$Å^{-1}$ | -.22% | .16$Å^{-1}$ | -.25% | .14$Å^{-1}$ | -.26% | .20$Å^{-1}$ |
| $<I_{a,o}>$ | -.23% | .18 | -.35% | .18 | -.83% | .17 | -.15% | .22 |
| $<I_{b,o}>$ | -.23% | .18 | -.16% | .18 | -.16% | .15 | -5.0% | .21 |

For $\sin\theta/\lambda$ greater than .8$Å^{-1}$, the non-rigid effects are completely negligible (that is, it is beyond our modeling capabilities). As expected, the anharmonic correction increases with the amplitude of vibration. For the electron-nuclear interference terms, the larger anharmonic correction is for the center with smaller nuclear charge. For all systems these corrections are small, except for the proton center in the FH electron-nuclear interference term.

## AVERAGES OF $F_{mol}$-"ALICE IN WONDERLAND" ?

We now turn our attention to $I_{Bragg}$ as given by (22). In this case we imagine an experiment that can energy analyse the scattered X-Ray intensity to within the energy of a vibron. This would be an energy resolution of $\sim 10^{-5}$.

$$I_{Bragg}(\underset{\sim}{S}) = |F_{Bragg}|^2 \tag{46}$$

$$F_{Bragg}(\underset{\sim}{S}) = <F(\underset{\sim}{S}\ ;\ R)> \tag{47}$$

It was shown in earlier work[11] that generalized X-Ray scattering factors, which represent static charge properties, cannot in general be extracted from (47). For our specialized study here, a zero-point vibrational average of an oriented diatomic molecule, we will model the properties of $F_{Bragg}(\underset{\sim}{S})$ with non-rigid pseudoatoms and with an anharmonic potential.

At the [2|2] level, we have :

$$F(\underset{\sim}{S}\ ;\ R) = \sum_{j=0}^{2} i^j P_j(\eta) f_{a,j}(S,R)\exp(-i\underset{\sim}{S}.\underset{\sim}{R}/2) + \sum_{k=0}^{2} i^k P_k(\eta) f_{b,k}(S,R)\exp(i\underset{\sim}{S}.\underset{\sim}{R}/2) \qquad (48)$$

See equation (6) of the lecture on Multipolar Expansions of the One-Electron Densities.

We let $\underset{\sim}{u}_a$ be the displacement of nucleus a from equilibrium and $\underset{\sim}{u}_b$ the displacement vector of b. We also use (24) for the expansion of $\{f_a\}$ and $\{f_b\}$ and truncate at the second derivative terms. Then for an internal vibration, the average of (48) is :

$$\begin{aligned} F_{Bragg}(\underset{\sim}{S}) = {} & \exp(-i\underset{\sim}{S}.\underset{\sim}{R}_e/2) \sum_{j=0}^{2} i^j P_j(\eta) \\ & \times \Big[ f_{a,j}(S\ ;\ R_e)\langle\exp(i\underset{\sim}{S}.\underset{\sim}{u}_a)\rangle + \Big(\frac{\partial f_{a,j}}{\partial R}\Big)_{R_e} \langle \Delta R \exp(i\underset{\sim}{S}.\underset{\sim}{u}_a)\rangle \\ & + \frac{1}{2}\Big(\frac{\partial^2 f_{a,j}}{\partial R^2}\Big)_{R_e} \langle \Delta R^2 \exp(i\underset{\sim}{S}.\underset{\sim}{u}_a)\rangle \Big] \\ & + \exp(i\underset{\sim}{S}.\underset{\sim}{R}_e/2) \sum_{k=0}^{2} i^k P_k(\eta) \\ & \times \Big[ f_{b,k}(S\ ;\ R_e)\langle\exp(i\underset{\sim}{S}.\underset{\sim}{u}_b)\rangle + \Big(\frac{\partial f_{b,k}}{\partial R}\Big)_{R_e} \langle \Delta R \exp(i\underset{\sim}{S}.\underset{\sim}{u}_b)\rangle \\ & + \frac{1}{2}\Big(\frac{\partial^2 f_{b,k}}{\partial R^2}\Big)_{R_e} \langle \Delta R^2 \exp(i\underset{\sim}{S}.\underset{\sim}{u}_b)\rangle \Big] \end{aligned} \qquad (49)$$

where $\Delta R = R - R_e$. With normal coordinates $\xi$ ,

$$\underset{\sim}{u}_a = -\ \xi(\mu^{1/2}/m_a)\ \hat{\underset{\sim}{R}} \qquad (50)$$

$$\underset{\sim}{u}_b = \xi(\mu^{1/2}/m_b)\ \hat{\underset{\sim}{R}} \qquad (51)$$

$$\Delta R = \xi\mu^{-1/2} \qquad (52)$$

and $\mu$ is the reduced mass. The vibrational averages in (49) can be done by integration with respect to $\xi$ after (50), (51) and (52) are substituted in (49). We then get for pseudoatom a :

$$\langle \exp(i\underset{\sim}{S}.\underset{\sim}{u}_a)\rangle = \exp(-\alpha^2\eta^2/2)\ [1 - iC_1(2\alpha\eta) + iC_3(\tfrac{2}{3})^{1/2}\alpha^3\eta^3] \qquad (53)$$

$$\langle \Delta R \exp(i\underset{\sim}{S}.\underset{\sim}{u}_a)\rangle = \langle X_{12}^2\rangle \exp(-\alpha^2\eta^2/2)\,[-i\alpha\eta + 2C_1(1-\alpha^2\eta^2) - (\tfrac{2}{3})^{1/2}C_3\alpha^2\eta^2(3-\alpha^2\eta^2)] \quad (54)$$

$$\langle \Delta R^2 \exp(i\underset{\sim}{S}.\underset{\sim}{u}_a)\rangle = \langle X_{12}^2\rangle \exp(-\alpha^2\eta^2/2)\,[1 - \alpha^2\eta^2 - 2iC_1\alpha\eta(3-\alpha^2\eta^2) - i(\tfrac{2}{3})^{1/2}C_3(\alpha^5\eta^5 - 7\alpha^3\eta^3+6\alpha\eta)] \quad (55)$$

$C_1$ and $C_3$ are given by (28).
In the above expressions :

$$\alpha = S\ (\mu/m_a)\ \langle X_{12}^2\rangle^{1/2}$$

and $\eta$ is the direction cosine between $\underset{\sim}{S}$ and $\underset{\sim}{R}$. Expressions (53)-(55) also hold for pseudoatom b, except that $\alpha$ is replaced by :

$$\beta = S\ (\mu/m_b)\ \langle X_{12}^2\rangle^{1/2}$$

Substitution of (53)-(55) and the corresponding b-centered averages into (49) is an explicit representation for $F_{Bragg}(\underset{\sim}{S})$. For rigid pseudoatoms,

$$(\partial f_{a,j}/\partial R)_{R_e} = (\partial f_{b,k}/\partial R)_{R_e} = (\partial^2 f_{a,j}/\partial R^2)_{R_e} = (\partial^2 f_{b,k}/\partial R^2)_{R_e}.$$

For an harmonic oscillator $C_1 = C_3 = 0$.

We now pose the following deconvolution problem. We assume (49) is an observable and ask ourselves to what extent can $\{f_a\}$ and $\{f_b\}$ be recovered with a rigid pseudoatom model. We define :

$$F_{Bragg}^{rigid}(\underset{\sim}{S}) = \exp(-i\underset{\sim}{S}.\underset{\sim}{R}_e/2)\exp(-\alpha^2\eta^2/2)\,[1-iC_1(2\alpha\eta) + iC_3(\tfrac{2}{3})^{1/2}\alpha^3\eta^3\,]\sum_{j=0}^{2} i^j g_{a,j}(S)P_j(\eta)$$
$$+ \exp(i\underset{\sim}{S}.\underset{\sim}{R}_e/2)\exp(-\beta^2\eta^2/2)\,[1+iC_1(2\beta\eta) - iC_3(\tfrac{2}{3})^{1/2}\beta^3\eta^3\,]\sum_{k=0}^{2} i^k g_{b,k}(S)P_k(\eta) \quad (56)$$

Note in (56) that pseudoatom a, $\{g_a\}$ is assigned the a center vibrational parameters and pseudoatom b, $\{g_b\}$ is given b center vibrational parameters. We form the mean square function :

$$\varepsilon(S) = \int_{-1}^{1} \int_{0}^{2\pi} |F_{Bragg}(\underset{\sim}{S}) - F^{rigid}_{Bragg}(\underset{\sim}{S})|^2 \, d\eta d\phi_S \tag{57}$$

and for each S minimize $\varepsilon(S)$ with respect to $g_{a,j}(S)$ and $g_{b,k}(S)$ :

$$\partial\varepsilon/\partial g_{a,j} = 0 \quad j=0,1,2 \qquad \partial\varepsilon/\partial g_{b,k} = 0 \quad k=0,1,2 \tag{58}$$

We then compare $\{g_a(S)\}$ to $\{f_a(S\ ;\ R_e)\}$ and $\{g_b(S)\}$ to $\{f_b(S\ ;\ R_e)\}$. If the g's are close to f's we have successfully deconvoluted $F_{Bragg}(\underset{\sim}{S})$ to a static charge picture.

We divide the solutions to (58) into two studies. In the first $C_1=C_3=0$ in (56) so taht an harmonic potential is assumed for $F^{rigid}_{Bragg}(\underset{\sim}{S})$. The $\{g_a\}$ and $\{g_b\}$ determined from (58) in this case are denoted $g^H_{a,j}(S)$ and $g^H_{b,k}(S)$. The second set of solutions include the anharmonic terms, so that we denote the functions simply as $g_{a,j}(S)$ and $g_{b,k}(S)$.

The results we have are for $N_2$, CO, BF, FH. With the exception of the hydrogen pseudoatom in FH, the monopole scattering factors $g^H_{p,o}$ and $g_{p,o}$ differ from $f_{p,o}(S\ ;\ R_e)$ by less than $2\ 10^{-3}$. For $g^H_{H,o}$ the maximum difference is 2% at $\sin\theta/\lambda=.04Å^{-1}$. For $g_{H,o}(S)$ the max. relative difference from $f_{H,o}(S\ ;\ R_e)$ is .16% at $\sin\theta/\lambda = .28Å^{-1}$. In sharp contrast to the monopoles, $g^H_{p,1}$ for all the pseudoatoms differ appreciably from $f_{p,1}(S\ ;\ R_e)$. For B,N,C,O and for F pseudo-atoms, the $g^H_{p,1}$ give a contracted dipole component of opposite polarity, and larger magnitude, than the pseudoatom at $R_e$. On the other hand the H pseudoatom dipole scattering factor $g^H_{H,1}$ does not have such an artificial sharp dipole. The $g^H_{p,1}$ from (58) contain components due to the coupling of the monopole scattering factors and the anharmonic terms in $F_{Bragg}(\underset{\sim}{S})$. The dominant dipole component for pseudoatom a projected from $F_{Bragg}(\underset{\sim}{S})$ into $g^H_{a,1}$ is :

$$f_{a,1}(S\ ;\ R_e) - 2(\mu/m_a)C_1\langle X^2_{12}\rangle^{1/2} S\ f_{a,o}(S\ ;\ R_e)$$

A similar term occurs for projection of $F_{Bragg}(\underset{\sim}{S})$ into $f_{b,1}$. The maximum discrepancy between $g^H_{a,1}(S)$ and $f_{a,1}(S\ ;\ R_e)$ occurs where $Sf_{a,o}(S\ ;\ R_e)$ is a maximum. For all "heavy" pseudoatoms the $g_{p,1}(S)$ are negligibly different from $f_{p,1}(S\ ;\ R_e)$. Thus for a proper set of vibrational parameters (inclusion of anharmonic terms) a deconvolution of $F_{Bragg}(\underset{\sim}{S})$ to $F(\underset{\sim}{S}\ ;\ R_e)$ appears to be successful. For H in FH, $g^H_{H,1}$ does not differ appreciably from $f_{H,1}(S\ ;\ R_e)$, but $g_{H,1}$

does at $.5\text{Å}^{-1}$ in $\sin\theta/\lambda$. The apparent agreement between $f_{H,1}(S\ ;\ R_e)$ and $g^H_{H,1}(S)$ is apparently due to the cancellation of the anharmonic and non-rigid effects, which are neglected in the rigid harmonic model. The quadrupole scattering factors, $g^H_{p,2}(S)$, for the heavy atoms are in closer agreement to $f_{p,2}(S\ ;\ R_e)$ than the dipole scattering factors. The $g_{p,2}(S)$ indicate successful deconvolution. On the other hand the H-pseudoatom $g^H_{H,2}(S)$ is closer to $f_{H,2}(S\ ;\ R_e)$ than is $g_{H,2}(S)$. The maximum relative difference from the anharmonic rigid model is 3% at $\sin\theta/\lambda$ of $.30\text{Å}^{-1}$. As for the H dipole scattering factor it appears that non rigid effects cancel the anharmonic effect. Thus we seem to learn from this particular study that a rigid pseudoatom model can recover $F(\underset{\sim}{S}\ ;\ \underset{\sim}{R}_e)$ from $F_{Bragg}(\underset{\sim}{S})$ provided that anharmonic contributions are included in the model. The largest non-rigid effects are found for the H-pseudoatom in FH with a maximum relative error of 6% for the quadrupole scattering factor.

## VIBRATIONAL FORCE CONSTANTS

Polyatomic and crystal force constants can be related to charge densities and field gradients in a simple way[12,13]. For example,

$$k_{\alpha\beta} = \frac{\partial^2 \Phi}{\partial A_p \partial B_p} = Z_p\left[\ q_{\alpha\beta} + \frac{4\pi}{3}\ \delta_{\alpha\beta}\rho(\underset{\sim}{R}_p) - \int \alpha_p \frac{\partial\rho}{\partial B_p}\ \frac{d\underset{\sim}{r}}{r_p^3}\ \right] \quad (59)$$

$$\alpha,\beta = x,y,z \qquad A,B = X,Y,Z$$

In (59) $q_{\alpha\beta}$ are the components of the electric field gradient at nucleus p, $\rho(\underset{\sim}{R}_p)$ is the electron density on nucleus p and $\partial\rho/\partial B_p$ is the change in $\rho$ with the change of one of the nuclear coordinates of nucleus p. If $\rho$ can be represented with rigid pseudoatoms, then

$$k^{rigid}_{\alpha\beta} = Z_p\left[\ q'_{\alpha\beta} + \frac{4\pi}{3}\delta_{\alpha\beta}\sum_{p'\neq p} \rho_{p'}(\underset{\sim}{R}_p)\ \right] \quad (60)$$

where $q'_{\alpha\beta}$ is the contribution to the electric field gradient from all p' p pseudoatoms and their associated nuclei.

For the present we explore the applicability of (60) to the diatomics $N_2$, CO, BF and FH for which $f_a(S\ ;\ R)$ and $f_b(S\ ;\ R)$ at the [2|2] expansion level. In these studies the near Hartree-Fock wavefunctions do not rigorously obey the Hellmann-Feynman theorem[14]. Nevertheless the pseudoatoms at [2|2] must satisfy the internal forces on the nuclei as determined by $\rho_{mol}(\underset{\sim}{r}\ ;\ \underset{\sim}{R})$, even though there is a net translational force. We start with

$$k_e = -\,(dF_{int}(R)/dR)_{R=R_e} \tag{61}$$

where $R_e$ is the internuclear distance for which $F_{in}(R) = 0$. (For these $\rho_{mol}(\underset{\sim}{r}\,;\,R)$ this is not the same $R_e$ at which the electronic energy is a minimum). For the $\rho_{mol}(\underset{\sim}{r}\,;\,R)$ with which we work

$$F_{int}(R) = \frac{Z_aZ_b}{R^2} - \frac{1}{2} Z_a \int \rho_{mol}(\underset{\sim}{r}\,;\,R)P_1(\cos\theta_a)r_a^{-2}\,d\underset{\sim}{r}$$

$$+ \frac{1}{2} Z_b \int \rho_{mol}(\underset{\sim}{r}\,;\,R)P_1(\cos\theta_b)r_b^{-2}\,d\underset{\sim}{r} \tag{62}$$

($Z_a$ and $Z_b$ are the nuclear charges)
We recast (62) in the equivalent form, given the definition (37) (see also re. 2)

$$F_{int}(R) = \frac{Z_aZ_b}{R^2} - \frac{Z_a}{\pi}\int_0^\infty I_{a,1}(S\,;\,R)SdS + \frac{Z_b}{\pi}\int_0^\infty I_{b,1}(S\,;\,R)SdS \tag{63}$$

At [2|2] the $\{f_a\}$ and $\{f_b\}$ satisfy $I_{a,1}$ and $I_{b,1}$. From (61) and (63)

$$k_e = \frac{2Z_aZ_b}{R^3} + \frac{Z_a}{\pi}\int_0^\infty \left(\frac{\partial I_{a,1}}{\partial R}\right)_{R_e} SdS - \frac{Z_b}{\pi}\int_0^\infty \left(\frac{\partial I_{b,1}}{\partial R}\right)_{R_e} SdS \tag{64}$$

For [2|2] pseudoatoms, (64) is :

$$\begin{aligned} k_e = {} & \frac{2Z_aZ_b}{R^3} + \frac{2\pi}{3}[\rho_b(a)+\rho_b(b)] - Z_a\langle r_a^{-3}P_2(\cos\theta_a)\rangle_b \\ & - Z_b\langle r_b^{-3}P_2(\cos\theta_b)\rangle_a + \frac{1}{3\pi}\left[Z_a\int_0^\infty\left(\frac{\partial f_{a,1}}{\partial R}\right)_{R_e}SdS - Z_b\int_0^\infty\left(\frac{\partial f_{b,1}}{\partial R}\right)_{R_e}SdS\right] \\ & + \frac{Z_a}{\pi}\sum_{k=0}^{2}\frac{(-1)^k}{2k+1}\int_0^\infty\left(\frac{\partial f_{b,k}}{\partial R}\right)_{R_e}[(k+1)j_{k+1}(SR_e) - kj_{k-1}(SR_e)]SdS \\ & + \frac{Z_b}{\pi}\sum_{n=0}^{2}\frac{1}{2n+1}\int_0^\infty\left(\frac{\partial f_{a,n}}{\partial R}\right)_{R_e}[(n+1)j_{n+1}(SR_e) - nj_{n-1}(SR_e)]SdS \end{aligned} \tag{65}$$

In (65), $\rho_b(a)$ and $\langle r_a^{-3}P_2(\cos\theta_a)\rangle_b$ are the contributions of pseudoatom b to the charge density on nucleus a and the electric field gradient about a, respectively. The term

$$(Z_a/3\pi)\int_0^\infty(\partial f_{a,1}/\partial R)_{R_e}\,SdS$$

is the derivative of the force on nucleus a from pseudoatom a.

Equation (65) in terms of $\{f_a\}$ and $\{f_b\}$ and the first derivatives must give us the same result as that from (61), where we determine the slope of $F_{int}$ at $R_e$ directly from $\rho_{mol}(\tilde{r}\ ;\ R)$ via eq. (62). We use this as an internal check for our methods of integration for evaluation of (65). For the rigid pseudoatom model we evaluate (65) with

$$(\partial f_{a,n}/\partial R)_{R_e} = (\partial f_{b,k}/\partial R)_{R_e} = 0$$

For an diatomic molecule in terms of pseudoatoms :

$$\rho_{mol}(\tilde{r}\ ;\ \tilde{R}) = (2\pi)^{-3} \int [f_a(S\ ;\ \tilde{R}) \exp(-i\tilde{S}.\tilde{R}/2)\, d\tilde{S} + f_b(S\ ;\ R) \exp(i\tilde{S}.\tilde{R}/2)] \quad (66)$$

Stewart[16] made the erroneous assumption that the pseudoatoms are rigid with respect to the space-fixed coordinate system : that is

$$(\partial f_p(S\ ;\ R)/\partial X_a) = (\partial f_p(S\ ;\ R)/\partial Y_a) = (\partial f_p(S\ ;\ R)/\partial Z_a) = 0 \quad (67)$$

(67) is not a rotationally invariant condition. The pseudoatoms are only rigid with respect to the internal molecular coordinate R. The generalized X-Ray scattering factors for pseudoatom a are :

$$f_a(S\ ;\ R) = \sum_{j=0}^{J} i^j P_j(\eta)\, f_{a,j}(S\ ;\ R) \quad (68)$$

Then

$$\frac{\partial f_a}{\partial X_a}(S\ ;\ R) = \sum_{j=0}^{J} i^j [\frac{\partial P_j}{\partial X_a} f_{a,j} + P_j \frac{\partial f_{a,j}}{\partial X_a}] \quad (69)$$

The first term is the $X_a$ component of the angular relaxation and the second is for radial relaxation. Recall that $\eta$ is the direction cosine between $\tilde{S}$ and $\tilde{R}$. (69) can be explicitely written :

$$\frac{\partial f_a}{\partial X_a}(S\ ;\ R) = \sum_{j=0}^{J} i^j f_{a,j}(S\ ;\ R)[\frac{dP_j(\eta)}{d\eta}(\frac{\Delta X}{R^2} - \frac{1}{R}P_1'(\eta_S)\cos\theta_S) - P_j(\eta)\frac{\partial f_{a,j}}{\partial R}\frac{\Delta X}{R^2}] \quad (70)$$

In (70) $\eta_S = \cos\theta_S$ and $\phi_S$ are the polar coordinates for $\tilde{S}$ with respect to the space fixed coordinate system. $\partial f_a/\partial Y_a$ and $\partial f_a/\partial Z_a$ are similar to (70). We now choose the condition : $R=R_e$, $\Delta Z=R_e$, $\Delta X=\Delta Y=0, \eta_S=\eta, \phi_S=\phi$.

$$\partial f_a/\partial X_a = -\frac{1}{R}\sum_{j=0}^{J} i^j f_{a,j}(S\ ;\ R)\, P_j'(\eta)\cos\phi \quad (71)$$

$$\partial f_a/\partial Y_a = -\frac{1}{R} \sum_{j=0}^{J} i^j f_{a,j}(S\ ;\ R)\ P_j'(\eta)\ \sin\phi \tag{72}$$

$$\partial f_a/\partial Z_a = -\frac{1}{R} \sum_{j=0}^{J} i^j\ \partial f_{a,j}/\partial R\ P_j(\eta) \tag{73}$$

In general (71) and (72) are not zero. When (71) and (72) are included in $\partial\rho/\partial X_a$ and $\partial\rho/\partial Y_a$ expressions (eq.16 in ref. 15) and integration over S is carried out, then

$$k_e^{rigid} = \frac{2Z_aZ_b}{R^3} + \frac{2\pi}{3}[Z_a\rho_b(a) + Z_b\rho_a(b)] - Z_a\langle r_a^{-3}P_2(\cos\theta_a)\rangle_b - Z_b\langle r_b^{-3}P_2(\cos\theta_b)\rangle_a \tag{74}$$

The non rigid correction, $(\partial f_a/\partial R)_{Re}$ terms from (73) can be included to get (65). In terms of f's

$$\rho_b(a) = \frac{2}{\pi} \sum_{k=0}^{2}(-1)^k \int_0^\infty f_{b,k}(S\ ;\ R_e) j_k(SR_e) S^2 dS \tag{75}$$

$$\langle r_a^{-3}P_2(\cos\theta_a)\rangle_b = \frac{2}{3\pi} \sum_{k=0}^{2} \sum_{n=|2-k|}^{2+k} (-1)^{(n+k-2)/2}\ b_n(2,k) \int_0^\infty f_{b,k}(S\ ;\ R_e) j_n(SR_e) S^2 dS \tag{76}$$

and corresponding terms. It is important to note that (74) depends on the partitioning of the pseudoatoms. Thus a [J|K] dependence for $k_e^{rigid}$ can occur in the same way as other pseudoatom expectation values. (see table 1 of the lecture on multipole expansion of one-electron densities). On the other hand if $\partial f_a/\partial R$ and $\partial f_b/\partial R$ are known, then $k_e$ is exactly calculated by (65).

Results for $N_2$, CO, BF and FH from (74) using a [2|2] expansion are given in table 4. Also included are the radial relaxation terms and $k_e$ from (64). Also tabulated is $k_e$ from (61) by determining $F_{int}$ directly from $\rho_{mol}(\underset{\sim}{r}\ ;\ R)$ at several R about $R_e$. In all cases studied the 2 2 rigid pseudoatoms predict a force that is too large. For FH the value is 24% off and the other constants are too large by over 50%. In all cases the radial relaxation terms are negative, but it is not clear why this is the case. Our evaluation of (64) agrees to within 1% of the value determined from (61).

Acknowledgment.
This research was supported by NSF grant CHE-77-09649

Table 4. Contributions to $k_e$ from [2|2] pseudoatoms

| molecule | $N_2$ | CO | BF | FH |
|---|---|---|---|---|
| $k_e^{rigid}$(au) from (74) | 2.83 | 2.73 | 1.25 | 1.07 |
| radial relaxation terms | -1.04 | -.99 | -.43 | -.21 |
| $k_e$(au) from (64) | 1.79 | 1.74 | .83 | .86 |
| $k_e = -\left(\frac{dF_{int}}{dR}\right)_{R_e}$ | 1.81 | 1.76 | .83 | .86 |

References

1. M. Born, Rep. Prog. Phys. 9, 294-333 (1942-1943)
2. R.F. Stewart, J. Bentley, B. Goodman, J. Chem. Phys. 63, 3786-3793 (1975)
3. J. Bentley, R.F. Stewart, J. Chem. Phys. 63, 3794-3803 (1975)
4. P.E. Cade, A.C. Wahl, At. Data and Nucl. Data Tables 13, 371 (1974)
5. P. E. Cade, K.D. Sales, A.C. Wahl, J. Chem. Phys. 44, 1973-2003 (1966)
6. W. Huo, J. Chem. Phys. 43, 624-647 (1965)
7. A.D. McLean, M. Yoshimine, IBM J. Res. Dev. 12, (1976) Suppl. pp 20-26
8. P.E. Cade, W. Huo, J. Chem. Phys. 47, 614-648 (1967)
9. J. Bentley, R.F Stewart, J. Comp. Phys. 11, 127-145 (1973)
10. P. Debye, J. Chem. Phys. 9, 55-60 (1941)
11. R.F. Stewart, Israel J. Chem. 16, 137-143 (1977)
12. A.F. Anderson, N.C. Handy, R.G. Parr, J. Chem. Phys. 50, 3634-3635 (1969)
13. R.J. Bartlett, R.G. Parr, J. Chem. Phys. 67, 5828-5837 (1977)
14. P.P. Feynman, Phys. Rev. 56, 340-343 (1939)
15. R.F. Stewart, Chem. Phys. Lett. 26, 121-125 (1974)

FORCE CONSTANTS IN DIATOMIC MOLECULES, FROM CHARGE DENSITIES

(from the work of Parr and coworkers)

Pierre BECKER

Following Parr, we represent the charge density of a molecule :

$$\rho(\underset{\sim}{r}) = \sum_{\eta} \rho_{\eta}(\underset{\sim}{r}_{\eta}) + \rho_{npf}(\underset{\sim}{r},\underset{\sim}{R})$$

where $\rho_{\eta}(\underset{\sim}{r}_{\eta})$ is a perfectly following pseudoatom, which follows the motion of nucleus $\underset{\sim}{R}_{\eta}$. $\rho_{npf}$ is the non perfectly following part of the density.

1. Show that :

$$\nabla^2_{\alpha} W = 4\pi Z_{\alpha} \sum_{\eta\neq\alpha} \rho_{\eta}(\alpha) - Z_{\alpha} \int \underset{\sim}{\nabla}_{\alpha}\rho_{npf} \cdot \underset{\sim}{\nabla}_{\alpha}\left(\frac{1}{r_{\alpha}}\right) dv$$

where $\rho_{\eta}(\alpha)$ is the contribution at point $\underset{\sim}{R}_{\alpha}$ of pseudoatom

We recall that :

$$\nabla^2\left(\frac{1}{r}\right) = -4\pi\delta(\underset{\sim}{r})$$

W is the energy of the system.

2. Show that if $\rho_{npf}$ can be assumed to be a sum of terms that follow points as :

$$K_{\alpha\beta}\underset{\sim}{R}_{\alpha} + (1-K_{\alpha\beta})\underset{\sim}{R}_{\beta}$$

its contribution to $\nabla^2 W$ is zero. In the following, we assume that :

$$\nabla^2 W = 4\pi Z_{\alpha} \sum_{\eta\neq\alpha} \rho_{\eta}(\alpha)$$

and moreover we shall restrict ourselves to the diatomic case $(\alpha,\beta)$

3. Show that the harmonic force constant is :

$$k_e = 4\pi Z_{\alpha}[\rho_{\beta}(\alpha)]_e = 4\pi Z_{\beta}[\rho_{\alpha}(\beta)]_e$$

e meaning the equilibrium distance. We shall assume that variation

of $\rho_\alpha$ or $\rho_\beta$ with the distance is exponential. Discuss.
Show that we get :

$$k_e = 4\pi Z_\alpha Z_\beta C \exp(-\zeta R_e)$$

How can one verify if this expression is in correct agreement with experimental force constants ?

4. Show, as a function of geometry, that :

$$d^2W/dR^2 + \frac{2}{R}\, dW/dR = 4\pi Z_\alpha \rho_\beta(\alpha) = F(R)$$

where F(R) acts as an effective density function. Integrate this equation to retrieve W(R) as a function of $W(R_e)$ and F(R). What is the result if we use the function F(R) introduced earlier ? We define $\chi = \zeta R$, $\chi_e = \zeta R_e$. Calculate the harmonic $k_e$, cubic $l_e$ and quartic $m_e$ force constants as a function of $\zeta$, $R_e$. Calculate the predicted dissociation energy.

5. Other F(R) can be assumed :

$$F(R) = C \exp(-\zeta R)/R$$

$$F(R) = C/R^3.$$

$k_e$, $l_e$, $m_e$ in CGS

| | $R_e$(au) | $k_e 10^5$ | $-l_e 10^{13}$ | $m_e 10^{21}$ | $D_e$(eV) |
|---|---|---|---|---|---|
| BeH | 2.54 | 2.26 | 10. | 38.4 | 2.3 |
| BH | 2.33 | 3.04 | 15.8 | 70.3 | 3.6 |
| CH | 2.12 | 4.51 | 26.7 | 136. | 3.6 |
| OH | 1.83 | 7.79 | 54.5 | 337. | 4.6 |
| FH | 1.73 | 9.66 | 69.9 | 446. | 6.5 |
| MgH | 3.27 | 1.28 | 4.92 | 13.3 | 2.5 |
| LiF | 2.92 | 2.48 | 12.4 | 55.1 | 6.5 |
| BO | 2.27 | 13.7 | 89.7 | 481. | 9.2 |

Hint. The expressions to be derived are given in the text. See the papers by Parr and coworkers, references in the notes on kinematical theory, part III.

# Part IV
# Related Techniques

# COMPTON SCATTERING

R. J. Weiss

Army Materials and Mechanics Research Center

Watertown, Massachusetts 02172 U.S.A.

## 1. Momentum Space

The Schrödinger equation in momentum space is an integral equation and appears too cumbersome to solve directly. As such theoreticians rely on the Dirac transformation to obtain the momentum wave function $\chi(\vec{p})$ from their calculated position wave function $\psi(\vec{r})$

$$\chi(\vec{p}) = (2\pi)^{-3/2} \int \psi(\vec{r}_i) \prod_i \exp(-i\vec{p}_i \cdot \vec{r}_i)\, d\vec{r}_i \tag{1}$$

For a one electron wave function separable into radial and hydrogenic angular variables $R(r)\ \Theta(\theta)\Phi(\phi)$. The one electron momentum density becomes

$$|\chi_\ell|^2 = \frac{2}{\pi} |\Theta(\theta')|^2 |\Phi(\varphi')|^2 \left| \int R(r)\, j_\ell(pr)\, r^2 dr \right|^2 \tag{2}$$

where $\theta'$ and $\phi'$ are the angular variables in momentum space, $j_\ell(x)$ is the spherical Bessel function ($\ell$ = 1,2,3, etc. for s, p, d, and f functions). Table I lists some position and momentum wave functions.

TABLE I

| Electron | Position Wavefunction | Momentum Wavefunction |
|---|---|---|
| 1s | $2b^{3/2} \exp(-br)/\sqrt{4\pi}$ | $4b^{5/2}/\pi\sqrt{2}(b^2+p^2)^2$ |
| 2p ($m_\ell = \pm 1$) | $(2\pi)^{-1/2} b^{5/2}\, r \exp(-br) \sin\theta \exp(\pm i\varphi)$ | $\dfrac{8 i b^{7/2} (\sin\theta\cos\varphi \pm i \sin\theta \sin\varphi) p}{\pi (b^2+p^2)^3}$ |
| 2p ($m_\ell = 0$) | $b^{5/2}\, r \cos\theta \exp(-br)/\sqrt{\pi}$ | $\dfrac{8 i b^{7/2} (\cos\theta) p}{\pi\sqrt{2}(b^2+p^2)^3}$ |

## 2. Inelastic Scattering from Free Electrons

A photon of momentum $\vec{k}_0$+ energy $E_o$ scattered through an angle $2\Theta$ by a free electron initially at rest suffers an energy loss $\Delta E$ and a wave length increase $\Delta\lambda$ given by

$$\Delta E = |\vec{k}-\vec{k}_o|^2/2m = \frac{(2E_o/mc^2)\sin^2\theta}{[1+(2E_o\sin^2\theta/mc^2)]} \tag{3}$$

$$\Delta\lambda = (2h/mc)\sin^2\theta \tag{4}$$

($2h/mc$ = 0.0485262 Å; $mc^2$ = 510.964 KeV) where $\vec{k}$ is the final momentum. The final energy of the photon $E_c$ is given by

$$E_c = E_o/[1+(2E_o\sin^2\theta/mc^2)] \tag{5}$$

Table II lists values of $\Delta E$ imparted to the electron for some typical photon energies $E_o$ employed in Compton scattering ($2\Theta$=180°).

TABLE II

| $E_o$(eV) | $\Delta E$(eV) | |
|---|---|---|
| 17,374 | 1,106 | (MoKα X-ray) |
| 59,570 | 11,400 | (Americium γ-ray source) |
| 159,000 | 60,994 | (Tellurium γ-ray source) |

If the free electron has an initial component of momentum z along the photon scattering vector $|\vec{k}-\vec{k}_o|$ then the photon energy loss is given by

$$\Delta E = z|\vec{k}-\vec{k}_o|/m + |\vec{k}-\vec{k}_o|^2/2m \tag{6}$$

Thus a symmetric distribution of initial electron momenta leads to a distribution in energy losses centered around the energy loss for an electron at rest.

The differential cross section into solid angle $\Omega$ for such a process is given by

$$\frac{d\sigma}{d\Omega\, d(\Delta E)} = \left(\frac{e^2}{mc^2}\right)^2 K^2 \frac{E}{E_o} \sum_f \sum_o \left|\langle\psi_f|e^{i(\vec{k}-\vec{k}_o)\cdot\vec{r}}|\psi_o\rangle\right|^2 \delta(E-E_o-\Delta E) \tag{7}$$

where K is the usual polarization factor, $\psi_f$ and $\psi_o$ are the final and initial electron wave functions and the delta function signifies conservation of energy. Since the final state electron wave function must be a plane wave $e^{i\vec{p}\cdot\vec{r}}$ and since momentum is conserved the matrix elements in eq. 7 reduce with the help of eq 1 to an integral over the initial state momentum density i.e.

$$\frac{d\sigma}{d\Omega\, dz} = \left(\frac{e^2}{mc^2}\right)^2 K^2 \frac{E}{E_o} \int_{|z|}^{\infty} |\chi(\vec{p}_o)|^2\, dp_x\, dp_y \tag{8}$$

where the integration is over the xy plane in momentum space at a distance $|z|$ (eq.6) from the origin. By convention the integral in eq. 8 is called the Compton profile J (z)

$$J(z) = \int_{|z|}^{\infty} |\chi(\vec{p}_o)|^2 \, dp_x dp_y \tag{9}$$

If $|\chi(\vec{p}_o)|^2$ is spherically symmetric then

$$J(z) = 2\pi \int_{|z|}^{\infty} |\chi(p_o)|^2 p_o \, dp_o \tag{10}$$

## 3. Inelastic Scattering From Bound Electrons

For electrons bound on atoms the differential cross section eq. 7 is only soluble in closed form for the hydrogenic atom (single electron). While the initial state wave function $\psi_o$ is fairly simple for the hydrogenic atom the final continuum state wave functions $\psi_f$ involve hypergeometric functions. For a many electron atom a series of approximations must be made by the theoretician in order to calculate the matrix elements in eq. 7.

For the ground state $\psi_o$, approximations such as H.F. functions for the free atom, molecular orbitals of various sorts for molecules, and assorted band or cluster calculations for solids are traditionally part of the literature and have been discussed by Professors Smith and March. It is the final state wave functions for the many electron system which are the problem. Theoreticians have employed screened hydrogenic-like solutions and numerical solutions but in the main the problem has been swept under the rug by selecting experimental conditions such that the final state electron is imparted an energy large compared to its binding energy. This is called the impulse approximation and permits us to approximate the final state wave function as a plane wave so that eq. 7 reduces to eq. 8. Comparison of the exact solution of eq. 7 for the hydrogenic atom with the impulse approximation of eq. 8 indicates that for a photon energy loss four to five times the electron binding energy the impulse approximation is accurate to better than 1%. Thus eq. 8 has come to be the standard operational equation for Compton scattering.

## 4. Measurement Techniques

While X-ray sources were initially employed for Compton profile measurements monochromatic $\gamma$-ray sources such as radioactive Americium and Tellurium (Table III) are the most popular and clearly satisfy the impulse approximation for elements up to the first transition group. The source is housed in a lead shield and the $\gamma$-rays scattered through an angle of $\sim 160^o$ are detected with a non-dispersive pure Ge detector and a multichannel recorder. The resolution of a Ge detector is about 350 eV at 60 KeV and $\sim$550 eV at 159 KeV.

After making background, absorption, multiple scattering and resolution corrections the data is converted to J(z) by application of eq. 8 and eq. 6. The latter is more conveniently written

$$z \cong \left(\frac{mc}{2}\right)\frac{(E_0/E) - \left(1 + \frac{2E_0}{mc^2}\sin^2\theta\right)}{(E_0/E)^{1/2}\,\delta\,\sin\theta} \tag{11}$$

where $\delta$ is small a correction term

$$\delta = \left\{1 + \left(\sum_{n=2}^{\infty}(\Delta E/E_0)^n\right)(1+\sin^2\theta)\right\}^{1/2} \tag{12}$$

Measurements are often made on single crystals since the momentum density like the charge density must reflect the crystal symmetry (eq. 2).

## 5. Results of Measurements

Hundreds of substances have been measured but we shall select just a few to illustrate the salient features of the electron momentum distribution.

### a. Li Metal

Differentiation of the measured Compton profile (eq. 9) gives the radial momentum density

$$I(p) \equiv 4\pi|\chi(p)|^2 p^2 = \left|2z\frac{dJ}{dz}\right| \tag{13}$$

and this is shown in Figure 1 for Li after subtracting the $is^2$ core electron contribution. Clearly shown are the discontinuity at the Fermi momentum $p_F$ = 0.59 a.u. which in free electron theory is related to the density of electrons

$$P_F = \pi\left(\frac{3N}{\pi}\right)^{1/3} \quad \text{a.u.} \tag{14}$$

where N is the number of valence electrons per unit volume in Bohr units (1 Bohr unit of length is 0.529 Å = $\hbar^2/me^2$). The shaded area gives the uncertainty in the experimental results while the dashed curve is the result of a theoretical calculation for the interacting electron gas. Above the Fermi momentum there are states due to the Coulomb repulsion between the the valence electrons as well as due to orthogonalization with the core electron and interaction with the crystal potential.

### b. Diamond

The so-called forbidden (222) reflection in diamond is a direct consequence of the anisotropy in the charge density i.e. the electron charge points toward the four nearest neighbors tetrahedrally deployed around each atom. The anisotropy in the momentum density is quite large (~10%) and is shown in Figure 2 for the [100] and [111] directions together with 2 theoretical calculations. The momentum density anisotropy can not be described in terms of a simple tetrahedrally pointed model but is significantly more complicated.

c. Transition Metals

Measurements of the paired reflections 330-411 and 600-442 (X-ray reflections in cubic substances with identical $\sin\theta/\lambda$) in b.c.c. vanadium and iron metal indicated a significant anisotropy in the charge density with lobes pointing toward the eight nearest neighbors along the body diagonals. It initially came as a surprise that the momentum density has a very small anisotropy. It was a further surprise that the theoretical band calculations predicted just the reverse, a large momentum density anisotropy. For the f.c.c. close-packed transition metal nickel theory and experiment do agree. An understanding of the discrepancy in vanadium and iron may help in understanding why all the early transition metals are b.c.c and the latter transition metals close packed.

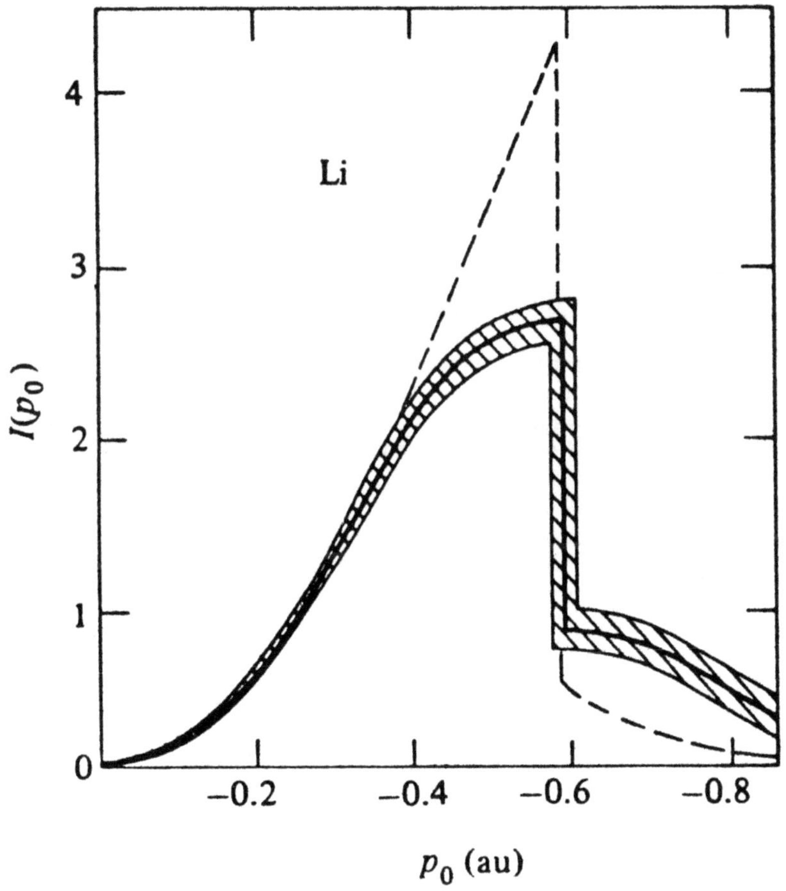


Fig. 1 - $I_{(p_o)}$ for lithium metal

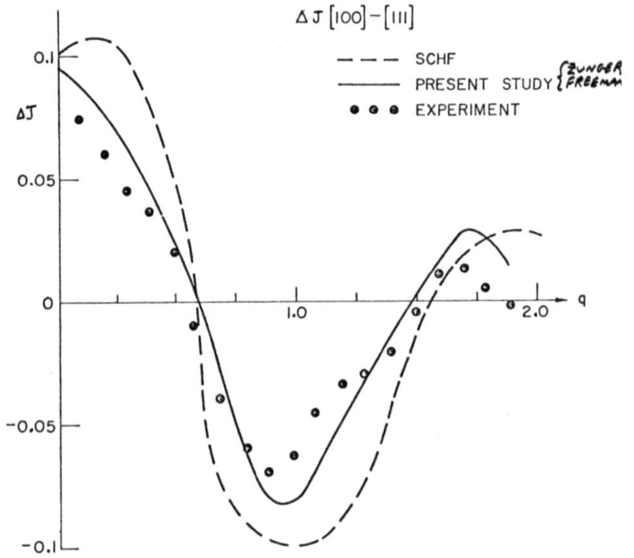


FIG 2 - T (100)-(111) for diamond

## Problems

1. Calculate the momentum wave functions for the **hydogenic 3d** orbital and demonstrate their orthogonality for $m_\ell$ = ±2, ±1 and 0. Show that the charge and momentum density point in the same direction. Calculate the X-ray scattering factor and Compton profiles for these orbitals.

2. Calculate the structure factors and Compton profile for the valence electrons in Be metal in the free electron approximation. Compare with experiment and free atom theory.

3. Study the paper of Phillips and Weiss on Charge, Spin and Momentum Density in Iron, Phys. Rev B 6, 4213, 1972. What does one learn about the relationship between these densities?

## Answers

1.

Position space :

$$\Psi = \frac{2}{3}\left(\frac{2}{5}\right)^{1/2}\alpha^{1/2} r^2 \exp(-\alpha r)\left\{\frac{5(2-m)!}{4\ (2+m)!}\right\}^{1/2} P_1^m(\cos\theta)\exp(im\phi)$$

| m | $P_2^m(\cos\theta)$ |
|---|---|
| 2,-2 | 3 |
| 1,-1 | $3\cos\theta$ |
| 0 | $1/2(3\cos^2\theta - 1)$ |

Momentum space :

$$\chi = -\left(\frac{\alpha}{5\pi}\right)^{1/2} \frac{64\alpha^4 p^2}{(\alpha^2+p^2)^4} (-1)^m\left\{\frac{5(2-m)!}{4\ (2+m)!}\right\}^{1/2} P_2^m(\cos\Theta)\exp(im\Phi)$$

$$f(S) = \int \Psi\Psi^* \exp(i\underset{\sim}{S}.\underset{\sim}{r})\ d\underset{\sim}{r}$$ eqn 1-34 Azoroff... X-Ray Diffraction 1974 Mc Graw Hill

$$J(q) = \int_{q_z}^{\infty} \chi\chi^* dp_x dp_y$$ eqn 1-96

The following integrals are useful

$$\int_0^\infty r^4 \exp(-\alpha r) j_2(pr) dr = \frac{48\alpha p^2}{(\alpha^2+p^2)^4}$$

$$\int_q^\infty \frac{p^n}{(z^2+p^2)^r} dp = \frac{q^{n-1}}{2(r-1)(z^2+q^2)^{r-1}} + \frac{(n-1)q^{n-2}}{2(r-1)(r-2)(z^2+q^2)^{r-2}}$$

$$\ldots\ldots \frac{(n-1)(n-2)\ldots(2)q^0}{2(r-1)(r-2)\ldots(r-n+2)(z^2+q^2)^{r-n+2}} \qquad n \text{ odd}$$

2.

Structure factors : zero

Compton profile :

$$J(q) = \frac{3n(p_F^2 - q^2)}{4\ p_F^2}$$

$n = 2$

$p_F$ around 1.0

## DETERMINATION OF CHARGE DENSITIES AND RELATED QUANTITIES BY USE OF HIGH ENERGY ELECTRON SCATTERING


R. A. Bonham and M. Fink

Chem. Dept., Indiana Univ. Bloomington, Ind., 47401

Phys. Dept., Univ. of Texas Austin, Texas, 78712


### I. INTRODUCTION

According to the theory of scattering developed in my first lecture the intensity expressions for electron scattering can be written as

$$I_{EL} = \frac{4I_0}{R^2K^4}[Z - F(K)]^2 \tag{1.0}$$

for elastic scattering from atoms where $I_0$ is a constant depending on the experimental parameters and

$$I_{TOTAL} = \frac{4I_0}{R^2K^4}[Z^2 - 2ZF(K) + N + \langle\Psi_0|e^{iK\cdot r_{12}}|\Psi_0\rangle] \tag{1.1}$$

Averaging over all angles of orientation of the target yields the results in terms of charge densities as

$$I_{EL} = \frac{4I_0}{R^2K^4}[Z^2-2Z\int d\vec{r}\rho(\vec{r})j_0(Kr)+\int d\vec{r}\rho(r)\int d\vec{r}'\rho(\vec{r}')j_0(K|\vec{r}-\vec{r}'|)] \tag{1.2}$$

and

$$I_{TOTAL} = \frac{4I_0}{R^2K^4}[Z^2-2Z\int d\vec{r}\rho(\vec{r})j_0(Kr) + N + \int d\vec{r}\rho_c(\vec{r})j_0(Kr)] \tag{1.3}$$

We can easily show that

$$\int_0^\infty dK\left[\left(\frac{R^2K^4}{4I_0}\right)I_{TOTAL} - Z^2-N\right] = \frac{2}{\pi}(\overline{V}_{en} + \overline{V}_{ee}) = \frac{1}{\pi}E \tag{1.4}$$

where E is the total electronic energy of the atom in its ground state and[1,2]

$$\int_0^\infty dK\left[\left(\frac{R^2K^4}{4I_0}\right)I_{EL} - Z^2\right] = \frac{2}{\pi}\left(\bar{V}_{en} + \bar{V}_{ee}^{\,Coul.}\right) \tag{1.5}$$

Comparing these results with the x-ray results shows that the contributions to the average electronic potential energy $\bar{V}_{en}$, $\bar{V}_{ee}$, $\bar{V}_{ee}^{\,Coul.}$ and $\bar{V}_{ee}^{\,Ex}$ are all accessible to experimental measurement.

What happens in the molecular case? The main differences are that the electron density is no longer spherically symmetric and more than one nuclear center exists. The scattering amplitude operator can now be written as

$$\Psi_{sc} = \frac{2e^{ikR}\left[\sum_{n=1}^{M} Z_n e^{i\vec{K}\cdot\vec{R}_n} - \sum_{i=1}^{N} e^{i\vec{K}\cdot\vec{r}_i}\right]}{RK^2} \tag{1.6}$$

where $Z_n$ is the charge on the <u>nth</u> of M nuclei and $\vec{R}_n$, $\vec{r}_i$ are the instantaneous positions of the <u>nth</u> nucleus and the <u>ith</u> electron with respect to the center of mass of the molecule. Note that the x-ray scattering amplitude operator is obtained from (1.6) by letting all the $Z_n$'s vanish (i.e. no nuclear scattering).

From Eq. (1.6) we can calculate the elastic scattering at electronic resolution (all translational, rotational and vibrational inelasticity included) as

$$I_{EL}^{\,e} = \frac{4I_0}{R^2K^4}\Bigg\{\sum_{n=1}^{M} Z_n^2 + \sum_{n\neq m=1}^{M}\sum^{M} \langle Z_n Z_m j_0(KR_{nm})\rangle_{vib} \tag{1.7}$$

$$- 2\sum_{n=1}^{M} Z_n \left\langle \int d\vec{r}\rho(\vec{r}) j_0(K|\vec{R}_n - \vec{r}|)\right\rangle_{vib}$$

$$+ \left\langle \int d\vec{r}\rho(\vec{r}) \int d\vec{r}\,'\rho(\vec{r}\,') j_0(K|\vec{r}-\vec{r}\,'|)\right\rangle_{vib}\Bigg\}$$

and for the total scattering as

$$I^{e}_{TOTAL} = \frac{4I_0}{R^2K^4}\Big\{\sum_{n=1}^{M} Z_n^{\,2} + \sum_{n\neq m=1}^{M}\sum^{M} \langle Z_n Z_m j_0(KR_{nm})\rangle_{vib} \tag{1.8}$$

$$- 2\sum_{n=1}^{M} Z_n \langle \int d\vec{r}\rho_c(\vec{r}) j_0(K|\vec{R}_n - \vec{r}|)\rangle_{vib}$$

$$+ N + \langle \int d\vec{r}\rho_c(\vec{r}) j_0(Kr)\rangle_{vib}\Big\}$$

The last term to the right in Eqs. (1.7) and (1.8) corresponds to the x-ray scattering cases of elastic and total scattering respectively. We may still define energy relationships in the same way as in the atomic case. That is

$$\frac{2}{\pi}\bar{V} = \frac{2}{\pi}[\bar{V}_{nn} + \bar{V}_{ne} + \bar{V}_{ee}] \tag{1.9}$$

$$= \int_0^\infty dK\left[\left(\frac{R^2K^4\, I^{e}_{TOTAL}}{4I_0}\right) - \sum_{n=1}^{M} Z_n^{\,2} - N\right]$$

and

$$\frac{2}{\pi}[\bar{V}_{nn} + \bar{V}_{ne} + \bar{V}_{ee}^{\,Coul.}] \tag{1.10}$$

$$= \int_0^\infty dK\left[\left(\frac{R^2K^4\, I^{e}_{EL}}{4I_0}\right) - \sum_{n=1}^{M} Z_n^{\,2}\right]$$

There is one difference in the molecular case which shows up in the relationship between total average potential energy and the total electronic energy of the molecular system. This is due to the fact that the molecular virial theorem is

$$E(\vec{R}_i) + \frac{1}{2}\sum_{j=1}^{M} \vec{R}_j\cdot\vec{\nabla}_j E(\vec{R}_i) = \bar{V}(\vec{R}_i) \tag{1.11}$$

where $E(\vec{R}_i)$ is the total electronic energy of the molecule with all nuclei frozen into the positions $\vec{R}_i$, $i = 1, \ldots M$. If all the $E(\vec{R}_i)$ in Eq. (1.11) are expanded in normal coordinates about the equilibrium configuration and if the vibrational average is carried out assuming simple harmonic motion then the total electronic energy for the equilibrium configuration of the molecule can be shown to be approximately

$$E(\vec{R}_i^{\,e}) = \frac{1}{2}\langle\overline{V}\rangle_{vib} - \langle E_{vib}\rangle \qquad (1.12)$$

with $\langle E_{vib}\rangle \cong \sum_i \hbar\omega_i \coth(\hbar\omega_i/2k_BT)$ where $\omega_i$ is the frequency of the ith normal mode, $k_B$ is the Boltzman factor, T is the absolute temperature and $\langle E_{vib}\rangle$ is the thermally averaged total vibrational energy of the system.

## II. ADVANTAGES OF THE ELECTRON SCATTERING METHOD

Perhaps the major advantage of electron scattering is the large cross section which makes possible the attainment of high accuracy for intensity measurements. This advantage for electron scattering from single molecules turns into somewhat of a disadvantage for electron scattering by crystals. In this case because of the large single molecule scattering, multiple scattering tends to dominate and any interpretation of the scattering in terms of molecular electron charge densities becomes rather complicated.[3] Here we will limit ourselves to a discussion of electron molecule scattering in the gas phase.

A second advantage somewhat related to the first is the fact that single scattering conditions can be relatively easily obtained. In addition only vibrational corrections for intra molecular motions need be made. These motions make a relatively small correction to the experimental data in the region of momentum transfer ($.5 \leq K \leq 5$ (a.u.)) most sensitive to the charge distribution. The mean square amplitudes needed to make vibrational corrections can be obtained from the electron scattering data by analysis of the scattered intensity at values of K beyond the charge sensitive region. At the same time such large angle data[4] can be analyzed for bonded and non bonded internuclear distances. This data combined with the mean amplitude data can be used to evaluate the nuclear-nuclear scattering term so it can be subtracted from the experimental intensity. This allows us to define the three experimentally accessible intensity functions.

$$N_{TOTAL}(K) = -2\sum_{n=1}^{M} Z_n\langle\int d\vec{r}\,\rho(\vec{r})j_0(K|\vec{r}-\vec{R}_n|)\rangle_{vib} \qquad (2.0)$$

$$+\ \langle\int_0^\infty dr\ P_0(r)j_0(Kr)\rangle_{vib}$$

where $P_0(r)$ is the spherically averaged electron pair correlation function $4\pi r^2 \int d\Omega \rho_c(\vec{r})$,

$$N_{EL}(K) = -2 \sum_{n=1}^{M} Z_n \langle \int d\vec{r} \rho(\vec{r}) j_0(K|\vec{r}-\vec{R}_n|) \rangle_{vib} \tag{2.1}$$

$$+ \langle \int d\vec{r} \rho(\vec{r}) \int d\vec{r}\,' \rho(\vec{r}\,') j_0(K|\vec{r}-\vec{r}\,'|) \rangle_{vib}$$

and

$$N_{INEL}(K) = \langle \int_0^\infty dr P_0(r) j_0(Kr) \rangle_{vib} - \langle \int d\vec{r} \rho(\vec{r}) \int d\vec{r}\,' \rho(\vec{r}\,') j_0(K|\vec{r}-\vec{r}\,'|) \rangle_{vib} \tag{2.2}$$

The total scattering and the elastic scattering contain the interesting term $-2 \sum_{n=1}^{M} Z_n \int d\vec{r} \rho(\vec{r}) j_0(K|\vec{r}-\vec{R}_n|)$ which is linear in the charge density and has no counterpart in the x-ray scattering case. This term arises because the incident electrons see both the nuclei and planetary electrons of the target system with comparable scattering power. In addition the nebulous planetary electron cloud is viewed upon a relatively rigid nuclear framework. If $\rho(\vec{r})$ is expanded in spherical harmonics about a single center as

$\rho(\vec{r}) = \frac{1}{4\pi} \sum_{\ell=0}^{\infty} \sum_{m=-\ell}^{+\ell} \rho_\ell(r) Y_{\ell,m}(\theta,\varphi)$ then the electron-nuclear interference term can be written as

$$-2 \sum_{n=1}^{M} Z_n \int d\vec{r} \rho(\vec{r}) j_0(K|\vec{r}-\vec{R}_n|) \tag{2.3}$$

$$= \sum_{n=1}^{M} Z_n \sum_{p=0}^{\infty} \sum_{m=-p}^{+p} j_p(KR_n) Y_{p,m}(\theta_n,\varphi_n) \cdot \int_0^\infty dr r^2 \rho_p(r) j_p(Kr)$$

while the Coulomb term becomes

$$\int d\vec{r}\rho(\vec{r})\int d\vec{r}'\rho(\vec{r}')j_0(K|\vec{r}-\vec{r}'|) = \frac{1}{4\pi}\sum_{n=0}^{\infty}\left[\int^{\infty} drr^2\rho_n(r)j_n(Kr)\right]^2 \qquad (2.4)$$

where $R_n$, $\theta_n$, $\varphi_n$ define the instantaneous position of the nth nucleus with respect to the principle axis system of the molecule.[5] As we will see from the problem set the integral $\int_0^{\infty} drr^2\rho_n(r)j_n(Kr)$ which is a Fourier-Bessel transform of the nth radial component of the 3-D electron density, is a kind of generalized scattering factor. In actual practice we will find it more convenient to use simultaneous expansion of the 3-D density about each nuclear center in the molecule.[6] The main point here is the information content of the experiment. The experiment contains all the information necessary to reconstruct the 3-D one electron density (i.e. the $\rho_n(r)$'s) but only a single piece of information concerning the electron pair correlation density (i.e. $P_0(r)$). How much of this information can be extracted from the data is another matter which will be discussed in the next section.

Several additional advantages are worth mentioning before proceeding to a discussion of problems. Since we are dealing with gas scattering the scattered intensity is a continuous function of the scattering angle which means that we can measure it at as many points as we wish as long as the auto-correlation between adjacent data points is small. A further advantage is the fact that current experimental relative intensity measurement accuracy is at or below the 0.1% accuracy level. This is partly due to the larger scattering cross sections using electrons but has probably more to do with the hard work and ingenuity of the experimenters working in this field.[7] Let me simply point out that 0.1% involves not only a counting uncertainty of better than .1% but also ability to measure the scattering angle to better than 2 arc seconds!

## III. DISADVANTAGE OF THE ELECTRON SCATTERING METHOD

Because of the fact that the target is a dilute gas no molecular orientational effects are possible. This means the scattering information is one dimensional. Hence the number of possible observations in the region sensitive to the charge density is limited by correlation between adjacent data points to something like 100 independent observations. This can be compared to the thousands or tens of thousands which can be observed in x-ray single crystal work. Another consequence of one dimensional data is the fact that the transformation from the experimental intensities to the 3-D charge density is non unique.[2,6] In some favorable cases (probably limited to homo nuclear diatomics) it may be possible to infer a 3-D charge density from electron scattering data alone. One should not lose sight of the fact however that the transformation from a theoretical charge distribution to a scattered intensity is unique. The real value of such electron scattering measurements ultimately lies in their use as a very sensitive test of theoretical predictions about the nature of charge densities. As we shall show, current experimental accuracy is now comparable in sensitivity to the best wave function calculations.

There is a problem in using electrons which is the consequence of the quantum mechanical principal of the indistinguishability of identical particles. That is if we shoot an electron at a target and detect a scattered electron there is no assurance that the detected electron is the original projectile electron. It may in fact be an electron originally belonging to the target. Such an event is referred to as exchange scattering. Fortunately exchange scattering corrections are small in the limits of high incident electron energy and small scattering angles.[8,9]

In addition to the general disadvantages listed above which are peculiar to the electron scattering method there are a number of difficulties which have their counterpart in the x-ray method. In the case of measuring the total scattered electron intensity the elastic and inelastic scattering are of course recorded simultaneously. While both are recorded at the same scattering angle, unfortunately they are not exactly at the same momentum transfer which is a requirement of the simple theory we have presented in the previous sections. This necessitates the use of small but necessary theoretical corrections to the scattering data.[9] In addition the approximation that the scattered amplitude is given by the expression $2[Z-F(K)]/K^2$ for elastic scattering breaks down and the potential scattering problem must be solved by more rigorous means.[10] These corrections are of the order of 4% for Ar (Z=18) and smaller for the lighter elements. For very careful work corrections must be applied even to first row elements.[2] This is because 0.1% relative measurement accuracy is required to achieve an energy accuracy of the order of 0.2 eV - 0.5 eV by use of the energy theorems presented in previous sections. Such corrections are generally less than 0.5% of the total intensity.

There is also the problem that the experiment observes a vibrating molecule while most theoretical calculations neglect the effects of vibration. It has been established that vibrational effects in the momentum transfer region of greatest interest ($K = 1$ a.u. $\rightarrow 7$ a.u.) are usually small and corrections based on the perfectly following model are adequate.[11] In fact one can guess from the fact that the Debye-Waller type factor, $e^{-K^2\ell^2/2}$, has only changed from its value at $K = 0$, by 7% for $H_2$ and 1% for $N_2$ by $K = 5$ a.u. ($\frac{\sin\theta}{\lambda} = 0.75$ Å$^{-1}$) that the perfectly following approximation can be expected to be a reasonable approximation. Here $\ell^2$ is the mean square amplitude of vibration. In addition we saw from Tavard's energy theorem that the effect of vibration adds only a few hundredths of an eV to the total energy. This is a little misleading as the largest energy contribution from N(K) comes from the region $K \cong 1 - 2$ a.u. where vibrational corrections are very small. The region around $K = 3 - 4$ a.u. which is sensitive to the quadrupole nature of the field is also more sensitive to vibrations. Of course the uncertainty we are talking about here is not the magnitude of the factor $1-e^{-K^2\ell^2/2}$ but rather the error in the perfectly following approximation which is something much smaller.

## IV. EXPERIMENTAL METHODS

Examples of all three experimental types, total scattering, elastic scattering and total inelastic scattering exist in the literature. All experiments must have certain features in order to obtain relative intensity accuracies approaching 0.1% or better. Obviously $10^6$ or more signal events must be recorded. In addition any background scattering must also be characterized and recorded. An angular accuracy of $\pm .001°$ ($\sim 2$ arc seconds) must be obtained. The reason for this is that $N(K) \propto K^4 I$ which means that the error in $K(\sim k\theta)$ is multiplied by a factor of 4. In order to achieve this angular accuracy the magnetic field in the experimental chamber must be reduced to a few milli Gauss and must be homogeneous to better than 1 milli Gauss.

All experiments carried out to date for the purpose of measuring electron densities have employed the beam-beam scattering method.[2-12] The gas beam is normally a jet. In all cases incident electron energies range from 25 keV to 50 keV. In total scattering experiments a plastic scintillator coupled to a photomultiplier has been used.[2] In the case of elastic scattering and total inelastic scattering a Si surface barrier detector has normally been employed.

In the case of total scattering the detector is allowed to view the entire scattering volume in order to help obviate corrections for the variation of path length through the gas beam as a function of changing scattering angle. For total inelastic scattering the detector solid angle of acceptance is limited to $10^{-6}$ - $10^{-8}$ steradians to eliminate background noise. In the latter case a complete energy loss spectrum is obtained and the inelastic intensity is converted to a relative generalized oscillator strength by means of the formula

$$\frac{df(K,E)}{dE} = \frac{E|\langle 0|\sum_{n=1}^{N} e^{iK\cdot r_i}|n\rangle|^2}{K^2} = \frac{K^2A}{E} I_{Expt.}(K,E) \tag{4.0}$$

where K is the momentum transfer, A is a normalizing constant and E is the energy loss. The Bethe sum rule[8, 12]

$$\int_0^\infty dE \frac{df(K,E)}{dE} = N \tag{4.1}$$

is then employed to place the experimental data on an absolute scale. This places the elastic intensity on an absolute scale at the same time. The fact that the data are normally obtained for a fixed scattering angle over a finite range of energies while the sum rule calls for a fixed value of the momentum transfer and an infinite data range necessitates some small corrections.[13]

In the case of total scattering accurate relative intensities must be obtained up to values of K that are large enough to guarantee that a comparison theory which is independent of the details of the charge density can be used to place the data on an absolute scale. To do this the independent atom molecule (IAM) or pro molecule model is normally used.[2] The IAM consists of spherically averaged ground state atoms arranged in the equilibrium geometry. The IAM is usually allowed to vibrate according to the perfectly following approximation.

The equations for comparing the N(K) functions with the IAM are

$$\Delta N_{TOTAL}(K) = N_{TOTAL}^{EXPT}(K) - N_{TOTAL}^{IAM}(K) \tag{4.2}$$

$$= -2 \sum_n Z_n \int d\vec{r}\ \Delta\rho(\vec{r}) j_0(K|\vec{r}-\vec{R}_n|)$$

$$+ \int_0^\infty dr\ \Delta P_0(r) j_0(Kr)$$

$$\Delta N_{EL}(K) = N_{EL}^{EXPT}(K) - N_{EL}^{IAM}(K) \tag{4.3}$$

$$= -2 \sum_n Z_n \int d\vec{r}\, \Delta\rho(\vec{r}) j_0(K|\vec{r}-\vec{R}_n|)$$

$$+ \int d\vec{r}\, \rho(\vec{r}) \int d\vec{r}\,'\, \rho(\vec{r}\,') j_0(K|\vec{r}-\vec{r}\,'|)$$

$$- \sum_n \int d\vec{r}\, \rho_n(|\vec{r}-\vec{R}_n|) \sum_m \int d\vec{r}\,'\, \rho_m(|\vec{r}\,'-\vec{R}_m|) j_0(K|\vec{r}-\vec{r}\,'|)$$

and

$$\Delta N_{INEL}(K) = N_{INEL}^{EXPT}(K) - N_{INEL}^{IAM}(K) \tag{4.4}$$

$$= \int_0^\infty dr\, \Delta P_0(r) j_0(Kr) - \int d\vec{r}\, \rho(\vec{r}) \int d\vec{r}\,'\, \rho(\vec{r}\,') j_0(K|\vec{r}-\vec{r}\,'|)$$

$$+ \sum_n \int d\vec{r}\, \rho_n(|\vec{r}-\vec{R}_n|) \sum_m \int d\vec{r}\,'\, \rho_m(|\vec{r}\,'-\vec{R}_m|) j_0(K|\vec{r}-\vec{r}\,'|)$$

where

$$\Delta\rho(\vec{r}) = \rho(\vec{r}) - \sum_n \rho_n(|\vec{r}-\vec{R}_n|) \tag{4.5}$$

and

$$\Delta P_0(r) = P_0(r) - \sum_n P_0^{\,n}(r) \tag{4.6}$$

with $\rho_n(|\vec{r}|)$ the one electron density of the nth atom, and $P_0^{\,n}(r)$ the electron pair correlation function for the nth atom.

Perhaps the major reason for using a comparison model is now apparent. The densities are referenced to a well defined comparison density which means that small deviations between the comparison density and the experiment can be displayed in a very sensitive manner. Examples of actual experimental data will be given in the next section.

As mentioned in Section III the simple scattering theory is never realized in actual practice and a number of corrections must be applied to the data before it can be interpreted in terms of Eqs. (4.2 - 4.4). We must ask how good are these corrections. We note that the quantities in Eqs. (4.2 - 4.4) are functions only of the momentum transfer K. Hence a necessary condition that our procedure is correct is the demonstration that corrected intensity data for the same target system collected at different incident electron energies, usually at 30, 40 and 50 keV, are identical within the experimental uncertainties. Such a procedure is also sufficient providing that no omitted theoretical corrections are functions only of K. To the best of our present knowledge no such terms exist in electron scattering.

## V. A SURVEY OF RECENT EXPERIMENTAL RESULTS

The first accurate demonstration of electron correlation effects on electron scattering was made in the case of Ne.[14] The results are shown in Fig. 1. The first attempt to separate elastic and total inelastic scattering for the express purpose of obtaining $\Delta N(K)$ curves was made by Ulsh et al.[15] The results of these investigators shown in Figs. 2 and 3 must be regarded as a first crude attempt.

Fink and coworkers have perfected total scattering experiment measurements and recently reported a first attempt at obtaining a 3-D one dimensional electron density map for $N_2$[18a] and the study of electron correlation in atoms.[18b] To obtain a 3-D density map from total scattering data requires assuming a theoretical model for the total inelastic scattering and subtracting it out. Some recent theoretical work by Stewart appears to lend some credence to this procedure.[19] In Figs. 4 - 6 we see the results of the latest attempt to invert the electron scattering data for $N_2$ by Fink's group at the University of Texas. In Fig. 4 we see the experimental density difference map where the IAM model is based on spherically averaged ground state nitrogen atoms which include essentially all of the electron correlation effects. In Fig. 5 the difference map based on Hartree-Fock (H.F.) theory is shown with the IAM based on spherically averaged H.F. atoms. Note that the two maps are pleasingly similar at a qualitative level. If the assumptions employed in obtaining these plots are warranted then we should be able to obtain a direct measure of correlation effects on the molecular density. Pierre Becker[20] has calculated the effect of electron correlation on the 3-D density in $N_2$ and the magnitude of the effect is in excellent agreement with the difference between the experiment and H.F. molecular theory as shown in Fig. 6.

It must be pointed out that in order to obtain a 3-D density map from 1-D experimental data a simple analytical model with a minimum of parameters must be employed. In the present case the model contained 3 one center monopole terms and a single quadrupole term for a total of 8 parameters which had to be obtained by least squares fitting of the experimental data. Quantitative comparison with the correlation map of Becker[20] cannot be achieved with such a

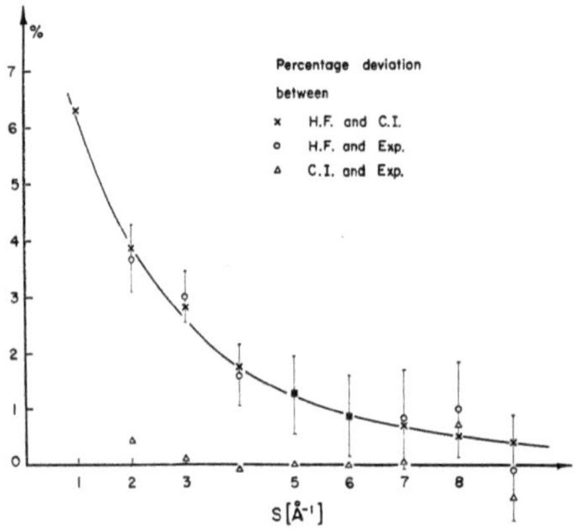


Fig. 1. Comparison between HF and CI calculated scattered intensities and HF and the experimental intensity.
X: $100[I_{total}(HF)-I_{total}(CI)/I_{total}(HF)]$;
O: $100[I_{total}(HF)-I_{total}(expt)]/I_{total}(HF)$;
Δ: $100[I_{total}(CI)-I_{total}(expt)]/I_{total}(HF)$.

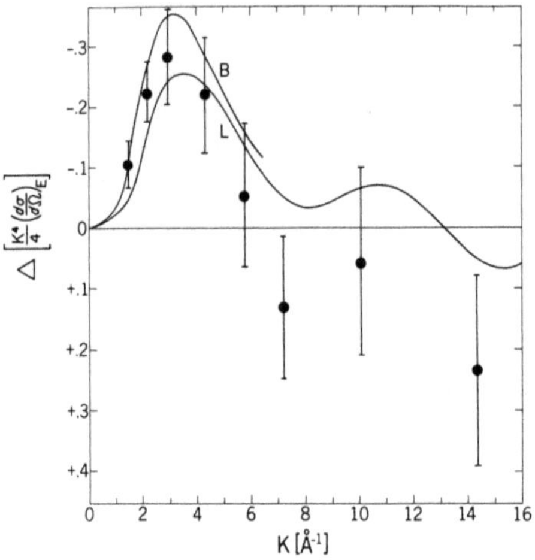


Fig. 2. Elastic scattering, $\Delta N_{EL}(K)$, for $H_2$ obtained with 25 keV electrons. The upper solid line labeled B is the theory of Ford and Browne[16] and the lower solid line labeled L is due to Liu.[17]

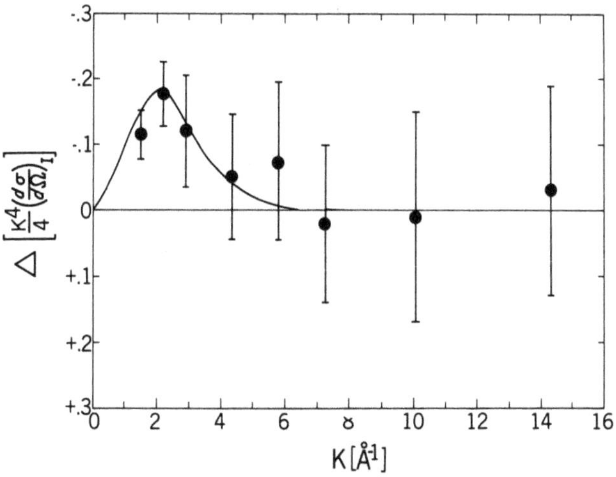


Fig. 3. The total inelastic scattering, $\Delta N_{INEL}(K)$, for $H_2$ obtained with 25 keV electrons. The solid line is theory due to Liu.[17]

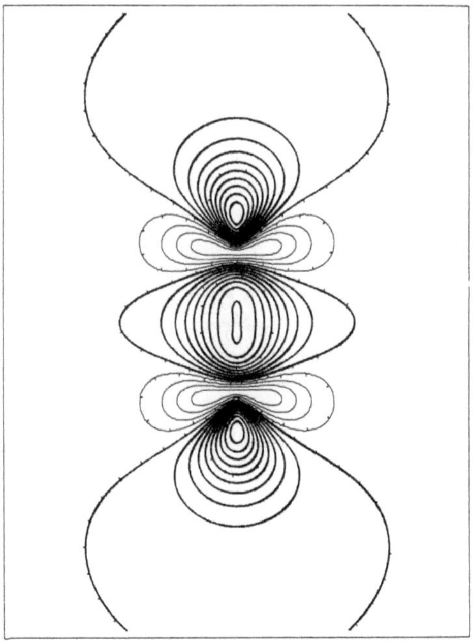

Fig. 4. Density difference map for $N_2$, $\rho_{expt.}(\vec{r}) - \rho_{IAM}(\vec{r})$, in units of 0.01 $e^-/a_0{}^3$. The light contours represent charge loss and the dark charge gain. The model includes 2 monopole, 2 dipole and 1 quadrupole terms.

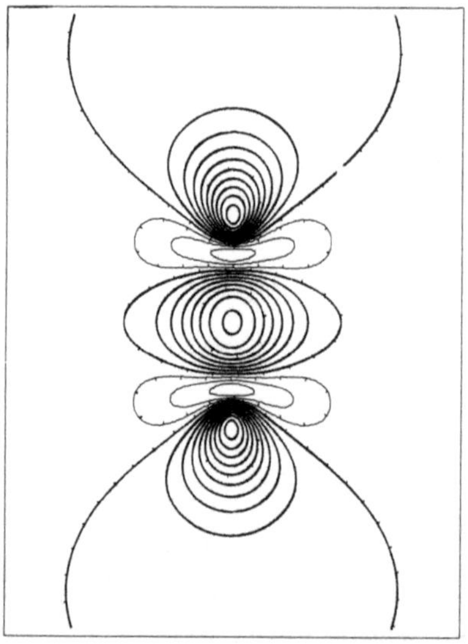

Fig. 5. Density difference map for $N_2$, $\rho_{H.F.}(\vec{r}) - \rho_{IAM}(\vec{r})$, in units of 0.01 $e^-/a_0^3$. The light contours represent charge loss and the dark charge gain. The model includes 2 monopole, 2 dipole and 1 quadrupole terms.

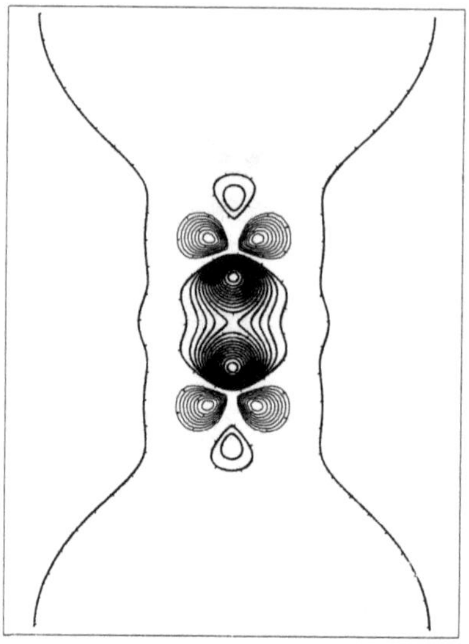

Fig. 6. Density difference map for $N_2$, $\rho_{H.F.}(\vec{r}) - \rho_{expt.}(\vec{r})$, in units of 0.002 $e^-/a_0^3$. The light contours represent less charge in the Hartree-Fock theory and the dark more. The model includes 2 monopole, 2 dipole and 1 quadrupole terms. Note that this map gives a qualitative representation of the correlation contribution to the chemical binding.

model since the shape of Becker's map dictates the need for inclusion of at least dipole and possibly octapole and/or hexadecapole terms in the analytical model.

As a further test of Fink's density map it is of interest to compare the value of properties derived from the map to the corresponding theoretical or experimental value. It is thought that the largest source of uncertainty in the calculation of properties comes from the restrictions in the choice of a model. In order to ascertain the magnitude of this uncertainty a least squares fit of the H.F. molecular 3-D density with the same model was obtained (column 3 in Table I). Properties were then derived from this H.F. model fit which could be compared with values computed directly from the H.F. molecular wave function (column 4 in Table I). In this way an estimate of the model dependent contribution to the uncertainty could be obtained. From Table I it can be noted that certain quantities such as the field gradient at the nucleus, quadrupole moment and force on the nucleus are particularly sensitive to the experimental data. This can be ascertained by noting the contribution to a derived property from the IAM model which is shown in column 1. In general the agreement between theory and experiment is excellent. While this latest study still contains a number of assumptions which need further verification the results are encouraging.

The most recent determination of a total inelastic $\Delta N(K)$ function has been carried out by Wellenstein and coworkers[21] at Brandeis University for $H_2$ with an incident electron energy of 35 keV. In Fig. 7 the results are shown and compared to rigorous theory as well as Hartree Fock theory. This is perhaps the most striking demonstration of the ability of the electron scattering method to yield sensitive information on charge densities comparable to the very best theoretical calculations.

Some recent studies of elastic scattering have been undertaken in Professor Rouault's laboratory at the Universite de Paris Sud in Orsay. There Duguet, Wellenstein and Bennanni have obtained preliminary results[24] for $NH_3$ which agrees well with the theory of Ostlund[25] for $0.5 < K < 2$ a.u. but shows definite disagreement for $K > 3$ a.u.. These preliminary results are shown in Fig. 8. Work is currently underway to improve the experiment and to locate the source of the large angle disagreement.

Table I. Experimental and calculated values for some selected molecular properties of $N_2$ in a.u.'s.

| Property (a.u.) | IAM Contribution | Fink Experiment | H.F. Model Fit of Stewart's Data | H.F. Direct Calculation | Experiment Direct Observation |
|---|---|---|---|---|---|
| Charge | 14.0182 | 24.0606 | 14.0178 | 14.000 | 14.000 |
| Force on nucleus | -1.192546 | -0.8141 | -0.8570 | +0.0796[c]<br>-0.8696[a] | -0.00126[b] |
| $\bar{V}_{ne}$ | -303.85 | -305.46 | -304.15 | -303.21[c] | -303.38[d] |
| $q_{zz}$ (Field gradient at nucleus) | +0.439 | -0.96 | -1.17 | -1.37[c] | -1.24 ± .54[e],<br>-1.40 ± .62[e]<br>(- 1.233)[d] |
| $Q_{zz}$ (Quadrupole moment) | 0 | -1.13 | -0.86 | -.95[c] | -1.13[k]<br>(±) 1.04 ± .07[f]<br>(-1.243)[d] |
| $\langle x^2 \rangle$ (⊥ to bond) | 8.156 | 7.644 | 7.831 | 7.701[c] | 7.5[g]<br>(7.487)[d] |
| $\langle z^2 \rangle$ (‖ to bond) | 23.269 | 23.887 | 23.803 | 23.616[c] | 23.6[g]<br>(23.854)[d] |
| $H_D$ (Hexadecapole moment) | 0 | -7.29 | -5.57 | -6.843[c] | ± 8.0 ± 2.7[ℓ,m] |
| $\langle k_e \rangle$ (Quadratic force constant) | 1.927 | 1.871 | 1.906 | 1.974[i]<br>1.805[j] | 1.58[h] |

a. Neglect of dipole term. Private Communication, R. F. Stewart.

b. The vibrational force was computed using the values $\langle Q \rangle_{vib}$ = 0.008 a.u., $\langle Q^2 \rangle_{vib}$ = 0.004 a.u. and $\langle Q^3 \rangle_{vib} = 7 \times 10^{-5}$ a.u.

c. R. F. Stewart, Private Communication.

d. T. H. Dunning, Jr., D. C. Cartwright, W. J. Hunt, P. J. Hay and F. W. Bobrowicz, J. Chem. Phys. 64, 4755 (1976).

e. R. F. Stewart, Private Communication. Calculated from data given in C. C. Lin, Phys. Rev. 119, 1026 (1960) and I. A. Scott, J. Chem. Phys. 36, 1459 (1962).

e′. R. F. Stewart, Private Communication. Calculated from data given in C. C. Lin, Phys. Rev. 119, 1026 (1960) and J. A. Scott and T. J. Haigh, J. Chem. Phys. 38, 117 (1963). Note this result is supposed to be free of librational effects.

f. A. D. Buckingham, R. L. Disch and D. A. Dunmur, J. Amer. Chem. Soc. 90, 3104 (1968).

g. W. H. Flygare, R. L. Shoemaker and W. Hüttner, J. Chem. Phys. 50, 2414 (1969).

h. Calculated from the values given in b.) above.

i. P. Cade, Private Communication.

j. R. F. Stewart, Private Communication.

k. D. E. Stogryn and A. P. Stogryn, Mol. Phys. 11, 371 (1966).

ℓ. J. E. Gready, G. B. Bacskay and N. S. Hush, Chem. Phys. 31, 467 (1978).

m. G. Birnbaum and E. R. Cohen, Mol. Phys. 32, 161 (1976).

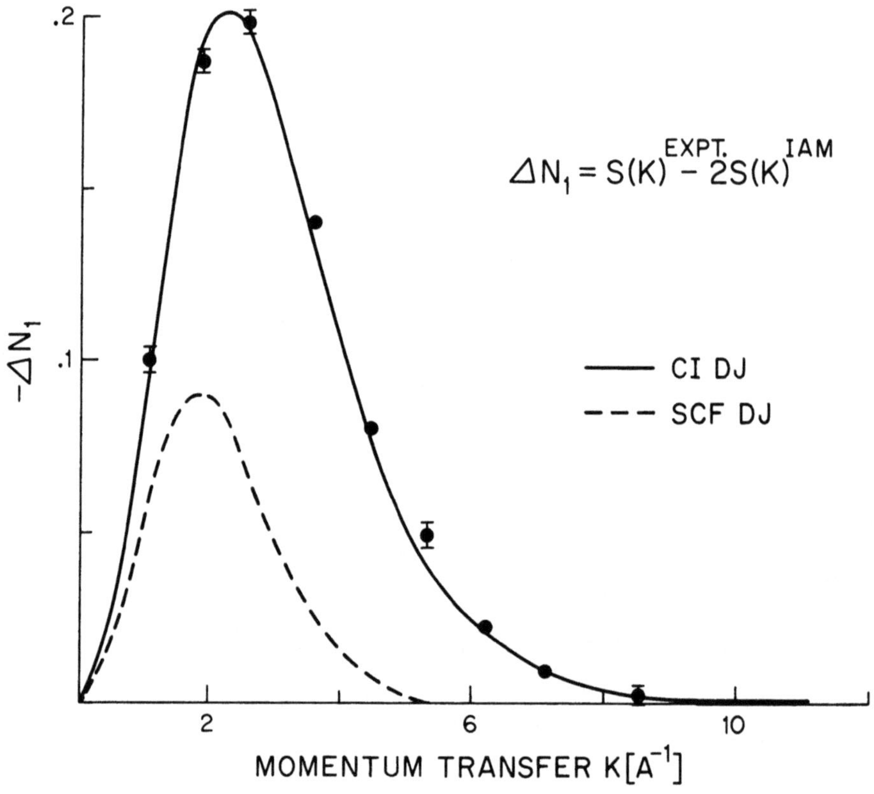


Fig. 7. The $\Delta N_{INEL}(K)$ for $H_2$ obtained with 25 keV electrons. The upper solid curve is due to Liu and Smith[22] employing a CI wave function by Davidson and Jones.[23] The lower dotted curve is also from Liu and Smith[22] using the H.F. limit part of the Davidson and Jones[23] wave function (courtesy of H. Wellenstein).

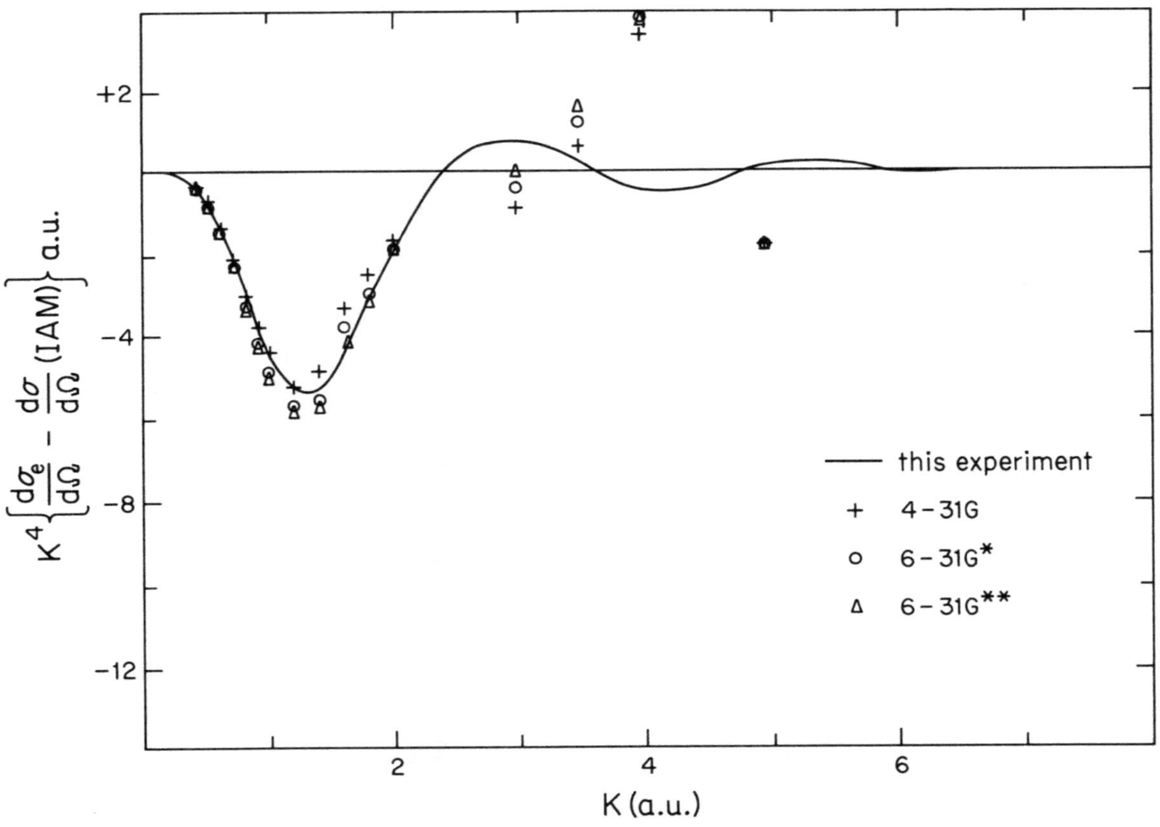


Fig. 8. The preliminary $\Delta N_{EL}(K)$ for $NH_3$ (solid line) compared to the calculations of Ostlund.[25] The Δ's represent what should be the best theoretical estimate (courtesy A. Duguet and H. Wellenstein).

## Bibliography


1. C. Tavard and M. Roux, Compt. Rend. 260, 4933 (1965). C. Tavard, M. Rouault and M. Roux, J. Chim. Phys. 62, 1410 (1965).
2. R. A. Bonham and M. Fink, "High Energy Electron Scattering", (Van Nostrand Reinhold Co., New York, 1974).
3. B. K. Jap and R. M. Glaeser, Acta Cryst. A34, 94 (1978).
4. In a typical electron diffraction experiment with 40 keV electrons we have $\lambda = 0.05\text{Å}$, $\theta_{max} = 20°$, $\sin(\theta_{max})/\lambda = 3.74\text{Å}^{-1}$ and $S = K = 43.6\text{Å}^{-1}$. Some measurements have been made to $S = K = 70\text{Å}^{-1}$.
5. E. B. Wilson, J. C. Decius and P. C. Cross, "Molecular Vibrations" (McGraw-Hill Book Co., New York, 1955).
6. c.f. D. A. Kohl and L. S. Bartell, J. Chem. Phys. 51, 2891; 2905 (1969); R. F. Stewart, J. J. Bently and B. Goodman, J. Chem. Phys. 63, 3786 (1975).
7. The details of the most recent techniques are available at present only in the form of Ph.D. theses. c.f. Dilek Barlas, Ph.D. thesis, Department of Physics, Brandeis University, 1977. Wolfgang Brueckner, Ph.D. thesis, Department of Physics, Brandeis University, 1977. P. G. Moore, Ph.D. thesis, Department of Physics, The University of Texas at Austin, 1976.
8. R. A. Bonham, J. Chem. Phys. 36, 3260 (1962). R. A. Bonham and C. Tavard, J. Chem. Phys. 59, 4691 (1973).
9. R. A. Bonham, Chem. Phys. Letters 52, 305 (1977).
10. c.f. R. A. Bonham and L. Schäfer, "International Tables for X-Ray Crystallography, Vol. IV". (The Kynoch Press, Birmingham, England, 1974) Sec. 2.5, pp. 176.
11. J. W. Liu and R. A. Bonham, Chem. Phys. Letters 14, 346 (1972).
12. H. F. Wellenstein, H. Schmoranzer, R. A. Bonham, T. C. Wong and J. S. Lee, Rev. Sci. Instr. 46, 92 (1975). H. Schmoranzer, H. F. Wellenstein and R. A. Bonham, Rev. Sci. Instr. 46, 89 (1975).
13. R. A. Bonham, Chem. Phys. Letters, 31, 559 (1975).
14. M. Fink and R. A. Bonham, Phys. Rev. 187, 114 (1969).
15. R. C. Ulsh, H. F. Wellenstein and R. A. Bonham, J. Chem. Phys. 60, 103 (1974).
16. A. L. Ford and J. C. Browne, Chem. Phys. Letters 20, 284 (1973).
17. J. W. Liu, J. Chem. Phys. 59, 1988 (1973).
18. a. M. Fink, D. Gregory and P. G. Moore, Phys. Rev. Letters 37, 15 (1976). b. M. Fink and P. G. Moore, Phys. Rev. A15, 112 (1977).
19. R. F. Stewart, Private Communication.
20. P. Becker, Private Communication.
21. H. F. Wellenstein, Private Communication.
22. J. W. Liu and V. H. Smith, Jr., Chem. Phys. Letters 45, 59 (1977).
23. E. R. Davidson and L. L. Jones, J. Chem. Phys. 37, 2966 (1962).
24. A. Duguet and H. F. Wellenstein, Private Communication.
25. H. F. Wellenstein, Private Communication.

Problem 1 : R.A. Bonham

SUBJECT AREA: PROPERTIES OF CHARGE DENSITIES OBTAINED FROM ELECTRON SCATTERING FROM NITROGEN

## INTRODUCTION

The charge density may be partitioned into separate contributions from each of the atoms by the method of Stewart. For a diatomic molecule this means that

$$\rho(\vec{r}) = \rho_A(\vec{r}-\vec{R}_A) + \rho_B(\vec{r}-\vec{R}_B)$$

$$= \rho_A(\vec{r}_A) + \rho_B(\vec{r}_B)$$

where $\vec{R}_A$ is a vector from the center of mass of the molecule to nucleus A, $\vec{R}_B$ is a vector from the center of mass of the molecule to nucleus B and $\vec{r}$ is a vector from the center of mass to a point in the electron charge distribution. The density about each center can be expanded as

$$\rho_A(\vec{r}-\vec{R}_A) = \rho_A(\vec{r}_A) = \frac{1}{4\pi} \sum_{n=0}^{\infty} \rho_{An}(r_A) P_n(\cos\theta_A)$$

Further the $\rho_{An}(r_A)$ may be parameterized as

$$\rho_{An}(r_A) = \sum_{k=0}^{M} A_{nk}\, r^{n+K}\, e^{-\lambda_{nk} r}$$

Such a parameterized model has been used by Fink to fit electron scattering data.

The procedure is as follows. The total density is approximated by

$$\rho(\vec{r})_{Expt} \simeq \rho_A{}^{IAM}(\vec{r}_A) + \rho_B{}^{IAM}(\vec{r}_B) + \Delta\rho_A(\vec{r}_A) + \Delta\rho_B(\vec{r}_B)$$

where IAM stands for the independent atom model and is simply a ground state atomic density, or an approximation to it, averaged over all orientations in space. Fink employed atoms in his IAM model which were based on theory and were of sufficiently high quality that the wave functions used reproduced the total energy of the nitrogen atom almost exactly. To use such a description forces one to employ numerical tables. For our purposes here we will employ simplified analytic representations of Hartree-Fock nitrogen atoms with the density

$$\rho^{IAM}(r) = \frac{Z}{4\pi} \sum_{i=1}^{M} \gamma_i \lambda_i^2 e^{-\lambda_i r}/r$$

The necessary parameters for the IAM and to approximately represent Fink's data are shown in Table I.

The total electron scattered intensity measured by Fink can be related to charge density quantities by use of the Born approximation of scattering theory as

$$I_{TOTAL} = \frac{A}{K^4} \{Z_A^2 + Z_B^2 + 2Z_A Z_B j_0(KR_{AB}) e^{-\frac{\ell_{AB}^2 K^2}{2}}$$

$$- 2Z_A \langle \int d\vec{r}\rho(\vec{r}) j_0(Kr_A)\rangle_{VIB} - 2Z_B \langle \int d\vec{r}\rho(\vec{r}) j_0(Kr_B)\rangle_{VIB}$$

$$+ N + 4\pi \langle \int_0^\infty dr_{12} r_{12}^2 \rho_c^{\,0}(r_{12}) j_0(Kr_{12})\rangle_{VIB}\}$$

where $\ell_{AB}^2$ is the mean amplitude of vibration, $R_{AB}$ is the bond length $\rho_c^{\,0}(r_{12})$ is the electron pair correlation density averaged over all angles and $\langle \ \rangle_{VIB}$ denotes vibrational averaging. The nuclear charges $Z_A$, $Z_B$ and the total number of electrons in the system N are known and can be subtracted out. The data extend from K = .5 a.u. to almost K = 30 a.u. but the region in K sensitive to $\Delta\rho$ only extends to about K = 10 a.u.. This means that the data from 10 to 30 a.u. can be used to characterize the parameters $R_{AB}$ and $\ell_{AB}$ allowing subtraction of the structure dependent term. If the approximation that $\rho_c^{\,0}(r_{12}) \simeq 2\rho_c^{\,A}(r_{12})$ is used where $\rho_c^{\,A}(r_{12})$ is the electron pair density for a Hartree-Fock atom averaged over all orientations in space then the experimental intensity for a homonuclear diatomic can be written as

$$\frac{K^4}{A} \Delta I_{TOTAL}(K) \cong [I_{TOTAL} - 2Z^2 - 2Z^2 j_0(KR_{AB}) e^{-\frac{\ell_{AB}^2 K^2}{2}} - N$$

$$-4\pi\, 2\int dr_{12} r_{12}^2 \rho_c^{\,A}(r_{12}) j_0(Kr_{12})]$$

$$= -2Z\langle \int d\vec{r}\rho(\vec{r}) j_0(Kr_A)\rangle_{VIB} - 2Z\langle \int d\vec{r}\rho(\vec{r}) j_0(Kr_B)\rangle_{VIB}$$

Note that the most serious approximation made so far is that

$\rho_c^{\,O}(r_{12}) = 2\rho_c^{\,A}(r_{12})$. There is some evidence as noted by Stewart that this approximation may be valid. Another point to note is that $e^{-\ell_{AB}K^2/2}$ remains very close to unity over the range $0 \leq K$ (a.u.) $\leq 10$ for a molecule such as $N_2$ ($\ell_{AB} = 0.032$Å).

So far we have not discussed the normalizing parameter A. The data is matched at large K (K > 10 a.u.) to the independent atom model (IAM) given as

$$K^4\Delta^{IAM}(K) = -2Z\int d\vec{r}\,\rho^{IAM}(\vec{r})j_0(Kr_A)$$

$$-2Z\int d\vec{r}\,\rho^{IAM}(\vec{r})j_0(Kr_B)$$

and the difference

$$\Delta\sigma(K) = \frac{K^4}{A}\,[\Delta I_{TOTAL}(K) - \Delta I^{IAM}(K)]$$

least squares fitted by the $\Delta\rho(r)$ model introduced previously. This is a slightly over simplified view as in actual practice corrections for the failure of the Born approximation are also applied to the experimental data.

The questions in this problem will deal with application of the Stewart type model to actual experimental data obtained by Fink and H. F. theoretical data to characterize various properties of the one electron density.

Problem:

1.) Show that for a homonuclear diatomic that $\rho_{An}(r_A) = (-1)^n\rho_{Bn}(r_A)$.

2.) Show that for a homonuclear diatomic that

$$\Delta N(K) = -4Zf_{A,0}(K) - 4Z\sum_n(-1)^n j_n(KR)f_{A,n}(K)$$

where $f_{A,n}(K) = \int_0^\infty drr^2\rho_{A,n}(r)j_n(Kr)$.

3.) Calculate the total electronic charge for the experimental model and the H.F. model. $N = \int d\vec{r}\rho(\vec{r})$ [see Table I for the parameters].

4.) Calculate the force acting on one of the nuclei. The force balance equation can be written as

$$\langle 0|n_R\cdot\vec{\nabla}_A V|0\rangle = n_R\cdot\vec{\nabla}_A E_e$$

where $\hat{n}_R$ is a unit vector parallel to the bond direction, $\vec{\nabla}_A$ is the gradient operator with respect to $\vec{R}_A$ and the potential energy V is given as

$$V = \frac{Z_A Z_B}{R_{AB}} - Z_A \sum_i \frac{1}{|\vec{r}_i - \vec{R}_A|} - Z_B \sum_j \frac{1}{|\vec{r}_j - \vec{R}_B|} + \sum_{i \neq j} \sum + \frac{1}{r_{ij}} .$$

5.) Calculate the electron-nuclear attractive energy

$$V_{ne} = -Z_A \int \frac{d\vec{r}\rho(\vec{r})}{|\vec{r}-\vec{R}_A|} - Z_B \int \frac{d\vec{r}\rho(\vec{r})}{|\vec{r}-\vec{R}_B|}$$

where $\rho(\vec{r}) = \rho^{IAM}(\vec{r}) + \Delta\rho(\vec{r})$

6.) Calculate the field gradient at the nucleus

$$q^A_{zz} = Z_B \left[ \frac{3(z_Z - z_B)^2}{R_{AB}^5} - \frac{1}{R_{AB}^3} \right]$$

$$- \int dr\,\rho(r) \left[ \frac{e(z-z_A)^2}{|\vec{r}-\vec{R}_A|^5} - \frac{1}{|\vec{r}-\vec{R}_A|^3} \right]$$

where $z_A$ and $z_B$ are the projections of $R_A$ and $R_B$ on the inter nuclear axis.

7.) Calculate the quadrupole moment

$$Q = \sum_\alpha \frac{Z_\alpha}{2} (3z_\alpha^2 - R_\alpha^2) - \int d\vec{r}\rho(\vec{r}) r^2 P_2(\cos\theta)$$

8.) Calculate the expectation values

$$\langle x^2 \rangle = \int d\vec{r}\rho(\vec{r}) x^2$$

and

$$\langle z^2 \rangle = \int d\vec{r}\rho(\vec{r}) z^2$$

Table 1. Parameters in the model in a.u.

$R = 2.078$

| | | | | | |
|---|---|---|---|---|---|
| $\gamma_1^{IAM}$ | = | 1.3678 | $\lambda_1^{IAM}$ | = | 2.0273 |
| $\gamma_2^{IAM}$ | = | -0.2380 | $\lambda_2^{IAM}$ | = | 16.5256 |
| $\gamma_3^{IAM}$ | = | 2.8625 | $\lambda_3^{IAM}$ | = | 5.9953 |
| $\gamma_4^{IAM}$ | = | 1.9402 | $\lambda_4^{IAM}$ | = | 9.9313 |
| $\gamma_5^{IAM}$ | = | -3.0396 | $\lambda_5^{IAM}$ | = | 5.0882 |
| $\gamma_6^{IAM}$ | = | -1.8916 | $\lambda_6^{IAM}$ | = | 9.0839 |
| | | | | | |
| $a_{00}^{ex}$ | = | -3.6464 | $\lambda_{00}$ | = | 2.5889 |
| $a_{01}^{ex}$ | = | 59.0651 | $\lambda_{01}$ | = | 4.5592 |
| $a_{02}^{ex}$ | = | -244.1168 | $\lambda_{02}$ | = | 6.8848 |
| $a_{20}^{ex}$ | = | 49.3469 | $\lambda_{20}$ | = | 3.8538 |
| | | | | | |
| $a_{00}^{HF}$ | = | -3.9078 | $\lambda_{00}$ | = | 2.8401 |
| $a_{01}^{HF}$ | = | 58.5597 | $\lambda_{01}$ | = | 4.5904 |
| $a_{02}^{HF}$ | = | -251.6363 | $\lambda_{02}$ | = | 6.6910 |
| $a_{20}^{HF}$ | = | 68.3479 | $\lambda_{20}$ | = | 4.1959 |

## A. Introduction: The Model

The theoretical model of interest here for the interpretation of electron scattering charge density data consists in representing molecular one electron charge densities as a sum of three dimensional atomic densities. In effect the density about each atom is expanded in a spherical harmonic series and the radial functions obtained by least squares fitting of the model to theoretical or experimental molecular charge densities. This idea was first applied to molecules in the context of electron scattering by Kohl and Bartell[1,2] although the model had been used to analyze charge densities in x-ray scattering somewhat earlier.[3,4] In fact the success of the approach in the x-ray case[5,8] eventually led to the reapplication of the method to electron scattering by Stewart and coworkers.[9,10] We shall hereafter refer to the model as the Kohl-Bartell-Stewart model (KBS).

The fundamental quantity in this study is the molecular one electron charge density $\rho(\vec{r})$ where $\vec{r}$ specifies a point, with the center of mass of the molecules taken as the origin, at which the electron density has the value $\rho(\vec{r})$. The electron density will be subdivided into charge densities located on each atom which for the diatomic case can be written as

$$\rho(\vec{r}) = \rho_A(\vec{r}-\vec{R}_A) + \rho_B(\vec{r}-\vec{R}_B) \tag{1}$$

where unless otherwise noted the units used will be Hartree atomic units. The "atomic" densities can be expanded in spherical harmonics as for example

$$\rho_A(\vec{r}-\vec{R}_A) = \rho_A(\vec{r}_A) = \frac{1}{4\pi}\sum_{n=0}^{\infty} \rho_{An}(r_A)P_n(\cos\theta_A) \tag{2}$$

where $\vec{r}_A = \vec{r}-\vec{R}_A$ with $\vec{R}_A$ the vector from the center of mass to the $A\underline{th}$ nucleus. The use of such density functions is generally straightforward with the exception of the homonuclear case. Here one must be careful about sign conventions as we will show. The problem is that the function on center B must be identical to that on A (except that it is inverted) for the homonuclear case. To see how one converts from B to A consider what happens at a point in the plane perpendicular to the bond axis and containing the center of mass. It is obvious that for points on this plane the contribution from $\rho_A$ and $\rho_B$ must be equal in both sign and magnitude. Hence we have as shown in Fig. 1 the coordinate axis for this example. Our condition is that

$$\rho_A(\vec{r}_A) = \sum_n \rho_{An}(r_A)\, P_n(\cos\theta_A) = \rho_B(\vec{r}_B) = \sum_n \rho_{Bn}(r_B)\, P_n(\cos\theta_B) \tag{3}$$

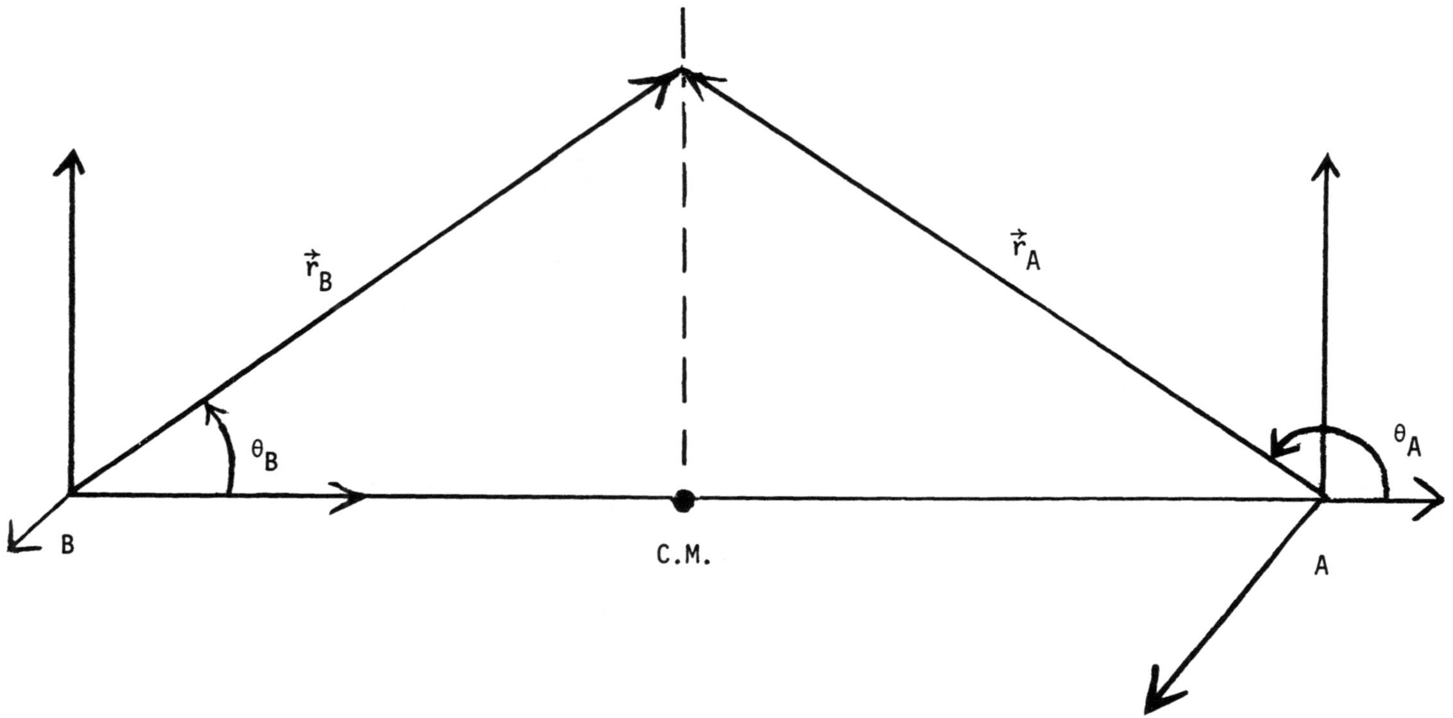


Fig. 1. The coordinate system for the model expansion showing a point on the plane $\perp$ to the bond axis and containing the center of mass (C.M.). Note that the $X_A$, $Y_A$, $Z_A$ axes is taken in the same directional sense as the $X_B$, $Y_B$, $Z_B$ axes.

where since $\cos\theta_B = -\cos\theta_A$ we have

$$\sum_n \rho_{An}(r_A)\, P_n(\cos\theta_A) = \sum_n \rho_{Bn}(r_B)(-1)^n P_n(\cos\theta_A) \qquad (4)$$

or since in the plane $r_A = r_B$ we see that

$$\rho_{An}(r_A) = (-1)^n \rho_{Bn}(r_A) \qquad (5)$$

In other words the sign of all odd radial components of the spherical harmonic expansion terms change sign when converted to the function on the other center. We will derive all formulas here first in terms of the two centers A and B so that the extension of the results of this study to heteronuclear diatomics will be simpler. The final results for the homonuclear case will then be obtained by the use of Eq. 5.

The great virtue of the model is that least squares fits of realistic theoretical charge densities show excellent convergence using no more than four terms in the spherical harmonic expansions about any one center.[6-8] In such an expansion the nomenclature "monopole" is applied to the n = 0 term, "dipole" to the n = 1 term, "quadrupole" to the n = 2 term and "octupole" to the n = 3 term. Fortunately for charge density studies the relative contributions to the total density from each term is very different for each molecule.[1,2] In addition the monopole term is quite close to the independent atom spherically averaged density. All this means that experimental information can be analyzed in terms of dipole, quadrupole and octupole contributions plus small perturbations to model contributions. The availability of accurate theoretical atomic spherically averaged one electron charge densities is a very important aid to the analysis of diffraction data as we shall see.

It will also prove useful to develop relationships between the Fourier transform of the charge density, the coherent x-ray scattering factor, and the charge density model. The x-ray coherent scattering factor $F(\vec{K})$ is defined as

$$F(\vec{K}) = \int d\vec{r}\; \rho(r)\; e^{i\vec{K}\cdot\vec{r}} \qquad (6)$$

By use of Eq. 1 we obtain

$$F(\vec{K}) = e^{i\vec{K}\cdot\vec{R}_A} \int d\vec{x}\; \rho_A(\vec{x}) e^{i\vec{K}\cdot\vec{x}} + e^{i\vec{K}\cdot\vec{R}_B} \int d\vec{x}\; \rho_B(\vec{x}) e^{i\vec{K}\cdot\vec{x}} \qquad (7)$$

which from Eq. 2 reduces to

$$F(\vec{K}) = e^{i\vec{K}\cdot\vec{R}_A} \sum_n f_{An}(K)P_n(\cos\theta_A) + e^{i\vec{K}\cdot\vec{R}_B} \sum_n f_{Bn}(K)P_n(\cos\theta_B) \quad (8)$$

where

$$f_{An}(K) = \int_0^\infty drr^2 \rho_{An}(r)j_n(Kr) \quad (9)$$

Stewart has termed the little f's "Generalized atomic scattering factors". They are in fact central to the analysis of both electron and x-ray data. Note that Eq. 7 can be inverted as

$$\rho(\vec{r}) = \frac{1}{(2\pi)^3}\int d\vec{K} e^{-i\vec{K}\cdot(\vec{r}-\vec{R}_A)} f_A(\vec{K}) + \frac{1}{(2\pi)^3}\int d\vec{K} e^{-i\vec{K}\cdot(\vec{r}-\vec{R}_B)} f_B(\vec{K}) \quad (10)$$

by use of the Fourier integral theorem where

$$f_A(\vec{K}) = \sum_n f_{An}(K)P_n(\cos\theta_A) \quad (11)$$

## Application of the Model to Electron Scattering

In the Born theory of electron scattering the total scattered molecular intensity can be written approximately as

$$I = \frac{A}{4\pi K^4}\int d\Omega_{ROT}\langle\Psi_e\Big|\sum_\alpha\sum_\beta Z_\alpha Z_\beta e^{iK\cdot R_{\alpha\beta}} - 2\sum_\alpha\sum_i Z_\alpha e^{i\vec{K}\cdot(\vec{r}_i-\vec{R}_\alpha)} \quad (12)$$

$$+ \sum_i\sum_j e^{iK\cdot r_{ij}}\Big|\Psi_e\rangle$$

where K is the momentum transfer on scattering, $\vec{r}_i$ is the instantaneous position of the $i\underline{th}$ of N target electrons, $\vec{R}_\alpha$ is the instantaneous position of the $\alpha\underline{th}$ of M nuclei with charge $Z_\alpha$, $\Psi_e$ is the molecular electronic wave function and $\int d\Omega_{ROT}$ is a classical rotational average. Note that the right hand side of the equation must also be averaged over the molecular vibrational motion and A stands for a normalization constant to put the data on an absolute scale. Eq. 12 can be written in terms of molecular densities as

$$I = \frac{A}{K^4}\left\{\sum_{\alpha} Z_\alpha^2 + \sum_{\alpha\neq\beta}\sum Z_\alpha Z_\beta j_0(KR_{\alpha\beta}) - 2\sum_{\alpha} Z_\alpha \int d\vec{r}\, \rho(\vec{r}) j_0(K|\vec{r}-\vec{R}_\alpha| \right. \qquad (13)$$

$$\left. + N + 4\pi\int_0^\infty drr^2\rho_c^{\,0}(r) j_0(Kr)\right\}$$

where $4\pi r^2\rho_c^{\,0}(r)$ is the spherically averaged molecular electron pair correlation function.

Two approaches have been suggested for analyzing experimental data in terms of Eq. 13. Either the correlation term $N + 4\pi\int_0^\infty drr^2\rho_c^{\,0}(r)j_0(Kr)$ is computed and subtracted out using a theoretical model or else the term $\frac{1}{4\pi}\int d\Omega\, F(\vec{K})^2$ is added and subtracted to Eq. 13. In the latter case one obtains

$$I = \frac{A}{K^4}\left\{\sum_{\alpha} Z_\alpha^2 + \sum_{\alpha\neq\beta}\sum Z_\alpha Z_\beta j_0(KR_{\alpha\beta}) - 2\sum_{\alpha} Z_\alpha \int d\vec{r}\, \rho(\vec{r}) j_0(K|\vec{r}-\vec{R}_\alpha|) \right. \qquad (14)$$

$$\left. + \frac{1}{4\pi}\int d\Omega_{ROT}\, F(\vec{K})^2 + S(K)\right\}$$

where the x-ray incoherent scattering factor S(K) is given as

$$S(K) = N + 4\pi\int_0^\infty drr^2\rho_c^{\,0}(r)j_0(Kr) - \frac{1}{4\pi}\int d\Omega_{ROT}\, F(\vec{K})^2 \qquad (15)$$

This last case is attractive because theoretical calculations by Stewart indicate that S(K) is given extremely well by a sum of spherical atom x-ray incoherent scattering factors. In addition the quantities remaining constitute the elastic scattering intensity which can in principle be measured directly. There are some problems in separating inelastic scattering contributions from the total but these appear to be manageable.[11]

In the case where the term $N + 4\pi\int_0^\infty drr^2\rho_c^{\,0}(r)j_0(Kr)$ is subtracted out the remaining experimental intensity is compared to the term

$$-2\sum_{\alpha} Z_{\alpha} \int d\vec{r}\; \rho(\vec{r}) j_0(K|\vec{r}-\vec{R}_{\alpha}|)$$

providing the purely structure dependent terms are also subtracted out. For the diatomic case this only remaining density dependent term reduces to

$$-2Z_A \int d\vec{x} \rho_A(\vec{x}) j_0(Kx) \; -2Z_A \int d\vec{x} \rho_B(\vec{x}) j_0(K|\vec{x}-\vec{R}_{AB}|)$$

$$-2Z_B \int d\vec{x} \rho_A(\vec{x}) j_0(K|\vec{x} + \vec{R}_{AB}|) \; -2Z_B \int d\vec{x} \rho_B(\vec{x}) j_0(Kx)$$

which in terms of the expansion in Eq. 2 becomes

$$-2Z_A \int_0^{\infty} dx x^2 \rho_{A,0}(x) j_0(Kx) \; -2Z_B \int_0^{\infty} dx x^2 \rho_{B,0}(x) j_0(Kx)$$

$$-2Z_A \sum_n j_n(KR_{AB}) \int_0^{\infty} dx x^2 \rho_{Bn}(x) j_n(Kx) - 2Z_B \sum_n (-1)^n j_n(KR_{AB}) \int_0^{\infty} dx x^2 \rho_{An}(x) j_n(Kx)$$

This result simplifies for the homonuclear case to

$$-4Z \int_0^{\infty} dx x^2 \rho_{A,0}(x) j_0(Kx) - 4Z \sum_n (-1)^n j_n(KR) \int_0^{\infty} dx x^2 \rho_{A,n}(x) j_n(Kx)$$

$$= -4Z\; f_{A,0}(K) - 4Z \sum_n (-1)^n j_n(KR)\; f_{A,n}(K)$$

B. Use of the KBS Model to Infer Properties of the Charge Distribution

We categorize the properties of the charge distribution as general theoretical properties (i.e. those enforced by charge conservation, the stationary property of a pure quantum state and results enforced by symmetry) and experimental properties (i.e. those measurable by other experimental methods). In the latter category are the moments of the charge distribution which we will further categorize into inner moments, $\langle r^{-n} \rangle$, with $n = 1,2...$ and outer moments, $\langle r^n \rangle$ with $n = 1,2...$ In addition to these moments and properties there exist some properties such as force constants which can be only approximately calculated from the charge density and will not be considered here.

General theoretical properties of the charge density.
a.) <u>Total charge in the molecule.</u>

Conservation of total charge in the molecule can be written as

$$N = \langle \int d\vec{r} \rho(\vec{r}) \rangle = \langle \int d\vec{r} [\rho_A(\vec{r}-\vec{R}_A) + \rho_B(\vec{r}-\vec{R}_B)] \rangle \tag{16}$$

where N is the number of electrons in the molecule and the brackets $\langle\ \rangle$, indicate rotational and vibrational averaging. Transforming each integral to its own center (i.e. let $\vec{x} = \vec{r}-\vec{R}_A$) and using Eq. 2 we obtain

$$N = \langle \int_0^\infty dx x^2 \rho_{A,0}(x) \rangle_{vib} + \langle \int_0^\infty dx x^2 \rho_{B,0}(x) \rangle_{vib} \tag{17}$$

where only the vibrational averaging remains to be done. For the homonuclear case Eq. (17) reduces to

$$N = 2 \int_0^\infty dx x^2 \langle \rho_{AO}(x) \rangle_{vib} \tag{18}$$

It is often convenient to adopt a comparison independent atom model (IAM) model in which case

$$\rho(\vec{r}) = \sum_n [\rho_n(\vec{r})^{IAM} + \Delta\rho_n(\vec{r})] \tag{19}$$

where the IAM atomic density is usually normalized to the number of electrons on the neutral atom and $\Delta\rho_n(r)$ is the correction density. If spherically averaged atomic densities are employed then the $\rho_n^{IAM}(r)$ are spherically symmetric about their nuclear centers and only the n = 0 term exists in the expansion. Eq. 17 then reads

$$N = \langle \int_0^\infty dx x^2 \rho_{AO}^{IAM}(x) \rangle_{vib} + \langle \int_0^\infty dx x^2 \rho_{BO}^{IAM}(x) \rangle_{vib} \tag{20}$$

$$+ \langle \int_0^\infty dx x^2 \Delta\rho_{AO}(x) \rangle_{vib} + \langle \int_0^\infty dx x^2 \Delta\rho_{BO}(x) \rangle_{vib}$$

or for the homonuclear case

$$N = 2\langle \int_0^\infty dx x^2 \rho_{AO}^{IAM}(x) \rangle_{vib} + 2 \int_0^\infty dx x^2 \Delta\rho_{AO}(x) \rangle_{vib} \quad (21$$

or if the IAM model is normalized to the atomic charge density one has

$$\boxed{\left\langle \int_0^\infty dx x^2 \Delta\rho_{AO}(x) \right\rangle_{vib} = 0} \tag{22}$$

which is a strong constraint on the adopted model. If the IAM model is not normalized to Z then since the IAM x-ray coherent scattering factor exhibits the property

$$\lim_{K\to 0} F^{IAM}(K) = \lim_{K\to 0} \int_0^\infty drr^2 \rho_{AO}^{IAM}(r) j_0(Kr) = \int_0^\infty drr^2 \rho_{AO}^{IAM}(r) \tag{23}$$

Eq. 21 can be written as

$$\int_0^\infty dx x^2 \Delta\rho_{AO}(x) = \frac{N}{2} - F(0). \tag{24}$$

Ehrenfests theorem or force balance in a stationary state.

This is a fundamental theorem[14] and can be stated in many ways. For example

$$\langle 0 | [\sum_\alpha \nabla_\alpha + \sum_i \nabla_i, H] | 0 \rangle = 0 \tag{25}$$

is perhaps the most general statement where $\alpha$ refers to nuclear coordinates, i to electronic and H is the total Hamiltonian with $|0\rangle$ the total wave function. Why Eq. 25? Because the commutator is $\vec{\nabla}_\alpha V + \vec{\nabla}_i V$ which is the sum of all forces acting on the system. For our purposes however it will prove more convenient to consider the force acting on a particular nucleus for a non vibrating molecule. Vibrational averaging can be accomplished later. We start by considering the commutator

$$\langle 0 | [\vec{\nabla}_A, H] | 0 \rangle = \langle 0 | \vec{\nabla}_A V | 0 \rangle \tag{26}$$

which is clearly the force on nucleus A. But does the average of this force vanish as in Eq. 25? The answer is clearly no because $\vec{\nabla}_A$ does not commute with the electronic energy since the electronic energy still depends on the internuclear separation in the Born-Openheimer approximation. Thus we have[15]

$$\hat{n}_R \cdot \langle 0 | \vec{\nabla}_A V | 0 \rangle = \hat{n}_R \cdot \vec{\nabla}_A E_e \tag{27}$$

where $\hat{n}_R \cdot \vec{\nabla}_A$ is the projection of the force on nucleus A along the bond axis. Eq. 27 reduces to

$$\vec{n}_R \cdot \vec{\nabla}_A E_e = -\vec{n}_R \cdot \frac{Z_A Z_B}{R_{AB}^3} (\vec{R}_A - \vec{R}_B) - Z_A \int dr \frac{\rho(\vec{r}) \vec{n}_R \cdot (\vec{r} - \vec{R}_A)}{|\vec{r} - \vec{R}_A|^3}$$

$$= -\frac{Z_A Z_B}{R_{AB}^2} - \frac{Z_A}{3} \int_0^\infty dx \rho_{A1}(x) - Z_A \int dx \frac{\rho_B(x)(x\cos\theta - R_{AB})}{(x^2 + R_{AB}^2 - 2xR_{AB}\cos\theta)^{3/2}}$$

$$= -\frac{Z_A Z_B}{R_{AB}^2} - \frac{Z_A}{3} \int_0^\infty dx \rho_{A1}(x) - Z_A \sum_n \frac{1}{(2n+1)} \frac{d}{dR_{AB}} \left[ \int_0^{R_{AB}} \frac{dx x^{n+2}}{R_{AB}^{n+1}} \rho_{Bn}(x) \right.$$

$$\left. + R_{AB}^n \int_{R_{AB}}^\infty dx x^{1-n} \rho_{Bn}(x) \right]$$

or

$$\frac{dE_e}{dR_A} = -\frac{Z_A Z_B}{R_{AB}^2} - \frac{Z_A}{3} \int_0^\infty dx \rho_{A1}(x) + Z_A \sum_n \frac{(n+1)}{(2n+1) R_{AB}^{n+2}} \int_0^{R_{AB}} dx x^{n+2} \rho_{Bn}(x) \quad (28)$$

$$- Z_A \sum_n \frac{n R_{AB}^{n-1}}{(2n+1)} \int_{R_{AB}}^\infty dx x^{1-n} \rho_{Bn}(x)$$

plus a similar relationship for the force balance on nucleus B. For a homonuclear diatomic Eq. 28 simplifies to

$$\frac{2dE}{dR} = -\frac{Z^2}{R^2} - \frac{Z}{3} \int dx \rho_{A1}(x) + Z \sum_n \left[ \frac{(n+1)(-1)^n}{(2n+1) R^{n+2}} \int_0^R dx x^{n+2} \rho_{An}(x) \right. \quad (29)$$

$$\left. - \frac{n(-1)^n R^{n-1}}{(2n+1)} \int_R^\infty dx x^{1-n} \rho_{An}(x) \right]$$

If an IAM model is adopted then Eq. (29) can be written as

$$\frac{2dE}{dR} = -\frac{Z}{R^2}\int_R^\infty dx x^2 \rho_{AO}^{IAM}(x) + \frac{Z}{R^2}\int_0^R dx x^2 \Delta\rho_{AO}(x) \tag{30}$$

$$-\frac{Z}{3}\int_0^R dx \Delta\rho_{A1}(x) - \frac{2Z}{3R^3}\int_0^R dx x^3 \Delta\rho_{A1}(x)$$

$$+ Z\sum_{n=2}\frac{(n+1)(-1)^n}{(2n+1)R^{n+2}}\int_0^R dx x^{n+2}\Delta\rho_{An}(x) - Z\sum_{n=2}\frac{nR^{n-1}(-1)^n}{(2n+1)}\int_R^\infty dx x^{1-n}\Delta\rho_{An}(x)$$

In order to make use of Eq. (30) $\frac{dE}{dR}$ must be evaluated by some independent means. We note that we can write E(R) as

$$E(R) = \frac{k_e}{2}(R-R_e)^2 + \frac{\ell_e}{6}(R-R_e)^3 + \frac{m_e}{24}(R-R_e)^4 + \ldots \tag{31}$$

where $k_e$, $\ell_e$ and $m_e$ are the quadratic, cubic and quartic force constants and $R_e$ is the equilibrium internuclear distance. Designating $R-R_e$ as Q and carrying out vibrational averages on both sides with the neglect of small terms we have

$$2k_e\langle Q\rangle_{vib} + \ell_e\langle Q^2\rangle_{vib} + \frac{m_e}{3}\langle Q^3\rangle_{vib} + \ldots \tag{32}$$

$$= \langle -\frac{Z}{R^2}\int_R^\infty dx x^2 \rho_{AO}^{IAM}(x) + \frac{Z}{R^2}\int_0^R dx x^2 \Delta\rho_{AO}(x)$$

$$-\frac{Z}{3}\int_0^R dx \Delta\rho_{A1}(x)\left[1 + \frac{2x^3}{R^3}\right]$$

$$+ 2\sum_{n=2}\frac{(-1)^n(n+1)}{(2n+1)R^{n+2}}\int_0^R dx x^{n+2}\Delta\rho_{An}(x) - Z\sum_{n=2}\frac{(-1)^n nR^{n-1}}{(2n+1)}\int_R^\infty dx x^{1-n}\Delta\rho_{An}(x)\rangle_{vib}$$

We note that the average vibrational force just balances the average electronic force and presumably the left hand side of Eq. (32) can be computed from experimentally available vibrational spectroscopic data from a standard source such as Herzberg's Spectra of Diatomic Molecules. As in the case of charge balance (Eq. 22) we can expect Eq. (32) to provide a very sensitive test of the experimental data since the left hand side will be small and the IAM term large so that the

$\Delta\rho$ contribution will have to largely cancel the IAM contribution.
c.) The electron-nuclear attractive energy.

The electron nuclear attractive energy $V_{ne}$ is accessible by theoretical calculation and indirectly from experiment by use of Tavard's theorem[14] and is defined as

$$V_{ne} = -Z_A \int \frac{d\vec{r}\rho(\vec{r})}{|\vec{r}-\vec{R}_A|} - Z_B \int \frac{d\vec{r}\rho(\vec{r})}{|\vec{r}-\vec{R}_B|} \tag{33}$$

$$= -Z_A \int \frac{d\vec{x}\rho_A(\vec{x})}{x} - Z_A \int \frac{d\vec{x}\rho_B(\vec{x})}{|\vec{x}-\vec{R}_{AB}|}$$

$$-Z_B \int \frac{d\vec{x}\rho_A(\vec{x})}{|\vec{x}+\vec{R}_{AB}|} - Z_B \int \frac{d\vec{x}\rho_B(\vec{x})}{x}$$

$$= -Z_A \int_0^\infty dxx\rho_{AO}(x) - Z_B \int_0^\infty dxx\rho_{BO}(x)$$

$$-Z_A \sum_n (2n+1) \frac{1}{k^{n+1}} \int_0^R dxx^{n+2}\rho_{Bn}(x) - Z_A \sum_n \frac{R^n}{(2n+1)} \int_R^\infty dxx^{1-n}\rho_{Bn}(x)$$

$$-Z_B \sum_n (2n+1) \frac{(-1)^n}{R^{n+1}} \int_0^R dxx^{n+2}\rho_{An}(x) - Z_B \sum_n \frac{(-1)^n R^n}{(2n+1)} \int_R^\infty dxx^{1-n}\rho_{An}(x)$$

which reduces for the homonuclear case to

$$V_{ne} = -2Z\int_0^\infty dxx\rho_{AO}(x) - 2Z\sum_n \frac{(-1)^n}{(2n+1)^{Rn+1}} \int_0^R dxx^{n+2}\rho_{An}(x) \tag{34}$$

$$-2Z\sum_n \frac{(-1)^n R^n}{(2n+1)} \int_R^\infty dxx^{1-n}\rho_{An}(x)$$

Insertion of the IAM model in Eq. (34) coupled with some rearrangement yields

$$V_{ne} = -2Z\int_0^R dx\,x\,\rho_{AO}^{IAM}(x)(1+\frac{x}{R}) - 4Z\int_R^\infty dx\,x\,\rho_{AO}^{IAM}(x)$$
$$-2Z\int_0^R dx\,x\,\Delta\rho_{AO}(x)(1+\frac{x}{R}) - 4Z\int_R^\infty dx\,x\,\Delta\rho_{AO}(x)$$
$$-2Z\sum_{n=1}\frac{(-1)^n}{(2n+1)^{Rn+1}}\int_0^R dx\,x^{n+2}\Delta\rho_{An}(x) - 2Z\sum_{n=1}\frac{(-1)^n R^n}{(2n+1)}\int_R^\infty dx\,x^{1-n}\Delta\rho_{An}(x) \quad (35)$$

Eq. (35) would not be expected to be terribly sensitive to the $\Delta\rho$ part of the model since the IAM contribution to $V_{ne}$ will be a major one. However the difference between experiment and a good CI calculation may be very sensitive.

2. Experimentally observable properties of the charge density.

a.) The inner moments of the charge density.

The field gradient at the nucleus[16-17] and the potential energy of the molecule are the two most notable examples of inner moments. The electron nuclear potential energy actually belongs in the previous section since it is only a part of the total energy and not one that can be experimentally measured directly. Hence in this section we will focus exclusively on the field gradient at the nucleus which for nucleus A is defined as[17]

$$q_{ZZ}^{A} = Z_B\left(\frac{3(z_A-z_B)^2}{R_{AB}^5} - \frac{1}{R_{AB}^3}\right) - \int d\vec{r}\rho(\vec{r})\left[\frac{3(z-z_A)^2}{|\vec{r}-\vec{R}_A|^5} - \frac{1}{|\vec{r}-\vec{R}_A|^3}\right] \quad (36)$$
$$= \frac{2Z_B}{R_{AB}^3} - 2\int d\vec{x}\rho_A(\vec{x})\frac{P_2(\cos\theta)}{x^3} - \int d\vec{x}\rho_B(\vec{x})\left[\frac{3(x\cos\theta-R_{AB})^2}{|\vec{r}-\vec{R}_{AB}|^5} - \frac{1}{|\vec{r}-\vec{R}_{AB}|^3}\right]$$
$$= \frac{2Z_B}{R_{AB}^3} - 2\int dx\rho_A(x)\frac{P_2(\cos\theta)}{x^3} - \left(\frac{d}{dR}\right)^2\int dx\rho_B(\vec{x})\frac{1}{|\vec{x}-\vec{R}_{AB}|}$$

Eq. (36) can be further reduced by reference to Eq.(28) where one of the $\frac{d}{dR_{AB}}$ derivatives has already been taken. The result is

$$q_{ZZ}^{A} = \frac{2Z_B}{R_{AB}^3} - \frac{2}{5}\int_0^\infty \frac{dx\,\rho_{A2}(x)}{x} + \sum_n \rho_{Bn}(R_{AB})$$

$$- \sum_n \frac{(n+1)(n+2)}{(2n+1)R_{AB}^{n+3}} \int_0^{R_{AB}} dx\,x^{n+2}\rho_{Bn}(x)$$

$$- \sum_n \frac{n(n-1)R_{AB}^{n-2}}{(2n+1)} \int_{R_{AB}}^\infty dx\,x^{1-n}\rho_{Bn}(x)$$

which reduces for the homonuclear case to

$$q_{ZZ} = \frac{2Z}{R^3} - \frac{2}{5}\int_0^\infty \frac{dx\,\rho_{A2}(x)}{x} + \sum_n (-1)^n\rho_{An}(R) \qquad (37)$$

$$-\sum_n \frac{(-1)^n(n+1)(n+2)}{(2n+1)R^{n+3}} \int_0^R dx\,x^{n+2}\rho_{An}(x) - \sum_n \frac{(-1)^n n(n-1)R^{n-2}}{(2n+1)} \int_R^\infty dx\,x^{1-n}\rho_{An}(x)$$

which with the insertion of the IAM model one obtains

$$q_{ZZ} = -\frac{2}{5}\int_0^R \frac{dx\,\Delta\rho_{A2}(x)}{x}\left[1 + 6\left(\frac{x}{R}\right)^5\right] - \frac{4}{5}\int_R^\infty \frac{dx\,\Delta\rho_{A2}(x)}{x} + \Delta\rho_{A2}(R) \qquad (38)$$

$$+ \frac{2}{R^3}\int_R^\infty dx\,x^2\rho_{A0}^{IAM}(x) + \rho_{A0}^{IAM}(R) + \Delta\rho_{A0}(R) - \frac{2}{R^3}\int_0^\infty dx\,x^2\Delta\rho_{A0}(x) - \Delta\rho_{A1}(R)$$

$$+ \frac{2}{R^4}\int_0^{o} dx\,x^3\Delta\rho_{A1}(x) - \sum_{n=3} \frac{(-1)^n(n+1)(n+2)}{(2n+1)R^{n+3}} \int_0^R dx\,x^{n+2}\Delta\rho_{An}(x)$$

$$- \sum_{n=3} \frac{(-1)^n n(n-1)R^{n-2}}{(2n+1)} \int_R^\infty dx\,x^{1-n}\Delta\rho_{An}(x)$$

where it has been assumed that $\int_0^\infty dx\,x^2\rho_{A0}^{IAM}(x) = Z_A$. Eq. (39) can be expected to be a fairly sensitive test of the $\Delta\rho$ terms since IAM contributions will not be terribly severe.

One might wonder about the other two field gradient components but since $q^A_{xx} + q^A_{yy} + q^A_{zz} = 0$ it is a simple matter to show that $q^A_{xx} = q^A_{yy} = -\frac{1}{2} q^A_{zz}$ and no new information is obtained. One could look at the non diagonal components of the field gradient tensor such as $q^A_{xy}$ but it is not clear whether or not any observations to compare with are available. In fact at present it appears that no direct gas phase measurements of $q_{zz}$ are available for $N_2$ although a value inferred from solid state measurements is available.[16]

b.) The outer moments of the charge density.

The outer moments are defined as $\langle r^n \rangle$. Among the non zero moments for a homonuclear diatomic are the various components of the quadrupole tensor. The total quadrupole moment $Q = Q_{zz}$ is given by[18]

$$Q = \sum_\alpha Z_\alpha \frac{1}{2}(3z^2_\alpha - R_\alpha{}^2) - \langle 0|\sum_i \frac{1}{2}(3z_i{}^2 - r_i{}^2)|0\rangle \tag{39}$$

$$= Z_A R_A{}^2 + Z_B R_B{}^2 - \frac{1}{2}\int d\vec{x}\rho_A(\vec{x})[3(x\cos\theta + R_A)^2 - |\vec{x}+\vec{R}_A|^2]$$

$$- \frac{1}{2}\int d\vec{x}\rho_B(\vec{x})[3(x\cos\theta + R_B)^2 - |\vec{x}+\vec{R}_B|^2]$$

$$= Z_A R_A{}^2 + Z_B R_B{}^2 - \int d\vec{x}\rho_A(\vec{x})[x^2 P_2(\cos\theta) + 2xR_A\cos\theta + R_A{}^2]$$

$$- \int dx\rho_B(x)[x^2 P_2(\cos\theta) + 2xR_B(\cos\theta) + R_B{}^2]$$

$$= Z_A R_A{}^2 + Z_B R_B{}^2 - \frac{1}{5}\int_0^\infty dx x^4 \rho_{A2}(x) - \frac{2}{3} R_A \int_0^\infty dx x^3 \rho_{A1}(x)$$

$$- \frac{1}{5}\int_0^\infty dx x^4 \rho_{B2}(x) - \frac{2}{3} R_B \int_0^\infty dx x^3 \rho_{B1}(x) - Z_A R_A{}^2 - Z_B R_B{}^2$$

or

$$\boxed{Q = -\frac{2}{5}\int_0^\infty dx x^4 \Delta\rho_{A2}(x) - \frac{2}{3} R \int_0^\infty dx x^3 \Delta\rho_{A1}(x)} \tag{40}$$

which is very sensitive to the details of the charge density since it shows no dependence on the IAM model.

Flygare et al[18-20] have succeeded in measuring the electronic moments

$$\langle 0|\sum_i x_i^2|0\rangle \text{ and } \langle 0|\sum_i z_i^2|0\rangle$$

for a number of diatomic and polyatomic molecules by means of the molecular Zeeman effect. These moments are given as

$$\langle 0|\begin{bmatrix} x_i^2 \\ z_i^2 \end{bmatrix}|0\rangle = \int dx\,\rho_A(x)\begin{bmatrix} (x\sin\theta\cos\varphi)^2 \\ (x\cos\theta+R_A)^2 \end{bmatrix} \tag{41}$$

$$+ \int dx\,\rho_B(x)\begin{bmatrix} (x\sin\theta\cos\varphi)^2 \\ (x\cos\theta+R_B)^2 \end{bmatrix}$$

or

$$\langle 0|x^2|0\rangle = \frac{1}{3}\int_0^\infty dx\,x^4\rho_{A0}(x) - \frac{1}{15}\int_0^\infty dx\,x^4\rho_{A2}(x) \tag{42}$$

$$\frac{1}{3}\int_0^\infty dx\,x^4\rho_{B0}(x) - \frac{1}{15}\int_0^\infty dx\,x^4\rho_{B2}(x)$$

and

$$\langle 0|z^2|0\rangle = Z_A R_A^2 + Z_B R_B + \frac{1}{3}\int_0^\infty dx\,x^4[\rho_{A0} + \rho_{B0}(x)] \tag{43}$$

$$+ \frac{2}{3}\int_0^\infty dx\,x^3[R_A\rho_{A1}(x) + R_B\rho_{B1}(x)]$$

$$+ \frac{2}{15}\int_0^\infty dx\,x^4[\rho_{A2}(x) + \rho_{B2}(x)]$$

Eqs. (42) and (43) reduce for the homonuclear case to

$$\langle x^2\rangle = \frac{2}{3}\int_0^\infty dx\,x^4\rho_{A0}^{IAM}(x) + \frac{2}{3}\int_0^\infty dx\,x^4\Delta\rho_{A0}(x) \tag{44}$$

$$- \frac{2}{15}\int_0^\infty dx\,x^4\Delta\rho_{A2}(x)$$

and

$$\langle z^2\rangle = \frac{ZR^2}{2} + \frac{2}{3}\int_0^\infty dxx^4\rho_{AO}^{IAM}(x) + \frac{2}{3}\int_0^\infty dxx^4\Delta\rho_{AO}(x) + \frac{2}{3}\int_0^\infty dxx^3\Delta\rho_{A1}(x) + \frac{4}{15}\int_0^\infty dxx^4\Delta\rho_{A2}(x) \tag{45}$$

It appears highly desirable to eliminate any dependence on the IAM model hence the difference $\langle z^2\rangle - \langle x^2\rangle - \frac{ZR^2}{2}$ seems like a suitable quantity to compute. It is unfortunately

$$\langle z^2\rangle - \langle x^2\rangle - \frac{ZR^2}{2} = \frac{2}{3} R\int_0^\infty dxx^3\Delta\rho_{A1}(x) + \frac{2}{5}\int_0^\infty dxx^4\Delta\rho_{A2}(x) = -Q_{zz} \tag{46}$$

and contains no new information other than a consistency check on the various electrical measurements.

In the case where $\rho_{A1}(x)$ is small or completely missing we have

$$\frac{2}{3}\int_0^\infty dxx^4\Delta\rho_{AO}(x) = \langle x^2\rangle - \frac{1}{3} Q_{zz} - \frac{2}{3}\langle r^2\rangle^{IAM} \tag{47}$$

or

$$\frac{2}{3}\int_0^\infty dxx^4\Delta\rho_{AO}(x) = \langle z^2\rangle + \frac{2}{3} Q_{zz} - \frac{2}{3}\langle r^2\rangle^{IAM} - \frac{ZR^2}{2} \tag{48}$$

or

$$2\int_0^\infty dxx^4\Delta\rho_{AO}(x) = \langle z^2\rangle + 2\langle x^2\rangle - 2\langle r^2\rangle^{IAM} - \frac{ZR^2}{2} \tag{49}$$

Assuming the mean square radius of the IAM model is well defined it may be possible to use Eqs. (47) - (49) as further checks on the monopole correction density. Eqs. (44) and (45) used firectly do not provide a sensitive test of the density correction since well over 90% of $\langle x^2\rangle$ and $\langle z^2\rangle$ come from the IAM model.

It is not known yet whether values for the next higher non-vanishing moment, the hexadecapole moment, $H_D$, are available for comparison. Some values have been given for octahedral molecules where the hexadecapole moment is the first nonvanishing moment available.[21] The moment is given as

$$H_D = \sum_\alpha Z_\alpha R_\alpha{}^4 P_4(\cos\theta_\alpha) - \langle \sum_i r_i{}^4 P_4(\cos\theta_i)\rangle$$

$$= Z_A R_A{}^4 + Z_B R_B{}^4 - \frac{1}{8}\int dr\,\rho(r)[35z^4 - 30Z^2 r^2 + 3r^4]$$

$$= Z_A R_A{}^4 + A_B R_B{}^4 - \int dx\,\rho_A(x)[x^4 P_4(\cos\theta) + 4R_A x^3 P_3(\cos\theta)$$

$$+ 6R_A{}^2 x^2 P_2(\cos\theta) + 4R_A{}^3 x P_1(\cos\theta) + R_A{}^4]$$

$$- \int dx\,\rho_B(x)[x^4 P_4(\cos\theta) + 4R_B x^3 P_3(\cos\theta) + 6R_B{}^2 x^2 P_2(\cos\theta)$$

$$+ 4R_B{}^3 x P_1(\cos\theta) + R_B{}^4]$$

$$= -\Big[\frac{1}{9}\int_0^\infty dx x^6 \rho_{A4}(x) + \frac{4R_A}{7}\int_0^\infty dx x^5 P_{A3}(x) + \frac{6}{5} R_A{}^2 \int_0^\infty dx x^4 \rho_{A2}(x)$$

$$+ \frac{4R_A{}^3}{3}\int_0^\infty dx x^3 \rho_{A1}(x)$$

$$+ \frac{1}{9}\int_0^\infty dx x^6 \rho_{B4}(x) + \frac{4R_B}{7}\int_0^\infty dx x^5 \rho_{B3}(x) + \frac{6R_A{}^2}{5}\int_0^\infty dx x^4 \rho_{B2}(x)$$

$$+ \frac{4R_A{}^3}{3}\int_0^\infty dx x^3 \rho_{B1}(x)\Big]$$

which for the homonuclear case with the IAM model becomes

$$H_D = -\frac{2}{9}\int_0^\infty dx x^6 \Delta\rho_{A4}(x) - \frac{4R}{7}\int_0^\infty dx x^5 \Delta\rho_{A3}(x) - \frac{3}{5}R^2\int_0^\infty dx x^4 \Delta\rho_{A2}(x) \quad (50)$$

$$- \frac{R^3}{3}\int_0^\infty dx x^3 \Delta\rho_{A1}(x)$$

C. Analysis of Fink's Electron Scattering Data for $N_2$ in Terms of Properties of the Charge Distribution.

1.) The model.

The Fink fits for experimental data and Hartree Fock data consist of the model with the radial density components $\rho_{AO}^{IAM}(r)$, $\Delta\rho_{AO}(r)$ and $\Delta\rho_{A2}(r)$ which are given as

$$\Delta\rho_{AO}(r) = \sum_{n=0}^{2} a_{on} r^{n} e^{-\lambda_{on} r} \tag{51}$$

and

$$\Delta\rho_{A2}(r) = a_{2O} r^{2} e^{-\lambda_{2O} r} \tag{52}$$

The $\rho_{AO}^{IAM}(r)$ was defined in terms of a nearly exact spherically averaged nitrogen molecule. Since this is essentially a numerical representation we will approximate it here by use of the expression

$$\rho_{AO}^{IAM}(r) = 7 \sum_{i=1}^{6} \gamma_i \lambda_i^2 \frac{e^{-\lambda_i r}}{r} \tag{53}$$

where all parameters needed in Equations (51) - (53) are given in Table I. In addition parameters for fitting Eqs. (51) and (52) to molecular Hartree Fock data were obtained and are listed in Table I.

2.) The Results

The basic equations are those outlined in rectangular blocks in the preceding text. A summary of these equations is given in Table II. Inspection of the equations in Table II show that only a limited number of integrals are required to obtain values for the indicated quantities. These integrals and the analytical results of their evaluation are presented in Table III. In Table IV the quantities given in Table II are reexpressed in terms of the basic $I_n$ integrals given in Table III. In Table V numerical values of the integrals are given. In Table VI the values of the force constants and vibrational moments in a.u. are given. These values are based on a Morse model for the vibrational motion.[26]

Finally in Table VII the final values for the calculated quantities are assembled and compared with experiment. The IAM contribution is listed separately for convenience.

We note that the agreement for most derived quantities is quite acceptable at this stage in the development of the analysis. Some of the numbers, depending strongly on the IAM model, could change with the introduction of improved estimates of the actual IAM contribution. Comparison of the results in columns 4 and 5 in Table VII shows that the error in the use of the model is of the order of 5 - 15% depending on the quantity in question. The agreement between values calculated from the experimental charge distribution and direct experimental measurements of the same quantities appears to be within this uncertainty.

It is interesting to note that the lack of charge balance is due to a failure of the monopole term. It is also interesting to note that the quadrupole terms furnish about 80% of the force balance which suggests the monopole terms may be deficient here also. It would be interesting to know if forcing the charge and force balance conditions upon the monopole fit would lead to an improved picture of the energy. It would also be interesting to know if the addition of a dipole term in the charge expansion would improve things. Perusual of the Eqs. in the first part of this report shows that dipole terms will contribute to many quantities especially the force balance. On the other hand the model does seem to provide an adequate fit of the Hartree Fock density (10-15%) thus showing that the experimental data is fundamentally different from the H.F. data.

Table II. Summary of Basic Equations for Computing Properties of the Charge Density for the Fink Model of $N_2$

1.) $\int_0^\infty dx\ x^2 \Delta\rho_{AO}(x) = 0$ [Charge Conservation]

2.) $2k_e \langle Q\rangle_{vib} + \ell_e \langle Q^2\rangle_{vib} + \frac{me}{3}\langle Q^3\rangle_{vib} = \frac{-Z}{R^2}\int_R^\infty dx x^2 \rho_{AO}^{IAM}(x)$

$+ \frac{Z}{R^2}\int_0^R dx x^2\ \Delta\rho_{AO}(x)$

$+ \frac{3Z}{5R^4}\int_0^R dx x^4 \Delta\rho_{A2}(x) - \frac{2ZR}{5}\int_R^\infty dx\ \frac{\Delta\rho_{A2}(x)}{x}$ [Force Balance]

3.) $\overline{V}_{ne} = -2Z\int_0^R dx x \rho_{AO}^{IAM}(x)[1 + \frac{x}{R}]\int_0^\infty dx x \rho_{AO}^{IAM}(x)$

$-2Z\int_0^R dx x \Delta\rho_{AO}(x)[1 + \frac{x}{R}] - 4Z\int_R^\infty dx x \Delta\rho_{AO}(x)$

$- \frac{2Z}{5R^3}\int_0^R dx x^4 \Delta\rho_{A2}(x) - \frac{2ZR^2}{5}\int_R^\infty dx\ \frac{\Delta\rho_{A2}(x)}{x}$ [Electron-Nuclear Attractive Energy]

4.) $q_{zz} = \frac{2}{R^3}\int_R^\infty dx x^2 \rho_{AO}^{IAM}(R) - \frac{2}{R^3}\int_0^\infty dx x^2 \Delta\rho_{AO}(x) + \Delta\rho_{AO}(R) + \Delta\rho_{A2}(R)$

$- \frac{2}{5}\int_0^R \frac{dx \Delta\rho_{A2}(x)}{x}[1 + 6(\frac{x}{R})^5] - \frac{4}{5}\int_R^\infty \frac{dx \Delta\rho_{A2}(x)}{x} \Delta\rho_{A2}(R)$ [Field Gradient At Necleus]

5.) $Q_{zz} = -\frac{2}{5}\int_0^\infty dx x^4 \Delta\rho_{A2}(x)$ [Quadrupole Moment]

6.) $H_D = -\frac{3}{5} R^2 \int_0^\infty dx x^4 \rho_{A2}(x)$ [Hexadecapole Moment]

7.) $\langle x^2\rangle = \frac{2}{3}\int_0^\infty dx x^4 \rho_{AO}^{IAM}(x) + \frac{2}{3}\int_0^\infty dx x^4 \Delta\rho_{AO}(x) + \frac{Q_{zz}}{3}$

8.) $\langle z^2\rangle = \frac{ZR^2}{2} + \langle x^2\rangle - Q_{zz}$

Table III. Integrals Involved in Computing Moments of the Charge Density.

$$I_1 = \int_0^\infty dx\, x\, \rho_{A0}^{IAM}(x) = Z \sum_i \gamma_i \lambda_i$$

$$I_2 = \int_R^\infty dx\, x\, \rho_{A0}^{IAM}(x) = Z \sum_i \gamma_i \lambda_i e^{-\lambda_i R}$$

$$I_3 = \int_0^\infty dx\, x^2 \rho_{A0}^{IAM}(x) = Z$$

$$I_4 = \int_R^\infty dx\, x^2 \rho_{A0}^{IAM}(x) = Z \sum_i \gamma_i e^{-\lambda_i R}(1 + \lambda_i R)$$

$$I_5 = \rho_{A0}^{IAM}(R) = Z \sum_i \gamma_i \lambda_i^2 \frac{e^{-\lambda_i R}}{R}$$

$$I_6 = \int_0^\infty dx\, x\, \Delta\rho_{A0}(x) = \frac{a_{00}}{\lambda_{00}^2} + \frac{2a_{01}}{\lambda_{01}^3} + \frac{6a_{02}}{\lambda_{02}^4}$$

$$I_7 = \int_R^\infty dx\, x\, \Delta\rho_{A0}(x) = \frac{a_{00}(1+\lambda_{00}R)e^{-\lambda_{00}R}}{\lambda_{00}^2} + \frac{2a_{01}(1+\lambda_{01}R+\frac{\lambda_{01}^2 R^2}{2})e^{-\lambda_{01}R}}{\lambda_{01}^3}$$

$$+ \frac{6a_{02}(1+\lambda_{02}R+\frac{\lambda_{02}^2 R^2}{2}+\frac{\lambda_{02}^3 R^3}{6})e^{-\lambda_{02}R}}{\lambda_{02}^4}$$

$$I_8 = \int_0^\infty dx\, x^2 \Delta\rho_{A0}(x) = \frac{2a_{00}}{\lambda_{00}^3} + \frac{6a_{01}}{\lambda_{01}^4} + \frac{24a_{02}}{\lambda_{02}^5}$$

$$I_9 = \int_R^\infty dx\, x^2 \Delta\rho_{A0}(x) = \frac{2a_{00}}{\lambda_{00}^3}(1 + \lambda_{00}R + \frac{\lambda_{00}^2 R^2}{2})e^{-\lambda_{00}R}$$

$$+ \frac{6a_{01}}{\lambda_{01}^4}(1 + \lambda_{01}R + \frac{\lambda_{01}^2 R^2}{2} + \frac{\lambda_{01}^3 R^3}{6})e^{-\lambda_{01}R}$$

$$+ \frac{24a_{02}}{\lambda_{02}^5}(1 + \lambda_{02}R + \frac{\lambda_{02}^2 R^2}{2} + \frac{\lambda_{02}^3 R^3}{6} + \frac{\lambda_{02}^4 R^4}{24})e^{-\lambda_{02}R}$$

$$I_{10} = \Delta\rho A_0(R) = a_{00}e^{-\lambda_{00}R} + a_{01}Re^{-\lambda_{01}R} + a_{02}R^2 e^{-\lambda_{02}R}$$

$$I_{11} = \int_0^\infty \frac{dx\, \Delta\rho_{A2}(x)}{x} = \frac{a_{20}}{\lambda_{20}^2}$$

$$I_{12} = \int_R^\infty \frac{dx\, \rho_{A2}(x)}{x} = \frac{a_{20}}{\lambda_{20}^2} e^{-\lambda_{20}R}(1 + \lambda_{20}R)$$

$$I_{13} = \int_0^\infty dxx^4\Delta\rho_{A2}(x) = \frac{a_{20}(6!)}{\lambda_{20}^7}$$

$$I_{14} = \int_R^\infty dxx^4\Delta\rho_{A2}(x) = \frac{a_{20}(6!)}{\lambda_{20}^7}[1 + \lambda_{20}R + \frac{\lambda_{20}^2R^2}{2} + \frac{\lambda_{20}^3R^3}{6} + \frac{\lambda_{20}^4R^4}{24} + \frac{\lambda_{20}^5R^5}{120} + \frac{\lambda_{20}^6R^6}{720}]e^{-\lambda_{20}R}$$

$$I_{15} = \int_0^\infty dxx^4\rho_{AO}^{IAM}(x) = 6Z\sum_i \frac{\gamma_i}{\lambda_i^2}$$

$$I_{16} = \int_0^\infty dxx^4\Delta\rho_{AO}(x) = \frac{24a_{00}}{\lambda_{00}^5} + \frac{120a_{01}}{\lambda_{01}^6} + \frac{720a_{02}}{\lambda_{02}^7}$$

$$I_{17} = \Delta\rho_{A2}(R) = a_{20}R^2e^{-\lambda_{20}R}$$

Table IV. Charge Density Properties Expressed in Terms of the Basic Integrals, $I_n$ from Table III.

---

1.) $I_8 = 0$

2.) $F_{vib} = 2k_e\langle Q\rangle_{vib} + \ell_e\langle Q^2\rangle_{vib} + \frac{me}{3}\langle Q^3\rangle_{vib} = -\frac{Z}{R^2}[I_4-(I_8-I_9)] + \frac{3Z}{5R^4}[I_{13}-I_{14}] - \frac{1ZR}{5} I_{12}$

3.) $\bar{V}_{ne} = -2Z[I_1-I_2] - \frac{2Z}{R}[I_3-I_4] - 4ZI_2 - 2Z[I_6-I_7] - \frac{2Z}{R}[I_8-I_9] - 4ZI_7 - \frac{2Z}{5R^3}[I_{13}-I_{14}] - \frac{2ZR^2}{5} I_{12}$

4.) $q_{zz} = \frac{2}{R^3} I_4+I_5 - \frac{2}{R^3} I_8+I_{10} - \frac{2}{5} I_{11} - \frac{2}{5} I_{12} - \frac{12}{5R^5}[I_{13}-I_{14}]+I_{17}$

5.) $Q_{zz} = -\frac{2}{5} I_{13}$

6.) $H_D = -\frac{3}{5} R^2I_{13}$

7.) $\langle x^2\rangle = \frac{Q_{zz}}{3} + \frac{2}{3}[I_{15}+I_{16}]$

8.) $\langle z^2\rangle + 2\langle x^2\rangle = \langle r^2\rangle = \frac{ZR^2}{2} + 2[I_{15}+I_{16}] - \frac{Q_{zz}}{3}$

Table V. Values of the Integrals $I_n$ for the IAM Density, $I_n^{IAM}$, the experimental density, $I_n^{Ex}$, and the Hartree-Fock density, $I_n^{HF}$.

| | | | | | |
|---|---|---|---|---|---|
| $I_1^{IAM}$ | = | + 18.3703 | 18.2859 (Stewart) | | |
| $I_1^{IAM}$ | = | + 0.28546 | | | |
| $I_3^{IAM}$ | = | + 7.0091 | | | |
| $I_4^{IAM}$ | = | + 0.73469 | | | |
| $I_5^{IAM}$ | = | + 0.275298 | | | |
| $I_{15}^{IAM}$ | = | + 12.23418 | | | |
| $I_6^{ex}$ | = | + 0.0505599 | $I_6$ | = | - 0.026942 |
| $I_7^{ex}$ | = | - 0.0109384 | $I_7$ | = | - 0.0046449 |
| $I_8^{ex}$ | = | + 0.02118 | $I_8$ | = | - 0.000178 |
| $I_9^{ex}$ | = | - 0.028517 | $I_9$ | = | - 0.028374 |
| $I_{10}^{ex}$ | = | - 0.008022 | $I_{10}$ | = | - 0.002920 |
| $I_{16}^{ex}$ | = | - 0.203012 | $I_{16}$ | = | - 0.058247 |
| $I_{11}^{ex}$ | = | 3.32263 | $I_{11}$ | = | 3.88217 |
| $I_{12}^{ex}$ | = | + 0.009959 | $I_{12}$ | = | + 0.006167 |
| $I_{13}^{ex}$ | = | 2.81429 | $I_{13}$ | = | 2.14922 |
| $I_{14}^{ex}$ | = | + 0.879112 | $I_{14}$ | = | + 0.502013 |
| $I_{17}^{ex}$ | = | + 0.0709 | $I_{17}$ | = | + 0.0482 |

Table VI. Values of Vibrational Quantities for $N_2$ in a.u.

$$\omega_e = 2359.61\ \text{cm}^{-1}$$

$$\mu_A = 7.00377$$

$$D_0{}^0 = 9.756\ \text{eV}$$

$$a = 2.710\text{Å}^{-1} = 1.3557 \times 10^{-3}\ \omega_e \left(\frac{\mu_A}{D_0{}^0}\right)^{1/2} \text{Å}^{-1}$$

| | |
|---|---|
| $k_e = 2a^2 D_0{}^0 = 1.476$ a.u. | $\langle Q \rangle^a_{vib} = 0.0041\text{Å} = 0.00775$ a.u. |
| $\ell_e = -3ak_e = -7.098$ a.u. | $\langle Q^2 \rangle^a_{vib} = 0.00102\text{Å}^2 = 0.00364$ a.u. |
| $m_e = 7a^2 k_e = 75.902$ a.u. | $\langle Q^3 \rangle^b_{vib} \simeq 1 \times 10^{-5}\text{Å}^3 \simeq 6.8 \times 10^{-5}$ a.u. |

a.) Value was taken from: R. A. Bonham and J. L. Peacher, J. Chem. Phys. 38, 3219 (1963).

b.) Value was taken from: E. W. Ng, L. S. Su and R. A. Bonham, J. Chem. Phys. 50, 2038 (1969). The value is for $I_2$ at 0°K. It is anticipated that the value for $N_2$ will be even smaller.

Table VII. Experimental and Calculated Values for Some Selected Molecular Properties of $N_2$ in a.u.'s.

| Property (a.u.) | IAM Contribution | Fink Experiment | H.F. Model Fit of Stewarts Date | H.F. Direct Calculation | Experiment Direct Observation |
|---|---|---|---|---|---|
| Charge | 14.0182 | 14.0606 | 14.0178 | 14.000 | 14.000 |
| Force on nucleus | -1.192546 | -0.8141 | -0.8570 | +0.0796[c]<br>-0.8696[a] | -0.00126[b] |
| $\overline{V}_{ne}$ | -303.85 | -305.46 | -304.15 | -303.21[c] | -303.38[d] |
| $q_{zz}$ | +0.439 | -0.96 | -1.17 | -1.37[c] | -1.24 ± .54[e],<br>-1.40 ± .62[e]<br>(± 1.233)[d] |
| $Q_{zz}$ | 0 | -1.13 | -0.86 | -.95[c] -.91[ℓ]<br>-1.1576 (SCF)[ℓ]<br>-1.3455 (CI)[ℓ] | -1.13[k] -1.0 ± 0.1[ℓ]<br>(±)1.04 ± .07[f] -1.1[ℓ]<br>(-1.243)[d] -0.90 ± .07[ℓ] |
| $\langle x^2 \rangle$ | 8.156 | 7.644 | 7.831 | 7.701[c] | 7.5[g]<br>(7.487)[d] |
| $\langle z^2 \rangle$ | 23.269 | 23.887 | 23.803 | 23.616[c] | 23.6[g]<br>(23.854)[d] |
| $H_D$ | 0 | -7.29 | -5.57 | -8.8583[ℓ]<br>-6.843[c]<br>-8.4203 (CI)[ℓ] | ± 8.0 ± 2.7[ℓ] |
| $\langle k_e \rangle$ | 1.927 | 1.871 | 1.906 | 1.974[i]<br>1.805[j] | 1.58[h] |

a.) Neglect of dipole term. Private communication, R. F. Stewart.
b.) The vibrational force was computed using the values given in Table VI.
c.) R. F. Stewart, Private Communication.
d.) T. H. Dunning, Jr., D. C. Cartwright, W. J. Hunt, P. J. Hay and F. W. Bobrowicz, J. Chem. Phys. 64, 4755 (1976).
e.) R. F. Stewart, Private Communication. Calculated from data given in C. C. Lin, Phys. Rev. 119, 1026 (1960) and I. A. Scott, J. Chem. Phys. 36, 1459 (1962).
e′.) R. F. Stewart, Private Communication. Calculated from data given in C. C. Lin, Phys. Rev. 119, 1026 (1960) and J. A. Scott and T. J. Haigh, J. Chem. Phys. 38, 117 (1963). Note this result is supposed to be free of liberational effects.
f.) A. D. Buckingham, R. L. Disch and D. A. Dunmur, J. Amer. Chem. Soc. 90, 3104 (1969).
g.) W. H. Flygare, R. L. Shoemaker and W. Hüttner, J. Chem. Phys. 50, 2414 (1969).
h.) Calculated from the values given in Table VI.
i.) P. Cade, Private Communication.
j.) R. F. Stewart, Private Communication.
k.) D. E. Stogryn and A. P. Stogryn, Mol. Phys. 11, 371 (1966).
ℓ.) J. E. Gready, G. B. Bacskay and N. S. Hush, Chem. Phys. 31, 467 (1978).

# MAGNETIC RESONANCE AND RELATED TECHNIQUES

J. MARUANI

C.M.O.A. du C.N.R.S. and UNIVERSITE PARIS VI

23, rue du Maroc, 75019 PARIS, FRANCE

## 1. INTRODUCTION

In this chapter we shall describe the principles associated with magnetic resonance and related techniques as applied to the determination of electronic spin and charge distributions in molecular systems. These techniques have more or less overlapping fields and therefore can complement each other as well as the various scattering techniques. They have in common several related features which distinguish them from the latter and also from the other spectroscopic techniques.

In the first place, they involve interactions between the molecular electronic cloud and the intrinsic moments of internal or surrounding nuclei, not incident particles. One then measures the effect of these interactions on the pattern of stationary states of the molecule, not on that of trajectories of diffracted particles. As the energies relative to the higher-order molecular interactions are rather small, one can monitor the splittings with an external magnetic field, the scanning of which brings the pairs of levels successively at resonance with a fixed radiofrequency or microwave. A related feature is that the probability for spontaneous emission is negligible at these frequencies, which permits the use of saturation effects in some magnetic resonance techniques. Finally, it should be noted that any other phenomenon involving very sharp frequencies, such as recoilless nuclear emission or laser molecular excitation, can also be used to study these interactions.

This chapter is divided into four sections. In the following section, we recall the origin and shape of the various electric and magnetic interactions within molecular systems and between molecules and external fields and the way effective spin Hamiltonians can be constructed by the use of angular-momentum-symmetry and order-of-magnitude considerations. In the third section, we show how the characteristic coupling tensors occurring in these Hamiltonians are related to the one and two-electron density and transition matrices in the different degrees of approximation. In the last section, we outline the basic principles of electron paramagnetic resonance (EPR) for molecular systems, nuclear magnetic resonance (NMR) for paramagnetic molecules, electron-nuclear and electron-electron double resonance (ENDOR and ELDOR), nuclear quadrupole resonance (NQR) and Mössbauer spectroscopy (MS), define their application fields as regard to the determination of electronic spin and charge distributions in molecular systems, and compare their possibilities and limitations with the help of a few examples.

## 2. SPIN HAMILTONIANS

The interpretation of experimental results obtained with magnetic resonance and related techniques requires the use of effective operators and functions involving electronic and nuclear spins. However, while in the theory of chemical bonding electron-spin operators and functions come into play only through their symmetry properties, in the spectroscopic domain with which we are concerned various spin operators also occur in additional interaction terms of the basic Hamiltonian operator. It may be worthwhile to recall first how these latter terms arise from the combination of special-relativity and quantum-theory requirements for the relative motions of atomic elementary particles.

### 2.1. Basic Hamiltonians

The classical Hamiltonian function for a free particle with rest mass $m_e$ and electric charge $-e$ : $H = \underline{p}^2/2m_e$, becomes in relativity theory : $H = c\,(m_e^2 c^2 + \underline{p}^2)^{1/2}$. The corresponding quantum Hamiltonian operator gives rise to the wave-equation : $\{\hat{H} - c\,(m_e^2 c^2 + \hat{\underline{p}}^2)^{1/2}\}\ \psi\ (\underline{r},t) = 0$, where $\hat{H} = i\hbar\partial/\partial t$ and $\hat{\underline{p}} = -i\hbar\underline{\nabla}$. The solutions of the latter also obey the quadratic (Klein-Gordon) equation : $\{\hat{p}_0^2 - (m_e^2 c^2 + \hat{\underline{p}}^2)\}\ \psi = 0$, since $\hat{p}_0 \equiv \hat{H}/c$ and $\hat{p}_1 \equiv \hat{p}_x$, $\hat{p}_2 \equiv \hat{p}_y$, $\hat{p}_3 \equiv \hat{p}_z$ commute. The Klein-Gordon operator can be rewritten as the factor-product operator : $(\hat{p}_0 - \hat{\alpha}_1\hat{p}_1 - \hat{\alpha}_2\hat{p}_2 - \hat{\alpha}_3\hat{p}_3 - \hat{\alpha}_4 m_e c)(\hat{p}_0 + \hat{\alpha}_1\hat{p}_1 + \hat{\alpha}_2\hat{p}_2 + \hat{\alpha}_3\hat{p}_3 + \hat{\alpha}_4 m_e c)$, provided that the $\hat{\alpha}_a$'s commute with the $\hat{p}_\mu$'s and that they satisfy the anticommutation relations : $\hat{\alpha}_a\hat{\alpha}_b + \hat{\alpha}_b\hat{\alpha}_a = 2\delta_{ab}$. The above equation will then be fulfilled by the solutions of the linear (Dirac) equation : $(\hat{p}_0 + \hat{\underline{\alpha}}.\hat{\underline{p}} + \hat{\alpha}_4 m_e c)\ \psi = 0$, which can be seen to satisfy simultaneously the quantum-theory requirement of linearity in $\hat{p}_0$ and the special-relativity requirement of invariance under Lorentz transformations. In the presence of a classical electromagnetic field with the scalar and vector potentials $A_0(\underline{r},t)$ and

$\underline{A}(\underline{r},t)$, $\hat{p}_0$ and $\hat{\underline{p}}$ will be replaced by $\hat{\pi}_0 = \hat{p}_0 + eA_o/c$ and $\hat{\underline{\pi}} = \hat{\underline{p}} + e\underline{A}/c$, giving[1] :

$$(\hat{\pi}_0 + \hat{\underline{\alpha}}.\hat{\underline{\pi}} + \hat{\alpha}_4 m_e c)\ \psi\ (\underline{r},t) = 0. \tag{1}$$

For the four operators $\hat{\alpha}_a$ to obey the above anticommutation relations, four-dimensional matrix representations are required, e.g. :

$$\alpha_i = \begin{pmatrix} 0 & \sigma_i \\ \sigma_i & 0 \end{pmatrix} (i = 1, 2, 3),\ \alpha_4 = \begin{pmatrix} 1 & 0 \\ 0 & -1 \end{pmatrix}, \tag{2}$$

where the $\sigma_i$'s are the two-dimensional Pauli matrices. This implies that the wave-function $\psi\ (\underline{r},t)$ also depends on some internal variables inducing a four-dimensional vector representation $\psi$. These can be shown[1] to be the rest-mass energy $(\pm m_e c^2)$ and a spin-momentum component $(\pm \hbar/2)$. Writing $\psi = \exp.(-im_e c^2 t/\hbar)\ \bar{\psi}$ to eliminate the former from the positive energies and substituting $\bar{\psi} = (\Phi,\Psi)^t$ with Eqns. (2) in Eqn. (1), one obtains :

$$(\hat{\pi}_0 + 2m_e c)\ \Phi + \hat{\underline{\pi}}.\sigma\ \Psi = 0,\ \hat{\pi}_0\ \Psi + \hat{\underline{\pi}}.\sigma\ \Phi = 0,$$

$$\int_{R^3} \{|\Psi|^2 + |\Phi|^2\}\ dv = 1, \tag{3}$$

where the occurrence of $2m_e c$ with $\hat{\pi}_0$ as a coefficient of $\Phi$ makes it much smaller than $\Psi$. Solving for the larger components and writing the explicit form for $c\hat{\pi}_0$, one obtains[2] :

$$\{(\hat{\underline{\pi}}.\sigma)(\hat{f}\hat{\underline{\pi}}.\sigma)/2m_e - eA_0\}\ \Psi\ (\underline{r},t) = i\hbar\partial\ \Psi\ (\underline{r},t)/\partial t, \tag{4}$$

where $\hat{f} \equiv (1 + \hat{\pi}_0/2m_e c)^{-1}$. One has $\hat{f} \simeq 1$ everywhere except in the neighbourhood of a source or a sink of electric field, $\zeta e/r$, where it behaves as $r/\zeta r_0$, $r_0 \equiv e^2/2m_e c^2 \simeq 1.4$ Fermi.

The spin terms can be separated by applying to the first operator in Eqn. (4) the identity : $(\hat{\underline{a}}.\sigma)(\hat{\underline{b}}.\sigma) = \hat{\underline{a}}.\hat{\underline{b}} + i\sigma.(\hat{\underline{a}}\Lambda\hat{\underline{b}})$, which results from the commutation properties of the $\sigma_i$'s. This gives the kinetic-energy-like term :

$$\hat{T} = (\hat{\underline{\pi}}\hat{f}.\hat{\underline{\pi}})/2m_e = (\hat{f}\hat{\underline{\pi}}^2 + \hat{\underline{\pi}}^2\hat{f})/4m_e - i\beta_e(\hat{f}\underline{E}\hat{f}\hat{\underline{\pi}} - \hat{\underline{\pi}}\hat{f}\underline{E}\hat{f})/4m_e c, \tag{5}$$

and the spin-energy additional term :

$$\hat{S} = i\sigma.(\hat{\underline{\pi}}\hat{f}\Lambda\hat{\underline{\pi}})/2m_e = 2\beta_e \underline{s}.(\hat{f}\underline{B} + \hat{f}\underline{E}\Lambda\hat{f}\hat{\underline{\pi}}/2m_e c), \tag{6}$$

where we have used the relations :

$$\hat{\underline{\pi}}\hat{f} - \hat{f}\hat{\underline{\pi}} = \hat{f}\ \{\hat{\pi}_0,\hat{\underline{\pi}}\}\ \hat{f}/2m_e c,$$

$$\{\hat{\pi}_0,\hat{\underline{\pi}}\} = (i\hbar e/c)\ (1/c.\partial\underline{A}/\partial t + \underline{\nabla}A_0) = -(i\hbar e/c)\ \underline{E}(\underline{r},t),$$

$$\{\hat{\underline{\pi}}\wedge\hat{\underline{\pi}}\} = -(i\hbar e/c)\ (\underline{\nabla}\wedge\underline{A}) = -(i\hbar e/c)\ \underline{B}(\underline{r},t),$$

and introduced the Bohr magneton : $\beta_e \equiv e\lambda_e \equiv e\hbar/2m_ec$, and the spin-matrix vector : $\underset{\approx}{s} \equiv \underset{\approx}{\sigma}/2$. The operator factor $\hat{f}$, which also contains $\hat{\pi}_0$, helps to introduce relativistic corrections or remove energy singularities wherever this may be needed(2).

The first term in Eqn. (6) expresses the interaction of the effective magnetic flux density $\hat{f}\underline{B}$ with the intrinsic magnetic moment $\underset{\approx}{\mu}_e = -2\beta_e\underset{\approx}{s} \equiv \gamma_e\hbar\underset{\approx}{s}$, where $\gamma_e$ is the magnetogyric ratio of $\underset{\approx}{\mu}_e$ by the spin-momentum vector $\underset{\approx}{m}_e \equiv \hbar\underset{\approx}{s}$. Further refinements introduced by the quantization of the electromagnetic field and polarization of the vacuum lead to the exact experimental value for the electron : $g_0 = 2.0023193$, instead of the factor 2 multiplying $\beta_e\underset{\approx}{s}$. The theory also holds well for the other known charged leptons (for the muon, which is 207 times heavier than the electron with a lifetime of 2.2 μs, the measured and calculated values of $g_0$ both equal 2.00233). However it breaks down for the more complex baryons (for the proton, which is 1836 times heavier than the electron with an infinite lifetime, $g_p = (-)5.58569$). For atomic nuclei, which are made up of nucleons bound by strong nuclear forces, one can define, in addition to the isotopic rest mass M(Z,A) and total electric charge +Ze, an effective spin-momentum $\hbar\underset{\approx}{I}$ (which may take on values different from $\pm\hbar/2$) and an effective magnetogyric ratio $\gamma_n$, as well as higher-order moments (such as the quadrupolar electric moment) if the distribution of charges and currents is not spherical. These moments (and the effective size of the nucleus) may actually depend on whether the nucleus is in the ground or in an excited state.

We shall now see with a simple classical example how the other terms arising from Eqns. (5) and (6) can be interpreted(3). The scalar and vector potentials induced by a fixed point nucleus can be written : $A_0 = Ze/r$, $\underline{A} = \gamma_n\hbar\underline{I}\wedge\underline{r}/r^3$, giving : $\hat{f}\underline{E} = -\hat{f}\underline{\nabla}A_0 = Ze\hat{f}\underline{r}/r^3$, $\hat{f}\underline{B} = \hat{f}(\underline{\nabla}\wedge\underline{A}) = \gamma_n\hbar\hat{f}\{(\underline{\nabla}.\underline{r}/r^3)\underline{I} - (\underline{I}.\underline{\nabla})\underline{r}/r^3\} = \gamma_n\hbar\hat{f}\{4\pi\delta(r)\underline{I} - \underline{I}/r^3 + 3(\underline{I}.\underline{r})\underline{r}/r^5\} = -\gamma_n\hbar\hat{f}\{\underline{I}/r^3 - 3(\underline{I}.\underline{r})\underline{r}/r^5\}$ since $\hat{f}\sim r/Zr_0$ when $r\sim 0$ and $r\delta(r) = 0$. Then, the first term of Eqn. (6) becomes the dipolar-interaction operator :

$$\hat{\underset{\sim}{S}}_d = \gamma_e\gamma_n\hbar^2\hat{f}\{\underset{\approx}{s}.\underline{I}/r^3 - 3(\underline{I}.\underline{r})(\underset{\approx}{s}.\underline{r})r^5\}, \tag{7}$$

with the allowed singularity $r^{-2}$. The second term can be split into a spin-orbit contribution :

$$\hat{\underset{\sim}{S}}_o = (g_0\beta_e/2m_ec)\underset{\approx}{s}.(\hat{f}\underline{E}\wedge\hat{f}\hat{\underline{p}}) = g_0Z\beta_e^2\hat{f}^2\underset{\approx}{s}.\hat{\underline{\ell}}/r^3, \tag{8}$$

where we have introduced the orbital angular momentum $\hbar\hat{\underline{\ell}} \equiv \underline{r}\wedge\hat{\underline{p}}$ ; and a Fermi contact contribution :

$$\hat{S}_c = (g_0\beta_e e/2m_e c^2)\underset{\approx}{s}.(\hat{f}\underline{E}\wedge\hat{f}\underline{A}) = g_0\beta_e Z r_0 \gamma_n \hbar \hat{f}^2 \underset{\approx}{s}.\{(\underline{r}.\underline{r})\underline{I} - (\underline{I}.\underline{r})\underline{r}\}/r^6.$$

For a stationary state, the operator $\hat{f}$ can be replaced by the function : $f(r) = \{1 + \{W + eA_0(r)\}/2m_e c^2\}^{-1}$, the derivative of which is : $f'(r) = Zr_0 f^2(r)/r^2$. Then $\hat{S}_c$ takes the familiar form :

$$\hat{S}_c = -\gamma_e\gamma_n\hbar^2 \{f'(r)/r^2\} \{\underset{\approx}{s}.\underline{I} - s_r I_r\} = -\gamma_e\gamma_n\hbar^2 \times 4\pi k\delta(r) \times (2/3)\underset{\approx}{s}.\underline{I} = -(8\pi/3)\gamma_e\gamma_n\hbar^2 k\delta(r)\underset{\approx}{s}.\underline{I}, \tag{9}$$

where we have used the fact that only isotropic components of the electronic distribution will give non-zero expectation values for this operator, and where we have taken note of the fact that $f'(r)/r^2$ decreases as $r^{-4}$ when $r >> Zr_0$ and diverges as $r^{-2}$ for $r << Zr_0$ while the integral :

$$\int_{R^3} \{f'(r)/r^2\}\, dv = \int_0^\infty f'(r)\, 4\pi dr = 4\pi\{f(\infty) - f(0)\} = 4\pi/(1 + W/2m_e c^2) \equiv 4\pi k,$$

converges approximately to $4\pi$ (the $\delta$ function is taken to simulate the real behaviour of the integrand in space). The second term of Eqn. (5) gives rise to the Darwin correction to the Coulomb potential in the two-component wave-equation :

$$\hat{T}_D = -(i\beta_e/4m_e c)\{\hat{f}^2\underline{E}, \hat{\underline{p}}\} = (Z\beta_e^2/2)\{(\underline{\nabla}\hat{f}^2).\underline{r}/r^3 + (\underline{\nabla}.\underline{r}/r^3)\hat{f}^2\} = (Z\beta_e^2/2)\{\underline{\nabla}(1/k + Zr_0/r)^{-2}.\underline{r}/r^3\} \text{ \{in a stationary state and since } (\underline{\nabla}.\underline{r}/r^3).f^2(r) \sim \delta(r).r^2 = 0\} = Z\beta_e^2 Z r_0/r(r/k + Zr_0)^3 = 2\pi Z\beta_e^2 k^2\delta(r), \tag{10}$$

using similar arguments as above. The first term of Eqn. (5) can be split into the three following contributions : the proper kinetic energy (including relativistic corrections) :

$$\hat{T}_K = (\hat{f}\hat{\underline{p}}^2 + \hat{\underline{p}}^2\hat{f})/4m_e = \hat{\underline{p}}^2/2m_e - \hat{\underline{p}}^4/8m_e^3c^2 + \ldots, \tag{11}$$

where we have used $\hat{f} = \{1 + (\hat{H} + eA_0)/2m_e c^2\}^{-1} \simeq 1 - (\hat{H} + eA_0)/2m_e c^2 \simeq 1 - \hat{\underline{p}}^2/4m_e^2c^2 + \ldots$ ; the coupling energy of the electron's orbital motion with the nuclear magnetic field :

$$\hat{T}_C = (e/2m_e c)(\hat{f}\underline{A}.\hat{\underline{p}} + \underline{A}.\hat{\underline{p}}\hat{f}) = \beta_e\gamma_n\hbar\underline{I}.\{\hat{f}(\underline{\ell}/r^3) + (\underline{\ell}/r^3)\hat{f}\} \text{ ;} \tag{12}$$

and the screening diamagnetic energy :

$$\hat{T}_M = (e^2/2m_e c^2)(\hat{f}\underline{A}^2 + \underline{A}^2\hat{f})/2 = r_0\hat{f}\underline{A}^2. \tag{13}$$

If we also have an external constant magnetic field $\underline{B}_Z$ ($\underline{A}_Z = \underline{B}_Z \Lambda \underline{r}/2$ if this field is uniform), we have, from Eqns. (5) and (6), the additional Zeeman terms :

$$\hat{T}_Z = (e/2m_e c)(\hat{f}\underline{A}_Z.\hat{\underline{p}} + \underline{A}_Z.\hat{\underline{p}}\hat{f}) = \beta_e(\hat{f}\underline{B}_Z.\underline{\ell} + \underline{\ell}.\hat{f}\underline{B}_Z)/2, \quad (14)$$

$$\hat{S}_Z = g_0\beta_e \underline{s}.\hat{f}\underline{B}_Z \ ; \quad (15)$$

the corresponding contact term can be shown to be zero (noticing that there is no $r^{-3}$ singularity in $\underline{A}_Z$). But one must not forget the nuclear Zeeman term :

$$I_Z = -\gamma_n \hbar \underline{I}.\underline{B}_Z, \quad (16)$$

and also the contribution of $\underline{A}_Z$ to $\hat{T}_M$. Gathering the operators from Eqns. (7-16), one can rewrite Eqn. (4) as :

$$(\hat{T}_K + \hat{T}_C + \hat{T}_Z + \hat{T}_M + \hat{T}_D + \hat{S}_d + \hat{S}_o + \hat{S}_c + \hat{S}_Z + I_Z - Ze^2/r)\Psi(\underline{r},t)$$
$$= i\hbar\partial\Psi(\underline{r},t)/\partial t. \quad (17)$$

This equation, and the corresponding one for the smaller components, can be solved to any meaningful degree of accuracy(3). It can also be extended to include explicitly quantum-electrodynamic effects and the nuclear structure and motion.

For many-particle systems, a double problem arises : (i) how to define a consistent, relativistically invariant, wave-equation ; and (ii) how to solve, with appropriate accuracy, whatever approximation has been chosen for this equation, with relevance to a particular class of effects. The first problem stems mainly from the difficulty of incorporating the Coulomb interactions, $q_j q_k/r_{jk}$, in such an equation. In the previous paragraph, the nucleus was considered just as a classical source of fields. For a quantum-relativistic two-particle system, Breit(4) was the first to set up a 16-component analog of the 4-component Dirac equation correct to the order of $v_1 v_2/c^2$, i.e. :

$$\{\hat{\pi}_0 + (\hat{\underline{\alpha}}^1.\hat{\underline{\pi}}^1 + \hat{\underline{\alpha}}^2.\hat{\underline{\pi}}^2) + (\hat{\alpha}_4^1 m^1 c + \hat{\alpha}_4^2 m^2 c) + (q^1 q^2/c) \times$$
$$\{-1/r^{12} + \hat{\underline{\alpha}}^1.\hat{\underline{\alpha}}^2/2r^{12} + (\hat{\underline{\alpha}}^1.\underline{r}^{12})(\hat{\underline{\alpha}}^2.\underline{r}^{12})/2(r^{12})^3\}\}\Psi(\underline{r}^1, \underline{r}^2, t) = 0, \quad (18)$$

where : $\hat{\pi}_0 = \{i\hbar\partial/\partial t - q^1 A_0(\underline{r}^1, t) - q^2 A_0(\underline{r}^2, t)\}/c$, $\hat{\underline{\pi}}^i = \hat{\underline{p}}^i - q^i \underline{A}(\underline{r}^i, t)/c$ (i = 1, 2), $\hat{\underline{\alpha}}^i$ and $\hat{\alpha}_4^i$ (i = 1, 2) are analogs of $\hat{\underline{\alpha}}$ and $\hat{\alpha}_4$ in Eqn. (1), and $\underline{r}^{12} = \underline{r}^2 - \underline{r}^1$. The last two terms of the operator in Eqn. (18) have to be understood as a first-order diagonal perturbation to the eigenstates of its other terms (4). That the above Breit equation is not fully covariant is obvious from the fact that it involves a single time variable for both particles, which places time and position on different footings. Using similar procedures as before, it was again possible[3] to obtain a Pauli-type effective Hamiltonian, $\hat{H}_e$, acting on the 4 larger components, $\Psi(\underline{r}_1, \underline{r}_2, t)$, of the wave-function to give $i\hbar\partial\Psi/\partial t$, to the order of $v_1v_2/c^2$. In addition to one-particle terms analogous to those gathered in Eqn. (17), $\hat{H}_e$ contains orbit-orbit, spin-spin (dipolar and contact) and spin-other orbit magnetic interactions, as well as the corrected Coulomb interaction between the two particles. For a many-electron system, a quantum-electrodynamic treatment has given[5a], to a similar degree of accuracy, the extended sums :

$$\hat{H}_e = \sum_i (\hat{T}_i + V_i + \hat{S}_i) + \sum_{j<k}\sum (\hat{T}_{jk} + V_{jk} + \hat{S}_{jk}) ; \tag{19}$$

$$\hat{T}_i = \hat{\underline{p}}_i^2/2m_e - \hat{\underline{p}}_i^4/8m_e^3c^2 + (e/m_ec)\underline{A}(\underline{r}_i).\hat{\underline{p}}_i + r_0\underline{A}(\underline{r}_i)^2, \tag{20a}$$

$$V_i = -e\{A_0(\underline{r}_i) - 2\pi\lambda_e^2\rho(\underline{r}_i)\}, \tag{20b}$$

$$\hat{S}_i = g_0\beta_e\underline{s}_i \cdot \{\underline{B}(\underline{r}_i) + \underline{E}(\underline{r}_i)\wedge(\hat{\underline{p}}_i + e\underline{A}(\underline{r}_i)/c)/2m_ec\} ; \tag{20c}$$

$$\hat{T}_{jk} = -(r_0/r_{kj})\{\hat{\underline{p}}_j.\hat{\underline{p}}_k + (\underline{r}_{kj}.\hat{\underline{p}}_j)(\underline{r}_{kj}.\hat{\underline{p}}_k)/r_{kj}^2\}/m_e, \tag{21a}$$

$$V_{jk} = e^2\{1/r_{kj} - 4\pi\lambda_e^2\delta(r_{kj})\}, \tag{21b}$$

$$\hat{S}_{jk} = -g_0\beta_e^2\{2(\underline{s}_j.\hat{\underline{\ell}}_{k/j} + \underline{s}_k.\hat{\underline{\ell}}_{j/k}) + (\underline{s}_j.\hat{\underline{\ell}}_{j/k} + \underline{s}_k.\hat{\underline{\ell}}_{k/j})\}/r_{kj}^3 + g_0^2\beta_e^2\{\underline{s}_j.\underline{s}_k/r_{kj}^3 - 3(\underline{s}_k.\underline{r}_{kj})(\underline{s}_j.\underline{r}_{kj})/r_{kj}^5 - (8\pi/3)\delta(r_{kj})\underline{s}_j.\underline{s}_k\}, \tag{21c}$$

where $\hbar\hat{\underline{\ell}}_{k/j} \equiv \underline{r}_{jk}\wedge\hat{\underline{p}}_k$ is the orbital angular momentum of electron k with respect to electron j. The Darwin and Fermi corrections have been gathered together with the corresponding electrostatic and magnetic-dipolar interactions, and the exact $g_0$ value has been substituted for the factors 2 multiplying $\beta_e\underline{s}_i$[5b]. For atomic, molecular or crystalline systems, one has to substitute, in Eqns. (20a-c), the expressions :

$$\rho(\underline{r}_i) = \sum_\lambda Z_\lambda e\delta(r_{\lambda i}), \tag{22a}$$

$$A_0(\underline{r}_i) = \sum_\lambda Z_\lambda e/r_{\lambda i} + P_0(\underline{r}_i), \tag{22b}$$

$$\underline{A}(\underline{r}_i) = \sum_\lambda \gamma_\lambda\hbar\underline{I}_\lambda\wedge\underline{r}_{\lambda i}/r_{\lambda i}^3 + \underline{P}(\underline{r}_i), \tag{22c}$$

where the summations extend over all nuclei and $P_0$ and $\underline{P}$ are the scalar and vector potentials of any additional, corrective or applied, electromagnetic fields. This gives rise to as many terms of types given by Eqns. (7-12) as there are electrons and nuclei, and also to some others, induced by $P_0$ and $\underline{P}$. If, instead of electrons, we had considered nuclei, most of the corrective terms in Eqns. (19-21), including those involving the orbital angular momenta, would be negligible, because of the large rest masses and of the slow and restricted motions. There would remain :

$$\hat{H}_n = \sum_\lambda \{\hat{\underline{p}}_\lambda^2/2M_\lambda + Z_\lambda e A_0'(\underline{r}_\lambda) - \gamma_\lambda \hbar \underline{\underline{I}}_\lambda . \underline{B}'(\underline{r}_\lambda)\} + \sum_{\mu<\nu}\sum \{Z_\mu Z_\nu e^2/r_{\nu\mu} + \gamma_\mu\gamma_\nu\hbar^2 \underline{\underline{I}}_\mu . \underline{\underline{\mathcal{D}}}(\underline{r}_{\nu\mu}) . \underline{\underline{I}}_\nu\}, \quad (23)$$

where $\mathcal{D}(\underline{r}_{\nu\mu})$ is the dipolar-coupling tensor given in Eqn. (21c), with no contact part. In an effective-Hamiltonian formulation, $A_0'$ and $\underline{B}'$ would include the fields induced by the electronic cloud at the locations of the nuclei.

## 2.2. Effective Hamiltonians

The Hamiltonian obtained by gathering all the terms expressed through Eqns. (19-23) contains the usual, zeroth-order Schrödinger Hamiltonian :

$$\hat{H}_0 = \sum_i \hat{\underline{p}}_i^2/2m_e - \sum_{i\lambda}\sum Z_\lambda e^2/r_{\lambda i} + \sum_{j<k}\sum e^2/r_{kj} + \sum_\lambda \hat{\underline{p}}_\lambda^2/2M_\lambda + \sum_{\mu<\nu}\sum Z_\mu Z_\nu e^2/r_{\nu\mu}, \quad (24)$$

the last two terms of which are generally treated through the Born-Oppenheimer approximation (which helps to define parametrized effective Hamiltonians acting on the electronic or nuclear wave-functions separately) and the third term through the self-consistent-field orbital approximation (with its so-called semi-empirical, ab-initio and numerical varieties and configuration-interaction extensions). It also contains non-magnetic relativistic corrections of the order of $\lambda_e^2$ :

$$\hat{H}_R = \lambda_e^2 \{\sum_i \{-\hbar^2 \underline{\nabla}_i^4/2m_e + \sum_\lambda Z_\lambda e^2 2\pi\delta(r_{\lambda i})\} + \sum_{j<k}\sum e^2\{2r_{kj}^{-1}\underline{\nabla}_j . (\underline{\underline{1}} + \underline{r}_{kj} x \underline{r}_{kj}/r_{kj}^2) . \underline{\nabla}_k - 4\pi\delta(r_{kj})\}\}, \quad (25)$$

which may become significant if the system contains heavy atoms (that is, for at least the core-electrons of about 80% of the elements). In a quantum-relativistic calculation, one has to consider also the smaller components of the electronic orbitals in the determination of the mean values of observables, in accordance with the normalization condition of Eqn. (3). A survey of theoretical studies on relativistic effects in Quantum Chemistry has been given recently[6]. In the following, we shall assume $H_R$ to be negligible and that the wave-function made up from the larger components of the electronic orbitals is normalized. Finally, $\underset{\sim}{H}_e + \underset{\sim}{H}_n$ also contains magnetic terms which can be written, making use of Eqns. (7-9) and (12-14) :

$$\hat{\mathcal{H}}_M = \hat{\mathcal{H}}_{hf} + \hat{\mathcal{H}}_{eZ} + \hat{\mathcal{H}}_{SO} + \mathcal{H}_{SS} + \mathcal{H}_{nZ} + \mathcal{H}_{nd} + \mathcal{H}_{dia} + \hat{\mathcal{H}}_{add}, \quad (26)$$

$$\hat{\mathcal{H}}_{hf} = \sum_i \beta_e \sum_\lambda \gamma_\lambda \hbar \{2\hat{\underline{\ell}}_{i/\lambda} \cdot \underline{I}_\lambda / r_{\lambda i}^3 - g_0 \underline{s}_i \cdot \{\underline{\underline{\mathcal{D}}}(\underline{r}_{\lambda i}) - (8\pi/3)\delta(r_{\lambda i})\} \cdot \underline{I}_\lambda\}, \quad (27)$$

$$\hat{\mathcal{H}}_{eZ} = \sum_i \beta_e (\hat{\underline{\ell}}_i + g_0 \underline{s}_i) \cdot \underline{B}_Z, \quad (28)$$

$$\hat{\mathcal{H}}_{SO} = g_0 \beta_e^2 \sum_i \underline{s}_i \cdot \{\sum_\lambda Z_\lambda \hat{\underline{\ell}}_{i/\lambda} / r_{\lambda i}^3 - \sum_j{}' (2\hat{\underline{\ell}}_{j/i} + \hat{\underline{\ell}}_{i/j}) / r_{ji}^3\}, \quad (29)$$

$$\mathcal{H}_{SS} = g_0^2 \beta_e^2 \sum_{j<k} \sum \underline{s}_j \cdot \{\underline{\underline{\mathcal{D}}}(\underline{r}_{kj}) - (8\pi/3)\delta(r_{kj})\} \cdot \underline{s}_k, \quad (30)$$

$$\mathcal{H}_{nZ} = -\sum_\lambda \gamma_\lambda \hbar \underline{I}_\lambda \cdot \underline{B}_Z, \quad (31)$$

$$\mathcal{H}_{nd} = \sum_{\mu<\nu} \sum \gamma_\mu \gamma_\nu \hbar^2 \underline{I}_\mu \cdot \underline{\underline{\mathcal{D}}}(\underline{r}_{\nu\mu}) \cdot \underline{I}_\nu, \quad (32)$$

$$\mathcal{H}_{dia} = r_0 \sum_i \underline{A}(\underline{r}_i)^2, \quad (33)$$

where $\underline{\underline{\mathcal{D}}}(\underline{r}) \equiv \underline{\underline{1}}/r^3 - 3(\underline{r}\times\underline{r})/r^5$. Here we have assumed that $A_0' \equiv P_0 \equiv 0$ and $\underline{B}' \equiv \underline{B}_Z$ with $\underline{P}(\underline{r}) = \underline{B}_Z \Lambda \underline{r}/2$. But the inclusion of applied or effective (crystalline or molecular) electric fields can be made through the provisional term $\mathcal{H}_{add}$. The magnetic corrective terms will split the levels determined by $\hat{\mathcal{H}}_0$, whether or not account is taken of $\mathcal{H}_R$. This will give rise to fine and hyperfine structure in the high-frequency spectra (including recoilless γ-ray lines) and to microwave and radiofrequency transitions between the induced sublevels (in magnetic resonance techniques).

An important contribution to $P_0(\underline{r}_i)$ may come from the nuclei themselves. We have already pointed out that an atomic nucleus is generally the seat of a complex distribution of charges and currents which may give rise to higher multipole components than the point electric charge $Z_\lambda e$ and dipole magnetic moment $\gamma_\lambda \hbar \underline{I}_\lambda$. The assumption that nuclear states have definite parity (that is, are either symmetric or antisymmetric with respect to the inversion of all nuclear coordinates) gives a theoretical justification to the experimental fact that nuclear operators with odd parity (such as electric dipole and octupole as well as magnetic monopole and quadrupole moments) have vanishing expectation values in a defined nuclear state. Among the remaining multipoles, only electric quadrupoles have been of some experimental significance, even though magnetic octupoles have also been detected in a few heavy nuclei. The expansion of the electrostatic interaction between the constituent protons p and surrounding electrons i of a nucleus with a center of mass and charge λ can be written :

$$-e^2\sum_{ip}\sum|\underline{r}_{\lambda i} - \underline{r}_{\lambda p}|^{-1} = -e^2\sum_{ip}\sum\{1 - \underline{r}_{\lambda p}\cdot\underline{\nabla}_{\lambda i} + (\underline{r}_{\lambda p}\cdot\underline{\nabla}_{\lambda i})^2/2 + \ldots\}|\underline{r}_{\lambda i}|^{-1} = -e^2\sum_{ip}\sum\{1/r_{\lambda i} + \underline{r}_{\lambda p}\cdot\underline{r}_{\lambda i}/r^3_{\lambda i} - \{\underline{r}^2_{\lambda p}/r^3_{\lambda i} - 3(\underline{r}_{\lambda p}\cdot\underline{r}_{\lambda i})^2/r^5_{\lambda i} + (4\pi/3)\underline{r}^2_{\lambda p}\delta(r_{\lambda i})\}/2 + \ldots\}.$$

The p summation on the first terms gives $-Z_\lambda e^2\sum_i 1/r_{\lambda i}$, which is part of $\hat{H}_0$ in Eqn. (24), while the p summation on the second terms gives 0, since λ is the center of nuclear charge. The third set of terms can be rewritten as a sum of quadrupole-splitting and isotope/isomer-shift operators :

$$\hat{H}_{qp,\lambda} = (1/6) \sum_{\alpha,\beta=x,y,z}\sum Q^{\alpha\beta}_\lambda V^{\alpha\beta}_\lambda,$$
$$Q^{\alpha\beta}_\lambda = e\sum_p(3r^\alpha_{\lambda p}r^\beta_{\lambda p} - \delta^{\alpha\beta}r^2_{\lambda p}),\ V^{\alpha\beta}_\lambda = -e\sum_i(3r^\alpha_{\lambda i}r^\beta_{\lambda i} - \delta^{\alpha\beta}r^2_{\lambda i})/r^5_{\lambda i}\ ; \qquad (34a,b,c)$$

$$\hat{H}_{is,\lambda} = -(1/6)S_\lambda C_\lambda,\ S_\lambda = e\sum_p \underline{r}^2_{\lambda p},\ C_\lambda = -e\sum_i 4\pi\delta(r_{\lambda i}). \qquad (35a,b,c)$$

$\hat{H}_{qp,\lambda}$ appears as the scalar product of two traceless, symmetric, second-rank Cartesian tensors : the nuclear quadrupole-moment operator $\underline{\underline{Q}}_\lambda$, which will shortly be given a more practical expression, and the electric field-gradient operator $\underline{\underline{V}}_\lambda$, whose components can also be written : $-e\sum_i\nabla^\alpha\nabla^\beta(1/r_{\lambda i})$ ; the complementary term $\hat{H}_{is,\lambda}$ is the product of a nuclear size factor $S_\lambda$ and an electronic contact-density operator $C_\lambda$. Since each of the two tensors $\underline{\underline{Q}}_\lambda$ and $\underline{\underline{V}}_\lambda$ can be completely defined by five independent parameters, $\hat{H}_{qp,\lambda}$ can be rewritten as the scalar product of two irreducible, second-rank spherical tensors :

$$\hat{H}_{qp,\lambda} = \sum_{m=-2}^{+2} Q^m_{2,\lambda}V^{-m}_{2,\lambda}(-1)^m. \qquad (36)$$

The usual expression of $\underline{\underline{Q}}_\lambda$ is obtained by making use of the Wigner-Eckart theorem :

$$\langle\eta JM|T^m_\ell|\eta'J'M'\rangle = C(J'\ell J,\ M'mM)\ \langle\eta J||\underline{T}_\ell||\eta'J'\rangle, \qquad (37)$$

where η stands for all quantum numbers other than the expectation values, J and M, of the total angular momentum and its z component, and $T^m_\ell(-\ell \leq m \leq +\ell)$ is one of the standard components of the irreducible tensor operator of rank ℓ, $\underline{T}_\ell$, which obey the commutation relations : $[J^z, T^m_\ell] = mT^m_\ell$, $[J^\pm, T^m_\ell] = \{\ell(\ell+1)-m(m\pm1)\}^{1/2}T^{m\pm1}_\ell$. The C's are Clebsch-Gordan coefficients for the coupling of angular momenta : they are the same for all $T^m_\ell$'s with a given ℓ and m and are zero for all values except those satisfying m = M − M' and $|J'-J| \leq \ell \leq |J'+J|$ ; the reduced matrix element $\langle\eta J||\underline{T}_\ell||\eta'J'\rangle$ does not depend on M, M', or m but may depend on the particular $\underline{T}_\ell$. Examples of $T^m_\ell$'s are $\mp J^\pm/\sqrt{2}$, $J^z$ for ℓ = 1 and the spherical harmonics,

$Y_\ell^m(\theta,\phi)$, for $\ell = 0, 1, 2, \ldots$ . If $T_\ell^m(q)$ and $S_\ell^m(p)$ are two sets of functions of the operators of the system with same $\ell$ which obey the above commutation relations, and if one defines $G_\ell(q) \equiv \sum_m a_m T_\ell^m(q)$ and $F_\ell(p) \equiv \sum_m a_m S_\ell^m(p)$, then Eqn. (37) entails :

$$\langle\eta JM|G_\ell(q)|\eta'J'M'\rangle/\langle\eta J||\underline{T}_\ell(q)||\eta'J'\rangle = \langle\eta JM|F_\ell(p)|\eta'J'M'\rangle/\langle\eta J||\underline{S}_\ell(p)||\eta'J'\rangle. \tag{38}$$

Next we notice that $Q_\lambda^{\alpha\beta}$ for each proton is a linear combination of second-rank harmonic polynomials, $\rho_{\lambda p}^2 Y_2^m(\theta_{\lambda p},\phi_{\lambda p})$, and that replacement in these polynomials of $x_{\lambda p}$, $y_{\lambda p}$, $z_{\lambda p}$ by the components $I_\lambda^x$, $I_\lambda^y$, $I_\lambda^z$ of the total angular momentum gives irreducible tensor operators of the same form because of the similarity of the commutation relations of the components of the two vectors, e.g. : $\{I_\lambda^x, I_\lambda^y\} = iI_\lambda^z \sim \{I_\lambda^x, y_{\lambda p}\} = iz_{\lambda p}$, where $\underline{I}_\lambda \equiv \sum_p(\underline{\ell}_{\lambda p} + \underline{s}_{\lambda p})$ can now be understood to include both protons and neutrons. Substituting in Eqn. (38) for a nuclear state defined by the quantum numbers $\eta$ and I (since we are concerned only with the spatial reorientation of the nucleus, which is defined by a change of M), we obtain :

$$\langle\eta IM|Q_\lambda^{\alpha\beta}|\eta IM'\rangle = K_\lambda\langle\eta IM|3(I_\lambda^\alpha I_\lambda^\beta + I_\lambda^\beta I_\lambda^\alpha)/2 - \delta^{\alpha\beta}\underline{I}_\lambda^2|\eta IM'\rangle, \tag{39}$$

where $K_\lambda$ does not depend on M, M' or $\alpha$ and $\beta$. The constant $K_\lambda$ can be obtained by setting M = M' = I and $\alpha = \beta = z$ :

$$K_\lambda = \langle\eta II|e\sum_p(3z_{\lambda p}^2 - r_{\lambda p}^2)|\eta II\rangle/\langle\eta II|3(I_\lambda^z)^2 - \underline{I}_\lambda^2|\eta II\rangle = eQ_\lambda/I_\lambda(2I_\lambda-1), \tag{40}$$

where we introduce the quadrupole-moment constant $eQ_\lambda$. Eqns. (34a) and (39) show that, within the nuclear manifold $|\eta IM\rangle$ $(-I \leqslant M \leqslant +I)$, $\hat{H}_{qp,\lambda}$ can be replaced by the equivalent operator :

$$\hat{H}_{qp,\lambda} = (K_\lambda/6) \sum_{\alpha,\beta=x,y,z}\sum \{3(I_\lambda^\alpha I_\lambda^\beta + I_\lambda^\beta I_\lambda^\alpha)/2 - \delta^{\alpha\beta}\underline{I}_\lambda^2\} V_\lambda^{\alpha\beta} =$$

$$(K_\lambda/4) \sum_{\alpha,\beta=x,y,z}\sum (I_\lambda^\alpha I_\lambda^\beta + I_\lambda^\beta I_\lambda^\alpha) V_\lambda^{\alpha\beta} = (K_\lambda/2)\, \underline{I}_\lambda\cdot\underline{V}_\lambda\cdot\underline{I}_\lambda =$$

$$\{e^2Q_\lambda/2I_\lambda(2I_\lambda-1)\}\, \underline{I}_\lambda\cdot\sum_i\underline{D}(\underline{r}_{\lambda i})\cdot\underline{I}_\lambda, \tag{41}$$

where we have used $\sum_\alpha V_\lambda^{\alpha\alpha} = 0$ and $V_\lambda^{\alpha\beta} = V_\lambda^{\beta\alpha}$ and noticed the similarity between the field-gradient and dipolar-coupling tensors in Eqn. (34c) and applied Eqn. (40). Similarly, $\hat{H}_{is,\lambda}$ {Eqns. (35a-c)} can be replaced by :

$$\hat{H}_{is,\lambda} = (e^2/6)\sum_i 4\pi\delta(r_{\lambda i})\langle\eta II|\sum_p r_{\lambda p}^2|\eta II\rangle \underline{I}_\lambda^2/I_\lambda(I_\lambda+1) =$$

$$(2\pi e^2/3)\ Z_\lambda R_\lambda^2 \sum_i \delta(r_{\lambda i}), \tag{42}$$

where we introduce the effective nuclear radius $R_\lambda$ and replace $\underset{\approx}{I}_\lambda^2$ by its expectation value in the nuclear manifold $^\lambda|\eta IM\rangle$. One can now write the provisional term as :

$$\hat{\underset{\sim}{H}}_{add} = \sum_\lambda (\hat{\underset{\sim}{H}}_{qp,\lambda} + \hat{\underset{\sim}{H}}_{is,\lambda}). \tag{43}$$

The selection rule $\ell \leq 2I$ restricts the existence of quadrupolar terms ($2^\ell = 4$) to nuclei with $I_\lambda \geq 1$.

In the last two paragraphs, we have met some examples of effective Hamiltonian operators. To zeroth-order in $\lambda_e$, the Schrödinger Hamiltonian $\hat{H}_0$, operating on the larger-component wave-function $\underset{\sim}{\Psi}$, generates the manifold of quantum states with positive total energy for all electrons. These are subject to further splittings when one includes magnetic corrections from $\hat{\underset{\sim}{H}}_M$. Similarly, the magnetic-like correction $\hat{\underset{\sim}{H}}_{qp,\lambda}$, within the manifold of nuclear states with total angular momentum $\underset{\approx}{I}_\lambda$, gives the same matrix elements as the quadrupole-coupling operator $\hat{H}_{qp,\lambda}$. Frequently, an effective Hamiltonian is introduced through semi-phenomenological considerations before it is formally shown to be equivalent to a more basic Hamiltonian, to a given order of perturbation and within some manifold of states. Since the early work of Van Vleck on the canonical-transformation treatment of the perturbations of degenerate states[7], which has received wide applications in the determination of effective Hamiltonians for various physical systems[8], a number of different schemes have been elaborated[9], most of which are recalled and compared in recent studies[10]. Of particular interest to us are effective Hamiltonians operating on pure spin spaces[11], although in some specific cases the subspace of interest may also involve phonon, rotation or other variables[12]. We shall now see with a simple classical example[8c] how an effective spin Hamiltonian can be constructed. For a free atom or ion with zero nuclear spin, Eqns. (24) and (26) become, in the center-of-mass frame :

$$\hat{H}_0 = \sum_i (\hat{\underline{p}}_i^2/2m_e - Ze^2/r_i) + \sum_{j<k}\sum e^2/r_{kj}, \quad \hat{\underset{\sim}{H}}_M = \hat{\underset{\sim}{H}}_{eZ} + \hat{\underset{\sim}{H}}_{SO} + \underset{\sim}{H}_{SS} + H_{dia}.$$

For transition-metal ions in crystalline electric fields, $H_{dia}$ is negligible, $\hat{\underset{\sim}{H}}_{SS}$ gives much smaller effects than $\underset{\sim}{H}_{SO}$ and the latter, together with $\underset{\sim}{H}_{eZ}$ and the crystalline-potential extra term : $\hat{H}_C \equiv -e\sum_i P_0(\underline{r}_i)$, is usually much smaller than $\hat{H}_0$. Since both $\hat{\underline{L}} \equiv \sum_i \hat{\underline{l}}_i$ and $\underset{\approx}{S} \equiv \sum_i \underset{\approx}{s}_i$ commute with the main operator $\hat{H}_0$, it is convenient to use simultaneous eigenkets of $\hat{H}_0$ and of $\hat{\underline{L}}^2$, $\hat{L}^z$, $\underset{\approx}{S}^2$ and $\underset{\sim}{S}^z$, i.e. $|\eta SLM_SM_L\rangle$, as a basis for a perturbation treatment (Russell-Sanders coupling scheme). Due to the antisymmetry of the total electronic wave-function, the allowed energies usually depend strongly on the eigenvalues S and L, while they are independent of $M_S$ and $M_L$ in this approximation. Using similar arguments as in our previous treatment of the nuclear quadrupole-coupling operator, one can then see that, within an electronic manifold (or spectroscopic term) $|\eta SLM_S M_L>$ $(-S \leqslant M_S \leqslant +S,\ -L \leqslant M_L \leqslant +L)$, $\hat{H}_{SO}$ given by Eqn. (29) has the same matrix elements as the operator $\lambda \underline{L}.\underline{S}$, where $\lambda$ is a constant characteristic of the system and of the manifold. The corresponding Hamiltonian can then be replaced by the approximate equivalent operator :

$$\hat{H}' = \hat{H}_0' + \hat{H}_M',\quad \hat{H}_0' = \hat{H}_0 + \hat{H}_C,\quad \hat{H}_M' = \beta_e(\hat{\underline{L}} + g_0\underline{S}).\underline{B}_Z + \lambda\hat{\underline{L}}.\underline{S}. \qquad (44a,b,c)$$

For the iron-group ions, $\hat{H}_M' \ll \hat{H}_C \ll \hat{H}_0$ ; therefore, one can first remove the orbital degeneracy by introducing $H_C$, the crystal-site symmetry of which, always lower than the spherical symmetry of the free ion, induces rotations of the eigenkets of $\hat{H}_0$ into symmetry-adapted, real eigenkets with slightly different energies, $E^o$, ..., $E^n$, ..., then construct an effective Hamiltonian acting on the lowest (2S+1)-fold degenerate level of the term, $|oM_S>$, including the coupling brought in by $(\beta_e\hat{\underline{L}}.\underline{B}_Z + \lambda\hat{\underline{L}}.\underline{S})$ with the other levels of the same term, $|nM_S>$ (the admixtures from other terms, very distant in energy, are usually neglected). The projection-operator partitioning of the eigenvalue problem proposed by Pryce(8c) can be given a matrix form similar to that used by Löwdin(2) in the separation of the Dirac equation (§2.1). Any eigenket $|\ >$ of $\hat{H}'$ can be expanded in the form : $|\ > = \sum_k c_k|k>$, where the $|k>$'s are the orthonormalized eigenkets $|oM_S>$, ..., $|nM_S'>$, ... of $\hat{H}_0'$ $(-S \leqslant M_S \leqslant +S)$, and the $c_k$'s form a vector $\underline{c}$. Introducing the matrix elements $H_{k\ell}' \equiv <k|\hat{H}'|\ell>$ and partitioning the basis $\{|k>\}$ into two parts : $\{|oM_S>\}$ and $\{|\nu M_S'>\}$ $\nu = \{1, ..., n, ...\}$, one obtains the following partitioned-matrix form of the eigenvalue equation :

$$\begin{pmatrix} \underline{\underline{H}}'_{oo} & \underline{\underline{H}}'_{o\nu} \\ \underline{\underline{H}}'_{\nu o} & \underline{\underline{H}}'_{\nu\nu} \end{pmatrix}\begin{pmatrix} \underline{c}_o \\ \underline{c}_\nu \end{pmatrix} = E\begin{pmatrix} \underline{c}_o \\ \underline{c}_\nu \end{pmatrix}.$$

This is equivalent to two matrix equations and, solving $\underline{c}_\nu$ from the second and substituting into the first, one obtains :

$$\tilde{\underline{\underline{H}}}_{oo}\underline{c}_o = E\underline{c}_o,\quad \tilde{\underline{\underline{H}}}_{oo} = \underline{\underline{H}}'_{oo} + \underline{\underline{H}}'_{o\nu}\,(E\underline{\underline{1}}_{\nu\nu} - \underline{\underline{H}}'_{\nu\nu})^{-1}\,\underline{\underline{H}}'_{\nu o}. \qquad (45a,b)$$

The decoupling of the two subspaces has thus lead to a reduced equation giving the same eigenvalues as the original equation in the subspace $\{|oM_S>\}$. To the contracted matrix $\tilde{\underline{\underline{H}}}_{oo}$ (which contains implicitly the unknown eigenvalue) corresponds an effective operator $\tilde{H}$, the matrix elements of which are obtained by substituting Eqn. (44a) into Eqn. (45b) and letting $\hat{H}_0'$ operate on its eigenkets. Replacing $E\underline{\underline{1}}_{\nu\nu}$ and $\underline{\underline{H}}'_{\nu\nu}$ by their zeroth-order equivalents $E^o\underline{\underline{1}}_{\nu\nu}$ and $E^\nu\underline{\underline{1}}_{\nu\nu}$, one obtains explicit matrix elements correct to the second order :

$$<oM_S|\tilde{H}|oM_S'> = E^o\delta_{M_S M_S'} + <oM_S|\hat{H}_M'|oM_S'> -$$

$$- \sum_{\nu(\neq o)M''_S} \langle oM_S|\hat{\mathcal{H}}'_M|\nu M''_S\rangle (E^\nu - E^o)^{-1} \langle \nu M''_S|\hat{\mathcal{H}}'_M|oM'_S\rangle + \dots . \qquad (46)$$

The first term on the right may be dropped if one measures all energies in the lowest level relatively to $E^o$. Using Eqn. (44c) and noticing that the matrix elements of $\hat{L} \equiv -i\Sigma r_i \wedge \nabla_i$ between our real eigenkets are purely imaginary (and, since $\hat{L}$ is Hermitian, vanish within one level - quenching of the orbital angular momentum - and are antisymmetrical between two levels), one sees that the second term on the right reduces to the spin Zeeman splitting : $\langle oM_S|g_0\beta_e \underset{\approx}{S}.\underline{B}_Z|oM'_S\rangle$. As $\langle oM_S|\underset{\approx}{S}.\underline{B}_Z|\nu M''_S\rangle = \sum_{\alpha=x,y,z} \langle oM_S|S^\alpha|\nu M''_S\rangle B_Z^\alpha = 0$ since the levels o and $\nu$ are orthogonal, the following sum reduces to :

$$- \sum_{\alpha,\beta=x,y,z}\sum \langle oM_S|\beta_e^2\hat{\Lambda}^{\alpha\beta}B_Z^\alpha B_Z^\beta + \lambda^2 S^\alpha\hat{\Lambda}^{\alpha\beta}S^\beta + \beta_e\lambda B_Z^\alpha(\hat{\Lambda}^{\alpha\beta}S^\beta + S^\beta\hat{\Lambda}^{\beta\alpha})|oM'_S\rangle,$$

$$\hat{\Lambda}^{\alpha\beta} = \sum_{\nu(\neq o)M''_S} \hat{L}^\alpha|\nu M''_S\rangle (E^\nu - E^o)^{-1}\langle \nu M''_S|\hat{L}^\beta.$$

Because $\hat{\underline{\Lambda}}$ and $\underset{\approx}{S}$ act on different variables of the kets, the result will be the same if one first replaces $\hat{\Lambda}^{\alpha\beta}$ by $\Lambda^{\alpha\beta} \equiv \langle oM_S|\hat{\Lambda}^{\alpha\beta}|oM'_S\rangle$, then uses the above-mentioned properties of $\langle oM_S|\hat{L}^\alpha|\nu M''_S\rangle$ and $\langle \nu M''_S|\hat{L}^\beta|oM'_S\rangle$ to show that $\Lambda^{\alpha\beta}$ is real and symmetrical (and positive definite if $E^o$ is the lowest eigenvalue). Then the equivalent operator $\tilde{\mathcal{H}}$ within the manifold $|oM_S\rangle$ can be written, in tensorial form :

$$\tilde{\mathcal{H}} = g_0\beta_e \underset{\approx}{S}.\underline{B}_Z - \beta_e^2 \underline{B}_Z.\underline{\underline{\Lambda}}.\underline{B}_Z - \lambda^2 \underset{\approx}{S}.\underline{\underline{\Lambda}}.\underset{\approx}{S} - 2\beta_e\lambda \underline{B}_Z.\underline{\underline{\Lambda}}.\underset{\approx}{S}. \qquad (47)$$

This allows one to define an effective spin Zeeman term : $\beta_e \underset{\approx}{S}.\underline{\underline{g}}.\underline{B}_Z$, $\underline{\underline{g}} = g_0\underline{\underline{1}} - 2\lambda\underline{\underline{\Lambda}}$, an effective spin-spin coupling, and an effective diamagnetic shielding : all of which are second-order effects from $(\lambda\hat{\underline{L}}.\underset{\approx}{S} + \beta_e\hat{\underline{L}}.\underline{B}_Z)$. Using angular-momentum-symmetry considerations, it has been shown(11c) that Eqn. (47) holds in more general cases than the one considered above. Inclusion of the hitherto neglected terms $\mathcal{H}_{SS}$ and $\mathcal{H}_{dia}$ just adds small first-order contributions to the terms quadratic in $\underset{\approx}{S}$ and $\underline{B}_Z$, respectively. Introduction of the hyperfine (magnetic dipolar and electric quadrupolar) interactions given by Eqns. (27) and (43) brings in terms bilinear in ($\underset{\approx}{S}$, $\underset{\approx}{I}$) and ($\underset{\approx}{I}$, $\underset{\approx}{I}$)(11a) (the electronic orbital momentum being again incorporated in the effective electron spin), to which one naturally adds the nuclear Zeeman term, bilinear in ($\underline{B}_Z$, $\underset{\approx}{I}$). As we shall see in the next section, an effective Hamiltonian, bilinear in all combinations of ($\underset{\approx}{S}$, $\underset{\approx}{I}_\mu$, $\underset{\approx}{I}_\nu$, $\underline{B}_Z$), can also be used with polyatomic systems(11b,d,e). However, it must be noted that when $S \geq 3/2$, higher-power terms in S may occur(11c). In addition, to third order in the perturbation treatment, the tensors in the bilinear terms may no longer be symmetrical(9l).

## 3. EFFECTIVE SPIN-HAMILTONIAN PARAMETERS AND MOLECULAR ELECTRONIC STRUCTURE

In this section we will essentially show how the characteristic coupling tensors occurring in the effective spin Hamiltonians of molecular systems can be related to their electronic structure through the first and second-order density and transition matrices in the different degrees of approximation. We shall devote the first part of this section to the introduction of the definitions and notations needed together with a review of some properties relevant to our subject.

### 3.1. Density and Transition Matrices

The use of density matrices in the study of quantum-mechanical pure states was first introduced by Dirac[(13)], and later generalized by Löwdin[(14)], McWeeny[(15)] and others[(16-21)]. The relation of quantum-mechanical to statistical density matrices was analyzed by Ter Haar[(16a)], who also gave an extensive survey of earlier studies on density matrices and related matters[(16b)]. The use of the former in the study of electron correlation and molecular structure has been reviewed by various authors[(17)]. The permutational, angular-momentum and point-group symmetry properties of density matrices have been extensively studied[(18)]. Under the impulsion of Coleman[(19)], the N-representability problem has received particular attention[(20d-u, 21a,c)]. This problem arises when one wishes to compute reduced density matrices directly with the equation derived from that governing the needlessly complicated, total wave-function : it consists essentially in the search for a set of tractable, necessary and sufficient conditions to impose on the solutions to ensure their derivability from an N-particle wave-function having the proper permutational symmetry (e.g., the Pauli principle for Fermions) ; in spite of the use of the most advanced mathematical techniques, the solutions found so far are incomplete or approximate or apply to special cases only. Further references on the properties and applications of reduced density matrices can be found in recent books and reviews[(21)]. In the following, we shall basically adopt McWeeny's notations and conventions and closely follow his developments[(15e, 11d, 21b)].

The Pauli multi-component representation of an N-electron wave-function, $\underset{\sim}{\Psi}(\underline{r}_1, \underline{r}_2, \ldots, \underline{r}_N, t)$, will be replaced by the equivalent one-component representation : $\Psi(\underline{x}_1, \underline{x}_2, \ldots, \underline{x}_N, t)$, where $\underline{x}_i \equiv \{\underline{r}_i, s_i\}$ combines the continuous spatial coordinates $\underline{r}_i$ and the two-valued spin coordinate $s_i$ for Electron i. Accordingly, the matrix representations $\underset{\sim}{s}_i$, etc., occurring in the various terms of Eqn. (26) will be replaced by the corresponding operator symbols $\hat{\underline{s}}_i$, etc. If $\hat{H}_M$ is treated as a perturbation, we shall make use of eigenfunctions of the spinless Hamiltonian $\hat{H}_0$ characterized by well-defined eigenvalues of the total spin operators $\hat{\underline{S}}^2$ and $\hat{S}^z$, which we shall single out from all other quantum numbers labelling the states.

Let us first consider a single electron in a stationary orbital A with a defined spin component + 1/2. The wave-function factorizes into the form : $\Psi(\underline{x}) = A(\underline{r})\alpha(s)$, omitting the phase factor exp.$(-iEt/\hbar)$. The probability of finding the electron in (space-spin) volume element $d\underline{x} = d\underline{r}ds$ around (space-spin) point $\underline{x}$ is proportional to $d\underline{x}$ and to the probability (or electron) density function at point $\underline{x}$ : $\rho(\underline{x}) = |\Psi(\underline{x})|^2 = |A(\underline{r})|^2|\alpha(s)|^2$. The probability of finding the electron in (space) volume element $d\underline{r}$ around (space) point $\underline{r}$ with any spin is proportional to $d\underline{r}$ and to the density function obtained by integrating $\rho(\underline{x})$ over the spin variable s, i.e. : $P(\underline{r}) = \int\rho(\underline{x})ds = |A(\underline{r})|^2$. These two functions can easily be extended to an N-electron system, using the following formulae :

$$\rho_1(\underline{x}_1) = N \int \Psi(\underline{x}_1, \underline{x}_2, \ldots, \underline{x}_N)\Psi^*(\underline{x}_1, \underline{x}_2, \ldots, \underline{x}_N)d\underline{x}_2\ldots d\underline{x}_N,$$

$$P_1(\underline{r}_1) = \int \rho_1(\underline{x}_1)ds_1. \qquad (48a,b)$$

$\rho_1(\underline{x}_1)d\underline{x}_1$ is the probability of finding any of the N-electrons in $d\underline{x}_1$ around $\underline{x}_1$, the other electrons being anywhere. $P_1(\underline{r}_1)$ is the ordinary electron density function measured by crystallographers. As in the theory of liquids, it is also possible to define probability densities for different configurations of any number of particles. Thus :

$$\rho_2(\underline{x}_1, \underline{x}_2) = N(N-1) \int \Psi(\underline{x}_1, \underline{x}_2, \ldots, \underline{x}_N)\Psi^*(\underline{x}_1, \underline{x}_2, \ldots, \underline{x}_N)$$

$$d\underline{x}_3\ldots d\underline{x}_N \qquad (49a)$$

determines the probability of any two electrons being found simultaneously at (space-spin) points $\underline{x}_1$, $\underline{x}_2$, while :

$$P_2(\underline{r}_1, \underline{r}_2) = \int \rho_2(\underline{x}_1, \underline{x}_2)ds_1ds_2 \qquad (49b)$$

determines the probability of finding them at (space) points $\underline{r}_1$, $\underline{r}_2$, with any combination of spins. These pair density functions tell us how the motions of two electrons are correlated as a result of their mutual interactions. In general, two particles may also interact in a way dependent upon their interactions with other particles, and such many-body interactions may be important in nuclear physics. However, since the many-electron Hamiltonian given by Eqn. (19) contains only pairwise interactions, we have no need to consider distribution functions of order higher than the second here.

A slight generalization of the above density functions will make it possible to express the expectation values of all physical quantities as averages of the corresponding operators over them. If we define the one-electron and electron-pair (continuous) density matrices :

$$\rho_1(\underline{x}_1\ ;\ \underline{x}_1') = N \int \Psi(\underline{x}_1, \underline{x}_2, \ldots, \underline{x}_N)\Psi^*(\underline{x}_1', \underline{x}_2, \ldots, \underline{x}_N)$$

$$d\underline{x}_2 \ldots d\underline{x}_N,\ P_1(\underline{r}_1\ ;\ \underline{r}_1') = \int_{s_1'=s_1} \rho_1(\underline{x}_1\ ;\ \underline{x}_1')ds_1\ ; \qquad (50a,b)$$

$$\rho_2(\underline{x}_1, \underline{x}_2\ ;\ \underline{x}_1', \underline{x}_2') = N(N-1) \int \Psi(\underline{x}_1, \underline{x}_2, \ldots, \underline{x}_N)$$

$$\Psi^*(\underline{x}_1', \underline{x}_2', \ldots, \underline{x}_N)d\underline{x}_3 \ldots d\underline{x}_N,\ P_2(\underline{r}_1, \underline{r}_2\ ;\ \underline{r}_1', \underline{r}_2') =$$

$$\int_{s_1'=s_1,\ s_2'=s_2} \rho_2(\underline{x}_1, \underline{x}_2\ ;\ \underline{x}_1', \underline{x}_2')ds_1 ds_2, \qquad (51a,b)$$

the density functions considered in Eqns. (48a,b) and (49a,b) are the corresponding diagonal elements : $\rho_1(\underline{x}_1) \equiv \rho_1(\underline{x}_1\ ;\ \underline{x}_1)$; $\rho_2(\underline{x}_1, \underline{x}_2) \equiv \rho_2(\underline{x}_1, \underline{x}_2\ ;\ \underline{x}_1, \underline{x}_2)$, etc.. A physical quantity F will be represented by a Hermitian operator $\hat{F}$ symmetrical in the indices of identical particles, e.g., for electrons :

$$\hat{F} = f(0) + \sum_i \hat{h}(i) + \sum_{j<k}\sum \hat{g}(j,k)\ ; \qquad (52)$$

for instance, f(0) may represent the last two terms of Eqn. (24) or those given by Eqns. (31) and (32), etc.. Consequently :

$$\langle\hat{F}\rangle = \int \Psi^*(\underline{x}_1, \underline{x}_2, \ldots, \underline{x}_N)\hat{F}\Psi(\underline{x}_1, \underline{x}_2, \ldots, \underline{x}_N)d\underline{x}_1 d\underline{x}_2 \ldots d\underline{x}_N =$$

$$f(0) + N \int_{\underline{x}_1'=\underline{x}_1} \Psi^*(\underline{x}_1', \underline{x}_2, \ldots, \underline{x}_N)\hat{h}(1)\Psi(\underline{x}_1, \underline{x}_2, \ldots, \underline{x}_N)d\underline{x}_1 d\underline{x}_2 \ldots d\underline{x}_N$$

$$+ \{N(N-1)/2\} \int_{\underline{x}_1'=\underline{x}_1,\ \underline{x}_2'=\underline{x}_2} \Psi^*(\underline{x}_1', \underline{x}_2', \ldots, \underline{x}_N)\hat{g}(1,2)$$

$$\Psi(\underline{x}_1, \underline{x}_2, \ldots, \underline{x}_N)d\underline{x}_1 d\underline{x}_2 \ldots d\underline{x}_N = f(0) + \int_{\underline{x}_1'=\underline{x}_1} \hat{h}(1)$$

$$\rho_1(\underline{x}_1\ ;\ \underline{x}_1')d\underline{x}_1 + (1/2) \int_{\underline{x}_1'=\underline{x}_1,\ \underline{x}_2'=\underline{x}_2} \hat{g}(1,2)\rho_2(\underline{x}_1, \underline{x}_2\ ;\ \underline{x}_1', \underline{x}_2')d\underline{x}_1 d\underline{x}_2, \qquad (53)$$

where we have used the symmetry of $\Psi^*\Psi$ to replace the summations by multiplications of the result for the first term of the sums by their number of terms, and introduced primes to protect $\Psi^*$ from the action of the operators when it appears to the right of $\Psi$ ; of course the primes must be removed before integrating over the density-matrix variables, once the operations have been effected. If the operators are just multipliers (e.g., functions of coordinates), only diagonal matrix-elements are needed, for then :

$$<\hat{F}> = f(0) + \int h(1)\rho_1(\underline{x}_1)d\underline{x}_1 + (1/2) \int g(1,2)\rho_2(\underline{x}_1, \underline{x}_2)d\underline{x}_1 d\underline{x}_2 \; ; \tag{54}$$

while if the operators involve non-commutative operations (such as differentiations or integrations), off-diagonal elements are also necessary. If the operators do not contain spins, $\rho$ and $\underline{x}$ can be replaced by the corresponding P and $\underline{r}$ in the above expressions. It follows from the definitions that successive reduced matrices are related, e.g. :

$$(N-1)\rho_1(\underline{x}_1 \; ; \underline{x}_1') = \int \rho_2(\underline{x}_1, \underline{x}_2 \; ; \underline{x}_1', \underline{x}_2)d\underline{x}_2,$$

$$(N-1)P_1(\underline{r}_1 \; ; \underline{r}_1') = \int P_2(\underline{r}_1, \underline{r}_2 \; ; \underline{r}_1', \underline{r}_2)d\underline{r}_2,$$

so that all electronic properties can actually be discussed in terms of the single reduced 2-matrix $\rho_2$ . If $\Psi$ is a single-determinantal wave-function :

$$\Psi(\underline{x}_1, \underline{x}_2, \ldots, \underline{x}_N) = (1/\sqrt{N!})\det|\psi_A(\underline{x}_1)\psi_B(\underline{x}_2)\ldots\psi_X(\underline{x}_N)|, \tag{55}$$

then it can be shown[13] that the 2-matrix and eventually the wave-function itself factorize in terms of the reduced 1-matrix $\rho_1$ :

$$\rho_2(\underline{x}_1, \underline{x}_2 \; ; \underline{x}_1', \underline{x}_2') = (1-P_{12})\rho_1(\underline{x}_1 \; ; \underline{x}_1')\rho_1(\underline{x}_2 \; ; \underline{x}_2'), \tag{56}$$

where $P_{12}$ interchanges the unprimed variables, $\underline{x}_1$, $\underline{x}_2$. One also has the following diagonal form for the matrix representation of $\rho_1$ in the basis of spin-orbitals $\psi_R$ :

$$\rho_1(\underline{x}_1 \; ; \underline{x}_1') = \sum_{R=A}^{X} \psi_R(\underline{x}_1)\psi_R^*(\underline{x}_1'). \tag{57}$$

Substituting $\psi_R(\underline{x}) = R(\underline{r})\sigma(s)$ leads to :

$$\rho_1(\underline{x}_1 \; ; \underline{x}_1') = P_1^{\alpha,\alpha}(\underline{r}_1 \; ; \underline{r}_1')\alpha(s_1)\alpha^*(s_1') + P_1^{\beta,\beta}(\underline{r}_1 \; ; \underline{r}_1')\beta(s_1)\beta^*(s_1'),$$

$$P_1^{\alpha,\alpha}(\underline{r}_1 \; ; \underline{r}_1') = \sum_{R(\alpha)} R(\underline{r}_1)R^*(\underline{r}_1'), \quad P_1^{\beta,\beta}(\underline{r}_1 \; ; \underline{r}_1') = \sum_{R(\beta)} R(\underline{r}_1)R^*(\underline{r}_1'). \tag{58c,a,b}$$

On integrating Eqn. (58c) over spin we obtain :

$$P_1(\underline{r}_1 \; ; \underline{r}_1') = P_1^{\alpha,\alpha}(\underline{r}_1 \; ; \underline{r}_1') + P_1^{\beta,\beta}(\underline{r}_1 \; ; \underline{r}_1'), \tag{58d}$$

which gives (putting $\underline{r}_1' = \underline{r}_1$) the regular charge density $P_1(\underline{r}_1)$ as a sum of up-spin and down-spin contributions. We shall also need the difference :

$$Q_1(\underline{r}_1 \; ; \underline{r}_1') = P_1^{\alpha,\alpha}(\underline{r}_1 \; ; \underline{r}_1') - P_1^{\beta,\beta}(\underline{r}_1 \; ; \underline{r}_1'), \tag{58e}$$

which reduces (apart from normalization) to the spin density intro-

represented by an operator of the form given by Eqn. (52) will have the off-diagonal matrix elements :

$$\langle \Psi_L | \hat{F} | \Psi_K \rangle = f(0)\delta_{KL} + \int_{\underline{x}_1'=\underline{x}_1} \hat{h}(1)\rho_1(KL|\underline{x}_1 \; ; \underline{x}_1')d\underline{x}_1 + $$
$$(1/2) \int_{\underline{x}_1'=\underline{x}_1,\ \underline{x}_2'=\underline{x}_2} \hat{g}(1,2)\rho_2(KL|\underline{x}_1, \underline{x}_2 \; ; \underline{x}_1', \underline{x}_2')d\underline{x}_1 d\underline{x}_2. \quad (63)$$

The same remarks can be made as with Eqn. (53). If both $\Psi_K$ and $\Psi_L$ are single determinants, the expressions of $\rho_1$ and $\rho_2$ in terms of the spin-orbitals can be obtained by applying Slater's rules. In the particular case where K = L (all spin-orbitals being the same), $\rho_1(KK|\underline{x}_1 \; ; \underline{x}_1')$ is given by Eqn. (57), and $\rho_2(KK|\underline{x}_1, \underline{x}_2 \; ; \underline{x}_1', x_2')$ by Eqn. (56). If $\Psi_K$ and $\Psi_L$ differ by one spin-orbital, R :

$$\rho_1(KL|\underline{x}_1 \; ; \underline{x}_1') = \psi_R(\underline{x}_1)\psi_R'^{*}(\underline{x}_1'), \quad (57')$$

$$\rho_2(KL|\underline{x}_1, \underline{x}_2 \; ; \underline{x}_1', \underline{x}_2') = (1-P_{12}) \sum_{S(\neq R)} \{\psi_R(\underline{x}_1)\psi_S(\underline{x}_2)$$
$$\psi_R'^{*}(\underline{x}_1')\psi_S^{*}(\underline{x}_2') + \psi_S(\underline{x}_1)\psi_R(\underline{x}_2)\psi_S^{*}(\underline{x}_1')\psi_R'^{*}(\underline{x}_2')\} \; ; \quad (56')$$

and if they differ by two spin-orbitals, R and S :

$$\rho_1(KL|\underline{x}_1 \; ; \underline{x}_1') = 0, \quad (57'')$$

$$\rho_2(KL|\underline{x}_1, \underline{x}_2 \; ; \underline{x}_1', \underline{x}_2') = (1-P_{12})\{\psi_R(\underline{x}_1)\psi_S(\underline{x}_2)\psi_R'^{*}(\underline{x}_1')\psi_S'^{*}(\underline{x}_2') +$$
$$\psi_S(\underline{x}_1)\psi_R(\underline{x}_2)\psi_S'^{*}(\underline{x}_1')\psi_R'^{*}(\underline{x}_2')\}. \quad (56'')$$

If $\Psi_K$ and $\Psi_L$ differ by more than two spin-orbitals, both $\rho_1$ and $\rho_2 = 0$. Slightly more complicated expressions for both $\rho_1$ and $\rho_2$ are obtained when the $\psi$'s are expressed in terms of the non-orthogonal basis functions $\phi_R$. Similar expressions have also been derived when $\Psi_K$ and $\Psi_L$ are orthogonal antisymmetrized products of group functions (15c) (e.g., geminals or core shells, valence bonds, lone pairs and, more generally, loges or even weakly interacting molecular systems), which reduce to Slater determinants in the particular case where each group consists of a single electron described by a spin-orbital. If $\Psi_K$ and $\Psi_L$ are expanded as the forms :

$$\Psi_K = \sum_{\kappa} c_{K\kappa}\Phi_{\kappa}, \quad \Psi_L = \sum_{\lambda} c_{L\lambda}\Phi_{\lambda},$$

where the $\Phi$'s may be single determinants or other analytical or numerical functions of the N electron coordinates, one obtains for $\rho_1$ and $\rho_2$ the following weighted sums :

$$\rho_1(KL|\underline{x}_1 \; ; \underline{x}_1') = \sum_{\kappa,} \sum_{\lambda} c_{K\kappa}c_{L\lambda}^{*}\rho_1(\kappa\lambda|\underline{x}_1 \; ; \underline{x}_1'),$$

duced by McConnell[22] on putting $\underline{r}_1' = \underline{r}_1$. One could of course replace $P_1^{\alpha,\alpha}$ and $P_1^{\beta,\beta}$ in Eqn. (58c) by the appropriate combinations of $P_1$ and $Q_1$. If the spin-orbitals $\psi_R$ (which may be orthonormal MO's or VB's) are expressed as linear combinations of some basis functions $\phi_R$ (which may be non-orthogonal AO's or other MO's) through the (not necessarily unitary) transformation : $\underline{\psi} = \underline{\underline{T}}\underline{\phi}$ (giving the overlap matrix $\underline{\underline{S}}$ for the $\phi$'s), Eqn. (56) continues to hold but Eqn. (57) takes the more general form :

$$\rho_1(\underline{x}_1 \ ; \ \underline{x}_1') = \sum_{R,S=A-X}\sum (\underline{\underline{S}}^{-1})_{RS}\phi_R(\underline{x}_1)\phi_S^*(\underline{x}_1'). \tag{59}$$

This can again be written in the form of Eqn. (58c) but Eqns. (58a,b) now become :

$$P_1^{\alpha,\alpha}(\underline{r}_1 \ ; \ r_1') = \sum_{R(\alpha)}\sum_{S(\alpha)} (\underline{\underline{S}}^{-1})_{RS}R(\underline{r}_1)S^*(\underline{r}_1'), \tag{60a}$$

$$P_1^{\beta,\beta}(\underline{r}_1 \ ; \ \underline{r}_1') = \sum_{R(\beta)}\sum_{S(\beta)} (\underline{\underline{S}}^{-1})_{RS}R(\underline{r}_1)S^*(\underline{r}_1'), \tag{60b}$$

where account has been taken of the fact that $(\underline{\underline{S}}^{-1})_{RS}$ vanishes if $\phi_R$ and $\phi_S$ differ in spin factors. The physical interpretation of the terms of $P_1$ and $Q_1$ (charge and spin population analysis) may be easier in the new basis.

When one is performing CI between SCF wave-functions such as that given by Eqn. (55) or, more generally, calculating the effects of stationary perturbations on the exact or approximate eigenstates of a given Hamiltonian, and also when one is calculating the probabilities of the transitions induced by time-dependent perturbations between different stationary states, one is led to extend the above definitions and to introduce spin-including and spinless, one-electron and electron-pair, reduced <u>transition matrices</u> and functions:

$$\rho_1(KL|\underline{x}_1 \ ; \ \underline{x}_1') = N\int \Psi_K(\underline{x}_1, \underline{x}_2, \ldots, \underline{x}_N)\Psi_L^*(\underline{x}_1', \underline{x}_2, \ldots, \underline{x}_N) d\underline{x}_2\ldots d\underline{x}_N, \quad P_1(KL|\underline{r}_1 \ ; \ \underline{r}_1') = \int_{s_1'=s_1} \rho_1(KL|\underline{x}_1 \ ; \ \underline{x}_1')ds_1 \ ; \tag{61a,b}$$

$$\rho_2(KL|\underline{x}_1, \underline{x}_2 \ ; \ \underline{x}_1', \underline{x}_2') = N(N-1)\int \Psi_K(\underline{x}_1, \underline{x}_2, \ldots, \underline{x}_N)\Psi_L^*(\underline{x}_1', \underline{x}_2', \ldots, \underline{x}_N)d\underline{x}_3\ldots d\underline{x}_N, \quad P_2(KL|\underline{r}_1, \underline{r}_2 \ ; \ \underline{r}_1', \underline{r}_2') = \int_{s_1'=s_1,\ s_2'=s_2} \rho_2(KL|\underline{x}_1, \underline{x}_2 \ ; \ \underline{x}_1', \underline{x}_2')ds_1ds_2, \tag{62a,b}$$

which for K = L reduce to Eqns. (50a,b) and (51a,b) (where the unnecessary labels are implied) ; as usual, we write the diagonal elements as $\rho_1(KL|\underline{x}_1)$; $\rho_2(KL|\underline{x}_1, \underline{x}_2)$, etc.. A physical quantity F

$$\rho_2(KL|\underline{x}_1, \underline{x}_2 ; \underline{x}'_1, \underline{x}'_2) = \sum_{\kappa, \lambda} \sum c_{K\kappa} c^{*}_{L\lambda} \rho_2(\kappa\lambda|\underline{x}_1, \underline{x}_2 ; \underline{x}'_1, \underline{x}'_2),$$

where $\rho_1(\kappa\lambda|\underline{x}_1 ; \underline{x}'_1)$ and $\rho_2(\kappa\lambda|\underline{x}_1, \underline{x}_2 ; \underline{x}'_1, \underline{x}'_2)$ are defined as in Eqns. (61a) and (62a), and similar expressions for $P_1$ and $P_2$. When $\Psi_K$ and $\Psi_L$ are arbitrary functions, Eqn. (58c) takes the more general form :

$$\rho_1(KL|\underline{x}_1 ; \underline{x}'_1) = P_1^{\alpha,\alpha}(KL|\underline{r}_1 ; \underline{r}'_1)\alpha(s_1)\alpha^{*}(s'_1) + P_1^{\alpha,\beta}(KL|\underline{r}_1 ; \underline{r}'_1)\alpha(s_1)\beta^{*}(s'_1) + P_1^{\beta,\alpha}(KL|\underline{r}_1 ; \underline{r}'_1)\beta(s_1)\alpha^{*}(s'_1) + P_1^{\beta,\beta}(KL|\underline{r}_1 ; \underline{r}'_1)\beta(s_1)\beta^{*}(s'_1), \qquad (64)$$

and similarly $\rho_2$, which no longer is factorizable in terms of the 1-matrix, can be expanded as a multilinear form in the complete orthonormal spin basis :

$$\rho_2(KL|\underline{x}_1, \underline{x}_2 ; \underline{x}'_1, \underline{x}'_2) = \sum_{\sigma_1, \sigma_2 = \alpha, \beta} \sum \sum_{\sigma'_1, \sigma'_2 = \alpha^{*}, \beta^{*}} \sum P_2^{\sigma_1\sigma_2,\sigma'_1\sigma'_2}(KL|\underline{r}_1, \underline{r}_2 ; \underline{r}'_1, \underline{r}'_2)\sigma_1(s_1)\sigma_2(s_2)\sigma'_1(s'_1)\sigma'_2(s'_2). \qquad (65)$$

Now it can be shown[(15e)] that if $\Psi_K$ and $\Psi_L$ are characterized by well-defined eigenvalues of the operators $\hat{S}^2$ and $\hat{S}^z$ : $\Psi_K = |\eta' S' M'\rangle$, $\Psi_L = |\eta\ S\ M\rangle$, $\eta$ standing for all other labels of the $(2S+1)$-fold degenerate level and M designating a particular state within the multiplet, at most two $P_1$'s in Eqn. (64) and at most six $P_2$'s in Eqn. (65) do not vanish, which can be expressed in terms of at most three independent functions with direct physical interpretation. More precisely, all $P_1$'s vanish unless $\Delta S = 0, \pm 1$ and $\Delta M = 0, \pm 1$ and similarly for the $P_2$'s with $\Delta S = 0, \pm 1, \pm 2$ and $\Delta M = 0, \pm 1, \pm 2$, the numbers of $P_1$'s and $P_2$'s increasing to their maximum values when $|\Delta M|$ decreases to 0, and the numbers of terms in these spinless components also increasing when either $|\Delta M|$ or $|\Delta S|$ decreases ; the expressions of these terms and their coefficients in the expansions of the $P_1$'s and $P_2$'s have been tabulated[(15e)]. The demonstration of the above-stated results consists basically in writing formally the $\rho$'s as matrix elements of appropriate operators $\hat{O}$, e.g. : $\rho_1(KL|\underline{x}_1 ; \underline{x}'_1) = N \langle\Psi_L|\hat{O}^r(1)\hat{O}^s(1)|\Psi_K\rangle$, then showing that the spinless components in Eqns. (64) and (65) can be expanded in terms of the matrix elements of irreducible tensor operators of spin, e.g. : $P^{\alpha,\beta}(KL|\underline{r}_1 ; \underline{r}'_1) = N \langle\Psi_L|\hat{O}^r(1)\hat{S}^-(1)|\Psi_K\rangle = \sqrt{2}\ N \langle\eta\ S\ M|\hat{O}^r(1)T^{-1}_1|\eta' S' M'\rangle$, and finally applying the Wigner-Eckart theorem, cf. Eqn. (37), to express these matrix elements as products of Clebsch-Gordan coefficient-ratios with standard reduced matrix elements ; for $\rho_1$ these latter define charge and spin transition matrices very similar to the density matrices, $P_1$ and $Q_1$, introduced in Eqns. (58d,e), while for $\rho_2$ they define three independent functions, $P_2$, $Q_{SO}$ and $Q_{SS}$, to which we shall return later.

To end this first part we shall recall that all density matrices defined above can be regarded as kernel representations of abstract density operators, e.g. :

$$\int \rho_1(KL|\underline{x}_1\ ;\ \underline{x}_1')\psi(\underline{x}_1')d\underline{x}_1' \equiv \hat{\rho}_1^{(KL)}\psi(\underline{x}_1) = \psi'(\underline{x}_1). \tag{66}$$

If the kernel can be expanded as a multilinear form in a complete orthonormal one-particle basis set, e.g. :

$$\rho_1(KL|\underline{x}_1\ ;\ \underline{x}_1') = \sum_{R,}\ \sum_{S} \rho_{1,RS}^{(KL)}\ \psi_R(\underline{x}_1)\psi_S^*(\underline{x}_1') \tag{67}$$

{which is the more general form of Eqn. (57)}, it follows from the definition that the matrix elements of the operator in this basis are the coefficients of the expansion :

$$\langle\psi_R|\hat{\rho}_1^{(KL)}|\psi_S\rangle = \rho_{1,RS}^{(KL)}, \tag{67'}$$

this providing a (discrete) matrix representation for the operator. When the basis is chosen so as to give a diagonal form to $\underline{\underline{\rho}}_1^{(KK)}$, one refers to its components as the natural states of $\Psi_K$ (e.g., natural spin-orbitals). Properties of these operators and of their matrix representations follow from those derived for their kernels. Finally, it may be useful to have the current notations and conventions (14,15,19) related :

$$\gamma_{LK}(\underline{x}_1'|\underline{x}_1) \equiv \rho_1(KL|\underline{x}_1\ ;\ \underline{x}_1') \equiv ND_{KL}^1(\underline{x}_1\ ;\ \underline{x}_1'),$$

$$\Gamma_{LK}(\underline{x}_1'\underline{x}_2'|\underline{x}_1\underline{x}_2) \equiv (1/2)\rho_2(KL|\underline{x}_1,\ \underline{x}_2\ ;\ \underline{x}_1',\ \underline{x}_2') \equiv$$

$$\{N(N-1)/2\}D_{KL}^2(\underline{x}_1,\ \underline{x}_2\ ;\ \underline{x}_1',\ \underline{x}_2').$$

When no confusion is possible, one often abbreviates further the sequence of (space, spin or space-spin) coordinates as 1'2'...N or, even more briefly, x'y (distinguishing so the primed and unprimed variables).

### 3.2. Characteristic Coupling Tensors

When we discussed the resolution of $\rho_1$ and $\rho_2$ into the spinless components defined in Eqns. (64) and (65), we mentioned that in the determination of the expressions of these components one is led to evaluate the matrix elements of appropriate products of space and spin operators between the state functions. Such matrix elements do actually occur in the calculation of the effects of the magnetic perturbation $\hat{H}_M$ on the eigenstates of the Schrödinger Hamiltonian $H_0$. We shall now see how the specific functions entering the expressions of the $P_1$'s and $P_2$'s can be utilized.

In Eqn. (26) one may basically distinguish five different kinds of terms : (i) those which do not depend on any electron coordinate, but may depend on nuclear spin operators and position vectors, i.e. $\hat{H}_{nZ}$ and $\hat{H}_{nd}$ : these can be written in the form f(0) ; (ii) those which may also depend on one-electron position vectors and orbital angular momenta (and two-electron spatial coordinates if one includes the relativistic corrections $\hat{H}_R$), but not on the electron spins, i.e. the first parts of $\hat{H}_{hf}$ and $\hat{H}_{eZ}$ and also $H_{dia}$ and $H_{add}$ as given by Eqns. (41-43) : these will be written in the form $\Sigma \hat{h}(i)$ ; (iii) those which involve one-electron spin operators multiplying axial vectors which may be functions of the same electron position vectors and orbital angular momenta, i.e. the second parts of $\hat{H}_{hf}$ and $\hat{H}_{eZ}$ and the first part of $\hat{H}_{SO}$ ; (iv) those which involve one-electron spin operators multiplying functions of the same and of another electron spatial operators, i.e. the second part of $\hat{H}_{SO}$ ; (v) those involving two-electron spin operators, i.e. the terms of $H_{SS}$. To express the terms mentioned in (iii)-(v) in convenient form, one may make use of the spherical components $\hat{S}_1^m$ of vector operators (which may be $\hat{s}_i$'s, $\ell_i$'s, etc.) with Cartesian components $\hat{S}_1^u$, i.e. : $\hat{S}_1^{+1} = -(\hat{S}_1^x + i\hat{S}_1^y)/\sqrt{2}$, $\hat{S}_1^o = \hat{S}_1^z$, $\hat{S}_1^{-1} = +(\hat{S}_1^x - i\hat{S}_1^y)/\sqrt{2}$ (which have the same transformation properties under a rotation of the axis of quantization as a set of spin states with quantum number S = 1), and of the well-known group-theoretical property(11d) that from a product of two vector operators $\underline{S}_1(i)$ and $\underline{S}_1(j)$ one can construct the nine irreducible tensor-operator components $\hat{S}_\ell^m(i,j)$ ($\ell$ = 0, 1, 2 ; $-\ell \leq m \leq +\ell$). Then one can write :

$$\begin{aligned}\hat{H}_M &= f(0) + \sum_i \hat{h}(i) + \sum_i \underline{\hat{s}}_i\cdot\{\underline{\hat{h}}(i) + \sum_j{}' \underline{\hat{g}}(i,j)\} + \sum_{j<k} \sum \underline{\hat{s}}_j\cdot\underline{\underline{g}}(j,k)\cdot\underline{\hat{s}}_k \\ &= f(0) + \sum_i \hat{h}(i) + \sum_i\sum_m \hat{S}_1^m(i)\{\hat{H}_1^{-m}(i) + \sum_j{}' G_1^{-m}(i,j)\}(-1)^m + \\ &\sum_{j<k}\sum\sum_{\ell,m}\sum \hat{S}_\ell^m(j,k)\hat{G}_\ell^{-m}(j,k)(-1)^m, \end{aligned} \quad (68)$$

where the coefficients of the $\hat{S}_\ell^m$'s can be deduced from Eqns. (27-30).

The preceding analysis, taken with Eqn. (63), shows that $\hat{H}_{nd}$ and $\hat{H}_{nZ}$ simply retain their forms or vanish when one takes the matrix elements of $H_M$ between the eigenfunctions of $\hat{H}_0$. The first sum of Eqn. (68) will give the matrix elements :

$$\int_{\underline{x}_1'=\underline{x}_1} \hat{h}(1)\rho_1(KL|\underline{x}_1 ; \underline{x}_1')d\underline{x}_1 = \int_{\underline{r}_1'=\underline{r}_1} \hat{h}(1)P_1(KL|\underline{r}_1 ; \underline{r}_1')d\underline{r}_1, \quad (69)$$

where $P_1$ is defined in Eqn. (61b), since h(1) does not involve the electron spin. Similarly, the symmetric sum of two-electron spinless operators would involve the matrix $P_2$ defined in Eqn. (62b). The definitions themselves yield the result that $P_1$ and $P_2$ are identically zero if they correspond to functions with different **M** values. The other sums of Eqn. (68) will give matrix elements with the following forms :

$$\int_{\underline{x}_1'=\underline{x}_1} (-1)^m \hat{H}_1^{-m}(1)\hat{S}_1^m(1)\rho_1(KL|\underline{x}_1 ; \underline{x}_1')d\underline{x}_1 = \int_{\underline{r}_1'=\underline{r}_1} (-1)^m \hat{H}_1^{-m}(1)$$

$$Q_S(S'M'SM|\underline{r}_1 ; \underline{r}_1')_1^m d\underline{r}_1, \qquad (70a)$$

$$Q_S(S'M'SM|\underline{r}_1 ; \underline{r}_1')_1^m = \int_{s_1'=s_1} \hat{S}_1^m(1)\rho_1(KL|\underline{x}_1 ; \underline{x}_1')ds_1 ; \qquad (70b)$$

$$\int_{\underline{x}_1'=\underline{x}_1,\ \underline{x}_2'=\underline{x}_2} (-1)^m \hat{G}_1^{-m}(1,2)\hat{S}_1^m(1)\rho_2(KL|\underline{x}_1, \underline{x}_2 ; \underline{x}_1', \underline{x}_2')d\underline{x}_1 d\underline{x}_2 =$$

$$\int_{\underline{r}_1'=\underline{r}_1,\ \underline{r}_2'=\underline{r}_2} (-1)^m \hat{G}_1^{-m}(1,2)Q_{SO}(S'M'SM|\underline{r}_1, \underline{r}_2 ; \underline{r}_1', \underline{r}_2')_1^m d\underline{r}_1 d\underline{r}_2, \qquad (71a)$$

$$Q_{SO}(S'M'SM|\underline{r}_1, \underline{r}_2 ; \underline{r}_1', \underline{r}_2')_1^m = \int_{s_1'=s_1,\ s_2'=s_2} \hat{S}_1^m(1)\rho_2(KL|\underline{x}_1, \underline{x}_2 ;$$

$$\underline{x}_1', \underline{x}_2')ds_1 ds_2 ; \qquad (71b)$$

$$(1/2) \int_{\underline{x}_1'=\underline{x}_1,\ \underline{x}_2'=\underline{x}_2} (-1)^m \hat{G}_\ell^{-m}(1,2)\hat{S}_\ell^m(1,2)\rho_2(KL|\underline{x}_1, \underline{x}_2 ; \underline{x}_1', \underline{x}_2')$$

$$d\underline{x}_1 d\underline{x}_2 = (1/2) \int_{\underline{r}_1'=\underline{r}_1,\ \underline{r}_2'=\underline{r}_2} (-1)^m \hat{G}_\ell^{-m}(1,2)Q_{SS}(S'M'SM|\underline{r}_1, \underline{r}_2 ;$$

$$\underline{r}_1', \underline{r}_2')_\ell^m d\underline{r}_1 d\underline{r}_2, \qquad (72a)$$

$$Q_{SS}(S'M'SM|\underline{r}_1, \underline{r}_2 ; \underline{r}_1', \underline{r}_2')_\ell^m = \int_{s_1'=s_1,\ s_2'=s_2} \hat{S}_\ell^m(1,2)$$

$$\rho_2(KL|\underline{x}_1, \underline{x}_2 ; \underline{x}_1', \underline{x}_2')ds_1 ds_2, \qquad (72b)$$

where the spin quantum numbers associated with the state labels have been singled out. Now from the definitions and the Wigner-Eckart theorem, cf. Eqn. (37), one can write :

$$Q_S(S'M'SM|\underline{r}_1 ; \underline{r}_1')_1^m = C(S'1S, M'mM)\ Q_S^1(S'S|\underline{r}_1 ; \underline{r}_1'), \qquad (73)$$

and similar expressions for $Q_{SO}$ and $Q_{SS}$. The functions $Q_S^1$, $Q_{SO}^1$ and $Q_{SS}^\ell$, if used in Eqns. (70a), (71a) and (72a), respectively, would give the so-called reduced matrix elements of the tensor operators, from which all other matrix elements (with various values of M, M' and m) may be obtained by multiplying by the appropriate Clebsch-Gordan coefficient C. To determine $Q_S^1$, for instance, one may make any one convenient choice of M and M', take m = M − M' (since the C's are zero otherwise) and evaluate the corresponding matrix element, called standard matrix element, using Eqn. (70b). If one chooses M = S, M' = S', and accordingly writes : m = S − S', $\Psi_K = |\eta' S' S'\rangle$, $\Psi_L = |\eta\ S\ S\rangle$, then $Q_S^1$ can be obtained from :

$$Q_S^1(S'S \mid \underline{r}_1 \ ; \ \underline{r}_1') = \overline{Q}_S(S'S'SS|\underline{r}_1 \ ; \ \underline{r}_1')_1^{\overline{m}}/C(S'1S,\ S'\overline{m}S), \qquad (74)$$

and $Q_{SO}^1$ and $Q_{SS}^{\ell}$, from similar expressions. Defining the scaled coefficients (which are non - zero only for $m = M - M'$, $|S'- S| \leq \ell \leq |S'+ S|$) :

$$C_\ell(S'S,\ M'M) = C(S'\ell S,\ M'mM)/C(S'\ell S,\ S'\overline{m}S), \qquad (75)$$

and combining Eqns. (73) and (74), one obtains :

$$Q_S(S'M'SM|\underline{r}_1 \ ; \ \underline{r}_1')_1^m = C_1(S'S,\ M'M)\overline{Q}_S(S'S|\underline{r}_1 \ ; \ \underline{r}_1')_1^{\overline{m}}, \qquad (70c)$$

where the first factor can be found in tables[11d] and the second factor can be calculated from Eqn. (70b) by making the appropriate substitutions. Similarly, one can write :

$$Q_{SO}(S'M'SM|\underline{r}_1,\ \underline{r}_2 \ ; \ \underline{r}_1',\ \underline{r}_2')_1^m = C_1(S'S,\ M'M)\overline{Q}_{SO}(S'S|\underline{r}_1,\ \underline{r}_2 \ ; \ \underline{r}_1',\ \underline{r}_2')_1^{\overline{m}} \ ; \qquad (71c)$$

$$Q_{SS}(S'M'SM|\underline{r}_1,\ \underline{r}_2 \ ; \ \underline{r}_1',\ \underline{r}_2')_\ell^m = C_\ell(S'S,\ M'M)\overline{Q}_{SS}(S'S|\underline{r}_1,\ \underline{r}_2 \ ; \ \underline{r}_1',\ \underline{r}_2')_\ell^{\overline{m}} \ . \qquad (72c)$$

The physical interpretation of the above functions can be made by setting the corresponding spatial operators in Eqn. (68) equal to unity (thus obtaining the diagonal elements in the coordinates) and considering only states with the same S and M (thus obtaining diagonal elements in the states). Using Eqns. (70a,c) (with m = 0), one obtains :

$$\langle\Psi_L|\sum_i \hat{S}_1^o(i)|\Psi_K\rangle = \langle\eta SM|\hat{S}^z|\eta' SM\rangle = M =$$
$$\int Q_S(SMSM|\underline{r}_1 \ ; \ \underline{r}_1)_1^o d\underline{r}_1 = (M/S) \int \overline{Q}_S(SS|\underline{r}_1 \ ; \ \underline{r}_1)_1^o d\underline{r}_1,$$

so that the normalized quantity :

$$D_S(S|\underline{r}_1) = (1/S)\overline{Q}_S(SS|\underline{r}_1 \ ; \ \underline{r}_1)_1^o, \qquad (70d)$$

can be identified with the spin density introduced by McConnell[22]. Hence the name spin transition matrix for $Q_S$. One can obtain $\overline{Q}_S$ in terms of the spinless components of $\rho_1$ by substituting Eqns. (64) into (70b) for K = L (m = 0) and operating ; the result :

$$\overline{Q}_S(SS|\underline{r}_1 \ ; \ \underline{r}_1)_1^o = (1/2)\{P_1^\alpha(KK|\underline{r}_1) - P_1^\beta(KK|\underline{r}_1)\}, \qquad (70e)$$

thus appears as the excess density of spin-up over spin-down electrons times the magnitude of their spin z-component (in ℏ units). Similarly, one can define the normalized quantity :

$$D_{SO}(S|\underline{r}_1, \underline{r}_2) = (1/S)\overline{Q}_{SO}(SS|\underline{r}_1, \underline{r}_2 ; \underline{r}_1, \underline{r}_2)^o_1, \quad (71d)$$

and show that, in terms of the spinless components of $\rho_2$ introduced in Eqn. (65) :

$$\overline{Q}_{SO}(SS|\underline{r}_1, \underline{r}_2 ; \underline{r}_1, \underline{r}_2)^o_1 = (1/2)\{P_2^{\alpha\alpha}(KK|\underline{r}_1, \underline{r}_2) - P_2^{\beta\alpha}(KK|\underline{r}_1, \underline{r}_2) + P_2^{\alpha\beta}(KK|\underline{r}_1, \underline{r}_2) - P_2^{\beta\beta}(KK|\underline{r}_1, \underline{r}_2)\} \quad (71e)$$

behaves as a <u>conditional spin density</u> for an electron at point 1 with a second electron at point 2. Finally, one can also write (for $S > 1/2$) :

$$<\Psi_L|\sum_{j<k}\Sigma\hat{S}^o_2(j,k)|\Psi_K> = <\eta SM|(1/2)\{3(\hat{S}^z)^2 - \underline{\hat{S}}^2\}|\eta' SM> = \{3M^2 - S(S+1)\}/2 = (1/2) \int Q_{SS}(SMSM|\underline{r}_1, \underline{r}_2 ; \underline{r}_1, \underline{r}_2)^o_2 d\underline{r}_1 d\underline{r}_2$$

$$= \{\{3M^2 - S(S+1)\} / 2S(2S-1)\} \int \overline{Q}_{SS}(SS|\underline{r}_1, \underline{r}_2 ; \underline{r}_1, \underline{r}_2)^o_2 d\underline{r}_1 d\underline{r}_2,$$

and introduce the normalized quantity :

$$D_{SS}(S|\underline{r}_1, \underline{r}_2) = \{1/S(2S-1)\}\overline{Q}_{SS}(SS|\underline{r}_1, \underline{r}_2 ; \underline{r}_1, \underline{r}_2)^o_2, \quad (72d)$$

where $\overline{Q}_{SS}$ can be written, in terms of the spinless components of $\rho_2$ :

$$\overline{Q}_{SS}(SS|\underline{r}_1, \underline{r}_2 ; \underline{r}_1, \underline{r}_2)^o_2 = (1/2)\{P_2^{\alpha\alpha,\alpha\alpha} + P_2^{\beta\beta,\beta\beta} - P_2^{\alpha\beta,\alpha\beta} - P_2^{\beta\alpha,\beta\alpha} - P_2^{\alpha\beta,\beta\alpha} - P_2^{\beta\alpha,\alpha\beta}\}, \quad (72e)$$

which behaves as a <u>spin correlation density</u>. We shall now see how the functions $P_2$ and $Q_{SO}$ (from which $P_1$ and $Q_S$, respectively, can be deduced by integration over $\underline{r}_2$) and the function $Q_{SS}$ determine completely the characteristic coupling tensors of effective spin Hamiltonians. It may be worth noting that when the state function is represented by a single determinant, Eqn. (56) entails a similar factorization of $P_2$, $Q_{SO}$ and $Q_{SS}$ in terms of only $P_1$ and $Q_S$.

The derivation of a general spin Hamiltonian for molecular systems follows the same lines as that given in §2.2 for paramagnetic ions in crystals. Any eigenket of $\hat{H}_0 + H_M$ can be expanded in terms of the orthonormal eigenkets of $\hat{H}_0$ : $|oM_SM_I>$, ..., $|nM'_SM'_I>$, ..., where o designates the unperturbed level of interest and n any one other, $M_S$ the electron spin label and $M_I$ the collection of nuclear spin labels characterizing the state within the multiplet. Partitioning the secular equation to define an effective Hamiltonian $\hat{H}$ operating within the subspace $\{|oM_SM_I>\}$, as in Eqns. (45a,b), then retaining the leading terms in the matrix elements of $H$ lead to an expression for the latter similar to Eqn. (46), with $M_S$ replaced everywhere by $M_SM_I$. One then looks for a spin operator $H_S$, containing operators for the total electron spin and nuclear spins, so defined that :

$$H_S = H'_S + H''_S + \ldots, \tag{76}$$

$$\langle M_S M_I | H'_S | M'_S M'_I \rangle = \langle o M_S M_I | \hat{H}_M | o M'_S M'_I \rangle, \tag{76'}$$

$$\langle M_S M_I | H''_S | M'_S M'_I \rangle = \sum_{\nu(\neq o) M''_S M''_I} \langle o M_S M_I | \hat{H}_M | \nu M''_S M''_I \rangle \, (E^o - E^\nu)^{-1}$$

$$\langle \nu M''_S M''_I | \hat{H}_M | o M'_S M'_I \rangle, \ \ldots \ . \tag{76''}$$

Such equivalent operators may be obtained under rather general conditions[11c], which then allow reduction of the full secular problem to a pure spin problem. The various terms in $H_M$ clearly make additive contributions to $H'_S$, while in $H''_S$ a large variety of cross-terms appear. The operators $H'_S$ and $H''_S$ depend upon the electronic wave-functions through the characteristic coupling tensors which appear when one applies Eqns. (76''"), as we shall now see in some examples.

For reasons similar to those leading to the quenching of orbital angular momenta for paramagnetic ions in crystals (see p. 14), the first parts of $\hat{H}_{hf}$ and $H_{eZ}$ as well as both parts of $\hat{H}_{SO}$ will give vanishing first-order contributions to the effective spin Hamiltonian. The second part of $\hat{H}_{hf}$ is a sum of dipolar and contact terms, $\hat{H}_{hfd}(\lambda)$ and $H_{hfc}(\lambda)$ respectively, and similarly for $\hat{H}_{SS}$ : the contributions of these operators to $H'_S$ will be investigated shortly. The second part of $\hat{H}_{eZ}$ (with the total electron-spin operator) as well as $H_{nZ}$ and $\hat{H}_{nd}$ (with one and two nuclear-spin operators) are already in spin-Hamiltonian form. Averaging $H_{dia}$ over the electronic wave-function will give a uniform diamagnetic shift to all levels. Finally, averaging $H_{add}$ as given by Eqns. (41-43) will give a sum of single nuclear-spin operators and nuclear-induced shifts, which we shall consider first.

a) Isotope/Isomer-Shift Terms. Substituting Eqn. (42) into Eqn. (76') and using Eqn. (69), one obtains :

$$\langle o M_S M_I | \hat{H}_{is,\lambda} | o M'_S M'_I \rangle = (2\pi e^2/3) Z_\lambda R_\lambda^2 \delta_{M_I M'_I} \int \delta(r_{\lambda 1}) P_1(KL|\underline{r}_1) d\underline{r}_1 =$$

$$(2\pi e^2/3) Z_\lambda R_\lambda^2 \delta_{M_I M'_I} \delta_{M_S M'_S} P_1(\underline{r}_\lambda) = \langle M_S M_I | (2\pi e^2/3) Z_\lambda R_\lambda^2 P_1(\underline{r}_\lambda) | M'_S M'_I \rangle;$$

whence the (spinless) spin-Hamiltonian term :

$$H'_{is,\lambda} = (2\pi e^2/3) Z_\lambda R_\lambda^2 P_1(\underline{r}_\lambda), \tag{77}$$

which is proportional to the effective cross-section of the nucleus and to the charge-density value at its center (induced by all "s-electrons"). It will later be shown how, for suitable nuclei, the Möss-bauer effect permits the determination of the difference, between two nuclear states, of either the effective cross-sections or the contact charge-densities when one knows the other[23].

b) Quadrupole-Splitting Constants. Substituting Eqn. (41) into Eqn. (76') and using Eqn. (69), one obtains :

$$<oM_SM_I|\hat{H}_{qp,\lambda}|oM_S'M_I'> = \{e^2Q_\lambda/2I_\lambda(2I_\lambda-1)\}<M_I|\hat{\underline{I}}_\lambda.\{\int \underline{\underline{D}}(\underline{r}_{\lambda 1}) P_1(KL|\underline{r}_1)d\underline{r}_1\}.\hat{\underline{I}}_\lambda|M_I'> = <M_I|\hat{\underline{I}}_\lambda.\underline{\underline{Q}}_\lambda\delta_{M_SM_S'}.\hat{\underline{I}}_\lambda|M_I'> = <M_SM_I|\hat{\underline{I}}_\lambda.\underline{\underline{Q}}_\lambda.\hat{\underline{I}}_\lambda|M_S'M_I'>, \quad (78')$$

$$\underline{\underline{Q}}_\lambda = \{e^2Q_\lambda/2I_\lambda(2I_\lambda-1)\}\int \underline{\underline{D}}(\underline{r}_{\lambda 1})P_1(\underline{r}_1)d\underline{r}_1 = \{eQ_\lambda/2I_\lambda(2I_\lambda-1)\}\underline{\underline{V}}_\lambda \;;$$

whence the (single nuclear-spin) spin-Hamiltonian term :

$$H'_{qp,\lambda} = \hat{\underline{I}}_\lambda.\underline{\underline{Q}}_\lambda.\hat{\underline{I}}_\lambda, \quad (78)$$

whose coupling tensor, given by Eqn. (78'), is proportional to the effective quadrupole-moment constant $eQ_\lambda$ and to the charge-density-averaged dipolar-coupling tensor $\underline{\underline{D}}(\underline{r}_{\lambda 1})$ (which involves all "p, d or f-electrons"). In the principal-axis system of the traceless field-gradient tensor $\underline{\underline{V}}_\lambda$, one usually defines the axial constant $-eq_\lambda = V_\lambda^{zz}$ and the asymmetry parameter $\eta_\lambda = (V_\lambda^{xx} - V_\lambda^{yy})/V_\lambda^{zz}$ (with the convention $|V_\lambda^{zz}| \geq |V_\lambda^{xx}| \geq |V_\lambda^{yy}|$ entailing $0 \leq \eta_\lambda \leq 1$). The above operator then takes the familiar form :

$$H'_{qp,\lambda} = \{-e^2q_\lambda Q_\lambda/4I_\lambda(2I_\lambda-1)\}\{\{3(\hat{I}_\lambda^z)^2 - \hat{\underline{I}}^2\} + \eta_\lambda\{(\hat{I}_\lambda^x)^2 - (\hat{I}_\lambda^y)^2\}\}. \quad (78'')$$

The product $-e^2q_\lambda Q_\lambda$ and the parameter $\eta_\lambda$ can, in favorable cases, be determined again by using the Mössbauer effect for excited nuclear states[23] and for ground nuclear states by using a variety of techniques, including NQR or NMR for diamagnetic systems and EPR or ENDOR for paramagnetic systems[24-27].

c) Contact Hyperfine Couplings. Substituting the contact terms of Eqn. (27) into Eqn. (76') and using successively Eqns. (70a,c), one obtains for each nucleus :

$$<oM_SM_I|\hat{H}_{hfc}(\lambda)|oM_S'M_I'> = (8\pi/3)g_0\beta_e\gamma_\lambda\hbar\sum_m(-1)^m<M_I|\hat{I}_1^{-m}(\lambda)|M_I'> \int_{\underline{x}_1'=\underline{x}_1} \delta(r_{\lambda 1})\hat{S}_1^m(1)\rho_1(KL|\underline{x}_1\;;\underline{x}_1')d\underline{x}_1 = (8\pi/3)g_0\beta_e\gamma_\lambda\hbar\sum_m(-1)^m <M_I|\hat{I}_1^{-m}(\lambda)|M_I'>C_1(SS,\;M'M)\overline{Q}_S(SS|\underline{r}_\lambda\;;\underline{r}_\lambda)_1^o.$$

If one then uses Eqn. (70d) and the following relation, deduced from Eqns. (75) and (37) :

$$C_1(SS, M'M) = C(S1S, M'mM)/C(S1S, SOS) = \langle SM|\hat{S}_1^m|SM'\rangle/\langle SS|\hat{S}_1^o|SS\rangle$$
$$= \langle M_S|\hat{S}_1^m|M_S'\rangle/S, \quad (79)$$

and introduces the hyperfine-coupling constant :

$$a_\lambda = (8\pi/3)g_0\beta_e\gamma_\lambda\hbar D_S(S|\underline{r}_\lambda), \quad (80')$$

one obtains :

$$\langle oM_SM_I|\hat{H}_{hfc}(\lambda)|oM_S'M_I'\rangle = a_\lambda\sum_m(-1)^m\langle M_I|\hat{I}_1^{-m}(\lambda)|M_I'\rangle\langle M_S|\hat{S}_1^m|M_S'\rangle =$$
$$\langle M_SM_I|a_\lambda\sum_m(-1)^m\hat{I}_1^{-m}(\lambda)\hat{S}_1^m|M_S'M_I'\rangle ;$$

whence the spin-Hamiltonian term :

$$H'_{hfc}(\lambda) = a_\lambda\underline{\hat{S}}.\underline{\hat{I}}_\lambda, \quad (80)$$

whose coupling constant, given by Eqn. (80'), is proportional to the magnetogyric ratio $\gamma_\lambda$ and to the spin-density value at the nucleus ( induced by "open-shell" and "spin-polarized" s-electrons). Unlike the isotope/isomer-shift term, this term will combine with the dipolar hyperfine coupling to contribute to the splitting of the multiplet level, and will in fact be responsible for the only splitting observed in liquids where rapid tumbling motions of the molecules average out the dipolar-coupling tensor(24).

d) Dipolar Hyperfine Couplings. The dipolar terms of Eqn. (27) can be written for each nucleus, cf. Eqn. (36) :

$$\hat{H}_{hfd}(\lambda) = -g_0\beta_e\gamma_\lambda\hbar\sum_i\sum_{m=-2}^{+2}\hat{S}_2^m(i,\lambda)\mathcal{D}_2^{-m}(i,\lambda)(-1)^m,$$

where $\hat{S}_2^m(i,\lambda)$ is constructed from $\underline{\hat{s}}_i x\underline{\hat{I}}_\lambda$ and $\mathcal{D}_2^{-m}(i,\lambda)$ from $\underline{r}_{\lambda i}x\underline{r}_{\lambda i}$, e.g. :

$$\hat{S}_2^o = (\hat{S}_1^{-1}\hat{I}_1^{+1} + 2\hat{S}_1^o\hat{I}_1^o + \hat{S}_1^{+1}\hat{I}_1^{-1}),\ \mathcal{D}_2^o = (3z^2 - r^2)/r^5,$$

etc.(11d). Proceeding as in §c and introducing a Cartesian frame, one obtains the spin-Hamiltonian term :

$$H'_{hfd}(\lambda) = \underline{\hat{S}}.\underline{\underline{b}}_\lambda.\underline{\hat{I}}_\lambda, \quad (81)$$

$$\underline{\underline{b}}_\lambda = -g_0\beta_e\gamma_\lambda\hbar\int\underline{\underline{\mathcal{D}}}(\underline{r}_{\lambda 1})D_S(S|\underline{r}_1)d\underline{r}_1 = g_0\beta_e\gamma_\lambda\hbar\underline{\underline{d}}_\lambda. \quad (81')$$

For instance, substituting the terms of Eqn. (27) involving the z components of $\underline{\hat{s}}_i$ and $\underline{\hat{I}}_\lambda$ into Eqn.(76') and using as in §c Eqns. (70a, c and d) and (79), one obtains in a straightforward manner :

$$\langle oM_SM_I|-\sum_i\hat{s}_i^z(r_{\lambda i}^2 - 3z_{\lambda i}^2)/r_{\lambda i}^5\hat{I}_\lambda^z|oM_S'M_I'\rangle = \langle M_I|\hat{I}_1^o(\lambda)|M_I'\rangle\ x$$

$$<M_S|\hat{S}_1^o|M_S'> \int \{(3z_{\lambda 1}^2 - r_{\lambda 1}^2)/r_{\lambda 1}^5\} D_S(S|\underline{r}_1)d\underline{r}_1 = <M_S M_I|d_\lambda^{zz}\hat{S}^z\hat{I}_\lambda^z|M_S'M_I'>.$$

The coupling tensor given by Eqn. (81') is proportional to the magnetogyric ratio $\gamma_\lambda$ and to the spin-density-averaged dipolar-coupling tensor $\underline{\underline{\mathcal{D}}}(\underline{r}_{\lambda 1})$ (which involves "open-shell" and "spin-polarized" p, d or f-electrons). For nearly axial couplings, one may again define axial and asymmetry constants, $b_\lambda$ and $c_\lambda$, which, together with the isotropic constant $a_\lambda$, completely determine the principal values of the resulting tensor $\underline{\underline{c}}_\lambda$ :

$$\underline{\underline{c}}_\lambda = a_\lambda \underline{\underline{u}} + \underline{\underline{b}}_\lambda. \tag{82}$$

Besides the Mössbauer effect when appropriate(23), such tensors are currently measured by EPR(25) or, when relaxation times permit, NMR (26), ENDOR or ELDOR(27).

Summarizing §§a - d, we have met averages of : (i) different operators (contact and dipolar) over the same density function (charge or spin), and (ii) the same operator over different density functions. The use of appropriate nuclear probes can thus provide detailed information on local values and distribution shapes of charge and spin densities. The interpretation of the measured constants in terms of electronic structure and molecular configuration is, in some cases, made easier by the use of simple relations, the most famous of which relate, in π-electron hydrocarbon radicals : (i) the hf coupling tensor of an α-proton to the spin-density population on the adjacent carbon-atom $2p^\pi$ orbital(28a), and (ii) the hf coupling constant of a β-proton to the previous population and to the local torsional angle(28b). We shall now turn to the last first-order spin-Hamiltonian term, which corresponds to a non-local property.

e) Electron Spin-Spin Coupling. The treatment of the contact and dipolar terms of Eqn. (30) is similar to that of the corresponding terms of Eqn. (27), except that now one has to use Eqns. (72a, c and d) and the following relation, deduced in a similar way as Eqn. (79) :

$$C_2(SS, M'M) = <SM|\hat{S}_2^m|SM'>/<SS|\hat{S}_2^o|SS> = <M_S|\hat{S}_2^m|M_S'>/S(2S-1). \tag{83}$$

One obtains :

$$H_{SS}' = \underline{\hat{S}}.\underline{\underline{\mathcal{D}}}.\underline{\hat{S}}, \tag{84}$$

$$\underline{\underline{\mathcal{D}}} = (g_0^2\beta_e^2/2) \int \{\underline{\underline{\mathcal{D}}}(\underline{r}_{21}) - (8\pi/3)\delta(r_{21})\} D_{SS}(S|\underline{r}_1, \underline{r}_2)d\underline{r}_1 d\underline{r}_2. \tag{84'}$$

The δ-term produces a uniform shift of the levels which, due to the Pauli principle, would be exactly zero if the δ-approximation were rigorous, cf. Eqn. (9) and Ref. (2). The remaining dipolar-coupling term can be written, in the principal-axis system of $\underline{\underline{\mathcal{D}}}$ :

$$\mathcal{H}'_{SS} = X(\hat{S}^x)^2 + Y(\hat{S}^y)^2 + Z(\hat{S}^z)^2 \sim D(\hat{S}^z)^2 + E\{(\hat{S}^x)^2 - (\hat{S}^y)^2\}, \quad (84'')$$

where we have introduced the axial and asymmetry constants D = 3Z/2 and E = (X - Y)/2, used the fact that X + Y + Z = 0 and dropped a term $-D\,\hat{S}^2/3$ which shifts uniformly the levels. The constants D and E, which can provide gross information on the <u>spin correlation density</u> in paramagnetic systems with spin S ≥ 1, are currently measured by EPR in solids[25].

f) <u>Second-Order Effects</u>. Such effects can be defined by substituting any one group of terms of $\hat{\mathcal{H}}_M$ into the first matrix elements of Eqn. (76") and any other - or the same - group of terms into the second matrix elements, provided that some of the products in the sum take on significant values. The spin-Hamiltonian terms thus obtained add up to the first-order terms of similar form, though different origin, described above and may in some cases outweigh them. For instance, in the case of paramagnetic ions in crystals considered in §2.2, the effective spin-spin coupling $\hat{\mathcal{H}}''_{SS}$ arising from the combination $(\lambda\underline{L}.\underline{S}) \times (\lambda\underline{L}.\underline{S})$ dominates the dipolar spin-spin coupling $\mathcal{H}'_{SS}$ just described. In high-resolution NMR experiments, where $\mathcal{H}_{nd}$ is averaged out by fast molecular reorientations or selective-averaging procedures, only second-order effects are observed, which arise principally from the combinations[11b] $\mathcal{H}_{hfo}(\lambda) \times \hat{\mathcal{H}}_{eZo}$ ($\underline{\underline{\sigma}}_\lambda$ shifts) and $\hat{\mathcal{H}}_{hfc}(\mu) \times \hat{\mathcal{H}}_{hfc}(\nu)$ ($J_{\nu\mu}$ splittings). The following enumeration summarizes the second-order effects originating from the various terms of Eqns. (27-29).

$\hat{\mathcal{H}}_{hf} \times \hat{\mathcal{H}}_{hf}$ : nuclear pseudo-quadrupolar and internuclear pseudo-dipolar splittings (non-traceless and non-symmetric contributions to $\underline{\underline{Q}}_\lambda$ and $\underline{\underline{\mathcal{D}}}_{\nu\mu}$, particularly $J_{\nu\mu}$).

$\hat{\mathcal{H}}_{hf} \times \hat{\mathcal{H}}_{eZ}$ : nuclear pseudo-Zeeman, "chemical" and "paramagnetic" shifts (non-traceless and generally non-symmetric $\underline{\underline{\sigma}}_\lambda$ tensors).

$\hat{\mathcal{H}}_{hf} \times \hat{\mathcal{H}}_{SO}$ : pseudo-hyperfine couplings[29].

$\hat{\mathcal{H}}_{eZ} \times \hat{\mathcal{H}}_{eZ}$ : second-order diamagnetism.

$\hat{\mathcal{H}}_{eZ} \times \hat{\mathcal{H}}_{SO}$ : electronic pseudo-Zeeman, "$\underline{\underline{g}}$-tensor" shift[30].

$\hat{\mathcal{H}}_{SO} \times \hat{\mathcal{H}}_{SO}$ : electronic pseudo-dipolar coupling (the generally non-zero trace of the generally symmetric $\underline{\underline{\mathcal{D}}}$ contribution uniformly shifts the levels).

As the complicated expressions of these terms as functions of the transition matrices $P_1$, $Q_S$ and $Q_{SO}$ with the closest excited states of appropriate symmetry cannot yield direct information on the density functions of interest, we shall refer to the literature for them(21b).

In the previous derivations, we have made use of first and second-order density and transition functions and matrices involving only electron coordinates : the nuclei, implicitly assumed to be fixed in space, entered the total wave-functions as single spin-function factors. If now one takes into account the fourth term of Eqn. (24) in the determination of the eigenkets of $\hat{H}_0$, the symbols o and ν in Eqns. (76'’") will stand for quantum numbers relative to both electrons and nuclei and thus refer to different vibronic states. Since the corresponding wave-functions depend upon electronic and nuclear coordinates alike, the density matrices defined in Eqns. (48-51,61-62,64-65,70-72) will contain all nuclear coordinates : $\underline{X} \equiv \{\underline{X}_1, \underline{X}_2, \ldots, \underline{X}_n\}$ with $\underline{X}_\lambda \equiv \{\underline{R}_\lambda, I_\lambda\}$, as variables, e.g. : $\underline{P}_1(\underline{r}_1 : \underline{X})$, $\mathcal{D}_S(S|\underline{r}_1 : \underline{X})$, $\mathcal{D}_{SS}(S|\underline{r}_1, \underline{r}_2 : \underline{X})$. The derivation of effective spin-Hamiltonian terms from Eqns. (76'’"), using the factorization of nuclear spin functions as above, consequently involves an additional integration over $\underline{R} \equiv \{\underline{R}_1, \underline{R}_2, \ldots, \underline{R}_n\}$ in the subsequent expressions, e.g. Eqns. (77,78',80',81',84'), while Eqn. (32) gives rise to :

$$H'_{nd} = \sum_{\mu<\nu}\sum \gamma_\mu \gamma_\nu \hbar^2 \int_{\underline{X}'=\underline{X}} \{\hat{\underline{I}}_\mu \cdot \underline{\underline{\mathcal{D}}}(\underline{r}_{\nu\mu}) \cdot \hat{\underline{I}}_\nu\} \rho_0(KL|\underline{X} ; \underline{X}')d\underline{X},$$

$$\rho_0(KL|\underline{X} ; \underline{X}') = \int_{\underline{x}} \Psi_{oM_SM_I}(\underline{x} : \underline{X}) \Psi^*_{oM'_SM'_I}(\underline{x} : \underline{X}')d\underline{x}.$$

If there is a temperature-dependent, statistical mixture of vibronic states, one has to consider a thermal average of the previous expectation values. We may also mention for the record that additional terms in $H'_S$ may arise from $\hat{H}_{hfo}(\lambda)$, $\hat{H}_{eZo}$ and $\hat{H}_{SO}$ when there is an orbital degeneracy of electronic states, e.g., with free atoms or linear molecules.

Gathering the various terms of the effective spin Hamiltonian, one can write, in the general standard case :

$$H_S = g_0\beta_e \hat{\underline{S}} \cdot (\underline{\underline{u}} + \underline{\underline{G}}) \cdot \underline{B}_Z + \hat{\underline{S}} \cdot \underline{\underline{\mathcal{D}}} \cdot \hat{\underline{S}} + \sum_\lambda \hat{\underline{S}} \cdot (a_\lambda \underline{\underline{u}} + \underline{\underline{b}}_\lambda) \cdot \hat{\underline{I}}_\lambda - \sum_\lambda \gamma_\lambda \hbar \hat{\underline{I}}_\lambda \cdot (\underline{\underline{u}} - \underline{\underline{\sigma}}_\lambda) \cdot \underline{B}_Z + \sum_\lambda \hat{\underline{I}}_\lambda \cdot (P_\lambda \underline{\underline{u}} + \underline{\underline{Q}}_\lambda) \cdot \hat{\underline{I}}_\lambda + \sum_{\mu<\nu}\sum \hat{\underline{I}}_\mu \cdot (J_{\nu\mu} \underline{\underline{u}} + \underline{\underline{\mathcal{D}}}_{\nu\mu}) \cdot \hat{\underline{I}}_\nu - \underline{B}_Z \cdot \underline{\underline{\chi}} \cdot \underline{B}_Z. \qquad (85)$$

Here, terms linear in $\hat{\underline{S}}$ or $\hat{\underline{I}}_\lambda$ occur only if $\hat{\underline{S}}$ or $\hat{\underline{I}}_\lambda \geq 1/2$, and terms quadratic in $\hat{\underline{S}}$ or $\hat{\underline{I}}_\lambda$ only if $\hat{\underline{S}}$ or $\hat{\underline{I}}_\lambda \geq 1$, respectively. The terms are written in the order of decreasing magnitude in which they appear in most magnetic-resonance experiments. Most of these terms contain an isotropic and an anisotropic part, both of which may involve a first and a second-order contribution : for instance, the traces of the tensors $\underline{\underline{G}}$ and $\underline{\underline{\sigma}}_\lambda$, which relate to second-order contributions as we have seen, show up by shifting the free-electron or nucleus value $g_0$ or $\gamma_\lambda$ ; on the other hand, the anisotropic part of the second-order contribution $\underline{\underline{J}}_{\nu\mu}$ has been incorporated into $\underline{\underline{\mathcal{D}}}_{\nu\mu}$. The isotropic part of the term quadratic in $\underline{S}$, which is mostly from second-order origin, is not observable, while that of the term quadratic in $\hat{\underline{I}}_\lambda$, $P_\lambda = (2\pi e^2/3)Z_\lambda R_\lambda^2 P_1(\underline{r}_\lambda)/I_\lambda(I_\lambda + 1)$, can be measured for instance by Mössbauer spectroscopy. Finally, both the isotropic and anisotropic parts of the term quadratic in $\underline{B}_Z$ can be obtained through magnetic-polarization measurements.

## 4. MEASUREMENT TECHNIQUES

This section is not intended to provide even a gross account of the various instrumental devices or numerous application fields of the magnetic resonance and related techniques. Several textbooks and monographs are available on these matters, a few of them being quoted in the bibliography(23-27,31-33). The aim of this section is simply to outline the very basic principles of the most standard techniques, in order to compare their respective application fields as regard to the determination of the various structural parameters introduced in the preceding section. A simple example(24) will help us understand EPR, NMR, ENDOR and ELDOR, which are then considered in the more general paramagnetic systems studied, NQR and MS being introduced separately. Comparison of the possibilities and limitations of these techniques will be made, with some typical examples.

### 4.1. The Hydrogen Atom in its Ground State

For this one-proton, one-electron, spherical system with no nuclear motion and no orbital degeneracy, Eqn. (85) reduces to :

$$\mathcal{H}_S = g\beta_e\hat{\underline{S}}.\underline{B}_Z + a\hat{\underline{S}}.\hat{\underline{I}} - \gamma\hbar\hat{\underline{I}}.\underline{B}_Z = g\beta_e B_Z\hat{S}^z - \gamma\hbar B_Z\hat{I}^z + a\{\hat{S}^z\hat{I}^z + (\hat{S}^+\hat{I}^- + \hat{S}^-\hat{I}^+)/2\}, \quad (86)$$

where $g \simeq g_0$, $\gamma \simeq \gamma_p$, and the diamagnetic shifting term has been omitted. If $|M_S\rangle$ and $|M_I\rangle$ ($M_S$, $M_I = \pm 1/2$) are the eigenkets of $\hat{S}^z$ and $\hat{I}^z$, respectively, with z directed along $\underline{B}_Z$, the ket products $|M_S\rangle|M_I\rangle \equiv |M_SM_I\rangle$ can be used as a complete orthonormal basis set for writing Eqn. (86) in matrix form : the resulting matrix is quasi-diagonal, only the kets $|+-\rangle$ and $|-+\rangle$ being coupled through the weak cross-terms $(a/2)(\hat{S}^+\hat{I}^- + \hat{S}^-\hat{I}^+)$. The eigenvalue equation $\mathcal{H}_S|n\rangle = E_n|n\rangle$ then yields the solutions in closed form :

$$E_{1,4} = \pm (Z_e - Z_p)/2 + a/4, \quad (87a)$$

$$E_{2,3} = - a/4 \pm \{(Z_e + Z_p)^2 + a^2\}^{1/2}/2, \quad (87b)$$

where $Z_e \equiv g\beta_e B_Z >> Z_p \equiv \gamma\hbar B_Z$ since $m_e << M_p$. For $B_Z = 0$, $E_{1,2,4} =$ $a/4$ and $E_3 = - 3a/4$ ; for $(Z_e + Z_p) >> a$ (strong-field case):

$$E_{2,3} = \pm (Z_e + Z_p)/2 - a/4. \quad (87b')$$

The orthonormal eigenkets corresponding to the degenerate, zero-field case are : $|1\rangle = |{+}{+}\rangle$, $|2\rangle = (|{+}{-}\rangle + |{-}{+}\rangle)/\sqrt{2}$, $|4\rangle = |{-}{-}\rangle$ and $|3\rangle = (|{+}{-}\rangle - |{-}{+}\rangle)/\sqrt{2}$ ; the zeroth-order eigenkets corresponding to the first-order energies given in Eqns. (87a,b') are : $|1\rangle = |{+}{+}\rangle$, $|2\rangle = |{+}{-}\rangle$, $|3\rangle = |{-}{+}\rangle$, $|4\rangle = |{-}{-}\rangle$. The "branching diagram" ($E_n/h$ versus $B_Z$) for the hydrogen atom (where $a/h = 1420.4$ Mc/s) is presented in Fig. (1a) (there, $\alpha$ stands for $|+\rangle$ and $\beta$ stands for $|-\rangle$).

*Fig. 1 - The hydrogen atom in its ground state. (a) Branching diagram. (b) Strong-field levels and transitions. (c) An idealized EPR spectrum.*

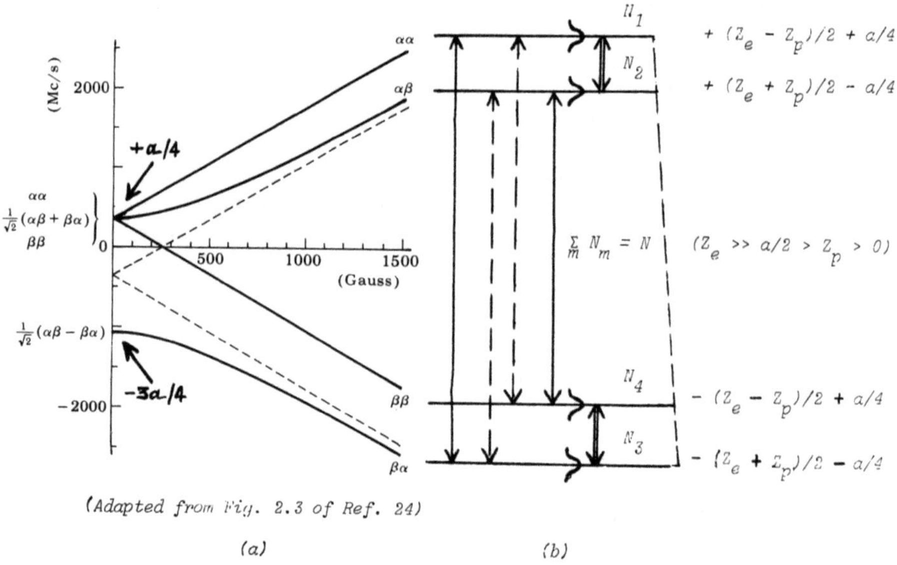


(c)
(Adapted from Fig. 11a of : K.A. McLAUCHLAN, "Magnetic Resonance" (Clarendon Press, Oxford, 1972))

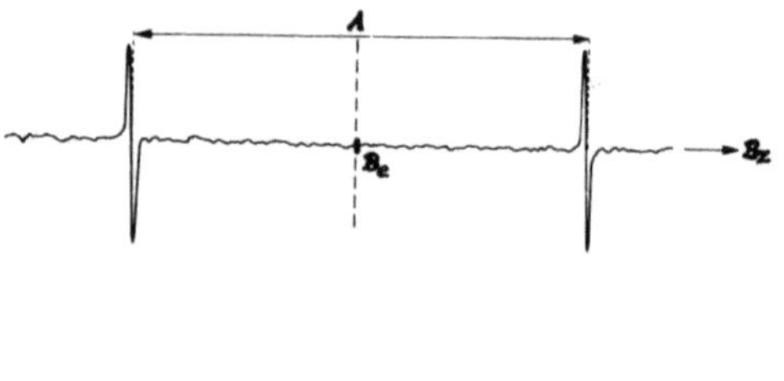

If there is a large number of systems described by Eqn. (86) with fixed values of $B_Z$, g, $\gamma$ and a across the sample, and if these systems interact weakly with each other and with a lattice reservoir at temperature $T_L$, then the stationary populations in the levels will be given, at not too low a temperature, by the Boltzmann distribution law :

$$p_n \equiv N_n/N = \exp.(-E_n/kT_L)/\sum_m \exp.(-E_m/kT_L) \simeq (1 - E_n/kT_L)/Z, \tag{88}$$

where N and Z are the total numbers of spins and states, respectively. These populations are represented in Fig. (1b) by the lengths of horizontal bars drawn at the locations of the energy levels : they are shown to increase slightly towards lower levels because in usual magnetic-resonance circumstances $kT_L >> E_n > 0$. The very mechanisms which help maintain a uniform temperature throughout the sample also induce finite lifetimes of the systems in the levels and, consequently, a so-called homogeneous broadening of the levels. An additional broadening, of inhomogenous nature, may also arise from a dispersion across the sample of the values of $B_Z$, g, $\gamma$ and a and, if the system were anisotropic, of the orientations of its characteristic tensors. The resulting overall broadenings are represented in Fig. (1b) by the widths of bell-shaped curves drawn across the level bars. As a matter of fact, the hyperfine splittings themselves may be considered as a resolved form of inhomogeneous broadening.

Magnetic dipolar transitions can be induced between the above-described levels by a radiofrequency or microwave magnetic field $B_1 e^{i\omega t}\underline{u}$, provided that the angular frequency $\omega$ and polarization vector $\underline{u}$ satisfy the two resonance conditions :

$$\hbar\omega = E_m - E_n, \tag{89}$$

$$w^M_{mn} \equiv |\langle m|\hat{\underline{\mu}}.\underline{u}|n\rangle|^2 \neq 0, \tag{90}$$

where $\hat{\underline{\mu}}$ is the total magnetic moment of the system. Eqn. (89) can be achieved by sweeping either $\omega$ or, which is technically easier, the static magnetic field $B_Z$ upon which $E_m$ and $E_n$ depend[31a]. Even though the probabilities $W^M_{mn}$ and $W^M_{nm}$ for induced absorption and emission are equal, there is a net absorbed power due to the difference in the populations of states $|m\rangle$ and $|n\rangle$, i.e. :

$$P_{mn} = \hbar\omega(N_n - N_m)W^M_{mn}. \tag{91}$$

With a standard magnetic-resonance spectrometer[31a], the absorption signal $V_{mn}$ for a transition $m \leftarrow n$ is approximately given by the following product :

$$V_{mn} = KS_{mn}N_{mn}(\omega_{mn}/2kT_L)w^M_{mn}\pi g_{mn}(\omega - \omega_{mn})B_1, \tag{92}$$

where K is a constant depending on the units and on the instrument parameters ; $S_{mn}$ is defined as $(N_n - N_m)/(N_n^o - N_m^o)$, where $N_n^o$ and $N_m^o$ are the values taken by $N_n$ and $N_m$ when the spins are in thermal equilibrium with the lattice, and depends upon the relative efficiencies of those mechanisms disturbing and maintaining the equilibrium, whose respective probabilities we write : $W_{jk}^M = W_{kj}^M$ and $W_{jk}^L = W_{kj}^L.\exp.(\Delta E_{jk}/kT_L)$ ; $N_{mn} = N_m + N_n$ ; $\omega_{mn}$ is a solution of Eqn. (89) in angular-frequency units ; $w_{mn}^M$ is defined in Eqn. (90), and $g_{mn}(\omega - \omega_{mn})$ is the density of transitions between states $|m\rangle$ and $|n\rangle$ at angular frequency $\omega$. When the oscillating-field amplitude $B_1$ increases, the signal intensity first increases as $B_1$ and then decreases together with the saturation factor $S_{mn}$. Before saturation, its shape is defined by the function $g_{mn}$, which is a convolution of the broadening shapes of the two levels. The other factors explain why NMR requires much larger numbers of spins than EPR (i.e., both larger samples and concentrations, since $N_{mn}$ has to make up for both $\omega_{mn}$ and $w_{mn}^M$).

In principle, 6 transitions could be induced between the 4 levels of Fig. (1b), whose energies are represented by the lengths of vertical arrows drawn between the level bars, i.e. :

$$\Delta E_{13} = Z_e + a/2,\ \Delta E_{24} = Z_e - a/2\ (\Delta M_S = +1,\ \Delta M_I = 0), \quad (93a)$$

$$\Delta E_{23} = Z_e + Z_p,\ \ \Delta E_{14} = Z_e - Z_p\ (\Delta M_S = \pm 1,\ \Delta M_I = \mp 1), \quad (93b)$$

$$\Delta E_{12} = |a/2 - Z_p|,\ \ \Delta E_{43} = |a/2 + Z_p|\ (\Delta M_S = 0,\ \Delta M_I = \pm 1). \quad (93c)$$

As $Z_e \gg Z_p$ and a/2, the first four transitions would occur at frequencies much higher than the last two. Consequently, different technical devices are required to produce, direct, and measure the electromagnetic waves in the two ranges. The order of magnitude, $Z_e$ or $Z_p$, of the transition energy determines the type of spectrometer, EPR or NMR, to be used. The maxima of the transition probability factors : $w_{mn}^M = |\langle M_S M_I|g\beta_e \hat{S}_u - \gamma\hbar\hat{I}_u|M_S' M_I'\rangle|^2$, are attained by using an oscillating field perpendicular to the static field : $\underline{u}(\underline{x}) \perp \underline{z}$. They are readily calculated to be :

$$w_{13}^M = w_{24}^M = g^2\beta_e^2/4, \quad (94a)$$

$$w_{23}^M = w_{14}^M = 0, \quad (94b)$$

$$w_{12}^M = w_{43}^M = \gamma^2\hbar^2/4. \quad (94c)$$

As a result of these formulae, only two EPR transitions, those represented by full arrows in Fig. (1b), are observed in the strong-field case. In a fixed-frequency, swept-field experiment, the corresponding resonance fields are obtained by combining Eqns. (93a) and (89), i.e. :

$$B_{13} = (\hbar\omega - a/2)/g\beta_e,\ B_{24} = (\hbar\omega + a/2)/g\beta_e. \tag{93a'}$$

Fig. (1c) shows the strong-field EPR spectrum of atomic hydrogen in a solid matrix : it displays two lines centered at $B_e \equiv \hbar\omega/g\beta_e$ with a splitting $A \equiv |a|/g\beta_e$. Most EPR spectrometers use static-field modulation and phase-sensitive detection to improve the signal-to-noise ratio, which results in a derivative of the absorption signal as the output(31a). In zero field, as can be seen from Fig. (1a), a strong single line would be observed, at frequency a/h. The values measured for a/h and g vary between 1401 and 1437 MHz and 1.9997 and 2.0066, respectively, depending on the trapping matrix(31d). Accurate determinations made in gaseous samples have given : a/h = 1420.405749 MHz and g = 2.002296, while use of Eqn. (80') with $\mathcal{D}_S(S|\underline{r}_\lambda) = |1s_H(0)|^2$ gives a slightly larger value : a/h = 1422.8 MHz, due to the various approximations involved in this formula.

It has been recalled earlier that NMR requires much larger concentrations than EPR, the $w^M_{mn}$'s given by Eqns. (94a,c) being much weaker for NMR than for EPR. Under such circumstances, the frequency of the electron-spin flip-flops induced by the exchange interactions between the unpaired electrons grows much larger than the splitting of the NMR doublet given by Eqn. (93c), that is : either a/h or $2Z_p/h$, whichever is smaller. Then it can be shown(26) that the doublet lines coalesce into a broad single line centered at the statistical average of their positions, i.e. :

$$\hbar\omega = P_{12}\Delta E_{12} + P_{34}\Delta E_{43}. \tag{95}$$

Here the $P_{mn}$'s are obtained by substituting Eqns. (87a,b') into Eqn. (88) :

$$P_{12} \equiv p_1 + p_2 \simeq (1 - Z_e/2kT_L)/2, \tag{96a}$$

$$P_{34} \equiv p_3 + p_4 \simeq (1 + Z_e/2kT_L)/2. \tag{96b}$$

For the hydrogen atom, $a/2 \gg Z_p$ at the magnetic fields commonly used in magnetic resonance. However, the high reactivity of atomic hydrogen prevents the required concentrations for NMR to be reached. Moreover, the broadening of NMR lines, which increases as $a^2$ due to specific spin-relaxation processes, may make them undetectable unless $a/2 \ll Z_p$(26). Both requirements of high concentration and small hyperfine coupling are usually fulfilled by the protons of large conjugated radicals and paramagnetic metal complexes. Then Eqn. (95) gives, by substitution of Eqns. (93c) and (96a,b) :

$$\hbar\omega = Z_p + aZ_e/4kT_L \quad (a/2 \leqslant Z_p). \tag{97}$$

In a fixed-frequency, swept-field experiment, one obtains, using $Z_e/2 \ll kT_L$ and introducing $B_p \equiv \hbar\omega/\gamma\hbar$, the corresponding resonance field :

$$B_r = B_p\ (1 - g\beta_e a/4\gamma\hbar kT_L). \qquad (97')$$

The so-called paramagnetic shift : $\Delta B_r \equiv B_r - B_p$, is proportional to $-aB_p/T_L$, while the diamagnetic or chemical shift $\sigma B_p$ does not depend directly on $T_L$. Typically, for $g \simeq g_0$, $\gamma \simeq \gamma_p$, $a/h \simeq$ 3MHz, $B_p \simeq$ 3300G and $T_L \simeq$ 300K, $\Delta B_r \simeq -0.26$G. Comparison of **Eqns.** (93a') and (97') shows that NMR can provide additional information over EPR, that is, the absolute sign of the hyperfine coupling a and hence, through Eqn. (80'), of the spin density $\mathcal{D}_S$.

The levels described in Fig. (1b) will finally help us introduce the double-resonance techniques[27]. Most generally, a double-resonance experiment consists of the observation of the variation of intensity of one transition while saturating another transition. Such an effect can be observed because the saturation factor $S_{mn}$ in Eqn. (92) can be affected not only by a decrease of the difference $N_n - N_m$ induced by an increase of the field $B_1$ at angular frequency $\omega_{mn}$, but also by independent variations of $N_n$ and $N_m$ induced by the interplay of pumping and relaxation processes at other angular frequencies. If the observing frequency is in the EPR range and the pumping frequency in the NMR range {4 x 2 possibilities on Fig. (1b)}, one is performing electron-nuclear double resonance (ENDOR). If both the observing and pumping frequencies are in the EPR range {4 x 3 possibilities on Fig. (1b)}, one is performing electron-electron double resonance (ELDOR). Because of the exchange-induced electron-spin flip-flops occurring at the large concentrations required for NMR, nuclear-nuclear double resonance (INDOR) cannot be performed in practice with paramagnetic systems.

To perform an ENDOR experiment, one first positions the magnetic field on one of the EPR lines observed with a partly-saturating microwave power having frequency $\omega_o$, then sweeps a pumping radiofrequency $\omega_p$ over the expected range for NMR transitions while scanning the recorder x-axis. The ENDOR response consists of the enhancement (sometimes reduction) of the EPR signal occurring when an NMR transition gets saturated. With the energy levels of Fig. (1b), ENDOR lines occur, according to Eqns. (89), (93c) and (93a'), at frequencies given by :

$$\hbar\omega_p = |a/2 \mp Z_p|\ (Z_p = \gamma\hbar(\hbar\omega_o \mp a/2)/g\beta_e). \qquad (98)$$

For $|a|/2 >> |Z_p|$, there appears two ENDOR lines centered at $|a|/2\hbar$, with a splitting of $2|Z_p|/\hbar$ (which is slightly larger if the higher-field EPR line is monitored). For $|a|/2 << |Z_p|$, there appears two ENDOR lines centered at $|Z_p|/\hbar$ (which is slightly shifted to larger frequency if the higher-field EPR line is monitored), with a splitting of $|a|/\hbar$. If $|a|/2 = |Z_p|$, there appears only one ENDOR line, at the position $|a|/2 + |Z_p|$. It is seen that ENDOR can provide additional information over EPR, that is, the effective magnitude of the magnetogyric ratio $\gamma$, through $|Z_p|$.

To perform an ELDOR experiment, one usually fixes the observing microwave frequency $\omega_o$ and sweeps either, as in ENDOR, the pumping microwave frequency $\omega_p$ (above or below $\omega_o$ depending on whether the static magnetic field $B_z$ is positioned on a high or low-field EPR line) or, as in EPR, the field $B_z$ ($\omega_p$ being set to a predetermined value above or below $\omega_o$). The ELDOR response consists of the reduction (sometimes enhancement) of the observed EPR-line intensity occurring when another EPR transition gets saturated. With the energy levels of Fig. (1b), an "allowed-allowed" ELDOR line occurs when the difference $\hbar|\omega_p - \omega_o|$ equals the difference of the two energy spacings given by Eqn. (93a), i.e. :

$$\hbar|\omega_p - \omega_o| = |a|. \tag{99}$$

Forbidden-allowed, allowed-forbidden and forbidden-forbidden ELDOR lines may also be observed, at $\hbar|\omega_p - \omega_o| = |a/2 \mp Z_p|$ and $2|Z_p|$.

Both ENDOR and ELDOR signals usually appear as a derivative of the EPR absorption variation, a proper magnetic-field or microwave-frequency modulation being used to improve the signal-to-noise ratio[27]. However, contrary to both EPR and NMR, ENDOR and ELDOR intensities do not depend in a simple way either on the number of spins or degeneracy of the lines, because they are intimately related to the relative efficiencies of the spin-relaxation processes which determine $S_{mn}$ ; in particular, corresponding low and high-field/frequency resonance lines need not have equal intensities. The intensity anomalies, if properly interpreted, can provide additional information on the relaxation processes involved[27]. The main advantage of both ENDOR and ELDOR in measuring hyperfine couplings (preferably small for the former and large for the latter) on systems more complex than H· is in their improved resolution and accuracy over EPR, accompanied by a much better sensitivity than in the case of NMR[27].

### 4.2. Polynuclear and Anisotropic Paramagnetic Systems the Case of Diphenylpicrylhydrazyl

The considerations developed in the preceding section are immediately applicable to all mononuclear and isotropic paramagnetic systems with effective electronic and nuclear spins $S = 1/2$ and $I = 1/2$, changing only the values of g, $\gamma$ and a for a many-electron cloud and many-nucleon nucleus. The expression doublet state will be used for any neutral or charged atomic or molecular system with spin $S = 1/2$, including trapped electrons and holes, stable and transient free radicals, transition-metal and rare-earth complexes, etc.. Examples of nuclei with spin $I = 1/2$ include, in addition to the proton $^1H$, the naturally present $^{13}C$, $^{15}N$, $^{19}F$, $^{29}Si$ and $^{31}P$. The terms mononuclear and isotropic must be understood in an effective sense. The first one means that there is only one magnetic nucleus interacting with the electronic distribution, even though the system may also contain nuclei with zero spin, such as $^{12}C$, $^{16}O$, $^{28}Si$ and $^{32}S$. The second term means that if the system has no spherical symmetry by itself, like a free atom, it shows an effective isotropy which may be induced, for instance, by random tumbling motions at a characteristic rate considerably larger than the greatest anisotropic contributions to the spin Hamiltonian expressed in frequency units : $\tau_c^{-1} >> \Delta\nu_{anis.}^{max.}$ (24). These latter motions can be an important source of spin-lattice relaxation and homogeneous line-broadening in the liquid and gas phases. The expression triplet state will be used for a system in its ground or in an excited state with spin S = 1, like the phosphorescent state of an aromatic molecule. In such a case, the analogue of Fig. (1b) will display three sets of strong-field levels instead of two, one for each value of $M_S$ : -1, 0, +1. When the anisotropic part of the term quadratic in $\hat{\underline{S}}$ in Eqn. (85) is averaged out by rapid random reorientations in a fluid isotropic medium, these three sets of levels are equidistant and, as a result, a single distinct set of components can be observed by EPR, with the selection rule $\Delta M_S = 1$, the resonance fields being again given by Eqn. (93a'). Similarly, it can be shown that when the characteristic rate for electron-spin flip-flops is much larger than the hyperfine splitting : $T_e^{-1} >> a/h$, a single line can be observed by NMR, resulting from a statistical average on the $M_S$ states, the resonance field being given by Eqn. (97') where 1/4 has to be replaced by S(S + 1)/3. It can also be shown that now ENDOR lines occur at the 2S + 1 positions given by : $\hbar\omega_p = |Z_p + M_S a|$, instead of Eqn. (98), whereas the principal ELDOR line is still given by Eqn. (99). If now the interacting nucleus has spin I = 1, as is the case with $^2D$, $^6Li$ or $^{14}N$, each set of strong-field levels in the analogue of Fig. (1b) displays three levels instead of two, one for each value of $M_I$ : -1, 0, +1. When the anisotropic part of the term quadratic in $\hat{\underline{I}}$ in Eqn. (85) is averaged out by rapid random reorientations in a fluid isotropic medium, these three levels are equidistant, and the three distinct components which can be observed by EPR, one for each value of $M_I$ with the selection rule $\Delta M_I = 0$, are given by Eqn. (93a') where $\mp$ 1/2 has to be replaced by the 2I + 1 values of $M_I$. The expression for the single NMR-line position remains unchanged. The same holds true for the principal ELDOR-line position, but the above substitution has to be made in the expression of $Z_p$ which enters the ENDOR-line positions. Generalizations to systems with spin S > 1 (quadruplets, quintuplets, etc.) or spin I > 1 (3/2 for $^7Li$, $^{23}Na$, $^{33}S$, $^{35,37}Cl$, 5/2 for $^{17}O$, etc.) are implicit in the expressions given above. It must be recalled that these expressions are correct only to first order in $a/Z_p$, while second-order corrections may be required especially for such accurate determinations as those made by ENDOR and ELDOR.

When one goes over to polynuclear, isotropic-like, paramagnetic systems, the number of levels, Y, in each of the 2S + 1 sets in the analogue of Fig. (1b) increases exponentially : $Y = \prod_{\lambda}(2I_{\lambda} + 1)$. In the first order, if the level sets are equidistant, the number of distinct allowed components in the EPR spectrum is also equal to Y, according to the selection rules : $\Delta M_S = 1$, $\Delta M_{I_\lambda} = 0$ for all $\lambda$'s. The integrated intensities of these components, obtained from Eqn. (92), are very nearly equal, and their average positions appear as the simple sums of contributions from the different nuclei generalizing Eqn. (93a'), i.e. :

$$B_{mn} = (\hbar\omega + \sum_{\lambda} M_{I_\lambda} a_{\lambda})/g\beta_e, \quad M_{I_\lambda} = -I_{\lambda}, \ldots, +I_{\lambda}. \tag{100}$$

If there are equivalences (i.e., equality of spin and coupling constant) among the nuclei, some hyperfine components overlap and, in the first order, appear as a single line, with an intensity proportional to the degeneracy of the involved levels. For instance, a paramagnetic system with n equivalent protons, such as the free radical $\cdot CH_3$ or the radical ion $C_6H_6^-$, displays n + 1 EPR lines centered at $B_e$, with a uniform splitting A and with intensities given by the binomial distribution, as a result of superpositions among the $2^n$ components. The NMR spectrum of a polynuclear system more simply consists of as many distinct lines as there are groups of equivalent nuclei (e.g., a single line in the examples cited above), the intensities being proportional to the numbers of nuclei in the groups and the positions given by a generalization of Eqn. (97'), i.e. :

$$\Delta B_r = -\,(g\beta_e/\gamma_\lambda\hbar)S(S + 1)a_\lambda B_\lambda/3kT_L, \quad B_\lambda = \omega/\gamma_\lambda. \tag{101}$$

Similarly, each group of equivalent nuclei is characterized by 2S + 1 ENDOR lines with first-order positions given by the generalization of Eqn. (98) :

$$\hbar\omega_p = |Z_\lambda + M_S a_\lambda|, \tag{102}$$

the intensities being now principally determined by the spin-relaxation paths involved. Finally, each group of equivalent nuclei produces a single "primary" allowed-allowed ELDOR line with a first-order position again given by Eqn. (99), but now one may also observe combination ELDOR lines with first-order positions given by :

$$\hbar|\omega_p - \omega_o| = |\sum_{\lambda} M_\lambda a_\lambda|, \tag{103}$$

the $M_\lambda$'s being small integers limited by the $I_\lambda$'s ; the intensities of these lines, which are commonly reduced with respect to those of the primary lines, depend on the setting of the observing resonance frequency and field and relative direction of sweep of the pumping frequency. One can see that the spectral density (i.e., the number of distinct lines per unit of spectral width) increases more rapidly with the number of nuclei in EPR than in NMR, ENDOR or ELDOR. In addition, the ratio of NMR to EPR widths is usually smaller than $10^{-2}$

and the ENDOR and ELDOR widths close to the EPR spin-packet widths. For these reasons, NMR, ENDOR and ELDOR generally offer better effective resolutions than EPR and, as a consequence, can provide more accurate hyperfine couplings(25-27). In this respect, the ENDOR and ELDOR techniques are more complementary than competitive : ENDOR is more appropriate for measuring small couplings, and ELDOR for measuring large couplings, with accuracy ; this is because the former requires long, and the latter short, nuclear relaxation times, the inverses of which tend to increase with increasing hyperfine coupling, among other variables. A similar comparison can be drawn between the NMR and EPR techniques, the former requiring shorter electron relaxation times than the latter to be gainfully applied. The price for the above-mentioned gain in resolution is a significant loss in sensitivity : typically, good-quality high-resolution EPR spectra can be obtained with spin concentrations as low as $10^{-6}$M per MHz of spectral width ; the limiting concentrations are 10 to 100 times larger for ELDOR and ENDOR, respectively, and $10^{-3}$M per kHz of spectral width for conventional NMR, with commonly larger samples and stronger fields. The use of Fourier-transform averaging techniques, especially for nuclei in low natural abundance such as $^{2}$D, $^{13}$C, $^{15}$N or $^{17}$O, or with short spin-relaxation times due to strong magnetic-dipolar or electric-quadrupolar interactions with the electronic distribution, can improve the sensitivity of NMR by one or two orders of magnitude (33b). For the phosphorescent states of some aromatic molecules, it may be appropriate to use a combination of simple or double magnetic resonance, with or without external field, and optical or laser spectroscopy, for example to achieve the optical detection of magnetic resonance (ODMR), which involves optical and microwave or radiofrequency pumpings and depends upon the interplay of the population and depopulation rates from and to the neighbouring singlet levels and spin-lattice relaxation rates within the triplet sublevels(33b).

If the medium is not perfectly fluid and if, as an effect, the condition $\tau_c^{-1} \gg \Delta\nu_{anis.}$ is not met for some specific term of $\mathcal{H}_S$, the anisotropic part of this term shows up in magnetic resonance spectra to a degree which will increase with $\tau_c$ until the limit $\tau_c^{-1} \ll \Delta\nu_{anis.}$ is reached for this term. If the medium is more fluid along some particular directions, as with liquid crystals or membranes, the manifestation of the anisotropy will depend upon characteristic ordering parameters. In every case, the determining factor is the degree and form of effective local rigidity : for instance, a small radical trapped in a clathrate powder may behave as fully isotropic, and a large radical trapped in a crystal or a glass may undergo variable internal motions. EPR and NMR line-shapes will strongly depend upon the rate processes in ranges of correlation times where $\tau_c^{-1} \simeq \Delta\nu_{anis.}$ for some significant term of $\mathcal{H}_S$, and ENDOR and ELDOR intensities and widths in other ranges as well due to their sensitivity to the induced spin-relaxation rates. The general interpretation of magnetic resonance spectra in terms of coupling tensors, ordering parameters, correlation times, spin-relaxation rates and inhomogeneous line-widths is very complicated, and can be handled only in simple cases. However, in many instances, the correlation rates are either much smaller or much larger than the anisotropic part of every individual term of $H_S$, so that one can define an effective static spin Hamiltonian with some coupling tensors presenting a less anisotropic, axial or spherical form. The determination of the eigenvalues and eigenfunctions of a static spin Hamiltonian can be achieved by using standard variation or perturbation procedures, the choice of an appropriate spin basis being facilitated by the introduction of effective fields. EPR resonance fields and transition probabilities are calculated by applying Eqns. (89) and (90). If the first term in Eqn. (85) is much larger than all other terms, the center of gravity of the EPR spectrum is located at the field : $B_e(\Omega) = \hbar\omega/g(\Omega)\beta_e$, where $g(\Omega)$ is the effective value of the tensor $\underline{\underline{g}} = g_0(\underline{\underline{u}} + \underline{\underline{G}})$ along the direction $\Omega = (\Theta,\Phi)$ of the field $\underline{B_Z}$, i.e. : $g(\Omega) = (\sum_u g_u^2\ell_u^2)^{1/2}$, $g_u$ being a principal value of $\underline{\underline{g}}$ and $\ell_u$ the direction cosine of $\underline{B_Z}$ on the corresponding principal axis. The second term in Eqn. (85) gives rise to a fine structure which begins at zero field and consists in the strong-field case, in the first order, of (2S) $\Delta M_S = 1$ components symmetrically disposed around $B_e$ with a uniform splitting which can be written, if the tensors $\underline{\underline{g}}$ and $\underline{\underline{D}}$ have coincident axes : $\Delta B_f(\Omega) = 3\sum_u D_u g_u^2\ell_u^2/g^3(\Omega)\beta_e$, $D_u$ being a principal value of $\underline{\underline{D}}$ and $\ell_u$ the direction cosine of $\underline{B_Z}$ on the corresponding principal axis. The next three sums in Eqn. (85) give rise to a hyperfine structure which consists for each $\Delta M_S = 1$ component, in the first order, of $(Y^2)$ $\Delta M'_{I_\lambda} = 0$ or $\neq 0$ subcomponents with splittings and intensities resulting from additive and multiplicative contributions, respectively, from the nuclei. The expressions of these contributions involve in general the principal values and axes of the various tensors and the magnitude and orientation of the applied field. The prime on $M'_{I_\lambda}$ accounts for the fact that when its tensors are anisotropic, the nuclear spin $\underline{\hat{I}_\lambda}$ is no longer quantized along the external field but, if the quadrupole term is small, along an effective field formed by adding to $\underline{B_Z}$ (corrected for $\underline{\underline{\sigma}}_\lambda$) the internal field exerted, through the hyperfine coupling, by the electronic spin in one of its states of quantization along $\underline{B_Z}$ (corrected for $\underline{\underline{G}}$). The resulting mixture of nuclear states entails that the previously forbidden $\Delta M'_{I_\lambda} = \pm 1$ transitions become more intense as $\underline{\underline{c}}_\lambda$ becomes more anisotropic and comparable in magnitude to $Z_\lambda$. If the quadrupole term is large, $\Delta M'_{I_\lambda} = \pm 2$ transitions can also become intense. Similarly, if the fine-structure term is not too small with respect to the electronic Zeeman term, $\Delta M_S = 2$ transitions can be observed. Finally, the last double sum in Eqn. (85) contributes to the inhomogeneous broadening of resolved EPR lines, together with neglected hyperfine couplings, deviations of larger couplings from their statistical averages and exchange and dipolar coupling terms between the electron spins.

The dependence of the positions, intensities and widths of the lines on the direction of $\underline{B_Z}$ can in principle be used to measure all the relevant tensors. There are two limiting cases of particular interest. In a single crystal, only a small number of symmetry-related orientations are possible in the unit cell, and one can follow and ascribe the lines when the crystal is rotated around appropriate reference axes. In a polycrystalline powder, glass, or other amorphous sample, all orientations are equally probable, and the anisotropies result in a more or less structured broadening : every line from the microcrystalline stick diagram is spread over its range of resonance fields, and the turning points of these as a function of field orientation result in singularities of the absorption (discontinuities of its derivative) as a function of the field magnitude. As the resonance field surfaces $B_r(\Theta,\Phi)$ have the topology of the sphere and due to magnetic-field-reversal symmetry, the numbers of non-degenerate singularities of increasing-step type (i), decreasing-step type (d) and logarithmic type ($\ell$) occur in related numbers in polycrystalline unbroadened patterns, i.e. : $i + d - \ell = 1$, the degenerate singularities being counted as the merging ones taken together. There are always more or less intense polycrystalline singularities corresponding to field directions parallel or perpendicular to the symmetry axes and symmetry planes of the investigated system, such as the $C_3$ axis and the perpendiculars to the three $\sigma_v$ planes in the non-planar $\cdot CF_3$ radical, or the principal axes common to all coupling tensors in simpler systems, but such singularities may also occur along non-symmetry directions, like the $B_{min}$ singularity in a triplet $\Delta M_S = 2$ spectrum. However, as a result of the increased overlap of the lines and of the other sources of inhomogeneous broadening, most polycrystalline singularities are habitually smeared out, except when the coupling tensors are few in number and strongly anisotropic.

What we have said for EPR can easily be extended to NMR, ENDOR and ELDOR, which, as we have seen earlier, give simpler spectra. For anisotropic even more than for isotropic systems, the interplay of sensitivity and resolution is critical in choosing the appropriate method. However, whereas EPR spectra contain in principle enough information to permit the extraction of all tensors by the disentanglement of their effects, NMR spectra can only provide an intricate combination of them for each group of equivalent nuclei, with the additional signs of the shifts, due to the unicity of the dynamically-averaged lines. ENDOR and ELDOR, for their part, present distinct advantages : ENDOR offers the possibility to unravel the inhomogeneous broadening due to very small hyperfine couplings, particularly from matrix nuclei, and to measure accurately quadrupole couplings, and ELDOR proves useful in separating overlapping spectra, particularly from different orientations within single crystals, and can also be used to measure zero-field splittings. Measurements can be performed under observing or pumping conditions where the allowed and forbidden lines are clearly differentiated. Single-crystal-like spectra can be obtained from polycrystalline samples by monitoring an appropriate singularity while sweeping the pumping frequency.

We shall now see with an illustrative example how the complementary applications of these and other techniques can lead to improved knowledge of the electronic distributions in paramagnetic systems.

The stable conjugated free radical 1,1-diphenyl-2-picryl-hydrazyl (DPPH), Fig. (2a), was prepared in crystalline form a long time ago [34a], and was among the first paramagnetic species investigated by EPR : the polycrystalline powder at 300 K gives an exchange-narrowed line of width $\lambda \simeq 1.4$ G centered at $g \simeq 2.0036$ [34b], but g and $\lambda$ have been found to be orientation-dependent in a single crystal and

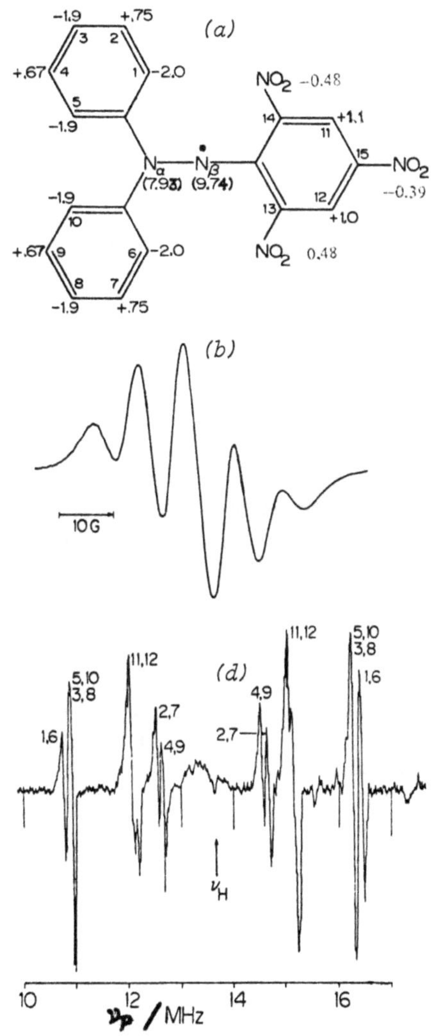


*Fig. 2 - The stable conjugated free radical DPPH. (a) Planar formula and hyperfine coupling constants/G. (b) Dilute liquid EPR spectrum. (c) Concentrate solid NMR spectrum. (d) $10^{-2}$M liquid ENDOR spectrum. (e) $10^{-3}$M liquid ELDOR spectrum. See text.*

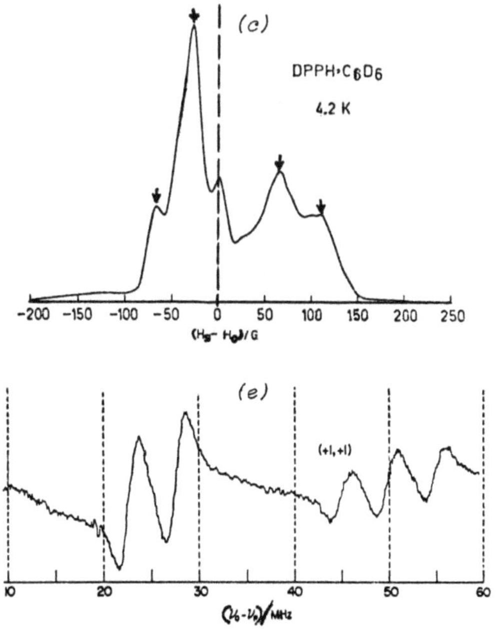

to increase with decreasing temperature[34c]. The remarkable chemical stability and EPR-line narrowness of this radical have made it a wide-spread standard for measuring unpaired-spin concentrations and effective magnetogyric ratios[34d]. Early crystallographic data have been reported[35a] on one of its best-grown crystal forms, the DPPH-$C_6H_6$ complex. Complete determination of the crystal structure from three-dimensional diffraction data has shown[35b] that the hydrazyl backbone consists of two quasi-trigonal nitrogens whose planes are twisted by about 28°, the phenyl and picryl rings being further inclined at angles of 22°, 49° and 33° and the ortho-nitro groups shielding the picryl-hydrazyl nitrogen. Early static susceptibility measurements have been performed in both high[36a] and low[36b] temperature regions. Subsequent extensions and refinements[36c-h] have confirmed the existence of several Curie points around $\theta_H \simeq -23$ K and $\theta_L \simeq -0.4$ K. The results suggest[36d,e] that the exchange coupling in DPPH crystals is antiferromagnetic.

The first observation of the hyperfine structure of DPPH was made on a dilute liquid solution which gave five EPR lines of widths $\lambda_{hf} \simeq 10$ G and splittings and intensities characteristic of two equivalent nitrogens with isotropic coupling constants $a_{N1,N2} \simeq 10$ G[37a], Fig. (2b)[39b]. Analysis of the line-shape later showed that the nitrogen couplings are in the exact ratio of 0.82[37b]. The first accurate determination of the couplings yielded the following values : $a_{N1} \simeq 7.8$ G, $a_{N2} \simeq 9.4$ G, and $\lambda_{hf} \simeq 6.7$ G[37c]. The smaller and larger couplings were assigned to the diphenyl and picryl-hydrazyl nitrogens, respectively, using $^{15}$N isotopic substitution[37d]. It was also shown that these couplings depend slightly upon the solvent[37e]. There was early evidence for a strong anisotropy of these in the marked broadening and modification of the EPR spectra when going from liquid to plastic solvents[37f]. The complete coupling tensors $\underline{\underline{c}}_{N1,N2}$ were measured, and assigned using $^{15}$N isotopic substitution, on a dilute mixed crystal of DPPH in the parent diamagnetic hydrazine ($DPPH_2$)[37g], and later refined on a similar mixed crystal[37h] : the $\underline{\underline{c}}_N$'s were found to be almost axial with axes making an angle of about 13°, but their traces appeared inconsistent with the $a_N$'s measured in liquids. With the help of some simplifying assumptions, including constancy of the $a_N$'s when going from liquid to glass solvents and axial $\underline{\underline{c}}_N$'s with parallel axes, several authors[37c-e] were able to simulate the glass line-shapes with the axial coupling constants : $b_{N1} \simeq 6$ G, $b_{N2} \simeq 7$ G, and to show that for both nitrogens $a_N$ and $b_N$ have the same sign. The marked differences between the tensors obtained from glasses and from single crystals were confirmed by the study of polycrystalline line-shapes[37i] and interpreted as resulting from different conformations in different media. Recently it has been shown[37j], using line-shape simulations free from the above assumptions and including the anisotropy of g, that the $a_N$'s may greatly differ even from liquids to glasses.

There is already a hint of superhyperfine structure from the ring protons and the picryl nitrogens in the wide Gaussian-shaped broadening of the low-resolution hyperfine components[37c], and of anisotropy of this structure in the increasing component width with increasing viscosity[37d]. However, the first direct evidence of coupling of the unpaired electron with ring protons came as an Overhauser enhancement of the strong-field proton resonance of cooled solid samples[38a,b]. In many respects, the Overhauser effect is to NMR what ENDOR is to EPR, the variation of intensity of an NMR transition observed while saturating an EPR transition[24]. The superhyperfine structure of the EPR spectrum was eventually resolved by using carefully desoxygenated liquid solutions[38c]. There was a first attempt to analyse it using line-shape simulations with trial couplings[38d], but it was completely interpreted by applying a best-fit procedure to the Fourier transform of the line-shape[38e]. In the meanwhile, several measurements had been made of the magnitudes and signs of the hyperfine couplings of the ring protons by strong-field NMR of cooled solid samples. As expected from the radical structure, Fig. (2a), and from Eqn. (101), the wide triangular single line observed at room temperature[38f] splits up into a dissymmetric doublet around liquid-nitrogen temperature[38f,g] and further into a well-resolved quartet below liquid-helium temperature[38h,i], Fig. (2c)[38j]. The upward and downward shifts and integrated intensities of the NMR lines appear roughly consistent with the partitioning of the ring protons into four groups, with the hyperfine couplings $a_\lambda/g\beta_e \simeq -2.0$ G (4 ortho), +0.7 G (4 meta), -1.3 G (2 para) and +1.5 G (2 picryl)[38h,i]. The assignments in parentheses are made by comparison with the results of spin-polarization-including quantum-mechanical calculations[38g] and by the use of selective ring deuterations[38h]. There is also an unshifted line for both $C_6H_6$-containing DPPH samples[38h] and samples obtained from solutions of DPPH in $CCl_4$ or $CS_2$[38i], which disappears with $C_6H_6$ deuteration but increases with decreasing temperature for the solvent-free samples. The explanation of this anomaly came with the use of samples prepared with variable amounts of $C_6D_6$, which permitted to demonstrate that there is a diamagnetic pairing of DPPH molecules which increases with decreasing temperature and interstitial-molecule content[38j]. The proton couplings measured by NMR on powder samples[38g-j] appear slightly different from those measured by EPR on liquid solutions[38d,e], partly because the media are very different, and partly because the former are intermediate principal values and the latter trace-averages of the hyperfine coupling tensors. Single-crystal NMR spectra have also been reported[38h] and their detailed interpretation has shown[38k] that the two phenyl groups are not equivalent, in accordance with the x-ray data[35b], and that the proton hfc tensors fall in fact into five groups : 2 groups of 3 protons (the 2 ortho and 1 para of each phenyl), 2 groups of 2 protons (the 2 meta of each phenyl) and 1 group of 2 protons (the 2 meta of the picryl), at variance with the previous assignments[38i]. The proton NMR spectra of concentrated liquid solutions exhibited only two broad lines corresponding to the hfc constants $a_\lambda/g\beta_e \simeq -1.8$ G and +0.75 G[38ℓ], additional equivalences being induced by dynamical conformational changes. Recently NMR measurements of the picryl-nitrogen couplings have also been made on liquids[39c], yielding values comparable to those obtained from high-resolution EPR spectra[38e] but with negative signs and inversed assignments.

The most accurate determinations of the hyperfine couplings of DPPH in liquid solutions have been achieved by ELDOR for the hydrazyl nitrogens[39a], ENDOR for the ring protons[39b], and NMR for the picryl nitrogens[39c]. An ENDOR spectrum of a $10^{-2}$M solution of DPPH in mineral oil at 300 K is given in Fig. (2d)[39b]. It displays six pairs of lines around the free-proton resonance frequency $\nu_H$, which correspond to six groups of equivalent protons whose couplings, obtained using Eqn. (102) corrected to second order, are given in Fig. (2a), the assignments being made with the help of the magnitudes and signs provided by NMR[38g-ℓ] and by quantum-mechanical calculations[40]. An ELDOR spectrum of a $10^{-3}$M solution of DPPH in toluene at 290 K is given in Fig. (2e)[39a]. It displays two intense lines which correspond to the two hydrazyl nitrogens whose couplings, obtained using Eqn. (103) corrected to second order, are given in Fig. (2a), the assignments being made with the help of $^{15}$N isotopic substitution[39a] and proving consistent the results of quantum-mechanical calculations[40]. The top-right label indicates the $M_I$ values of the observed EPR component, and the weaker lines to the right are particular combination lines. The picryl-nitrogen couplings measured by strong-field NMR on a saturated solution of DPPH in $CH_2Cl_2$ around 300 K[39c] are also given in Fig. (2a). ELDOR measurements have also been performed on a DPPH-containing glass[39a], but so far no ENDOR or ELDOR experiments have been reported on DPPH-containing crystals. The use of $^{13}$C isotopic substitutions could shed some new light on the spin-density distribution in DPPH. The direct determination of the spin-density distribution away from the nuclei from the neutron-diffraction data would bring an outstanding contribution to the understanding of this remarkable system.

### 4.3. Quadrupole Splittings and Isotope/Isomer Shifts

Until now we have restricted our attention to the techniques of measurement of spin-coupling parameters in paramagnetic systems in condensed phases under strong magnetic fields inducing resonances. We have mentioned that in such circumstances quadrupole couplings can be determined practically, together with other couplings but without the absolute signs, by EPR or, more accurately, by ENDOR, if concentrations and relaxation times permit, in either single crystals or liquid crystals or, if the couplings are large enough, powders or glasses, using appropriate curve-fitting procedures. We shall now briefly consider the techniques of measurement of quadrupole couplings in other circumstances. We have also mentioned earlier that the isotropic parts of the terms quadratic in $\hat{\underline{S}}$ or $\hat{\underline{I}}_\lambda$ in Eqn. (85), which shift uniformly the levels, are not observable by magnetic resonance techniques. We shall thus also consider the techniques by which the latter, which are directly related to the charge densities at the nuclei, can in specific circumstances be measured.

The argument according to which the anisotropic parts of spin-coupling terms may be averaged out by rapid random reorientations in a fluid isotropic medium is not valid in empty space, where the systems can exist in well-defined quantum states and the averages of the traceless terms over the quantized states are usually not zero. As an effect, the electronic spectra of atoms and rotational spectra of molecules in low-pressure gases can exhibit fine or hyperfine structure originating from anisotropic terms. For atoms or simple linear molecules in low-pressure gases or particle beams, magnetic resonance or, even better, electric resonance, if the molecules are polar, can yield more accurate determinations of these terms, using the properly adapted effective Hamiltonians. These techniques have been applied to obtain the sign and magnitude of quadrupole couplings in a number of isolated systems(32b), where they appear slightly larger than for the same systems in solids. In liquids, approximate estimates of the anisotropic parts of spin-coupling terms can often be made, under particular assumptions about the spin-relaxation mechanisms and dynamics of the molecules in solution, from the measured homogeneous broadenings and spin-lattice relaxation rates. This method can be used to obtain quadrupole couplings from broadened NMR lines of diamagnetic liquid solutions(32b), and permits easy acquisition of comparable data for series of related molecules of similar sizes.

NMR and NQR are the most generally applicable techniques for measuring quadrupole couplings in diamagnetic solid samples(32). For such systems, all the terms involving $\underline{S}$ in Eqn. (85) vanish, and the remaining terms for each nucleus are similar to the first two terms relative to the electron in Eqn. (85), while the last double sum contributes to the inhomogeneous broadening of the resolved lines. NMR experiments are performed under strong magnetic fields, ranging from about 6 kG to 24 kG with conventional electromagnets and 59 kG to 94 kG with supraconducting magnets, and for this latter value one of the largest Larmor frequencies for nuclei possessing quadrupolar moments (that of $^7$Li) is about 156 MHz, while the measured quadrupole couplings range from a few hundred kHz (for $^2$D in hydrocarbons) to several thousand MHz (for $^{127}$I in polyhalides). Now magnetic resonance transitions can be produced only if the Larmor frequency is much larger than the quadrupole coupling, and under such conditions a treatment parallel to that outlined p. 43 for the electronic-Zeeman and fine-structure terms can be developed. Here again, one can study either single crystals, bigger now than in EPR for the reasons recalled earlier, or amorphous samples, usually easier to prepare. As an example, Fig. (3) shows the NMR absorption line-shape for deuterium in a polycrystalline clathrate hydrate[41] : one can easily discern the i, $\ell$ and d-type singularities corresponding to the two overlapping $\Delta M_I = 1$ components for the deuterons with the magnetic field quasi-parallel to the three principal axes of the quadrupole-coupling tensor, as marked on the figure. If one neglects the slight dissymmetry produced by a possible chemical-shift anisotropy but mostly the second-order correction, one can obtain the quadrupole-coupling constant $-e^2qQ$ and asymmetry parameter $\eta$ introduced in Eqn. (78") directly from the separations between the singularities by using the simple formulae :

$$Z'Z'' = 2\nu_Q, \quad Y''Y' = \nu_Q(1 + \eta), \quad X''X' = \nu_Q(1 - \eta),$$

$$\nu_Q = (3/4)\,|-e^2qQ|/h.$$

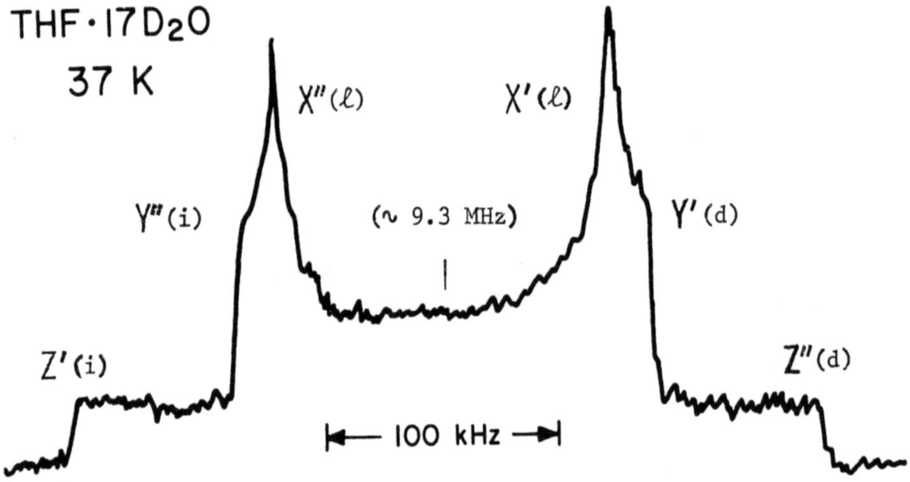


*Fig. 3 - NMR absorption line-shape for deuterium in a polycrystalline clathrate hydrate (Reproduced from Ref. 41, marks added).*

NQR must be utilized when the quadrupole couplings are so large that one cannot bring the levels into resonance with the available magnetic fields. It consists of swept-frequency observation of the transitions induced between zero-field levels (pure quadrupole resonance) or weak-field levels (Zeeman quadrupole resonance). The simplest case is that of pure quadrupole resonance with an axial coupling tensor, where the spin Hamiltonian reduces to : $H'_{qp} = A\{3(\hat{I}^z)^2 - \hat{\underline{I}}^2\}$, see Eqn. (78"). This operator is diagonal in an $|M_I\rangle$ representation and gives rise to (I + 1) or (I + 1/2) $\pm M_I$-degenerate levels. $\Delta M_I = \pm 1$ transitions can be induced by the perpendicular component of the oscillating magnetic field at angular frequencies given by :

$$\hbar\omega_r = 3|A|(2|M_I| + 1), \qquad |M_I| = I-1, I-2, \ldots, 1/2 \text{ or } 0. \quad (104)$$

When the quadrupole-coupling tensor does not have axial symmetry, the term involving $\eta$ in Eqn. (78") mixes states differing by $\Delta M_I = \pm 2$ and the eigenstates of $H'_{qp}$ can be obtained by diagonalization of its matrix in the $|M_I\rangle$ representation. Except for I = 1 and I = 3/2 the secular equations must be solved numerically. For I = 1 three resonance frequencies are obtained :

$$\hbar\omega_r = 3|A|(1 \pm \eta/3) \text{ or } 2|A|\eta, \quad (105')$$

and for I = 3/2 only one resonance frequency :

$$\hbar\omega_r = 6|A|(1 + \eta^2/3)^{1/2}, \quad (105")$$

with the consequence in the latter case that pure quadrupole resonance cannot yield separable values for $|A|$ and $\eta$. When pure quadrupole resonance is performed on a single crystal, the tensor's principal axes can in favorable cases be obtained from the orientation dependence of the intensities of the lines. The application of a weak external magnetic field ($B_Z$ < 100 G) mixes states differing by $\Delta M_I = \pm 1$, lifts the $\pm M_I$-degeneracy and further splits the lines, producing an additional orientation dependence of the positions and intensities of the components. These features can be utilized to determine the asymmetry parameter for spin-3/2 nuclei in either single crystals or amorphous samples, and the tensor's principal axes in single crystals when pure quadrupole resonance does not give appropriate results. We mention for the record that double-resonance techniques have also been devised in the NMR and NQR ranges, and in some cases have been used to determine the quadrupole-coupling signs.

The spectroscopic technique derived from the Mössbauer effect[23] allows the determination of all hyperfine interactions for radioactive nuclei imbedded in rigid lattices, including the isotropic part of the term quadratic in $\hat{\underline{I}}$ in Eqn. (85). Effects due to this isotropic part have been observed in atomic spectroscopy even before Mössbauer's discovery[42a], either on isotopic mixtures of heavy atoms (such as Cd, which has 8 natural isotopes), or on isomeric mixtures of atoms with the same nuclei in the ground and an excited state (such as artificial radioactive $^{197}$Hg). This is possible because the energy of a given atomic transition depends on the nuclear size if the two atomic states involved in the transition are affected differently by the nuclear effective radius. In Mössbauer spectroscopy (MS), the energy of a given nuclear transition depends on the electronic wave-function if the two nuclear states involved in the transition are affected diffe-

*(a)*
*(Adapted from Fig. V.1 of Ref. 23)*

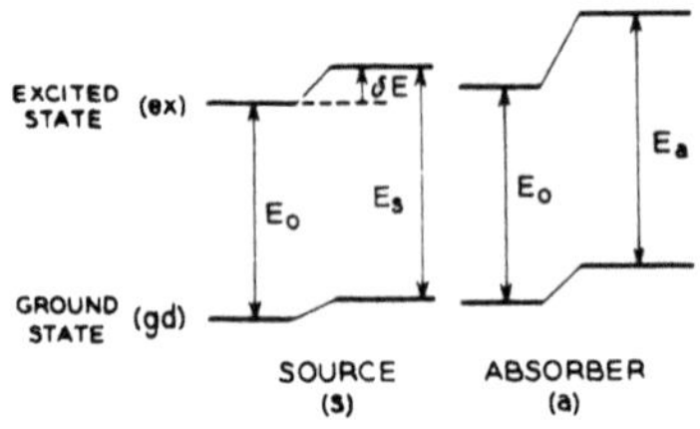


$$\delta E = \frac{2\pi}{3} Ze^2 |\psi(0)|^2 R^2$$

$$E_s = E_0 + \frac{2\pi}{3} Ze^2 |\psi_s(0)|^2 \left[R^2_{ex} - R^2_{gd}\right]$$

$$E_a = E_0 + \frac{2\pi}{3} Ze^2 |\psi_a(0)|^2 \left[R^2_{ex} - R^2_{gd}\right]$$

$$\Delta E = \frac{2\pi}{3} Ze^2 \left[|\psi_a(0)|^2 - |\psi_s(0)|^2\right]\left[R^2_{ex} - R^2_{gd}\right]$$

*(b)*
*(Reproduced from Ref. 42b)*

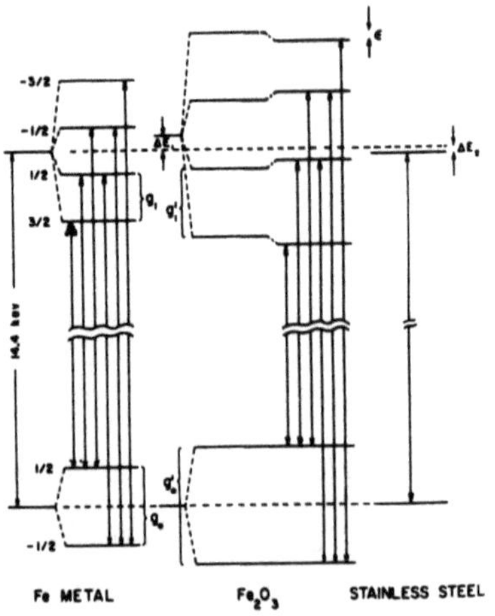


*(c) (Reproduced from Ref. 42b)*

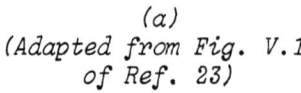

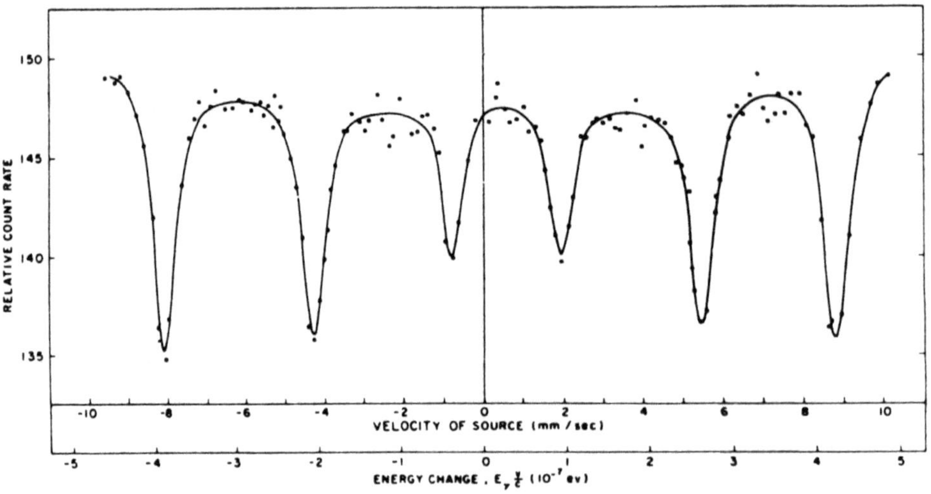


*Fig. 4 – Hyperfine structure in Mössbauer spectroscopy. (a) Definition of the relative isomer shift. (b) Isomer shift and quadrupole and magnetic splittings for $^{57}Fe$ bound in various environments. (c) Absorption by $^{57}Fe$ bound in $Fe_2O_3$ of the 14.4-keV γ ray emitted in the decay of $^{57}Fe^*$ bound in stainless steel.*

rently by the contact charge density. However, whereas in optical spectroscopy Doppler and Heisenberg broadenings are commonly a few orders of magnitude larger than the recoil energy of the emitting or absorbing atoms and resonance presents no problem, in $\gamma$-ray spectroscopy the recoil energy of the emitting or absorbing nuclei may be so large as to reduce appreciably the overlap of the respective lines. More precisely, an isolated system at rest which decays from an excited state with an energy $E_0$ by emitting a photon of energy $E_\gamma$ and associated linear momentum $p_\gamma = E_\gamma/c$ receives a recoil momentum $p_R = -p_\gamma$ with an associated recoil energy $E_R = p_R^2/2M$ (neglecting the small mass change during the emission) $= E_\gamma^2/2Mc^2$ ($\ll E_\gamma$). Energy conservation requires that $E_0$ be shared between $E_\gamma$ and $E_R$, so that : $E_\gamma = E_0 - E_R$. Similarly, for the absorbing system, one must have : $E_\gamma = E_0 + E_R$. For a given difference of the recoil shifts, resonance will be the more effective as the widths of the absorption and emission lines will be larger. It will thus tend to decrease with decreasing temperature of either the source or the absorber, until one comes to the point where no quantized lattice mode can be excited by the nuclear recoil energy. More precisely, if $E_R$ is larger than the binding energy of the atom in the lattice ($\sim$ 20 eV), the atom will be dislodged from its site ; if $E_R$ is smaller than that but larger than the characteristic energy of the lattice vibrations ($\sim 10^{-2}$ eV), there will be thermal dissipation of the recoil energy, and the widths of the $\gamma$-ray lines will be of the same order as the phonon energies ; finally, if $E_R$ is even smaller than these latter, there will be recoil-free emission or absorption for a fraction of the transitions, and the recoil-free components will be in perfect resonance and have their natural widths ($\lambda_H \sim 10^{-8}$ eV for $\tau_{1/2} \sim 10^{-7}$ s). This is the Mössbauer effect, first discovered on $^{191}$Ir and now observed on several scores of nuclides, particularly $^{57}$Fe for both reasons of convenience and importance. The recoilless events bear some similarity with the elastic processes in the scattering of x rays or neutrons by atoms, and it can be shown that when such events take place the nuclear recoil momentum is transferred to the whole sample while the first and second moments of the energy spectrum are not affected. Because Mössbauer lines are so narrow, the resonance of the emitting and absorbing nuclei may again be destroyed by different shifts or splittings produced by hyperfine interactions. Resonance can then be restored by supplying to the source or the absorber a Doppler energy shift with an appropriate moving device : therefore MS hyperfine shifts and splittings are usually given in velocity units.

The diagram and formulae presented in Fig. (4a)[23] help understand how the quantity defined in Eqn. (77) is related to the MS isomer shift : the decay and excitation energies of the source and absorber nuclei, respectively, would be both equal to $E_0$ if there were no differential perturbations of the electronic environments ; if the isomer-shift term $\delta E$ depends on both nuclear state and chemical binding, then an energy difference $\Delta E = E_a - E_s$ can be observed.

The more complex case where the ground and excited states of the nuclei in the source and absorber are also split by quadrupole and magnetic interactions is presented in Fig. (4b)[42b], which refers specifically to $^{57}Fe$ : this nucleus has a spin I = 1/2 in its ground state, which gives rise to magnetic splittings in Fe metal and antiferromagnetic $Fe_2O_3$, $g_0$ and $g_0'$ respectively, but not in stainless steel ; it has a spin I = 3/2 in its first excited state ($E_0 \simeq 14.4$ keV, $\tau_{1/2} \simeq 1.4 \times 10^{-7}$ s, obtained from electron capture and step decay of $^{57}Co$), which gives rise again to magnetic splittings in Fe metal and $Fe_2O_3$, $g_1$ and $g_1'$ respectively, and in addition to a small quadrupole splitting, ε, in non-cubically-symmetric $Fe_2O_3$ (note the $\pm M_I$-degeneracy of this splitting). The relative isomer shifts $\Delta E_1 (> 0)$ and $\Delta E_2 (< 0)$ in $Fe_2O_3$ and stainless steel are defined with respect to that in Fe metal. The six allowed transitions ($\Delta M_I = 0, \pm 1$) are indicated. Fig. (4c) gives an example of a Mössbauer spectrum[42b] : the relative count rate measured by the detector placed behind the absorber decreases when the relative velocity of the source provides the Doppler energy balance for two hyperfine levels in the source and absorber to enter into resonance ; the double scale allows conversion of velocity into energy units. The displacement of the center of the spectrum from zero velocity is a measure of the relative isomer shift $\Delta E_1 - \Delta E_2$, while the splittings are associated with the internal magnetic field and electric-field gradient in the $Fe_2O_3$ sample.

In these lectures we have tried, in the first place, to show exactly what are the quantities measured in magnetic resonance and related techniques, with particular stress on those most directly related to the charge and spin densities at and around the nuclei, and in the second place to give a bird's view on how these quantities are measured. Magnetic resonance and related techniques appear complementary to x-ray and neutron scattering techniques, which provide charge and spin densities away from the nuclei.

We wish to thank the copyright owners of the original spectra and diagrams presented in Fig. (1-4) for allowing their reproduction in this paper.

REFERENCES


(1) P.A.M. DIRAC, Proc. Roy. Soc. (London) 117, 610 (1928) ; "The Principles of Quantum Mechanics" revised 4th edition (Clarendon Press, Oxford, 1967), ch. 11.

(2) P.-O. LÖWDIN, J. Mol. Spectry. 14, 131 (1964). See also R.G. WOOLLEY, Mol. Phys. 30, 649 (1975).

(3) H.A. BETHE and E.E. SALPETER, "Quantum Mechanics of One- and Two-Electron Atoms" (Springer Verlag, Berlin, 1957).

(4) G. BREIT, Phys. Rev. 34, 553 (1929) ; 39, 616 (1932).

(5) (a) T. ITOH, Rev. Mod. Phys. 37, 159 (1965). (b) A. RICH and J.C. WESLEY, Rev. Mod. Phys. 44, 250 (1972).

(6) P. PYYKKÖ, "Relativistic Quantum Chemistry", submitted to "Advances in Quantum Chemistry" (1978).

(7) (a) E.C. KEMBLE, "The Fundamental Principles of Quantum Mechanics" (Mc Graw- Hill, New York, 1937), ch. XI, § 48. (b) J.H. VAN VLECK, Phys. Rev. 33, 467 (1929). (c) O.M. JORDAHL, Phys. Rev. 45, 87 (1934).

(8) (a) E.B. WILSON Jr. and J.B. HOWARD, J. Chem. Phys. 4, 260 (1936). (b) L.L. FOLDY and S.A. WOUTHUYSEN, Phys. Rev. 78, 29 (1950). (c) M.H.L. PRYCE, Proc. Phys. Soc. A63, 25 (1950). (d) R.A. HARRIS, J. Chem. Phys. 47, 3967 (1967).

(9) (a) T. KATO, Progr. Theor. Phys. 4, 514 (1949). (b) P.-O. LÖWDIN, J. Chem. Phys. 19, 1396 (1951). (c) C. BLOCH, Nucl. Phys. 6, 329 (1958). (d) J. DESCLOISEAUX, Nucl. Phys. 20, 321 (1960). (e) H. PRIMAS, Helv. Phys. Acta 34, 331 (1961). (f) P.-O. LÖWDIN, J. Math. Phys. 3, 969 (1962). (g) J.O. HIRSCHFELDER, J. Chem. Phys. 39, 2099 (1963). (h) T. MORITA, Progr. Theor. Phys. 29, 351 (1963). (i) P.-O. LÖWDIN, in "Perturbation Theory and its Applications in Quantum Mechanics" (John Wiley, New York, 1966), p. 255. (j) L.N. BULAEVSKIY, Soviet Physics - J.E.T.P. 24, 154 (1967). (k) B. KIRTMAN, J. Chem. Phys. 49, 3890 (1968). (l) C.E. SOLIVEREZ, J. Phys. C2, 2161 (1969).

(10) (a) P.R. CERTAIN and J.O. HIRSCHFELDER, J. Chem. Phys. 52, 5977 (1970). (b) P.-O. LÖWDIN and O. GOSCINSKI, Int. Jl. Quant. Chem. 5, 685 (1971). (c) D.J. KLEIN, J. Chem. Phys. 61, 786 (1974). (d) F. JØRGENSEN, Mol. Phys. 29, 1137 (1975). (e) B.H. BRANDOW, Adv. Quant. Chem. 10, 187 (1977).

(11) (a) A. ABRAGAM and M.H.L. PRYCE, Proc. Roy. Soc. (London) A205,

135 (1951). (b) N.F. RAMSEY, Phys. Rev. 78, 699 (1950) ; ibid. 91, 303 (1953). (c) J.S. GRIFFITH, Mol. Phys. 3, 79 (1960). (d) R. McWEENY, J. Chem. Phys. 42, 1717 (1965). (e) R. CALVO, J. Magn. Res. 26, 445 (1977), and references therein.

(12) (a) R.A. FROSCH and H.M. FOLEY, Phys. Rev. 88, 1337 (1952). (b) M. TINKHAM and M.W.P. STRANDBERG, Phys. Rev. 97, 937 (1955). (c) R.D. MATTUCK and M.W.P. STRANDBERG, Phys. Rev. 119, 1204 (1960). (d) K.F. FREED, J. Chem. Phys. 45, 4214 (1966). (e) T.A. MILLER, Mol. Phys. 16, 105 (1969).

(13) P.A.M. DIRAC, Proc. Cambridge Phil. Soc. 26, 376 (1930) ; ibid. 27, 240 (1931). See also : V. FOCK, Z. Phys. 61, 126 (1930) ; J.E. LENNARD-JONES, Proc. Cambridge Phil. Soc. 27, 469 (1931).

(14) P.-O. LÖWDIN, (a) Phys. Rev. 97, 1474, 1490, 1509 (1955) ; (b) Adv. Phys. 5, 1 (1956) ; (c) Adv. Chem. Phys. 2, 207 (1959) ; (d) Rev. Mod. Phys. 32, 328 (1960).

(15) R. McWEENY, (a) Proc. Roy. Soc. (London) A232, 114 (1955) ; (b) ibid. A235, 496 (1956) ; A237, 355 (1956) ; A241, 239 (1957); (c) ibid. A253, 242 (1959) ; (d) Rev. Mod. Phys. 32, 355 (1960) ; (e) R. McWEENY and Y. MIZUNO, Proc. Roy. Soc. (London) A259, 554 (1961) ; (f) R. McWEENY and W. KUTZELNIGG, Int. Jl. Quant. Chem. 2, 187 (1968).

(16) D. TER HAAR, (a) Physica 26, 1041 (1960) ; (b) Repts. Progr. Phys. 24, 304 (1961).

(17) (a) J.E. LENNARD-JONES, J. Chem. Phys. 20, 1024 (1952). (b) K. RUEDENBERG, Rev. Mod. Phys. 34, 326 (1962). (c) R. McWEENY, Int. Jl. Quant. Chem. 1S, 351 (1967). (d) E.R. DAVIDSON, Rev. Mod. Phys. 44, 451 (1972).

(18) (a) W.A. BINGEL, J. Chem. Phys. 32, 1522 (1960) ; 34, 1066 (1961) ; 36, 2842 (1962) ; W. KUTZELNIGG, Z. Naturforsch. 18a, 1058 (1963). (b) D.A. MICHA, J. Chem. Phys. 41, 3648 (1964) ; D.W. SMITH, ibid. 43, S258 (1965) ; Phys. Rev. 147, 896 (1966) ; K.F. FREED, J. Chem. Phys. 47, 3907 (1967). (c) J.E. HARRIMAN, J. Chem. Phys. 40, 2827 (1964) ; A. HARDISSON and J.E. HARRIMAN, ibid. 46, 3639 (1967). (d) R.D. POSHUTA and F.A. MATSEN, J. Math. Phys. 7, 711 (1966) ; R.D. POSHUTA, ibid. 8, 955 (1967). (e) W. KUTZELNIGG and V.H. SMITH Jr., Int. Jl. Quant. Chem. 2, 531 (1968). (f) W.A. BINGEL and W. KUTZELNIGG, Adv. Quant. Chem. 5, 201 (1970). (g) D.J. KLEIN and R.W. KRAMLING, Int. Jl. Quant. Chem. 3S, 661 (1970) ; D.J. KLEIN, ibid. 3S, 675 (1970). (h) G.G. DYADYUSHA and E.S. KRYACHKO, Preprints of the Institute for Theoretical Physics of the Ukrainian Aca-

demy of Sciences (Kiev, U.S.S.R.) 95E and 98E (1975) ; 116E (1976) ; E.S. KRYACHKO, ibid. 72E (1975) ; 137E (1977).

(19) A.J. COLEMAN, (a) Can. Math. Bull. 4, 209 (1961) ; (b) Rev. Mod. Phys. 35, 668 (1963) ; (c) J. Math. Phys. 6, 1425 (1965) ; (d) Int. Jl. Quant. Chem. 1S, 457 (1967) ; (e) J. Math. Phys. 13, 214 (1972) ; Repts. Math. Phys. 4, 113 (1973) ; (f) Int. Jl. Quant. Chem. 13, 67 (1978).

(20) (a) K. HUSIMI, Proc. Phys.-Math. Soc. Japan 22, 264 (1940). (b) J.E. MAYER, Phys. Rev. 100, 1579 (1955). (c) F. BOPP, Z. Phys. 156, 348 (1959). (d) H.W. KUHN, Proc. Symp. Appl. Math. 10, 141 (1960). (e) C.N. YANG, Rev. Mod. Phys. 34, 694 (1962). (f) T. ANDO, Rev. Mod. Phys. 35, 690 (1963). (g) C. GARROD and J.K. PERCUS, J. Math. Phys. 5, 1756 (1964) ; L.J. KIJEWSKI and J.K. PERCUS, Phys. Rev. 164, 228 (1967) ; ibid. A2, 1659 (1970). (h) T.B. GRIMLEY and F.D. PEAT, Proc. Phys. Soc. 86, 249 (1965) ; F.D. PEAT and R.J.C. BROWN, Int. Jl. Quant. Chem. 1S, 465 (1967). (i) J.J. BRANDSTATTER and C.P. PELTZER, J. Math. Anal. Appl. 16, 123, 472 (1966) ; 19, 179 (1967) ; 33, 263, 500 (1971) ; 34, 1 (1971). (j) F. WEINHOLD and E.B. WILSON Jr., J. Chem. Phys. 46, 2752 (1967) ; 47, 2298 (1967). (k) H. KUMMER, J. Math. Phys. 8, 2063 (1967) ; 11, 449 (1970). (l) M.B. RUSKAI and J.E. HARRIMAN, Phys. Rev. 169, 101 (1968) ; M.B. RUSKAI, ibid. 183, 129 (1969) ; A5, 1236 (1972) ; J. Math. Phys. 11, 3218 (1970) ; J. SIMONS and J.E. HARRIMAN, Phys. Rev. A2, 1034 (1970) ; J. SIMONS, Chem. Phys. Lett. 10, 94 (1971). (m) W.L. CLINTON et al, Phys. Rev. 177, 1, 7, 13, 19, 27 (1969) ; Int. Jl. Quant. Chem. 6, 519 (1972). (n) M.V. MIHAILOVIĆ and M. ROSINA, Nucl. Phys. A130, 386 (1969) ; A237, 229 (1975). (o) E.R. DAVIDSON, J. Math. Phys. 10, 725 (1969). (p) R.M. ERDAHL, J. Math. Phys. 13, 1608 (1972). (q) E.G. LARSON, Int. Jl. Quant. Chem. 7, 853 (1973). (r) F.D. PEAT, Int. Jl. Quant. Chem. 8S, 313 (1974). (s) R. CONSTANCIEL, Phys. Rev. A11, 395 (1975). (t) H. NAKATSUJI, Phys. Rev. A14, 41 (1976). (u) H. KUMMER, I. ABSAR and A.J. COLEMAN, J. Math. Phys. 18, 329 (1977).

(21) (a) A.J. COLEMAN and R.M. ERDAHL, eds., "Reduced Density Matrices with Applications to Physical and Chemical Systems" (Queen's University, Kingston, Ontario, 1968). (b) R. McWEENY and B.T. SUTCLIFFE, "Methods of Molecular Quantum Mechanics" (Academic Press, New York, 1969), esp. chs. 4 and 8. (c) A.J. COLEMAN and R.M. ERDAHL, eds., "Proceedings of Density Matrix Conference" (Queen's University, Kingston, Ontario, 1974). (d) E.S. KRYACHKO and Yu.A. KRUGLYAK, Fiz. Mol. 1, 1 (1975). (e) E.R. DAVIDSON, "Reduced Density Matrices in Quantum Chemistry" (Academic Press, New York, 1976). (f) M.M. MESTECHKIN, "Density Matrix Method in Quantum Chemistry" (Naukova Dumka, Kiev,

U.S.S.R., 1977).

(22) H.M. McCONNELL, J. Chem. Phys. 28, 1188 (1958) {Erratum 30, 328 (1960)}.

(23) G.K. WERTHEIM, "Mössbauer Effect" (Academic Press, New York, 1964).

(24) A. CARRINGTON and A.D. McLACHLAN, "Introduction to Magnetic Resonance" (Harper and Row, New York, 1967).

(25) N.M. ATHERTON, "Electron Spin Resonance" (Halsted Press, New York, 1973).

(26) G.N. LA MAR, W. DE W. HORROCKS Jr. and R.H. HOLM, eds., "NMR of Paramagnetic Molecules" (Academic Press, New York, 1973).

(27) L. KEVAN and L.D. KISPERT, "Electron Spin Double Resonance Spectroscopy" (John Wiley, New York, 1976).

(28) (a) H.M. McCONNELL and D.B. CHESNUT, J. Chem. Phys. 28, 107 (1958) ; H.M. McCONNELL and J. STRATHDEE, Mol. Phys. 2, 129 (1959). (b) C. HELLER and H.M. McCONNELL, J. Chem. Phys. 32, 1535 (1960) ; D. BAHIER and J. MARUANI, C.R. Acad. Sci. (Paris) C275, 257 (1972).

(29) F.K. KNEUBÜHL, Phys. Kond. Mat. 1, 410 (1963) ; R. LEFEBVRE, Mol. Phys. 12, 417 (1967).

(30) A.J. STONE, Proc. Roy. Soc. (London) A271, 424 (1963) ; Mol. Phys. 7, 311 (1964) ; S.H. GLARUM, J. Chem. Phys. 39, 3141 (1963) ; T.N. CASSELMAN and J.J. MARKHAM, ibid. 42, 4176 (1965).

(31) (a) C.P. POOLE Jr., "Electron Spin Resonance : a Comprehensive Treatise on Experimental Techniques" (John Wiley, New York, 1967). (b) J.E. WERTZ and J.R. BOLTON, "Electron Spin Resonance : Elementary Theory and Practical Applications" (McGraw-Hill, New York, 1972). (c) H.M. ASSENHEIM, ed., "Monographs on Electron Spin Resonance" (Adam Hilger, London, introductory volume in 1966). (d) LANDOLT-BÖRNSTEIN new tables, "Magnetic Properties of Free Radicals (vol. II.1, 1965) and Transition-Element Compounds (vol. II.2, 1966)" (Springer Verlag, Berlin).

(32) (a) T.P. DAS and E.L. HAHN, "Nuclear Quadrupole Resonance Spectroscopy" (Academic Press, New York, 1958). (b) E.A.C. LUCKEN, "Nuclear Quadrupole Coupling Constants" (Academic Press, New York, 1969).

(33) (a) F. SEITZ and D. TURNBULL, eds., "Solid State Physics" (Academic Press, New York, 32 issues and 14 supplements

since 1955, several of which contain reviews on magnetic resonance). (b) J.S. WAUGH, ed., "Advances in Magnetic Resonance" (Academic Press, New York, 9 issues and 1 supplement since 1965).

(34) (a) S. GOLDSCHMIDT and K. RENN, Ber. Deut. Chem. Ges. 55, 628 (1922). (b) A.N. HOLDEN, C. KITTEL, F.R. MERRITT and W.A. YAGER, Phys. Rev. 77, 147 (1950) ; C.H. TOWNES and J. TURKEVICH, ibid. 77, 148 (1950). (c) L.S. SINGER and C. KIKUCHI, J. Chem. Phys. 23, 1738 (1955), and references therein. (d) A.L. BUCHACHENKO, "Stable Radicals" (Consultants Bureau, New York, 1965), esp. §IV.1.a.

(35) (a) M. STERNBERG, C.R. Acad. Sci. (Paris) 240, 990 (1955). (b) D.E. WILLIAMS, J. Am. Chem. Soc. 89, 4280 (1967).

(36) (a) H. KATZ, Z. Phys. 87, 238 (1933). (b) H.J. GERRITSEN, R. OKKES, H.M. GIJSMAN and J. VAN DEN HANDEL, Physica 20, 13 (1954). (c) J.H. BURGESS, R.S. RHODES, M. MANDEL and A.S. EDELSTEIN, J. Appl. Phys. 33S, 1352 (1962). (d) A.M. PROKHOROV and V.B. FEDOROV, Soviet Physics-J.E.T.P. 16, 1489 (1963). (e) A. VAN ITTERBEEK and M. LABRO, Physica 30, 157 (1964). (f) W. DUFFY Jr. and D.L. STRANDBURG, J. Chem. Phys. 46, 456 (1967). (g) T. FUJITO, T. ENOKI, H. OHYA-NISHIGUCHI and Y. DEGUCHI, Chem. Lett. 7, 557 (1972). (h) P. GROBET, L. VAN GERVEN, A. VAN DEN BOSCH and J. VAN SUMMEREN, Physica 86-88B, 1132 (1977).

(37) (a) C.A. HUTCHISON, R.C. PASTOR and A.G. KOWALSKY, J. Chem. Phys. 20, 534 (1952). (b) R.M. DEAL and W.S. KOSKI, J. Chem. Phys. 31, 1138 (1959). (c) N.W. LORD and S.M. BLINDER, J. Chem. Phys. 34, 1693 (1961). (d) M.M. CHEN, K.V. SANE, R.I. WALTER and J.A. WEIL, J. Phys. Chem. 65, 713 (1961) ; K.V. SANE and J.A. WEIL, Proc. XIth. Coll. AMPERE (Eindhoven, 1962), p. 431. (e) N.S. GARIFJANOV, A.V. IL'YASOV and Yu.V. YABLOKOV, Proc. Acad. Sci. U.S.S.R., Phys. Chem. Sect. 149, 280 (1963). (f) E.E. SCHNEIDER, Disc. Faraday Soc. 19, 158 (1955). (g) R.W. HOLMBERG, R. LIVINGSTON and W.T. SMITH Jr., J. Chem. Phys. 33, 541 (1960). (h) K. GAMO, K. MASUDA, J. YAMAGUCHI and T. KAKITANI, J. Phys. Soc. Japan 20, 1730 (1965). (i) R. LEFEBVRE, J. MARUANI and R. MARX, J. Chem. Phys. 41, 585 (1964). (j) V.A. GUBANOV, V.I. KORYAKOV and A.K. CHIRKOV, J. Magn. Res. 9, 263 (1973), and references therein.

(38) (a) H.G. BELJERS, L. VAN DER KINT and J.S. VAN WIERINGEN, Phys. Rev. 95, 1683 (1954). (b) Y.H. TCHAO and J. HERVE, C.R. Acad. Sci. (Paris) 250, 700 (1960). (c) Y. DEGUCHI, J. Chem. Phys. 32, 1584 (1960). (d) Z. HANIOTIS and Hs. H. GÜNTHARD, Helv. Chim. Acta 51, 561 (1968). (e) V.A. GUBANOV, V.I. KORYAKOV and A.K. CHIRKOV, J. Magn. Res. 11, 326 (1973). (f) G. BERTHET and

R. REIMANN, C.R. Acad. Sci. (Paris) 246, 1830 (1958). (g) H.S. GUTOWSKY, H. KUZUMOTO, T.H. BROWN and D.H. ANDERSON, J. Chem. Phys. 30, 860 (1959). (h) M.E. ANDERSON, G.E. PAKE and T.R. TUTTLE Jr., J. Chem. Phys. 33, 1581 (1960). (i) J. HERVE, R. REIMANN and R.D. SPENCE, Proc. IXth. Coll. AMPERE (Pisa, 1960), p. 396. (j) R. VERLINDEN, P. GROBET and L. VAN GERVEN, Chem. Phys. Lett. 27, 535 (1974). (k) T. YOSHIOKA, H. OHYA-NISHIGUCHI and Y. DEGUCHI, Bull. Chem. Soc. Japan 47, 430 (1974). (1) R.Z. SAGDEEV, Yu. N. MOLIN, V.I. KORYAKOV, A.K. CHIRKOV and R.O. MATEVOSYAN, Org. Magn. Res. 4, 365 (1972).

(39) (a) J.S. HYDE, R.C. SNEED Jr. and G.H. RIST, J. Chem. Phys. 51, 1404 (1969). (b) N.S. DALAL, D.E. KENNEDY and C.A. McDOWELL, J. Chem. Phys. 59, 3403 (1973). (c) N.S. DALAL, J.A. RIPMEESTER and A.H. REDDOCH, J. Magn. Res. 31, 471 (1978).

(40) (a) R.I. WALTER, J. Am. Chem. Soc. 88, 1930 (1966), and references therein. (b) V.A. GUBANOV, V.I. KORYAKOV, A.K. CHIRKOV and R.O. MATEVOSYAN, J. Struct. Chem. 11, 941 (1970) ; V.A. GUBANOV and A.K. CHIRKOV, Acta Phys. Pol. A43, 361 (1973), and references therein.

(41) D.W. DAVIDSON, S.K. GARG and J.A. RIPMEESTER, J. Magn. Res. 31, 399 (1978).

(42) (a) G. BREIT, Rev. Mod. Phys. 30, 507 (1958). (b) O.C. KISTNER and A.W. SUNYAR, Phys. Rev. Lett. 4, 229 (1960).

# Part V
# Going to The Real World

# ELECTRONIC DENSITIES IN MOLECULES, CHEMICAL

# BONDS AND CHEMICAL REACTIONS

R. Daudel
Sorbonne, Académie des Sciences and Centre de Mécanique Ondulatoire Appliquée du CNRS, 23 Rue du Maroc, Paris, France

## FIRST PART : THE NATURE OF THE CHEMICAL BOND

### What is a Chemical Bond ?

This paper is mainly devoted to the discussion of a central problem in molecular physics and chemistry : the relations between electronic distribution, chemical bonds and chemical reactions.

In the first part, we shall study the nature of the chemical bond from the viewpoint of the wave mechanics.

Let us recall the rigourous mathematical framework of the wave-mechanics, as very often this is not well done.

The important point is to built a system containing a topological vectorial space and an operator domain in such a manner that each operator possesses a complete spectrum on that space.

Gelfan has shown that this can be done in the following way.

The topological vectorial space L is made of three parts :

a) A nuclear space K, that is to say, a space with a particular topology containing a set of related scalar products. This space can be realized by S, the space of rapidly decreasing functions.

b) An Hilbert space H which is the completed space of K with respect to the classical scalar product. This space can be realized by $L^2$, the space of functions for which the square of the modulus can be integrated throughout the configuration space following Lebesgue definition.

c) The dual $K^{x}$ of K.
It is easy to see that :

$$K \subset_i H \subset_i K^{x}$$

if $\subset_i$ denotes an isomorphic inclusion. If K is realized by S, $K^{x}$ is a space of distributions.

The operator domain $\Omega$ is made of the self-adjoint (or hermitian) operators acting on the vectors of the topological vectorial space L.

In this framework an observable is represented by a self-adjoint operator.

From this viewpoint, a chemical bond is not an observable because, as far as I now, no self-adjoint operator has been accepted to represent a given chemical bond. This is why the concept of chemical bond is not well defined from the wave mechanical viewpoint. From this point of view we do not really know what is a chemical bond and therefore there is not a unique approach to the concept. We are led to describe various facets of the chemical bond.

The Energetic Facet.

First of all we can try to understand the energetic nature of the chemical bond. Let us consider the simplest molecule : the hydrogen molecule-ion $H_2^+$. It is often said that the bond results from the fact that when a proton approaches an hydrogen atom a region appears between the two nuclei in which a favorable potential occurs because the electron is attracted there by both nuclei.

This statement is not convincing because, in fact, the potential in this region remains obviously less favorable than the one which occurs near each nucleus. The statement has to be completed by considering the velocity of the electrons in the corresponding regions.
As we know the momentum is related to the derivatives of the wave function with respect to the electron coordinates. Near the nuclei the wave function changes rapidly with the coordinates because the potential do so, therefore the velocity of an electron is this region is high. There, the potential is favorable but the electron does not remain a long time.

Between the nuclei it is easy to see that the potential does not change rapidly with the coordinates. Therefore the wave function does the same and the velocity of the electron remains small. This is why we can say that the electron visits the bonding region during an appreciable amount of time. The potential is there less favorable than near the nuclei but the kinetic energy being smaller is more

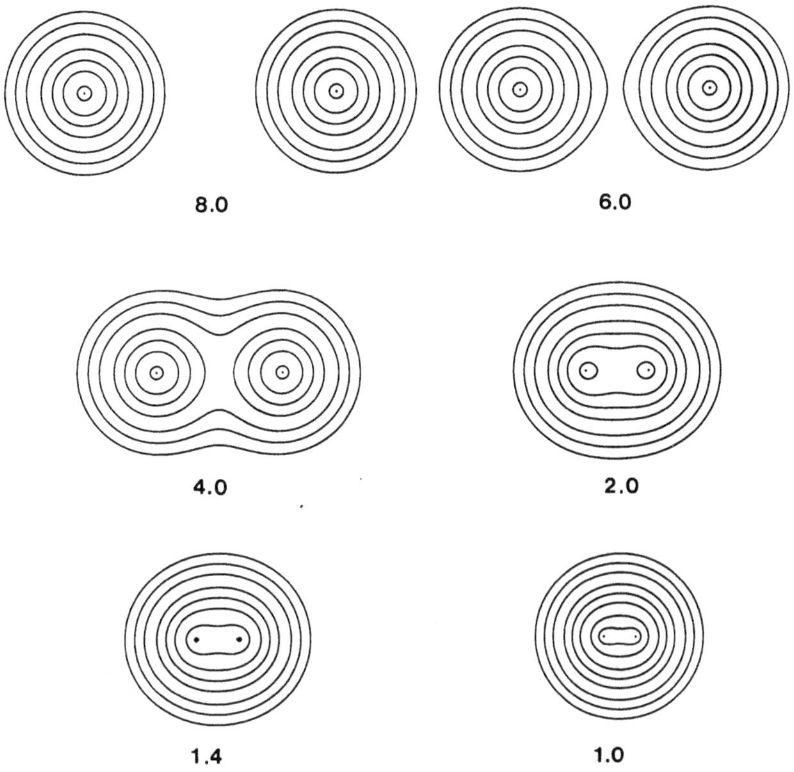


Figure 1 - Electronic density change during the formation of the hydrogen molecule. (Reproduced by permission from R.F.W. BADER, An introduction to the electronic structure of atoms and molecules, Clarke, Irwin and Cy. Lim. TORONTO, 1970)

favorable. Such would be the main source of bond energy.

The other source seems to be a consequence of the first one. The decrease of the kinetic energy in the binding region produces a decrease of the kinetic resistance against the nuclear suction. This result is a closer attachment of the electronic cloud to the nuclei with a concomitant lowering of the potential energy and increase in the corresponding kinetic energy. This is why both the modulus of the potential energy and the modulus of the kinetic energy increase simultaneously as required by the virial theorem.

For a more sophisticated analysis of the problem see Ruedenberg[(1)].

When a two electron bond is considered (as in the hydrogen molecule) it is sometimes said that a part of the bond energy is due to a certain exchange energy.

Let us recall that Lennard Jones has shown that the exchange energy has no physical meaning.
This term corresponds to a part of the total energy which has only a mathematical meaning in the framework of certain approximate calculations and which is not unvarying through certain unitary transforms which do not alter the wave function.

Furthermore as the virial theorem applies, the total energy E of a molecule satisfies the equation

$$E = \langle\psi H\psi\rangle = \langle\psi V\psi\rangle + \langle\psi T\psi\rangle = 1/2 \langle\psi V\psi\rangle$$

if V represents the potential energy operator and T the kineti energy operator. Therefore the bond energy can be understood i terms of potential energy mean value.

## The Electronic Facet

The analysis of the various density matrices is another interesting way to explore the nature of the wave function. The study of the electronic density itself $\alpha(M)$ is the more natural. This density can be measured or calculated. Figure 1 shows how the density maps of two hydrogen atoms change when their internuclear distance decreases

(1) K. RUEDENBERG in Localization and Delocalization, CHALVET, DINER, MALRIEU ed, REIDEL Pub. (1975) p. 223.

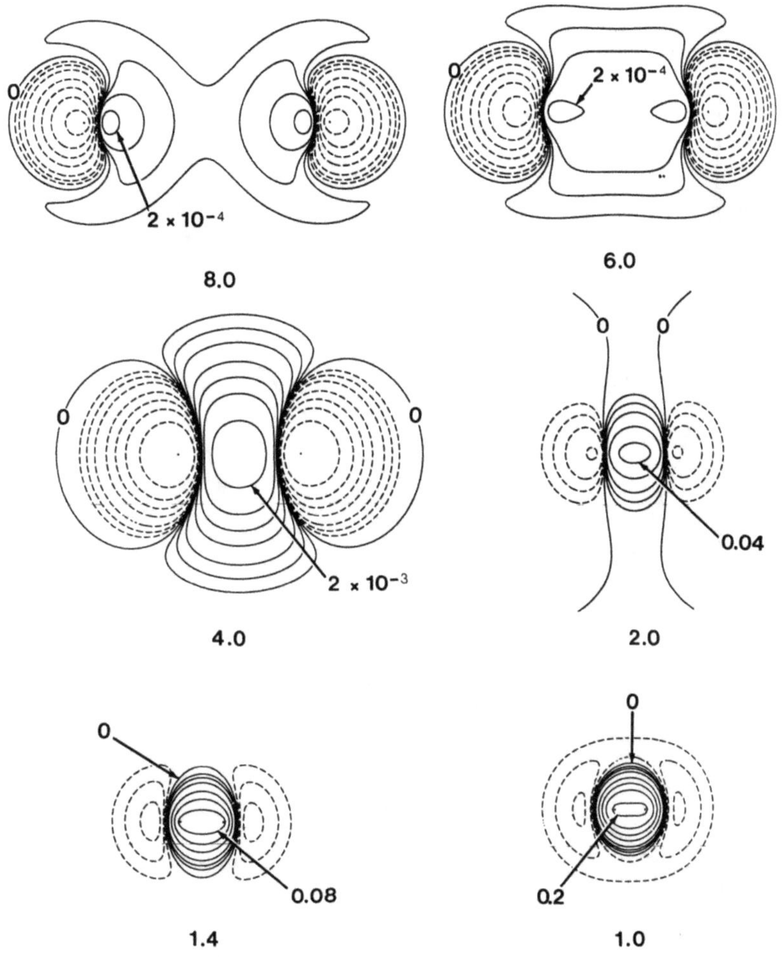


Figure 2 - Density difference function during the formation of the hydrogen molecule (after BADER, loc. cit.) Dashed lines corresponds to negative values, solide contours to positive values.

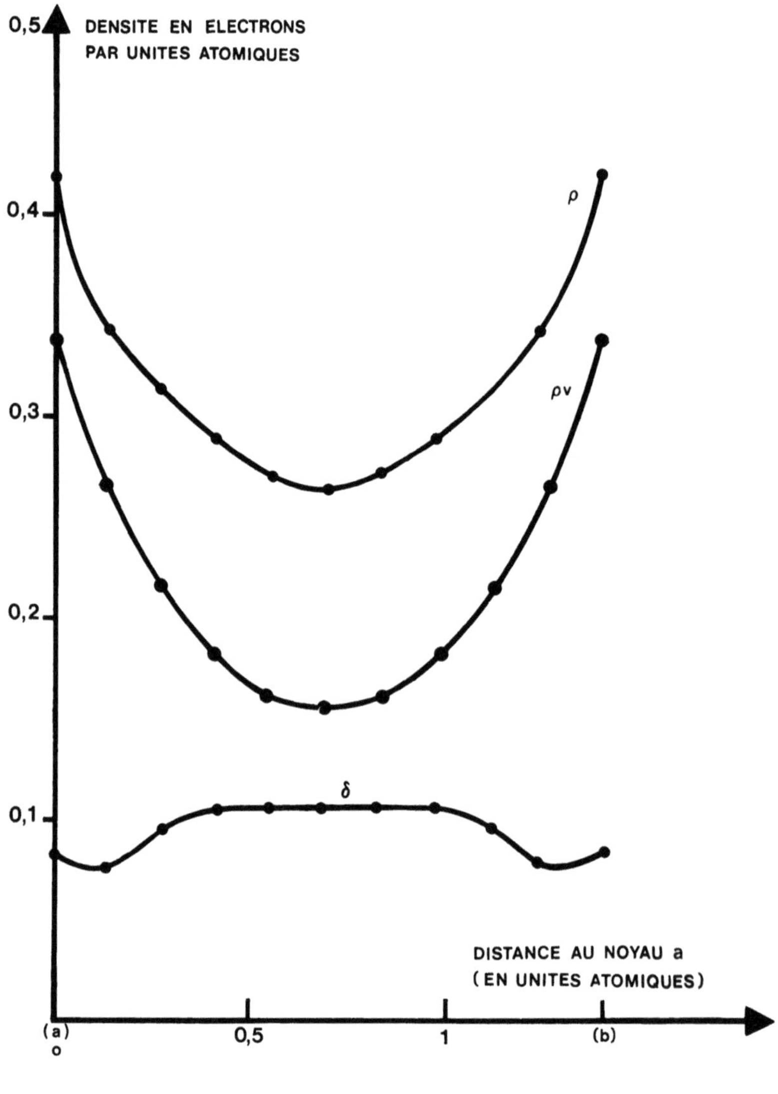


Figure 3 - Density difference function along the line of the nuclei in the hydrogen molecule.

The result is rather disappointing. The figure shows a kind of polarization which rapidly occurs between the two atoms : an increase of electronic density appears on each atom pointing in the direction of the other. For a shorter distance the two atomic electronic clouds are fused and a density saddle appears between the nuclei, but no specific region appears which could be associated with the idea of bond.

This is why I introduced the concept of density difference function $\delta(M)$ (1) which is defined as the difference between the actual density $\rho(M)$ which occurs at point M in a molecule and the fictivous density $\rho^f(M)$ which would simply result from the addition of the density in the free atoms.

$$\delta(M) = \rho(M) - \rho^f(M)$$

In a point where $\delta(M)$ is positive the formation of the molecule from free atoms leads to an increase in the electronic density. The contrary is true in a point where $\delta(M)$ is negative. It is readily seen that :

$$\int_{R^3} \delta(M)\, dv = 0$$

Figure 2 shows the density difference maps during the approach of two hydrogen atoms. We see clearly now that between the nuclei appears a region in which the electronic density has been increased during the formation of the bond. On the other hand outside of the central region of the molecule regions appear in which the formation of the bond has led to a decrease of the density. The formation of bond leads to a transfer of electron between the two nuclei as suggested by chemical intuition. The advantage of wave mechanics is to give a quantitative evaluation of the phenomenon and to give the frontiers of the region in which it occurs. It is seen that the phenomenon is important as for the equilibrium distance of the bond (1.4 $a_o$). $\delta(M)$ reaches a positive value of about one tenth of electron per cubic atomic unit on a volume of about one atomic unit.

Figure 3 gives an other representation of the density difference function along the line of the nuclei in the hydrogen molecule.

The figure 4 is concerned with one homonuclear molecule ($N_2$) and an heteronuclear one (FLi)
In the case of $N_2$ the density difference function is positive between the nuclei and also at the extremities of the molecule in a region where classical chemists see electron lone pairs. In the case of

(1) R. DAUDEL, C.R. Acad. Sci. 225, 886 (1952).
M. ROUX et R. DAUDEL, C.R. Acad. Sci. 240, 90 (1955).

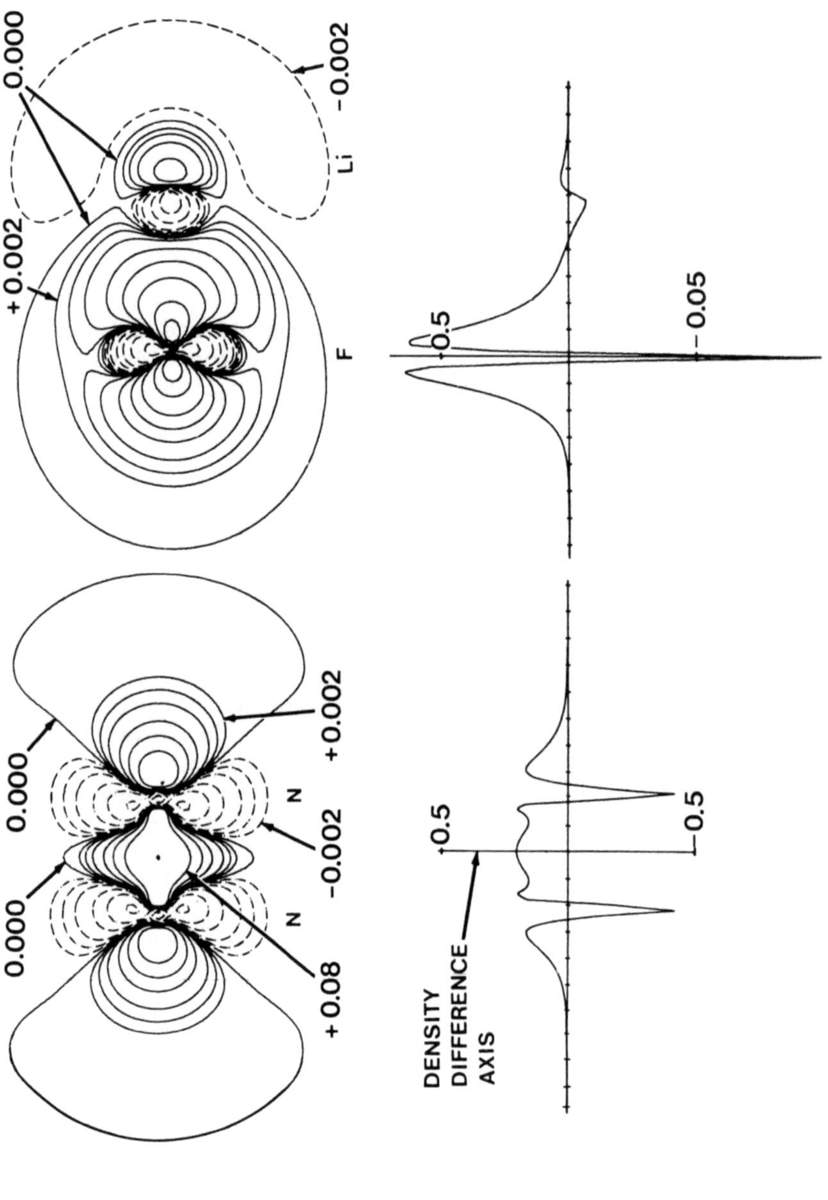


Figure 4 - Density difference function in $N_2$ and FLi (after BADER loc. cit.)

FLi the most interesting feature is the charge transfer which is produced between the lithium and the fluorine atoms during the formation of the bond as between the two nuclei the function $\delta(M)$ is negative on the lithium side and positive near the fluorine side.

Finally the figure 5 describes the $\delta(M)$ function along an hydrogen bond. It is seen that the main phenomenon is a decrease of the electronic density near the hydrogen making the bond and an increase of that density along the NH bond. Professor OLOVSSON has shown how it is in some cases distinguished between the part of the phenomenon which is really due to the formation of the hydrogen bond and the part which already exists in the two molecules before the formation of the bond.

All results presented until now come from wave-mechanical calculations. In fact, it is now well known that the agreement between theoretical calculations and experimental measurements is good when wave functions are calculated at least at the self consistent field level by using a sufficiently large basis set of gaussian functions. The main discrepancy appears near the nuclei and is due to the nuclear movement which occurs in the crystal and which is usely not taken into account in the theoretical calculations.[1] Usually the corrections introduced by the configuration interaction are not very important taking account of the uncertainties of the measurements.

An other method of using the electronic density to analyze the nature of the chemical bond has been introduced by Kurki-Suonio[2]. It consists of plotting the radial density $4\pi r^2 \rho(M)$ as a function of the distance r of M from the nucleus of an important atom of a molecule.

Figure 6 shows the result of such a study in BaO when the oxygen nucleus is taken as a center. The value of r which corresponds to the minimum of the radial density can be considered to be the frontier of the oxygen domain. The amount of charge which is then found in this domain corresponds approximately to $O^{-2}$ showing that in such a compound the bond is essentially ionic.

To end this section concerned with the electronic density let us say that they are some interesting relations between the various density matrices. We shall present only one result related to this topic. If we consider the total electronic charge Q in a certain

(1) See for example : P. BECKER, Physica Scripta, 15, 119 (1977)
(2) K. Kurki-Suonio, Acta Cryst. A24, 379 (1978).

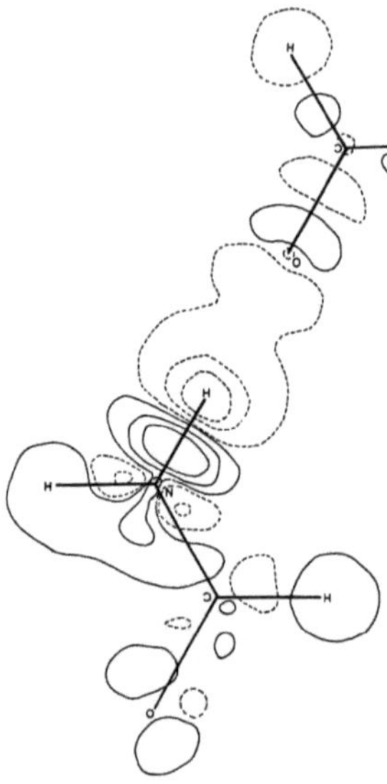


Figure 5 - Density difference function in the formamide dimer (after A. PULLMAN in R. DAUDEL et A. PULLMAN, Aspects de la Chimie Contemporaine, C.N.R.S, Paris, 1971) reproduced by permission.

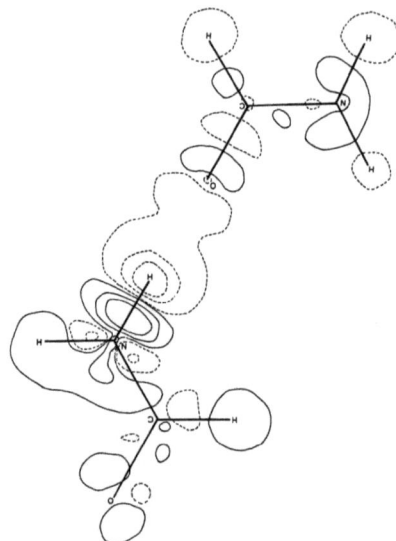

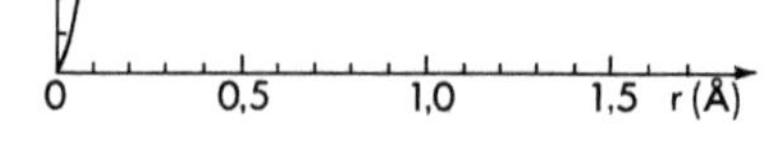

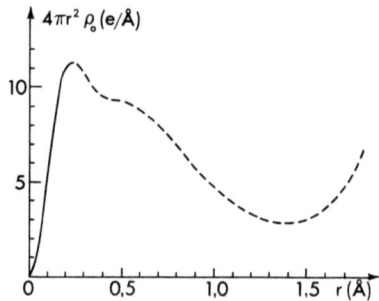

Figure 6

volume V for a system containing say four electrons it is obviously the sum of :

a) the probability $P_1$ of finding one electron and one only in V

b) twice the probability $P_2$ of finding two electrons and two only in V

c) three times the probability $P_3$ of finding three electrons and three only in V

d) four times the probability $P_4$ of finding four electrons and four only in V.

$$Q = P_1 + 2P_2 + 3P_3 + 4P_4$$

This formula is readily generalized and shows that two identical charges can have different fine structure and therefore different behaviour in chemical reactivity.

A completely different process to define chemical bonds from electron behaviour is the loge theory.(1) This theory provides a procedure of analyzing the localizability of the electrons of an atom or a molecule. A loge is a volume of the electronic system in which there is a high probability of finding a certain number n of electrons and only this number. The best partition of an atom of a molecule in loges in obtained by varying the frontiers of the loges to minimize the corresponding missing information function in such a way to obtain the maximum amount of information about the localizability of the electrons.

The example of the BH molecule(2), a six-electron problem, will help to understand the theory. Consider a sphere of arbitrary radius r centered at the boron nucleus. Seven "electronic events" are possible. We can find 0, 1, 2, 3, 4, 5 or 6 electrons in that sphere. The probability $p_i$ of occurrence of the electronic events corresponding to the finding of i electrons can be easily calculated from the wave-function. For example if V denotes the volume of the sphere :

$$p_1 = 6 \int_V dv_a \int_{R^3-V} dv_b dv_c dv_d dv_e dv_f |\Psi (M_a, M_b, M_c, M_d, M_e, M_f|^2$$

(1) R. Daudel, C.R. Acad. Sci. 237, 601 (1953) ; S. Odiot Cah. Phys. 81, 1 (1957) ; C. Aslangul B272, 1 (1971).
Localization and Delocalization Quantum Chemistry, Chalvet, Daudel, Diner and Malrieu éd. Reidel pub. Vol. I (1975) Vol. II (1976).

(2) R. Daudel, R.F.W. Bader, M.E. Stephens and D.S. Borrett, J. Can. Chem. 52, 1310 (1974).

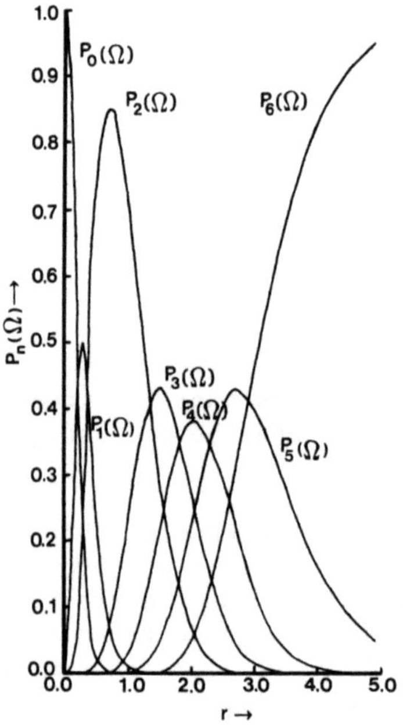


Fig 7

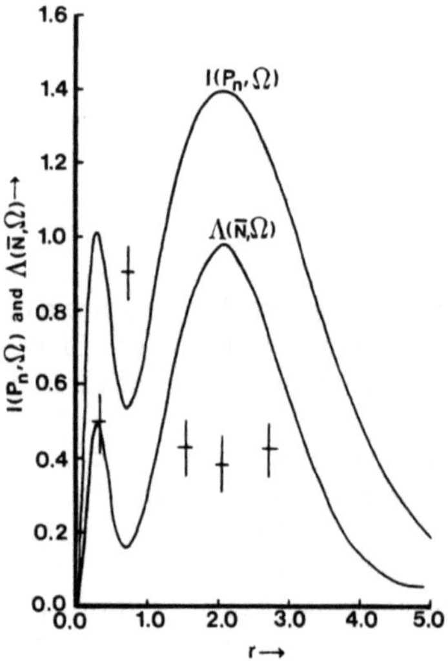


Fig 8

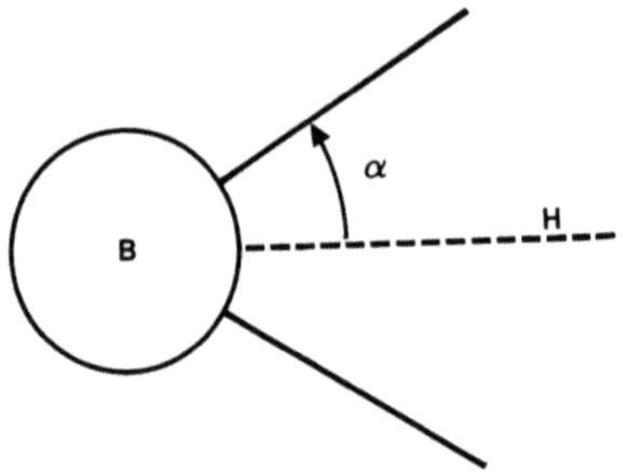


Fig 9

It is seen that the calculation of the $p_i$'s depends on the 6 order density matrix.

Fig. 7 shows the variation with r of the various probabilities $p_i$. It is seen that of all the $p_i$'s only $p_2$ reaches an important maximum value (>0.8) for a value of r different from zero or infinity. The probability of finding a pair of electrons in the sphere is the only event which can produce a good loge and this occurs for :

$$r = 0.7 \text{ a.u.}$$

If now we consider the missing information function I associated with that distribution of probabilities :

$$I = \sum_i p_i \log_2 p_i^{-1}$$

it is readily seen that it reaches a minimum value for

$$r = 0.7 \text{ a.u.}$$

(figure 8). Therefore the sphere possessing this radius produces the best partition in two loges of the molecule and we can call <u>core loge</u> that sphere because it has approximately the same radius that in the free boron atom.

To go further we can consider now a three loges partitioning containing the foregoing sphere and a portion of cone of angle $\alpha$ possessing the BH line as an axis (figure 9). The new information function reaches its minimum value for :

$$r = 0.7 \text{ a.u.}$$
$$\alpha = 73°$$

and the leading event that is to say the event which occurs with the highest probability corresponds to a finding of two electrons in each loge. We can say that the inside of the cone is the <u>bond loge</u> BH and the outside the <u>lone pair loge</u>.

The loges show many other important features. First of all, the fluctuation $\Lambda$ of the number N of electron in a loge :

$$\Lambda = \overline{N^2} - (\overline{N})^2$$

is minimized in the same time that the missing information function (figure 8). Therefore a good loge is a region of the space in which the number of electrons does not fluctuate much. It is important to remark that the fluctuation only depends on the 2-order density matrix.

Furthermore there is a relation between loge and electron

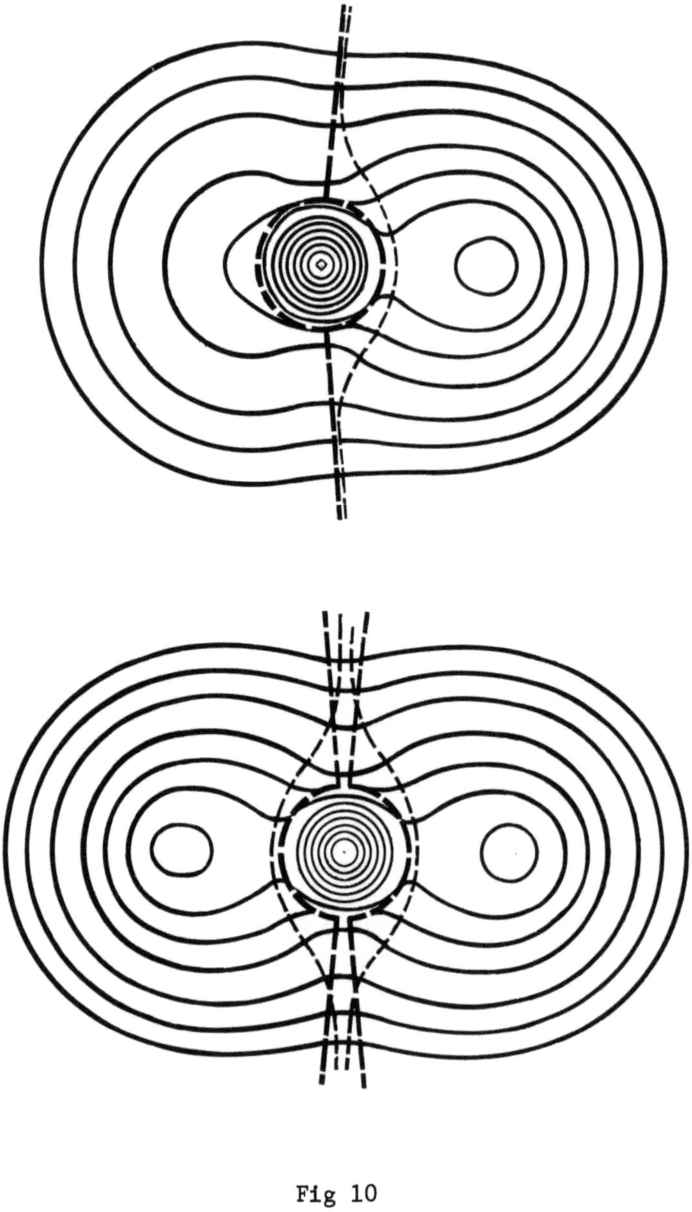

Fig 10

correlation.(1)
Let $P_2(M_a, M_b)$ be the probability density of finding one electron at point $M_a$ and another one at point $M_b$. If $P_1(M_a)$ is the probability density of finding one electron at point $M_a$ and $P_1(M_b)$ the probability density of finding one electron at point $M_b$. If the two events were independant we should have :

$$P_2(M_a, M_b) = P_1(M_a)\, P_1(M_b)$$

If a correlation occurs between the two electrons

$$P_2(M_a, M_b) \neq P_1(M_a)\, P_1(M_b)$$

and following McWeeny we can write

$$P_2(M_a, M_b) = P_1(M_a)\, P_1(M_b)\, [1 + f(M_a, M_b)]$$

where f is called the correlation factor. This factor has usually a negative value and measure the "Fermi hole".

Bader has shown that the fluctuation of the number N of electrons in a loge V can be written as :

$$\Lambda = \overline{N} + \int_V P_1(M_a)\, P_1(M_b)\, f(M_a, M_b)\, dv_a dv_b$$

Therefore $\Lambda$ is minimized when the integral reaches its highest negative value that is to say when the correlation between the electrons of the loge reaches its maximum value.

By a consequence the best partition into loges corresponds to a maximization of the correlation in the loges and therefore a minimization of the correlation between the loges.

An other important behaviour of the loges are their transferability from molecule to molecule.

Figure 10 permits the comparison between the BeH bond loge in BeH and the corresponding one in HBeH. It is seen that they are very similar in size and shape. Furthermore their "fine structures" (that is to say the probabilities of the various electronic events) are approximately the same.

It is possible to demonstrate rigorously(2) that any expecta-

(1) R.F.W. Bader in Localization and Delocalization in Quantum Chemistry, Chalver, Daudel, Diner et Malrieu ed, Reidel Pub. (1975) p. 15.
(2) See for example R. Daudel in Localization and Delocalization in Quantum Chemistry (loc.cit) Vol. II (1976) p. 72.

tion value $<\psi|\Omega|\psi>$ associated with a two-electron (or less) operator $\Omega$ can be expressed as a sum of loge contribution $\Omega_{\ell}$ and of pair loge contributions $\Omega_{\ell\ell'}$.

$$<\psi|\Omega|\psi> = \sum_{\ell} \Omega_{\ell} + \sum_{\ell,\ell'} \Omega_{\ell\ell'}$$

For some monoelectronic operator (as dipole moment) the $\Omega_{\ell\ell'}$'s vanish. In the case where the loges are transferable the modulus associated with a given kind of loge $\ell$ remains approximately the same in the family of molecules concerned. When the $\Omega_{\ell\ell'}$'s vanish the molecular property becomes a simple sum of the loge moduli. An additive law is observed. This is, for example, the case of the Faraday effect[(1)].

When the $\Omega_{\ell\ell'}$'s do not vanish they yield deviation to the additivity law. These terms are, for example, responsible for isomerization energies.[(2)]

The loge theory can be used as the starting point of a classification of chemical bonds.

First of all, it gives a good criterion to distinguish between localized and delocalized bonds. A bond can be considered to be localized when its loge lies between two adjacent core loges. The bond is delocalized when the loge is extended over more than two cores.

The loge theory has been also used to distinguish between a dative bond and a covalent bond. The main difference does not lie in the electronic distribution in the bond loge. It lies on the difference of its environment. A typical covalent bond loge (the CN loge in $H_3C - NH_2$ for example) lies between two groups of loges possessing exactly the charge + e for the leading event.
A typical dative bond (the BN loge in $H_3B \leftarrow NH_3$) lies between two groups of loges possessing respectively the charges 0 and + 2e for the leading event.[(3)]

Finally Bader and Stephens[(4)] have used the loge theory to characterize each "two electron bond" by a number measuring its distance from a pure two electron bond that is to say a loge in which the probability to find two electrons and two only is 1. They define a percentage of localization in the loge which is derived from the relative fluctuation number of the electrons in the loge.

(1) R. Daudel, F. Gallais and F. Smet, Int. J. Quantum Chemistry 1, 873 (1967).
(2) R. Daudel, C.R. Acad. Sci. 270, 929 (1970).
(3) R. Daudel and A. Veillard, Nature et Propriétés des Liaisons de Coordination C.N.R.S. pub. 1970 p. 15.
(4) R.F.W. Bader and M.E. Stephens, J. Am. Chem. Soc. 97, 739 (1975).

Table 1 gives a few examples.

Table 1
Properties of two electron bond

| Molecule | percentage localization |
|---|---|
| LiH | 95.5 |
| $BeH_2$ | 92.8 |
| $CH_4$ | 68.9 |
| $F_2$ | 17.4 |

It is seen that in LiH the bond is really near an ideal two electron bond but that in $F_2$ the bond is far from beeing so : the number of electrons in the loge largely fluctuates between 0, 1, 2, 3 and 4.

## The Functional Facet

Very often a molecular wave-function $\Psi$ can be written approximatively as :

$$\Psi \approx \mathcal{a}.\ K_A\ (M_a, M_b)\ K_B\ (M_c, M_d)\ \ldots\ P_{AB}\ (M_j, M_k)$$

$$\ell'_{AB}\ (M_\ell\ M_m)\ L_{CDE}\ (M_n, M_o, M_p)\ \ldots\ \sigma \qquad (1)$$

where $\mathcal{a}$ is an antisymmetrizer and $\sigma$ the spin function.

The various space functions may depends of one, two, three or more points.
The K's are assumed to be mainly important near one of the nucleus, A, B, C... . They are called core functions. The $\ell$'s and the L's are mainly important between two, three or more cores. They can be considered as bond functions. When a function only depends on one point it is called orbital.

The difference between the concept of loge and that functional description lies mainly in the fact that the functions are not localized in separate volumes. For example, as Coulson has shown, $\sigma$ and $\pi$ orbitals may have important values in the same region of the molecular space.

However very often the "factorization" of a function $\Psi$ as in equation 1 is not unique. Many equivalent "factorizations" are possible. This is, for example, the case if the $\Psi$ is written as a simple Slater determinant of spin-orbitals. It is well known that in this

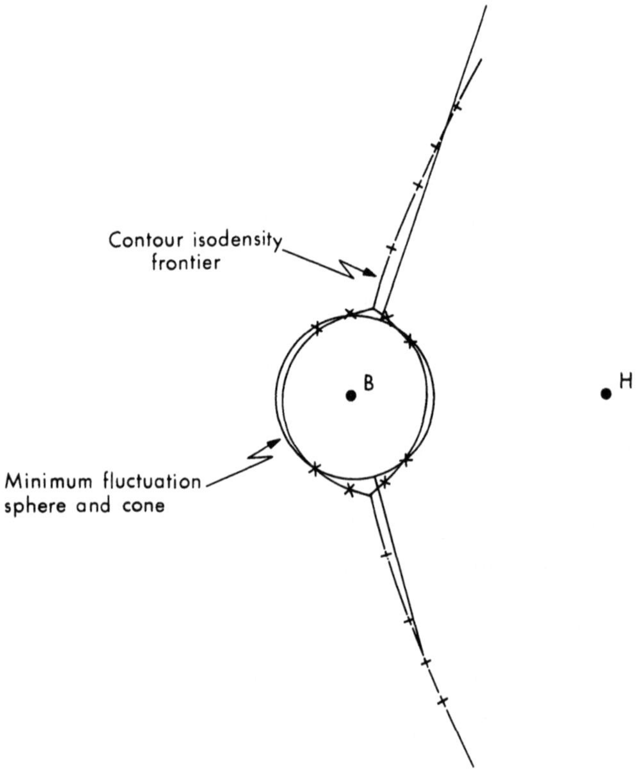
Contour isodensity
frontier
B
H
Minimum fluctuation
sphere and cone

Fig 11

case it is possible to replace a set of orbitals by another set resulting in the action on the formers ones of a unitary transform: the determinant and therefore the $\Psi$ does not change under the effect of this operation.

When this happens it is intersting to select the "most localized bond functions" and various criterions have been proposed to define and to calculate the most localized orbitals or to built more elaborate localized bond functions.

Usually localized orbitals and more generally localized bond function show properties analogous to those of loges. Their is a kind of isomorphism between loges and localized functions. This fact is very useful because it is more easy to calculate localized functions than to built loges.

Therefore we shall go further into some details.
First of all, the localized functions can be used to estimate the frontier of the loges.
To do so it is only necessary to plot the iso-density contours of the various localized functions. The intersections of contours with the same density values lie very near the loges. The figure 11 permits the comparison of the boundaries obtained by this procedure and the frontiers of the loges in the case of BH[1].

In fact, other procedure can be used to obtain an estimation of the frontier of the loges. Odiot[2] has shown that for an atom the radii of the frontiers of the loge correspond approximately to the minima of the radial density. Therefore the Kurki-Suonio method produces boundaries analogous to loge frontiers. Furthermore, as it is easy to see on figure 10, some frontiers of bond loges correspond approximately to steepest descent lines on the electronic density surface.

An other important point is the transferability of localized functions. The transferability of localized orbitals has been observed by Leroy and Peeters[3]. The transferability of more elaborate bond functions (group or loge functions) has been observed by Sanchez et al.[4]

A third interesting problem is related to the size and the shape of loges and localized functions.

(1) R. Daudel, M.E. Stephens, L.A. Burke and G. Leroy, Chem. Phys. Letters, 52, 426 (1977).
(2) S. Odiot, Cah. Phys. 81, 1 (1957).
(3) G. Leroy and D. Peeters, Localization and Delocalization (loc. cit.) Vol. I, p. 207.
(4) M. Sanchez, R. Daudel, P.D. Dacre, R. McWeeny, S. Kwun and C. Valdemoro, Int. J. Quantum Chemistry, 11, 415 (1977).

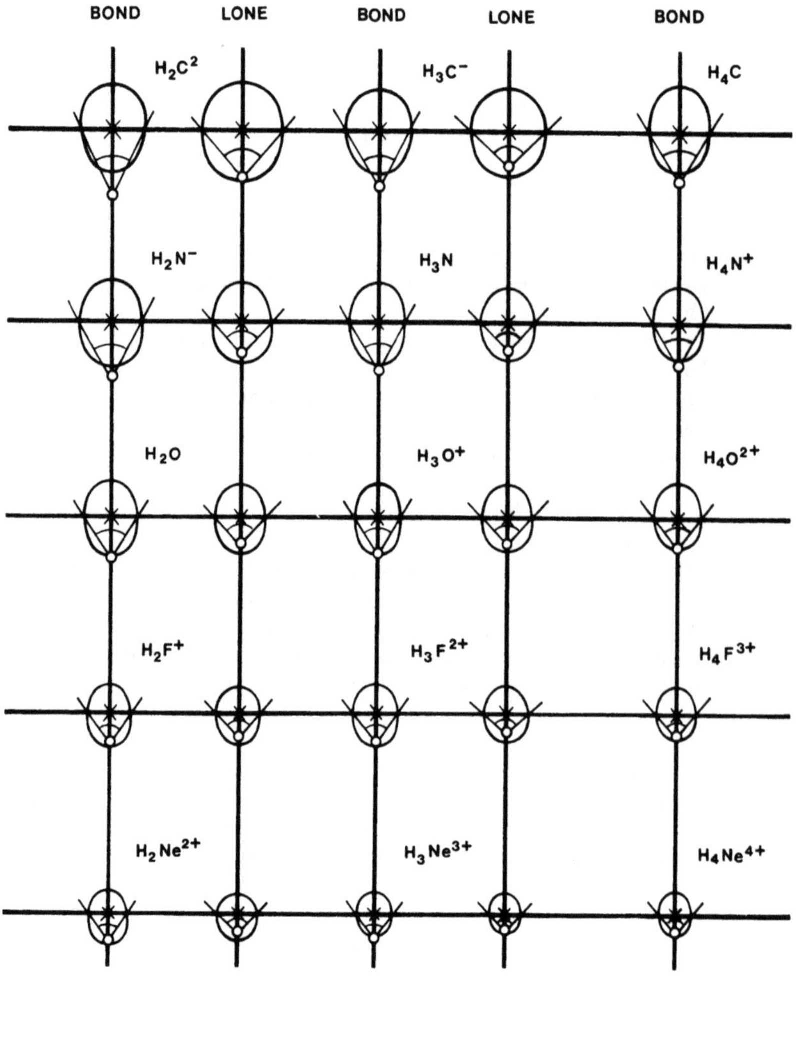
BOND
LONE
BOND
LONE
BOND
H2C2
H3C−
H4C
H2N−
H3N
H4N+
H2O
H3O+
H4O2+
H2F+
H3F2+
H4F3+
H2Ne2+
H3Ne3+
H4Ne4+

Figure 12

For atom a kind of Boyle Mariotte law has been found[1] between the volume v which an electron occupies when it visits a loge and the mean value p of the electric potential acting on that electron during the visit. More precisely the following relation applies for all loges of all atoms :

$$p^{3/2}.v = \text{cste} \qquad (2)$$

A relation of that kind has been also observed for molecules by using localized orbitals to measure the volume occupied by electrons.

The "size" of a localized orbital can be defined by considering the three quadratic moment of the orbital : $<x^2>$, $<y^2>$ and $<z^2>$ the origin of the coordinates being the centroid of charge.
The expression :

$$\sqrt{<x^2> <y^2> <z^2>}$$

can be taken as a measure of the size of the orbital.

An equation analogous is obtained[2] for molecule if v is replaced by this expression and p by the energy associated with the orbital.

It is also possible to represent the orbital by an ellipsoid with haf-axes $<x^2>^{1/2}$, $<y^2>^{1/2}$ and $<z^2>^{1/2}$. The figure 12 shows the result of some calculations of that kind.[3]

It is seen that contrarilly to Gillespie postulate a lone pair can be less bulky than a bond. This is, for example, the case in $NH_3$. However, the angles contained by the rays between the end points of the maximum width of the ellipsoid (as seen from the centroid of charge) provides a measure of the angular size of the various localized orbital. From figure 12 it appears that the angular size of a lone pair is always greater than the angular size of an adjacent bond. This is perhaps the reason of the success of the VSEPR method[4].

Finally it must be emphasized that the consideration of the most localized orbitals can be confusing. Because in certain cases the most localized orbitals are not really localized.

(1) S. Odiot and R. Daudel, C.R. Acad. Sci. 238, 1384 (1954).
(2) R. Daudel, J.D. Goddard and I.G. Csizmadia, Int. J. Quantum Chemistry
(3) R. Daudel, M.E. Stephens, I.G. Csizmadia, C. Kozmutza, E. Kapuy and J.D. Goddard, Int. J. Quantum Chemistry 11, 665 (1977).
(4) R.J. Gillespie, Molecular Geometry, (Van Nostrand, N.Y. (1972)).

To measure the degree of localizability of two orbitals f and g it is convenient to calculate the integral :

$$S' = \int |f|g|dv$$

If the relative localization is complete the integral vanishes. If the orbitals completely overlap the integral is unity.
Between two localized bond CH orbital is methane

$$S' = 0.404^{(1)}$$

We see that even on this typical example of covalent bond the localization of the most localized orbitals is far from beeing complete.

## SECOND PART : QUANTUM THEORY OF CHEMICAL EQUILIBRA

### Equilibrium Constant

Let us consider a reversible equilibrium of the simplest type, namely

$$A \underset{}{\overset{K}{\rightleftharpoons}} B$$

like tautomeric equilibria.
If we suppose that the law of mass action applies, we will get

$$K = \frac{[B]}{[A]}$$

if the square brackets show the concentration of the corresponding molecule. We saw that the law applies in the gaseous phase when the pressure is not too high.
Let us use $\varepsilon_{iA}$ to denote the possible energy levels of the molecules A, taking into account the conformations, rotation, vibration, and so on.
Boltzmann's law teach us that the number $N_{iA}$ of molecules possessing the energy $\varepsilon_{iA}$ is proportional to

$$p_{iA}\, e^{-(\varepsilon_{iA}/\chi T)}$$

if $\chi$ represents Boltzmann's constant and $p_{iA}$ the a priori probability of the energy $\varepsilon_{iA}$, that is to say the number of physically different states possessing the energy under consideration.

The total number $N_A$ of the molecules A may therefore be written

(1) R. Daudel, M.E. Stephens, E. Kapuy and C. Kozmutza, Chem. Phys. Letters.

$$N_A = b \sum_i p_{iA} e^{-(\varepsilon_{iA}/\chi T)}$$

in which b is a proportionality constant.

If these molecules occupy a volume V, one gets

$$[A] = \frac{N_A}{V} = \frac{b}{V} \sum_i p_{iA} e^{-(\varepsilon_{iA}/\chi T)}$$

whence, by applying the same formula to molecule B :

$$K = \frac{[B]}{[A]} = \frac{\sum_j p_{jB} e^{-(\varepsilon_{jB}/\chi T)}}{\sum_i p_{iA} e^{-(\varepsilon_{iA}/\chi T)}}$$

If $\varepsilon_{oA}$ and $\varepsilon_{oB}$ represent respectively the lowest possible energies for the species A and B, one may write :

$$K = \frac{\sum_j p_{jB} e^{-((\varepsilon_{jB}-\varepsilon_{oB})/\chi T)}}{\sum_i p_{iA} e^{-((\varepsilon_{iA}-\varepsilon_{oA})/\chi T)}} e^{-(\varepsilon_{oB}-\varepsilon_{oA})/\chi T)}$$

Putting

$$f_M = \sum_j p_{jM} e^{-((\varepsilon_{jM}-\varepsilon_{oM})/\chi T)}$$

a function, called partition function of molecules M, it is readily seen that :

$$K = \frac{f_B}{f_A} e^{-\Delta\varepsilon/\chi T}$$

if we put :

$$\Delta\varepsilon = \varepsilon_{oB} - \varepsilon_{oA}$$

This formula is easy to generalize for more complicated equilibrium.

Therefore an equilibrium constant in the gas phase depends on two main factors : the ratio of the partition function which depends on the temperature, the difference of the lowest energies of the molecules before and after reaction which does not depends on that temperature.

Very often we are interested to compare analogous reactions for example two protonation reactions :

$$A + H^+ \underset{}{\overset{K}{\rightleftharpoons}} AH^+$$

$$A' + H^+ \underset{}{\overset{K'}{\rightleftharpoons}} AH^+$$

The ratio $\frac{K'}{K}$ may be written as :

$$\frac{K'}{K} = \frac{f_{A'} f_{H^+} / f_{A'H^+}}{f_A f_{H^+} / f_{AH^+}} \, e^{\frac{-\Delta\varepsilon' - \Delta\varepsilon}{\chi T}}$$

Very often the ratio of the ratios of the partition functions is not far from unity and therefore :

$$\frac{K'}{K} \approx e^{-\frac{\Delta\varepsilon' - \Delta\varepsilon}{\chi T}}$$

In such a case the order of the equilibrium constant for the protonation reactions is given by the order of the protonation energies.

Obviously protonation energies can be directly calculated by using wave-mechanics. But at the beginning of quantum chemistry it was believed that the protonation energies will run parallel to the electronic charge of the atom on which the protonation occurs. And this idea establishes a bridge between some aspect of the chemical reactivity and the distribution of electronic density in molecules. But nowadays it is known that this idea is only very crude and that it is possible to do a better job by considering the electrostatic potential created by a molecule in the surrounding space.

If $\rho(M)$ represents the electronic density at point M of a molecule the electrostatic potential V(P) at a given point P of space is :

$$V(P) = \sum_\alpha \frac{Z_\alpha}{r_{\alpha P}} - \int \frac{\rho(M)\, dv_M}{r_{MP}}$$

if $r_{MP}$ denotes the distance between M and P and $r_{\alpha P}$ the distance between P and the nucleus $\alpha$ of atomic number $Z_\alpha$ (1).

The interest for us of this concept is mainly due to the fact that it can be calculated from both experimental or theoretical electronic density. Furthermore it is a global expression of the molecular reality that is in direct relation to what an approaching reagent sees when coming near the molecule. If a point charge q is placed in a point where the potential is V the quantity qV represents the first order approximation of the electrostatic energy

(1) R. Bonaccorsi, E. Scrocco and J. Tomasi, J. Chem. Phys. 52, 5270 (1970).

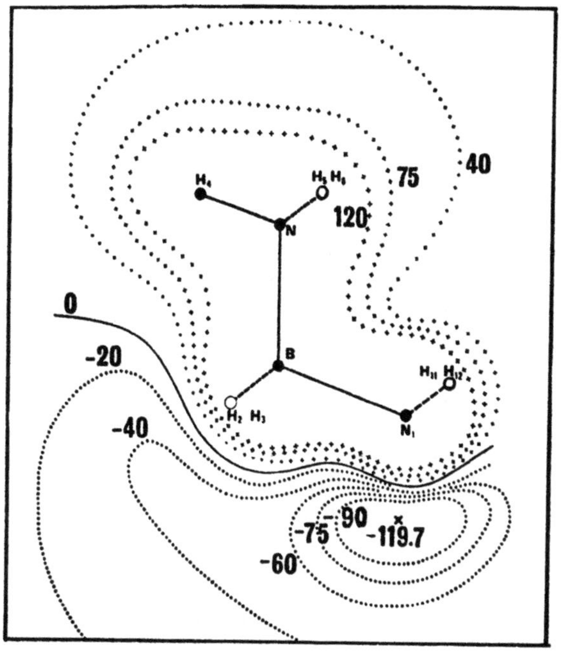


Fig 13

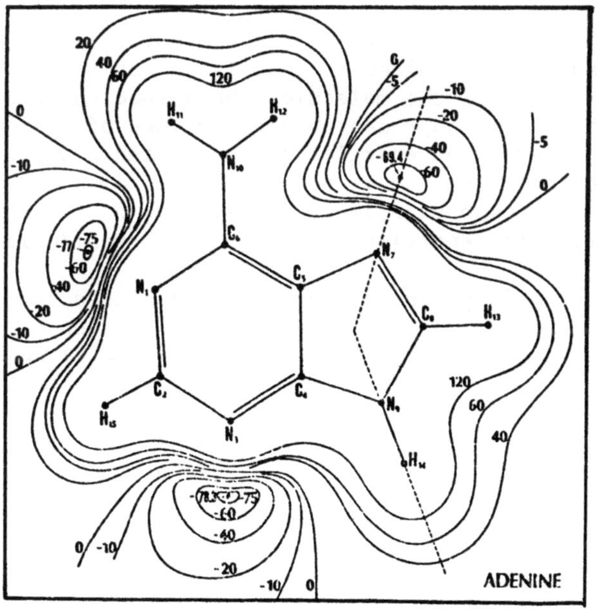


Fig 14

interaction. (the polarization effect is not included)

The best way to visualize the distribution of the potential V is to draw a map of isopotential contours.

Figure 13 shows for example the isopotential lines created by the molecule $NH_2$ $BH_2$ $NH_3$.[1] The most stricking feature is certainly, near the nitrogen bearing a lone pair, the presence of a potential well of about - 120 kcal/mole. Obviously this well is just prepared to receive a proton and rather often a relation is observed between the value of the potential well and the protonation energy.

Therefore for a molecule containing many potential wells the deepest must correspond to the prefered protonation site.

This method has been used to study the protonation of the base of the nucleic acids[2]. Figure 14 shows the isopotential contours of adenine. Three wells appear (near $N_1$, $N_3$ and $N_7$). The deepest are near $N_1$ and $N_3$ (-75 kcal/mole). Experimental evidence shows that $N_1$ and $N_3$ are the main protonation sites.

Recently the same procedure has been used to predict the position of the anion $Cl^-$ in the crystal of cytosine hydrochloride[3].

The anion is represented by a point negative charge in such a way that its electrostatic interaction with the protonated cytosine is -V, V being the electrostatic potential. B. Pullman[3] has calculated how this potential varies when the chlorine anion approaches as much as possible the molecule. That is to say remains at 3, 2 A from all heavy atoms of the molecule. The dashed curve on the figure 15 corresponds to this approach. Various values of the corresponding electrostatic interaction energies are indicated along the lines and a small circle indicates the experimental positions of the anions. It is seen that the two experimental positions correspond to the largest electrostatic energies (-104 and -87.7 kcal/mole).

The electrostatic potential maps have been used to study the pharmacological properties of various drugs. A group of pharmacologists asked me to study from this view point a set a derivatives of morpholine showing an analgesic effect. Is is known that in the brain

(1) R. Daudel, O. Chalvet, R. Constanciel and L. Esnault in Chemical and Biochemical Reactivity, E.D. Bergmann and B. Pullman ed. Israel Acad. Sci and Humanities, pub. (1974) p. 63.
(2) R. Bonaccorsi, A. Pullman, E. Scrocco and J. Tomasi, Theoretica Chimica Acta 24, 51 (1972).
(3) A. Goldblum and B. Pullman, Theoretica Chimica Acta 47, 345 (1978).

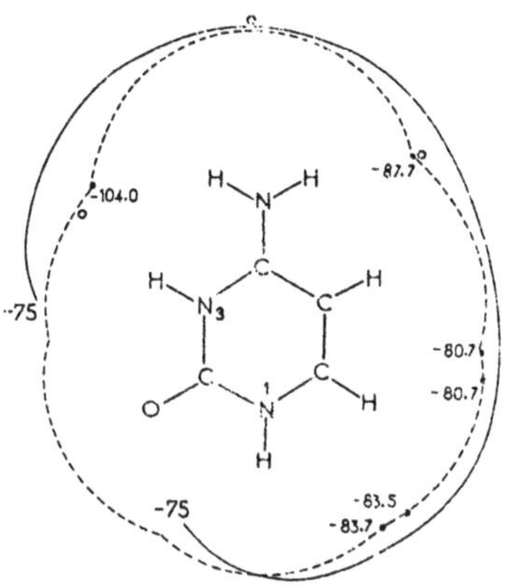

adjacent neurones are separated by gaps called synapsis. The nervous influx has to jump through those gaps. When the influx arrives at the end of a neurone transmitter molecules are produced which go though the gap and react on membrane receptors of a proximate neurone. It is therefore believed that analgesic drugs are able to add on membrane receptors but not able to circulate some specific nervous influx corresponding to some pains. Therefore, the analgesic molecules would blocked the corresponding receptors.

Effectively, the pharmacologists observed the fixation of morpholine derivatives on tryptaminergic receptors. On the other hand, calculations have shown that of more than two thirds of the surrounding space the potential distribution around the serotonine molecule greatly ressembles that for the most analgesic compound of the morpholine family. This help us to understand why this compound is able to react on tryptamineergic receptors. Calculations have also shown that a similarity appears between the potentiel map of this compounds and that of noradrenaline. As a consequence the pharmacologist were led to look for an interaction between morpholine derivatives and noradrenergic receptors. They observed a clear modification of the dose responre curve for noradrenaline when one of these morpholines is first left in contact with the convenient preparation.(1)

These observation permits to understand the antidepressant activity which has also been observed for these drugs.

(1) R. Daudel, L. Esnault, C. Labrid, N. Busch, J. Moleyre et J. Lambert, Eur. J. Med. Chem. 11, 443 (1976).

# SPIN, CHARGE, MOMENTUM DENSITIES : THEIR RELATIONSHIPS AMONGST THEMSELVES AND WITH OTHER PHYSICAL MEASUREMENTS

P.J. BROWN

Institut Laue-Langevin

Grenoble, France

## 1. INTRODUCTION

### a) Fundamental Position of the Wave Function

We are brought up to believe that if we were able to obtain an exact solution for the Schrodriger equation for any problem then we would be able to predict the result of any experiment which we could carry out on it. Unhappily of course we are never able to obtain an exact solution in the simplest possible cases. For crystalline solids which contain numbers in the order of $10^{22}$ atoms even the translational symmetry of the problem does not reduce it to a solvable size. One unit cell of even an extremely simple metal such as beryllium already contains 8 electrons. In order to attempt a calculation of the wave function in a solid some approximations must be made. Once the idea that an exact solution for the wave function can be calculated is abandoned then there is always a question as to whether the approximations made in obtaining a simpler solution are valid. From an exact solution one could predict with confidence any particular property, but an approximate solution may predict the result of one type of experiment accurately whilst giving a wildly incorrect result for another.

### b) Charge and Spin Density as Measures of the Wave-Function

The wave function for a many electron system can be written as an antisymmetric product of one-electron functions. This may be expressed as a Slater determinant -or more concisely as the sum

$$\Psi(\underline{R}) = \frac{1}{N!} \sum_{P} \epsilon_P P \psi_A(r_1 S_1) \psi_B(r_2 S_2) \cdots \psi_X(r_N S_N) \qquad (1)$$

where N is the number of electrons, P the permutation operator and $\varepsilon_p$ a constant which is $\pm 1$ depending on whether the number of permutations is odd or even. R is a vector of the space and spin coordinates of all N electrons.
The charge density as measured in an X-ray scattering experiment can then be written :

$$\rho(\underline{r}) = \sum_i \int \Psi^*(\underline{R}) \, \Psi(\underline{R}) \, dR^3_{-i} \qquad (2)$$

Here the sum is to be taken over all electrons i and the integration is over all spin co-ordinates and all space coordinates except those of the ith electron. The physical meaning of this equation is that it gives us the probability that any of the i electrons is in a volume element around the point $\underline{r}$ if we don't care what any of the other electrons are doing at the time ; this corresponds to the integration over all coordinates except $r_i$. The density is given by summing the result for each individual electron function over all electrons. This can be generalised by defining the electron density matrix which gives both the electron density and the pair correlation functions. Thus the element $\rho_{ij}$ in the electron density matrix

$$\rho_{ij}(\underline{r}_i \underline{r}_j) = \sum_{i \neq j} \int \Psi^*(\underline{R}) \, \Psi(\underline{R}) \, dR^3_{-i-j} \qquad (3)$$

where $\rho_{ij}$ is now the probability that there will be electrons simultaneously at $\underline{r}_i$ and at $\underline{r}_j$. A spin density function can be similarly defined

$$\underline{S}(\underline{r}) = \sum_i \int \Psi^*(\underline{R}) \, \tilde{S} \, \Psi(\underline{R}) \, dR^3_{-i} \qquad (4)$$

A note of caution must be introduced that it is not necessarily this function but the magnetisation density which may be measured in a neutron scattering experiment.

It is conjectured that since the electron energy is determined by pair-wise electron electron interactions the properties of an electronic system can be described completely by an appropriate density function. Such a function must include both space and spin functions so equation 3 above is generalised by replacing $r_i$ $r_j$ by $x_i$ $x_j$ so that $\underline{x}_i$ $\underline{x}_j$ are vectors describing both the space and spin coordinates of the electron, the integration is then over all coordinates except $x_i$ $x_j$. Both the charge and spin densities are one electron properties and are determined by the spatially diagonal elements. If we write $x_i = r_i S_i$ where $S_i = S^\alpha$ or $S^\beta$ for the two possible spin states then the charge density is given by :

$$\rho(\underline{r}) = \sum_i \left( \rho(r_i s^\alpha; r_i s^\alpha) + \rho(r_i s^\beta, r_i s^\beta) \right) \tag{5}$$

the z component of the spin density by

$$\langle S_z(r) \rangle = \sum_i \left( \rho(r_i s^\alpha; r_i s^\alpha) - \rho(r_i s^\beta; r_i s^\beta) \right) \tag{6}$$

and the x component by

$$\langle S_x(r) \rangle = 2 \sum_i \rho(r_i s^\alpha r_i s^\beta) \tag{7}$$

It is thus clear that the charge and spin densities give information which is rather closely related to the density functions. Most of the difficulties that arrive in deducing density functions from X-ray and neutron scattering measurements come rather from the limited accuracy and resolution of the data than from any conceptual difficulty in the transformation.

c) Physical Properties as Measures of the Wave-Function

Bulk physical properties such as specific heat, resistivity, optical absorption etc. whilst determined completely by the exact wave functions, cannot in general be used in a deductive way. The specific heat for instance comes from an average over accessible states of the densities of states within a particular energy range. In calculating the densities of states and in taking the averages much of the detailed information is lost. In such cases even through results may not be interpretable on an absolute basis comparative results between closely similar systems may be useful. For example low temperature specific heat measurements in metals can be analysed to give a measure of the density of states at the Fermi surface. Such measurements for a series of alloys between transition metals can be taken to show the way in which the density of states varies across the d-band in the metals.

Another difficulty with measurement of bulk physical properties is that they are very often strongly affected by specimen dependent phenomena such as dislocation densities impurity concentrations etc. To take a specific example the tensile strength of a piece of iron is very strongly dependent on the microcrystalline structure, and to a rather lesser extent by the precise form of the electron wave functions of the pure crystalline material. At the moment the range of model and computational techniques available limits the calculation of accurate wave functions either to perfectly periodic systems or to ones of limited physical extent. It is therefore more relevant to consider properties which can be modeled more easily in terms of simple wave-functions when considering comparisons with charge and spin density measurements.

### d) Studies of Conduction Electrons : - the Effect of Inaccurate Wave-Functions

There are a significant number of properties of metallic and semi-conducting systems which are sensitive to the density of states and the number of carriers at the Fermi surface. In such measurements such as conductivity, magnetic susceptibility etc. specimen dependent effects can often be eliminated either by measurements as a function of temperature or by very careful sample preparation. In the recent past numbers of techniques have become available which measure the dynamics of electrons at the Fermi surface and which can map the contours of equal energy in k space at the Fermi surface. For many pure metals and semi-conductors there is therefore a significant amount of data available which relates to the wave functions of electrons near the Fermi-surface. These data have been used to compare with predictions of wave function calculations leading to band structures for these metals, and in general a very good agreement is obtained. The point that becomes clear is that a wave function which gives good agreement with Fermi surface measurements will not necessarily give good agreement with resonance measurements or charge density measurements. The reasons are not hard to understand, Fermi surface measurements are sensitive to the relative energies of states near the surface and are insensitive to even quite large errors in the absolute energies of the states. Similarly they are more strongly influenced bvcehe choice of potential in the region remote from the atom cores Resonance measurements on the other hand require a good model of the wave function at the centre of the resonant nucleus and charge density measurements require that the eigen vectors as well as the relative eigen values should be approximately correct. Thus we expect different approximate methods for obtaining wave functions may give results in good agreement with some properties and not with others. Thus the APW (augmented plane wave technique) which gives a good description of the potential between atoms can give good agreement with Fermi surface measurements ; whereas the tight binding LCAO approach or the OPW techniques which deal more correctly with the atom cores, may be more favoured by resonance measurements.

## 2. RELATIONSHIPS BETWEEN CHARGE, SPIN AND MOMENTUM DENSITIES

### a) Charge and Spin Densities

We saw in the introductory section how the charge and spin densities are related to the elements of the electron density matrix. It is tempting to postulate that since the charge density gives the sum of the densities due to both spin states and the spin density that due to the difference, it ought to be possible to deduce the densities in each of the two spin states separately by combining the results of X-ray and magnetic scattering measurements. In fact this procedure is probably not possible in any general way. The problem being that magnetic scattering is sensitive to the magnetisation density not just to the spin density and measures only the projection of it perpendicular to the scattering vector. It is therefore not possible in principle to obtain complete three dimensional data on a single component of the spin density even if complications due to orbital moment are absent. In cases where orbital moment is present the current state of the theory enables its contribution to the magnetisation density to be calculated so long as the ground state wave function can be written in terms of non-overlapping atomic functions. At the present time there is little evidence as to whether the orbital moment associated with overlapping atomic charge distributions is important. The usual assumption made is that it is not, and that after subtraction of the orbital moment given by a model calculation the remaining magnetisation and in particular the delocalised magnetisation comes from spin only.

b) Complementarity of charge/spin and momentum densities

The complementary relationship which exists between the momentum and position representations of the wave function is well known and is expressed in the equation

$$\phi(\underline{k}) = \left(\frac{1}{\sqrt{2\pi}}\right)^3 \int \psi(\underline{R}) e^{i\underline{k}\cdot\underline{R}} dR^3$$

For a many electron system, in just the same way as we define the electron density matrix we can define a momentum density matrix

$$P_{ij}(k_i k_j) = \sum_{i \neq j} \Phi^*(K)\, \Phi(K)\, dK^3_{-i-j}$$

this gives the probability that there will exist simultaneously two electrons with momenta $k_i$ and $k_j$. The coordinates K contain the momenta of all the particles.

The possibility of measuring a momentum distribution rather than a charge distribution breaks one of the constraints which is implicit in measuring charge densities in crystalline materials. This constraint is due to the periodicity of the charge density which means that its Fourier transform is only finite at Bragg peaks. As a result, in simple structures, there are not many independent measurements contributing to the charge density in the region of low K where effects of interatomic interactions show themselves. However, although the electron density is periodic, the wave functions are not and although their amplitudes are periodic the phase is different at different lattice points. To carry out the Fourier transformation which leads from the space to the momentum wave-function it is necessary to know these phases which are not obtainable from the density matrix. Consequently measurement of momentum densities leading to a momentum density function will give information about wave functions complementary to that obtained by measuring the spatial density.

Two techniques are available for measuring momentum densities compton scattering, and positron anihilation. In compton scattering the momentum of the scattered photon is modified by the scattering process and the momentum change contains information about the momentum distribution of the scattering electrons. With present day equipment and photon fluxes the momentum resolution obtainable is not high and hence the method cannot achieve its full potential. By using polarised photons scattered from a magnetised sample the spin dependence of the momentum density can also be investigated. In positron anihilation experiments the angle between the paths of two $\gamma$-rays emitted when a position anihilates with an electron in the material is measured. Assuming that the positron is at rest when it anihilates, the angle measures the momentum of the electron. This technique is currently capable of much higher resolution than compton scattering and has the advantage, for some purposes, that because of the repulsion of the positron by the nucleus it samples the interatomic momentum density preferentially. The validity of results obtained is dependent on the accuracy with which the positron wave-function in the material can be calculated.

### c) Band Structure

As mentioned earlier the availability of a wave function from which a band structure in agreement with experiment has been calculated, does not necessarily mean that density functions in agreement with experiment can be calculated. This is particularly true in the case of magnetisation density studies on ferromagnetic metals. In this case it is usual to analyse the neutron scattering results in terms of a spin moment, an orbital moment and a diffuse moment. The symmetry of the band wave functions contributing to the spin moment can also be deduced. These results should correlate with measurements of the gyromagnetic ratio and with resonance measurements. They provide a particularly severe test for band structure calculations because of the central position of electron correlations in determining the magnetisation density.

Electron correlations play a less important rôle in relating magnetisation densities associated with aligned paramagnetic electrons in metallic systems to other band properties of the material. In such cases the band splitting induced by the applied field is very small compared to any measure of the band width and the spin density reflects simply the spatial properties of electron states at the Fermi surface. In such case the recent APW of Freeman and co-workers give results which are in relatively satisfactory agreement with the theory. It is emphasised by Freeman that while a variety of experimental methods yield information which can be related directly to the electronic band structure magnetic neutron scattering is one of the few experiments which yield direct information about the nature of the wave-functions.

## 3. OTHER MEASURES OF CHARGE AND MAGNETISATION DENSITIES

### a) Introduction

Before starting to inquire into the relationship between different techniques which measure particular features of charge and spin densities it is as well to remind ourselves of some of the basic properties of physical measurements which affect their comparability. One such feature is the time scale of a measurement. For Bragg scattering the effective time of the measurement is long and we measure that part of the time averaged distribution of charge or magnetisation density which is periodic. For diffuse scattering which is sensitive to the aperiodic part of the scattering density the time scale is measured by the time for passage of the scattered radiation across the scattering system. For X-ray this time is short compared with the velocities of all except the core electrons of very heavy elements. For neutrons on the other hand their passage across an atom takes $\sim 10^{13}$ secs, this is long compared with electronic fluctuations but of the same order of magnitude or less than the times associated with thermal fluctuations. It is for this reason that one does not expect to see magnetic diffuse scattering of neutrons from non magnetic atoms although each electron state in the one-electron representation is magnetic. Diffuse magnetic scattering from a paramagnet will only occur when the lifetime of a magnetised atomic state is comparable with or longer than $10^{-13}$ secs. In general for any physical measurement there is a characteristic time. A measuring technique will average some property of the system under investigation over a time of this order of magnitude. One can consider this time averaging as giving the time resolution of the technique. It may be related to the energy resolution by the uncertainty principle. In many techniques there is an additional spatial or momentum resolution imposed by the method of measurement. In charge and spin density studies it is well known that the spatial resolution is limited by the limit of data in K space which is gathered. In nuclear magnetic resonance the effective resolution length is the diameter of the nuclear moment, so that the resolution is very fine. On the other hand, the area of space over which information can be obtained is also limited to the nucleus and so is comparably small. The question of what information can be obtained from any measurement brings us to the third important property of any physical technique which is usually contained in the relevant matrix element.

It should be possible to relate any physical measurement to an appropriate average over matrix elements of the form

$$m_{KL} = \int \Psi_K^*(X)\, \tilde{T}\, \Psi(X)_L\, dX^3$$

where the operator $\tilde{T}$ is appropriate to the type of measurement.

b) Magnetisation

The relationship between magnetisation measurements and the magnetisation density is very simple.The magnetisation is directly equal to the integral of the magnetisation density over the crystal. It gives the term with K=0 in the Fourier expansion of the magnetisation density. One caution, the magnetisation density as determined by neutron Bragg scattering reflects only the periodic part of that density whereas the magnetisation is added up over the whole crystal. This can cause problems of interpretation where magnetic domains are present. It is the difference between the total moment determined by magnetisation measurements and the apparent moment determined by diffraction that leads to the idea that there is a non-localised contribution to the magnetic moment in ferromagnetic metals. The relationship is the same between the paramagnetic susceptibility and neutron Bragg scattering by aligned paramagnets. If we consider the generalised static susceptibility $\chi(q)$ then susceptibility measurements give $\chi(0)$ whereas Bragg scattering measurements give $\chi(\tau)$ where $\tau$ are reciprocal lattice vectors. The generalised static susceptibility is one part of the dynamic susceptibility $\chi(q\,\omega)$ which can be investigated in more detail using inelastic neutron scattering.

c) Resonance Techniques

Nuclear magnetic resonance and Mössbauer spectroscopy measure closely related quantities. In NMR the resonance frequency measures the Zeeman splitting under the action of a magnetic field between states with different spin parallel to the field direction. The field producing the splitting may be of external or internal origin, but it is the internal field which provides information about charge and spin density distributions. In Mössbauer spectroscopy the hyperfine splitting of the nuclear energy levels is observed directly by resonant absorption. In both cases the phenomena that cause the splitting are the same. Two major effects occur in one of which,the magnetic hyperfine interaction,the levels are split by the magnetic field of the surrounding electrons -the strength of the interaction depends directly on the magnetisation density at the nucleus. The second large effect is the electrostatic interaction which occurs between the nuclear and electronic charge distributions. The spherically symmetric term in this interaction causes a shift in the nuclear energy levels which is the same for all nuclear spin states. It therefore gives a shift (the isomer shift) in the Mössbauer spectrum but no change in the nuclear magnetic resonance. The dipolar term in the electrostatic interaction is zero since there is no nuclear dipole moment, but a quadropole interaction will exist if the nucleus has spin greater than 1/2. Quadrupole splitting occurs in both the Mössbauer and NMR spectra and measures the quadrupole term in the charge density expanded about the nucleus. The relationship between the magnetisation densities at the nucleus measured by neutron scattering and by hyperfine field measurements is not at all obvious because of the very different resolutions involved in the two techniques. The resolution of neutron scattering is of the order 0.25 Å which means that the density measured at the nucleus corresponds more nearly to the average magnetisation density in a sphere of radius $\sim$ .25 Å around the nucleus. The magnetisation density appropriate to the hyperfine field measurement corresponds to a sphere of radius more nearly $10^{-5}$ Å it is therefore very sensitive to small differences in the radial distributions of s electrons of up and down spins which have high densities at the nucleus.

There are other resonance techniques which give information about charge and spin densities such as electron spin resonance which can be used to obtain electronic susceptibilities and electron double resonance which gives indirect evidence of bonding electrons.

## d) Spectroscopic Measurements

There is no very real distinction between resonance spectroscopy and other types of spectroscopy and under this heading I am only going to mention very briefly some of the techniques which have recently created a kind of alphapet soup out of this subject. A few of the names in this soup which I have turned up on making a brief search of the recent literature are ESCA, EXAFS, XPS, UPS this list is by no means exhaustive and I have not included techniques such as Auger spectroscopy or electron scanning techniques which do not seem so relevant to charge and spin densities. If we add to the techniques above these of soft X-ray emission and absorption spectroscopy then the list covers practically all conversions of electrons to photons and vice versa.

Amongst the oldest established of these techniques are those of soft X-ray emission and absorption. In the emission technique a light atom target under investigation is bombarded with electrons of sufficient energy to excite the L or M radiation and the spectrum of the emitted X-rays is studied. In the emission process an electron from a closed shell of the atom is ejected by the bombarding electron and subsequently this hole is filled by an electron from the valence or conduction bands. The wavelength of the emitted X-ray gives the energy lost in the transition and the shape of the spectrum the density of states in the valence and conduction bends involved. The principle of soft X-ray absorption spectroscopy is very similar, the material is bombarded with photons of X-ray energies which cause electrons in the atomic core to be excited into the conduction band. The resultant absorption spectrum should reflect the density of empty states in the conduction band. The interpretation of such measurements in terms of the density of states is not without pit-falls one of the chief of these arises from the fact that by necessity either the initial or the final state of the atom involved in the process is an ionised one and the ion is of course positioned close to where all the action takes place. This problem -which is termed a many body problem has recently been tackled quite successfully theoretically and the spectra of pure materials now seem to be relatively well understood. Such is not the case for multicomponent systems such as alloys where the apparent band widths and shapes given by the individual atomic species in an alloy such as Al-Mg are found to be very different. This casts doubt on a simple picture of a common band a disordered alloy, but does not seem unexpected when it is recognised that the technique is such as to weight the measurements very heavily in terms of their local electronic environment which will undoubtedly be different for aluminium and magnesium.

A closely related technique is that of EXAFS (extended X-ray absorption fine structure measurements).In such measurements the oscillations in the X-ray absorption coefficient that occur close to an absorption edge are analysed. These fluctuations are supposed to be due to scattering of the ejected photoelectrons by atoms surrounding the absorbing atom. In the analysis of such measurements the total photo electron wave function including scattering is calculated from scattering theory and used to calculate the dipole transition matrix. It is the oscillatory part of this function which is needed. The technique can be used to obtain detailed data on the numbers and distances of different types of neighbour even in non-crystalline solids. It has the marked advantage over other scattering techniques that the species of the origin atom is labelled by the position of the K edge of which the oscillations are measured.

Two techniques in which the photon induced emission of electrons is measured are XPS and UPS : (X-ray and Ultra-Violet photoemission spectroscopy) they differ only in the energy of the exciting photons a distinction which is becoming difficult to make with the advant of synchrotron radiation sources. Photo-electron spectroscopy again gives information on valence bands. Monochromatic radiation is used to ionise a material and the energy and perhaps the angular distribution of the excited electrons are measured. The intensity of photo-emission depends on the square of the momentum matrix element between the initial and the final states. Because angular momentum must be conserved the direction of the photo-electron carries information about the wave-vector of its initial state and allows electron dispersion relationships to be deduced. In general X-ray photo-emission studies yield information on the initial state of the electrons, the final state being in the continuum whereas ultra violet photo-emission is more sensitive to the final states.

## 5. CONCLUSION

If one is to reach any conclusion from this survey of physical properties and techniques it is that there are not at present other techniques comparable to X-ray and neutron diffraction for determining charge and spin densities in crystals. It must be recognised however that the resolution and precision of these techniques is strongly limited. Where other techniques exist which give more precise information relevant to parts of these densities then such result should be incorporated into the interpretation of the densities and should enable more stringent tests of the validity of model calculations to be made.

ELECTRON DISTRIBUTIONS AND THERMODYNAMICS

Richard J. Weiss

Army Materials and Mechanics Research Center

Watertown, Massachusetts 02172

## 1. Kinetic and Potential Energies

If one integrates the spherically averaged Compton profile one can show that in the impulse approximation the total kinetic energy contained within a sphere of radius p in momentum space is

$$K.E.(p) = \left[3\int_0^p J(z)\, z^2\, dz\right] - p^3 J(p) \tag{1}$$

Thus the total ground state kinetic energy of the electrons is

$$\langle K.E.\rangle = 3\int_0^\infty J(z)\, z^2\, dz \tag{2}$$

Since it is difficult to obtain accurate data for large values of $z$ one can integrate $J(z)$ to perhaps $z = 3$ a. u. and assume that beyond $z = 3$ a. u. The Compton profile of a solid or liquid is given by a superposition of Hartree-Fock free atom profiles which are tabulated. In fact by taking a difference between the measured Compton profile and a superposition of H.F. free atom profiles one obtains the < K.E.> difference between free atoms and solid. Since both free atoms and solid satisfy the Virial Theorem <K.E.> = -1/2 <V> where <V> is the potential energy it follows that the K.E. is just minus the total energy.

$$\langle K.E.\rangle = -\langle E\rangle \tag{3}$$

Thus the <K.E.> energy difference between free atom and solid must just equal the cohesive energy of the solid.

It is more difficult to obtain the total potential energy from the charge density $\rho(\vec{r})$ due to the Coulomb repulsion between the electrons. It is possible to show that the attractive electron-nuclear potential energy is related to the structure factors

$$\langle V\rangle_{ne} = \frac{2e^2}{\pi}\langle Z\rangle\delta\sum_{hkl}\langle F(\vec{S})\rangle \qquad 4$$

where $\langle Z\rangle$ is the average nuclear charge in the unit cell, $\langle F(\vec{S})\rangle$ the spherically averaged structure factor and $\delta$ is the natural peak half width

$$\delta \simeq \frac{1.047}{4\pi\upsilon^{1/3}} \qquad 5$$

where $\upsilon$ is the volume of the unit cell. An approximate expression for the electron-electron coulomb repulsive energy is given by

$$\langle V\rangle_{ee} \simeq \frac{e^2\delta}{2\pi}\sum_{hkl}\langle F^2(\vec{S})\rangle \qquad 6$$

The total potential energy difference between a superposition of free atom scattering factors and the measured structure factors must again be related through the virial theorem to the cohesive energy.

## 2. Thermodynamics

The free energy G of any solid can be written as a sum of terms

$$G = E_0 + H(T) + PV - TS(T) \qquad 7$$

where $E_0$ is the cohesive energy at T=0 i.e. the energy difference between the solid and a superposition of free atom energies, H(T) the enthalpy i.e. the energy absorbed through lattice vibrations and electronic and magnetic excitations at any finite temperature T, P the pressure, V the volume, and S(T) the entropy at temp T. Eq 7 is most often employed for alloys where only differences in the terms relative to the pure elements in their solid state are considered.

## 3. Cohesive Energy

Eq 1 provides the most reliable technique at the moment in determining alloy heat of mixing $E_0$ i.e. one integrates the measured differences in the Compton profile between the pure elements and the alloy. For compounds it is also possible to take the difference between the measured Compton profile and the superposition of free atom profiles to obtain the cohesive energy. This is interesting in that it is a non-destructive technique compared to such measurements as acid solution, tin calorimetry or heats of vaporization.

While it is also possible to obtain this information from Eqs 4 and 6 this technique requires more experimental work.

## 4. Enthalpy H(T) and Entropy S(T)

The bulk of the energy and entropy absorbed by most crystals from T=0 to T excites the normal modes of vibration. These manifest themselves through the Debye-Waller factor exp(-2M) and diffuse background in the X-ray case and the Debye-Waller factor and inelastic scattering for neutrons. By employing a force constance model between atoms one can fit it to the neutron inelastic scattering data and calculate a normal mode frequency spectrum $q(\nu)$. Integration of $q(\nu)$ with the partition function yields H(T)

$$H(T) \cong (1 + 3\alpha) \int_0^{\infty} \frac{h\nu \, g(\nu) \, d\nu}{(\exp(h\nu/kT) - 1)} + \frac{1}{2} \int_0^{\infty} h\nu \, g(\nu) \, d\nu \tag{8}$$

where the term $(1 + 3\alpha)$ ($\alpha$ is the coefficient of linear expansion) is a correction due to anharmonicity. By measuring the absolute structure factors at large $\sin\theta/\lambda$ where the scattering factors approximate the Hartree-Fock free atom we can obtain the Debye-Waller factor exp (-2M) which is given by

$$2M \cong \frac{h}{3mN} (4\pi \sin\theta/\lambda)^2 \left[ (1+3\alpha) \int_0^{\infty} \frac{g(\nu)\, d\nu}{[\exp(h\nu/kT) - 1]} + \frac{1}{2} \int_0^{\infty} \frac{g(\nu)\, d\nu}{\nu} \right] \tag{9}$$

where m is the atomic mass and N the number of atoms/cc. The principal difference between Eqs 8 and 9 is the factor $1/\nu^2$ in the integrals. In most cases the Debye approximation $q(\nu) = 9N\nu^2/\nu_{max}^3$ is adequate since only the difference in H(T) or S(T) between the alloy and pure elements or between phases is required.

The Debye model has the advantage in that a single parameter $\nu_{max}$ for each element in an alloy enables one to determine H(T) and S(T) at any temperature. The Debye characteristic temperature $\Theta$ is related to $\nu_{max}$ by

$$h\nu_{max} = k\Theta$$

Another contribution to the entropy is the configurational entropy which can be determined from the X-ray order parameter in an alloy or the neutron determination of order in a magnetic system.

The only contribution to H(T) and S(T) not directly measurable by X-rays is due to the electronic specific heat. In principle an accurate measurement of J(q) as a function of T could reveal this contribution but in practice a measurement of the linear term in the specific heat may be more practicable.

## 5. PV Term

An X-ray or neutron determination of unit cell dimensions of an alloy and its pure elements gives a direct measure of the PV term. By loading a pressure cell with a mixture of alloy and pure elements one can determine small differences in V.

6. Water-Methanol

Some recent Compton profile work on water and methanol mixtures yielded reasonable agreement for the heat of mixing. It also showed that changes in the electron distributions occur in this mixture.

## Problems

1. Derive Eqns 1, 4, 6.

2. Assume a helium atom ground state wave function of the form

$$\psi = A e^{-ar_1} e^{-ar_2}$$

Calculate the ground state energy, scattering factor f(s) and Compton profile J(Z). Show that the following are true

$$\langle K.E.\rangle = 3\int_0^\infty Z^2 J(Z)\, dZ$$

$$\langle V\rangle_{ne} = -Ze^2 \int_0^\infty f(s)\, ds$$

$$\langle V\rangle_{ee} = e^2 \int_0^\infty f^2(s)\, ds$$

3. Look up the Debye $\Theta$ values, electronic specific heat coefficients, lattice parameters etc. for α and β Ti. Determine the difference in cohesive energies of the two phases and the latent heat of the α-β transformation. Estimate the phase diagram up to 100 kilobars. Calculate the Debye-Waller factors. How sensitive are the X-ray intensities to the value of $\Theta$; thus how sensitive are the free energies to the X-ray intensities within the Debye model?

Answers

1.

Equation 1. Kinetic energy within a sphere of radius p for $\chi(p)^2$ symmetric is in atomic units (m=1)

$$KE(p) = 4\pi \int_0^p |\chi^2| \frac{p^2}{2} p^2 dp.$$

Integrate by parts. Since $2\pi\int\chi^2 \, pdp = J(o) - J(p)$, get eqn 1

Equation 4. $F(\underset{\sim}{S})$ is, with nucleus 1 at origin :

$$F(\underset{\sim}{S}) = \int\rho(r_1)\exp(i\underset{\sim}{S}.\underset{\sim}{r}_1)r_1{}^2dr_1 + \exp(i\underset{\sim}{S}.\underset{\sim}{R}_2)\int\rho(r_2)\exp(i\underset{\sim}{S}.\underset{\sim}{r}_2)r_2^2dr_2 + \ldots$$

$$<F(S)> = \int\rho(r_1)j_o(Sr_1)r_1^2dr_1 + \int\rho(r_2)j_o[S(\underset{\sim}{r}_2+\underset{\sim}{R}_2)]r_2^2dr_2 + \ldots$$

$$\int <F(S)> dS = \frac{\pi}{2}\int\frac{\rho(r_1)}{r_1} r_1^2 \, dr_1 + \frac{\pi}{2}\int\frac{\rho(r_2)}{|\underset{\sim}{r}_2+\underset{\sim}{R}_2|} r_2^2dr_2 + \ldots$$

$$-Z_1e^2 \int<F(S)> dS = \frac{\pi}{2} <V_1>$$

Now if we use nucleus 2 as an origin, then F(S) is identical except for a phase factor $\exp(i\underset{\sim}{S}.\underset{\sim}{R}_2)$. Thus,

$$-Z_2e^2\int<F(S)> \exp(i\underset{\sim}{S}.\underset{\sim}{R}_2)dS = \frac{\pi}{2} <V_2>$$

$$- \quad Z_me^2 \int_0^\infty \exp(i\underset{\sim}{S}.\underset{\sim}{R}_m) \, dS <F(S)> = - \frac{\pi}{2} <V>$$

In a simple crystal such as bcc Fe, the Bragg peaks give

$$\exp(i\underset{\sim}{S}.\underset{\sim}{R}_m) = 1.$$

Thus replacing each Bragg peak with a peak height F(S) plus a width $\delta = 1.047/(4 \, V^{1/3})$, where V is the volume of unit cell :

$$-e^2Z\delta \sum_{hkl} <F(S)> = -\frac{\pi}{2} <V_{en}>$$

Equation 6. Consider two electron scattering factor

$$f_i(\underset{\sim}{S})f_j^*(\underset{\sim}{S}) = \iint\rho(\underset{\sim}{r}_i)\rho(\underset{\sim}{r}_j) \exp[i\underset{\sim}{S}.(\underset{\sim}{r}_i-\underset{\sim}{r}_j)] \, d\underset{\sim}{r}_i d\underset{\sim}{r}_j$$

The spherical average

$$<f_i(\underset{\sim}{S})f_j^*(\underset{\sim}{S})> = \iint<\rho(\underset{\sim}{r}_i)\rho(\underset{\sim}{r}_j)> r_i{}^2r_j{}^2dr_idr_j \, j_o(S|\underset{\sim}{r}_i-\underset{\sim}{r}_j|)$$

Integrating over S and multiplying by $e^2$ :

$$e^2\int_0^\infty <f_i(\underset{\sim}{S})f_j^*(\underset{\sim}{S})> dS = \frac{\pi e^2}{2}\iint\frac{<\rho(\underset{\sim}{r}_i)\rho(\underset{\sim}{r}_j)>}{|\underset{\sim}{r}_i-\underset{\sim}{r}_j|} dr_idr_jr_i{}^2r_j{}^2 =$$

$$\frac{\pi}{2} <V>_{ij}$$

which gives the Coulmb repulsion energy. For an atom the structure factor squared gives

$$1/2 \langle F^2(\underset{\sim}{S})\rangle = 1/2\langle(\sum_i f_i(\underset{\sim}{S}))^2\rangle = 1/2\langle \sum_i f_i^2(\underset{\sim}{S})\rangle + \sum_{i\neq j} \langle f_i(\underset{\sim}{S}) f_j(\underset{\sim}{S})\rangle$$

The last term yelding the required two electron scattering factors. For the deformation density in a crystal the first term cancels for the core electrons and the major contribution comes from the product of valence and core electrons to the last term, thus following the arguments of eqn 4 for a simple crystal we have :

$$\langle V_{ee}\rangle = 2e^2/\pi\ \delta/2 \sum_{hkl} \langle F^2(\underset{\sim}{S})\rangle/2$$

2.

Helium atom : ground state energy

$$\langle E\rangle = (Z-5/16)^2 = 2.84766 \text{ au} = 77.44 \text{ ev (exp : 78.6 ev)}$$

$$= \langle KE\rangle$$

$$f(S) = \frac{2}{(1 + S^2/4a^2)^2} \qquad S = 4\pi\sin\theta/\lambda \quad \text{(au)}$$

$$a = (Z-5/16)$$

$$J(q) = \frac{16}{3\pi a(1 + q^2/a^2)^3}$$

3.

| | $\alpha$ Ti (hcp) | $\beta$ Ti (bcc) |
|---|---|---|
| $\Theta$ | 365 | 300 |
| $\gamma$ | $8.25\ 10^{-4}$ cal/mol/$°^2$ | $5.7\ 10^{-4}$ |
| | $a_o$=2.950 | $a_o$ =3.33 |
| | $c_o$=4.686 | |

$T_F$ 1155°K ($\alpha\rightarrow\beta$)

$\Delta H = T_F \Delta S$ $(\Theta/T_F) = x$

at $T_F$ :

| | | |
|---|---|---|
| $S_{lattice}$ | 15.05 | 15.98 |
| $S_{electr.}$ | .95 | .66 |

$\Delta H = T\Delta S = 739$ cal/mol = latent heat

| | | |
|---|---|---|
| H | 2307.7 | 2924 |

Difference in cohesive energy : 739 - 617 = 122 cal/mole ($\alpha$ lower)

The volume per atom is $V = a_o^3/2$ for bcc = 18.463 Å$^3$

$V = .866\ a_o^2 c_o/2 = 17.658$ Å$^3$ (hcp is more stable)

$P\Delta V$ term is PX.805 $\text{Å}^3$ = P(kbars)11.582 cal/mole

Neglecting the difference in the coefficient of expansion we have for example at 1473°K

$$G(\alpha) = -122 + 3924 + 895 - 23921 - 1790 = -21014$$
$$G(\beta) = \quad + 4781 + 618 - 25748 - 1237 = -21586$$
$$\qquad U \quad \gamma T^2/2 \quad S_{latt} \quad S_{el}$$

$\Delta G = 572$ ($\beta$ lower) $= P(11.582)$ $P = 49.4$ kbars

This $\alpha\to\beta$ transformation rises from 1155°K at 0 pressure to 1473°K at 50 kbars

Debye Waller factor:

$$2M = \frac{22973}{A\,\Theta^2}\,T\,(\sin\theta/\lambda)^2\,\{1 + .026\,(\Theta/T)^2\} \qquad T = 295 \quad A = 47.90$$

$\alpha$ phase $\quad 2M = 1.1\ (\sin\theta/\lambda)^2$

$\beta$ phase $\quad 2M = 1.61(\sin\theta/\lambda)^2$

Thus for a reflection at $\sin\theta/\lambda=1$

$\alpha$ phase $\quad \exp(-2M) = .333$

$\beta$ phase $\quad \exp(-2M) = .200 \qquad$ Ratio : 1.67

FORM FACTORS AND THEIR APPLICATION TO

CALCULATING PROPERTIES

R. A. BONHAM

## INTRODUCTION

The coherent x-ray form factor is given as

$$F(\vec{K}) = \int dr\,\rho(r)e^{i\vec{K}\cdot\vec{r}}$$

where K is the momentum transfer, $\vec{K} = \vec{k}_i - \vec{k}_s$ and $\rho(\vec{r})$ is a three dimensional one electron density. The incoherent x-ray form factor is given as $S(K) = N + \int d\vec{r}_{12}\rho_c(\vec{r}_{12})e^{i\vec{K}\cdot\vec{r}_{12}} - |F(\vec{K})|^2$ where $\rho_c(\vec{r}_{12})$ is the electron pair correlation density.

1.) Calculate an analytic expression for the form factor for a hydrogenic atom in its ground state.

$$\varphi_{1s}(r) = \frac{1}{\sqrt{\pi}}\left(\frac{Z}{a_0}\right)^{3/2} \exp\left[-\frac{Zr}{a_0}\right].$$

2.) Show that the rotational average of $F(\vec{K})$ gives the same result obtained by averaging $\rho(\vec{r})$ over all orientations in space and then computing F(K) with the result.

3.) Show for a spherically symmetric charge distribution that for sufficiently small values of K, F(K) can be expanded in even positive integer power moments of the charge distribution.

4.) Derive an expansion for $F(\vec{K})$ when $\rho(\vec{r})$ is not spherically symmetric by expanding $\rho(\vec{r})$ in spherical harmonics as

$$\rho(r) = \frac{1}{\sqrt{4\pi}}\sum_{n=0}^{\infty}\sum_{\ell=-n}^{n}\frac{\rho_n(r)Y_{n\ell}(\theta,\varphi)}{\sqrt{2n+1}};\quad Y_{n\ell}(\theta,\varphi) = \sqrt{\frac{(2n+1)(n+\ell)!}{4\pi(n-\ell)!}}P_n^{\ell}(\cos\theta)e^{i\ell\varphi}$$

where $\theta$, $\varphi$ orient $\vec{r}$ with respect to some space fixed axis. Your answer should be expressed in terms of radial integrals of the form

$$f_n(K) = \int_0^\infty drr^2\rho_n(r)j_n(Kr).$$

5.) Derive an expansion for F(K), when $\rho(\vec{r})$ is spherically symmetric and also when K is large. You should be able to show that $\lim_{K\to\infty} F(K) = \frac{A}{K^4}$ where A is a constant. Show explicitly the relation between A and the density $\rho(\vec{r})$.

6.) In elastic or coherent x-ray scattering the cross section is proportional to $|F(\vec{K})|^2$. Average $|F(\vec{K})|^2$ over all angles of spatial orientation. Express the result in terms of the $f_n(K)$ functions in problem 4.

7.) Prove that the electron nuclear attractive energy, $\overline{V}_{ne}$, is given as

$$V_{ne} = -\frac{2Z}{\pi}\int_0^\infty dKf_0(K)$$

8.) Two possible analytic representations of the form factor are:

$$\text{a.) } F(K) = \sum_{i=1}^{M} \alpha_i e^{-\beta_i K^2} \quad \text{and} \quad \text{b.) } F(K) = \sum_{i=0}^{M} \alpha_i/[K^2 + \beta_i^2]^{2+i}$$

Indicate any desirable constraints that should be imposed on $\alpha_i$ and $\beta_i$ in each case and present arguments why one expansion might be favored over another.

9.) Calculate the form factors for a hydrogen atom in a 2s state, $2p_z$ state, $2p_x$ state and a $2p_y$ state.

$$\varphi_{2s}(r) = \frac{1}{4\sqrt{2\pi}}\left(\frac{Z}{a_0}\right)^{3/2}\left(2 - \frac{Zr}{a_0}\right)e^{-\frac{Zr}{2a_0}}$$

$$\varphi_{2p_z} = \frac{1}{4\sqrt{2\pi}}\left(\frac{Z}{a_0}\right)^{3/2}\frac{Zr}{a_0}e^{-\frac{Zr}{2a_0}}\cos\theta$$

$$\varphi_{2p_x} = \frac{1}{4\sqrt{2\pi}}\left(\frac{Z}{a_0}\right)^{3/2}\frac{Zr}{a_0}e^{-\frac{Zr}{2a_0}}\sin\theta\cos\varphi$$

$$\varphi_{2p_y} = \frac{1}{4\sqrt{2\pi}}\left(\frac{Z}{a_0}\right)^{3/2}\frac{Zr}{a_0}e^{-\frac{Zr}{2a_0}}\sin\theta\sin\varphi$$

Show that the sum of the scattering factors is equivalent to the scattering factor of the sum of the densities.

10.) Find the asymptotic expansions for $S(\vec{K})$, with both $\rho_c(\vec{r}_{12})$ and

$\rho(\vec{r})$ spherically symmetric, for both small and large K limits.

(Hint: remember that $\int d\vec{r}_{12}\rho_c(\vec{r}_{12}) = N(N-1)$, i.e. number of different electron-electron interactions in an N electron system.)

11.) Show that the Coulomb and exchange energies of a system are given as

$$\overline{V}^{Coul}_{ee} = \frac{1}{2\pi^2}\int_0^\infty \frac{d\vec{K}}{K^2}|F(\vec{K})|^2 \text{ and } \overline{V}^{Ex}_{ee} = \frac{1}{2\pi^2}\int_0^\infty \frac{d\vec{K}}{K^2}[S(\vec{K})-N]$$

12.) Calculate S(K) for an H atom in a 1S state and the $^1S$ ground state of He in the Hartree-Fock approximation.

PROBLEM 1

Key

1.) $$F(K) = \int dr\,\rho(r)e^{i\vec{K}\cdot\vec{r}} = 2\pi\int_0^\infty dr r^2\rho(r)\int_{-1}^{1} dx e^{iKrx}$$

$$= \frac{4}{K}\left(\frac{Z}{a_0}\right)^3\int_0^\infty dr r e^{-\frac{2Zr}{a_0}} \sin Kr = \frac{4}{K}\left(\frac{Z}{a_0}\right)^3\left(\frac{-d}{d\epsilon}\right)\int_0^\infty dr e^{-\epsilon r}\sin Kr$$

$$= \frac{4}{K}\left(\frac{Z}{a_0}\right)^3\frac{K}{\epsilon^2+K^2} = \frac{8(Z/a_0)^3\epsilon}{(\epsilon^2+K^2)^2}$$

$$\boxed{F(K) = \frac{16(Z/a_0)^4}{[4Z^2/a_0^2 + K^2]^2}}$$

Note $\lim_{K\to 0} F(K) = 1$ and $\lim_{K\to\infty} F(K) = \frac{16Z^4/a_0^4}{K^4}$

2.) $$\frac{1}{4\pi}\int d\Omega_{Rot}F(\vec{K}) = \frac{1}{4\pi}\int d\Omega_{Rot}\int d\vec{r}\rho(\vec{r})e^{i\vec{K}\cdot\vec{r}}$$

$$e^{i\vec{K}\cdot\vec{r}} = \sum_\ell i^\ell(2\ell+1)j_\ell(Kr)P_\ell(\cos\Omega)$$

$$\rho(\vec{r}) = \frac{1}{4\pi}\sum_n \rho_n(r)P_n(\cos\theta)$$

where $\cos\Omega = \cos(<\vec{K},\vec{r})$

Hence $P_\ell(\cos\Omega) = P_\ell(\cos\theta')P_\ell(\cos\theta) + 2\sum_{m=1}^{\ell}\frac{(\ell-m)!}{(\ell+m)!}P_\ell^m(\cos\theta')$

• $$P_\ell^m(\cos\theta)\cos m(\varphi-\varphi') = \frac{4\pi}{(2\ell+1)}\sum_{m=-\ell}^{\ell} Y_{\ell,m}(\theta,\varphi)Y_{\ell,m}^*(\theta',\varphi')$$

where $\theta$, $\varphi$ orient $\vec{r}$ w.r.t. a space fixed axis. Integrating over $d\Omega_{Rot}(\theta', \varphi')$ reduces the sum over $\ell$, m to a single term, $\ell = 0$, $m = 0$. Since $P_0^0(\cos\theta) = 1$ we are free to do the average over $\theta$, $\varphi$ which yields

$$\int d\Omega_{Rot} F(\vec{K}) = \int_0^\infty drr^2 \rho_0(r) j_0(Kr)$$

where $j_0(Kr) = \sin Kr/Kr$.
If we average $\rho(\vec{r})$ over all orientations first then we have

$$F(\vec{K}) = \int d\vec{r} \rho_0(r) e^{i\vec{K}\cdot\vec{r}}$$

since $\frac{1}{4\pi} \int d\Omega_{Rot} \rho(\vec{r}) = \frac{\rho_0(r)}{4\pi}$

or $$F(\vec{K}) = \int_0^\infty drr^2 \rho_0(r) j_0(Kr)$$

3.) Since $F(K) = \int_0^\infty drr^2 \rho_0(r) j_0(Kr)$

we have

$$F(K) = \int_0^\infty drr^2 \rho_0(r) - K^2 \frac{1}{6} \int_0^\infty drr^4 \rho_0(r)$$

$$+ K^4 \frac{1}{120} \int_0^\infty drr^6 \rho_0(r) - \ldots$$

$$F(K) = Z - \frac{K^2 \langle r^2 \rangle}{6} + \frac{K^4 \langle r^4 \rangle}{120}$$

4.) As in problem 2 we may write

$$F(\vec{K}) = \frac{1}{\sqrt{4\pi}} \int d\Omega_{\hat{r}} \sum_n \sum_p \int_0^\infty drr^2 \frac{\rho_n(r)}{\sqrt{2n+1}} Y_{n,p}(\theta,\varphi) \sum_\ell i^\ell(2\ell+1)$$

$$j_\ell(Kr)[P_\ell(\cos\theta')P_\ell(\cos\theta)+2\sum_{m=1}^{\ell} \frac{(\ell-m)!}{(\ell+m)!} P_\ell^m(\cos\theta)P_\ell^m(\cos\theta')\cos m(\varphi-\varphi')]$$

$$\text{or as } F(\vec{K}) = \sum_n \sum_p \int_0^\infty drr^2\rho_n(r)j_n(Kr)\; i^p Y_{n,p}{}^*(\theta',\varphi')$$

Let us define a generalized scattering factor as the Fourier-Bessel transform of $\rho_n(r)$ called $f_n(K)$. Then

$$\boxed{f_n(K) = \int_0^\infty drr^2\rho_n(r)j_n(Kr)}$$

and

$$\boxed{F(K) = \sum_n i^n f_n(K) \sum_{p=-n}^{n} Y_{n,p}{}^*(\theta',\varphi')}$$

5.) For K large and $\rho(\vec{r}) = \rho(r)$ we have

$$F(K) = -\frac{4\pi}{K}\int_0^\infty drr\rho(r)\,\frac{1}{K}\left(\frac{d}{dr}\right)\cos Kr$$

Integrating this last result by parts yields

$$-\frac{4\pi}{K^2}[r\rho(r)\cos Kr]\Big|_0^\infty + \frac{4\pi}{K^2}\int_0^\infty dr\left\{\frac{d}{dr}[r\rho(r)]\right\}\cos Kr$$

where the first term on the left vanishes at both limits. Continuing the procedure until a non vanishing term on the left is finally obtained we have

$$\frac{4\pi}{K^4}\left[\frac{2d\rho(r)}{dr}+\frac{rd^2\rho(r)}{dr^2}\right]\cos Kr\Bigg|_0^\infty = -\frac{8\pi}{K^4}\frac{d\rho(r)}{dr}\Bigg|_0$$

A well known theorem* due to Kato states that

*T. Kato, Commun. Pure Appl. Math. 10, 151 (1957).

$$\lim_{r\to 0}\left[\frac{\frac{d}{dr}\rho(r)}{\rho(r)}\right] = -2Z$$

Hence if both $\lim_{r\to 0}\frac{d}{dr}\rho(r)$ and $\lim_{r\to 0}\rho(r)$ are finite and different from zero we can write

$$\lim_{r\to 0}\frac{d\rho(r)}{dr} = -2Z\rho(0)$$

so that the leading term is

$$\boxed{\frac{16\pi}{K^4}Z\rho(0)}$$

For a more complete discussion see A. J. Thakkar and V. H. Smith, Jr., Chem. Phys. Letters 42, 476 (1976).

6.) $\frac{1}{4\pi}\int d\Omega_{Rot}|F(\vec{K})|^2=\frac{1}{4\pi}\int d\Omega_{Rot}\int d\vec{r}\rho(\vec{r})\int d\vec{r}\,'\rho(\vec{r}\,')e^{i\vec{K}\cdot(\vec{r}-\vec{r}\,')}$

Let $\rho(\vec{r}) = \frac{1}{\sqrt{4\pi}}\sum_n\sum_\ell \frac{\rho_n(r)Y_{n\ell}(\theta,\varphi)}{\sqrt{2n+1}}$

then $\frac{1}{4\pi}\int d\Omega_{Rot}|F(K)|^2 = \sum_{n=0}^{\infty}\left|\int_0^\infty drr^2\rho_n(r)j_n(Kr)\right|^2/(2n+1)$

$$\boxed{\frac{1}{4\pi}\int d\Omega_{Rot}|F(K)|^2 = \sum_{n=0}^{\infty}\frac{|f_n(K)|^2}{(2n+1)}}$$

7.) By definition $\bar{V}_{ne} = -Z\int dr \frac{\rho(\vec{r})}{r}$

Note that $\frac{1}{r} = \frac{1}{2\pi^2}\int \frac{d\vec{K}}{K^2} e^{i\vec{K}\cdot\vec{r}} = \frac{2}{\pi}\int_0^\infty dK \frac{\sin Kr}{K} = \frac{2}{\pi r}\int_0^\infty dx \frac{\sin x}{x} = \frac{1}{r}$

and

$$\bar{V}_{ne} = -\frac{Z}{2\pi^2}\int \frac{d\vec{K}\, F(\vec{K})}{K^2}$$

or

$$\boxed{\bar{V}_{ne} = -\frac{2Z}{\pi}\int_0^\infty dK f_0(K)}$$

8.) For either case F(0) = Z so that for

a.) we have

$$\sum_{i=1}^{M} \alpha_i = Z$$

and for

b.) we have

$$F(K) = \sum_{i=1}^{M} \frac{(\alpha_i/\beta_i{}^{4+2i})}{(1+\frac{K^2}{\beta_i{}^2})^{2+i}} = \sum_{i=1}^{M} \frac{\alpha_i{}'}{(1+\frac{K^2}{\beta_i{}^2})^{2+i}}$$

Hence $\sum_{i=1}^{M} \alpha_i{}' = Z$

For large K, $F(K) \rightarrow A/K^4$. We see immediately that form a.) cannot give the proper result here. For b.) we see that $\alpha_1{}'\beta_1{}^4$ should be constrained to be $16\pi Z\rho(0)$ from problem 5.

9.) 2s case

$$\rho_{2s}(r) = \frac{1}{32\pi}\,\alpha^3(4-4\alpha r + \alpha^2 r^2)e^{-\alpha r} \text{ with } \alpha = Z/a_o$$

and

$$\boxed{F_{2s}(K) = \frac{\alpha^4(\alpha^2-K^2)(\alpha^2-2K^2)}{(\alpha^2+K^2)^4}}$$

9.) $2p_z$ case

$$\rho_{2p_z}(\vec{r}) = \frac{1}{32\pi}\,\alpha^5\, r^2\, e^{-\alpha r}\,\cos^2\theta$$

and

$$F_{2p_z}(\vec{K}) = \frac{\sqrt{4\pi}\,\alpha^5}{24}\left[Y_{0,0}(\theta',\varphi')\int_0^\infty drr^4 e^{-\alpha r} j_0(Kr) - \frac{2}{\sqrt{5}}\,Y_{20}(\theta',\varphi')\int_0^\infty drr^4 e^{-\alpha r} j_2(Kr)\right]$$

9.) $2p_x$ case

$$\rho_{2p_x}(r) = \frac{1}{32\pi}\,\alpha^5\, r^2\, e^{-\alpha r}\,\sin^2\theta\,\cos^2\varphi$$

and

$$\begin{aligned} F_{2p_x}(\vec{K}) = \frac{\sqrt{4\pi}}{24}\,\alpha^5 [ & Y_{00}(\theta',\varphi')\int_0^\infty drr^4 e^{-\alpha r} j_0(Kr) \\ & + \frac{1}{\sqrt{5}}\,Y_{2,0}(\theta',\varphi')\int_0^\infty drr^4 e^{-\alpha r} j_2(Kr) \\ & - \frac{3}{\sqrt{30}}\,Y_{2,2}(\theta',\varphi\theta)\int_0^\infty drr^4 e^{-\alpha r} j_2(Kr) \\ & - \frac{3}{\sqrt{30}}\,Y_{2,-2}(\theta',\varphi')\int_0^\infty drr^4 e^{-\alpha r} j_2(Kr)] \end{aligned}$$

9.) $2p_y$ case

$$\rho_{2p_y}(\vec{r}) = \frac{1}{32\pi}\alpha^5 r^2 e^{-\alpha r} \sin^2\theta \sin^2\varphi$$

and

$$F_{2p_y}(\vec{K}) = \frac{\sqrt{4\pi}\,\alpha^5}{24}\Big[Y_{00}(\theta'\varphi')\int_0^\infty drr^4e^{-\alpha r}j_0(Kr) + \frac{1}{\sqrt{5}}Y_{20}(\theta',\varphi')\int_0^\infty drr^4e^{-\alpha r}j_2(Kr) + \frac{3}{\sqrt{30}}Y_{2,2}(\theta'\varphi')\int_0^\infty drr^4e^{-\alpha r}j_2(Kr) + \frac{3}{\sqrt{30}}Y_{2,-2}(\theta'\varphi')\int_0^\infty drr^4e^{-\alpha r}j_2(Kr)\Big]$$

In each case 2 integrals are important

$$I_1 = \int_0^\infty drr^4e^{-\alpha r}j_0(Kr) \text{ and } I_2 = \int_0^\infty drr^4e^{-\alpha r}j_2(Kr)$$

$$I_1 = (-\frac{d}{d\alpha})^3\frac{1}{K}\int_0^\infty dr\ e^{-\alpha r}\sin Kr$$

$$I_1 = \frac{24\alpha(\alpha^2-K^2)}{(K^2+\alpha^2)^4}$$

$$I_2 = K\frac{d}{dK}(\frac{1}{K}\frac{d}{DK})\frac{1}{K}\int_0^\infty drre^{-\alpha r}\sin Kr$$

$$I_2 = \frac{48\alpha K^2}{(\alpha^2+K^2)^4}$$

Summing up all the charge densities we obtain

$$\rho_{2s}(r) + \rho_{2p_z}(r) + \rho_{2p_x}(r) + \rho_{2p_y}(r) = \frac{\alpha^3 e^{-\alpha r}}{32\pi}[4 - 4\alpha r + \alpha^2 r^2 + \alpha^2 r^2 \cos^2\theta + \alpha^2 r^2 \sin^2\theta \cos^2\varphi + \alpha^2 r^2 \sin^2\theta \sin^2\varphi]$$

$$\rho_{2s}(r) + \rho_{2p_z}(r) + \rho_{2p_x}(r) + \rho_{2p_y}(r) = \frac{\alpha^3 e^{-\alpha r}}{32\pi}[4 - 4\alpha r + 2\alpha^2 r^2]$$

which is spherically symmetric. Note also that

$$\rho_{2p_x} + \rho_{2p_y} + \rho_{2p_z} = \frac{\alpha^5 r^2 e^{-\alpha r}}{32\pi}$$ is also spherically symmetric.

The x-ray form factor for $2s + 2_{px} + 2p_y + 2p_z$ is thus

$$F(K) = \frac{\alpha^3}{8K}\int_0^\infty drr\, e^{-\alpha r}\sin Kr[4 - 4\alpha r + 2\alpha^2 r^2]$$

$$\boxed{F(K) = \frac{2\alpha^4(K^2-\alpha^2)(K^2-2\alpha^2)}{(K^2+\alpha^2)^4}}$$

On the other hand adding up all the F(K) contributions we obtain

$$F(\vec{K}) = F_{2s}(\vec{K}) + F_{2p_z}(\vec{K}) + F_{2p_x}(\vec{K}) + F_{2p_y}(\vec{K})$$

$$F(K) = \frac{\alpha^4(\alpha^2-K^2)(\alpha^2-2K^2)}{(\alpha^2+K^2)^4} + \frac{3\alpha^6(\alpha^2-K^2)}{(\alpha^2+K^2)^4}$$

$$\boxed{F(K) = \frac{2\alpha^4(\alpha^2-K^2)(2\alpha^2-K^2)}{(\alpha^2+K^2)^4}}$$

We see that this result is identical to that at the bottom of page 15. In addition note that $\lim_{K\to 0} F(K) = N$ with N the number of orbitals in the sum and that $\lim_{K\to 0} F(K)$ is of order $1/K^4$ in all the examples studied.

10.) $$S(\vec{K}) = N + \int d\vec{r}_{12}\rho_c(\vec{r}_{12})e^{i\vec{K}\cdot\vec{r}_{12}} - |F(\vec{K})|^2$$

if both $\rho_c(\vec{r}_{12})$ and $\rho(\vec{r})$ are spherically symmetric then

$$S(K) = N + 4\pi\int_0^\infty dr_{12}r_{12}^2\rho_c(r_{12})j_0(Kr_{12}) - F(K)^2$$

We have already shown that $\lim_{K\to 0} F(K)$ is of the order of $1/K^4$ and that integrals of the type

$\int_0^\infty dr_{12}[4\pi r_{12}^2\rho_c(r_{12})]j_0(Kr_{12})$ behave as $1/K^4$ in the limit as $K\to\infty$. Hence the large K limit of S(K) behaves as N and S(K)-N as $1/K^4$. For small K we have

$$\int d\vec{r}_{12}\rho_c(\vec{r}_{12})e^{i\vec{K}\cdot\vec{r}_{12}} = \int d\vec{r}_{12}\rho_c(\vec{r}_{12}) + i\int d\vec{r}_{12}\vec{K}\cdot\vec{r}_{12}\rho_c(\vec{r}_{12})$$

$$- \frac{1}{2}\int d\vec{r}_{12}(\vec{K}\cdot\vec{r}_{12})^2\rho_c(\vec{r}_{12})$$

which for spherical symmetry yields

$$\int d\vec{r}_{12}\rho_c(r_{12})e^{iK\cdot\vec{r}_{12}} = N(N-1) - \frac{4\pi K^2}{6}\int_0^\infty dr_{12}\, r_{12}^4\rho_c(r_{12}) + \ldots .$$

$$\lim_{K\to 0} S(K) = \frac{K^2}{3}[N\langle r^2\rangle - \frac{1}{2}\langle r_{12}^2\rangle] + O(K^4)$$

11.) The classical Coulomb energy for a system is

$$V_{ee}^{Coul.} = \int d\vec{r}\rho(\vec{r})\int d\vec{r}'\rho(\vec{r}')\frac{1}{|\vec{r}-\vec{r}'|}$$

Using the transform $\frac{1}{|\vec{r}-\vec{r}'|} = \frac{1}{2\pi^2}\int\frac{d\vec{K}}{K^2}e^{i\vec{K}\cdot(\vec{r}-\vec{r}')}$

$$V_{ee}^{Coul.} = \frac{1}{2\pi^2}\int\frac{d\vec{K}}{K^2}F(\vec{K})F(\vec{K})^*$$

$$V_{ee}^{Coul.} = \frac{2}{\pi}\int_0^\infty dK\int d\vec{r}\rho(\vec{r})\int d\vec{r}'\rho(\vec{r}')j_0(K|\vec{r}-\vec{r}'|)$$

or alternatively

$$V_{ee}^{Coul.} = \frac{2}{\pi} \sum_{n=0}^{\infty} \int_0^{\infty} dK\, f_n(K) f_n^*(K)/(2n+1)$$

The total electron-electron repulsion energy is given as

$$V_{ee}^{Total} = \int d\vec{r}_{12} \frac{\rho_c(\vec{r}_{12})}{r_{12}} = \frac{1}{2\pi^2} \int \frac{d\vec{K}}{K^2} \int d\vec{r}_{12} \rho_c(\vec{r}_{12}) e^{i\vec{K}\cdot\vec{r}_{12}}$$

$$\text{But } S(\vec{K}) = N + \int d\vec{r}_{12} \rho_c(\vec{r}_{12}) e^{i\vec{K}\cdot\vec{r}_{12}} - |F(\vec{K})|^2$$

The exchange energy is defined as the difference between the total electron-electron repulsion energy and the classical Coulomb energy hence

$$V_{ee}^{Ex} = \frac{1}{2\pi^2} \int \frac{dK[S(K)-N]}{K^2}$$

If both $\rho_c(\vec{r}_{12})$ and $\rho(\vec{r})$ are spherically symmetric then

$$V_{ee}^{Coul.} = \frac{2}{\pi} \int_0^{\infty} dK\, F^2(K)$$

and

$$V_{ee}^{Ex} = \frac{2}{\pi} \int_0^{\infty} dK[S(K)-N]$$

12.) a.) Since there is only one electron in an H atom $\rho_c(\vec{r}_{12}) \equiv 0$ and we have

$$S(K) = N - F(K)^2$$

$$S(K) = 1 - \frac{256\alpha^4}{[4\alpha^2+K^2]^4} =$$

$$= K^2 \left\{ \frac{2(32\alpha^4 - 4K^2\alpha^2 + K^4)}{[4\alpha^2 + K^2]^3} - \frac{K^6}{[4\alpha^2 + K^2]^4} \right\}$$

Note that $\lim_{K\to 0} S(K)/K^2 = 1/\alpha^2$. Hence $\langle r^2\rangle = 1/\alpha^2 = \frac{a_0^2}{Z^2}$ .

b.) For the Hartree-Fock ground state of He

$$\Psi(\vec{r}_1, \vec{r}_2) = \varphi_{1s}(\vec{r}_1)\varphi_{1s}(r_2)[\alpha(1)\beta(2)-\alpha(2)\beta(1)]/2$$

$$\langle\Psi(\vec{r}_1, \vec{r}_2)|e^{i\vec{K}\cdot\vec{r}_{12}} + e^{i\vec{K}\cdot\vec{r}_{21}}|\Psi(\vec{r}_1, \vec{r}_2)\rangle$$

$$= 2\langle\varphi_{1s}(\vec{r}_1)|e^{i\vec{K}\cdot\vec{r}_1}|\varphi_{1s}(\vec{r}_1)\rangle\langle\varphi_{1s}(\vec{r}_2)|e^{-i\vec{K}\cdot\vec{r}_2}|\varphi_{1s}(\vec{r}_2)\rangle$$

and

$$F(\vec{K}) = \langle\Psi(\vec{r}_1, \vec{r}_2)|e^{i\vec{K}\cdot\vec{r}_1} + e^{i\vec{K}\cdot\vec{r}_2}|\Psi(r_1, r_2)\rangle$$

$$|F(\vec{K})|^2 = 4|\langle\varphi_{1s}(\vec{r}_1)|e^{i\vec{K}\cdot\vec{r}_1}|\varphi_{1s}(\vec{r}_1)\rangle|^2$$

For S(K) we have

$$S(K) = 2 + 2|\langle\varphi_{1s}(\vec{r}_1)|e^{i\vec{K}\cdot\vec{r}_1}|\varphi_{1s}(\vec{r}_1)\rangle|^2$$

$$- 4|\langle\varphi_{1s}(\vec{r}_1)|e^{i\vec{K}\cdot\vec{r}_1}|\varphi_{1s}(\vec{r}_1)\rangle|^2$$

or

$$\boxed{S(K) = 2[1 - |\langle\varphi_{1s}(\vec{r}_1)|e^{i\vec{K}\cdot\vec{r}_1}|\varphi_{1s}(\vec{r}_1)\rangle|^2]}$$

This result is similar to that for the H atom. Note however that for wave functions including electron correlation no such simple result will follow.

# CHEMICAL INTERPRETATION OF DEFORMATION DENSITIES

F.L. Hirshfeld

Department of Structural Chemistry

The Weizmann Institute of Science, Rehovot, Israel

## QUALITATIVE AND QUANTITATIVE INFORMATION

Before we can consider questions of chemical interpretation it is useful to have as complete as possible a description of our charge distribution. Such a description should ideally comprise

- a detailed quantitative specification of the molecular deformation density $\delta\rho(\boldsymbol{r})$;
- a reliable estimate of the accuracy of this information.

For complete specification of the function $\delta\rho(\boldsymbol{r})$ we clearly require something more than one or two contour diagrams. Such diagrams should be supplemented by some quantitative data on significant properties of the deformation density; unfortunately it is not easy to obtain a concensus on what these significant properties are.

Estimates of accuracy will take different forms for experimental and for theoretical deformation densities. Rees[1] has given general expressions for estimating the random errors in an experimental deformation density and in various derived quantities. But these expressions tend to be cumbersome to evaluate and actual numerical estimates are not easy to obtain. Non-random components of error, arising from systematic experimental error or from model deficiencies, are even harder to estimate.

The errors in a theoretical deformation density are almost entirely systematic. Often they can be roughly estimated by comparison of several calculations at different levels of sophistication (e.g. alternative basis sets, degree of optimization, etc.).

In many cases a qualitative interpretation of the deformation density is sufficient for our purpose. A contour map may then provide all the information we need. Such information may pertain, for example, to the position and shape of a peak or trough, the path of the zero contour, or the location of singular points, etc. As an example, the theoretical S.C.F. deformation density[2] of hydrazoic acid, I, shows several chemically interesting features (Fig. 1). The density peaks in the bonds from N(1) to its two neighbors lie slightly off the bond axes. This asymmetry confirms the expected bending of these bonds, attributed[3] to non-bonded repulsion between atoms H and N(2). An equally significant feature of this deformation map is the non-symmetrical shape of the lone-pair peak behind atom N(3). The out-of-plane section through this peak (Fig. 1b) is appreciably more extended than the in-plane section. This may be explained by resonance between structures $a$ and $b$:

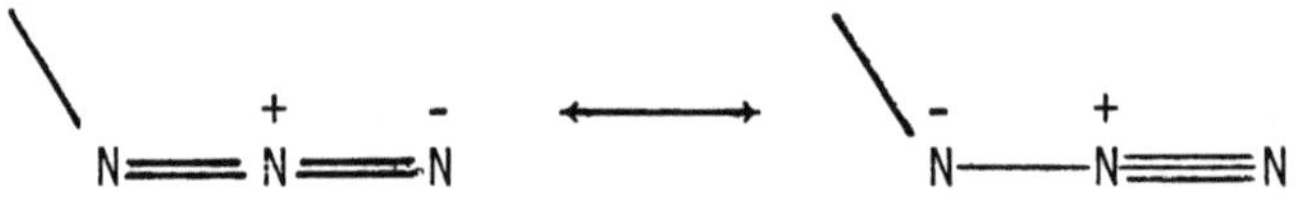

$$\diagdown N = N^+ = N^- \longleftrightarrow \diagdown N^- — N^+ \equiv N$$

$$a \qquad\qquad\qquad\qquad b$$

Structure $b$ alone would imply a symmetric lone-pair peak on the N(2)-N(3) axis behind atom N(3). The greater extension of this peak in the out-of-plane direction reflects the contribution of structure $a$, which assigns to N(3) two lone pairs, analogous to the "rabbit ear" lone-pair lobes of a carbonyl oxygen atom.

H\N(1)=N(2)=N(3)    N(4)≡C\N(1)=N(2)=N(3)    NH₂\C(—NH₂)(—N—C≡N)

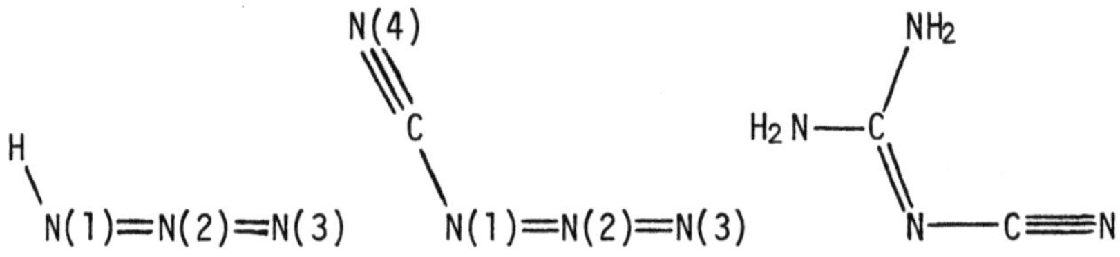

I          II          III

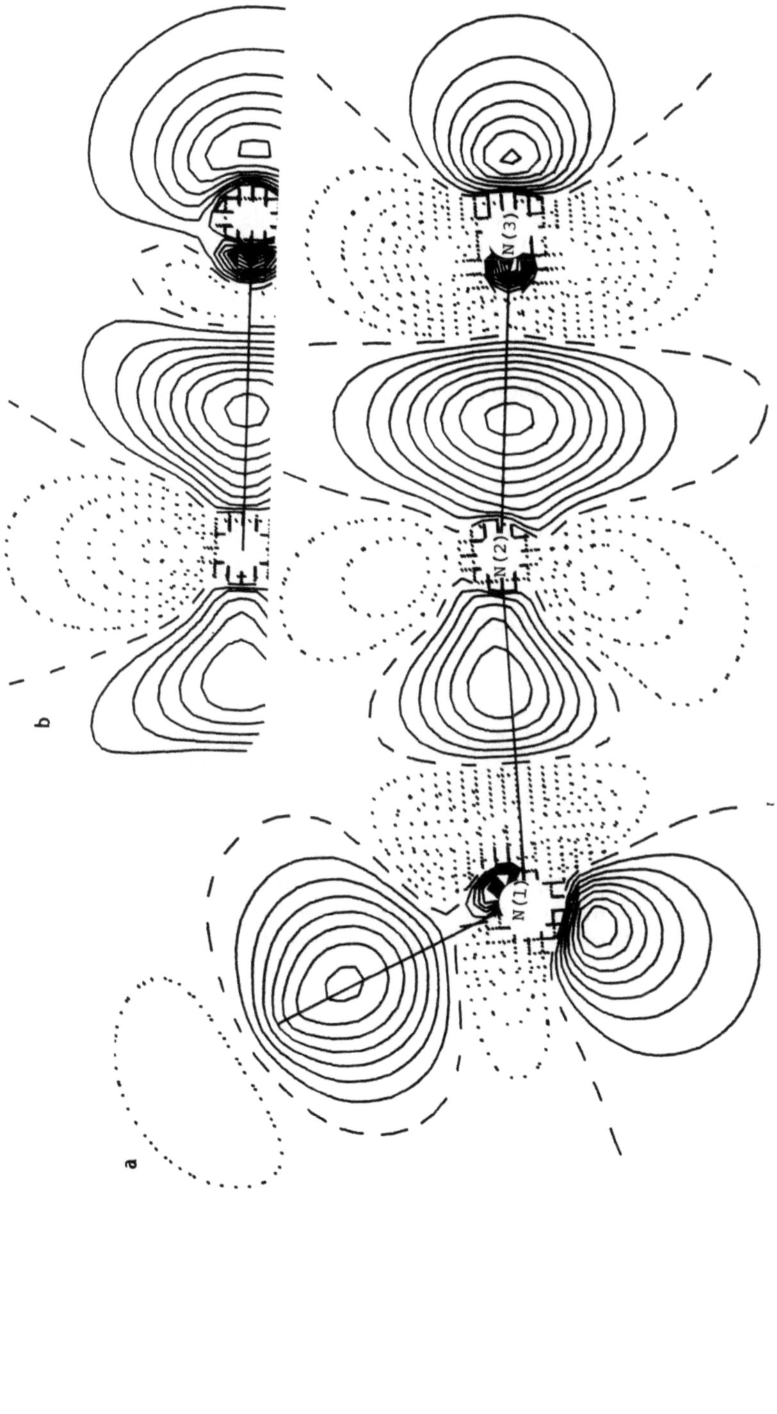


Fig. 1. $HN_3$: S.C.F. deformation density, contour interval 0.1 e/$A^3$

a. molecular plane
b. perpendicular section through N(2)-N(3) bond

The effects noted in Fig. 1 are closely reproduced in the S.C.F. deformation density[2] of cyanogen azide, II, and in the experimental deformation map[4] of 2-cyanoguanidine, III. Such agreement between two theoretical maps and one experimental map is encouraging evidence that the observed effects are genuine.

## QUANTITATIVE MEASURES OF DEFORMATION DENSITY

When we come to a quantitative specification of a molecular deformation density, there are no uniform standards of practice. Among X-ray crystallographers it is customary to present little or no information beyond what appears in contour maps in two or three selected sections through the molecule. Quantum chemists rarely present contour maps but they often list a few quantitative measures of the charge density, such as various inner and outer moments of the total density about particular centers, i.e. values of the integrals

$$\langle r^n \rangle = \int r^n \rho \, dv ,$$

where n is a positive or negative integer and the vector $r$ is measured from a specified center, e.g. an atomic nucleus or the molecular center of mass. In addition it is customary to present a list of Mulliken population indices[5], which are intended to indicate, in a poorly defined manner, the net charges on the several atoms and, sometimes, in the several bonds.

In a grossly overoptimistic attempt to promote some uniformity in the reporting of quantitative charge-density data, the author proposed [6] a conceptually simple recipe that is equally applicable to experimental and to theoretical density distributions. The crucial step is to fragment the molecular charge density into well defined atomic pieces. For this purpose we begin with the promolecule, defined as the sum of spherical free-atom charge densities, centered at the respective atomic positions in the molecule

$$\rho^{pro}(r) = \sum_\alpha \rho_\alpha^{at}(r).$$

For each atom we next define a sharing function

$$w_\alpha(r) = \rho_\alpha^{at}(r) \,/\, \rho^{pro}(r).$$

This specifies the fraction of the promolecule density at each point that is contributed by atom $\alpha$. The sum over $\alpha$ of $w_\alpha(r)$ at any point equals 1. We now define the bonded-atom fragment of the molecular charge density $\rho^{mol}(r)$ as

$$\rho_\alpha^{b.a.}(r) = w_\alpha(r)\, \rho^{mol}(r).$$

In this way the molecular density at every point is shared among all the atoms of the molecule in proportion to their respective shares in the promolecule density at the same point.

Just as the molecular density is closely approximated by the promolecule, so the bonded-atom density is not very different from the free atom. Thus it is well localized in the atomic vicinity, yet continuous, and approximately spherical. The difference between the bonded and the free atom is defined as the atomic deformation density

$$\delta\rho_\alpha = \rho_\alpha^{b.a.} - \rho_\alpha^{at}$$

But this may equally be regarded as the atomic share in the molecular deformation density

$$\delta\rho_\alpha(r) = w_\alpha(r) \quad \delta\rho(r) ,$$

where, as usual, $\delta\rho(r)$ represents the molecular deformation density.

The atomic deformation densities provide a very convenient basis for a quantitative description of the charge distribution. They allow us, for example, to define unambiguous net atomic charges $q_\alpha$ and multipole moments $\mu_\alpha$:

$$q_\alpha = -\int \delta\rho_\alpha(r)\, dv$$

$$\mu_{\alpha,i} = -\int x_i\, \delta\rho_\alpha(r)\, dv$$

$$\mu_{\alpha,ij} = -\int x_i\, x_j\, \delta\rho_\alpha(r)\, dv$$

etc. These quantities constitute a compact representation of some of the most significant properties of the molecular deformation density. In particular they permit the detailed evaluation of the external Coulomb potential of the molecule.

The usefulness of such an atomic decomposition of the molecular deformation density may be illustrated by three specific applications.

## A QUANTITATIVE COMPARISON OF TWO AZIDE MOLECULES

As noted above, the deformation density of $HN_3$ shows qualitative evidence of resonance between two canonical structures. For cyanogen azide we can write structures exactly analogous to these two, represented by *a* and *b* below, plus a third structure, *c*, with the negative charge localized on atom N(4) of the cyano group:

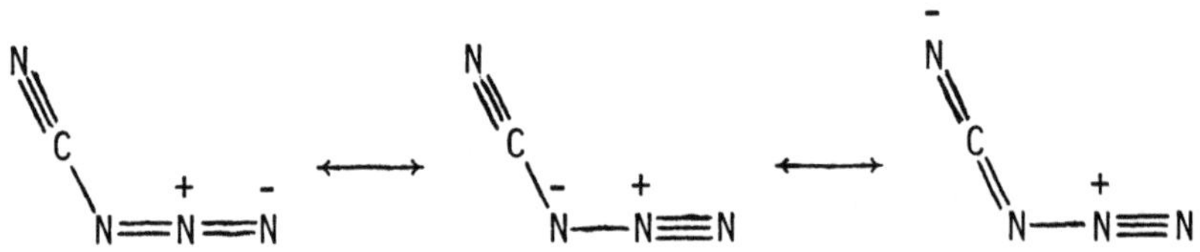

*a* *b* *c*

Thus we expect the difference between these two molecules to show itself in a transfer of negative charge from the azido group, especially from atoms N(1) and N(3), to atom N(4) in cyanogen azide. The net atomic charges $q_\alpha$ shown below confirm the expected behavior in detail:

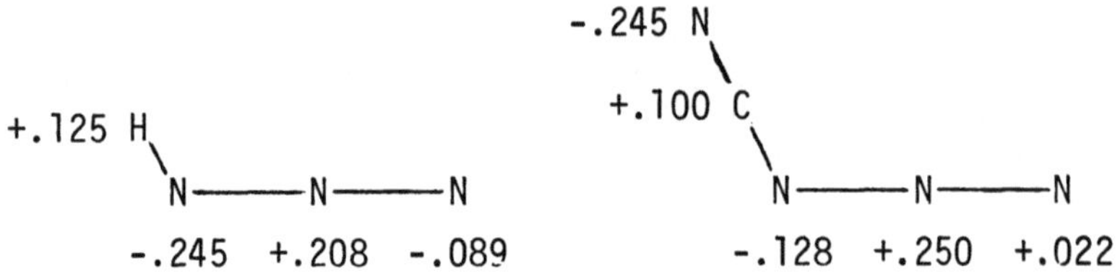


Further confirmation of this resonance picture may be found in the atomic dipole moments and second moments (Table 1). We see here that the differences between the two molecules are most pronounced in the out-of-plane $\pi$ regions. This is best seen in the second moments $\mu_{ii}$ on all three nitrogen atoms, of which the perpendicular components $\mu_{yy}$ consistently show the largest changes. These changes are in the positive sense, indicating a loss of negative charge from the $\pi$ regions of these atoms. The lesser weight assigned to structure *a* in cyanogen azide than in hydrazoic acid is shown not only by the reversed net charge on the

TABLE 1. Atomic net charges $q$ (e), dipole moments $\mu_i$ (eA), and second moments $\mu_{ij}$ ($eA^2$) in hydrazoic acid and cyanogen azide, from S.C.F. wavefunctions of M. Eisenstein, unpublished.

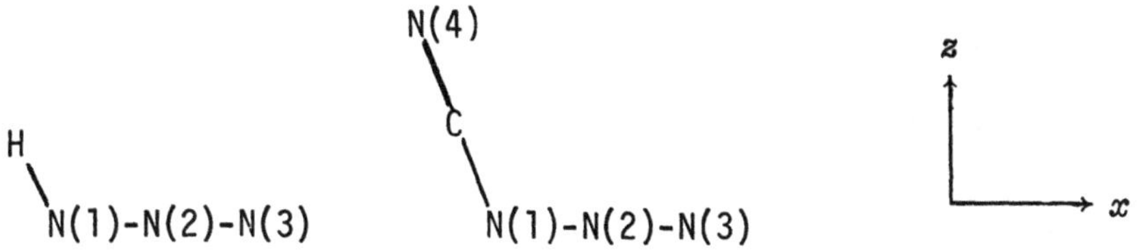


| $HN_3$ | H | N(1) | N(2) | N(3) |
|---|---|---|---|---|
| $q$ | .125 | -.245 | .208 | -.089 |
| $\mu_x$ | -.049 | .088 | -.042 | -.078 |
| $\mu_z$ | .094 | .129 | .006 | .002 |
| $\mu_{yy}$ | .051 | -.111 | .104 | -.044 |
| $\mu_{xx}$ | .076 | .018 | -.028 | -.048 |
| $\mu_{zz}$ | .061 | -.057 | .133 | .034 |
| $\mu_{xz}$ | -.002 | -.062 | -.003 | .010 |

| $NCN_3$ | N(4) | C | N(1) | N(2) | N(3) |
|---|---|---|---|---|---|
| $q$ | -.245 | .100 | -.128 | .250 | .022 |
| $\mu_x$ | .016 | .054 | .062 | -.056 | -.054 |
| $\mu_z$ | -.045 | -.134 | .116 | .017 | .002 |
| $\mu_{yy}$ | -.066 | .127 | -.032 | .142 | .026 |
| $\mu_{xx}$ | -.064 | .109 | .066 | .001 | -.013 |
| $\mu_{zz}$ | -.108 | -.008 | -.026 | .156 | .084 |
| $\mu_{xz}$ | .036 | .071 | -.037 | .005 | .019 |

terminal nitrogen but also by a decrease in its dipole moment $\mu_x$, corresponding to reduced lone-pair density on this atom. The anisotropy of this lone-pair density peak, seen in Fig. 1, is also reduced in cyanogen azide, as indicated by the much smaller difference $\mu_{zz} - \mu_{yy}$. Detailed examination of the data presented in Table 1 thus confirms and extends the qualitative deductions that can be drawn from contour diagrams.

SUBSTITUENT EFFECTS IN THREE CARBONYLS

Similar methods of analysis are illustrated by our second example. Theoretical deformation densities for the three molecules formaldehyde, IV, formic acid, V, and formamide, VI, have been derived from extended-basis S.C.F. wave functions.

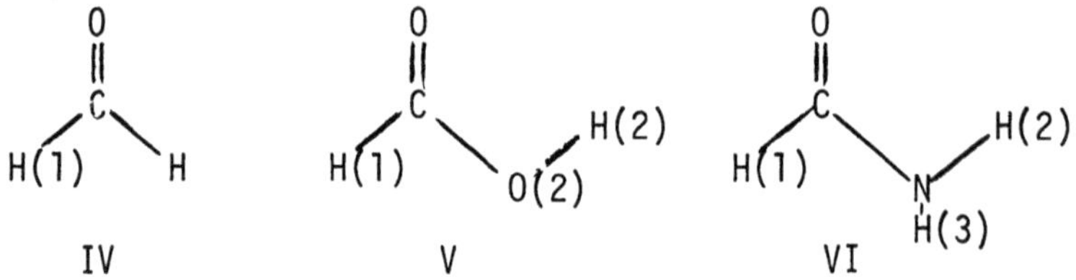


In IV the $\pi$ bonding density is well localized in the C=O double bond, but the other two molecules have alternative resonance structures in which the $\pi$ bond moves to the C-OH or C-$NH_2$ position:

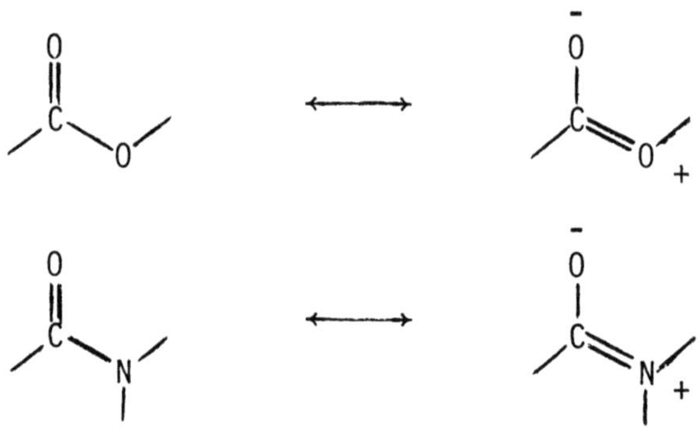


Table 2 lists the calculated net charges and moments of the atomic deformation densities in each of these three molecules. The atomic charges are seen to be consistently positive on carbon and hydrogen, negative on nitrogen and, even more, on oxygen, in accordance with the relative electronegativities of these atoms. In formaldehyde, for example, the excess charge on the oxygen atom is largely concentrated in its lone-pair lobes,

Table 2. Atomic net charges $q$ (e), dipole moments $\mu_i$ (eA), and second moments $\mu_{ij}$ ($eA^2$) in formaldehyde, IV, formic acid, V, and formamide, VI, from S.C.F. wavefunctions of M. Eisenstein, unpublished.

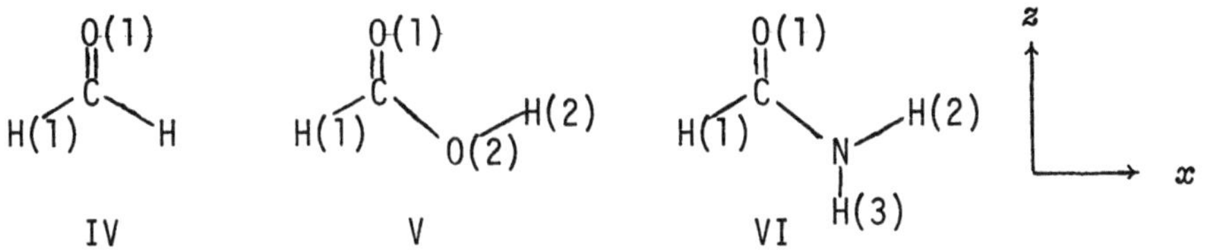


IV V VI

| $H_2CO$ | O(1) | C | H(1) | | | |
|---|---|---|---|---|---|---|
| $q$ | -.281 | .178 | .051 | | | |
| $\mu_z$ | -.055 | -.058 | -.036 | | | |
| $\mu_x$ | 0 | 0 | -.058 | | | |
| $\mu_{zz}$ | -.035 | .081 | .063 | | | |
| $\mu_{xx}$ | -.075 | .129 | .061 | | | |
| $\mu_{yy}$ | -.005 | .274 | .070 | | | |
| $\mu_{xz}$ | 0 | 0 | -.007 | | | |
| HCOOH | O(1) | C | H(1) | O(2) | H(2) | |
| $q$ | -.335 | .272 | .080 | -.219 | .200 | |
| $\mu_z$ | -.062 | -.059 | -.047 | .088 | .064 | |
| $\mu_x$ | .003 | .005 | -.062 | 0 | .096 | |
| $\mu_{zz}$ | -.035 | .048 | .069 | -.011 | .070 | |
| $\mu_{xx}$ | -.082 | .135 | .070 | .074 | .086 | |
| $\mu_{yy}$ | -.039 | .270 | .079 | -.025 | .078 | |
| $\mu_{xz}$ | -.002 | .022 | -.003 | .016 | 0 | |
| $HCONH_2$ | O(1) | C | H(1) | N | H(2) | H(3) |
| $q$ | -.386 | .204 | .054 | -.140 | .136 | .130 |
| $\mu_z$ | -.077 | -.041 | -.052 | -.004 | .043 | -.103 |
| $\mu_x$ | .008 | -.019 | -.047 | .014 | .093 | -.005 |
| $\mu_{zz}$ | -.047 | .033 | .065 | .080 | .078 | .072 |
| $\mu_{xx}$ | -.111 | .109 | .061 | .099 | .074 | .086 |
| $\mu_{yy}$ | -.075 | .217 | .071 | -.017 | .054 | .052 |
| $\mu_{xz}$ | 0 | .024 | -.005 | -.003 | -.003 | 0 |

which extend in the molecular plane, mainly in a direction perpendicular to the C=O bond (negative $\mu_{xx}$ and $\mu_{zz}$, the former more so) but strongly displaced away from the carbon atom (negative $\mu_z$).

The carbon atom is correspondingly deficient in charge (positive $q$) and is strongly polarized toward the double bond (negative $\mu_z$). Its largest second moment is in the out-of-plane direction (positive $\mu_{yy}$), showing that the loss of charge has occurred mainly from its $p_\pi$ orbital.

The hydrogen atom is slightly positive, is polarized into the C-H bond (negative $\mu_x$ and $\mu_z$), and is contracted isotropically (positive $\mu_{zz} \sim \mu_{xx} \sim \mu_{yy}$). This behavior is typical of covalently bonded hydrogen in general. Exactly the same pattern was found in hydrazoic acid (Table 1) and is duplicated by the two hydrogen atoms in formic acid and by all three hydrogens in formamide.

When one of the hydrogen atoms in formaldehyde is replaced by OH or $NH_2$, two kinds of change occur, an inductive and a conjugative effect. The former is mainly a local effect, being confined largely to the region of the bond between carbon and the oxygen or nitrogen substituent. This bond is polarized towards the electronegative substituent, causing the charge on carbon to decrease (more positive $q$), moderately when the substituent is nitrogen (with $q_N$ = -0.14 e), much more when it is oxygen ($q_O$ = -0.22 e). In either case a small portion of the extra positive charge on carbon spills over onto H(1).

The major effect in the region of the C=O bond arises from the conjugative influence of the polar resonance structures, in which the $\pi$ density is removed from the bond and concentrated in a third lone pair on the oxygen atom. This effect appears to be about twice as great in the amide, with the greater electron-donating tendency of the $NH_2$ group, as in the acid. It shows up in both these molecules in increasingly negative values of $q$ and $\mu_{yy}$ on O(1) along with appropriate changes in the atomic dipole moments of both C and O(1).

## LATTICE ENERGY CALCULATIONS: CYANOGEN

Our final example illustrates the evaluation, from the atomic net charges and multipole moments, of the electrostatic interaction energy of neighboring molecules in a crystal. Fig. 2 plots the calculated lattice energy of cyanogen, VII, whose charge distribution has been derived from the Hartree-Fock

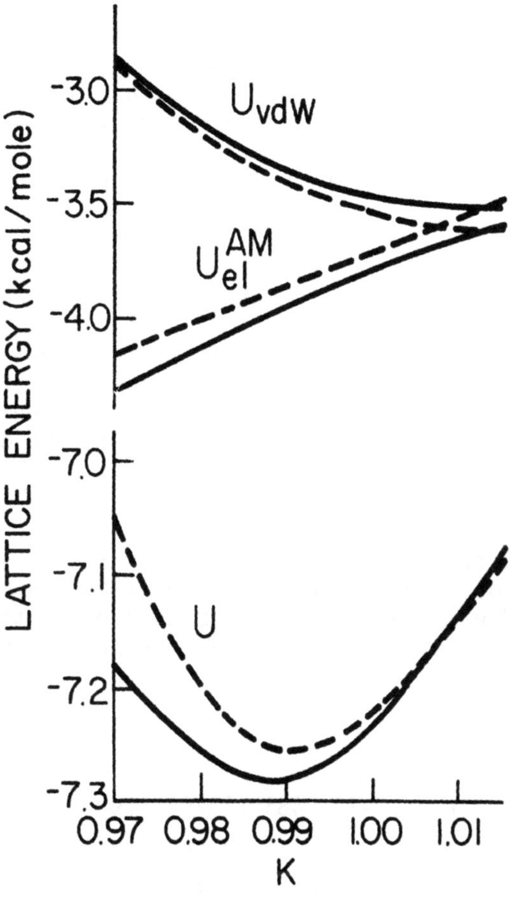


Fig. 2. Variation of calculated lattice energy $U$ of cyanogen, and of its van der Waals and electrostatic components $U_{vdW}$ and $U_{el}$, with a uniform expansion or contraction of the unit cell; scale factor $K$ multiplies observed cell edges. Full lines: observed structure, space group *Pcab*; broken lines: hypothetical structure in space group *Pa*3. Reproduced from Hirshfeld and Mirsky[8].

wavefunction computed by McLean and Yoshimine[7]. The abcissa in this plot is a scale factor $K$ that simultaneously scales all three

N≡C-C≡N

VII

unit-cell edges in this orthorhombic structure (space group $Pcab$). The value $K = 1$ corresponds to the observed cell dimensions at a temperature of 224 K. Since the calculation gives the theoretical lattice energy at $T = 0$, we expect the calculated minimum to occur at a unit-cell size slightly below that observed at elevated temperature. Indeed the total lattice energy $U$ has its minimum value at $K \sim 0.99$. But this represents a compromise between the van der Waals energy $U_{vdW}$, which favors a more expanded structure, and the electrostatic contribution $U_{el}$, which tends to compress the structure.

The broken lines in Fig. 2 show the corresponding plots for a hypothetical cubic structure, space group $Pa3$, such as is found in crystalline $CO_2$ and in the high-temperature form of acetylene. The van der Waals energy $U_{vdW}$ is seen to be lower for the cubic than for the observed orthorhombic structure and only the influence of the electrostatic energy, which is lower in the orthorhombic structure, accounts for the greater stability of the less symmetric structure.

These illustrations demonstrate the value of a quantitative description of molecular deformation densities, such as may be readily achieved via the proposed atomic decomposition, as a basis for chemical interpretation as well as for predictions of various molecular properties.

REFERENCES

1. B. Rees, Acta Cryst. A32, 483 (1976); Isr.J.Chem. 16, 180 (1977).
2. M. Eisenstein and F.L. Hirshfeld, Chemical Physics, in press.
3. F.L. Hirshfeld, Isr. J. Chem. 2, 87 (1964).
4. F.L. Hirshfeld and H. Hope, to be published
5. R.S. Mulliken, J.Chem.Phys. 3, 573 (1935); 23, 1833 (1955).
6. F.L. Hirshfeld, Theoret. Chim. Acta 44, 129 (1977).
7. A.D. McLean and M. Yoshimine, *Tables of Linear Molecule Wave Functions*, I.B.M. Corp., San Jose (1967).
8. F.L. Hirshfeld and K. Mirsky, Acta Cryst., in press.

DENSITY ANALYSIS FOR THE CUBIC MODIFICATION OF $C_2H_2$

A. VOS

A. Structural Data

a. The high temperature modification of $C_2H_2$ is cubic, space group Pa3, a (141°K) = 6.091 Å, Z = 4. Find the Wyckoff positions for the molecule. See Figure 1.

b. The intermolecular distances at 141°K are given in Figure 2. Do you expect small or large thermal motion.

B. Density analysis

We consider the possibility of a valence analysis with restricted radial density basis functions according to Stewart (1976). In this analysis the total dynamic density distribution $\rho(\underset{\sim}{r})$ is assumed to be the sum of the densities of the pseudoatoms p in the cell.

$$\rho(\underset{\sim}{r}) = \sum_p \rho_p(\underset{\sim}{r}_p)$$

with $\underset{\sim}{r}_p = \underset{\sim}{r} - \underset{\sim}{R}_p$, $\underset{\sim}{R}_p$ being the equilibrium position of atom p, and

$$\rho_p(\underset{\sim}{r}_p) = Pop_p^e(\text{core})\rho_p(\text{core},r_p) + \sum_{\ell,m=0} Pop_p^e(\ell,m)\rho_p(\ell,r_p)P^m(\cos\theta_p)\cos m\phi_p + \sum_{\ell,m=0} Pop_p^o(\ell,m)\rho_p(\ell,r_p)P^m(\cos\theta_p)\sin m\phi_p \qquad (1)$$

$\rho_p(\text{core},r_p)$ is a normalized SCF density function for p. A deviation of $Pop_p(\text{core})$ from the real core population, say by an amount $-q_p$, is not considered as a perturbation of the core, but as a modification of the monopole valence function $\rho_p(0,r_p)$, in this case with the amount $[q_p/Pop_p(0,0)]x[\rho_p(0,r_p)-\rho_p(\text{core},r_p)]$. The products of the radial density functions and the associated Legendre functions are normalized as follows. For monopole terms with $\ell=0$ the normalisation is such that $Pop_p(0,0)$ is the population of the monopole function, thus :

$$Pop_p(0,0) \int \rho_p(0,r_p) P_0^0(\cos\theta_p) d\underset{\sim}{r}_p = Pop_p(0,0)$$

For $\ell > 0$ where the terms in (1) integrate to zero, the normalisation is such that $Pop_p(\ell,m)$ gives the number of electrons tranferred from the total negative to the total positive part of the function, thus for $Pop^e(\ell,m)$

$$\int | \rho_p(\ell,r_p) P^m(\cos\theta_p) \cos(m\phi_p) | d\underset{\sim}{r}_p = 2$$

Scattering factors . The aspherical scattering factors are obtained by Fourier transformation of (1).

$$f_p(S) = Pop_p(\text{core}) f_p(\text{core},S) + \sum_{\ell,m} Pop_p^e(\ell,m) i^\ell f_p(\ell,S) P_\ell^m(\cos\theta_S)\cos(m\phi_S) \sum_{\ell,m} Pop_p^o(\ell,m) i^\ell f_p(\ell,S) P_\ell^m(\cos\theta_S)\sin(m\phi_S)$$

with $f_p(\ell,S)$ given by :

$$f_p(\ell,S) = 4\pi \int_o^\infty \rho_p(\ell,r_p)\ j_\ell(Sr_p) r_p^2 dr_p .$$

The questions below concentrate mainly on the C atoms. The scattering factor for the core is given in table 1, derived from the SCF $^3P$ wavefunction of Clementi. For the valence electrons two types of scattering factors will be considered :

1. The $\zeta$ scattering factors $f(\ell,\zeta)$, corresponding with the valence radial distribution functions :

$$\rho_n(r) = (4\pi)^{-1} [ (2\zeta)^{n+3}/(n+2)!]\ r^n \exp(-2\zeta r)$$

with $\zeta=1.72$, and n=2 for the monopole up to the quadrupole, and n=3 for the octupole function.

2. The $^3P$ scattering factor $f(\text{val};{}^3P)$ derived from the SCF function. Note that the total scattering factor 2f(core)+4f(val) is close to the Cromer-Mann scattering curve for C.

Questions.

Give the discussion for a local reference frame with z parallel to H-C-C-H and x and y perpendicular to the molecular axis.

1. Which independent parameters can be refined in a conventional (spherical) anisotropic refinement ?

2. Which multipole functions may, in view of the molecular symmetry, have populations different from zero (see table 2) ?

3. Make a rough estimate of the correlation between scale factor K and isotropic temperature factor B for the high order refinement. Data : Refinement on $F^2$ with $w_i = 1$ on reflections with $0.6\ A^{-1} < \sin\theta/\lambda < 0.8\ A^{-1}$. $B = 4\ Å^2$.

4. Show the influence of the assumed radial (monopole) function on scale, temperature factor B and core population for the following example : Take a single C atom with ζ scattering factor, K = 1, P(core) = 2, B =4. Make an estimate of the values of K, P(core) and B which would be obtained if the atom had been described by a $^3P$ valence scattering factor plus, of course, f(core). Reflection data available for $0.14 < \sin\theta/\lambda < 0.80$.

5. How do errors in z(C) and in the anisotropic thermal motion of C, show up in a difference map. Compare these features with the functional form of the dipole and quadrupole functions of question 2. Which correlations do you expect on the basis of this comparison ?

6. Try to find the correlations of question 5, but this time starting from the formula for the structure factor. Consider only two carbon atoms at distances +Z and -Z from the inversion center.

7. Find the best way to calculate the filtered and unfiltered deformation maps for the present density analysis. Discuss the standard deviation for positions of different type in the unfiltered deformation map. No absolute estimate is requested, but only relative values for the different types of positions you want to distinguish.

8. After the above seven questions have been answered, a correlation table of a X-Ray refinement for $C_2H_2$ will be provided ( $0.14 < \sin\theta/\lambda < 0.80$, least squares aspherical refinement on $F^2$ ). Check your answers with the correlation coefficients given in the table. Also some deformation maps can be available for inspection.

9. Can additional experiments be used to avoid the above difficulties ? Which accuracy would be required for such experiments ?

10. Which is the lowest molecular electric moment for acetylene ? Does the above density analysis give an accurate value for this moment ?

## ANSWERS

1. K, z(C), $U_{zz}$(C), $U_{xx}(C) = U_{yy}(C)$. Same for H

2. z is the main axis. Take $\Theta=\Theta_z$, z is a 3-fold axis.
Functions applicable : Q(0,0), Q(1,0), Q(2,0), Q(3,0), $Q^e(3,3)$, $Q^o(3,3)$
For cylindrical symmetry, $Q^e(3,3)$ and $Q^o(3,3)$ cannot be used

3. Single carbon at the origin, and spherical part of F.

$$|F|^2 = K^2 f_c^2 \exp(-2B\sin^2\theta/\lambda^2)$$

$$\frac{\partial |F|^2}{\partial K} = 2K f_c^2 \exp(-2B\sin^2\theta/\lambda^2)$$

$$\frac{\partial |F|^2}{\partial B} = -2(\sin^2\theta/\lambda^2)K^2 f_c^2 \exp(-2B\sin^2\theta/\lambda^2)$$

Elements $a_{KK}$, $a_{BB}$, $a_{BK}$ of normal matrix. Summation over reflections is replaced by integration and $f_c^2$ is considered as constant for the high region involved. Write $y = \sin\theta/\lambda$. Apart from a constant factor

$$a_{KK} = \int_{.6}^{.8} y^2 \exp(-4By^2)\, dy$$

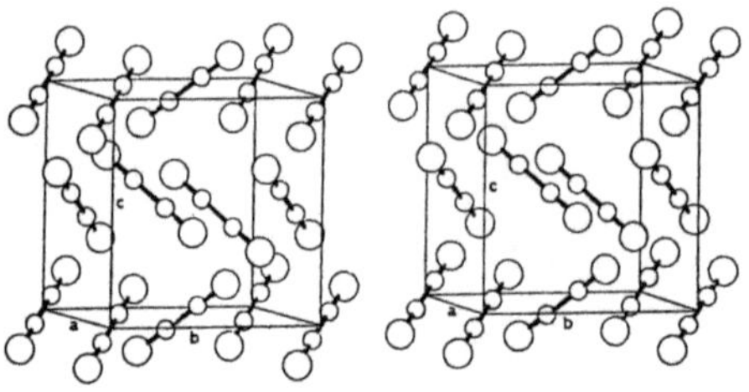


*Figure 1.*
*Stereopicture of the packing of the molecules in the cubic phase of acetylene.*

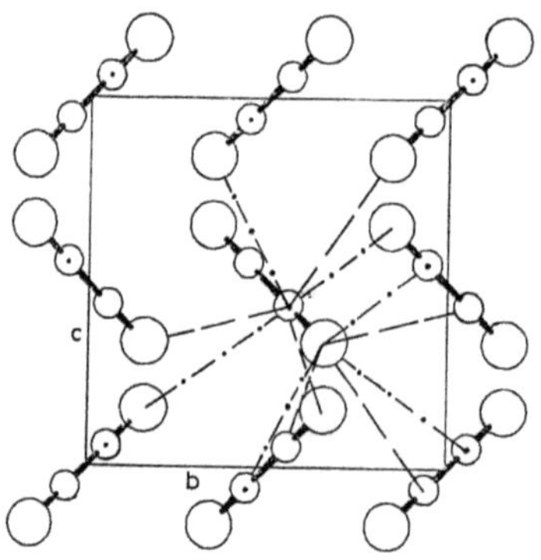


*Figure 2.*
*$C_2H_2$ structure along a.*
*Non bonding distances shorter than the sum of van der Waals radii (r(C) =1.7, r(H) = 1.2 ) plus 0.4 Å. Dashed lines 3.05 Å, chain dotted lines 3.26 Å.*

Table 1. Carbon scattering factors f(ℓ) per electron

$^3P$ scattering factors

| sinΘ/λ | f(0;core) | f(0;valence) | f(1;dipole) | f(2;quadrupole) |
|---|---|---|---|---|
| 0.0 | 1.000 | 1.000 | 0.000 | 0.000 |
| 0.1 | 0.993 | 0.785 | 0.291 | 0.080 |
| 0.2 | 0.972 | 0.410 | 0.333 | 0.183 |
| 0.3 | 0.939 | 0.156 | 0.233 | 0.197 |
| 0.4 | 0.895 | 0.040 | 0.129 | 0.158 |
| 0.5 | 0.843 | -0.001 | 0.060 | 0.109 |
| 0.6 | 0.786 | -0.009 | 0.021 | 0.069 |
| 0.7 | 0.726 | -0.007 | 0.002 | 0.041 |
| 0.8 | 0.665 | -0.002 | -0.007 | 0.024 |
| 0.9 | 0.605 | 0.002 | -0.010 | 0.013 |
| 1.0 | 0.547 | 0.005 | -0.011 | 0.007 |

ζ scattering factors for ζ = 1.72

| sinΘ/λ | f(0;valence) | f(1;dipole) | f(2;quadrupole) |
|---|---|---|---|
| 0.0 | 1.000 | 0.000 | 0.000 |
| 0.1 | 0.831 | 0.276 | 0.064 |
| 0.2 | 0.487 | 0.358 | 0.171 |
| 0.3 | 0.208 | 0.283 | 0.211 |
| 0.4 | 0.062 | 0.174 | 0.183 |
| 0.5 | 0.005 | 0.094 | 0.134 |
| 0.6 | -0.011 | 0.047 | 0.089 |
| 0.7 | -0.013 | 0.022 | 0.057 |
| 0.8 | -0.011 | 0.010 | 0.036 |
| 0.9 | -0.008 | 0.004 | 0.023 |
| 1.0 | -0.005 | 0.002 | 0.015 |

Table 2. Normalized multipole functions (see text).
$q_z$, $q_y$, $q_z$ are the direction cosini of a vector $\vec{r}$ relative to the chosen x, y and z axis.

Monopole

$Q(0,0) = P_o^o(\cos\vartheta) = 1$

Dipole

$Q(1,0) = P_1^o(\cos\vartheta) = 4q_z$

$Q^o(1,1) = P_1^1(\cos\vartheta)\sin\varphi = 4q_y$

$Q^e(1,1) = P_1^1(\cos\vartheta)\cos\varphi = 4q_x$

Quadrupole

$Q(2,0) = P_2^o(\cos\vartheta) = 9\sqrt{3}(q_z^2 - 1/3)/2$

$Q^o(2,1) = P_2^1(\cos\vartheta)\sin\varphi = 3\pi\, q_y\, q_z$

$Q^e(2,1) = P_2^1(\cos\vartheta)\cos\varphi = 3\pi\, q_x\, q_z$

$Q^o(2,2) = P_2^2(\cos\vartheta)\sin2\varphi = 3\pi\, q_x\, q_y$

$Q^e(2,2) = P_2^2(\cos\vartheta)\cos2\varphi = 3\pi\, (q_x^2 - q_y^2)/2$

Octupole

$Q(3,0) = P_3^o(\cos\vartheta) = 20(5q_z^2 - 3)q_z/7$

$Q^o(3,1) = P_3^1(\cos\vartheta)\sin\varphi = 4\pi(\tan^{-1}2 + 14/5 - \pi/4)\,(5q_z^2 - 1)q_y$

$Q^e(3,1) = P_3^1(\cos\vartheta)\cos\varphi = 4\pi(\tan^{-1}2 + 14/5 - \pi/4)\,(5q_z^2 - 1)q_x$

$Q^o(3,2) = P_3^2(\cos\vartheta)\sin2\varphi = 8\pi\, q_x\, q_y\, q_z$

$Q^e(3,2) = P_3^2(\cos\vartheta)\cos2\varphi = 4\pi(q_x^2 - q_y^2)q_z$

$Q^o(3,3) = P_3^3(\cos\vartheta)\sin3\varphi = 16(3q_x^2 - q_y^2)q_y/3$

$Q^e(3,3) = P_3^3(\cos\vartheta)\cos3\varphi = 16(q_x^2 - 3q_y^2)q_x/3$

$$a_{BB} = K^2 \int_{.6}^{.8} y^6 \exp(-4By^2)\, dy$$

$$a_{KB} = -K \int_{.6}^{.8} y^4 \exp(-4By^2)\, dy$$

Matrix : $\begin{pmatrix} 133.4 & -53.8\,K \\ -53.8\,K & 23.4\,K^2 \end{pmatrix}$ Calculate inverse matrix

Correlation coefficient : $r(K,B) = \dfrac{a^{KB}}{(a^{KK}a^{BB})^{1/2}} = 0.96$

4. a. Calculate the $\zeta$- Carbon atom.
   b. Calculate the $^3P$- Carbon atom for different core populations.
Try P(core) = 2, 1.9, 2.1

$$f^o(^3P) = P(core)f(core) + [6-P(core)]f(val)$$

c. $f^o(^3P)_{adjusted} = Kf^o(^3P)\exp(-\Delta B\sin^2\theta/\lambda^2)$ $f^o(\zeta)$

Make Wilson like plot : $\ln[f^o(\zeta)/f^o(^3P)] = \ln K - \Delta B\sin^2\theta/\lambda^2$

Calculation from Table 1 gives

| $\sin\Theta/\lambda$ | $f^o(\zeta;P=2)$ | $f^o(^3P;P=2)$ | $f^o(^3P;P=1.9)$ | $f^o(^3P;P=2.1)$ |
|---|---|---|---|---|
| 0.1 | 5.310 | 5.126 | 5.106 | 5.147 |
| 0.2 | 3.892 | 3.584 | 3.527 | 3.640 |
| 0.3 | 2.710 | 2.502 | 2.424 | 2.580 |
| 0.4 | 2.038 | 1.950 | 1.865 | 2.036 |
| 0.5 | 1.706 | 1.682 | 1,598 | 1.766 |
| 0.6 | 1.528 | 1.536 | 1.456 | 1.616 |
| 0.7 | 1.400 | 1.424 | 1.350 | 1.498 |
| 0.8 | 1.286 | 1.322 | 1.256 | 1.398 |

From the Wilson plots it is clear that P(core) = 1.9 gives the best agreement. For P(core) = 1.9 we find: K = 1.11, $\Delta B = 0.14$ Å$^2$.

(The least squares refinements gave P(core;$\zeta$) = 1,97, P(core;$^3$P) = 1.90 K($^3$P)/K($\zeta$) = 1.084, $\Delta B = 0.14$ Å$^2$).

5. Error in z. Slope at atomic position

Correlation between position and dipole

Error in anisotropy of thermal motion.

Difference Fourier  Correlation with quadrupole

6. Formula for structure factor.
Scale and monopole population fixed. Z = distance of C from inversion centre. $H_z$ = projection of H on z axis. $H_x, H_y$ = projections of H on x,y. $\underset{\sim}{r}\cdot\underset{\sim}{S} \doteq H_z Z$   $T(\underset{\sim}{H})$ = temperature factor.

$$F(\underset{\sim}{H}) = f(0)T(\underset{\sim}{H})\,[\exp(2\pi i H_z Z) + \exp(-2\pi i H_z Z)] +$$

$$iP(1,0)f(1)T(\underset{\sim}{H})Q(1,0)\,[\exp(2\pi i H_z Z) - \exp(-2\pi i H_z Z)] -$$

$$P(2,0)f(2)T(\underset{\sim}{H})Q(2,0)\,[\exp(2\pi i H_z Z) + \exp(-2\pi i H_z Z)]$$

$$T(\underset{\sim}{H}) = \exp[-2\pi^2(U_{xx}H_x^2 + U_{xx}H_y^2 + U_{zz}H_z^2)]$$

$H_x = Hq_x \ldots,$

$$T(\underset{\sim}{H}) = \exp[-2\pi^2(\frac{2U_{xx}+U_{zz}}{3})H^2]\ \exp[-2\pi^2(\frac{U_{zz}-U_{xx}}{3})(3q_z^2-1)H^2]$$

'isotropic part'   'ΔU'   $\frac{2\sqrt{3}}{9}Q(2,0)$

$$T(\underset{\sim}{H}) = \exp(-B\sin^2\theta/\lambda^2)\ \exp[-\frac{4\sqrt{3}}{9}\pi^2\Delta U Q(2,0)H^2]$$

Derivatives :

(a) $$\frac{\partial F(\underset{\sim}{H})}{\partial Z} = -4\pi f(0)T(\underset{\sim}{H})H_z\sin(2\pi H_z Z) = -4\pi f(0)T(\underset{\sim}{H})H\frac{Q(1,0)}{4}\sin(2\pi H_z Z)$$

(b) $$\frac{\partial F(\underset{\sim}{H})}{\partial P(1,0)} = 2f(1)T(\underset{\sim}{H})Q(1,0)\sin(2\pi H_z Z) \qquad (Z \text{ in } \mathring{A})$$

(a) and (b) have the same directional symmetry and the same goniometric dependence on $H_z Z$. The cross term

$$\sum_{\underset{\sim}{H}} \frac{\partial F(\underset{\sim}{H})}{\partial Z}\ \frac{\partial F(\underset{\sim}{H})}{\partial P(1,0)}$$

in the normal equations thus does not vanish and correlation is expected. The amount of correlation depends on the angular range considered, due to difference in radial function.

$$\frac{\partial F(\underset{\sim}{H})}{\partial U} = \frac{8\sqrt{3}}{9}\pi^2 f(0)H^2 T(\underset{\sim}{H})Q(2,0)\cos(2\pi H_z Z) \qquad \text{for monopole}$$

$$\frac{\partial F(\underset{\sim}{H})}{\partial P(2,0)} = -2f(0)T(\underset{\sim}{H})Q(2,0)\cos(2\pi H_z Z)$$

Correlation expected.

7. In general, amplitudes of deformation maps are defined by $[K^{-1}F_o-F_c(\text{mono})]$ with $F_c(\text{mono})$ based on scattering factor of a free spherical atom. This is impossible here as the correlation between K and B, type of scattering factor and monopole population is so large that independent determination of these quantities is impossible. The best approximation is to take for $F_c(\text{mono})$ the monopole terms (core+valence) of the refinement (including the refined population). Only higher multipole terms, which for the static density are equal to 0 at the atomic position, remain. The deviations from 0 at the atomic positions are small. They must be due to anisotropy in the thermal motion to restricted resolution and to multipole terms from neighbouring atoms.
Let the standard deviation at general position be p, the s.d. at the 3-fold axis is $p\sqrt{3}$, and at the origin $p\sqrt{6}$.

9. A. Neutron Diffraction.
Errors in positions and thermal parameters (especially in the $\Delta U$ value) will show up in changes of dipole and quadrupole moments. Errors in isotropic B's will show up in a change of K.

a. Positions and dipole moments. From (6) :

$$[2f(1)T(\underset{\sim}{H})Q(1,0)\sin(2\pi H_z Z)]\Delta P \simeq [\pi f(0)T(\underset{\sim}{H})HQ(1,0)\sin(2\pi H_z Z)]\Delta Z$$

giving for the least squares solution :

$$\Delta P = \frac{\pi}{2}\frac{\sum_{\underset{\sim}{H}} [f(0)T(\underset{\sim}{H})H][f(1)T(\underset{\sim}{H})]}{\sum_{\underset{\sim}{H}}[f(1)T(\underset{\sim}{H})]^2}$$

. Integration over $[Q(1,0)\sin(2\pi H_z Z)]^2$ gives the same for numerator and denominator. $\Delta Z$ is the error in Z. One gets :

$$\Delta P = -9.75\,\Delta Z \qquad (\,Z \text{ in } \mathring{A})$$

$\Delta Z = .001$ Å, thus $\Delta P = .01$. This is of the same order as the standard deviation in P of the $C_2H_2$ analysis ($\Delta\rho$ at the middle of the C-C bond is .024 eÅ$^{-3}$).

b. Anisotropy in thermal motion and quadrupole moment.

$$f(2)T(\underset{\sim}{H})\Delta P_2 \simeq [\frac{4\sqrt{3}}{9}\pi^2]f(0)H^2T(\underset{\sim}{H})\Delta U$$

with least squares solution :

$$\Delta P_2 = [\frac{4\sqrt{3}}{9}\pi^2]\frac{\sum_{\underset{\sim}{H}}[f(0)H^2T(\underset{\sim}{H})][f(2)T(\underset{\sim}{H})]}{\sum_{\underset{\sim}{H}}[f(2)T(\underset{\sim}{H})]^2}$$

Calculation gives $\Delta P_2 = 53.1\,\Delta U$

Say error in $\Delta U = .0005$ Å$^2$, thus error in $\Delta P_2 = .03$ (similar to X-ray experiment).

Conclusion. If the neutron work is not more accurate than the X-Ray experiment systematic errors may be avoided, but random errors larger than (or of the same order than) the st. d. in the X-Ray experiment will occur.

B. Molecular geometries from the gas phase : error discussion as above. Programs must be suitable for rigid body refinements as otherwise difficulties arise due to librational motion. Internal vibrations must be accounted for, so that these must be known from spectral data.

10. Quadrupole moment. Especially uncertainties in populations due to errors in radial distribution functions play a part.

Table 3. Correlation coefficients for K and C-atom parameters for $C_2H_2$. Details of the refinement : least-squares refinement on $F^2$, H constrained to C; $.14 < \sin\theta/\lambda < .80$. Weights obtained by comparison of intensities of equivalent reflections P(0,0) = valence monopole population, P(1,0) and P(2,0) dipole and quadrupole populations respectively.

| | K | ζ | Z | B | ΔU | P(0,0) | P(1,0) | P(2,0) |
|---|---|---|---|---|---|---|---|---|
| K | +1.00 | -0.84 | -0.47 | +0.98 | +0.23 | +0.64 | +0.28 | +0.32 |
| ζ | -0.84 | +1.00 | +0.52 | -0.73 | -0.33 | -0.91 | -0.40 | -0.26 |
| Z | -0.47 | +0.52 | +1.00 | -0.45 | -0.05 | -0.59 | -0.90 | -0.30 |
| B | +0.98 | -0.73 | -0.45 | +1.00 | +0.14 | +0.52 | +0.26 | +0.35 |
| ΔU | +0.23 | -0.33 | -0.05 | +0.14 | +1.00 | +0.26 | -0.15 | -0.75 |
| P(0,0) | +0.64 | -0.91 | -0.59 | +0.52 | +0.26 | +1.00 | +0.53 | +0.23 |
| P(1,0) | +0.28 | -0.40 | -0.90 | +0.26 | -0.15 | +0.53 | +1.00 | +0.48 |
| P(2,0) | +0.32 | -0.26 | -0.30 | +0.35 | -0.75 | +0.23 | +0.48 | +1.00 |

# APPLICATIONS OF ELECTRON DENSITY STUDIES TO COMPLEXES OF THE TRANSITION METALS

E.N. Maslen

Crystallography Centre

University of Western Australia, Nedlands 6009

## SIMPLE APPLICATIONS

It is important when applying electron density methods to remember the underlying simplicity of the experiment. An educated lay man, looking at an object is aware that light scattered by that object is recombined by the lens of the eye to form an image. For electron density studies, where we use X-rays of appropriate wavelength in place of the visible light for human vision, formation of an image (an inverse Fourier transformation) is done by computation.

The limitations of human vision due to restricted resolution, interference and aberrations in the lens forming the image are important to us only in so far as they obscure what we wish to observe. For simple applications of electron density studies the imperfections in the experiment may be regarded in the same way.

The scattering from the N-shell of a transition metal atom falls off relatively rapidly with Bragg angle, as shown in Fig. 1. This limits the accuracy to which gross electron populations may be measured, especially for structures with small cells. In such simple structures the 4s contribution to the scattering is appreciable for a limited number of inner reflections only. The poor statistics then limits the accuracy for gross electron populations on the transition metal atoms, since the contribution from the 4s density must be considered.

The 3d contribution to the scattering, on the other hand, persists at relatively large Bragg angles, especially for the

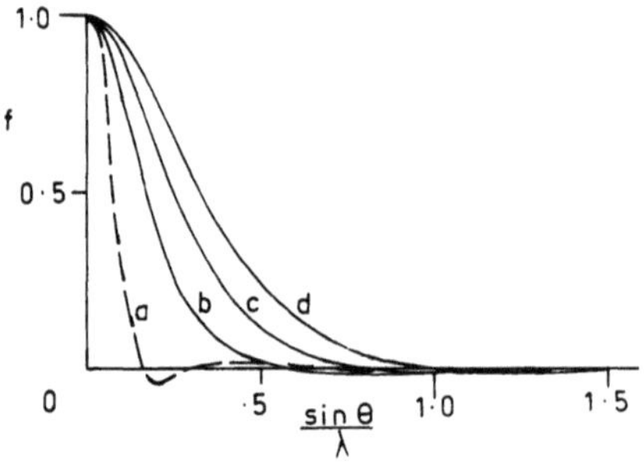


Figure 1. Variation of scattering amplitude per electron, f, with sin θ/λ for (a) the 4s density of scandium; (b), (c) and (d) the 3d density for scandium, manganese and copper respectively. θ is the Bragg and λ is the wavelength.

heavier members of the first transition series, as shown in Fig. 1. Thus we expect difference densities for transition metal complexes to show relatively detailed information on the 3d electron density. An example, taken from the work of Iwata & Saito (1973) on the complex $Co(NH_3)_6\ Co(CN)_6$ is shown in Fig. 2. The difference density is characterized by peaks along the body diagonals of a cube rather close to the metal nucleus (0.3 Å) with octahedrally disposed hollows 0.6 Å from the metal. Their appearance is correlated with the occupancies expected for $t_{2g}$ and $e_g$ orbitals.

Electron density methods may be used to study the change in the electron density with the number of 3d electrons. This is well illustrated by the studies on the isomorphous series $M_2SiO_4$, where M is Ni, Co or Fe, by Marumo, Isobe, Saito, Yagi and Akimoto (1974) and by Marumo, Isobe & Akimoto (1977).

Sections of the difference densities shown in Figs. 3, 4 and 5 show a progressive change in the disposition of the 3d density as the number of electrons is varied. For the nickel complex the density has near to ideal octahedral symmetry. For the iron complex the octahedral symmetry is strongly broken and there are subsidiary maxima in the features associated with the 3d electrons.

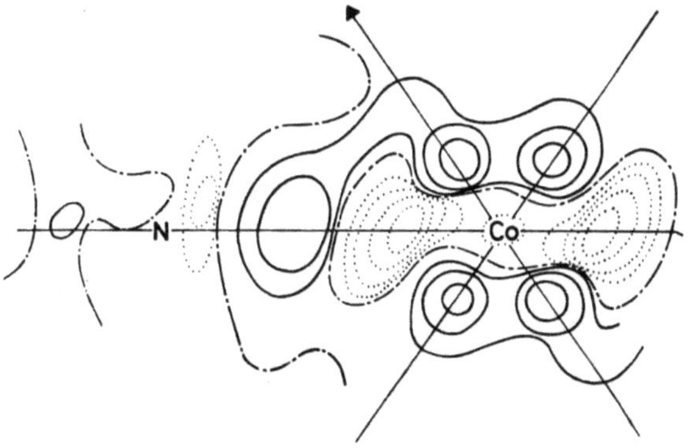


Figure 2. A section of the final difference Fourier synthesis through a Co-N bond and a threefold axis in $Co(NH_3)_6$ $Co(CN)_6$, taken from Iwata & Saito (1973). The arrow indicates the threefold axis. The solid contours are at intervals of 0.1 $e\mathring{A}^{-3}$. Negative contours are dotted, zero being chain-dotted.

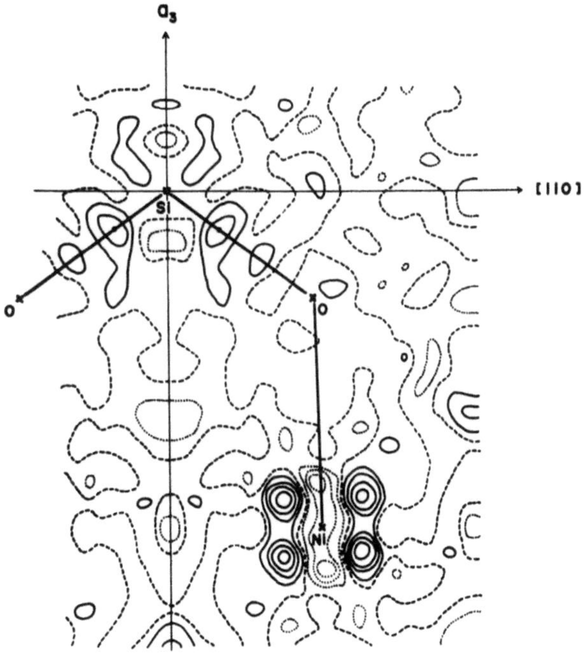


Figure 3. A section of the difference synthesis for $\gamma$-$Ni_2SiO_4$, taken from Marumo, Isobe, Saito, Yagi & Akimoto (1974). Contour interval 0.2 $e\mathring{A}^{-3}$. Zero contours are broken lines, and negative contours dotted.

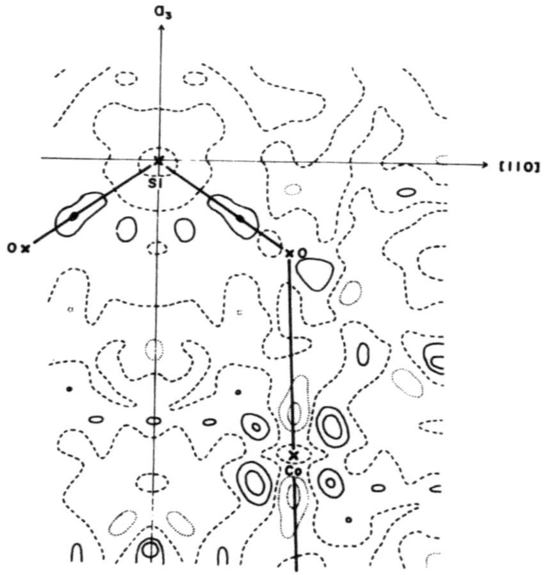


Figure 4. A section of the difference synthesis for $\gamma$-$Co_2SiO_4$, taken from Marumo, Isobe & Akimoto (1977). Contour interval 0.43 $Å^{-3}$. Zero contours are broken lines, and negative contours dotted.

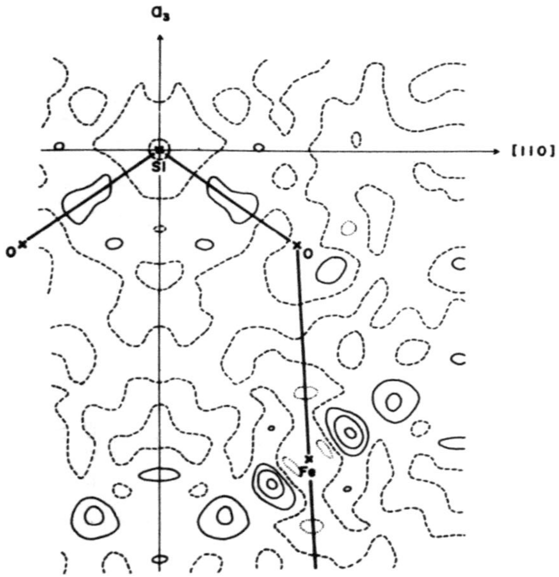


Figure 5. A section of the difference density for $\gamma$-$Fe_2SiO_4$, taken from Marumo, Isobe & Akimoto (1977). Contour interval 0.4 $eÅ^{-3}$. Zero contours are broken lines, and negative and negative contours dotted.

The map for the cobalt complex is intermediate in its appearance, as expected. The cobalt atom in this structure is in a divalent high spin state. The maxima are closer to and the minima further from the nucleus than the corresponding features in Fig. 2, which are for a low spin trivalent cobalt. These changes in the difference density are as predicted by physical reasoning.

The detailed features around the metal atoms in Figs 2 to 5 contrast with the slowly varying density surrounding titanium atom in rutile, $TiO_2$, studied by Shintani, Sato & Saito (1975). As shown in Fig. 6 the density around the titanium is rather featureless, which is expected since the $Ti^{4+}$ ion has no d electrons.

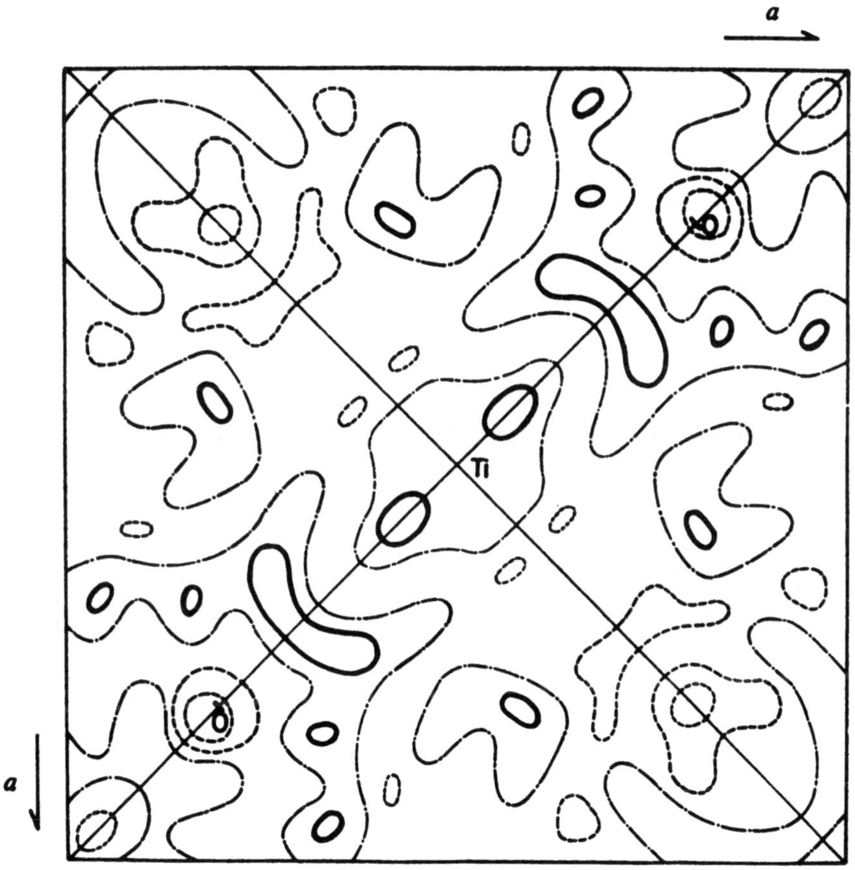


Figure 6. A section of the difference density through the titanium ion for rutile, taken from Shintani, Sato & Saito (1975). Contour interval 0.2 $e\AA^{-3}$. Negative contours are broken lines, zero being chain dotted.

## QUANTITATIVE STUDIES

The solution of many chemical problems requires more quantitative studies of the electron density. The use of the usual difference density in these cases is subject to conceptual and practical limitations. The difficulty is in allowing for the termination of the Fourier series used in forming the image at a maximum value determined by the experimental conditions. Naively it appears that the ideal data set should extend to infinite Bragg angle. If each term contributes appreciably to the variance the total error would be infinite. For the 'ideal' experiment we should be able to say nothing whatever about the problem!

We mav hvoothesize that some chemical problem could be resolved by an accurate determination of the electron density $\rho$ itself, or by some quantity $Q$ related by differentiation or integration to $\rho$. The true value of $Q$ will be related to a set of correct structure factors $\underline{F}$ by

$$Q = \frac{1}{V} \Sigma \underline{F} D(\underline{S}) \tag{1}$$

where V is the unit cell volume, and the summation extends over the complete set of structure factors. $\underline{S}$ is a reciprocal lattice vector with

$$|\underline{S}| = 4\pi \frac{\sin\theta}{\lambda} \tag{2}$$

where $\theta$ is the Bragg angle and $\lambda$ the radiation wavelength. For an integral $D(\underline{S})$ is given by

$$D(\underline{S}) = \int_{\nu} \exp(-i\underline{S}.\underline{r})\, d\underline{r}$$

where $\nu$ is the volume of interest. For $\rho$ and its derivatives $D(\underline{S})$ is obtained by successively applying the differential operator to $\int \exp(-i\underline{S}.r)\, d\underline{r}$.

The true structure factors $\underline{F}$ are unknown, but we have experimental estimates $\underline{F}_e$. We wish to determine the quantity

$$Q_M = \frac{1}{V} \Sigma M \underline{F}_e\, D(\underline{S}) \tag{3}$$

such that the variance

$$\langle\sigma^2(Q_M)\rangle_V = \frac{1}{V} \int \langle|Q_M - Q|^2\rangle\, d\underline{r} \tag{4}$$

is a minimum. The optimum form of $M$ is

$$M = \langle \underline{F}_e . \underline{F} \rangle / \langle \underline{F}_e . \underline{F}_e \rangle \qquad (5)$$

In a well conducted experiment $M$ is zero for all reflections beyond the maximum experimental Bragg angle. In practice it may be necessary to replace the ensemble averages in equations (5) with averages over regions of reciprocal space, such as small intervals of $|\underline{S}|$, as described by Davis, Varghese & Maslen (1978).

These methods have been used to study the deformation of the electron density associated with the electron density in copper sulphate pentahydrate. In this structure there are two crystallographically independent copper atoms for which the difference densities, shown in Figs. 7 and 8, and the density integrated over regions coinciding approximately with the 3d electrons, shown in Table 1, are significantly different.

Table 1. Integrated electron populations within a sphere of 1.1 Å of the copper atoms in copper sulphate. x, y and z correspond to the directions of two water ligands and the sulphate oxygen ligand respectively, taken from J.N. Varghese (1978).

| Atom | $P_{x^2-y^2}$ | $P_{z^2}$ | $P_{total}$ |
|---|---|---|---|
| $Cu_A$ | -0.72(1) | -0.29(1) | -1.01(2) |
| $Cu_B$ | -0.84(1) | -0.12(1) | -0.96(2) |

## MODELLING THE ELECTRON DENSITY

A model for the electron density which is optimal, in a least squares sense, may be obtained by adjusting variable parameters which minimise residuals related either to

$$R(F^2) = \Sigma\, \omega\, (F^2) \; \{|\underline{F}_e|^2 - |\underline{F}_c|^2\}^2 \qquad (6)$$

or to

$$R(F) = \Sigma\, \omega\, (F) \; \{\underline{F}_e - \underline{F}_c\}^{\frac{1}{2}} \qquad (7)*$$

---

*$R(F^2)$ is preferable on statistical grounds but $R(F)$, which in many cases is a linear function of the parameters, is frequently chosen to reduce the cost of the computations.

where $\omega$ is the reciprocal of the variance for the observation. $\underline{F}_c$, a calculated structure factor, is based on some model for the electron density which includes the variable parameters. For first row atom systems models based on one-centre multipole representations of the density, with radial dependence of atomic self consistent field, Gaussian or Slater-type functions have proved fruitful. Finding satisfactory representations of the densities around transition metal atoms is more difficult because of greater complexity. The idealized symmetries of the local density distributions are broken more strongly in many cases, and minor changes in the environment may produce marked changes in the radial dependence of the density. Such changes are particularly pronounced in the case of atoms affected by Jahn-Teller distortions, such as the copper atoms in copper sulphate pentahydrate, shown in Figs. 7 and 8.

So far the main use of models for the densities in transition complexes has been to estimate local properties of the density, such as gross atomic populations.

The subdivision of a continuous distribution of charge is necessarily somewhat arbitrary, and several procedures have been considered. In one method due to Varghese (1978) the radial densities for the multipole density functions are given by a series of moments of derivatives of the unperturbed density. Changing the radial forms of the terms in such an expansion has the effect of altering the relative weights for different regions of charge. Models which correspond to closely similar total distributions (and thus to similar residuals) may differ substantially in the subdivision of charge among different centres. In Table 2 results are given for the net electron populations for copper sulphate pentahydrate based on models, not described here in detail, but simply labelled a and b. These two models correspond roughly to reasonable extremes in the subdivision of charge. Although for a given model the results transfer moderately well for similar groups in other related structures, there are substantial differences between the results from the two models in the same sample. This is especially so for the hydrogen atoms. It is expected that further work will confine the optimum procedure to closer limits within these extremes.

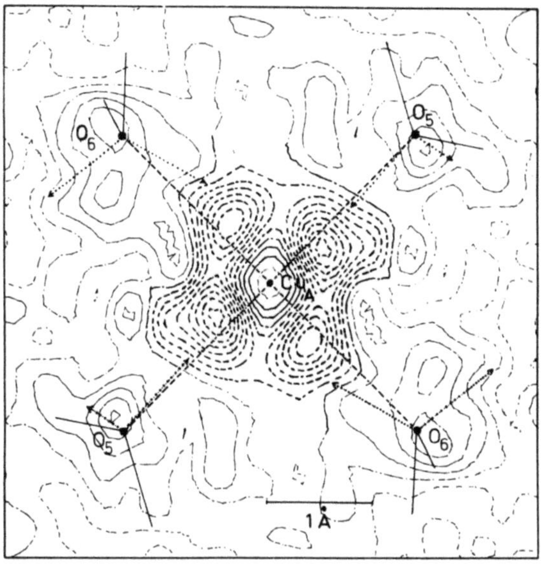


Figure 7. Section through the $Cu_A$ atom and two ligand water oxygens in copper sulphate pentahydrate. Contour interval 0.1 eÅ$^{-3}$. Negative contours are broken lines, zero being chain-dotted.

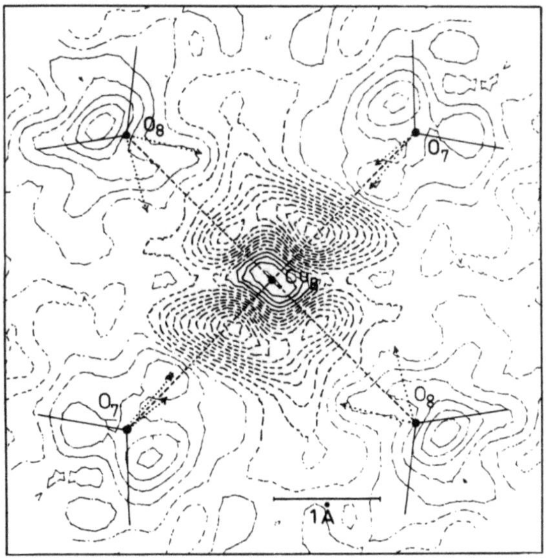


Figure 8. Section through the $Cu_B$ atom and two ligand water oxygens in copper sulphate pentahydrate. Contour interval 0.eÅ$^{-3}$. Negative contours are broken lines, zero being chain-dotted.

Table 2. Valence electron populations in electrons (with respect to neutral atoms as zero for copper sulphate pentahydrate, taken from J.N. Varghese (1978).

| Atom | Model a | Model b | Atom | Model a | Model b |
|---|---|---|---|---|---|
| Cu A | -.30 | -.14(3) | O(9) | .75 | .51(3) |
| Cu B | -.18 | -.06(3) | H(54) | -.98 | -.46(5) |
| S | -.55 | -.58(3) | H(59) | -.71 | -.32(5) |
| O(1) | .67 | .61(2) | H(62) | -.61 | -.29(4) |
| O(2) | .44 | .49(3) | H(69) | -.79 | -.44(4) |
| O(3) | .34 | .27(2) | H(74) | -.90 | -.49(5) |
| O(4) | .61 | .54(2) | H(73) | -.80 | -.44(4) |
| O(5) | .88 | .69(2) | H(83) | -.91 | -.50(5) |
| O(6) | .86 | .62(2) | H(84) | -.22 | -.10(5) |
| O(7) | 1.19 | .80(3) | H(91) | -.33 | -.11(5) |
| O(8) | .58 | .45(3) | H(92) | -.48 | -.38(5) |

| Group | | Model a | Model b |
|---|---|---|---|
| Sulphate | | 1.51 | 1.32(6) |
| Water | 5 | -.81 | -.09(7) |
| | 6 | -.53 | -.11(7) |
| | 7 | -.52 | -.13(7) |
| | 8 | -.54 | -.16(7) |
| | 9 | -.05 | .02(7) |

## REFERENCES

1. Davis, C.L., Maslen, E.N. & Varghese, J.N. (1978). Acta Cryst. A34, 371-377.

2. Iwata, M. & Saito, Y. (1973). Acta Cryst. B33, 59-70.

3. Marumo, F., Isobe, M. & Akimoto, S. (1977). Acta Cryst. B33, 713-716.

4. Marumo, F., Isobe, M., Saito, Y., Yagi, T. & Akimoto, S. (1974). Acta Cryst. B30, 1904-1906.

5. Shintani, H., Sato, S. & Saito, Y. (1975). Acta Cryst. B31, 1981-1982.

6. Varghese, J.N. (1978). Private Communication.

# THE CHEMICAL INTERPRETATION OF MAGNETIZATION DENSITY DISTRIBUTIONS

J.B. FORSYTH

Neutron Beam Research Unit, SRC Rutherford Laboratory

Chilton, Oxon, OX11 0QX, England

## 1. INTRODUCTION

The determination of accurate magnetisation density distributions through neutron diffraction experiments, which mainly involve the measurement of flipping ratios using the polarized beam technique, can give information on the ground state wavefunctions of magnetic ions and, in certain cases, can give quantitative information on the bonding between a magnetic ion and its ligands. Many of the results obtained so far by magnetic scattering have been discussed in terms of the molecular orbital (MO) model of transition metal complexes. This is useful both for relating trends between one ligand and another and for comparing neutron data with those obtained by magnetic resonance studies of the ligand hyperfine interactions (LHFI). Since the deviations from a simple ionic model introduced by covalent effects are normally quite small ($\sim$ 10%), it is clearly important that the metal ion ground state wavefunction be known before these data can be properly interpreted. For this reason, most covalency studies have been restricted to systems with spin-only ground states, but even in these cases the interpretation of the experimental observations is by no means trivial.

The LHFI was first observed by ESR for $^{35}Cl$ and $Ir^{4+}$ $[d^5, S=\frac{1}{2}]$ doped into $K_2PtCl_6$ by Owen and Stevens (1953). These observations gave direct evidence for the delocalisation of the metal d-electron density onto the ligand chlorine atoms. Many subsequent observations have been made by this method and also by NMR. The first discussion of the determination of covalency parameters from magnetic neutron scattering experiments was given by Hubbard and Marshall (1965). The timelag between these parallel approaches perhaps reflects the difficulties of making neutron measurements of the required accuracy, but the recent development of high flux neutron sources has improved the situation enormously and has made possible experiments on more complex materials than the simple oxides and halides which formed the bulk of the initial studies.

A number of articles and reviews have been written on the subject of neutron diffraction and covalency. (Rimmer, 1970; Jacobsen, 1973; Tofield, 1975). Tofield (1976) has also extended his survey of the results obtained by neutron scattering to include other magnetic techniques, such as EPR to measure the LHFI and ESR and Mossbauer spectroscopy to probe the d- and s- electron density at the metal ion nucleus.

The first sections of the present account will be devoted to a description of the MO model of covalency and its relation to magnetic neutron scattering. The next section will discuss its application to the detailed interpretation of three magnetisation density distributions. Section 5 contains examples of the application of neutron magnetic scattering to the determination of the ground states of actinide ions. Before the concluding remarks, an outline is given of some current investigations in the field of the chemical interpretation of magnetisation density distributions.

## 2. THE MOLECULAR ORBITAL DESCRIPTION OF COVALENCY

A simple example of the molecular orbital approach is provided by the analysis of the bonding in a simple octahedral complex of a 3d transition metal ion such as $(MnF_6)^{4-}$. The ionic wavefunctions which take part in the bonding are the metal 3d orbitals and the ligand 2s and 2p orbitals. $\sigma$ bonding can occur between the $3d_{x^2-y^2}$ and $3d_{z^2}(e_g)$ metal orbitals and the six ligand $p\sigma$ orbitals and $\pi$ bonds may be formed between the $3d_{xy,\,xz}$ and $_{yz}(t_{2g})$ orbitals and the twelve ligand $p\sigma$ orbitals. The filled inner shell orbitals may be neglected to first order in discussing neutron magnetic scattering. Figures 1 and 2 illustrate the overlap of the metal orbitals with the $2p\sigma$ and $2p\pi$ orbitals of the ligands.

Following the nomenclature of Tofield (1975) and others, we may describe the filled bonding orbitals of predominantly ligand character as:

$$\psi_S^B = N_S^B\ (\chi_{2s} + \gamma_s d_\sigma + \gamma_{s\sigma}\ \chi_{2p\sigma})$$

$$\psi_\sigma^B = N_\sigma^B\ (\chi_{2p\sigma} + \gamma_\sigma d_\sigma + \gamma_{\sigma s}\ \chi_{2s})$$

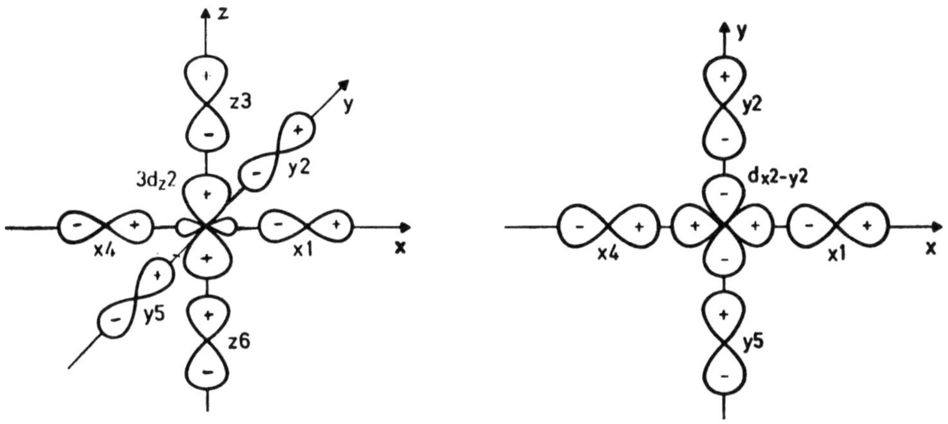


Figure 1 - Schematic illustration of the overlap of $3d_{z^2}$ and $3d_{x^2-y^2}$ metal orbitals with the $2p\sigma$ ligand orbitals. Ligands 1 and 4 are on the ± x axes, ligands 2 and 5 on ± y and ligands 3 and 6 on ± z.

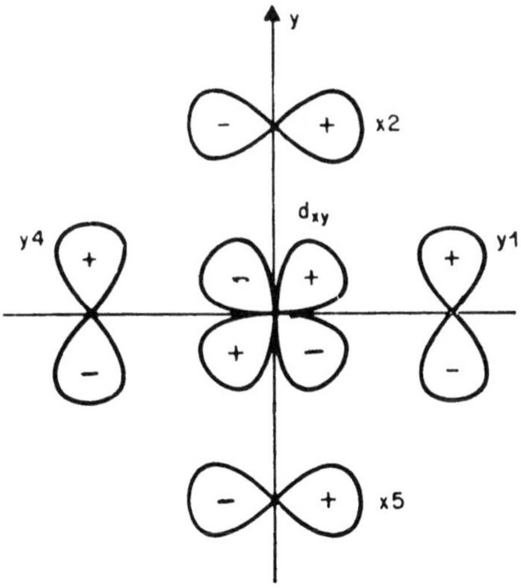


Figure 2 - The overlap of the metal $t_{2g}$ orbitals ($3d_{xy}$) with ligand $2p\pi$ orbitals. The ligands are numbered in the same way as in Figure 1.

$$\psi_\pi^B = N_\pi^B\ (\chi_{2p\pi} + \gamma_\pi\ d_\pi).$$

Here $N_s^B$, $N_\sigma^B$ and $N_\pi^B$ are normalization constants and $\chi_{2s}$, $\chi_{2p\sigma}$ and $\chi_{2p\pi}$ are the appropriate linear combinations of ligand 2s, 2pσ and 2pπ orbitals, respectively. $\gamma_s$ and $\gamma_\sigma$ are the admixture parameters describing the σ covalency between $d_\sigma$ and the ligand 2s and 2pσ orbitals respectively, and $\gamma_\pi$ measures the π covalency between $d_\pi$ and 2pπ orbitals.

Because the bonding orbitals are filled they do not directly contribute to the magnetic properties; the magnetic interactions associated with transition metal ions reflect the properties of the unpaired electrons in the antibonding orbitals, which have mainly d character. Orthogonal to the bonding orbitals, they may be written.

$$\psi_\sigma = N_\sigma\ (d_\sigma - \lambda_\sigma \chi_{2p\sigma} - \lambda_s \chi_{2s})$$

$$\psi_\pi = N_\pi\ (d_\pi - \lambda_\pi \chi_{2p\pi})$$

Expressing the antibonding orbitals explicitly we find:

$$\psi_{z^2} = N_\sigma \Big[ d_z^2 - \frac{1}{\sqrt{12}}\ \lambda_\sigma\ (-\ 2p_{z3} + 2p_{z6} + p_{x1} - p_{x4} + p_{y2} - p_{y5}) - \frac{1}{\sqrt{12}}\ \lambda_s\ (2s_3 + 2s_6 - s_1 - s_2 - s_4 - s_5) \Big]$$

$$\psi_{x^2-y^2} = N_\sigma \Big[ d_{x^2y^2} - \frac{1}{2}\lambda_\sigma\ (p_{x4} - p_{x1} + p_{y2} - p_{y5}) - \frac{1}{2}\lambda_s\ (s_1 + s_4 - s_2 - s_5) \Big]$$

$$\psi_{xy} = N_\pi \Big[ d_{xy} - \frac{1}{2}\lambda_\pi\ (p_{y1} - p_{y4} + p_{x2} - p_{x5}) \Big]$$

$$\psi_{xz} = N_\pi \Big[ d_{xz} - \frac{1}{2}\lambda_\pi\ (p_{z2} - p_{z5} + p_{y3} - p_{y6}) \Big]$$

$$\psi_{yz} = N_\pi \Big[ d_{yz} - \frac{1}{2}\lambda_\pi\ (p_{x3} - p_{x6} - p_{z1} - p_{z4}) \Big]$$

The normalization constants are defined by $\langle\psi|\psi\rangle = 1$ and

$$N_\sigma = (1 - 2\lambda_\sigma S_\sigma - 2\lambda_s S_s + \lambda_\sigma^2 + \lambda_s^2)^{-\frac{1}{2}}$$

$$N_\pi = (1 - 2\lambda_\pi S_\pi + \lambda_\pi^2)^{-\frac{1}{2}}$$

where the overlap integrals S are defined by $\langle d|\chi\rangle$ and

$$S_\sigma = 2\langle d_\sigma|p_{2p\sigma}\rangle \quad S_s = 2\langle d_\sigma|s_{2s}\rangle, \quad S_\pi = 2\langle d_\pi|p_{2p\pi}\rangle.$$

Because of the orthogonality relations

$$\langle\psi_\sigma|\psi_\sigma^B\rangle = \langle\psi_\sigma|\psi_s^B\rangle = \langle\psi_\pi|\psi_\pi^B\rangle = 0$$

it follows that, to first order in $\lambda$, $\gamma$ and S

$$\lambda_\sigma = \gamma_\sigma + S_\sigma$$
$$\lambda_s = \gamma_s + S_s$$
$$\lambda_\pi = \gamma_\pi + S_\pi$$

Experiments such as magnetic neutron diffraction and spin resonance which measure $\lambda$ are therefore sensitive to a combination of the covalency admixture parameter $\gamma$ and the overlap integral. Although it has become customary to treat $\lambda$ as a measure of the covalency, some recent neutron studies have been of sufficient accuracy to warrant explicit calculation of the overlap contribution.

Bonding also takes place between the ligand valence orbitals and the outer 4s and 4p orbitals of the transition metal ion, which are initially unoccupied. However, although charge transfer into the outer orbitals is interesting in the discussion of bonding and effective charges, for example, there is no involvement of unpaired electrons. Thus, to first order, magnetic neutron diffraction and spin resonance measurement of LHFI will not detect these interactions. This is not the case for other techniques such as nuclear quadrupole resonance and the Mössbauer effect isomer shift. To second order, where exchange effects and radial polarization are allowed for, magnetic moment densities may be induced in metal 4s orbitals and in nominally empty metal 3d orbitals.

A schematic molecular orbital diagram is shown in Figure 3 for bonding involving ligand 2s and 2p orbitals and metal 3d, 4s and 4p orbitals. The actual ordering of energy levels may be quite differ-

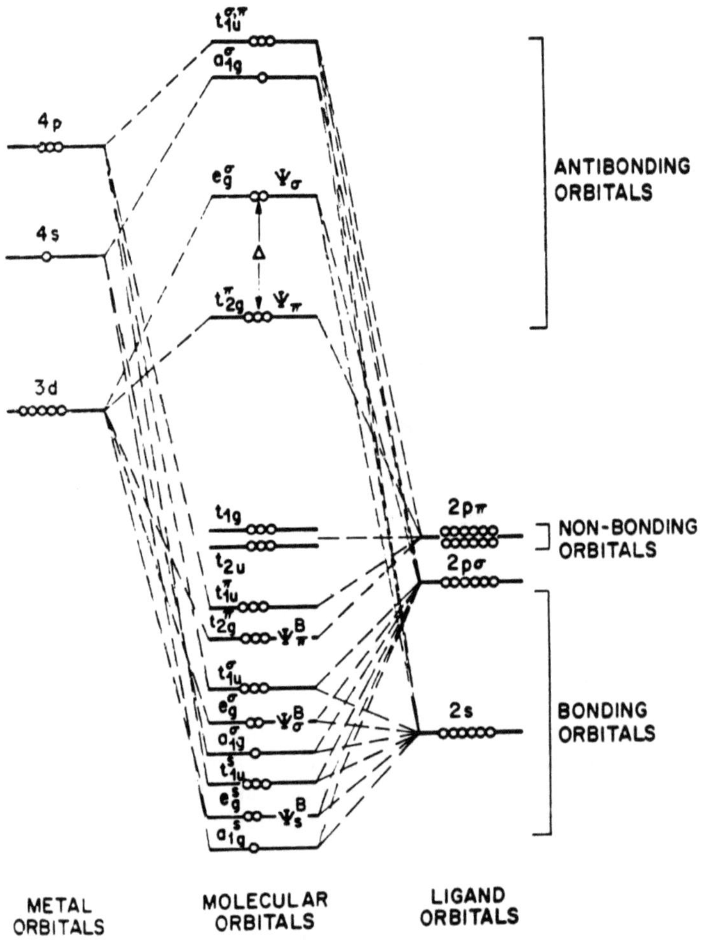


Figure 3 - Schematic MO diagram for bonding involving metal 3d, 4s and 4p orbitals with ligand 2s and 2p orbitals. The bonding and antibonding orbitals are indicated and also the ligand field splitting Δ. In $MnF_2$, the five unpaired electrons occupy the $e_g^\sigma$ and $t_{2g}^\pi$ antibonding orbitals.

ent in any real situation and polarization effects are not included. The convention has been followed that the σ and π antibonding orbitals, which contain the unpaired electrons in transition metal complexes, are less tightly bound than the mainly ligand bonding orbitals.

The ligand field splitting $\Delta$ is shown in Figure 3. From the simple MO model this is given by

$$\Delta = (\lambda_\sigma^2 - \lambda_\pi^2)\ (E_d - E_{2p}) + \lambda_s^2\ (E_d - E_{2s})$$

where $E_d$, $E_{2p}$ and $E_{2s}$ are the initial energies of the 3d, 2p and 2s orbitals. Although this is a considerable simplification, it is sufficient to make it clear that conclusions based on a simple crystal field splitting of the 3d levels, with no change in total energy, are of little significance. Site preference energies, Jahn-Teller splittings, and so forth are determined principally by the energies of the filled, mainly ligand, bonding orbitals. The calculation of $\Delta$, and of the magnetic moment distributions determined by neutron scattering and LHFI measurement remains a principal target of first principles calculations on simple transition metal complexes.

## 3. THE EFFECT OF COVALENCY ON THE MAGNETISATION DENSITY DISTRIBUTION

As has already been pointed out, the presence of a significant orbital contribution to the magnetisation complicates the interpretation of magnetic scattering data. For this reason most of the experiments which have given reliable estimates of the bonding parameters have been concerned with spin-only systems (for example, octahedral $d^3$, $d^5$ and $d^8$ ions) in which small orbital effects introduced by spin-orbit coupling can be accounted for more simply.

The magnetic moment distributions, $D(\underline{r})$, which correspond to $\sigma$ and $\pi$ antibonding orbitals containing one electron are, to second order:

$$D(\underline{r})_\sigma = |\psi_\sigma(\underline{r})|^2 = d_\sigma^2\ (1 + 2\lambda_\sigma S_\sigma + 2\lambda_s S_s - \lambda_\sigma^2 - \lambda_s^2)$$

$$- 2(\lambda_\sigma d_\sigma \chi_{2p\sigma} + \lambda_s d_\sigma \chi_{2s})$$

$$+ (\lambda_\sigma^2 \chi_{2p\sigma}^2 + \lambda_\sigma^2 \chi_{2s}^2\ ) + 2\lambda_\sigma \lambda_s \chi_{2s} \chi_{2p\sigma}$$

and

$$D(\underline{r})_\pi = |\psi_\pi(\underline{r})|^2 = d_\pi^2\ (1 + 2\lambda_\pi S_\pi - \lambda_\pi^2) - 2\lambda_\pi d_\pi \chi_{2p\pi}$$

$$+ \lambda_\pi^2 \chi_{2p\pi}^2$$

The explicit radial dependence of d and $\chi$ have been omitted from these expressions and Tofield's (1975) notation has been adopted. The origin for $D(\underline{r})$ is at the centre of the metal ion, but $\chi$ is ligand-centred. These two equations show that there are three contributions to $D(\underline{r})$: one of the form $d^2(1 + 2\lambda S - \lambda^2)$ associated with the metal ion, a term of the same sign associated with the ligand orbitals and a contribution of opposite sign given by the overlap between d and p orbitals. The measurement of magnetisation density by neutron scattering gives the resultant distribution, whereas the observation of LHFI by paramagnetic resonance in an isolated paramagnetic complex is only sensitive to the unpaired spin transferred from the paramagnetic ion to its ligands. Of course, the net charge transfer in covalent bonding is from the ligands to the metal ion, since there are more electrons occupying bonding orbitals (in which charge is transferred from ligands to metal) than there are electrons occupying antibonding orbitals, in which the transfer is in the other direction.

The magnetic form factor which corresponds to the moment distribution of a free ion will be modified by the presence of both the ligand and the overlap densities. The expression for $D(\underline{r})$ has a term in $d^2$ only which is $(1 - \lambda_\sigma^2 - \lambda_s^2)$ and this will be the value which multiplies the corresponding form factor, $f_{dd}$ at $\underline{k} = 0$. The overlap term is

$$2(\lambda_\sigma S_\sigma d_\sigma^2 + \lambda_s S_s d_\sigma^2 - \lambda_\sigma d_\sigma \chi_{2p\sigma} - \lambda_s d_\sigma \chi_{2s})$$

which gives a form factor, $f_{\ell d}$, which is zero at $\underline{k} = 0$. Finally, the ligand density contribution will have a form factor, $f_{\ell\ell}$, whose multiplier is $(\lambda_\sigma^2 + \lambda_s^2)$ at $\underline{k} = 0$.

The effect on the form factor for an ion such as $Ni^{2+}$ has been given by Hubbard and Marshall (1965) and is illustrated in Figure 4. The metal ion contribution is reduced by the factor $1 - \lambda_\sigma^2 - \lambda_s^2$. The ligand term has a very sharply peaked form factor, since it corresponds to spin density which is extended far from the metal ion site. A spherically-averaged $f_{\ell\ell}$ is approximately given by the expression

$$f_{\ell\ell}(\underline{k}) = \frac{\sin (kR)}{kR} \left| \lambda_\sigma^2 \int j_o(kr)p^2(r)dr + \lambda_s^2 \int j_o(kr)s^2(r)dr \right|$$

where R is the distance of the ligands from the metal ion, $j_o(kr)$ is the zero order spherical Bessel function and $p^2(\underline{r})$ and $s^2(\underline{r})$ are the charge densities of the 2p and 2s orbitals respectively, both of which are centred on the origin. The sin (kR)/kR multiplier arises from the change of origin for the ligand orbitals and ensures that the ligand contribution falls to zero at a very low scattering vector $\underline{k}$. Although a measurement of the form factor close to $\underline{k} = 0$ can therefore

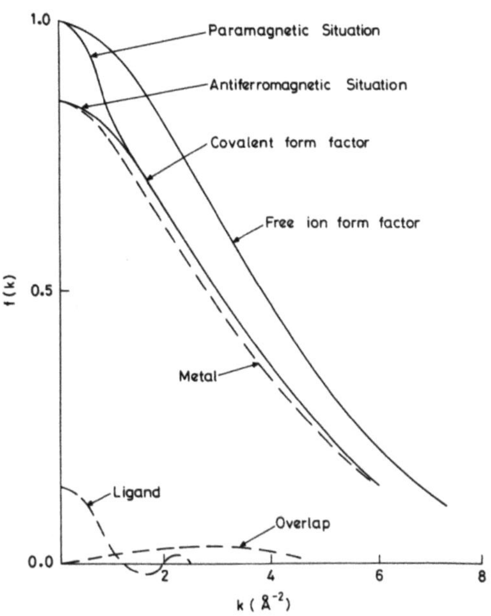


Figure 4 - Typical form factors for a 3d ion such as $Ni^{2+}$ based on simply MO theory. The free ion form factor and the covalent form factor for both paramagnetic and antiferromagnetic situations are shown. The three components of the covalent form factor due to the metal ion spin, the ligand spin and the overlap spin are shown by broken lines.

give a fairly direct measurement of λ, the practical difficulties associated with such measurements have restricted their application to one or two studies (see for example Tofield and Fender, 1971 and Hirakawa and Ikeda, 1973 and 1974).

If the form factor is determined from reflections which approach $\underline{k} = 0$, but do not encroach on the forward peak due to the ligand density, the effective moment will appear to be reduced by the factor $(1 - \lambda_\sigma^2 - \lambda_s^2)$. This had been the basis for the interpretation of numerous powder diffraction studies of magnetically ordered salts and oxides. The magnetic intensity of the lowest angle magnetic reflection is measured and placed on an absolute scale by comparison with the nuclear intensities. The factor $(1 - \lambda_\sigma^2 - \lambda_s^2)$ may then be determined without any detailed knowledge of the real form factor, since the difference between it and the calculated free-ion form factor is small in this region of $\underline{k}$ ($\sim 1.2$ Å$^{-1}$ for $4\pi \sin\theta/\lambda$). However, it must be emphasised that only by collecting full three-dimensional data on single crystals can the complete spin distribution be obtained by Fourier inversion methods and detailed information obtained on the form factor and the contributions from the ligand, overlap and metal regions.

Most inorganic salts and oxides which do order form antiferromagnetic structures: the net magnetisation over the magnetic unit cell is zero and the moments on neighbouring metal ions are generally antiparallel. In such situations, the ligand density is cancelled out if each ligand has an equal number of metal ion neighbours with spin up and with spin down, and the resulting form factor has been included in Figure 4. The expression for $D(\underline{r})_\sigma$ will then contain only two terms, $f_{dd}$ and $f_{\ell d}$. $f_{dd}$ is small in the region of the ligands, so in this region $D(\underline{r})$ is dominated by the overlap term which is negative if the sign of $\lambda$ is positive, as is generally the case. Since the magnetisation density is zero at the ligand nucleus by symmetry, we would expect every magnetic ion with positive magnetisation to be surrounded with a region of negative magnetisation and vice versa. The sign of $\lambda$ is likely to be negative with ligands such as (CN), but no neutron studies have yet been undertaken on these complexes.

As shown in Figure 4 the overlap form factor $f_{\ell\ell}(k)$ is zero at $k = 0$ and passes through a maximum as k increases. The simple molecular orbital model thus predicts a form factor for an antiferromagnet reduced from the free ion value, but somewhat expanded in shape because of the overlap moment. Many other factors such as the introduction of orbital effects via spin orbit coupling, the polarization of filled inner shells, the polarization of partially occupied outer orbitals and variations in the 3d radial functions need to be considered before a detailed form factor analysis can be attempted in any particular situation. However, NiO (Alperin, 1962) behaves qualitatively in the manner described above.

Tofield (1975, 1976) has published tables giving the combinations of covalency parameters for octahedral complexes which would be measured by the 'moment reduction' technique. These clearly depend on the number of d electrons present and Tofield also lists parameters which could be measured by LHFI or deduced from the quadrupole coupling constant determined in nuclear quadrupole resonance spectroscopy or from the quadrupole interactions observed in magnetic resonance. The two review papers summarise the results obtained for the spin transfer coefficients in octahedral complexes of $Ni^{2+}$, $Mn^{2+}$, $Fe^{3+}$, $V^{2+}$, $Cr^{3+}$ and $Mn^{4+}$ and also present the more limited amount of experimental data available on tetrahedrally coordinated ions.

## 4. SINGLE CRYSTAL NEUTRON STUDIES OF BONDING

This section will describe some experimental studies which have been sufficiently accurate to warrant a detailed analysis in terms of the picture of covalent bonding outlined in the previous two sections.

### 4.1 The $(CrF_6)^{3-}$ Group in $K_2NaCrF_6$

The cubic structure of $K_2NaCrF_6$ is illustrated in Figure 5. The $(CrF_6)^{3-}$ groups are effectively magnetically isolated by the intervening $Na^+$ and $K^+$ ions. The substance remains paramagnetic to at least 1.5K but it can be polarized to an appreciable extent by the application of an external field. Wedgwood (1976) has reported careful polarized beam measurements made in a field of 1.76T at 4.2K: in these conditions the aligned moment per Cr is 1.08 μB and it arises from spin alone, since the measured g value is 1.998 (Hukin, 1965). Some of the experimental scattering amplitudes $mf(\underline{k})$ per chromium atom are plotted in Figure 6. The radial dependence is significantly different along the [100] and [111] directions but has the same general form of the theoretical curves for these symmetry

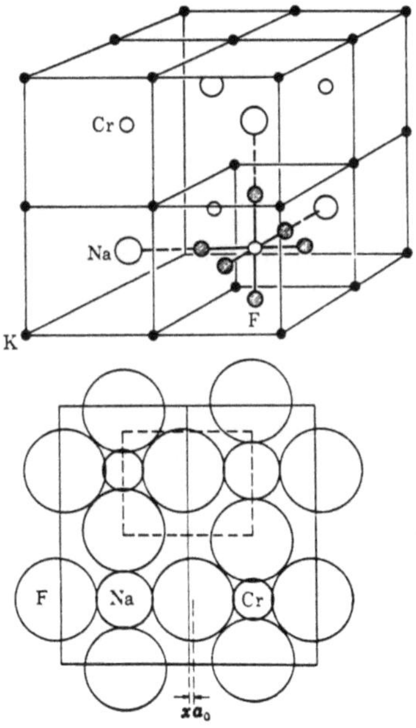


Figure 5 - The cubic structure of $K_2NaCrF_6$. The perspective drawing shows the alternating perovskite sub cells of $KNaF_3$ and $KCrF_3$ which are stacked in a rock-salt arrangement, space group Fd3m. The (001) section of the structure is illustrated above.

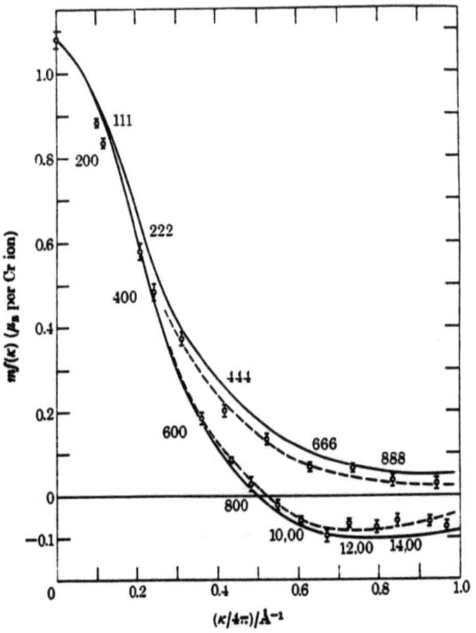


Figure 6 - Magnetic scattering amplitude mf ($\underline{k}$) per chromium atom in $K_2NaCrF_6$ at 4.2K and 1.76T. Only data of the form {h00} and {hhh} are plotted: the full lines correspond to theoretical free-ion functions and the broken lines are sketched through the data.

directions calculated by Watson and Freeman (1959) for a free $Cr^{3+}$ ion in the $t_{2g}^3$ state. Figure 7 illustrates the [001] section of the magnetisation density in the region of the chromium site. The data are complete to sin $\theta/\lambda$ (= $k/4\pi$) = 0.726 $Å^{-1}$ and the density has been averaged over a small cube of dimensions $\delta$ = 0.4 Å in order to minimise the effects of series termination. The $t_{2g}$ nature of the distribution is obvious, though the nodes along [$\bar{1}$00] and [001] are smeared out due to the finite resolution of the map. Wedgwood analysed these data in terms of the Hubbard Marshall model: the ground state of the cluster consists of three electrons, one in each of the $d_{xy}$, $d_{yz}$, $d_{xz}$ orbitals and with parallel spin. The form factor is given by the sum of the three one electron operators $e^{i\underline{k}\cdot\underline{r}}$ and it is simplified by the symmetry to

$$f(\underline{k}) = N_\pi^2 \; (f_{dd}(\underline{k}) + 4A_\pi^2 f_{pp}(k) - 8A_\pi S_\pi f_{pd}(k)),$$

where

$$f_{dd}(k) = \langle j_o \rangle_d - A_{hk\ell} \langle j_4 \rangle_d,$$

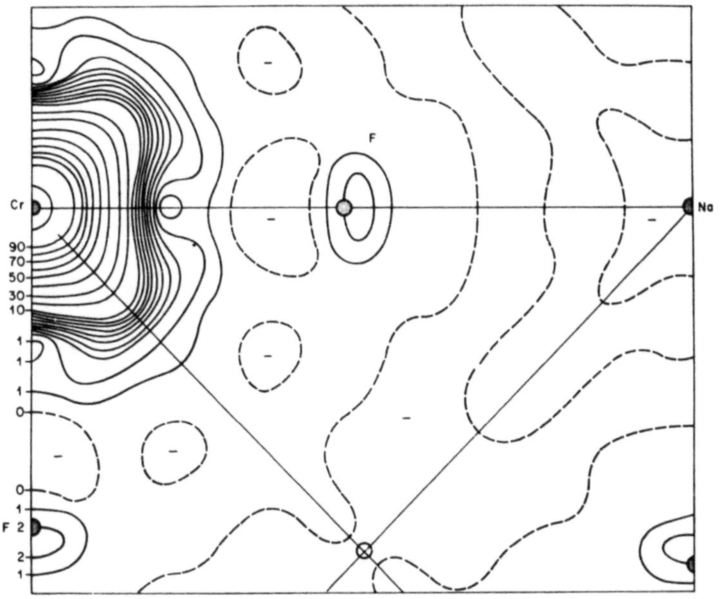


Figure 7 - Distribution of the magnetisation density in the (001) plane of $K_2NaCrF_6$ at z = 0. The contour intervals are in 0.01 μB/Å$^3$.

The ligand contribution arises solely from the 2p wavefunctions

$$f_{pp}(k) = \frac{1}{3}\langle j_o\rangle_p(\cos 2\pi hd/a + \cos 2\pi kd/a + \cos 2\pi \ell d/a)$$

$$+ \frac{1}{6}\langle j_2\rangle_p\left(\frac{3h^2 - s^2}{s^2}\cos 2\pi hd/a + \frac{3k^2 - s^2}{s^2}\cos 2\pi kd/a\right.$$

$$\left. + \frac{3\ell^2 - s^2}{s^2}\cos 2\pi \ell d/a\right),$$

where d is the distance from metal to ligand and a is the unit cell edge. $A_{hk\ell}$ and $\langle j_n\rangle$ have their usual meanings:

$$A_{hk\ell} = (h^4 + k^4 + \ell^4 - 3[h^2k^2 + k^2\ell^2 + \ell^2h^2])/s^4,$$

$$s^2 = h^2 + k^2 + \ell^2$$

$$<j_n> = \int_o^\infty R^2(r)j_n(kr)r^2dr,$$

where R(r) is the radial part of the wavefunction.

The quality $f_{pp}(k)$ can be expanded in cubic harmonics by expanding the cosine terms in spherical Bessel functions $j_n(kd)$. This gives

$$f_{pp}(k) = <j_o>_p j_o(kd) - \frac{1}{10} <j_2>_p j_2(kd)$$
$$+ A_{hk\ell} (\frac{7}{12} <j_o>_p j_4(kd) - \frac{3}{20} <j_2>_p j_2(kd)$$
$$+ \frac{5}{66} <j_2>_p j_4(kd)) + \dots .$$

Unlike $f_{dd}(k)$ this does not terminate at fourth order cubic harmonics but includes sixth and higher orders.

$f_{pd}(k)$ involves two-centre integrals which have to be integrated numerically.

The zero and fourth order terms of all three contributions to the form factor are shown in Figures 8 and 9. The main effect of the spin transfer is at small values of k where $f_{pp}(k)$ varies strongly. This causes a sharp peak in $f_o(k)$ at $k/4\pi \approx 0$ and a change of spin of $f_4(k)$ at $k/4\pi \approx 0.2$.

To test this model against the data, it is desirable to separate the date into spherical and aspherical parts. This can be done formally by a double Fourier transform method (Moss and Brown, 1972). The data can be written as

$$f(s) = f_o(s) + A_{hk\ell}f_4(s)$$
$$= \int_Q \rho(r)e^{-s.r}dr,$$

where the second Fourier transform is over a sphere of radius Q surrounding the chromium ion. The spherically separated form factors are now given by

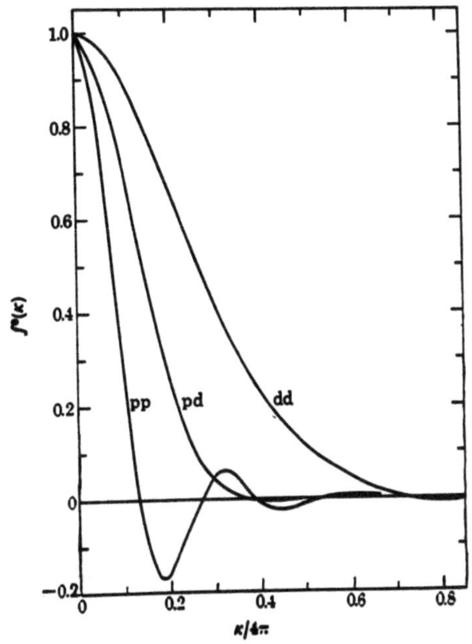


Figure 8 - Theoretical spherical form factors for the MO model of $K_2NaCrF_6$: $f^0_{dd}(k)$, $f^0_{pp}(k)$ and $f^0_{pd}(k)$.

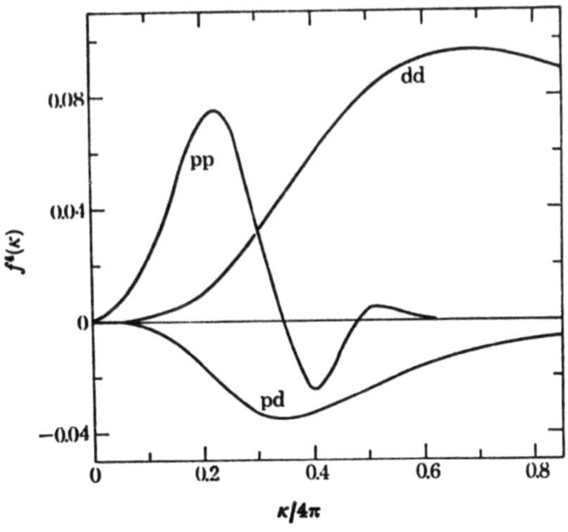


Figure 9 - Theoretical aspherical form factors for the MO model of $K_2NaCrF_6$: $f^4_{dd}(k)$, $f^4_{pp}(k)$ and $f^4_{pd}(k)$.

$$f_\ell(s) = \frac{16\pi^2 Q^2}{v} \sum_k \frac{f(k)}{k^{-2} - s^2} \left(k j_\ell(kQ) j_{\ell-1}(sQ)\right.$$

$$\left. - s j_\ell(sQ) j_{\ell-1}(kQ)\right),$$

which can be directly computed from the data $f(\underline{k})$.

$f_o(s)$ and $f_4(s)$ are plotted in Figures 10 and 11 for Q = 2.4 Å. $f_6(s)$ was also calculated but was found to be very small and to oscillate at large s due to the data cut-off at $k/4\pi$ = 0.726 $Å^{-1}$.

Molecular orbital theory curves for $A_\pi$ = 0 (free ion), 0.15 and 0.24 are also shown in Figures 10 and 11. $A_\pi$ = 0.15 appears to give the best fit for $f_o(s)$ apart from the region around $s/4\pi$ > 0.4. On the other hand $A_\pi$ ≙ 0.24 seems to give good agreement with the positive hump in $f_4(s)$ which is a strong feature of the theoretical curve. This inconsistency, together with the lack of agreement for large s is $f_4(s)$ is evidence that the one-parameter molecular field treatment is inadequate. It also points to the inadequacy of Fourier methods of data reduction. This is because the data cut-off at $k/4\pi$ > 0.726 $Å^{-1}$ causes oscillations in both $f_o(s)$ and $f_4(s)$ which are larger than the effects due to covalency, apart from the region s < 0.4 $Å^{-1}$.

### 4.2 The $(FeO_4)^{5-}$ Group in $Fe_3O_4$

The spinels can be represented by the formula $AB_2X_4$, where A and B denote the cations and X represents the anion, which in this case is oxygen. The structure is conveniently described as an fcc close-packed array of anions into which the smaller cations are located in two types of interstitial site, A and B. The 8 A-sites have tetrahedral and the 16 B-sites octahedral coordination to the neighbouring oxygen ligands. Magnetite contains both $Fe^{2+}$ and $Fe^{3+}$ ions and its formula can be written $Fe^{2+} Fe_2^{3+} O_4^{2-}$. The cation distribution corresponds to the fully inverted structure in which all the $M^{2+}$ ions are located on the B sites, which then contain equal numbers of $Fe^{3+}$ and $Fe^{2+}$ cations. Ferrimagnetic ordering occurs below 848K and at room temperature the structure is cubic with space group Fd3m. A second phase transition occurs at about 120K to a structure of lower symmetry and lower conductivity (by about two orders of magnitude) accompanied by some kind of charge reordering on the B sites. At room temperature an electron hopping mechanism between the $Fe^{2+}$ and $Fe^{3+}$ ions in the octahedral sites is generally presumed to be the origin of the higher conductivity of this phase.

The first single crystal spin density study of the room temperature phase was made by Nathans et al (1960), who noted some interesting features of the unpaired electron density distribution but drew no specific or quantitative inferences about the bonding. It

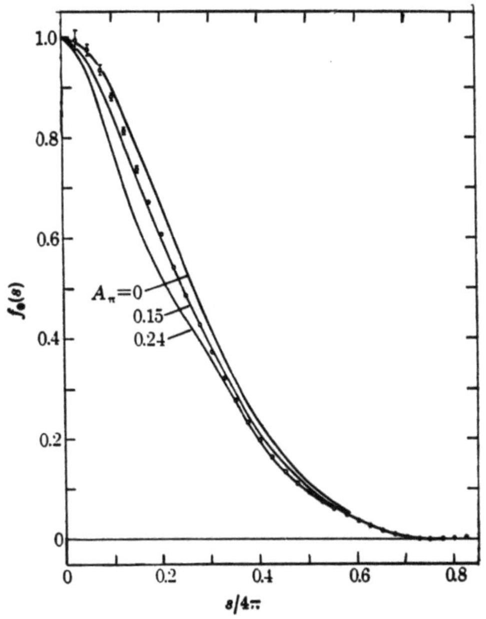


Figure 10 - The experimental values of $f_0(s)$ for $K_2NaCrF_6$ are plotted as open circles. The full lines correspond to a MO model with $A_\pi$ = 0, 0.15 and 0.24.

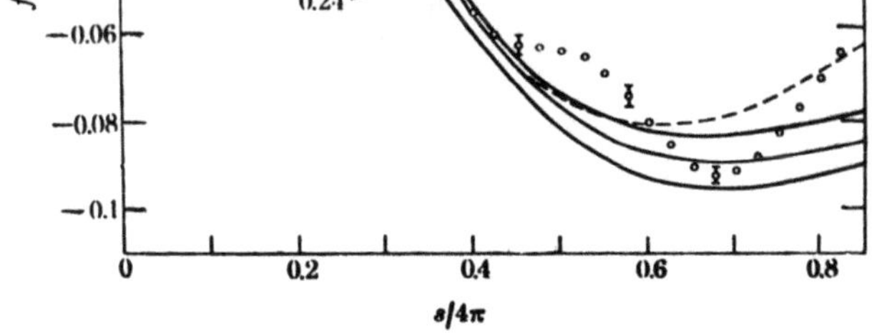

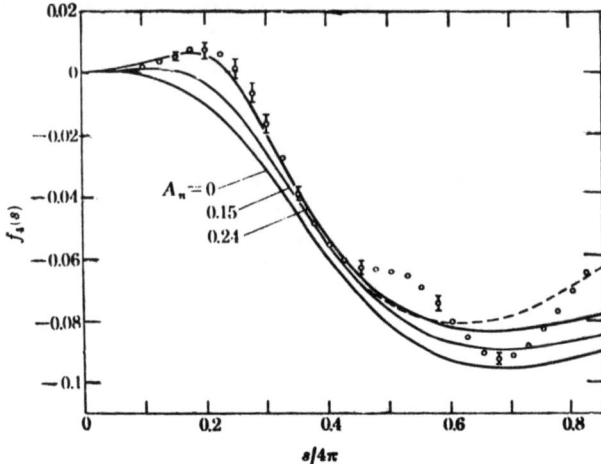

Figure 11 - The experimental values of $f_4(s)$ for $K_2NaCrF_6$ are plotted as open circles. The full lines correspond to the MO model with $A_\pi$ = 0, 0.15 and 0.24.

was for example, pointed out that there appeared to be a sizeable difference between the electron distribution on octahedral and tetrahedral sites, the moment density is less compact than in Fe metal and that there was moment density present on the oxygen or in the region between the cations and the oxygen anions. Recently, Rakhecha (1977) has completed a new polarized beam study which provides much more accurate values for the magnetic structure factors than some earlier work at Trombay (Shrinivasan et al 1974). The increased precision was mainly due to the use of thinner crystals which reduced the normally high extinction found in this class of materials.

Rakhecha has analysed his experimental data using the form factor approach and has concentrated his attention on the moment distribution of the A site group, which contains only a single species of transition metal cation. The tetrahedral crystal field again splits the d-levels into two groups, but with the E set lower than the $T_2$. In a pure crystal field situation, the moment density will remain centrosymmetric despite the absence of a centre of symmetry. However, when covalency is allowed for a density term $\rho_3(\underline{r})$ appear which in turn makes the form factor $f_3(\underline{k})$ non-zero in the total expression

$$f(\underline{k}) = f_o(\underline{k}) + iB(hk\ell)\, f_3(\underline{k}) + A(hk\ell)\, f_4(\underline{k})$$

Rakhecha has calculated the overlap form factors which correspond to a given overlap density using the Slater type orbitals for $Fe^{3+}$ given by Clementi (1965) and those for $O^{2-}$ given by Watson (1958). The admixture coefficients, which remain as adjustable parameters, then determine to what extent the overlap form factor for the E, $T_2^{\sigma}$ and $T_2^{\pi}$ states contribute to the net form factor for the A site density. A simple scaling of the experimental observations to a $Fe^{3+}$ free ion form factor gives a factor of 0.81, which suggests a rather strong covalency. A least-squares refinement of the parameters $A_{\sigma}^{T2}$, $A_{\pi}^{T2}$ and $A_{\pi}^{E}$ starting from the full temperature-corrected moment of 4.7 μB gave approximate values of 0.6, 0.0 and 0.4 respectively. These values lead to the conclusion that about 28% of the moment density is transferred from the A-site to the four neighbouring oxygens (ie about 7% each). The analysis of the B site magnetisation distribution in $Fe_3O_4$ is less satisfactory, since it corresponds to the static average of the $Fe^{2+}$ and $Fe^{3+}$ distributions. Furthermore, unlike the A sites, no group of reflections exist to which only the B sites contribute. No consideration was given to the possibility of unquenched orbital moment on the B sites. An excess (3-4%) $E_g$ population for the d-electrons was found together with an oxygen moment of between 0.1 and 0.3 μB.

Another detailed study of the magnetisation density associated with iron ions octahedrally and tetrahedrally coordinated by oxygen atoms has been carried out by Bonnet et al (1974, 1978), who have made polarized beam measurements on yttrium iron garnet. This material is also ferrimagnetic, but the unit cell is larger than that of $Fe_3O_4$ and there is a class of reflections, $K_{IV}$, to which the iron ions do not contribute. Figure 12 illustrates the difference between the total magnetisation density and that due to the iron ions. It can be seen that, in addition to a moment localised at the oxygen site which corresponds to some $0.032 \pm 0.005$ μB, there is direct evidence for significant density in the overlap region between the ligand and the metal ions. This is more pronounced in the case of the tetrahedral iron ion for which the metal-ligand interaction is stronger than in the octahedral case. The crystallographic residual R factor for some 50 $K_{IV}$ reflections was reduced from 0.61 to 0.19 by introducing a moment of $- 0.020 \pm 0.005$ μB midway between the oxygens and the tetrahedral iron ions, with a moment direction antiparallel to that of the iron moment. Bonnet et al are currently analysing their data to extract the admixture parameters.

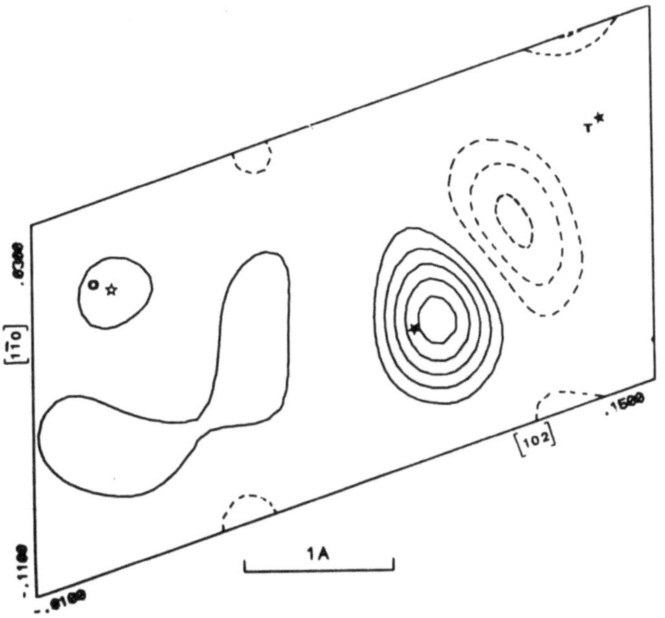


Figure 12 - Magnetisation difference density in YIG at room temperature. The atomic positions in the section are indicated by stars, O being the position of the octahedral iron ion and T that of the tetrahedral iron ion. The remaining star gives the oxygen position. The contours are at 0.02 μB/Å$^3$ intervals. (Bonnet et al, 1978). The tetrahedral iron moment is positive and the octahedral moment negative: the difference density located between the oxygen atom and the tetrahedral iron ion is therefore oppositely directed to the moment on the tetrahedrally co-ordinated iron site.

### 4.3 The $(IrCl_6)^{2-}$ Group in $K_2IrCl_6$

Lynn et al (1976) have made the first detailed form factor measurements on a material in which the 5d electrons order magnetically. Figure 13 illustrates the cubic antifluorite structure of $K_2IrCl_6$, which has space group Fm3m and a = 9.662 Å. Because of the symmetry of the ground state wavefunction, the magnetisation density differs radically from the cubic symmetry that the charge density possesses and this non-cubic symmetry reveals the covalent bonding to the coordinating ligands in a dramatic manner.

Like $K_2NaCrF_6$, the transition metal ions are magnetically isolated by the intervening alkali ions, but the indirect exchange paths Ir-Cl-Cl-Ir are sufficient to produce antiferromagnetic ordering below 3.05K. The low-lying, triplet $t_{2g}$ states produced by the octahedral crystalline electric field contain all five 5d electrons, since the field is strong enough to overcome Hund's rule. The low-spin $(t_{2g})^5$ state can conveniently be regarded as a single hole in the $\pi$ antibonding $t_{2g}$ state. Single-crystal integrated intensity measurements were made at 1.75K using 2.44 Å neutrons. The antiferromagnetic structure is of the 3A type, with a magnetic unit cell doubled along [001], which is the magnetic moment direction. The magnetic reflections are therefore completely separated from the nuclear intensity, so the polarized beam technique cannot be used.

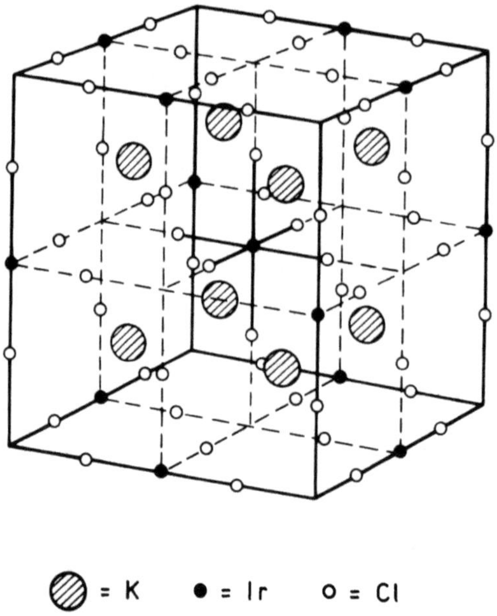


Figure 13 - The chemical structure of $K_2IrCl_6$. The unit cell is cubic space group Fd3m.

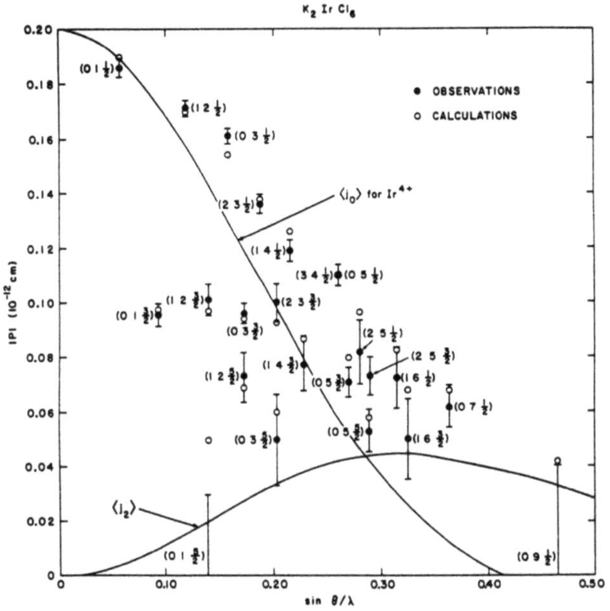


Figure 14 - The magnetic form factor for $K_2IrCl_6$ measured by Lynn et al (1976). The data not only differ sharply from spherical symmetry, but they also do not show cubic symmetry.

Figure 14 shows the observed structure factors and emphasises the difficulty of obtaining accurate values at high sin θ/λ from integrated intensities; the intensities are further weaked in the 3A structure, since only one third of the crystal contributes to any magnetic reflection. The structure factors fall off very quickly for the sequence (0 1 ½), (0 1 3/2) and (0 1 5/2), which lie close to the moment direction, compared to the slower variation for the (0 1 ½), (0 3 ½), (0 5 ½) and (0 7 ½) reflections which are roughly perpendicular to the moment direction. The Fourier section of the magnetisation density in the (010) plane is given in Figure 15, from which it can be seen that the distribution is elongated parallel to [001].

The cylindrically-symmetric moment density can be reproduced by a model in which the appropriate ground state wavefunction contains an admixture of p states from the chlorine ligands. The total multiplicity of the $^2T$ ground state, (2L + 1) (2S + 1), is six and spin-orbit coupling splits this into a spin doublet ground state and a higher quartet. The antiferromagnetic ordering further defines a singlet ground state which may be written as

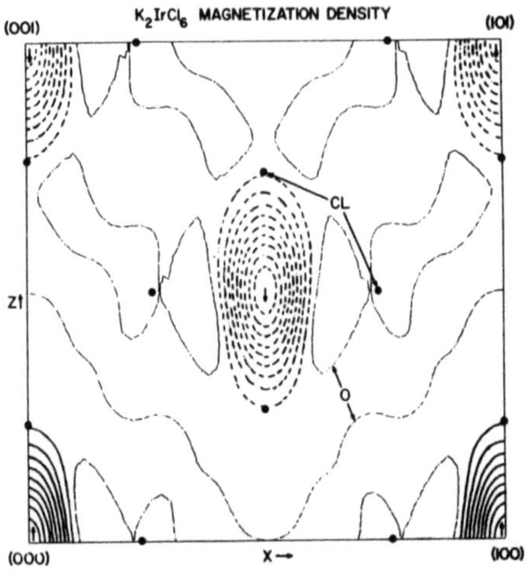


Figure 15 - The (010) section of the magnetisation density in $K_2IrC\ell_6$ at the level of the iridium ion (y = 0). There is a pronounced elongation of the density in the direction of the moment.

$$|+\rangle = \sqrt{2/3}\ |a,+\rangle + \sqrt{1/3}\ |c,-\rangle$$

where

$$|a\rangle = 1/\sqrt{2}\ \{|xz\rangle + i\ |yx\rangle\}$$

and

$$|c\rangle = |xy\rangle$$

It can be seen that this moment density will have the observed symmetry, since

$$\psi^* S_z \psi \propto x^2z^2 + y^2z^2 - x^2y^2$$

A detailed calculation shows that this asymmetry in the ground state wavefunction induced by the spin-orbit coupling is not sufficient to account for the measurements. However, if the $t_{2g}$ d orbitals are replaced by the molecular orbitals given in section 2, and the contributions to the various ligands are summed, it is found that they cancel for the chlorine ions located on the x and y axes. Thus, although the charge is transferred equally from all six ligands the magnetic moment transfer is only to the two ligands which are in the direction of the $Ir^{4+}$ moment. Lynn et al. (1976) were able to get a good fit to their observations with a simple model in which 28% of the total moment, μ, of 0.80 μB is on the two chlorine atoms at ± (0, 0, 0.24) and its distribution is described by a Gaussian with width parameter 0.06 a. These values are in good agreement with the estimates of μ = 0.89 μB with a 30% transfer obtained by Stevens (1953).

## 5. DETERMINATION OF THE GROUND STATE

The derivation of covalency parameters from magnetisation density distributions is dependent to a greater or lesser extent on the certainty with which the ground state of the magnetic ion is known. This is clearly crucial if the overall degree of covalency is to be deduced from moment reduction in an antiferromagnetic salt measured by the lowest angle magnetic lines from a powdered sample. However, the measurement of magnetic structure factors may provide a very important method for investigating the electronic structure of the more complex ions.

In actinide (5f) systems the ground state properties, and often even the ionization states, are unknown. Moreover, inelastic neutron experiments (Wedgwood, 1974) have failed to determine the crystal-field levels or the nature of the magnetic excitations. The first accurate magnetisation density determination was carried out by Wedgwood (1972) on ferromagnetic US using the polarized beam technique. He showed that the aspherical aspects of the magnetisation give a much more sensitive test of the fit between experiment and a number of different models for the ground state. Even so, he was unable to arrive at a unique determination of the ground state, though the $5f^2:\Gamma_1$ singlet configuration gave the best fit of those examined. The calculation of the magnetic structure factors corresponding to any given model requires the use of the tensor-operator method (Marshall and Lovesey, 1971) together with relativistic values for the one-electron radial integrals (Freeman et al, 1976).

There is considerably more anisotropy present in the magnetic form factor of USb compared to that in US. Figure 16 shows the data on USb obtained by Lander (1976). It can be seen that the overall fit with a $5f^2:\Gamma_8^{(1)}$ ground state is excellent which gives confidence in the radial integrals used. The anisotropy again allows a distinction to be made between a number of possible ground states. Figure 17 shows the differences between the observed anisotropy (open circles) and that calculated for two different states $\Gamma_8^{(1)}$ (upper) and $\Gamma_8^{(2)}$ (lower figure).

The ground state wavefunctions are:-

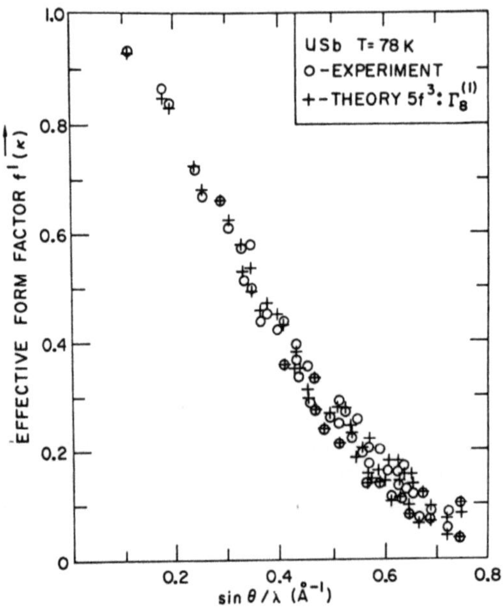


Figure 16 - Comparison between the observed and calculated effective form factor values for USb at 78K (Lander 1976).

$$\Gamma_8^{(2)} \quad \psi_e = 0.79\ |9/2\rangle - 0.59\ |1/2\rangle - 0.14\ |-7/2\rangle$$

$$\mu_{sat} = 2.12\ \mu B$$

$$\Gamma_8^{(1)} \quad \psi_e = 0.97\ |7/2\rangle - 0.25\ |-1/2\rangle - 0.01\ |-9/2\rangle$$

$$\mu_{sat} = 2.36\ \mu B$$

It is clear from Figure 17 that the $\Gamma_8^{(2)}$ state, which has a dominant $|9/2\rangle$ term, is clearly incorrect and the shape of the magnetisation density observed in USb is incompatible with both the $\Gamma_8^{(2)}$ or free-ion $|9/2\rangle$ states.

The neutron elastic magnetic cross-section has also been measured from ferromagnetic $^{242}$PuP at 4.2K by Lander and Lam (1976). As a result of the partial cancellation of S and L in the $f^5$ state, the cross-section is independent of angle for sin $\theta/\lambda < 0.35$ Å$^{-1}$. The choice of possible ground state is limited by the necessity of obtaining the correct value of the total moment and the need to have a large coefficient for the $\langle j_2\rangle$ radial integral to produce the

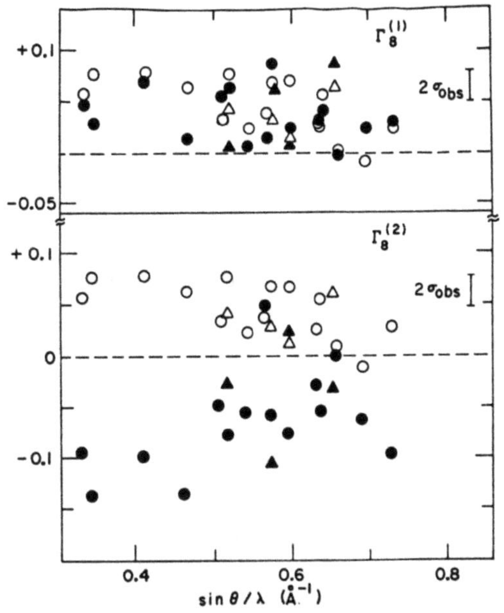


Figure 17 - Comparison of the observed aspherical components of the effective structure factor values for USb (open circles) with those calculated for the $\Gamma_8^{(1)}$ state (upper solid points) and the $\Gamma_8^{(2)}$ state (lower solid points), (Lander 1976).

essentially flat form factor at low values of sin θ/λ. The final choice for the ground state is based on $^6H_{5/2}$ with a small admixture (5%) of the J = 7/2 state.

## 6. SOME CURRENT STUDIES

The improvements in neutron fluxes and instrumentation provided by the high flux beam reactors have encouraged experimentalists to examine more complex structures and to make measurements on materials which are only weakly magnetic. Some examples of work which is still in progress will be given, since it gives an impression of the way the subject is enlarging, even though it must be admitted that much still remains to be done in the examination of simpler systems.

Potassium chloro-rhenate, $K_2ReCl_6$, is isomorphous with the chloro-iridate described in section 4.3. However, the ion $Re^{4+}$ is $d^3$, so to a first approximation the moment is spin only with three electrons in the $t_{2g}$ orbitals. Forsyth, Wedgwood and Brown have completed polarized beam measurements of the field-aligned paramagnetism at 30K, above the antiferromagnetic transition at 12K. Although the highest available field of 4.8T only gives an aligned moment of some 0.15 μB/Re, this is still sufficient to allow the magnetic structure factors to be quite accurately determined by the polarized beam technique. Figure 18 illustrates an (001) section of the magnetisation density through the centre of the rhenium ion, and the spin transfer to the four chlorine ligands in this plane can clearly be seen. Detailed interpretation of the measurements awaits a redetermination of the nuclear structure by powder profile refinement, since the material undergoes a phase transition between room temperature at 77K which leads to small displacements of the chlorine atoms (Smith and Bacon 1966). However, the major features of the density distribution are not likely to be significantly modified, since the magnetic scattering is confined to low values of sin θ/λ by the rapid fall-off of the 5d form factor and the structural modifications produce the largest changes in the nuclear structure factors at high values of momentum transfer.

The octahedral complex of $(M\ 6H_2O)^{2+}$ will be the subject of a series of investigations by Fender and Forsyth. The Tutton salts in the series $M^{2+}(NH_4)_2(SO_4)_2.6H_2O$ provide an excellent vehicle for this study, since M can be V, Cr, Mn, Fe, Co and Ni. The magnetic dilution is such that the transition metal complex is effectively isolated and the salts remain paramagnetic down to low temperature.

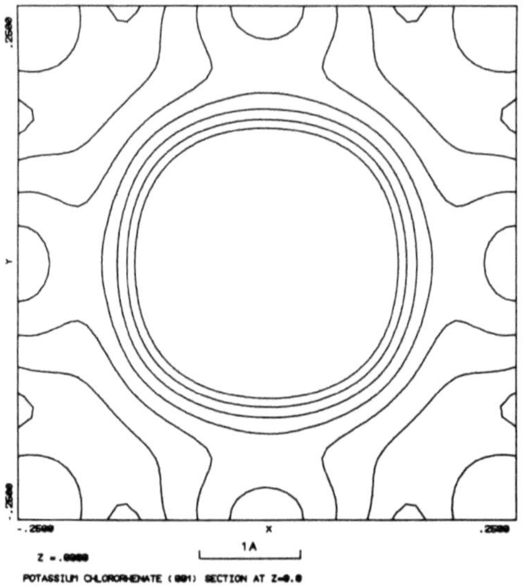


Figure 18 - The (001) section of the induced magnetisation density is $K_2ReCl_6$. The contours near the centre of the rhenium atoms have not been plotted.

A single crystal refinement of the nuclear structure of the manganese salt has been carried out and the first set of polarized beam measurements have just been made at 4.2K in an external field of 4.8T, which produces a moment in excess of 4 $\mu$B. It is hoped to extend these measurements to the $V^{2+}(d^3)$ and $Ni^{2+}(d^8)$ salts in the first place, since the interpretation of the resulting magnetisation distribution should be simplified by the need to consider only one type of orbital in a simple MO model, though the presence of spin polarization effects are likely to make the final model more complicated. There is a lack of neutron diffraction data on $V^{2+}$, but LHFI was observed by Davies et al (1972) using ENDOR on $V^{2+}$ in $KMgF_3$ and indicates a lower covalency than for $Cr^{3+}$. Tofield (1976) has suggested that the cubic perovskites $KVF_3$ and $RbVF_3$ (Cros et al, 1976) might also be suitable materials in which to study $V^{2+}$ covalency, but the high Néel temperatures of 120K and 130K respectively probably mean that the neutron data would have to be derived from integrated intensity measurements of the magnetic reflections at helium temperature.

Mason, Figgis and Williams have made measurements on both manganese (II) and cobalt (II) phthalocyanines (Pc). The transition metal ion is square - coordinated by four nitrogen atoms of the essentially planar molecule. MnPc becomes a canted ferromagnet below 8.6K (Miyoshi, 1974) and the magnetic susceptibility above this temperature implies that the ground state of the molecule is a quartet giving an intermediate spin value of S = 3/2. There are two molecules of $MnC_{32}H_{16}N_8$ on the monoclinic cell with a = 19.400, b = 4.761, c = 14.613, $\beta$ = 120.74°, space group $P2_1/a$. The planes of the two molecules are inclined to the b axis which is the direction of easy magnetisation by an angle of approximately 45°. Miyoshi finds that the saturation magnetisation is consistent with a total moment of 3 $\mu$B/Mn if the moment direction remains perpendicular to the plane of the nitrogen ligands. The polarized beam measurements of Mason et al have confirmed that the moments are indeed canted, since large magnetic contributions occur in reflections of the form {hkℓ}, {h3ℓ} for which there is no contribution from spherically-symmetric moment ferromagnetically aligned on the two Mn ions at (0 0 0) and (0 ½ 0) in the unit cell. The situation is simpler in CoPc which remains paramagnetic down to helium temperatures. The spin state has S = 1/2 and the neutron data have been measured to sufficient accuracy to enable a choice to be made amongst the possible cobalt magnetic configurations.

The crystal field analogue of an octahedral $d^3$ ion is a tetrahedral $d^7$ ion and tetrahedral $Co^{2+}$ occurs quite frequently in crystals and complexes. $Co_3O_4$ and $CoRh_2O_4$ are normal spinels with $Co^{2+}$ in the A sites and diamagnetic low spin $t_{2g}^6$ $Co^{3+}$ or $Rh^{3+}$ in the B sites. The Néel temperatures are 40K and 27K respectively and the reduction in the magnetic intensity observed by the powder method for $Co_3O_4$ is somewhat larger than for $CoRh_2O_4$ (Infante, 1975). Both reductions were, however, similar to that found in NiO by Fender et al (1968), which means that the total charge transfer must be greater for the tetrahedral $Co^{2+}$ case than for octahedral $Ni^{2+}$, since three electrons are involved in the former ion compared with two in the latter. A single crystal, polarized beam study of tetrahedral $Co^{2+}$ has recently been started by Williams, Figgis and Smith. The $(CoCl_4)^{2-}$ groups in $Cs_3CoCl_5$ are magnetically isolated, resulting in paramagnetism down to liquid helium temperatures. However, the simple MO picture of the ground state is complicated by the combined action of spin-orbit coupling and the tetragonal distortion of the chlorine tetrahedron in the tetragonal unit cell with a = 9.219 Å, c = 14.554 Å and space group I/mcm (Figgis et al 1964). Fortunately, only 5 positional and 19 thermal parameters are involved and a low-temperature structure refinement has already been carried out to an R factor of 0.03. Magnetisation measurements of Stapele et al (1966) indicate that almost complete alignment can be obtained at 4.2K with the available field of 4.8T parallel to the tetragonal axis. It is hoped that the ground state wavefunction can be defined with sufficient certainty so that the covalent contributions to the magnetisation can be isolated.

## 7. CONCLUSIONS

No attempt has been made to summarise all the magnetic neutron diffraction studies which have been made to measure chemical covalency or which have primarily determined the ground state wavefunction of a magnetic ion. It is hoped, however, that the more detailed discussion of selected investigations will have provided some insight into the difficulties of both the experiments and their interpretation. It is clear that the simple molecular orbital model, though capable of giving a semi-quantitative description of the bonding parameters in transition metal salts, is inadequate to explain the detailed three dimensional distributions of magnetisation which can now be measured by neutron diffraction. For example, Wedgwood (1976) showed conclusively that his data were not consistent with a single value of $A_\pi$ for the $d^3$ Cr ion, for which the simple MO model predicts only a $\pi$-bonding contribution to the moment redistribution.

The value of $A_\pi$ = 0.15, which gave a best fit to the spherical part of the form factor, was lower than the value 0.24 which fitted the aspherical part. The latter is much more sensitive to the ligand spin density and it is therefore not surprising that this value for $A_\pi$ is in close agreement with the value deduced by Shulman and Knox (1960) from their measurement of the transferred hyperfine interaction on the fluorine nucleus.

Tofield and Fender (1970) have suggested that the presence of some small parallel spin polarization of the empty $e_g$ orbitals could explain why the covalency parameter $A_\pi$ = 0.12, deduced from the moment reduction observed in a powder study of $LaCrO_3$, was less than that for $Cr^{3+}$ in the fluorides $KMgF_3$ or $K_2NaCrF_6$. Such a spin polarization would also lead to a lower value of the $^6A_\pi$ being found from the spherical part of the moment in $K_2NaCrF_6$ since this density is particularly sensitive to the metal spin. Although Fourier inversion of the observed magnetic structure factors is qualitatively revealing and can suggest what are likely to be the most important features for a model calculation, it is unlikely to be the best method for extracting the greatest information from a diffraction experiment. The presence of series termination errors due to the limited data set sets one limit to the technique, even when care is taken to average the density over a suitable volume (Shull and Mook, 1966). The presence of small non-collinear components in the otherwise collinear distribution of $K_2IrCl_6$ means that the transform shown in Figure 13 is not strictly accurate, as was indeed pointed out by its authors.

The most satisfactory final method of analysis must be a model calculation which either predicts the complete magnetisation distribution or which calculates its Fourier components directly. Ellis, Freeman and Ros (1968) were the first to perform a full Hartree-Fock calculation for all the electrons in an $(NiF_6)^{4-}$ cluster and derived form factors $f_0(\underline{k})$ and $f_4(\underline{k})$ by Fourier transformation. Their $f_4(\underline{k})$ curve shows a change in sign at $\sin\theta/\lambda \sim 0.2$ Å$^{-1}$ which is similar to that observed in $K_2NaCrF_6$ (Figure 11) but in an opposite sense since it relates to $e_g$ rather than $t_{2g}$ electrons. Brown and Burton (1970) and Larsson and Connolly (1974) have made spin-polarized Hartree-Fock calculations for $Cr^{3+}$ using different approximation schemes. Although the calculations give different estimates for the spin density coefficients $f_\sigma$ and $f_\pi$, they agree in predicting some $e_g$ polarization, in contradiction to the simple MO theory, but in line with the ideas of Tofield and Fender (1970) and the indications of the magnetisation density distribution in $K_2NaCrF_6$. In the case of $Fe_3O_4$, the closest Hartree-Fock calculation is the one made by Byron et al (1975) for the $(Fe_3O_5)^{5-}$ cluster of $Fe^{3+}$ in YIG, which has a similar Fe-O distance to that for the A sites in magnetite.

The examination of systems of more 'chemical' interest is still in its infancy. Most molecular crystals containing the requisite transition metal ion have large unit cells which make the determination of accurate magnetisation density distributions extremely difficult. The nuclear structure factors depend on a large number of positional and thermal parameters and this in turn limits the precision with which they are known. A large number of reflections must be measured to define the density in the asymmetric unit and the mean magnetisation density is usually low due to the small number of magnetic ions. Nevertheless, it seems profitable that such attempts should be made since it is clear that the diffraction technique, in conjunction with other magnetic measurements, provides unique information of relevance to the theories of the chemical bond.

## REFERENCES


Alperin H A (1962), J Phys Soc, Japan, Supl B III 17, 12.

Bonnet M, Delapalme A, Tcheou F and Fuess H (1974), Proc ICM (Moscow, 1973) IV, 251.

Bonnet M, Delapalme A, Becker P and Fuess H (1978), J Mag and Mag Materials, 7, 23.

Brown R D and Burton P G (1970), Theoret Chim Acta, 18, 309.

Byrom E, Freeman A J and Ellis D E (1975), AIP Conf Proc (1974), 210.

Clementi E (1965), Suppl to IBM J Res Develop, 9, 2.

Cros C, Feurer R and Pourchard M (1976), Mat Res Bull, 11, 117.

Davies J J, Smith S R P, Owen J and Hann B F (1972), J Phys C 5, 245.

Ellis D E, Freeman A J and Ros P (1968), Phys Rev, 176, 688.

Fender B E F, Jacobsen A J and Wedgwood F A (1968), J Chem Phys, 48, 990.

Figgis B N, Gerloch M and Mason R (1964), Acta Cryst, 17, 506.

Freeman A J, Desclaux J P, Lander G H and Faber J Jr (1976), Phys Rev B13, 1168.

Hirakawa K and Ikeda H (1973), J Phys Soc, Japan, 35, 1608.

Hirakawa K and Ikeda H (1974), Phys Rev Letters, 33, 374.

Hubbard J and Marshall W C (1965), Proc Phys Soc (London), 86, 561.

Hukin J (1965), M Sc Thesis, Clarendon Laboratory, Oxford University.

Infante C E (1975), D Phil Thesis, University of Oxford.

Jacobsen A J (1973) in Chemical Applications of thermal neutron scattering (Ed B T M Willis) London-Oxford University Press, p270.

Lander G H (1976), AIP Conference Proceedings 21st Annual Conference on Magnetism and Magnetic Materials, 29, 311.

Larsson S and Connolly J W D (1974), J Chem Phys, 60, 1514.

Lynn J W, Shirane G and Blume M (1976), Phys Rev Letters, 37 154.

Marshall W and Lovesey S W (1971), Theory of Thermal Neutron Scattering, London-Oxford University Press.

Miyoshi H (1974), J Phys Soc, Japan, 37, 50.

Moss J and Brown P J (1972), J Phys F, 2, 358.

Nathans R, Pickart S J and Alperin H A (1960), Bul Amer Phys Soc, 115, 455.

Owen J and Stevens K W H (1953), Nature 171, 836.

Rakhecha V R (1977), Theseis for PhD, University of Bombay, India.

Rimmer D E (1970) in Thermal Neutron Diffraction (Ed B T M Willis), London-Oxford University Press, p14.

Shrinivasan R, Rakhecha V C, Paranjpe S K, Begum R J, Madhav Rao L, Satya Murthy N S (1974), Proc ICM (Moscow 1973) IV, 246.

Shull C G and Mook H A (1966), Phys Rev Letters, 16, 184.

Smith H G and Bacon G E (1966), J Appl Phys, 37, 979.

van Stapele R P, Beljers H G, Bongers P F and Zijlstra H (1966), J Chem Phys, 44, 3719.

Stevens K W H (1953), Proc Roy Soc London, A226, 542.

Tofield B C (1975) in Structure and Bonding, 21, Springer-Verlag, Berlin.

Tofield B C (1976), J de Physique Coll C-6-539, Applications of the Mössbauer Effect.
Tofield B C and Fender B E F (1970), J Phys Chem Solids, 31, 2741.
Tofield B C and Fender B E F (1971), J Phys C, 4, 1279.
Watson R E (1958), Phys Rev, 111, 1108.
Watson R E and Freeman A J (1959), Acta Cryst, 14, 27.
Wedgwood F A (1972), J Phys C, 5, 2427.
Wedgwood F A (1974), J Phys C, 7, 3203.
Wedgwood F A (1976), Proc Roy Soc London A349, 447.

RELATIONSHIP OF MULTIPOLE POPULATIONS TO d-ORBITAL OCCUPANCIES OF A TRANSITION METAL ATOM IN A TETRAGONALLY DISTORTED OCTAHEDRAL FIELD.

Edwin D. Stevens

Chemistry Department
State University of New York at Buffalo
Buffalo, New York 14214

## I. Introduction

The $e_g$ and $t_{2g}$ orbitals of a transition metal atom are split by a tetragonal distortion of an octahedral field as follows:

$e_g$:

$$y_{22-} = d_{x^2-y^2} = R(r)(x^2-y^2)/r^2$$

$$y_{20} = d_{z^2} = R(r)(2z^2-x^2-y^2)/r^2$$

$t_{2g}$:

$$y_{22+} = d_{xy} = R(r)2xy$$

$$y_{21+}, y_{21-} = d_{xz}, d_{yz} = R(r)xz/r^2,\ R(r)yz/r^2$$

Let $P_1$, $P_2$, $P_3$, and $P_4$ be the occupancies of the $y_{22-}$, $y_{20}$, $y_{22+}$, and ($y_{21+}$, $y_{21-}$) orbitals respectively. Using the expressions and normalization constants given below,

(Part 1) Write an expression for the d-shell density as a function of the orbital occupancies ($P_1 \ldots P_4$).

(Part 2) If $P_{\ell m\pm}$ is the population of the $y_{\ell m\pm}$ density function obtained from a multipole refinement, derive expressions for $P_1 \ldots P_4$ as a function of the $P_{\ell m\pm}$.

Table 2. Selected Products of Spherical Harmonic Functions in Real Form

$$y_{20}y_{20} = (9/49\pi)^{1/2}\, y_{40} + (5/49\pi)^{1/2}\, y_{20} + (1/4\pi)^{1/2}\, y_{00}$$

$$y_{21\pm}y_{21\pm} = \pm(5/49\pi)^{1/2}\, y_{42+} \pm (15/196\pi)^{1/2}\, y_{22+} - (4/49\pi)^{1/2}\, y_{40} + (5/196\pi)^{1/2}\, y_{20} + (1/4\pi)^{1/2}\, y_{00}$$

$$Y_{22\pm}y_{22\pm} = \pm(5/28\pi)^{1/2}\, y_{44+} + (1/196\pi)^{1/2}\, y_{40} - (5/49\pi)^{1/2}\, y_{20} + (1/4\pi)^{1/2}\, y_{00}$$

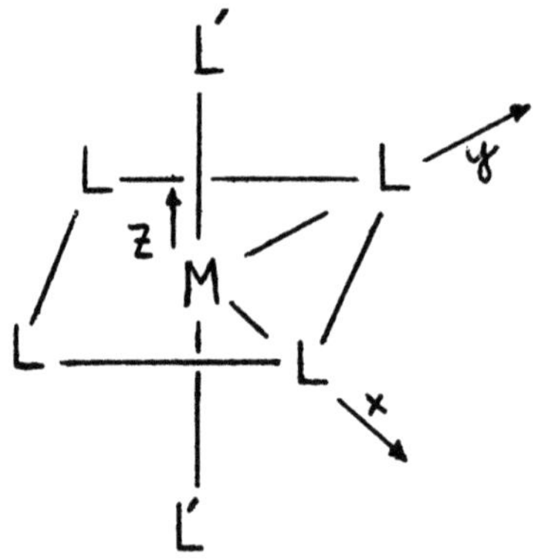

## II. Solution

$$\rho_{3d}\ (\underline{\underline{r}}) = (R_{3d}(r))^2\ \{P_1(y_{22-})^2 + P_2(y_{20})^2$$

$$+ P_3(y_{22+})^2 + \frac{1}{2}\ P_4(y_{21+}{}^2 + y_{21-}{}^2)\}$$

$$= (R_{3d}(r))^2\{y_{00}(1/4\pi)^{1/2}\ M_{00}\ (P_1 + P_2 + P_3 + P_4)$$

$$+\ y_{20}\ (5/196\pi)^{1/2}\ M_{20}\ (-2P_1 + 2P_2 - 2P_3 + P_4)$$

$$+\ y_{40}\ (1/196\pi)_{1/2}\ M_{40}\ (P_1 + 6P_2 + P_3 - 4P_4)$$

$$+\ y_{44+}\ (5/28\pi)^{1/2}\ M_{44}\ (-P_1 + P_3)\}$$

$$P_{00} = (1/4\pi)^{1/2}\ M_{00}\ (P_1 + P_2 + P_3 + P_4)$$

$$P_{20} = (5/196\pi)^{1/2}\ M_{20}\ (-2P_1 + 2P_2 - 2P_3 + P_4)$$

$$P_{40} = (1/196\pi)^{1/2}\ M_{40}\ (P_1 \div 6P_2 + P_3 - 4P_4)$$

$$P_{44+} = (5/28\pi)^{1/2}\ M_{44}\ (-P_1 + P_3)$$

Where $M_{\ell m} = N_{\ell m}$ (wave function)/$N_{\ell m}$ (density function)

$$P_1 = \frac{1}{5} P_{00} - \frac{1}{7} \left(\frac{196\pi}{5}\right)^{1/2} \left(\frac{16\pi}{5}\right)^{1/2} \left(\frac{3\sqrt{3}}{8\pi}\right) P_{20}$$

$$+ \frac{1}{70} \left(\frac{196\pi}{1}\right)^{1/2} \left(\frac{256\pi}{9}\right)^{1/2} ! \quad P_{40}$$

$$+ \left(-\frac{1}{2}\right)\left(\frac{28\pi}{5}\right)^{1/2} \left(\frac{32\pi}{315}\right)^{1/2} \left(\frac{15}{32}\right) P_{44+}$$

$$P_2 = \frac{1}{5} P_{00} + \frac{1}{7} \left(\frac{196\pi}{5}\right)^{1/2} \left(\frac{16\pi}{5}\right)^{1/2} \left(\frac{3\sqrt{3}}{8}\right) P_{20}$$

$$+ \frac{6}{70} \left(\frac{196\pi}{1}\right)^{1/2} \left(\frac{256\pi}{9}\right)^{1/2} ! \quad P_{40}$$

$$P_3 = \frac{1}{5} P_{00} - \frac{1}{7} \left(\frac{196\pi}{5}\right)^{1/2} \left(\frac{16\pi}{5}\right)^{1/2} \left(\frac{3\sqrt{3}}{8}\right) P_{20}$$

$$+ \frac{1}{70} \left(\frac{196\pi}{1}\right)^{1/2} \left(\frac{256\pi}{9}\right)^{1/2} ! \quad P_{40}$$

$$+ \frac{1}{2} \left(\frac{28\pi}{5}\right)^{1/2} \left(\frac{32\pi}{315}\right)^{1/2} \left(\frac{15}{32}\right) \quad P_{44+}$$

$$P_4 = \frac{2}{5} P_{00} + \frac{1}{7} \left(\frac{196\pi}{5}\right)^{1/2} \left(\frac{16\pi}{5}\right)^{1/2} \left(\frac{3\sqrt{3}}{8}\right) P_{20}$$

$$+ \left(-\frac{8}{70}\right)\left(\frac{196\pi}{1}\right)^{1/2} \left(\frac{256\pi}{9}\right)^{1/2} ! \; P_{40}$$

Where $! = \{2\pi(14A_-^{\,5} - 14A_+^{\,5} - 20A_-^{\,3} + 20A_+^{\,3} + 6A_- - 6A_+)\}^{-1}$

and $A_{\pm} = \left(\frac{30 \pm \sqrt{480}}{70}\right)^{1/2}$

# EXERCISE: CALCULATION OF THE MAGNETIC FORM FACTOR FOR THE SPIN DENSITY OF A 3d SHELL IN A GIVEN ENVIRONMENT

J.X. Boucherle

DRF/DN - Centre d'Etudes Nucléaires

85 X - 38041 Grenoble Cedex - France

The calculation of a spin structure factor can be performed for any local symmetry using the general formulae (1)

$$F(\vec{H}) = \mu \sum_K < j_K > \sum_Q C_Q^K \, Y_Q^{K^\star}(\theta,\phi)$$

$$C_Q^K = i^K(2\ell+1)\left[4\pi(2K+1)\right]^{1/2} \sum_{m,m'} (-1)^m \, a_m^\star \, a_{m'} \begin{pmatrix} \ell & K & \ell \\ 0 & 0 & 0 \end{pmatrix} \begin{pmatrix} \ell & K & \ell \\ -m & Q & m' \end{pmatrix}$$

where $\begin{pmatrix} j_1 & j_2 & j_3 \\ m_1 & m_2 & m_3 \end{pmatrix}$ are 3j coefficients, and $Y_Q^K(\theta,\phi)$ are spherical harmonics.

The coefficients $a_m$ define the wave functions :

$$\psi(\vec{r}) = R(r) \sum_m a_m \, Y_m^\ell \,(\Theta,\Phi)$$

and $<j_K>$ are the radial integrals :

$$<j_K\,(H)> = \int_0^\infty r^2 dr \, |R(r)|^2 \, J_K(Hr)$$

The aim of this exercise is to calculate magnetic form factor for a 3d shell considering different local symmetries.

1 - For a 3d shell the orbital quantum number is $\ell = 2$. Show, from the properties of the 3j coefficients (2), that only three different radial integrals are needed in the calculations.

2 - The 3d magnetic electrons correspond to a 5 fold degenerate level with the following wave functions :

| | |
|---|---|
| $d_{z^2}$ | $Y^2_0$ |
| $d_{x^2-y^2}$ | $1/\sqrt{2}\ (Y^2_2 + Y^2_{-2})$ |
| $d_{xy}$ | $-i/\sqrt{2}\ (Y^2_2 - Y^2_{-2})$ |
| $d_{xz}$ | $1/\sqrt{2}\ (Y^2_1 + Y^2_{-1})$ |
| $d_{yz}$ | $-i/\sqrt{2}\ (Y^2_1 - Y^2_{-1})$ |

In a crystal field of a low symmetry such as D2h, the degeneracy is completely removed. Calculate the structure factor for each of the orbitals above.

3 - In an hexagonal symmetry such as D3h, the decomposition involves only three levels : a singlet and two doublets.

A1g : $d_{z^2}$ E2g : $\begin{cases} d_{x^2-y^2} \\ d_{xy} \end{cases}$ E1g : $\begin{cases} d_{xz} \\ d_{yz} \end{cases}$

Starting from the preceding results, find the form factors corresponding to these levels. Show that the results are independant of the azimuthal angle $\phi$ and so present an axial symmetry.

4 - In a cubic environment, the decomposition involves one doublet and one triplet.

Eg : $\begin{cases} d_{z^2} \\ d_{x^2-y^2} \end{cases}$ Tg : $\begin{cases} d_{xy} \\ d_{xz} \\ d_{yz} \end{cases}$

Using the results of the second part, show that the form factor for these levels correspond to the formulae

$$F(Eg) = \langle j_0 \rangle + 3/2\ A\ \langle j_4 \rangle$$

$$F(Tg) = \langle j_0 \rangle - A\ \langle j_4 \rangle$$

$$A = 1/8(35 \cos^4\theta - 30 \cos^2\theta + 3) + 5/8 \sin^4\theta \cos 4\phi$$

$$\text{or} \quad A = \frac{h^4 + k^4 + l^4 - 3(h^2k^2 + h^2l^2 + k^2l^2)}{(h^2 + k^2 + l^2)^2}$$

Note that the second order is absent from these results.

References

(1) J. Schweizer, Interpretation of the spin densities in metals and alloys, this book and reference therein

(2) H. Appel, Numerical tables for 3 j symbols, Landolt-Börnstein, Vol. 3, Springer-Verlag (1968)

(3) F. Tasset, Thesis, Univ. Grenoble (1975) (A.O. 10916)

SOLUTION

Question 1

The 3j coefficient $\begin{pmatrix} \ell & K & \ell \\ 0 & 0 & 0 \end{pmatrix}$ imposes severe restriction on the values of K with $\ell = 2$ :

- The triangular condition ($K \leqslant 2\ell$) gives $K \leqslant 4$
- The relation

$$\begin{pmatrix} j_1 & j_2 & j_3 \\ m_1 & m_2 & m_3 \end{pmatrix} = (-1)^{j_1+j_2+j_3} \begin{pmatrix} j_1 & j_2 & j_3 \\ -m_1 & -m_2 & -m_3 \end{pmatrix}$$

shows that $K+2\ell$ must be even. K is also even. Only K = 0,2,4 are possible.

Question 2

For each level the form factor can be written as $F(\vec{H}) = \sum_{K=0,2,4} . B_K(\theta,\phi) \langle j_K \rangle$ . The values of $B_K$ are given in the table and in reference (3).

| | $\langle j_0 \rangle$ | $\langle j_2 \rangle$ | $\langle j_4 \rangle$ |
|---|---|---|---|
| $d_{z^2}$ | 1 | $- 5/7.\alpha$ | $9/28.\beta$ |
| $d_{x^2-y^2}$ | 1 | $+ 5/7.\alpha$ | $3/56.\beta+15/8\sin^4\theta\cos 4\phi$ |
| $d_{xy}$ | 1 | $+ 5/7.\alpha$ | $3/56.\beta-15/8\sin^4\theta\cos 4\phi$ |
| $d_{xz}$ | 1 | $- 5/14\alpha+15/14 \sin^2\theta\cos 2\phi$ | $-12/56\beta-15/14.\gamma.\sin^2\theta\cos 2\phi$ |
| $d_{yz}$ | 1 | $- 5/14\alpha-15/14 \sin^2\theta\cos 2\phi$ | $-12/56\beta+15/14.\gamma.\sin^2\theta\cos 2\phi$ |

$\alpha = 3 \cos^2\theta-1$ , $\beta = 35 \cos^4\theta-30 \cos^2\theta+3$ , $\gamma = 7 \cos^2\theta-1$

Questions 3 and 4

Using the table, the form factor is obtained for a degenerate level by averaging the form factors of the different constituent states. The results are given in reference (1).

# EFFECTS OF CRYSTAL FORCES AND HYDROGEN BONDING ON CHARGE DENSITY

Ivar Olovsson

Institute of Chemistry, University of Uppsala

Uppsala, Sweden

## ABSTRACT

The general features of the charge density in simple hydrogen-bonded compounds, as derived from quantum-mechanical calculations and from diffraction investigations, are compared in order to assess the relative influence of hydrogen-bond interactions and crystal forces on the electron distribution.

In hydrogen bonds of weak and intermediate strengths, the electron distribution may be considered to a first approximation simply as a superposition of the density of the undisturbed, constituent monomers. The experimental maps show very little variation in their general appearance as the bond is shortened from around 3.0 Å down to 2.7 Å. Theoretical work indicates, however, that there occurs as a second-order effect a definite electronic redistribution due to intermolecular interactions. The major effect of the environment is to increase the polarity of the functional groups that already exists in the isolated monomers.

In very short hydrogen bonds of type $[O\cdots H\cdots O]^-$, with oxygen-oxygen distances around 2.4 - 2.5 Å and where the proton interacts with two chemically very similar acceptor groups, the deformation density maps have a rather different appearance: there is a much less pronounced charge build-up in both the O···H and the H···O region.

## 1. INTRODUCTION

The majority of charge density studies has been concerned with bonding effects within molecules, ions or complexes of limited extension. The purpose has been to study such phenomena as covalency, charge transfer and related features. The effects of interaction with the surrounding environment have generally been neglected in comparing diffraction results with the results of theoretical calculations. These "intermolecular" effects are perhaps of lesser importance in the case of non-polar molecules or groups as they approach each other. On the other hand, the influence on the charge density is quite noticeable when the molecules interact more strongly as, for example, in the case of hydrogen bonding. The studies of intermolecular effects have so far been limited almost completely to hydrogen-bonded compounds. In the following, a discussion will be made of some general charge density features in such systems as derived from quantum-mechanical calculations and diffraction experiments. Although special attention is given to hydrogen-bonded systems the general approach may easily be extended to other types of interaction.

Hydrogen bonding may be seen in general terms as an interaction between polar groups. It has merited a special designation and treatment, however, through its many unique features. The hydrogen bond plays a fundamental role in many organic as well as inorganic compounds; the spectacular properties of water and ice are typical examples. The physical and chemical properties of water are quite different from those of most other liquids; this is a general consequence of the presence of hydrogen bonding in water. The hydrogen bond is thus also of fundamental importance in biological systems, since all living matter has evolved from and exists in an aqueous environment. It is generally believed that most biological processes at some stage involve the formation or breaking of hydrogen bonds.

## 2. CHARACTERIZATION AND NOMENCLATURE OF HYDROGEN BONDS

A hydrogen atom which is covalently bonded to an atom X may, under certain conditions, also be attracted to an atom Y to form a hydrogen bond X-H···Y. A general, empirical requirement for the formation of such a bond is that X and Y are both strongly electronegative. Employing a simplified overall picture, the electron density of X-H will then be deformed to give a certain net positive charge around H which is attracted by a negative charge distribution around Y. Accordingly, hydrogen bonds are generally formed between such atoms as oxygen, nitrogen, fluorine and chlorine. The bond energy is quite small, 5-10 kcal/mole (20-40 kJ/mole).

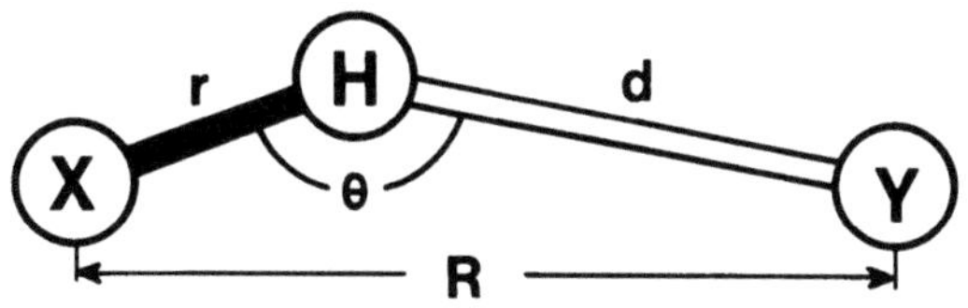


Fig. 1. Geometrical hydrogen-bond parameters.

An operational definition in common usage states that a hydrogen bond between two groups X-H and Y exists when (a) there is evidence of bonding X-H···Y and (b) there is evidence that this bonding involves the hydrogen in X-H.

A hydrogen bond can be characterized geometrically by the parameters r, d and θ as shown in Fig. 1. The most natural geometrical criterion for hydrogen bonding between the groups X-H and Y (cf. requirement (a) above is that the distance H···Y be shorter than the van der Waals approach, $d<\underline{w}_H+\underline{w}_Y$.

The most common acceptor atom (Y) is oxygen and here the van der Waals radius ($\underline{w}_O$) given by Pauling of 1.4 Å is generally employed. In the case of hydrogen, however, recent neutron diffraction measurements of H-H distances suggest 1.0 Å instead of the Pauling value 1.2 Å. This means that an H···O contact shorter than 2.4 Å can be taken as an indication of hydrogen bonding.

A general review of recent theoretical and experimental studies of hydrogen-bonded systems is given in Ref. 1. X-ray and neutron diffraction studies are here surveyed by Olovsson & Jönsson in chapter 8, and some of the topics discussed below are treated more extensively in this article.

Many of the features characteristic of hydrogen bonding can be explained by simplified models, some of these are presented in a later section. For a general, more physical interpretation of the charge redistribution effects associated with chemical bonding, attention should be drawn to a rather different approach outlined briefly below.

## 3. ELECTRON DENSITY AND CHEMICAL BONDING

Since electrons are ultimately responsible for the forces holding atoms together in a chemical system, it would seem desirable to discuss chemical bonding in terms of some experimentally observable quantity directly related to the behaviour of the electrons. From this point of view, many of the current concepts like σ- and π-bonding, covalency etc., are less satisfactory as they are intrinsically dependent on certain models utilized for their definition and, furthermore, they do not correspond to experimentally observable quantities. The possibility of interpreting the chemical bonding more directly in terms of the charge distribution itself is clearly of particular importance in the present context. Among the various quantum-mechanical theorems involving electron density, a special case of the general Hellmann-Feynman theorem [2,3] is of particular interest. According to this theorem, the force acting on each nucleus in a system is exactly that calculated by the principles of classical electrostatic theory from the charges and positions of the other nuclei and the electrons. The theorem requires that the net electrostatic field at each nucleus in the system must vanish at equilibrium. It thus provides a physical basis for an analysis of chemical bonding in terms of the one-electron distribution and the forces acting on each nucleus. It also offers a physically attractive way of investigating the relative influence of different details in the charge distribution on the total force acting on the different nuclei. The chemical bonding in diatomic molecules has been analyzed in these terms by Bader and coworkers [4] and by Hirshfeld & Rzotkiewisz [5]. The deformation density (cf. section 6.2) forms the basis for a partitioning of the total electronic force on the nuclei. The relative binding ability of the different orbitals has then been discussed. Several of the results in these studies have a bearing on the physical interpretation of charge redistributions associating with intermolecular interactions. As their approach has yet to be extended beyond first-row diatomic molecules, it is only possible to refer to these results in a very qualitative way in the later discussion. The interpretation of chemical bonding features in terms of the Hellmann-Feynman field is difficult in practice since there is no simple way of summarizing or visualizing the details of the charge distributions. At least for the present, it is thus perhaps necessary to employ standard concepts and simplified models in trying to relate bonding characteristics in similar systems. The following treatment of the nature of the hydrogen bond thus follows more or less traditional lines.

## 4. THE NATURE OF THE HYDROGEN BOND: ENERGY AND CHARGE ANALYSIS

### 4.1. Concepts

Several different concepts and models have been employed in the past to account for the characteristic properties of the hydrogen--bond interactions. Although simplified models have been found quite useful in interpreting many experimental results, it is clear that a general, quantitative analysis requires the use of non-empirical quantum mechanical calculations. Several years ago, Coulson argued that a better understanding of the details of hydrogen-bond behaviour may be obtained by considering the total interaction in terms of four components. He introduced the following concepts: electrostatic interaction, delocalization effects, repulsive forces and dispersion forces. For a discussion of these concepts, see Ref. 6, p. 339. Similar suggestions were brought forward independently by Tsubomura [7]. With certain modifications, analogous partitionings have later been employed by several authors (Dreyfus & Pullman [8], Kollman and Allen [9], Morokuma [10], Van Duijneveldt-van de Rijdt & Van Duijneveldt [11]). The energy decomposition scheme previously used by Morokuma [10] has recently been extended to include an electron distribution analysis (Yamabe & Morokuma [12], Kitaura & Morokuma [13]). Here, an analysis is also made of where the electron redistribution takes place in the interacting systems. We will return to a detailed description of this approach in the next section. Deviating slightly from the original concepts of Coulson, the following partitioning will be adopted, in accordance with that used by the majority of authors:

(a) Electrostatic (Coulomb) energy
(b) Polarization energy
(c) Charge transfer energy
(d) Exchange energy (repulsive)
(e) Dispersion energy

The sum of these five terms should thus represent the total intermolecular interaction energy $\Delta E_{HB}$, ie. the difference between the energy of the final hydrogen-bonded system at equilibrium and the total energy of the original, isolated molecules. The electrostatic contribution (a) corresponds to the energy change that would result if the free, constitutent molecules, A and B, were somehow brought together into the relative positions in which they appear in the hydrogen-bonded complex, without deforming the original monomer charge distributions and without any electron exchange taking place. The polarisation contribution (b) corresponds to the additional energy-gain on deforming the monomer charge distributions from the previous hypothetical situation to a state more closely resembling the final hydrogen-bond situation, but without there occurring any transfer of electrons between the original constituents. The charge transfer contribution (c) represents the energy change on also allowing electron transfer between the systems.

The "delocalization" effect of Coulson corresponds to the sum of the contributions (b) and (c). The concept of "covalency" in the hydrogen bond is also related to the charge transfer effect: an accumulation of charge density in the overlap region, where it is shared by the two original monomers, constitutes the traditional view of a covalency contribution.

The dispersion contribution (e) corresponds to the attraction between the systems due to the coordinated motion, or correlation, of the electrons in A and B (London dispersion forces). All contributions discussed so far are attractive. The two interacting systems are prevented from collapse by the repulsion term (d), denoted as the exchange energy contribution. ~~This represents the effect of~~ electron exchange between A and B, and corresponds more physically to the repulsion of the two electron systems when too many electrons are located within the same volume, thus violating the Pauli exclusion principle. (Note that the classical Coulomb repulsion between equal charges (electron-electron, nucleus-nucleus) is already included as part of the electrostatic contribution.) We will return to a critical examination of the physical significance of the various energy contributions in section 4.5.

## 4.2. Energy Partitioning within MO Theory

The above energy decomposition scheme has been used by Morokuma [10] in a more extensive analysis of the hydrogen-bond energy within the ab initio SCF-MO theory. A summary of his treatment follows.

Let the antisymmetrized (HF) wave functions of the two original, isolated molecules A and B be $A\Psi^o_A$ and $A\Psi^o_B$ (here $A$ is the antisymmetrizer of the electrons). These molecules are placed in the same relative positions as in the hydrogen-bonded complex. Assuming that the geometries of the original monomers are not changed, the following functions are defined for the total interacting system:

(1) The simple unsymmetrized product $\Phi_1$:

$$\Phi_1 = A\Psi^o_A \cdot A\Psi^o_B$$

(2) The unsymmetrized product $\Phi_2$ of the wave functions for each molecule optimized individually in the potential field of the other molecule:

$$\Phi_2 = A\Psi_A \cdot A\Psi_B$$

(3) The antisymmetrized product $\Phi_3$ of the original wave functions:

$$\Phi_3 = A(\Psi^o_A \, \Psi^o_B)$$

(4) The usual SCF wave function $\Phi_4$ for the entire system:

$$\Phi_4 = A(\Psi_{AB})$$

If the complete Hamiltonian of the interacting system AB is $H$, the expectation value of the energy is calculated for each of the four wave functions as

$$E_i = \int \Phi_i^* H \Phi_i \, d\tau.$$

If the sum of the energies of the original, isolated molecules is $E_o$ we may identify the various energy terms (a)-(d) with the following quantities:

(a) Electrostatic energy, $\Delta E_{els} = E_1 - E_o$

(b) Polarization energy, $\Delta E_{pol} = E_2 - E_1$

(d) Exchange energy, $\Delta E_{exc} = E_3 - E_1$

(e) Charge transfer energy, $\Delta E_{cht} = E_4 + E_1 - E_2 - E_3$

As no correlation effects were taken into account in the SCF calculations (single determinant wave functions), the dispersion energy contribution is not included in the total hydrogen-bond energy, $\Delta E_{HB} = E_4 - E_o$. The difference $E_4 - E_o$ - (a)-(b)-(d) = $E_4 + E_1 - E_2 - E_3$ would thus represent the only remaining energy contribution, due to electron transfer (c).

It has been found empirically that the results of energy partitioning vary considerably depending on the basis sets employed. The results from the calculations of Yamabe & Morokuma [12] on the water-water, formaldehyde-water and cyclopropenone-water dimer systems are given in Table 1 to give some impression of the relative order of magnitude of the different energy components. In all calculations the geometries were kept fixed to the values shown in Fig. 2.

Table 1. Hydrogen-bond energy and its components (kcal/mole) from ab initio SCF-MO-LCAO calculations with a 4-31 G basis set [12]. The geometry (fixed) is shown in Fig. 2.

| Dimer system | O···O | $\Delta E_{els}$ | $\Delta E_{pol}$ | $\Delta E_{exc}$ | $\Delta E_{cht}$ | $\Delta E_{HB}$ |
|---|---|---|---|---|---|---|
| Water-water | 2.98 Å | - 8.98 | -0.47 | 4.19 | -2.45 | -7.72 |
| Formaldehyde--water | 2.85 Å | - 9.49 | -0.77 | 6.95 | -2.78 | -6.09 |
| Cyclopropenone--water | 2.75 Å | -13.50 | -1.27 | 10.48 | -3.76 | -8.05 |

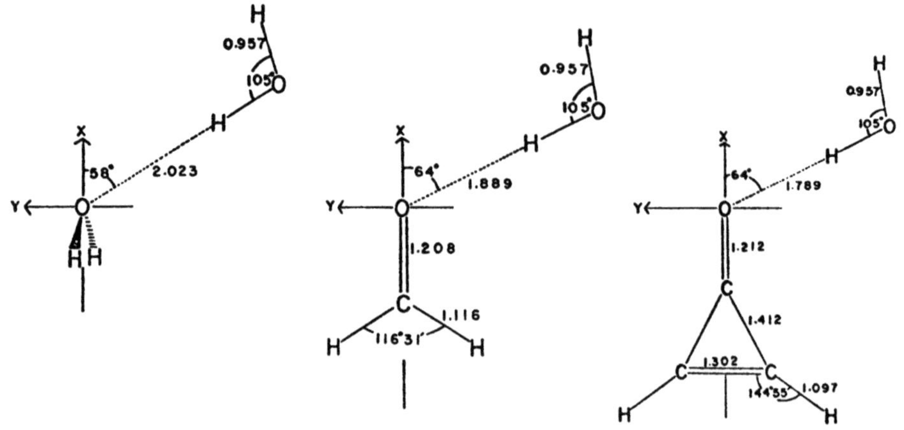


Fig. 2. Geometries used for electron density analyses of water--water, formaldehyde-water, and cyclopropenone-water dimer systems [12]. The distances are in Å units.

From recent large scale configuration interaction calculations (e.g. on the dimer of water [47] with O···O = 2.90 Å), we may estimate that the dispersion energy contribution is of the order of -1 kcal/mole.

In the case of weak and moderately strong hydrogen bonds, it is generally found that the polarization, exchange, charge transfer and dispersion contributions approximately cancel each other out (although this is only apparent for the water dimer in Table 1). Furthermore, the variation in electrostatic energy, follows the same trend as the total hydrogen-bond energy. Since each of the other contributions is less sensitive to the relative orientation of the monomers, this provides a crude basis for the empirical rule that most geometrical features of hydrogen bonding can be explained simply by electrostatic arguments. Thus, the orientation of water molecules in crystalline hydrates coincides with a minimum in the electrostatic energy [15]. The same is found to apply in many other cases [16].

It should be noted that all the decomposition schemes presently applied to hydrogen-bond interactions are based on the assumption that the geometry of the original monomers is not changed as the molecules interact. This seriously limits their usefulness in the case of strong interactions (cf. section 4.5). We should note, however, the modified procedure of Umeyama et al. [14] in which an energy decomposition analysis is made along the reaction coordinate. Since this treatment has not been applied to electron density features, it will not be discussed further in the present context.

### 4.3. Electron Distribution Analysis

As an extension of the energy decomposition analysis presented above, Yamabe & Morokuma [12] have studied the redistribution of the electron density associated with each modification of the wave functions $\Phi_1$ to $\Phi_4$.

The classical electrostatic interaction corresponding to (a) does not involve any change in the electron distribution. Accordingly, the one-electron density $\rho_1(1|1)$ associated with the wave function $\Phi_1$ represents the superposition of the electron density in the two original, undisturbed, molecules A and B (with A and B in the same relative positions as in the hydrogen-bonded complex).

The difference between the electron density $\rho_2$ $(1|1)$, associated with $\Phi_2$, and $\rho_1(1|1)$ clearly represents the electron redistribution due to polarization:

$$\rho_{pol}\ (1|1) = \rho_2\ (1|1) - \rho_1\ (1|1)$$

Analogously, the difference between the electron density $\rho_3(1|1)$ associated with $\Phi_3$, and $\rho_1(1|1)$ corresponds to the exchange effect:

$$\rho_{exc}\ (1|1) = \rho_3(1|1) - \rho_1(1|1)$$

The difference $\rho_4(1|1) - \rho_1(1|1)$ represents the total redistribution of the electron density due to hydrogen bonding. Referring to the previous comments concerning the dispersion interaction, we find by analogy with the expression for the energy, that the charge transfer effect corresponds to:

$$\rho_{cht}(1|1) = \rho_4(1|1) + \rho_1(1|1) - \rho_2(1|1) - \rho_3(1|1)$$

where $\rho_4(1|1)$ is the electron density associated with $\Phi_4$. Some approximations inherent in these expressions will be discussed in section 4.5.

The above partioning scheme will be illustrated for the water--water and formaldehyde-water dimers in section 10.2.

### 4.4. Perturbation Treatment of Hydrogen Bonds

The wave functions of the isolated monomers are also generally used as a starting point in the perturbation treatment of hydrogen-

-bond interactions. Partitioning schemes analogous to those used above have then been utilized to analyze the various contributions to the energy (cf. Van Duijneveldt-Van de Rijdt & Van Duinjeveldt [11]). The reader is referred to Schuster (ref. 1, p. 39) for a review of recent calculations. Qualitatively, the individual energy contributions are about the same in the MO and perturbation treatments, but the actual numbers differ considerably due to different model assumptions or basis sets.

### 4.5. General Comments

The above energy and charge decomposition schemes may be attractive at first sight. More detailed considerations reveal several difficulties, however:

The partitioning of the total hydrogen-bond interaction into various contributions is rather artificial; the definition of the individual components depends somewhat on the mathematical basis for the analysis. The detailed interpretation of the different concepts will also differ according to the various approximations made in, for example, current perturbation or MO treatments. Even the names of the different components may easily lead to misunderstanding. The term "electrostatic" or "Coulomb" energy appears particularly unfortunate since, strictly speaking, all the interactions discussed above are intrinsically electrostatic in nature: the only terms in the Hamiltonian considered here involve electron-nucleus, electron-electron or nucleus-nucleus interactions (spin-orbit, spin-spin and other magnetic effects are neglected).

It is conceptually difficult to partition the electron distribution when the systems A and B interact strongly. This is necessary in order to distinguish between polarization and charge transfer effects (or to estimate the degree of "covalency"). Analogously, serious difficulties also arise in attempting a step-wise partition of this type for the electron redistribution on hydrogen-bond formation.

The underlying reason for the above difficulties is clearly that the concepts are not based on physically observable quantities but on models which are often not clearly defined.

An approximation inherent in the above treatment is that the geometry (nuclear arrangement) of the isolated monomers A and B is unchanged at the different stages of the calculation. Only the electrons are thus allowed to redistribute, whereas the geometry should be modified successively under the influence of each separate interaction component. It is thus not possible to define a reference system in an altogether satisfactory way in calculating individual energy or charge contributions (see also the discussion of reference states in section 6.2). The assumption of constant geometry may be a reasonable one, however, for the medium or weak hydrogen bonding present in the actual calculations made [12] (O···O distances 2.75--2.98 Å for the cyclopropenone-water, formaldehyde-water and water--water dimer systems). This would appear to be too crude an approximation for strong hydrogen bonds. It is well known that hydrogen bonding affects the geometry of the entire participating molecule, in particular the X-H distance of the hydrogen-donor molecule. In the case of strong O-H···O bonds, the O-H distance may change by up to 0.25 Å. Furthermore, this change is concentrated to just that region where the electron density redistribution is of particular interest.

To summarize, then: a decomposition scheme as described above should be restricted to medium and weak hydrogen bond interactions. In the case of very strong hydrogen bonds, it is no longer possible to apply such energy or charge partitioning without introducing additional arbitrary assumptions. The modification to the scheme suggested by Umeyama et al. [14] and discussed in section 4.2. should be noted, however.

## 5. MODELS FOR THE HYDROGEN-BOND INTERACTION

Many simplified models have been suggested to describe the various phenomena associated with hydrogen bonding. Such models are often useful for qualitative comparisons of experimental results, or for prediction purposes in related systems. Even here, their range of application is limited, however.

### 5.1. Simple Point Charge Model

Early models tried to simulate the charge density in terms of fixed point charges assigned to the atoms of the individual monomers. The Coulomb energy was then computed for given relative positions corresponding to the hydrogen-bonded complex. Using this simple point charge model, an attractive electrostatic energy will usually be obtained provided the different charges have been set to reasonable values. The model correctly predicts hydrogen bonds involving charged monomers ($A^{+}\cdots B$ or $A\cdots B^{-}$) to be stronger than those between neutral molecules. It also favours linear or approximately linear hydrogen bonds.

Alternatively, the charge distribution within the monomers has been approximated by dipoles (and higher multipoles). In empirical calculations of molecular arrangement (*e.g.* the conformation of peptides and other biomolecules), the electrostatic interactions between the functional groups have been represented both as Coulomb forces between point charges at the atomic positions and as dipole-

-dipole interactions. As the dipole model does not give a good representation of the interaction at distances of the same order of magnitude as the dipole charge separation, the point charge model would appear to be preferable and simpler to use. One of the most serious problems remains, however: to assign suitable values for the partial charges. It is well known that the Mulliken population analysis gives net atomic charges which are strongly dependent on the basis sets used. This will later be illustrated for the water dimer. If a point charge model is used, it would be preferable to derive the partial charges directly from the observed charge density in some model compounds by suitable partitioning of the electron distribution between the atoms concerned. Several different partitioning schemes have been suggested. See, for example, the review of Smith [17]. On the other hand, if one is only interested in deriving the relative arrangement of the molecules, the actual values of the partial charges are not always crucial. This has been shown, for example, in Baur's method for calculating the orientation of water molecules in crystalline hydrates [15].

### 5.2. Simple Polarization Model

It is expected that only crude qualitative conclusions can be drawn from the model discussed in 5.1. As a natural extension of the simple point-charge model, it is reasonable to allow for charge redistributions within each of the constituent monomers as a result of the mutual electrostatic interaction. Such a polarization model is able to account for many of the features characteristic of hydrogen-bonded systems, e.g. IR intensity enhancement and the frequency shift of the X-H stretching band, NMR chemical shifts etc. Since polarization is a long-range phenomenon, this model again has its widest applications to weak and moderately strong hydrogen bonds ($-\Delta H$ = 5-10 kcal/mole).

### 5.3. Extended Point Charge Models

The above models have the advantage of permitting a discussion of hydrogen-bond parameters in terms of properties of the unperturbed constituent molecules. For the simple point charge model it is sufficient to have either an estimate of the formal charges on each atom or the charges and dipole moments of the molecules. The application of the polarization model requires, in addition, the polarizabilities of the monomers.

The simple point charge model can be further improved by introducing charges outside the nuclear positions to simulate lone pairs and other electronic features. Considerably extended models of this type have recently had some success in the calculation of molecular electrostatic potentials. In the case of the water molecule, the potential outside the molecular van der Waals surface was reproduced to an accuracy of about 1% by the use of 27 point charges (Tait & Hall [18]); only 13 point charges are sufficient to give a reasonable representation, however (Bonaccorsi et al. [16]).

Naturally, the point charge models can be improved to any desired level of accuracy by adding more charges around the nuclei to reproduce the total one-electron density of the individual monomers. However, such refined models are mainly useful for computational purposes and have lost the convenience of the original models for qualitative description and prediction purposes.

The approximations inherent in the models discussed above are not valid for the case of strong hydrogen bonds ($-\Delta H > 10$ kcal/mole):

It is no longer possible to make a quantitative analysis of the system in terms of monomer properties, so that models involving the complete hydrogen-bonded system must be used. This corresponds to the breakdown of the energy partitioning scheme discussed earlier.

### 5.4. Empirical Potential Energy Functions

Another class of models involves the introduction of analytical functions to approximate the total interaction energy. The parameters in these expressions are determined empirically in such a way that calculated quantities such as energies, structures and dipole moments, agree as closely as possible with the corresponding experimental results for several reference compounds. This empirical approach has been particularly popular for the calculation of the conformation and packing modes of large molecules containing various functional groups, e.g. biomolecules. Accurate quantum-mechanical calculations on such systems are clearly far beyond our present computational facilities.

The total interaction between two functional groups is then generally represented by the sum of several potential energy functions. The analytical form of each individual function is chosen to approximate to an energy contribution such as repulsion or electrostatic interaction as earlier described, or to express more directly the energy change when stretching the hydrogen bond, etc. An example of this approach is the Lippincott-Schroeder one-dimensional model [19] in which two of the terms resemble Morse functions for an X-H and H···Y bond, respectively, and the other two represent Coulomb attraction and repulsion contributions.

In an empirical model of this type it is not possible to identify each term with the earlier interaction components. The Morse potential contains, for example, contributions due to repulsion forces. In accordance with the above model, one often assumes that the

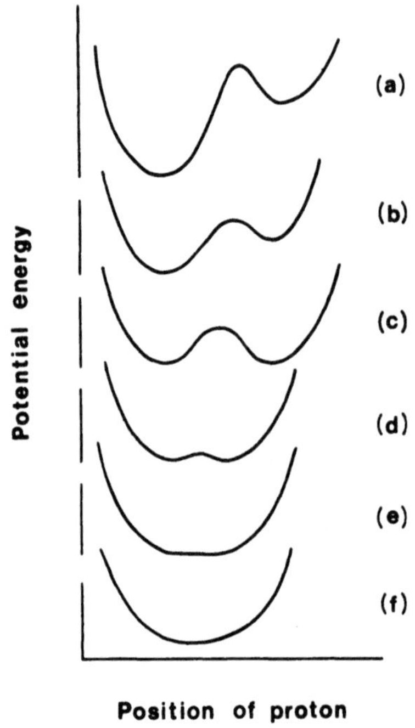


Fig. 3. Schematic illustration of potential functions assumed to be characteristic of hydrogen bonds of different lengths.

potential energy for hydrogen bonds of different strengths has the general form illustrated in Fig. 3.

In some recent calculations of the crystal packing modes of various amides, Hagler et al. [20] have employed simple classical expressions which are more closely related to earlier concepts. It has been found that the hydrogen-bond potential energy function can be represented quite effectively by an electrostatic potential based on the simple point charge model plus a Lennard-Jones 6-12 potential. The total hydrogen-bond interaction energy will thus be given by:

$$E_{HB} = \sum_{i} \sum_{j>i} \left( \frac{Q_i Q_j e^2}{R_{ij}} - \frac{A_{ij}}{R_{ij}^6} + \frac{B_{ij}}{R_{ij}^{12}} \right) \qquad \text{eq. 5.4}$$

The terms here thus correspond to electrostatic, dispersion and repulsion components, respectively. They found no significant contribution from "covalent bonding". Such a contribution is inherent in the Lippincott-Schroeder model in the term approximating to a Morse functional dependence of the H···O interaction. This result is perhaps not so unexpected since all hydrogen bonds considered here are rather weak (O···O and N···O > 2.8 Å).

## 6. CHARGE DENSITY OF THE ISOLATED MOLECULES

The relative arrangement of the molecules in a crystal is directly dependent on the charge distribution of the individual molecules. The interactions between the molecules as they crystallize will only result in small redistributions of these charges, and it will therefore be convenient to discuss first the electronic features of the monomers before proceeding to larger complexes.

### 6.1. Total Electron Density and the Overall Shape of Molecules

In the absence of a specific interaction between functional groups, the relative arrangement of the molecules is mainly determined by the overall shape and size of the molecules. This general form is expressed empirically by the molecular van der Waals surface, which gives the effective radius for the nonbonded interaction in various directions. The form should be directly related to the outer contours of the total molecular electron density. It should be noticed that the overall shape of the molecule at this level is very similar to that obtained by the superposition of the constituent spherical atoms. This is equally true for molecules containing functional groups, as demonstrated by ab initio MO-calculations for N-methylacetamide (AcNHMe), acetic acid, diketopiperazine and N-acetyl-N'-methylalanine (Hagler & Lapiccirella [21]). The two electron density maps for AcNHMe are compared in Fig. 4. The contour line C (0.0268 e/Å$^3$) corresponds approximately to the molecular van der Waals surface. As might be expected, the electron density is slightly contracted as the molecule is formed. However, the overall picture is very similar in the two cases; features associated with the formation of the molecule are hardly noticeable at this level. A quantitative estimate of the distortion of the spherical electron density on molecular formation was also made. For proposes of the later discussion of the role of lone pairs, it is particularly interesting to study the total electron density in various directions around the oxygen atom of C=O. In the case of AcNHMe, the contour of 0.0268 electrons/Å$^3$ is 1.52 Å from the oxygen centre at an angle of ~120° relative to the C=O direction (~lone pair direction), and 1.47 Å from the centre at an angle of 180°.

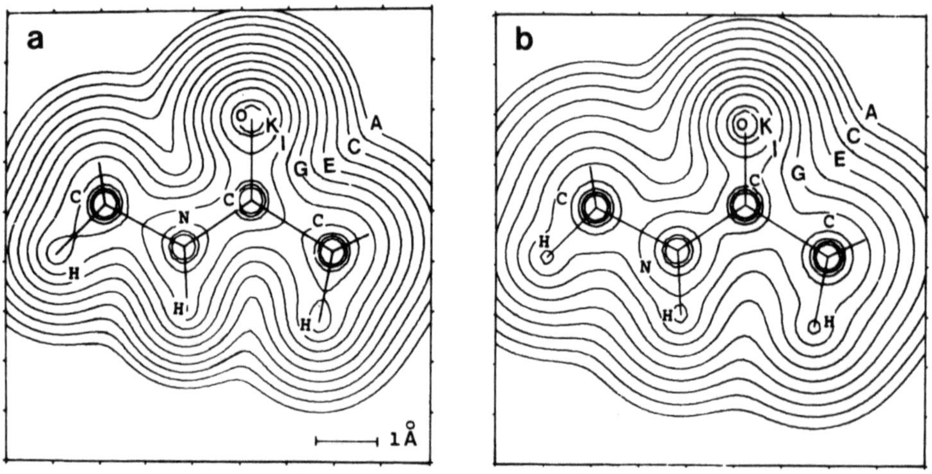


Fig. 4. Contour maps of the electron density in the amide plane of N-methylacetamide as calculated with an extended 6-31 G basis set [21]. (a): Total density from molecular wave function. (b): Total density from superposition of spherical atoms. Contour lines: A = 0.0067, B = 0.0134, C = 0.0268··· L = 13.4 e/Å$^3$.

In empirical calculations of molecular packing modes, non--bonded interactions have commonly been represented by atom-atom potentials (involved in the second and third terms in eq. 5.4. above). This implicitly assumes that the molecular van der Waals surface may be approximated by the superposition of spherical atoms. The close similarity between the shape of the electron density for the molecule and for the superposed spherical atoms (Fig. 4) indicates that this representation is a reasonable approximation [21].

It would also seem appropriate to illustrate the total electron density of the water molecule in the present context. The contour surfaces corresponding to five different electron densities are illustrated in Fig. 5 (Van Waser & Absar [22]). It is also interesting to compare the total density of the water molecule with that obtained by the superposition of spherical atoms in the same way as was done earlier for N-methylacetamide. The contour maps in three different planes are shown in Fig. 6 (Smith [23]). Particular attention should be paid to the bottom figures which illustrate the density in the plane of the lone pairs.

The previous illustrations of the total electron distribution in the lone pair region should serve as a reminder with reference to certain popular views. The "lone pair" density, sometimes imagined

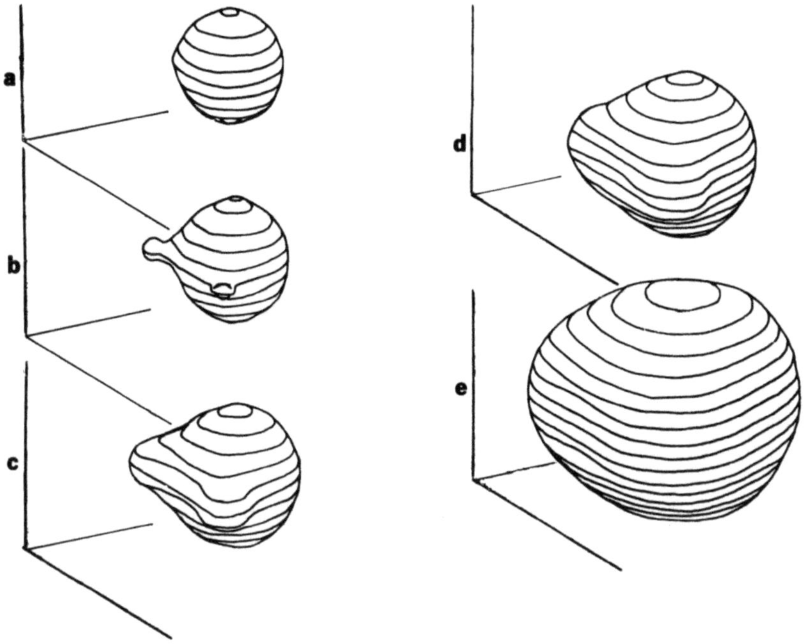


Fig. 5. Three-dimensional shape plots of the electron density of the free water molecule [22]. The contour surfaces correspond to 2.0, 1.7, 1.0, 0.6 and 0.07 e/Å$^3$, respectively.

as rabbit ears sticking out in the tetrahedral directions from the water molecule, corresponds to only part of the contribution to the total electron density in these regions (the electron density associated with the lone pair orbitals). If the electron densities associated with the bond orbitals are also added, the total density in this region will be very close to spherical, as illustrated. The small deviations from sphericity that do, in fact, exist will be discussed in some detail later.

### 6.2. Deformation Density and Reference States

Even if the deviation of the total molecular electron density from the superposition of spherical atomic densities is normally relatively small, it may have important consequences for the relative arrangement of the molecules as they approach each other. This charge migration on molecular formation is thus the only factor responsible for the polarity of the different functional groups of a molecule. It is well known that the specific interactions between the functional groups may result in a crystalline arrangement which is very

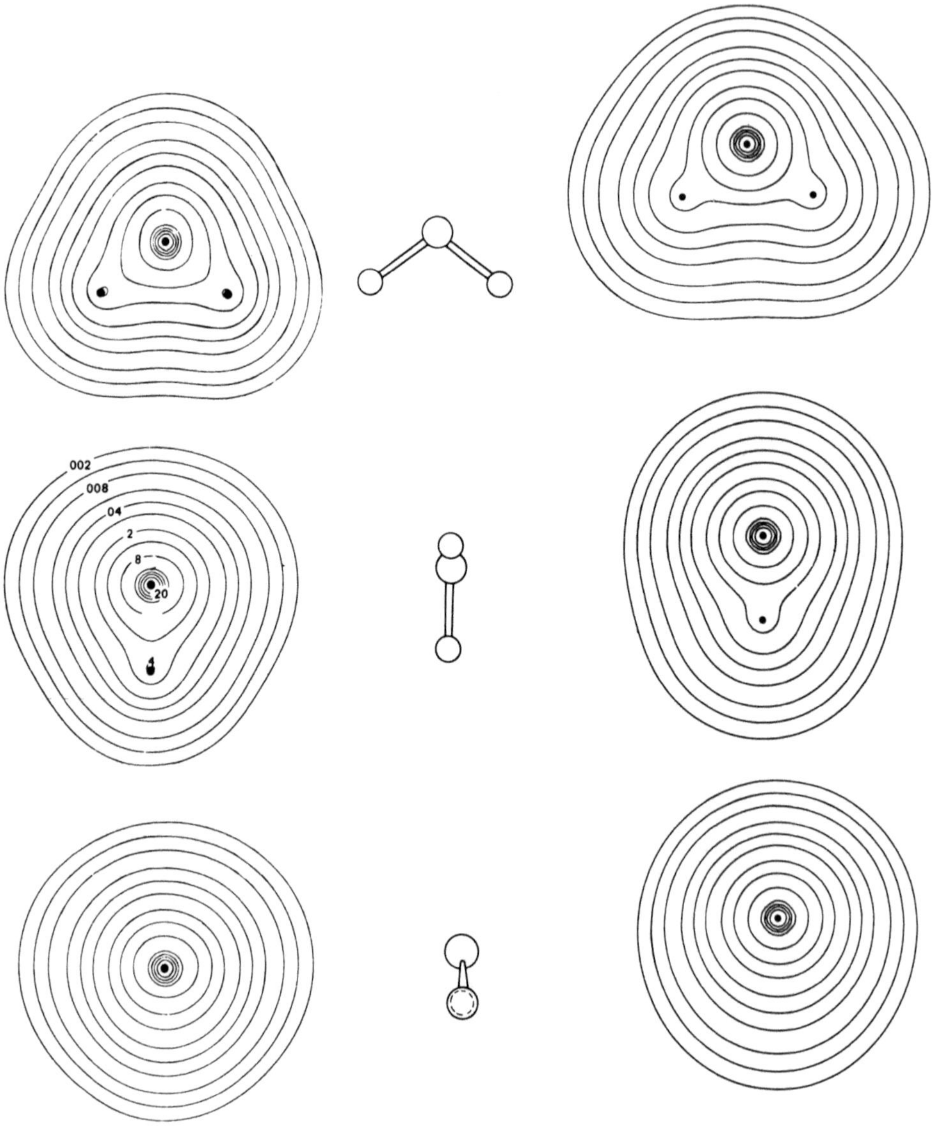


Fig. 6. Contour maps of the electron density in three different planes of the free water molecule as calculated with a DZP+ basis [23].
Left: Total density from molecular wave function.
Right: Total density from superposition of spherical atoms.
Top: In the H-O-H plane
Middle: ⊥H-O-H plane, containing O-H
Bottom: ⊥H-O-H plane, bisecting angle H-O-H

different from that expected from simple close-packing principles. The water molecule is the most spectacular example of this effect.

The total density map is clearly unsuitable to illustrate the finer details of the molecular electron distribution responsible for specific intermolecular interactions such as hydrogen bonding. It was earlier demonstrated (Figs. 4 and 6) that even the electron density change as a molecule is formed from its constituent atoms is barely noticeable in the total density maps. Nevertheless, these changes are in general considerably larger than those due to intermolecular interactions in crystals.

To study the details of the electron distribution in individual molecules, the current procedure is to form the difference

$$\Delta\rho(\underline{r}) = \rho(\underline{r}) - \rho_{ref}(\underline{r})$$

where $\rho(\underline{r})$ is the electron distribution in the actual molecule, and $\rho_{ref}(\underline{r})$ is the corresponding distribution in a suitable reference state. In principle, the choice of reference state could depend on the purpose of the investigation. To study the electron redistribution on successive substitutions in a benzene molecule, for example, it would seem natural to choose the free unsubstituted benzene molecule as a reference. Such a procedure is only possible in an approximate way, however, since the geometry of the benzene ring will also change slightly on making these substitutions. Furthermore, it is generally not so useful in practice to employ reference systems which are too specific to the system under investigation. Obvious difficulties can arise in comparing $\Delta\rho$-maps from different sources. The currently most commonly used reference state, the "promolecule" state, does not suffer from this complication. The reference state is here defined as the superposition of the spherical electron densities of the free constituent atoms arranged as they occur in the molecule. The free atom densities are then derived from quantum mechanical calculations. In this connection it is instructive to illustrate the "deformation" density $\Delta\rho$ for free N-methylacetamide and the free water molecule, Figs. 7 and 8. The amount of charge migrating into the different bonding and lone-pair regions on molecular formation is quite small, about 0.05-0.25 electrons. Particular attention should be paid to the appearance of the density in the lone-pair regions. It should be remembered, however, that these $\Delta\rho$ maps do not represent the total "valence" or "lone-pair" density but only the excess - or deficiency - relative the superposed spherical reference densities; see also section 6.1.

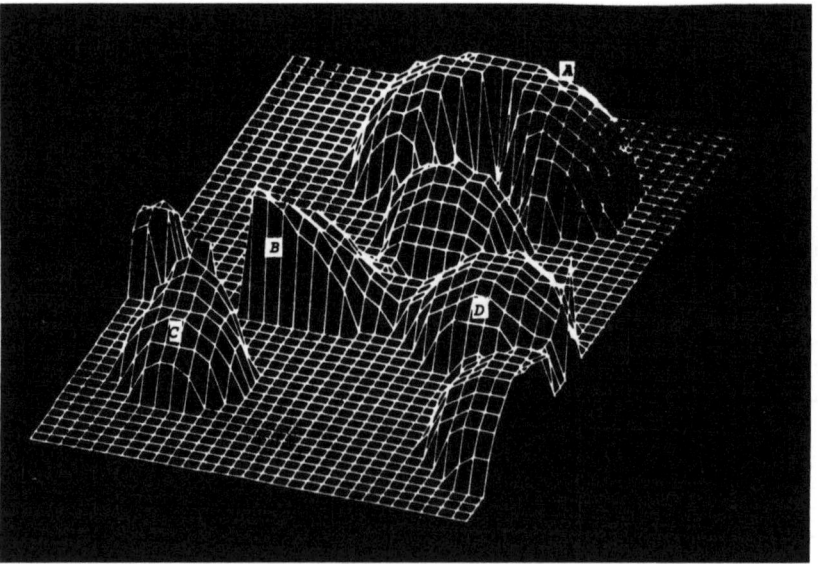

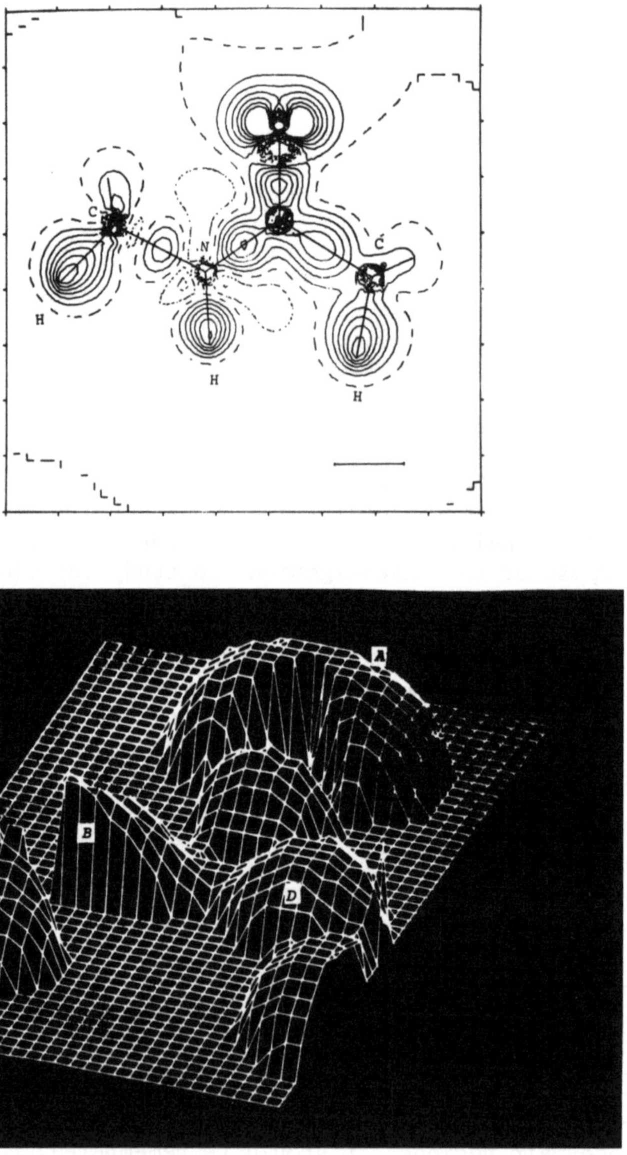


Fig. 7. Deformation density maps of N-methylacetamide in the plane of the amide group, as calculated with an extended 6-31 G basis set [21] (cf. Fig. 4). In the three dimensional shape plot the 0.1 e/$Å^3$ contour is illustrated; the lone pair region of oxygen is denoted A, the C-N bond density B, the N-H density C and the C-C' density D.

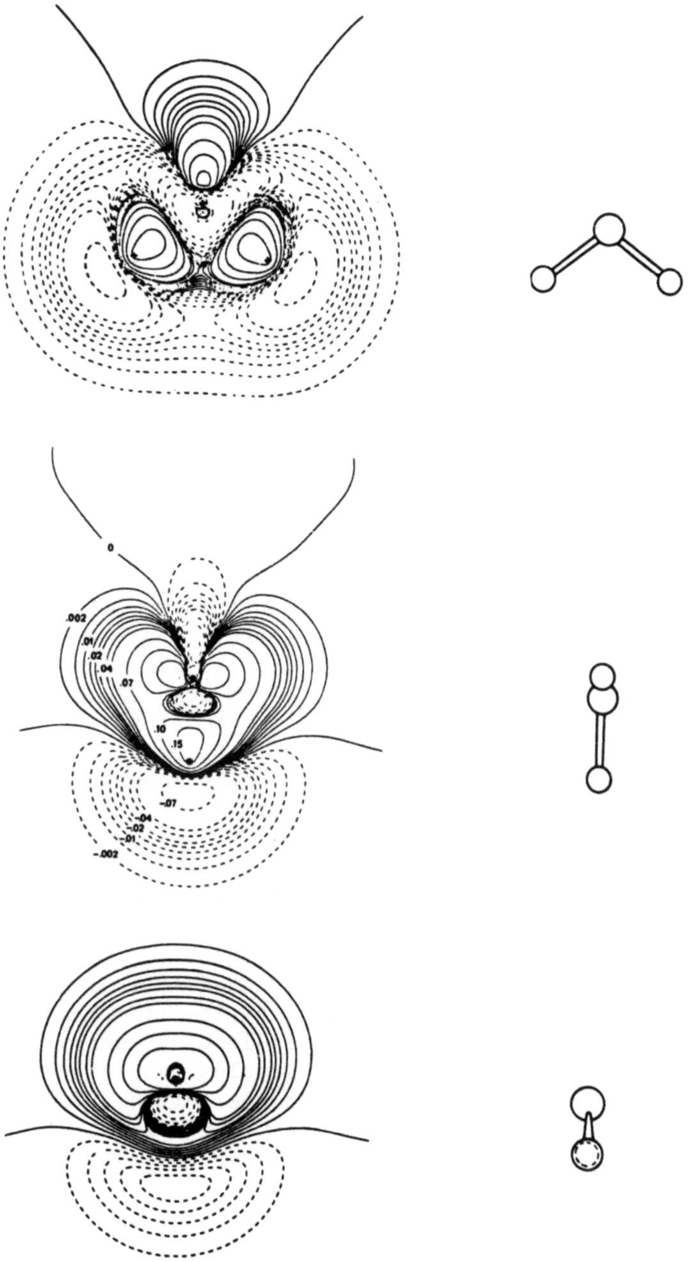


Fig. 8. Deformation density maps of the free water molecule as calculated with a DZP+ basis [23] (cf. Fig. 6).

## 7. CHARGE DENSITY IN HYDROGEN-BONDED COMPLEXES AND CRYSTALS: DEFORMATION DENSITY AND REFERENCE STATES

In an attempt to isolate the effect of intermolecular forces such as hydrogen bonding on the electron density of the molecules involved we might consider a procedure similar to that discussed above for substituted benzene molecules. We would then choose $\rho_{ref}(\underline{r}) = \rho_A^o(\underline{r}) + \rho_B^o(\underline{r})$, i.e. a superposition of the electron densities of the unperturbed monomers placed in the same relative positions as in the hydrogen-bonded complex. We have to rely here on accurate quantum-mechanical calculations of the charge densities for the free molecules. The accuracy needed in this context is currently attainable only for simple molecules, however. The objections raised earlier concerning the reference state may be raised again here: the nuclear arrangement in the constituent molecules is not, in practice, exactly the same in the free state and in the hydrogen-bonded complex; this is approximately true only in the case of weak hydrogen-bond interactions. Furthermore, it is likely that different references are chosen depending on the immediate purpose of the investigation.

To avoid this difference in nuclear arrangement for $\rho(\underline{r})$ and $\rho_{ref}(\underline{r})$, we could instead choose a reference state corresponding to charge density for the isolated molecules with the same (distorted) geometry as in the hydrogen-bonded complex. The difference density in this case does not reflect the total redistribution due to hydrogen bonding, however, as part of the redistribution has already been included automatically in calculating the electron densities of the <u>distorted</u> reference molecules.

So far we have only considered the interaction between two molecules. In a crystal, the charge density of the constituent molecules will be further distorted due to the influence of the total crystalline environment, the so-called crystal forces. An experimental deformation density with the isolated molecules as reference state would clearly include the effect of specific hydrogen-bond interactions between neighbouring molecules, as well as less specific interactions with more distant molecules (strictly speaking, we cannot differentiate between these interactions, but a certain verbal distinction is perhaps useful). The difficulties experienced in defining the reference state (a "procrystal" state) have already been discussed above. Furthermore, the electron distribution of the reference state cannot generally be determined with sufficient accuracy by ab initio quantum mechanical methods. In view of these difficulties, deformation maps referred to the "promolecule" state (superposition of independent spherical atoms) are, in general, the only real alternative in charge density studies by diffraction methods. However, it is clearly no simple matter to separate the different effects in such deformation maps as these represent the total redistribution due to molecular formation from the constituent atoms, hydrogen-bond interactions and crystal forces. If the primary purpose were to determine the redistribution associated with the formation of the molecules from the independent atoms, the effects due to the crystalline environment may be neglected to a first approximation. If the much smaller effects associated with hydrogen bonding are also to be studied, it is clearly more important to take the crystal forces into account. In deformation density maps currently being published, the effect of the crystalline environment is hardly ever considered.

To summarize, it would appear that a direct identification of the effects of crystal forces and hydrogen bonding is possible at present in only very simple systems involving weak interactions between the molecules, where the distorsion of the free monomer geometries may be neglected. In the energy and charge decomposition schemes earlier discussed, we recall that the quantum-mechanical expressions were also based on the assumption of fixed geometries for the constituent molecules.

## 8. THE RELATIVE EFFECTS OF HYDROGEN BONDING AND CRYSTAL FORCES ON GEOMETRY AND INTERACTION ENERGY

In order to distinguish the electron redistribution brought about by hydrogen bonding between adjacent molecules from that caused by the whole crystalline arrangement (crystal forces), we would need investigations of the same molecule as a free monomer, as an isolated hydrogen-bonded complex and in the crystal. However, such a wealth of experimental or theoretical data is not available for any single compound, at least not to an accuracy sufficient for the present purposes. Electron density data for the isolated monomer and the hydrogen-bonded complex are clearly only available from theoretical calculations.

Formic and acetic acid will be discussed in some detail below to illustrate the relative influence of hydrogen bonding and crystal forces on <u>geometry</u> and <u>interaction energy</u>. The electron distribution from ab initio calculations is not available for these compounds, but we will assume that there is a direct correlation between geometry/energy effects and electronic redistribution. The general conclusions drawn from these compounds will later be utilized in the discussion of electron redistribution in other systems, where no direct information exists concerning the relative importance of hydrogen bonding and crystal forces.

The gas-phase geometry of formic and acetic acid has been studied experimentally both in their monomeric and dimeric forms. In the crystalline state, these molecules are hydrogen bonded to form planar zig-zag chains, whereas propionic acid and higher homologues form cyclic dimers. A crystalline phase of pure acetic acid containing dimers has so far not been found. However, the 1:1 addition compound $H_3PO_4 \cdot HAc$ contains such dimers where, interestingly enough, the two monomers are crystallographically independent. Dimerization and crystallization induces significant changes in the monomer geometries; some data are given in Table 2.

It appears that the major changes are concentrated to the C-O and O-H bond lengths, whereas the C=O bond stays remarkably constant. It is also apparent that the major changes occur already as the monomers form the hydrogen-bonded dimer, whereas the further modification on crystallization is considerably less significant. This is a clear indication that a hydrogen-bond interaction between adjacent neighbours plays a more important role than the interaction with the remaining crystalline arrangement. This is well illustrated in a comparison of the acetic acid dimers in the gas phase and in the crystal of $H_3PO_4 \cdot HAc$.

It is also interesting to compare the hydrogen-bond distances O···O within the dimers and the chains. For formic acid, this distance is 2.70 and 2.63 Å, respectively. The discrepancy is similar for acetic acid, 2.68 and 2.63 Å. On the other hand, the distance is exactly the same, 2.68 Å, for the acetic acid dimer in the gas phase and in the $H_3PO_4 \cdot HAc$ crystal. This suggests that the significant difference between the dimers and the chains is not simply crystallization effect but is due rather to the difference in arrangement. The hydrogen bonds in the dimer are only affected indirectly as these dimers pack together in the crystal. In the chain structure, however, the hydrogen bonds within the zig-zag chain will influence each other much more directly. It thus appears likely that the shorter hydrogen-bond distance in the chain structure is a cooperative effect. This phenomenon will be further discussed below for the water molecule.

The previous qualitative discussion of the relative influence of nearest neighbours and the whole crystalline arrangement is in complete agreement with the electrostatic energy calculations made on formic acid by Smit et al.[31] First, ab initio MO-LCAO-SCF calculations were made on the isolated monomer for three different geometries, corresponding to that in the free monomer (M), in the free dimer (D) and in the chain of the crystal (C). The energies obtained with the largest basis set (DZP) are given in Table 2. The molecular distortion energies thus amount to: $E_D - E_M = 3.0$, $E_C - E_M = 3.4$ kcal/mole. Unfortunately, the old C=O and C-O values, 1.23 and 1.26 Å, were used for the monomer in the crystal, instead of the more recent values 1.221 and 1.311 Å. Accordingly, the distortion energy, $E_C - E_M$, may be somewhat overestimated. Furthermore, the O-H distance in the dimer geometry, 1.036 Å, appears too long: a value around 1.00 Å would seem more reasonable in view of carboxyl geometries in other studies (e.g. the acetic acid dimer in the

Table 2. Geometry of formic and acetic acid in the free monomer, dimer and crystal.

| | C=O(Å) | C-O(Å) | O-H(Å) | O-H···O(Å) | Monomer energy (a.u.)[g] |
|---|---|---|---|---|---|
| Formic acid: | | | | | |
| Monomer (gas)[a] | 1.217(3) | 1.361(3) | 0.972(5) | - | -188.8124 |
| Dimer (gas)[b] | 1.220(3) | 1.323(3) | 1.036(17) | 2.703 | -188.8076 |
| Chain (crystal)[c] | 1.221(2) | 1.311(2) | - | 2.624(2) | - |
| | (1.23) | (1.26) | (1.010) | | -188.8070 |
| Acetic acid: | | | | | |
| Monomer (gas)[d] | 1.214(3) | 1.364(3) | | - | |
| Dimer (gas)[d] | 1.231(3) | 1.334(4) | | 2.680(10) | |
| Dimer (crystal | 1.231(3) | 1.306(3) | 0.986(7) | 2.681(4) | |
| of $H_3PO_4 \cdot HAc$)[e] | 1.229(3) | 1.297(3) | 0.992(7) | 2.688(4) | |
| Chain (crystal)[f] | 1.216(5) | 1.320(5) | 1.011(15) | 2.627(5) | |

(a) C=O and C-O from gas electron diffraction (Almenningen et al.[24]), O-H from microwave spectroscopy (Kwei & Curl[25]).

(b) Almenningen et al.[24]

(c) X-ray diffraction at 98K (Nahringbauer[26]). The values in parentheses were used in the theoretical calculations [g].

(d) Gas electron diffraction (Derissen[27]).

(e) Neutron diffraction of $CH_3COOH \cdot H_3PO_4$ (Jönsson[28]).

(f) Average of X-ray and neutron diffraction referred to 133K (Nahringbauer[29], Jönsson[30]).

(g) Energy of formic acid monomers of different geometries from ab initio LCAO-MO-SCF calculations by Smit et al.[31], using a DZP basis.

$H_3PO_4 \cdot HAc$ crystal). If this is the case, $E_D - E_M$ is also somewhat too large. These modifications will then probably not alter the general conclusion that the energy of distortion due to hydrogen-bond interaction with the adjacent neighbour is much larger than that due to interaction with the remaining crystalline arrangement.

Smit et al.[31] also made a calculation of the energy of electrostatic interaction between a formic acid molecule and the surrounding molecules at various distances in the crystal. The charge distribution of each molecule (derived from the ab initio wave functions) was then approximated either by a simple point charge model or by multipoles up to sixth order. Separate calculations were made for each of the three monomer geometries M, D and C described above (cf. Table 2). As is generally the case, more attention had to be paid to the convergence behaviour of the calculation when employing the multipole expansion compared to the point-charge model. Due to the various approximations made in the models and calculations, the results are significantly different in the two cases. Below, we will use the numbers from the multipole model, but the general conclusions are the same if the point charge model is used instead. The energy contribution from different summation areas, employing the monomer geometry C, are given in Table 3. Since the electrostatic energy for a crystal is defined as half the sum of the pair interactions, the interaction with only half of the neighbours within each range of $R_{AB}$ around the reference molecule are included in the table; the contribution from the other half is equivalent by symmetry. Thus, within the radius R≤3.64 Å, there are six neighbours, but the energy -20.41 kcal/mole represents only the interaction with one of the hydrogen-bonded neighbours in the chain and with two other molecules in adjacent chains.

Table 3. Energy of electrostatic interaction between a formic acid molecule and surrounding molecules at different distances in the crystal. Molecular geometry C, multipoles and net charges from electron distribution employing the DZP basis.[31]

| Range of $R_{AB}$ (Å) | Multipole sum (to order six) kcal/mole*) | Point-charge sum, kcal/mole *) |
|---|---|---|
| 0 < R ≤ 3.64 | -20.41 | -10.70 |
| 3.64 < R ≤ 6 | 0.10 | 0.30 |
| 6 < R ≤ 10 | -0.54 | -0.26 |
| 10 < R ≤ 15 | 0.05 | 0.03 |
| 15 < R ≤ 20 | 0.00 | 0.01 |
| 20 < R ≤ 25 | 0.02 | 0.02 |
| 25 < R ≤ 35 | -0.02 | -0.01 |
| 35 < R ≤ ∞ | -0.38 | -0.25 |
| Total | -21.17 | -10.86 |

*) Only half of the neighbours around the reference molecule are included; see text.

The total electrostatic energy calculated for a crystal composed of monomers M and D was 2-3 kcal/mole smaller than that calculated with geometry C (the energy difference between M and D being much larger than between D and C). This energy difference is of the right order of magnitude to account for the molecular distortion energy discussed earlier. The best estimate of the electrostatic part of the lattice energy, corrected for molecular distortion, is thus -21.2 + 3.4 = -17.8 kcal/mole. Experimentally, the lattice energy of formic acid has been estimated to be -14.6 kcal/ /mole [32].

It is evident from Table 3 that practically the whole contribution to the electrostatic energy (-20.41 out of -21.17 kcal/mole) originates from the nearest neighbours ($R_{AB} \leq 3.64$ Å). A closer inspection of their data reveals, furthermore, that the dominant contribution within this shell is due only to the hydrogen-bonded neighbour in the same zig-zag chain: -19.5 out of -20.41 kcal/mole. After correction for molecular distortion, we may then estimate that the electrostatic energy component of the hydrogen-bond interaction between two formic acid molecules is around -19.5 + 3.4 = = -16.1 kcal/mole. Clementi et al. [33] obtained a value of -8.1 kcal/ /mole per bond for the cyclic dimer of formic acid (the geometry of the monomers was fixed to values somewhat different from the more recent ones given in Table 2, however).

As discussed in section 4, the electrostatic energy component reflects quite well the relative strengths of hydrogen bonds. Together with the above results, this suggests that the interaction between a molecule and its crystal environment may be well approximated to by the electrostatic part alone of the hydrogen-bond interaction with its immediate neighbours, the effect of more distant molecules is practically negligible.

The large molecular distortion in the present compounds due to hydrogen bonding and crystal forces emphasizes the difficulties of choosing isolated molecules as reference state in calculating charge deformation maps. This is true even for the case of only moderately strong hydrogen bonding (O···O distances in the range 2.62-2.70 Å). A deformation density map calculated with the distorted monomers C as reference state will certainly not reveal the whole charge redistribution due to hydrogen bonding and crystallization, since the molecular distortion energy, 3.4 kcal/mole, is hardly negligible compared to the total electrostatic interaction energy in the crystal, -17.8 kcal/mole (in fact, the relative importance of the molecular distortion energy is even greater since the calculated value for the lattice energy, -17.8, appears too large).

## 9. MOLECULAR COMPOUNDS WITH WEAK AND MODERATELY STRONG HYDROGEN BONDS

In the following sections we shall discuss the electron density distribution in some simple hydrogen-bonded compounds and, whenever possible, compare experimental and theoretical results. For several reasons (cf. section 7) weak and moderately strong hydrogen bonds will be treated separately from strong hydrogen bonds. Due to the unique features of the water molecule, the water polymers and crystalline hydrates are discussed in a separate section. A certain overlap between these main groups cannot be avoided, however.

The emphasis in the following presentation will be laid on general principles. Only a few selected examples will thus be treated in any detail; we regret that all results in the literature cannot be included for reasons of space.

### 9.1. α-Glycine

The charge distribution in the simplest amino acid, α-glycine, has been determined both experimentally (X-N) and theoretically (ab initio LCAO-MO-SCF, DZ basis) by Almlöf et al. [34] The experimental deformation maps in two planes are shown in Fig. 9a, b. An attempt to estimate the effect of the crystalline environment on the electron density was also made in the theoretical calculations. The simple point charge model was then employed to represent the charge distribution of the individual molecules in the crystal, with net charges derived from Mulliken populations. The electrostatic potential resulting from these charges was first evaluated at various points within the van der Waals surface of the reference molecule. The resulting potential was subsequently simulated by a set of ~50 point charges around the van der Waals surface, the magnitudes of the charges being derived by a least-squares fitting procedure. The effect of the crystal field could then be calculated by incorporating these charges into the Fock operator. The deformation density thus calculated is shown in Fig. 9c (a mirror plane was imposed on the molecule to simplify the theoretical calculations). The separate contribution to the deformation density due to the crystal field is illustrated in Fig. 11; this represents the deformation density calculated for the molecule in the crystal field with the isolated molecule of the same geometry as reference state. (Since the geometry of the isolated molecule used in calculating the maps shown in Figs. 9c and 11 was that observed in the crystal, the total charge redistribution due to hydrogen bonding and crystal forces does not appear in Fig. 11; cf. sections 6, 7 and 8.)

The qualitative agreement between the experimental and theoretical deformation maps is quite good (Figs. 9a, b and 9c) considering the medium-size basis employed in the theoretical calculations and the neglect of thermal motion (it seems likely that the effects of these deficiencies tend to some extent to cancel each other out). The maps show the commonly observed features: migration of electron density to the centre of the covalent bonds and to the lone-pair regions of the oxygen atoms. The X-N map through the strongest hydrogen bond (N-H1···O1 = 2.770Å) is typical of a medium strength hydrogen bond (Fig. 10): accumulation of charge in the donor N-H bond and in the H···O bond at some distance from the acceptor oxygen atom, together with a slight electron deficiency in the H···O region close to the hydrogen atom. However, all these features are already present qualitatively in the theoretical deformation map of the isolated molecule (Fig. 12). A distribution very similar to that found in Fig. 10 would thus be obtained by simply superposing the unmodified deformation maps of the isolated molecules. This is equivalent to stating that the electrostatic model gives the essential features of the hydrogen-bond interaction in the case of medium strength hydrogen bonds. The modification which actually takes place as the molecules interact with each other in the crystal constitutes only a second-order effect, as shown in Fig. 11. Although the significance of all details in this figure may be questioned, the overall picture is very simple: the major effect from the environement is to increase the polarity of the functional groups that already exists in the isolated monomer (note the similarity between the overall features in Fig. 11 and 12). Thus, the electron shift from hydrogen to nitrogen in the donor N-H bond and to the lone-pair regions of the acceptor oxygens is further reinforced. With reference to the earlier discussion of the relative influence of hydrogen bonding and crystal forces (section 8), we can attribute this general tendency simply to hydrogen bonding to nearest neighbours.

In Fig. 10, we notice that N-H is directed towards a distinct electron concentration in the oxygen lone-pair region. This may suggest a certain directional influence of this charge distribution. However, the isolated molecule (Fig. 12) shows a more extended region of electron density over the whole non-bonded region of the oxygen atom; the features appear very similar to those in Fig. 7. Furthermore, Fig. 11 suggests that the concentration of charge in the O···H-N direction is, to some extent, a secondary effect caused by the hydrogen-bond interaction. This conclusion is supported by the appearance of the electrostatic potential of the isolated molecule, shown in Fig. 13. The very extended form of the potential minima indicates that a positive charge would experience very little directional influence due to electrostatic interaction alone (see section 11). It is interesting to compare the electron density in the lone-pair regions of O1 and O2. In the case of O1, a hydrogen bond is accepted in the same plane as the carboxylate group and the polarization due to hydrogen bonding accentuates the excess charge already present in this region in the isolated molecule. O2 on the other hand accepts two somewhat weaker hydrogen bonds from out of
Experimental and theoretical deformation densities of α-glycine.[34]

a

b

c

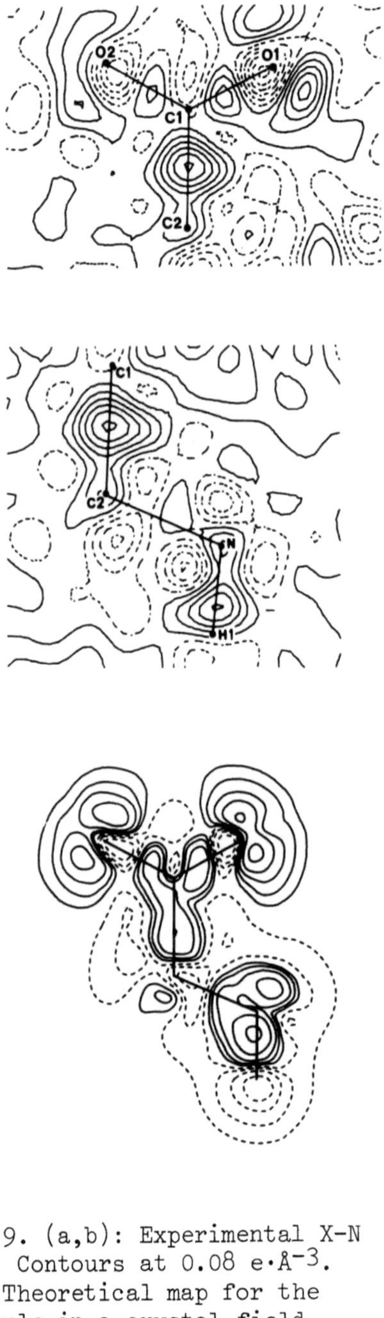


Fig. 9. (a,b): Experimental X-N maps. Contours at 0.08 $e\cdot Å^{-3}$. (c): Theoretical map for the molecule in a crystal field. Contours at 0.04, 0.10, 0.20, 0.40 and 0.80 $e\cdot Å^{-3}$.

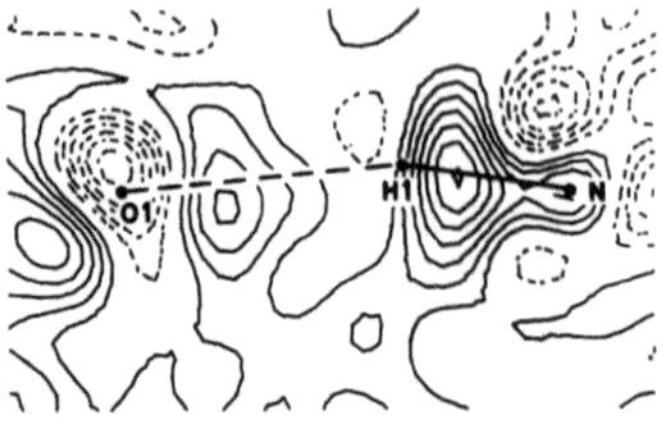


Fig. 10. X-N map through hydrogen bond N-H1···O1. Contours at 0.06 $e\cdot Å^{-3}$.

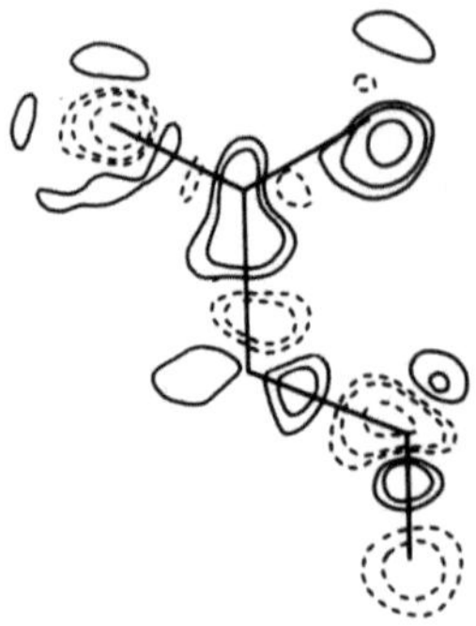

Fig. 11. Theoretical deformation density for a molecule in a crystal field relative to an isolated molecule. Contours at 0.02, 0.04 and 0.10 $e\cdot Å^{-3}$.

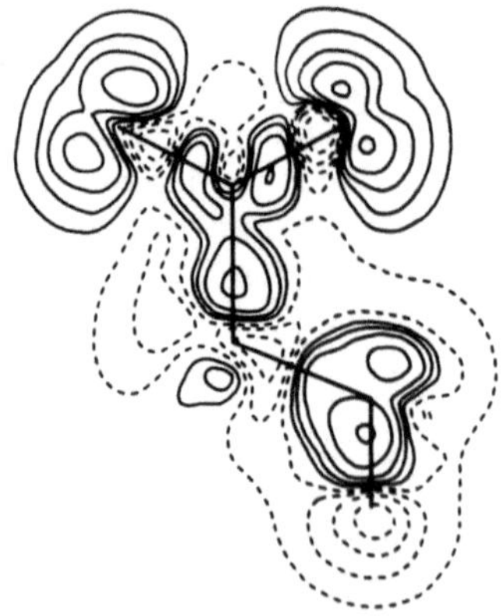

Fig. 12. Theoretical deformation density for the isolated molecule (priv.comm.). Contours as in Fig. 9c.

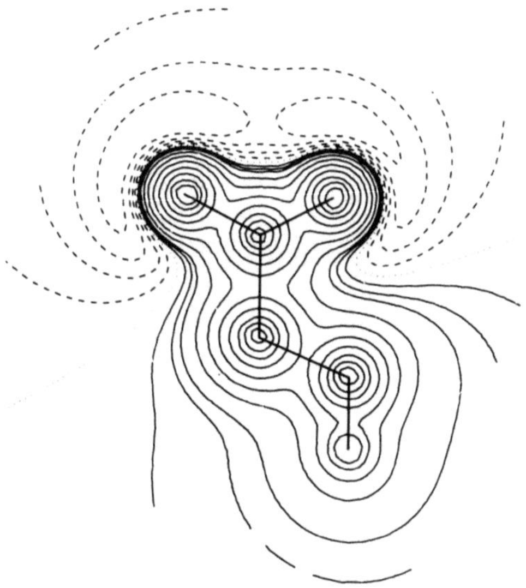

Fig. 13. Molecular electrostatic potential of α-glycine.[34] Contours at 20, 40, 80, 120, 250, 500, 1000, 2000 and 4000 kcal/mole.

this plane. The excess charge density around O2 is thus expected to be more extended.

### 9.2. Formamide

Hydrogen bonding in the cyclic as well as the linear dimer of formamide was studied by Dreyfus et al. [8,35] by ab initio LCAO-MO-SCF calculations; the largest Gaussian basis set used was (7,3|3) contracted to (2,1|1). No remarkable additional stability due to the cyclic arrangement was found: for the linear dimer $\Delta E = -7.95$ kcal/mole (at an optimized N-H···O distance of 2.85 Å) compared to -7 kcal/mole per hydrogen bond for the dimer (N-H···O distance fixed to 2.94 Å). The deformation density of the dimer relative to the monomers is shown in Fig. 14. This map thus illustrates the electron redistribution due to hydrogen bonding alone in contrast to the previous maps in Figs. 9 and 10 for α-glycine, which gave the total effect due to molecular formation from spherical atoms, hydrogen bonding and crystal forces. Assuming that the

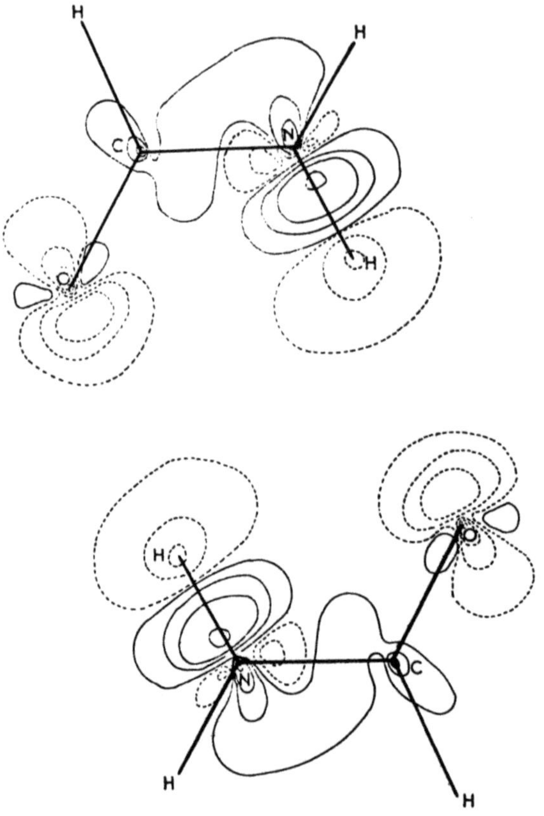


Fig. 14. Theoretical deformation density of the isolated cyclic dimer of formamide relative to the monomers.[35] Contours 0.001, 0.005, 0.01 and 0.03 e·Å$^{-3}$.

deformation of α-glycine in a crystalline environement illustrated in Fig. 11 is mainly due to hydrogen bonding, this map should be more directly comparable to Fig. 14. Focussing our attention only on the N-H···O bond, the main qualitative difference is that there is no charge accumulation in the H···O bond at some distance from the acceptor oxygen in Fig. 14. Comparing this with other accurate theoretical deformation maps with the monomers as reference (e.g. water dimer, section 10.2), it appears that such a charge accumulation is rather characteristic for hydrogen bonds in general, however. This feature is also typical for experimental densities in crystals, although it is clearly not an effect of hydrogen bonding alone, as just discussed for α-glycine. It is interesting to note that a very slight charge accumulation was actually observed when a smaller but less contracted basis set was used ((4,2|3) contracted to (2,1|2)). This may be particularly important for hydrogen.[35] It is remarkable that all the usual factors characteristic of hydrogen bonding were also present in the deformation maps calculated by Almlöf & Mårtensson [36] by the iterative extended Hückel approach for the dimers of formamide, formic and acetic acid.

The deformation density of solid formamide has recently been determined from X-ray diffraction data at 90K by Stevens ($X\text{-}X_{high}$ technique) [37]. The positional and anisotropic thermal parameters for the heavy atoms and hydrogen were then refined from data with $\sin\theta/\lambda > 0.85$ and $>0.60$ Å$^{-1}$, respectively, to reduce bias from the aspherical electron distribution. The deformation density in the least-squares plane of the molecule is shown in Fig. 15; the deviations from this plane are less than the estimated standard deviations for all atoms within the molecule. Two molecules are joined to form dimers by N-H2···O' bonds (2.948(3) Å) across a centre of symmetry. Alternate dimers are linked to form chains by hydrogen bonds N-H1···O'' (2.885(3) Å). Since the H1···O'' direction deviates considerably from the plane of the molecule (Fig. 15), the distribution in the hydrogen bonds is better shown in the plane defined by H1, H2 and O (Fig. 16). Note, however, that Fig. 15 should be consulted for the distribution in the N-H1 bond. It appears then that all features typical of medium to moderately strong hydrogen bonds are present in both hydrogen bonds: accumulation of charge in the donor N-H bond, and in the H···O bond at some distance from the acceptor oxygen atom, but a slight deficiency in the H···O region close to hydrogen. It should also be added that the true maxima around the oxygen lie above and below the molecular plane; perhaps an effect of the crystalline environment.

The corresponding theoretical deformation density of the free formamide molecule has been calculated by Stevens et al. [38] by ab initio LCAO-MO-SCF methods employing an extended Gaussian basis set (11,5,1|6,1) contracted to (4,3,1|4,1). The molecular geometry assumed was close to the final X-ray structure, with a mirror plane imposed on the molecule. The theoretical density was also thermally smeared, using the rigid body thermal parameters from the X-ray study (Fig. 17). The agreement between theory and experiment is very good, within twice the estimated experimental standard deviation in much of the molecule. A detailed comparison is made in Fig. 18, where the difference between the experimental and theoretical deformation density is plotted. As intermolecular interactions were neglected in the theoretical calculations, one might expect that some of the discrepancy was due to this omission. As discussed for α-glycine, we would expect hydrogen bonding to further accentuate

Deformation density of formamide [37,38]
(contours at 0.05 e.Å$^{-3}$)

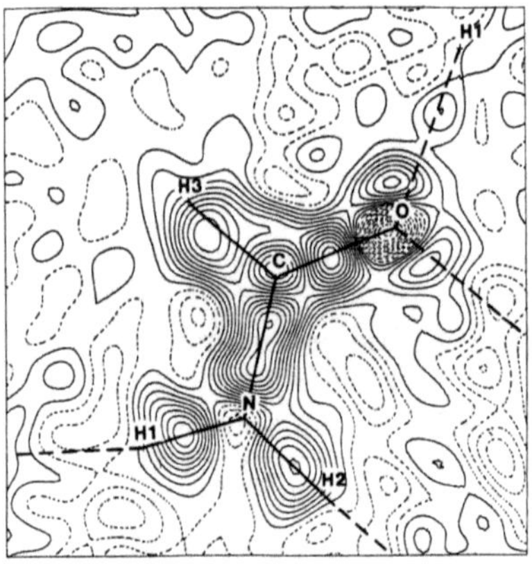


Fig. 15. Experimental X-$X_{high}$ map in the least-squares plane of the molecule.

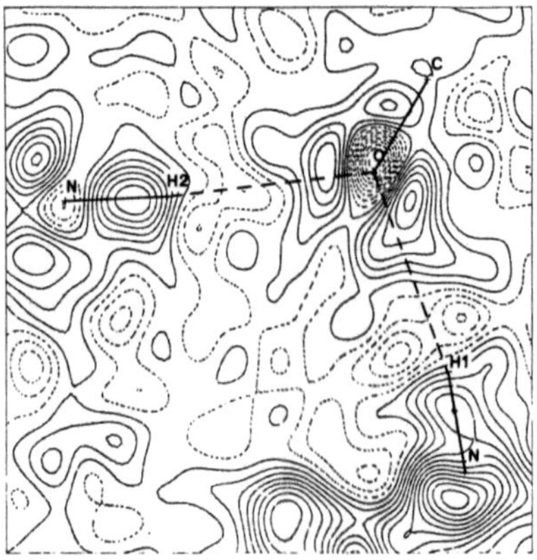


Fig. 16. Experimental X-$X_{high}$ map in the plane defined by H1, H2 and O.

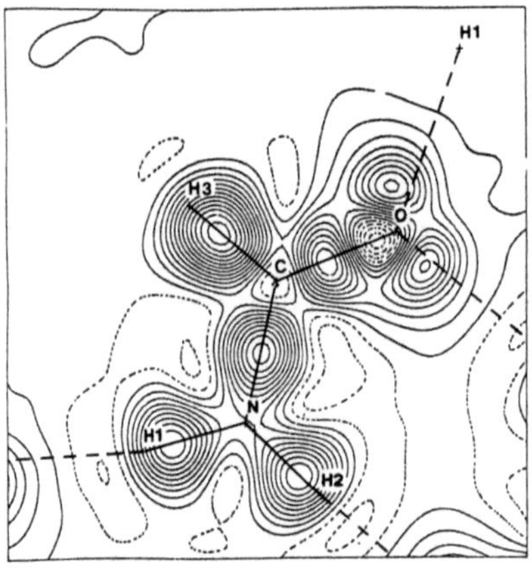


Fig. 17. Dynamic theoretical deformation density in the least-squares plane of the molecule.

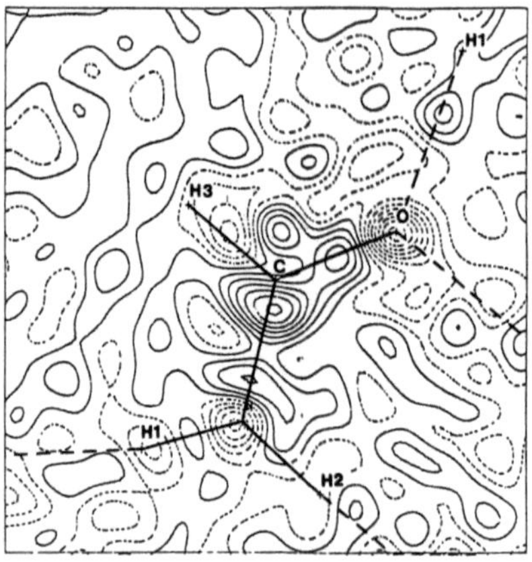


Fig. 18. Difference between the experimental and theoretical deformation densities in the molecular plane.

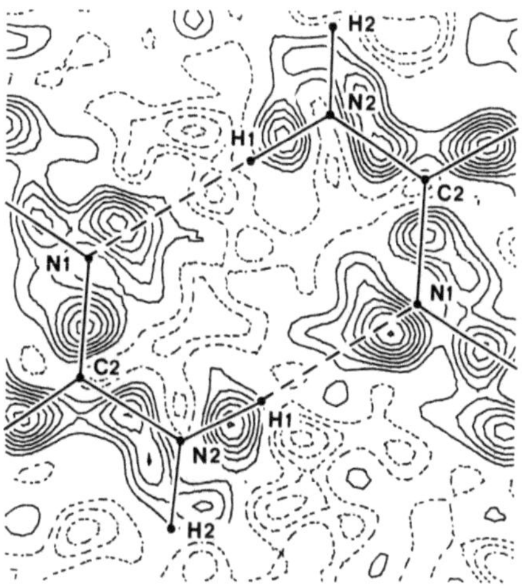


Fig. 19. Deformation density (X-N) of 2-amino-5-chloropyridine [39]. Contours of 0.06 e·Å$^{-3}$ intervals.

the polar features already existing in the isolated molecule. Such a rearrangement would result in worse agreement in the present case, so that other sources of error must be looked for.

### 9.3. 2-Amino-5-chloropyridine

The experimental deformation density of 2-amino-5-chloropyridine at 297K has recently been determined by the X-N technique, Fig. 19 (Kvick et al. [39]). Two molecules are linked to form (centrosymmetric) dimers by N2-H1···N1 hydrogen bonds of length 3.058(3) Å. Only one of the amino hydrogen atoms participates in hydrogen bonding, so that this compound can serve as an internal check of our previous conclusions concerning the redistribution effects caused by hydrogen bonding. As predicted, the migration of electrons is, in fact, more pronounced into the N2-H1 bond. The electron distribution in the N2-H1···N1 bond exhibits all the characteristics of weak hydrogen bonds.

## 9.4. Hydrogen peroxide

The experimental deformation density of hydrogen peroxide has recently been studied at 110K by the X-N technique (Savariault & Lehmann [40]). Each end of the $H_2O_2$ molecule donates one and accepts another hydrogen bond from a neighbouring molecule (O···O in both cases 2.80 Å). The deformation density in the O-H···O' bond is shown in Fig. 20, upper figure. The corresponding density in the hydrogen

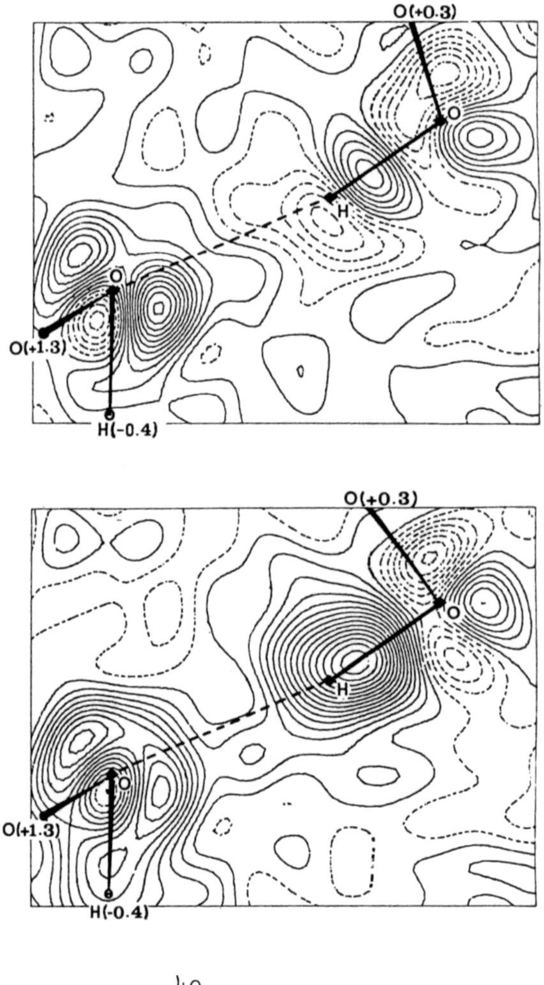


Fig. 20. Hydrogen peroxide.[40] X-N deformation density in the O-H···O bond.
Upper: Hydrogen subtracted. Lower: Hydrogen not subtracted. Contours at 0.10 e·Å$^{-3}$.

bond when hydrogen is not subtracted is also shown in Fig. 20, lower figure. As expected, the previous negative region close to H is now filled and there is, in fact, a slight electron excess in the whole H···O region.

## WATER POLYMERS AND CRYSTALLINE HYDRATES

### 10.1 General

The water molecule continues to hold its place as one of the most interesting substances in the hydrogen-bond field, constantly attracting the attention of crystallographers, spectroscopists and theoreticians alike. A water molecule can function in a crystal structure in a diversity of roles. It can act both as a donor and an acceptor of hydrogen bonds, as well as a ligand molecule, and in no case are the geometrical constraints severe. It is thus not surprising that crystalline hydrates are very common in nature. Many anhydrous ionic compounds which are difficult to prepare in a crystalline form are clearly stabilized by hydration.

Almost without exception, both hydrogen atoms of a water molecule are engaged in hydrogen bonds to neighbouring acceptors. The water molecule thus normally acts as a donor in two hydrogen bonds; on the other hand, the environment on the acceptor side can vary considerably. The coordination of water molecules is thus primarily one of two types: planar-trigonal or tetrahedral. In the former, a cation or hydrogen-bond donor is situated approximately on the two-fold axis of the water molecule. In the salt hydrates studied so far, such trigonal coordination occurs in about 50% of the cases in which a water molecule has a cation as its nearest neighbour. This coordination is almost always formed in the presence of highly charged cations (e.g. $M^{3+}$ and $M^{4+}$). In tetrahedral coordination, two cations or hydrogen-bond donors approach the water molecule approximately in the directions of the two lone-pairs. This coordination is preferred with monovalent (alkali) cations as nearest neighbours, and also in the presence of hydrogen-bond donors like NH or OH.

Besides the two main types of coordination described above, more complex situations often arise involving considerable distortion from the ideal geometrical configurations. The various geometries have sometimes been classified on the basis of the coordination of the lone pairs. However, the large fluctuation in the detail of the arrangements within each class and the diffuseness of the definition of the classes raise the question of the general usefulness of such detailed schemes. Furthermore, from the discussion in section 11, it is doubtful whether too much significance should be attached to the role of the lone-pair electrons as hydrogen-bond acceptors, at least for relatively weak hydrogen bonds.

For a review of recent X-ray and neutron studies of hydrates, see Olovsson & Jönsson [1]. A comprehensive treatise on experimental and theoretical aspects of water is given in Ref. 41.

## 10.2. Water polymers

In the molecules treated so far, the hydrogen-bond donor and acceptor groups are well separated from each other, and it is relatively easy to consider redistribution effects due to the intermolecular interactions in each group separately. The situation is much more complex, however, for the case of the water molecule since the same oxygen atom must here play the dual role of donor and acceptor. It is therefore appropriate to study first the behaviour of the water molecule in small, isolated hydrogen-bonded complexes, before proceeding to crystalline hydrates. The electron density distributions in such complexes are obviously only available from theoretical calculations.

The dimer of water has been studied in several extensive ab initio LCAO-MO-SCF calculations. The resulting equilibrium structures show good general agreement, cf. Kollman & Allen [42], Diercksen [43], Popkie et al. [44]. A review of the various calculations on water polymers is given by Rao in Ref. 41. The minimum energy occurs for a relative arrangement of the water molecules as illustrated in Fig. 21, with $\phi = 0°$ (trans form). ($\phi$ is the dihedral angle between the HOH plane of the donor molecule and the bisector plane of the acceptor molecule.)

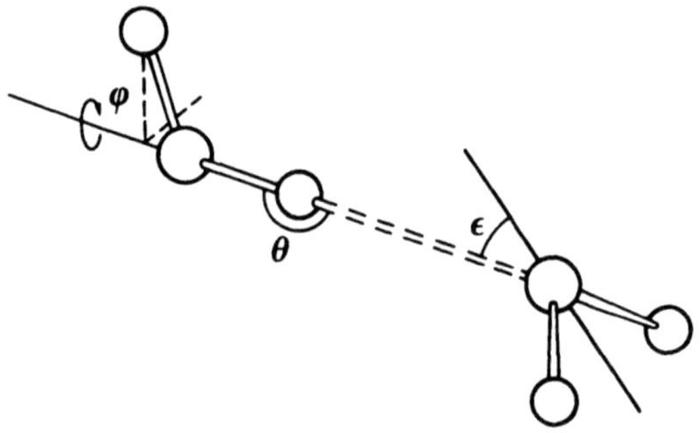


Fig. 21. Structure of the water dimer.

In the SCF calculations, the minimum energy is obtained for an O···O distance of about 3.0 Å and a bond energy between 4.5 and 6.5 kcal/mole. The energy curve for different $\varepsilon$ is very flat and values between 0 and 60° have been reported for the optimum $\varepsilon$. It should be noted, however, that a full geometry search has not been made of $\varepsilon$ while other paramters have been varied, e.g. when deviation from linearity is allowed. Rotation of the donor molecule around O-H···O will also affect the optimum value of $\varepsilon$: it may be concluded from ab initio calculations of Hankins et al. [45] (LCAO-MO-SCF with (531|21) Gaussian basis sets) that the repulsion between protons not involved in the hydrogen bond is responsible for the preference for an angle $\varepsilon \neq 0°$. If $\phi = 90°$ the energy minimum occurs for $\varepsilon = 0°$. The same conclusions were drawn from the extensive geometry search made by Kroon et al. [46], employing a minimal basis set. The collected evidence from this and other systems has emphasized that all interactions between the approaching molecules must be considered to determine the optimum geometry of the hydrogen-bonded complex.

Previous single determinant Hartree-Fock studies by Diercksen [43] have more recently been extended by large-scale configuration interaction (CI) calculations (Diercksen et al. [47]). As in the earlier calculations, the monomer geometry was kept fixed to that determined experimentally for the free water molecule, and a trans form was assumed as in Fig. 21. The equilibrium CI results give a slightly shorter O···O distance, 2.90 Å, and a correlation energy of -1.03 kcal/mole.

Several IR spectroscopic investigations have been made of matrix-isolated water at low temperature. Tursi & Nixon [48] interpreted their data in terms of an open-chain structure (it should also be noted that one unit of the dimer, the acceptor molecule, is only slightly perturbed compared to the free molecule). Recent microwave data by Dyke & Muenter [49] indicates a linear trans dimer as in Fig. 21 with O···O = 2.98(4) Å and $\varepsilon = 60(10)°$.

X-ray and neutron scattering of liquid water indicate that the majority of nearest neighbours are around 2.94 Å at 200°C and that this distance decreases to 2.84 Å at 4°C (cf. Narten in Ref. 41). As the average number of hydrogen bonds around each water molecule increases as the temperature is lowered, these data may be taken as an indication of a cooperative effect between the different hydrogen bonds. (We assume that the difference in distance is due to the difference in the number of hydrogen bonds at 200°C and 4°C; the possible thermal effect on the distances is neglected.) It is then interesting to note that in ice each molecule participates in four hydrogen bonds, each of which is even shorter, 2.76 Å.

The deformation density of the isolated water molecule relative to the superposed spherical atoms was shown in Fig. 8. The effect

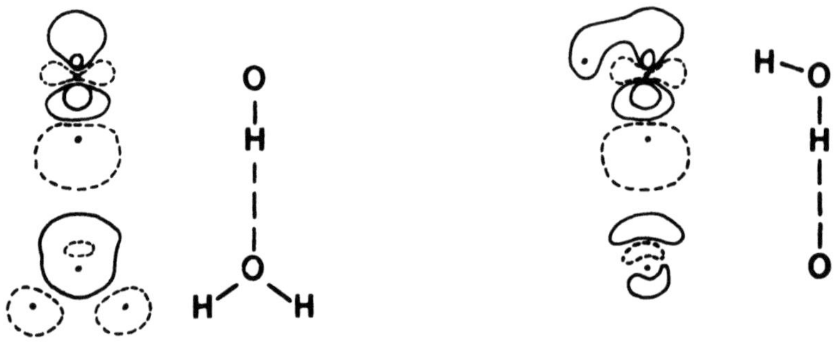


Fig. 22. Deformation density in the water dimer relative to the isolated water molecules. Contours at 0.001 and 0.01 $e \cdot (a.u.)^{-3}$ (the contours are taken from Diercksen's results [43], with the contour 0.0001 deleted).

of hydrogen bonding between two water molecules is illustrated in Fig. 22. This is a simplified drawing of the contour diagram due to Diercksen [43] (the dimer geometry corresponds to that of Fig. 21, with $\varepsilon = 0^{o}$).

This figure represents the difference between the electron distribution of the dimer and the two undistorted monomers at the equilibrium distance O···O = 3.00 Å. The overall features express the same tendency as earlier discussed for α-glycine: the hydrogen bond interaction reinforces the polarity already existing in the functional group. Thus, in the donor O-H group, the electron density is further shifted towards oxygen. A similar reinforcement is also observed at the acceptor oxygen. Note, however, that this is only a reliable trend for the groups directly participating in the hydrogen bond. In the donor water molecule, the hydrogen atom not participating in the hydrogen bond becomes less positive and will thus become less efficient as a donor to a second water molecule; the increased density around oxygen will make this water molecule a better acceptor of a second hydrogen bond, however. Similarly, in the acceptor molecule, the acceptance of one hydrogen bond makes this molecule a better donor of a second hydrogen bond. The changes in atomic charges, estimated by Mulliken population analysis are given in Fig. 23.

The above predictions are in complete agreement with ab initio calculations for water trimer aggregates[45] of the double donor and sequential trimer type, see Fig. 24a and c (cf. also Lentz & Scheraga [50]). From Fig. 23 one would also conclude that the oxygen atom of the acceptor molecule should become a better acceptor of a second hydrogen bond (due to the increased negative charge on the

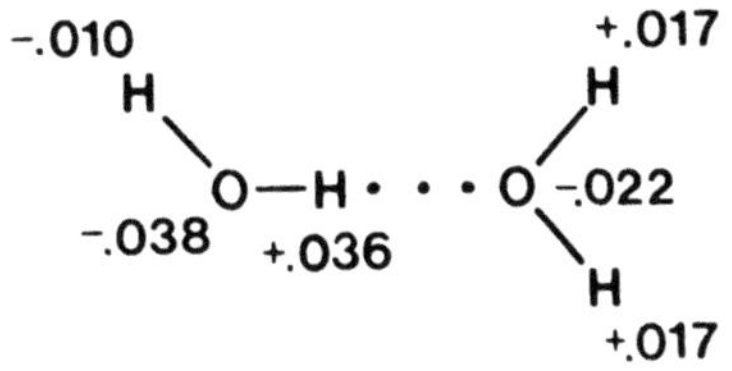


Fig. 23. Changes in atomic charges due to the formation of a hydrogen-bond dimer, derived from extended LCAO-MO-SCF calculations [45]. The dimer configuration corresponds to that of Fig. 21 with $\phi = 0$ and $\varepsilon = 40^{\circ}$.

acceptor oxygen atom). This conclusion does not agree with the energy calculated for the water trimer of the double acceptor type, however (Fig. 24b). Two explanations will be suggested here. Firstly, the detailed spatial distribution of the charges accumulated around the acceptor oxygen atom must be taken into account: when a water dimer geometry with $\varepsilon = 58^{\circ}$ is used in the calculations (Fig. 25) instead of $\varepsilon = 0^{\circ}$ (as in Fig. 22) it is evident that only little charge accumulation occurs just in the direction of acceptance of a second hydrogen bond which forms tetrahedral angles with the water molecule. Secondly, it is important to consider not only the hydrogen bonding, but also the interaction between all the other components of the trimer before any definite conclusions can be drawn about the magnitude of the inductive effect.

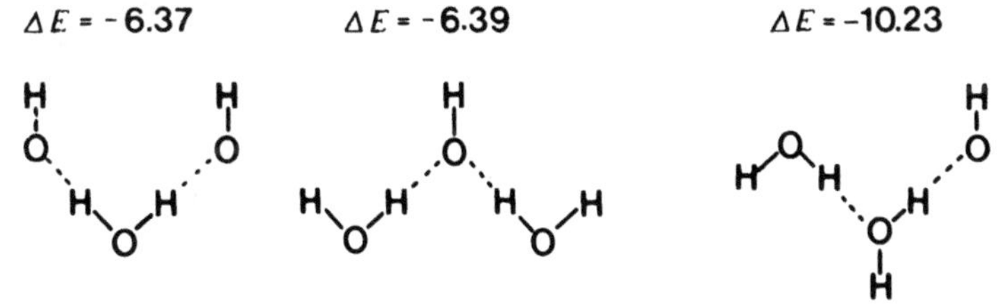


Fig. 24. Energy of trimerization (kcal/mole) for (a) double donor, (b) double acceptor and (c) sequential trimer of water, derived from extended LCAO-MO-SCF calculations [45].

It is appropriate at this stage to refer to the partitioning scheme discussed in section 4.3. The electron density change and its components for the water dimer are illustrated in Fig. 25 (Yamabe & Morokuma [12]). A 4-31 G basis was used, with a dimer geometry given in Fig. 2. The overall features of the total deformation

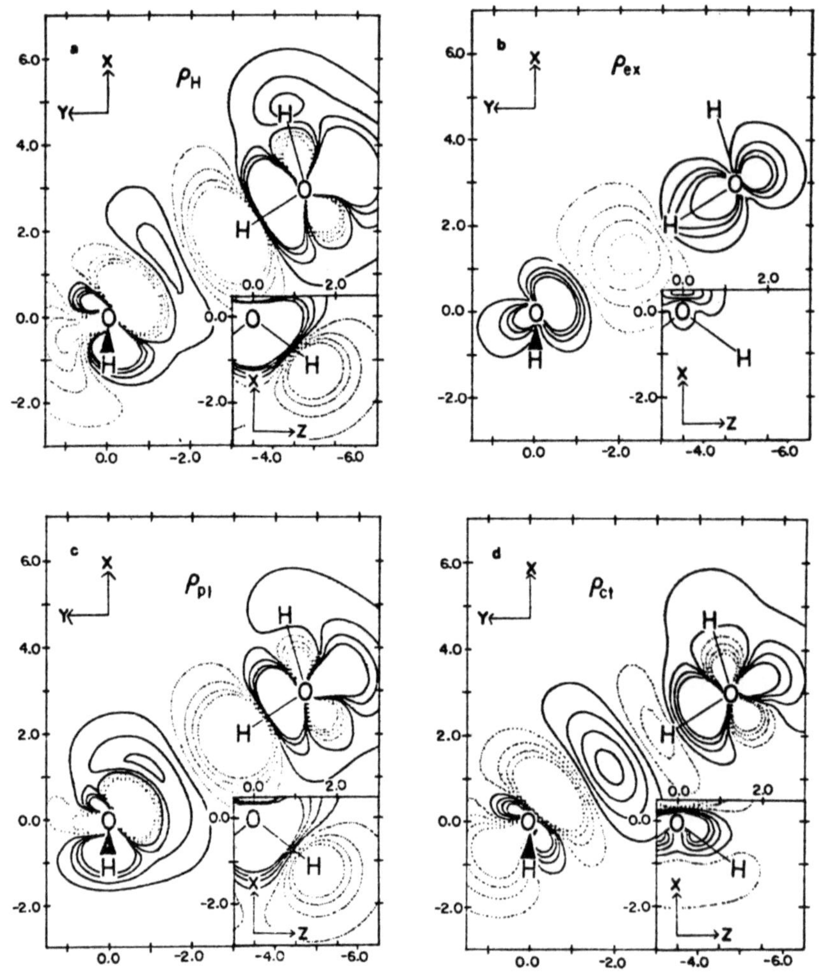


Fig. 25. Deformation density and its components in the water dimer, relative to the isolated monomers. [12] The fixed geometry is shown in Fig. 2.

density are very similar to those of Fig. 22. It should be noted that, whereas polarization and charge transfer modify the electron distribution of the whole system, the exchange effect is limited to the O-H···O bond and its extension. Furthermore, there is a remarkable similarity between the total and the polarization deformation densities. A comparison between the exchange and the charge transfer maps indicates that this similarity may be largely attributed to the appropriate cancellation of exchange and charge transfer effects (except around the oxygen atom of the donor molecule). The corresponding maps for the formaldehyde - water and cyclopropenone - - water systems are strikingly similar (cf. Fig. 26). It is particularly interesting to note the effect in the lone-pair region of formaldehyde: the density increase is most pronounced in the region which is directly involved in the hydrogen bond. This is in complete agreement with the earlier discussion of α-glycine (Fig. 11). Furthermore, the electron density in the whole system is affected by the hydrogen-bond interaction. This is true even for cyclopropenone, which may seem remarkable considering the size of the molecule.

To summarize the results for the water dimer: hydrogen bonding increases the polarity in the parts which directly participate in the hydrogen bond. The non-bonded hydrogen atom of the donor molecule becomes less positive, whereas the hydrogen atoms of the acceptor molecule become more positive. The partitioning analysis indicates that these modifications are largely polarization effects.

### 10.3. Crystalline Hydrates

As pointed out in section 10.1, the water molecule can participate in coordination and hydrogen bonding in many different ways. Since each of these can involve the same oxygen atom, the mutual influence of the various bonding effects may be considerable. Considering also the low electron density around hydrogen, the variation in O-H distances and the mobility of the molecule, we would expect a considerable variation in the electron distribution for different bonding situations. Inspection of the few accurate experimental electron density studies of the water molecule so far available confirms these expectations. It should be added, however, that the difficulties experienced in obtaining good electron densities are particularly severe for this type of compound. Density maps should thus be intrepreted with great caution at this stage.

$LiHCOO \cdot H_2O$. The crystal structure has been determined by neutron diffraction by Tellgren et al. [51] These data have subsequently been combined with X-ray data in two independent experiments to calculate X-N maps (Thomas et al. [52] and Harkema et al. [53]). This compound is particularly interesting as the two hydrogen bonds donated by the water molecule are unusually different, 2.714(2) and 2.896(2) Å. The corresponding O-H distances are: O-H1 = 0.976(3) and O-H2 =

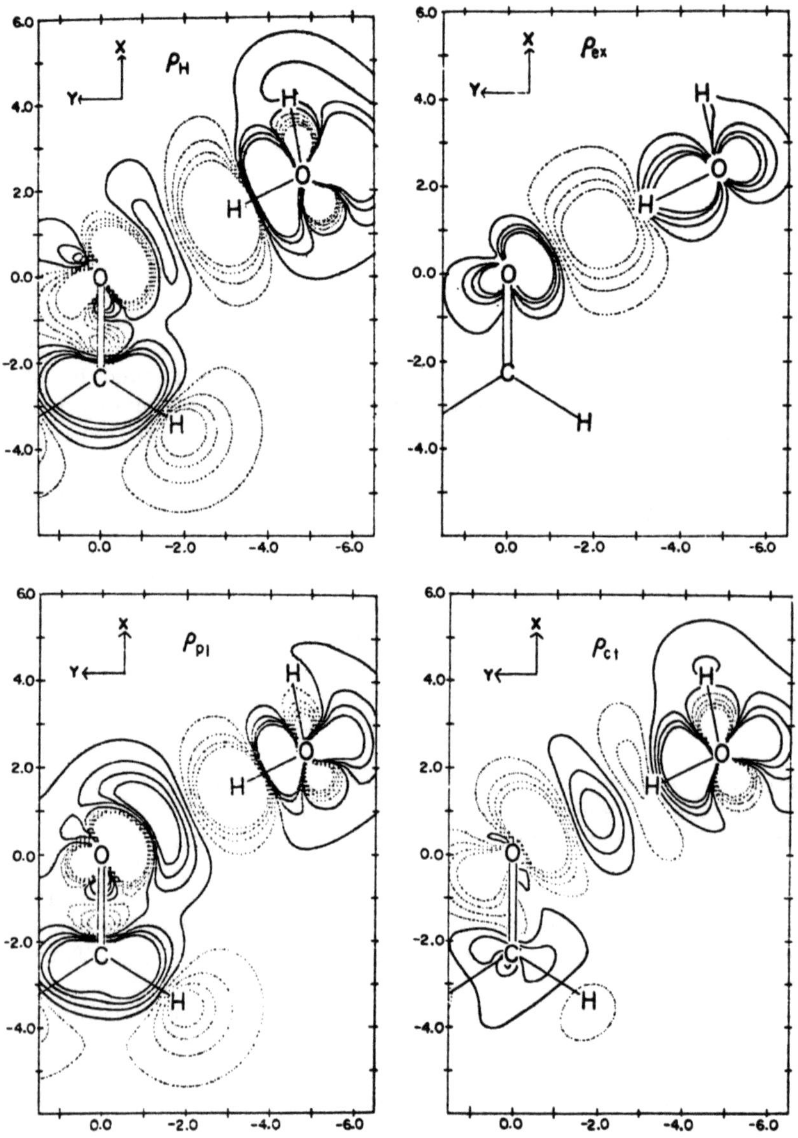


Fig. 26. Deformation density and its components in the formaldehyde--water dimer, relative to the isolated monomers [12]. The fixed geometry is shown in Fig. 2.

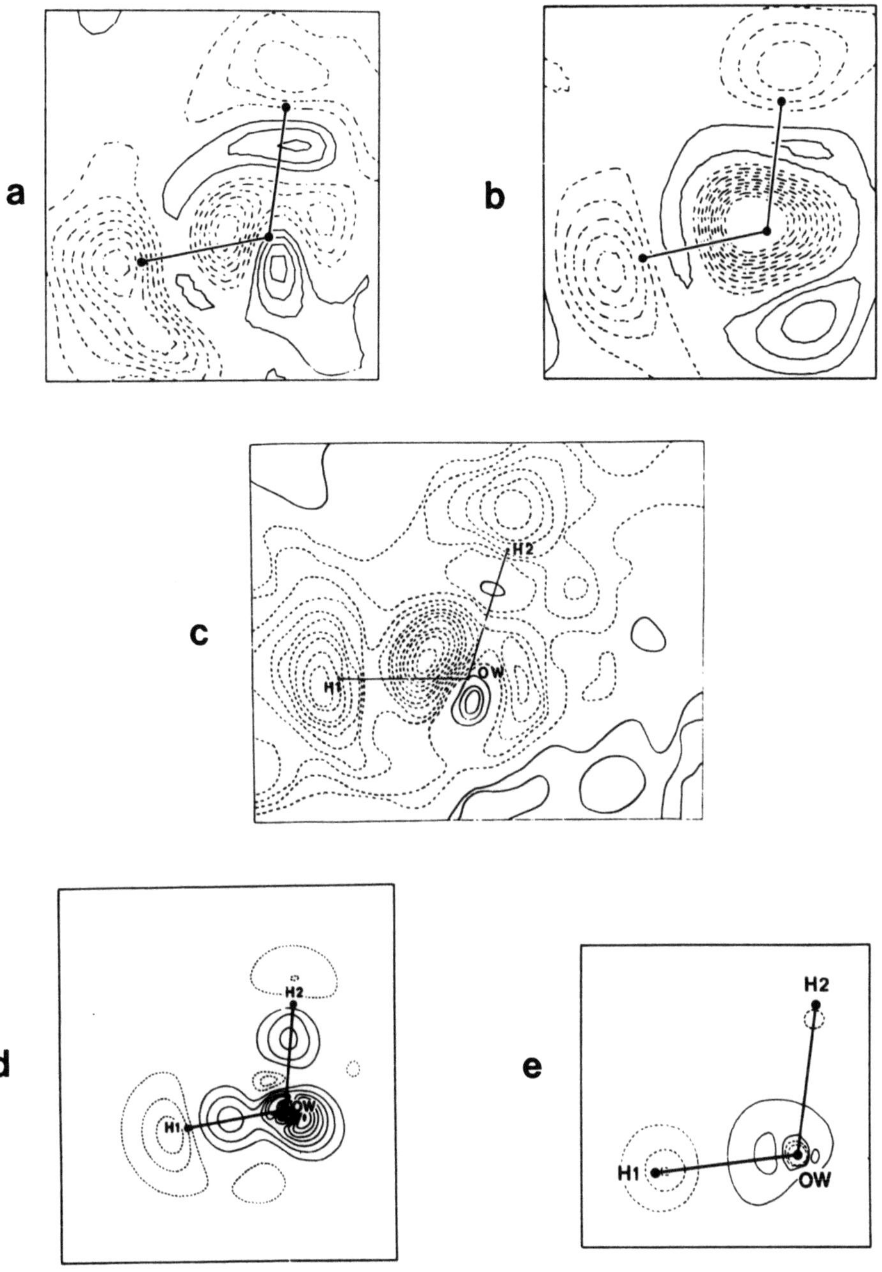


Fig. 27. LiHCOO·$H_2O$: Deformation density in the plane of the water molecule. (a) X-N [52]; (b) Multipole model [54]; (c) X-N [53]; (d) Theoretical, in crystal field [52]; (e) Theoretical, in crystal field relative to the free molecule [52].

=0.965(4) Å. The calculation of X-N maps is complicated by the non-centrosymmetric space group ($Pna2_1$). Thomas [54] has later proposed a procedure in which the phase of $F_{obs,X}$ is not evaluated in a conventional X-ray refinement using spherical free-atom form factors but, instead, in a refinement where multipole deformation density functions are also included in the model. Although this procedure gives a marked intensification of the peaks in the X-N maps for the $HCOO^-$ ion, especially in the lone-pair regions, the same cannot be said of the $H_2O$ molecule. The reason for this is not yet understood. The deformation density maps from the three investigations are illustrated in Fig. 27 a-c. The qualitative features are similar; the discrepancies are due partly to differences in the treatment of the data. As is generally expected, the plot of the deformation functions gives smoother features and less background noise. This map has at least some qualitative resemblance to that expected theoretically, cf. the earlier discussion of water polymers. A comparison with the theoretical deformation density of the water molecule in its crystalline environment can be made in Fig. 27d. (Thomas et al. [52]). The crystal field was simulated as in the case of α-glycine. The separate effect of the crystalline environment is shown in Fig. 27e, which represents the deformation density relative to that of the free water molecule. This map illustrates particularly well the

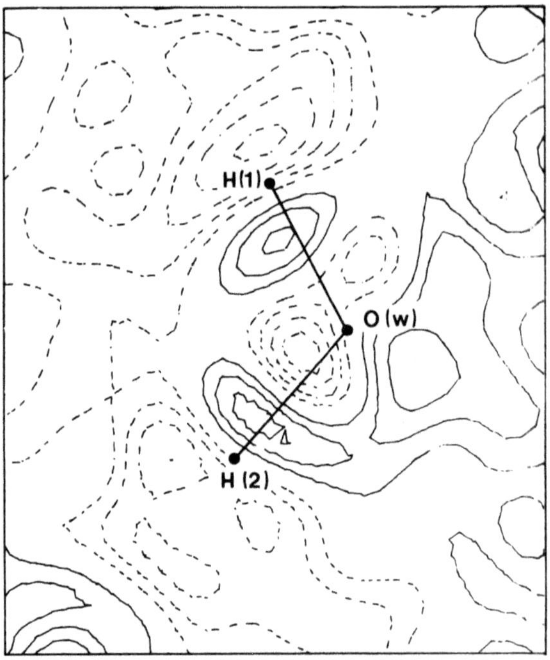


Fig. 28. X-N deformation density in the plane of the water molecule of $NaHC_2O_4 \cdot H_2O$ [56]. Contours at 0.05 $e \cdot Å^{-3}$.

difference in polarization in the O-H1 and O-H2 bonds; H1 is involved in a much stronger hydrogen bond than H2. Whereas the theoretical maps are in perfect agreement with our general expectation, the experimental maps indicate the opposite effect, O-H2 being more polarized than O-H1. The reasons for this discrepancy is again not clear.

$NaHC_2O_4 \cdot H_2O$. This compound has been studied both by X-ray and neutron diffraction (Tellgren & Olovsson [55], Tellgren et al. [56]). Its centrosymmetric space group ($P\bar{1}$) renders this compound more amenable to the calculation of X-N maps than the previous compound. The water molecule has a tetrahedral environment. It is coordinated to two $Na^+$ ions, and donates two hydrogen bonds of about equal length (2.806(1) and 2.824(1) Å). The X-N map in the plane of the water molecule is shown in Fig. 28. The general features are here in considerably better agreement with expectation. We will return later to this compound to illustrate the electron density in the short hydrogen bond between the hydrogen oxalate ions.

NaH Maleate$\cdot 3H_2O$. The electron density of this compound has recently been studied by X-ray and neutron diffraction at 120K (Olovsson et al. [59]). The hydrogen maleate ion will be discussed in section 10.4. The three water molecules are crystallographically independent and each molecule is approximately tetrahedrally surrounded by four neighbours. The X-N deformation density in the plane of the water molecules is shown in Fig. 29. As observed earlier, the deformation density differs considerably from that calculated for the free water molecule (Fig. 8). The corresponding deformation densities without subtraction of hydrogen are also shown in Fig. 29; as expected, the water molecules now appear more symmetric as the superposed spherical density of hydrogen tends to obscure the unsymmetric features.

## 11. STRONG HYDROGEN BONDS

The lower limit on the length of an O···O hydrogen bond is around 2.40 Å, and the proton may be located at the centre of the shortest hydrogen bonds. It can be noted from the survey made by Olovsson & Jönsson [1] (Vol. II, p. 426) that all crystallographically symmetric hydrogen bonds are found for O···O distances in the interval 2.40-2.50 Å, but also that asymmetric bonds are encountered equally frequently in this range. The feature common to all short hydrogen bonds is that the proton interacts with two chemically very similar acceptor groups X and Y, forming $[X\cdots H\cdots Y]^-$ or $[X\cdots H\cdots Y]^+$. In the first case, X and Y are negatively charged (generally carboxylate ions), whereas the second case normally involves neutral molecules (water, etc.). The atoms in the groups X and Y which are directly bonded to the proton are, in all cases, identical (oxygen atoms) but the structural details further away from the hydrogen

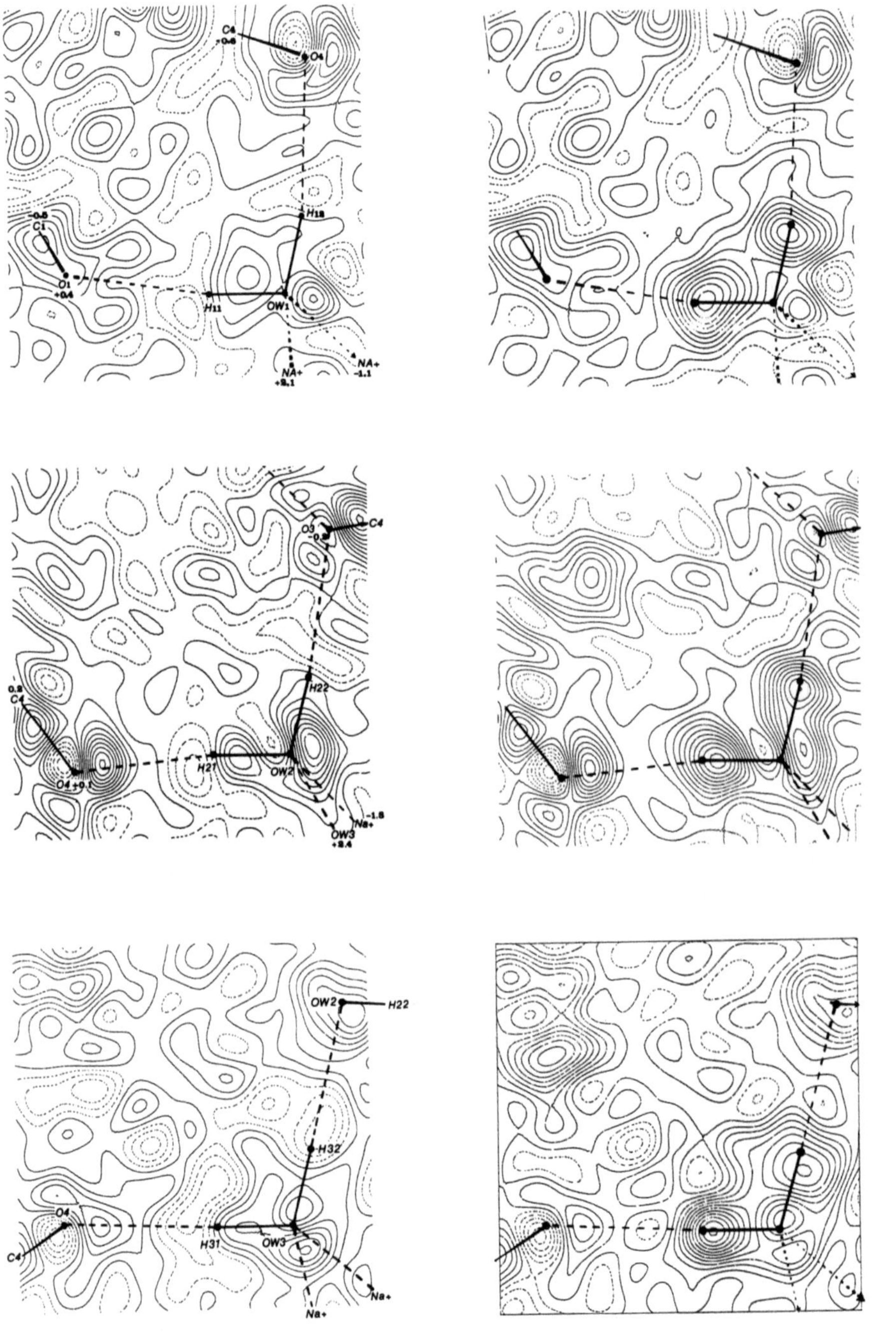


Fig. 29. NaH Maleate·3$H_2O$.[59] X-N deformation density in the planes of the $H_2O$ molecules. Contours at 0.10 e.Å$^{-3}$. Left: hydrogen subtracted; Right: hydrogen not subtracted.

bond need not be identical. A slightly asymmetric hydrogen bond environment can thus arise due to chemical differences between the X and Y groups, such as in pyridine-2,3-dicarboxylic acid (Kvick et al. [57]). However, even if the two groups X and Y are chemically identical, slight asymmetry in the environment of the hydrogen bond may be present due to differences in the structural details. A distortion will be expected in the potential functions shown in Fig. 3 c-e, if the arrangement with respect to the centre of the hydrogen bond is sufficiently asymmetric; Fig. 3f illustrates such a slightly asymmetric situation. The mean position of the proton may then deviate significantly from the centre, the magnitude of the deviation depending on the degree of asymmetry of the environment. The influence of an asymmetry in the proton environment is demonstrated by the structure of potassium hydrogen maleate and sodium hydrogen maleate (Peterson & Levy [58]; Olovsson et al. [59]). In the first case, there is a crystallographic mirror symmetry in the hydrogen bond and the proton is effectively centrally located. In the second case, there is no crystallographic symmetry in the hydrogen bond and the proton is markedly off-centred. As the two halves of the maleate ion are chemically identical, the asymmetry is clearly caused by the influence of more distant neighbours.

NaH Maleate·3$H_2O$. The data given below refer to X-ray and neutron diffraction results at 120K (Olovsson et al. [59]). The deformation density in the water molecules was discussed in section 10.3. The two carboxyl groups of the maleate ion form a very short intramoleculear hydrogen bond, O···O = 2.445 Å. The space group is centrosymmetric ($P\bar{1}$) but there is no crystallographic symmetry in the short intramolecular hydrogen bond and the proton is clearly off--centered: O2-H = 1.079 Å and H···O3 = 1.367 Å. The end containing O2 may thus be considered approximately as a carboxylic group, and the other end as a carboxylate ion. The X-N deformation density map in the plane of the maleate ion is shown in Fig. 30a (since the main feature of interest is the intramolecular hydrogen bond, the plane is defined by O2, O3 and (arbitrarily) C3; the other atoms of the maleate ion deviate by 0.02-0.08 Å from this plane). We notice that the deformation density in the intramoleculear hydrogen bond has a rather different appearance compared to previous maps (cf. Figs. 10, 16 and 20): there is a much less pronounced charge build up in the donor O2-H bond as well as in the acceptor region H···O3. The electron density is also more symmetrically distributed around the midpoint. The X-N map with the hydrogen atom not subtracted, is shown in Fig. 30b. It is then even more evident that O2-H still has a considerable character of a hydroxyl group. This is also apparent from the distribution around O2 and O3: the distribution in the second "lone pair" region, not participating in the intramolecular hydrogen bond, has a distinct peak for O2 but is more smeared out for O3 (geometrical considerations would suggest that O3 corresponds approximately to $O^-$). The low electron density near the hydrogen should also be noted. From Fig. 30 b it appears that the total densi-

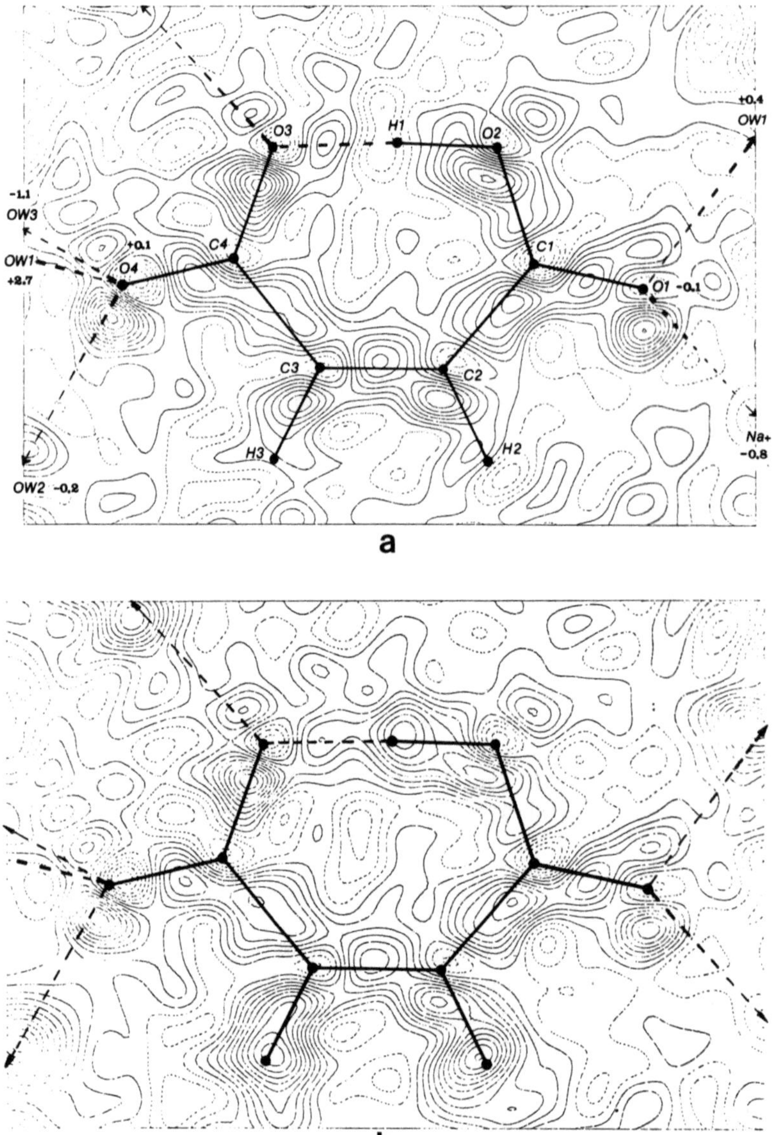


Fig. 30. NaH Maleate. $3H_2O$.[59] X-N deformation density in the hydrogen maleate ion. Contours at 0.10 e.Å$^{-3}$.
(a): hydrogen subtracted; (b): hydrogen not subtracted.

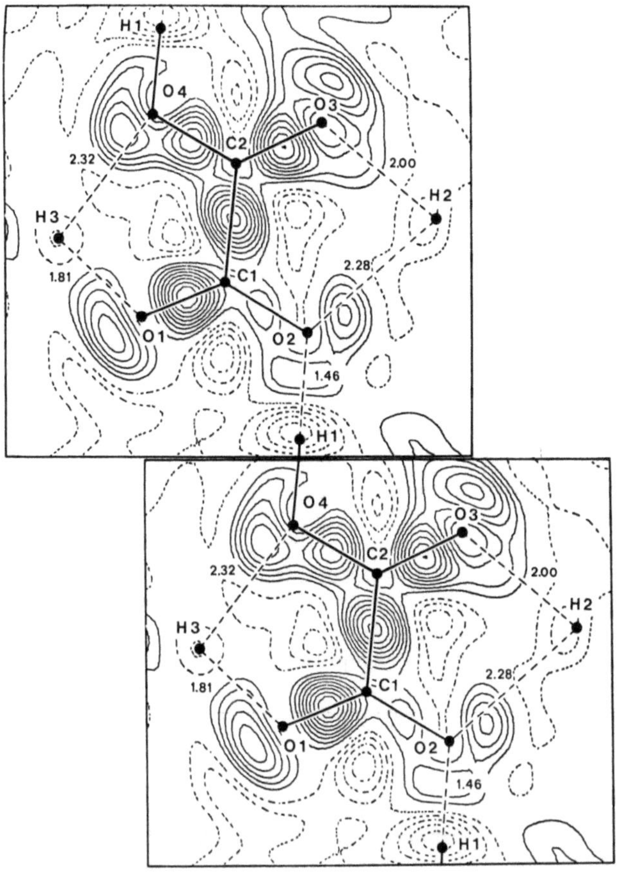


Fig. 31. X-N deformation density in the least-squares plane through the heavy atoms of the $HC_2O_4^-$ ion in $(CH_3)_2NH_2HC_2O_4$ [61]. Contours at 0.05 e·Å$^{-3}$.

ty around hydrogen is only of the same order of magnitude as the excess accumulation of electrons in a CO bond (excess relative to the superposed spherical atom contribution of carbon and oxygen).

The features in the lone-pair regions of O1 and O4 are perhaps also worth commenting. In the case of O1, the slightly higher peak may be attributed to a stronger polarization influence of $Na^+$ compared to that of the water molecule which approaches in the other lone-pair direction. A hydrogen bond is accepted by O4 from a water molecule close to the plane of the molecule in the direction of the pronounced peak; in contrast, two hydrogen bonds are accepted considerably out of the molecule plane in the other lone-pair region, with a smeared distribution in the map as a result.

$(CH_3)_2NH_2HC_2O_4$. Earlier X-ray data by Thomas & Pramatus [60] have been combined with neutron data by Thomas [61] to calculate the X-N deformation density. The structure comprises $HC_2O_4^-$ ions linked by short hydrogen bonds (O···O = 2.533(1) Å) to form infinite chains, Fig. 31. In the present case the bond length is just outside the limit for potentially symmetric hydrogen bonds with a very low barrier for the proton (Fig. 3 c-f). It is thus interesting to compare Fig. 31 with the previous map, Fig. 30a. A more distinct difference between the deformation density in the O4-H1 and H1···O2 regions is clearly noticeable in the present case (O4 - H1 = 1.073(4) Å), H1···O2 = 1.455(5) Å). There remains, for example, a distinct peak around O2 in the direction of O2···O4, although this is markedly lower than the other one directed towards H2.

$NaHC_2O_4 \cdot H_2O$. The deformation density of the water molecule was discussed in section 10.3. Here the $HC_2O_4^-$ ions are again linked end-to-end to form infinite chains by means of short asymmetric hydrogen bonds (O2···O4 = 2.571(1), O2 - H = 1.036(1), H···O4 = = 1.537(1) Å). A composite X-N map giving the deformation density in the different parts of the chain is shown in Fig. 32 (Tellgren et al. [56]). The overall features in the short hydrogen bond are quite similar to those found in $(CH_3)_2NH_2HC_2O_4$. The less pronounced build up in the H···O4 direction compared to the other "lone-pair" direction (in which O4 accepts a hydrogen bond from a water molecule, O···O = 2.806(1) Å), is equally noticeable in the present case. The same applies to the density around O(2).

A general comparison of deformation density features in hydrogen bonds of different lengths will be given in section 12.

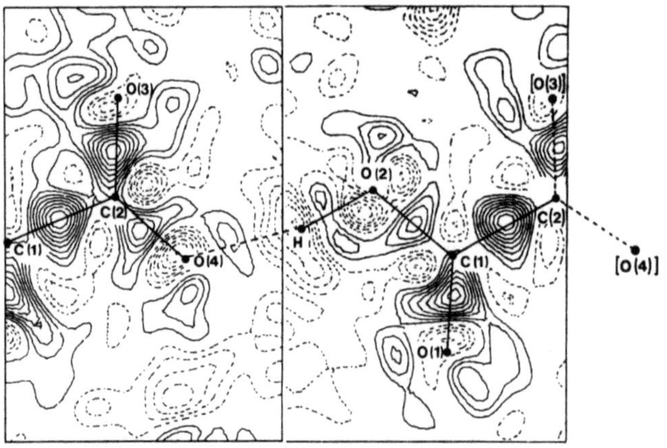


Fig. 32. X-N deformation density in the $HC_2O_4^-$ ion of $NaHC_2O_4 \cdot H_2O$ [56] (the two least-squares planes shown are twisted 12.92° with respect to each other). Contours at 0.05 $e \cdot Å^{-3}$.

## 12. THE ROLE OF THE LONE-PAIR ELECTRONS ON THE ACCEPTOR ATOM

The electron density in the non-bonded region has been illustrated in numerous compounds in the previous sections. It would here seem appropriate to consider in more detail the general role of the lone-pair electrons in the formation of hydrogen bonds.

### 12.1. General Considerations

In elementary discussions, the reciever of the hydrogen bond is considered to be a lone pair on the acceptor atom. Let us investigate if this simple model is fulfilled in practise. From empirical data it is found that, if a molecule contains "active" hydrogen atoms (i.e. contains groups normally forming hydrogen bonds), then all such hydrogens have a strong tendency to participate in hydrogen bonding. It is very seldom found, for example, that the hydrogen atoms of a water molecule are not engaged in hydrogen bonding. This fact has an important bearing on the arrangement of hydrogen bonds, as will be illustrated for the case of a molecule $AH_n$, containing n active hydrogen atoms. Suppose that we wish to build up a three-dimensional structure containing only $AH_n$ molecules, and assume that all these molecules have equivalent surroundings. Thus, if each molecule acts as a donor for n hydrogen bonds, Fig. 33a, each molecule must also act as an acceptor of the same number of hydrogen bonds, Fig. 33b. We will thus need n lone-pairs on each $AH_n$ molecule. However, one seldom finds molecules with the same number of hydrogen atoms as lone-pairs. Water is thus a rather unique molecule in this respect. The ideal situation is not found in most other compounds; ammonia is a typical example. Here, there is only one lone-pair available and it might therefore be concluded that only one of the three hydrogen atoms can be engaged in hydrogen bonding. The structure of solid ammonia is illustrated in Fig. 34 (Olovsson & Templeton [62]). We notice that all three hydrogen

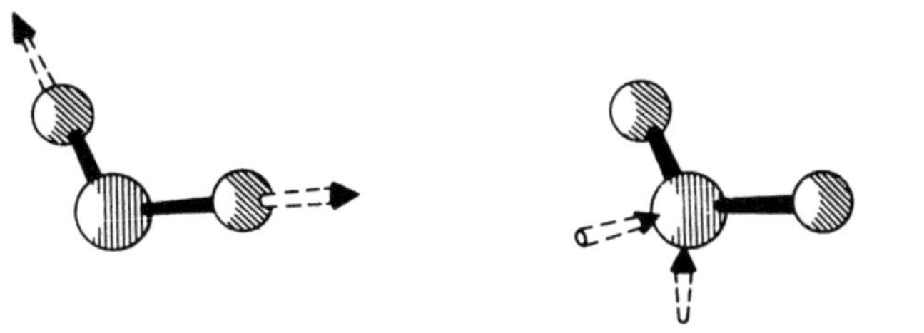


Fig. 33. Geometrical requirements in hydrogen bond formation.

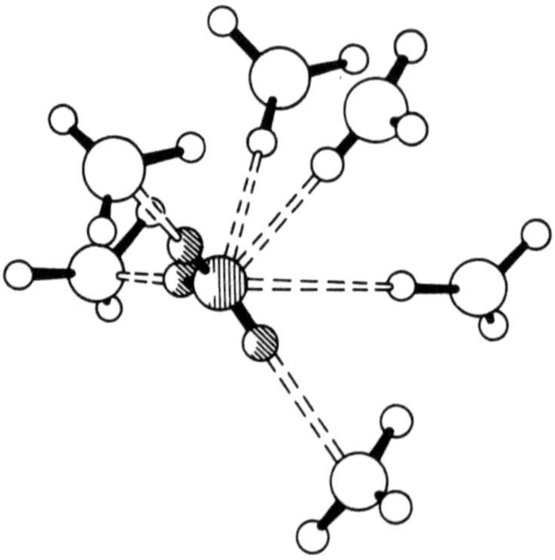

Fig. 34. Structure of solid ammonia [62].

atoms do, in fact, participate in hydrogen bonding, which means that the single lone pair has to accept no less than three hydrogen bonds. The simple model of one lone pair per hydrogen bond is clearly not particularly relevant. The above discussion may be extended to include cases with more than one type of molecule in the structure. If the number of lone pairs available is less than the number of active hydrogens, we may encounter situations like those described above. In many other cases there is an excess of available lone pairs as, for example, in many organic compounds. Here, either some of the lone pairs are not involved in hydrogen bonding, or the donor hydrogen atom is directed towards some point between several lone pairs.

From the above discussion, it is doubtful whether lone pairs should be regarded as the immediate receivers of hydrogen bonds. The details of the hydrogen-bond arrangement are often mainly determined by simple geometrical requirements. An inspection of the electron density features in the lone-pair regions further supports this point of view (see 12.2). Many examples may be taken to illustrate that it is important to take into account the whole electron <u>and</u> nuclear distribution in discussing the relative arrangement of interacting molecules. As long as the electrostatic component plays the dominant role in the intermolecular interaction, the total influence of the electron and nuclear distribution may be directly illustrated by the electrostatic potential energy surface. This represents the electrostatic energy of interaction between a molecule and a positive unit charge at different positions. Such diagrams have been used, for example, to discuss preferred protonation sites in molecules. In studying these maps, it should then be noted that the molecular charge distribution is assumed not to be modified by the presence of the positive test charge. This is precisely the same assumption which underlies our electrostatic model for the hydrogen-bond interaction. The electrostatic potential for α-glycine (Fig. 13) shows a minimum which extends over a major part of the non-bonded region around each carboxylate oxygen atom. This means that the major part of the non-bonded region should be approximately equally favourable to an approaching proton, as long as the latter does not cause a significant perturbation of charge distribution in the α-glycine molecule. Consideration of the Coulomb interaction with only the local electronic charges in the lone-pair regions can often be very misleading.

Several attempts have previously been made to study the directional influence of the lone-pair electrons on a carbonyl group when acting as a hydrogen-bond acceptor. A statistical study of the angle H···O = C in compounds accurately studied by neutron diffraction shows a certain accumulation around $120^{o}$, but there are many large deviations from this value (cf. Olovsson & Jönsson [1], Vol. II p. 417). Naturally, it is impossible to decide from structural data alone whether this effect is caused by a genuine directional influence of the lone-pair electrons or by other geometrical factors, e.g. the most favourable direction of approach considering the form of the molecule.

From a statistical analysis of 196 hydrogen bonds from 45 crystal structures of polyalcohols, saccarides and related ROH compounds studied by X-ray diffraction, there appears to be no distinct preference for acceptance along the lone-pair directions (Kroon et al. [46]), Fig. 35. The entire non-bonded region seems to be equally accessible for hydrogen bonding. The same conclusions were drawn from their extensive search of the optimum geometry for the water dimer from minimal basis ab initio calculations.

## 12.2. Electron Density in the Non-bonded Region

Even if the electron density in the lone-pair region does not alone determine the relative arrangement of the neighbours, it is interesting to compare the electron concentration in various parts of the non-bonded region of the isolated molecules. Furthermore, the electron density is expected to be modified due to polarization or charge transfer as hydrogen bonds are formed.

In studying deformation density maps, one should be reminded

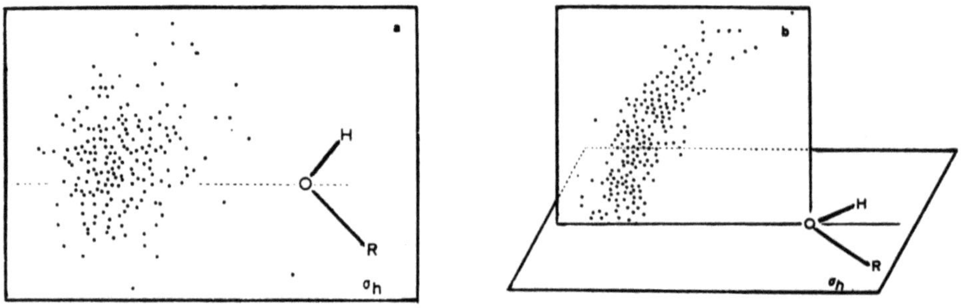


Fig. 35. Observed hydrogen positions in donors to ROH[46].

that these only reveal the deviation of the electron density from the superposed spherical atom distribution. The charge build up in the lone-pair region, which is sometimes observed in these maps, thus does not at all represent the complete lone-pair distribution. Similarly, the total electron distribution in a chosen region around an atom represents the total contribution from all the orbitals, bonding as well as non-bonding, including the inner shells. It is well known that the superposition of the four equivalent $sp^3$ orbital densities (for example) gives a perfectly spherical density. Even if these orbitals are not exactly equivalent when the atom considered is part of a molecule, the virtually spherical contribution from the inner shells leads to a total density in the non-bonded region which is very close to spherical. This was earlier illustrated for N-methylacetamide (Fig. 4) and for water (Figs. 5 and 6). Even if the electron distribution in the non-bonded region were solely responsible for the interaction with an approaching molecule, we would thus expect very little tendency for an approach precisely along the lone-pair direction.A closer inspection by means of deformation density maps also indicate the same tendency: Figs. 7 and 8 demonstrate that the electron concentration extends over a large part of the non-bonded region.

The general features of the deformation density in the non--bonded region are maintained as the molecules form moderately strong hydrogen bonds with each other, with a more pronounced polarization in the direction of acceptance of hydrogen bonds in experimental as well as theoretical maps (sections 9-11). The relative importance of the electron migration to the lone-pair regions is not easily estimated without detailed calculations. The theoretical calculations on water dimers discussed earlier (section 10.2) are particularly illustrative in this context. In several accurate ab initio calculations, the energy minimum is found for an angle $\epsilon$ which is different from zero (Fig. 20). These results might then suggest a clear directional influence of the lone pairs of the acceptor molecule. A more complete geometry search indicates, however, that the deviation of $\epsilon$ from zero is simply a result of the interaction between the non-bonded hydrogen atoms at the extremities of the dimer.

## 13. SUMMARY

The general features of the charge density in simple hydrogen-bonded compounds, as derived from quantum-mechanical calculations and diffraction investigations, are compared in an attempt to study the relative influence of hydrogen-bond interactions and crystal forces on the electron distribution.

In the absence of a specific interaction between functional groups, the relative arrangement of the molecules is mainly determined by their overall shape and size. The latter is directly related to the outer contours of the total molecular electron density, which is, in turn, very similar to that obtained by the superposition of the constituent spherical atoms. This is equally true for molecules containing functional groups, which normally form hydrogen bonds. It is particularly important to note that the total electron density in a lone-pair region is very close to spherical.

Even if the deviation of the total molecular electron density from the superposition of spherical atomic densities is relatively small, it is important in determining the relative arrangement of the molecules as they approach each other. This charge migration on molecular formation is thus the only factor responsible for the polarity of the different functional groups of a molecule. The specific interactions, like hydrogen bonding, between the functional groups will generally result in a crystalline arrangement which is very different from that expected from simple close-packing principles. In analyzing intermolecular effects, the total interaction may be decomposed into electrostatic, polarization, charge transfer, exchange and dispersion components. In the case of weak and moderately strong hydrogen bonds, the energy contributions of the last four components approximately cancel. Furthermore, the variation in electrostatic energy follows the same trend as the total hydrogen-bond energy. As each of the other four separate contributions is less sensitive to the relative orientation of the monomers, this roughly explains the empirical rule that most geometrical features of hydrogen bonding can be explained simply by an electrostatic model for the hydrogen bond. This corresponds to the classical Coulomb interaction between the undistorted constituent molecules as they are brought together into the relative positions corresponding to the hydrogen-bonded complex, without any deformation of the original monomer charge distributions and without any electron exchange.

To study the finer details of the electron distribution in a given system, the current procedure is to form the difference $\Delta\rho(\underline{r}) = \rho(\underline{r}) - \rho_{ref}(\underline{r})$, "the deformation density", where $\rho(\underline{r})$ is the electron density in the actual system and $\rho_{ref}(\underline{r})$ is the corresponding distribution in an appropriate reference state. In diffraction studies, the most common reference state is defined as the superposition of the spherical electron densities of the free constituent atoms arranged as in the actual system (the "promolecule" state). In theoretical studies of intermolecular interactions it is more common, however, to choose a reference state corresponding to the superposition of the electron densities of the unperturbed monomers, placed in the same relative positions as in the actual complex. In this case it is assumed that the geometry of the original monomers is not changed as these interact with each other. This may be approximately true for weak and moderately strong interactions, but is certainly not so for strong interactions. It is thus not possible to define the reference system in an altogether satisfactory way in this latter case. The same difficulty arises when one applies current interaction decomposition schemes to study the effects due to the separate components, as the geometry is assumed to be undistorted under the influence of each separate interaction contribution. This difficulty does not arise with the promolecule reference state but, on the other hand, it is then not possible to separate the effects due to molecular formation, hydrogen bonding, and crystal forces.

Theoretical calculations of the energy of the electrostatic interaction between a formic acid molecule and the surrounding molecules at various distances in the crystal demonstrate that practically the whole contribution to the electrostatic energy originates solely from the interaction with the two nearest neighbours within the hydrogen-bonded chain. Together with the previous observation that the electrostatic energy component reflects the relative strengths of hydrogen bonds, these results suggests that a good approximation to the interaction between a molecule and its crystal environment may be obtained by considering only the electrostatic part of the hydrogen-bond interaction to its immediate neighbours; the effect due to more distant neighbours are practically negligible. It should be recalled, however, that this electrostatic approximation is only applicable to weak and moderately strong interactions.

Since the electrostatic approximation implies that no electron redistribution occurs when the monomers interact with each other, we would expect that the electron density in a hydrogen-bonded crystal could be obtained approximately by simply superposing the unmodified electron density of the isolated monomers. This is in perfect agreement with the form of the electron density maps derived from diffraction experiments. The modification actually taking place, as the molecules interact with each other in the crystal, constitutes only a second-order effect.

The typical deformation density (referred to the superposed spherical atomic densities) for a weak or moderately strong hydrogen bond X-H···Y thus consists of an electron excess in the X-H bond, a slight electron deficiency in the H···Y bond close to hydrogen, and an electron excess closer to the acceptor atom Y. The major effect from the environment is to increase the polarity of the functional groups that already exists in the isolated monomers. Thus, the electron shift from H to X in the donor X-H bond and to the lone-pair region of the acceptor atom Y is reinforced. An interaction decomposition analysis, performed for a number of dimers, indicates that the polarization, exchange and charge transfer components all contribute significantly to this electron redistribution. A comparison between the exchange and the charge transfer density maps shows, however, that these two components to a large extent cancel (except around the atom X of the donor molecule). To a certain approximation, it may thus be stated that the electron redistribution due to weak or moderately strong interaction is largely a polarization effect.

In very short hydrogen bonds of type $[O\cdots H\cdots O]^-$, with oxygen-oxygen distances around 2.4 - 2.5 Å, and where the proton interacts with two chemically very similar acceptor groups, the deformation density maps have a rather different appearance: there is a much less pronounced charge build-up in the O···H as well as in the H···O region, and the charge density is also somewhat more symmetrically distributed around the midpoint.

As one proceeds from weak to very strong hydrogen bonds, we may thus present the following trend on the basis of theoretical and experimental results. In the weak and intermediate hydrogen bonds, the electron distribution may be considered simply as a superposition of the density of the undisturbed, constituent monomers, and the experimental maps show very little difference in general appearance as the bond is shortened from around 3.0 to 2.7 Å. The theoretical results indicate, however, that there is a definite redistribution due to the intermolecular interactions as a second-order effect, and we expect this electron migration to increase as the interaction gets stronger. For these bonds, the proton transfer across the hydrogen bond from the donor to the acceptor still has a very low probability. As the interaction becomes very strong, the tunnelling barrier is considerably reduced and there is now a much less pronounced electron excess in both the donor and the acceptor region. The electrostatic approximation is no longer so applicable for these bonds, and it becomes conceptually more difficult to partition the electron distribution between the two systems. The current interaction partitioning schemes break down, so that it is no longer possible for example to distinguish between the polarization and charge transfer effects without introducing further arbitrary assumptions.

The total electron density in the lone-pair region is close to spherical and the actual electron excess in this area, which is commonly observed in deformation density maps, only represents a very small fraction of the total density. As the whole electron and nuclear distribution must be taken into account when discussing the most favorable direction of approach, it then seems likely that there should be little tendency for an approach in just the lone--pair direction. The details of the hydrogen bond arrangement are often determined by simple geometrical requirements.

REMERCIEMENTS

Le présent travail a été effectué au Laboratoire de Cristallographie du Centre National de la Recherche Scientifique de Grenoble durant l'année 1977-78. J'exprime ma gratitude au Gouvernement Francais et à la Délégation Générale à la Recherche Scientifique et Technique pour leur soutien financier pendant ce séjour.

J'exprime ma respectueuse reconnaissance à Monsieur E.F. Bertaut pour m'avoir accueilli dans son laboratoire, en mettant à ma disposition tous les moyens nécessaires. Tout au long de ce séjour, j'ai bien apprécié l'assistance de l'ensemble du personnel du laboratoire.

I wish to express my gratitude to Dr Josh Thomas for valuable linguistic and other suggestions.

## 14. REFERENCES


1. The hydrogen bond. Recent developments in theory and experiments (1976). Vol. I-III. Eds.: P. Schuster, G. Zundel & C. Sandorfy. North Holland Publ. Co., Amsterdam.

2. Hellmann, H. (1937). Einführung in die Quantenchemie (Franz Deuticke, Leipzig, 1937), pp. 285 ff.

3. Feynman, R.P. (1939). Phys. Rev. 56, 340.

4. Cade, P.E., Bader, R.F.W., Henneker, W.H. & Keaveny, I. (1969). J. Chem. Phys. 50, 5313.

5. Hirshfeld, F.L. & Rzotkiewisz, S. (1974) Mol. Phys. 27, 1319.

6. The Hydrogen Bond. Proceedings of the First International Conference on Hydrogen Bonding, Ljubljana (1957). Eds. D. Hadẑi and H.W. Thompson. Pergamon Press, London.

7. Tsubomura, M. (1954) Bull. Chem. Soc. Japan 27, 445.

8. Dreyfus, M. & Pullman, A. (1970) Theoret. Chim. Acta 19, 20.

9. Kollman, P.A. & Allen, L.C. (1970) Theoret. Chim. Acta 18, 399.

10. Morokuma, K. (1971) J. Chem. Phys. 55, 1236.

11. Van Duijneveldt-van de Rijdt, J.G.C.M. & Van Duijneveldt, F.B. (1971) J. Amer. Chem. Soc. 93, 5644.

12. Yamabe, S. & Morokuma, K. (1975) J. Amer. Chem. Soc. 97, 4458.

13. Kitaura, K. & Morokuma, K. (1976). Int. J. Quantum Chem. 10 (1976) 325.

14. Umeyama, H., Kitaura, K. & Morokuma, K. (1975) Chem. Phys. Letters 36, 11.

15. Baur, W.H. (1965). Acta Cryst. 19, 909.

16. Bonaccorsi, R., Petrongolo, C., Scrocco, E. & Tomasi, J. (1971). Teor. Chim. Acta 20, 331.

17. Smith, V.H. (1977) Physica Scripta 15, 147.

18. Tait, A.D. & Hall, G.G. (1973) Theor. Chim. Acta 31, 311.

19. Lippincott, E.R. & Schroeder, R. (1955). J. Chem. Phys. 23, 1099.

20. Hagler, A.T., Huler, E. & Lifson, S. (1974). J. Amer. Chem. Soc. 96, 5319.

21. Hagler, A.T. & Lapiccirella, A. (1976). Biopolymers 15, 1167.

22. Van Waser, J.R. & Absar, I. (1975). Electron densities in molecules and molecular orbitals (Acad. Press, New York).

23. Smith, V.H. (1978). Priv. comm.

24. Almenningen, A., Bastiansen, O. & Motzfeldt, T. (1969). Acta Chem. Scand. 23, 2848.

25. Kwei, G.H. & Curl, R.F. (1960). J. Chem. Phys. 32 (1592).

26. Nahringbauer, I. (1978). Acta Cryst. B34, 315.

27. Derissen, J.L. (1971). J. Mol. Structure 7, 67.

28. Jönsson, P.G. (1972). Acta Chem. Scand. 26, 1599.

29. Nahringbauer, I. (1970). Acta Chem. Scand. 24, 453.

30. Jönsson, P.G. (1971). Acta Cryst. B27, 893.

31. Smit, P.H., Derissen, J.L. & van Duijneveldt, F.B. (1977), J. Chem. Phys. 67, 274.

32. Minicozzi, W.P. & Stroot, M.T. (1970). J. Comput. Phys. 6, 95.

33. Clementi, E., Mehl, J. & von Niessen, W. (1971). J. Chem. Phys. 54, 508.

34. Almlöf, J., Kvick, Å. & Thomas, J.O. (1973). J. Chem. Phys. 59, 3901.

35. Dreyfus, M., Maigret, B. & Pullman, A. (1970). Theoret. Chim. Acta 17, 109.

36. Almlöf, J. & Mårtensson, O. (1971). Acta Chem. Scand. 25, 355 & 1413.

37. Stevens, E.D. (1978). Acta Cryst. B34, 544.

38. Stevens, E.D., Rys, J. & Coppens, P. (1978). J. Am. Chem. Soc. 100, 2324.

39. Kvick, Å., Thomas, R. & Koetzle, F. (1976). Acta Cryst. B32, 224.

40. Savariault, J.-M. & Lehmann, M.S. (1978). To be published.

41. Water, a comprehensive treatise (1972-1975). Vol. I-V. Ed.: F. Frank. New York: Plenum Press.

42. Kollman, P.A. & Allen, L.C. (1969). J. Chem. Phys. 51, 3286.

43. Diercksen, G.H.F. (1971). Theoret. Chim. Acta 21, 335.

44. Popkie, H., Kistenmacher, H. & Clementi, E. (1973). J. Chem. Phys. 59, 1325.

45. Hankins, D., Moskowitz, J.W. & Stillinger, F.H. (1970). J. Chem. Phys. 53, 4544. Ibid. 59 (1973) 995.

46. Kroon, J., Kanters, J.A., Van Duijneveldt-van de Rijdt, J.G.C.M., van Duijneveldt, F.B. & Vliegenthart, J.A. (1975). J. Mol. Struct. 24, 1975.

47. Diercksen, G.H.F., Kraemer, W.P. & Roos, B.O. (1975). Theoret. Chim. Acta 36, 249.

48. Tursi, A.J. & Nixon, E.R. (1970). J. Chem. Phys. 53, 518.

49. Dyke, T.R. & Muenter, J.S. (1974). J. Chem. Phys. 60, 2929.

50. Lentz, B.R. & Scheraga, H.A. (1973). J. Chem. Phys. 58, 5296.

51. Tellgren, R., Ramanujam, P.S. & Liminga, R. (1973). Ferroelectrics 6, 191.

52. Thomas, J.O., Tellgren, R. & Almlöf, J. (1975). Acta Cryst. B31, 1946.

53. Harkema, S., deWith, G. & Keute, J.C. (1977). Acta Cryst. B33, 3971.

54. Thomas, J.O. (1978). Acta Cryst. A34, 819.

55. Tellgren, R. & Olovsson, I. (1971). J. Chem. Phys. 54, 127.

56. Tellgren, R., Thomas, J.O. & Olovsson, I. (1977). Acta Cryst. B33, 3500.

57. Kvick, Å., Koetzle, T.F., Thomas, R. & Takusagawa, F. (1974). J. Chem. Phys. 60, 3866.

58. Peterson, S.W. & Levy, H.A. (1958). J. Chem. Phys. 29, 948.

59. Olovsson, G., Kvick, Å., Lehmann, M.S. & Olovsson, I. (1978). To be published.

60. Thomas, J.O. & Pramatus, S. (1975). Acta Cryst. B31, 2159.

61. Thomas, J.O. (1977). Acta Cryst. B33, 2867.

62. Olovsson, I. & Templeton, D.H. (1959). Acta Cryst. 12, 832.

INDEX

$A_2$ molecules, 53ff
Ab Initio methods, 36
Absorption correction, 300, 359
Acetic acid, 855
Acetylene, 769
Acoustic modes, 176, 196
Accuracy of charge density, 784
A-H molecules, 51ff
2-Amino-5-chloropyridine, 865
Amorphous silicon, 68, 79
Analysis of density, 375ff
Analytical corrections for TDS, 207
Anharmonic corrections, 198ff, 523, 529, 555
Anharmonicity in CO, 524
Anharmonicity in p-terphenyl, 526
Anharmonicity in silicon, 525
Anisotropic Compton profile, 91ff, 121, 129
Anisotropic magnetization density 514
Anisotropy in actinides and rare earths, 517
Antiferromagnets, 260
Asymptotic charge density, 748
Atomic magnetization density, 115
Autocorrelation functions, 14, 111
Azide molecules, 762

Band theory, 64, 83, 121, 728
Basic interactions in electronic systems, 376
Basis set, 31ff
Basis set in reciprocal space, 455
Basis set for transition metals, 457
Be, 420, 448
Beam characteristics, 289
Beam inhomogeneity, 290
Becker-Coppens theory, 242
Bending vibrations, 539
BF, 556, 560, 568
BH, 705
Bloch functions, 64, 83, 123
Bond density in Si, 76
Born approximation, 213, 255

Born-Oppenheimer approx., 148, 550
Born-von Karman conditions, 174
Borrmann effect, 226
Bragg intensity, 553, 557
Brillouin's theorem, 30, 41

Canonical orbitals, 29
C-C bonds, 50
$CeAl_2$, 516
$CH_4$, 42, 711
$C_2H_2$, 42
$C_2H_4$, 42
$(CH_3)_2NH_2HC_2O_4$, 882
Charge and bond order matrix, 19, 390
Charge density, 63ff, 87, 95, 375ff, 405ff, 447ff, 521ff, 560, 569, 585, 601ff, 698, 723, 845
Charge density and reactivity, 718ff
Charges in chemical reactivity, 395ff
Charge transfer, 835
Chemical applications, 757
Chemical reactivity, 716ff
Chemisorption, 100ff
ClH, 41
Clusters, 119
CO, 34, 40, 42, 417, 556, 560, 568
Cobalt, 481
Cobalt phtalocyanines, 817
$Co_2SiO_4$, 782
Cohesive energy, 736
$Co(NH_3)_6Co(CN)_6$, 780
Complementarity between densities, 726
Compton profile, 16, 110, 575
Compton scattering, 573ff
Conduction electrons, 726
Configuration interaction, 39, 42
Co-Ni, 485
Contact hyperfine coupling, 660
$Co_3O_4$, 817
Copper sulfate, 787
Core polarization, 54
Core-valence partitioning, 427
$CoRh_2O_4$, 818
Correlation, 11, 39
Correlation in extinction, 245
Counting statistics, 356
Covalency and spin, 797
Cr, 80
Cr-acetate, 453
$Cr(CO)_6$, 118, 459
Cryostat, 314, 339
Crystal field, 459, 467, 479, 823, 827
Crystal forces, 831ff
Crystal suitability criterion, 299
Cubic crystal field, 483
Cumulant expansion, 528, 531
Cusp condition, 11, 18
Cyanogen, 764

Cyanuric acid, 535

Data processing, 344
Dawson's formalism, 199, 533
Debye approximation, 177
Debye-Waller factors, 186, 195ff, 561, 777
Deformation density, 34, 48, 54, 450ff, 590, 602ff, 831, 845
Deformed pseudoatom, 554
Delocalized density, 65
Delocalized spin density, 491ff
Densities as measure of wave-function, 723
Densities and thermodynamics, 735
Density functional, 19, 64, 83
Density localisation, 428
Density matrices, 7, 107, 647
Depolarisation, 347
Derived properties, 465ff
Diamond, 87ff, 576
Difference density, 42, 700
Diffracted wave, 137
Diffraction scans, 294
Diphenylpicrylhydrazyl, 671ff
Dipolar hyperfine coupling, 661
Dipole moment, 40, 465ff
Dirac equation, 634
Directional bonding, 76
Discrete variational method, 86
Dispersion, 836
Dispersion surface, 216
Domains, 263
Dynamic charge density, 375
Dynamical scattering, 213
Dynamical theory and band theory, 220

Effective Hamiltonian for magnetic resonance, 640, 664
Einstein model, 177
ELDOR, 665
Elastic average electron scattering 559, 581
Elastic scattering, 148
Elastic X-Ray scattering, 64, 156, 183ff
Elastic wave approximation, 179
Electric field, 49, 60
Electric field gradients, 414, 423
Electron density, 3, 40ff, 48
Electron-electron energy, 753
Electronic properties from charge densities, 611, 743, 763
Electron-nuclear energy, 617, 749
Electron scattering and $N_2$, 590, 601-631
Electron scattering experiment, 584, 586
Electrostatic description of binding, 47, 57, 58ff
Electrostatic potential, 718
Electrostatic potential maps, 720
Empirical potentials, 843
ENDOR, 665
Energy and chemical bond, 696

Energy and charges in Hydrogen bonding, 834
Energy and scattering, 583ff
EPR, 665
Equilibrium constants, 716
Error analysis, 355ff
Errors in corrections, 359ff
Errors in deformation density, 365
Errors in expectation values, 36
Errors in measurements, 357
Ewald construction, 291
Exact partitioning, 379
EXAFS, 732
Exchange-correlation, 64, 68, 72
Experiment control, 341
Experimental charges and dipoles, 467
Extinction, 213ff, 237ff, 307, 348, 351, 360
Extinction length, 214
Extinction and polarized neutron 235
Extinction and TDS, 228

$f^5$ ions, 507
Fe, 80, 481, 486, 489
$Fe_3O_4$, 806
$Fe_2SiO_4$, 782
Fermi surface, 74
Ferroelectricity and extinction, 233
Ferromagnet, 260
$FeS_2$, 461
$F_2$, 58, 711
FH, 34, 41, 51, 57, 417, 556, 560
Flipping efficiency, 346
Flipping ratio, 265, 326, 351, 506
Fluctuation of observables, 376, 392, 707
Fock matrix, 27
Force constants, 564, 569
Forces on nulei, 59, 614
Formaldehyde, 765
Formaldehyde-water, 875
Formamide, 765, 861
Form factor, 14, 167, 743ff
Form factor model, 443ff
Formic acid, 765, 855
Fourier transform of multipoles, 439, 443
Free atom, 49, 60
Free electron Compton scattering, 574
Friedel oscillations, 67
Functional factorisation of wavefunctions, 711
Functional partitioning, 393

Gadolinium, 489
Gamma ray diffraction, 231
Gamma ray sources, 575
Gaussian type orbitals, 31, 440
Generalized scattering factors, 407, 551, 554, 609, 798
Generalized temperature factor, 527

Generalized Willis formalism, 529, 532ff
Glycine, 858

$H_2$, 53, 60, 598
Half wavelength correction, 345, 352
Hamilton-Zachariasen approx.,239
Harmonic averaging of scattering, 552
Harmonic oscillator, 153
Hartree-Fock limit, 34
Hartree-Fock model, 7, 27ff, 64, 87
Hellman-Feynman theorem, 47, 58, 74, 375
High energy electron scattering, 581ff
High temperature limit of anharmonicity, 532
$H_2O$, 33, 42, 51, 851
Hohenberg-Kohn-Sham theory, 19, 65, 84ff, 375
Hydration of uracil, cytosine, 399
Hydrates, 867
Hydrogen atom, 665
Hydrogen bonding, 831ff
Hydrogen bond density, 852
Hydrogen peroxide, 866

Impulse approximation, 110
Incommensurable structures, 262
Independent particle model, 27, 38, 589, 601
Inelastic electron scattering, 585
Inelastic scattering, 149
Inner moments of $N_2$, 618
Integrated intensity, 185, 238, 249
Intensity propagation equations, 222
Intensity standard deviation, 356
Intermolecular density effects, 397
Intracule function, 10, 11
Ionic crystals, 65
Isotope-Isomer shift, 659, 680

J,M decomposition of magnetisation, 502ff

Kinematical theory, 173ff, 213, 238
Kinetic energy, 109, 735
$K_2IrCl_6$, 810
$K_2NaCrF_6$, 801
$K_2ReCl_6$, 815
Kohn-Sham theory, 69
Korringa, Kohn, Rostoker, 70

Lattice dynamics, 173
Lattice energy, 764
$Li_2$, 53, 60
LiF, 95ff, 97, 99
LiH, 40, 711
$LiHCO_2$, $H_2O$, 874
Linear combination of atomic orbitals, 31, 84, 406
Linear response, 72

Liquid structure factors, 66
Lithium metal, 67, 70, 576
Local density and surfaces, 100
Local potential, 21, 83, 112
Localized density, 65
Localized orbitals, 29, 713
Localized spin density, 479
Loge decomposition, 710
Loge theory, 386, 705
Lone pair density, 56
Lone pairs and hydrogen bonds, 884
Low frequency anharmonicity, 525
Lowdin's partitioning, 381

Magnetic coupling, 642ff
Magnetic diffraction by powders, 324
Magnetic form factor, 485
Magnetic ground state, 510, 813
Magnetic interaction vector, 258
Magnetic neutron scattering, 255
Magnetic resonance, 633ff, 730
Magnetic structures, 271ff
Magnetic structure factor, 323, 479ff
Magnetization density, 259, 501, 791
Many body theory, 64
Mean squares partitioning, 388ff
Melamine, 423
Metals and alloys, 479ff
Metal-metal bond, 450
Methylacetamide, 846, 850
Migration field, 49, 52, 56ff
Minimum basis set, 33
$M(NH_4)_2(SO_4)_2{,}6H_2O$, 816
Molecular densities, 695
Molecular orbital description of covalency, 792
Moller-Plesset theorem, 39
Momentum density, 13ff, 110, 123, 573, 723
Momentum wavefunction, 573
Mosaic distributions, 223
Mosaicity and experimentation, 294
$M_2SiO_4$, 780
Mulliken's partitioning, 380
Multiple reflection, 307, 344, 361
Multipolar expansion, 388, 405ff, 447ff, 473, 606ff, 769
Multipolar expansion of momentum densities, 121
Multipolar projection to a given order, 411ff
Multipoles, 410ff, 761
Multipoles and properties, 414ff

NaCl, 67
$NaHC_2O_4$, $H_2O$, 877, 883
NaH maleate, $3H_2O$, 879
Natural spin orbitals, 8, 13
Nature of chemical bond, 695
$NbO_2$, 119
$NbO_6$, 119
$Nd^{3+}$, 489

$NdAl_2$, 494, 507, 512
Neutron diffuse scattering, 189
Neutron dynamical diffraction, 214ff
Neutron-electron potential, 256
Neutron nuclear diffraction, 184
Neutron polarisation, 335
$NH_3$, 595
Nickel, 481, 799
$Ni(CO)_6$, 111
Ni-Cu, 485
NiO, 800
$Ni_2SiO_4$, 781
NMR, 665
$N_2$, 53, 60, 556, 560, 568, 591, 596, 623ff
Nomenclature of hydrogen bonds, 832
Non localized magnetism, 266
Non rigid pseudoatoms, 545
Nonuniqueness of charge definition, 377
Normal probability plots, 364
N-Representability, 7, 20
Nuclear polarisation, 268
Numerical TDS correction, 210

One body potential, 64, 68
$O_2$, 34, 52
One particle density matrix, 9, 390, 394, 647
One particle potential, 528
Optimal experimental conditions, 287ff
Orbital scattering, 267
Orbit and spin structure factors, 501
Outer moments for $N_2$, 619
Overlap, 38, 793

Pair distribution, 10, 146, 379
Pair potential, 67
Parameter's correlation, 457
Paraterphenyl, 539
Partitioning of density, 375, 839
Partitioning of molecular interactions, 836
Penetration field, 49, 52
Perfectly following approximation, 153, 407
Phonons, 72, 176
Physical properties and wavefunctions, 725
Point charge model, 841
Point source model of extinction, 225, 246
Poisson equation, 158
Polarisation, 53, 61, 835
Polarisation analysis, 331ff
Polarisation in $MnF_2$, 280ff
Polarisation model of Hhydrogen bonding, 842
Polarisation orbitals, 33, 36
Polarized beam diffractometer, 333, 337
Polarized beam technique, 328, 330

Polarized neutron cross section, 264, 801
Potential energy function, 522, 735
Primary extinction, 226
Profile analysis, 314
Projection coefficients, 409
Projection methods, 379, 390ff
Promolecule, 34, 48, 59, 740
Pseudoatoms, 413ff, 554ff, 563, 601, 761
Pu compounds, 509

Quadrupole splitting constants, 660, 680
Quantum anharmonicity, 534
Quasi harmonicity, 198

Radial electron density, 9
Radial moments in NH, 419
Radial nodes, 56
Radial refinement, 464ff
Rare gas crystals, 66
Real space Analysis, 447ff
Reciprocal space analysis, 447ff
Reduced density matrices, 3, 5ff, 652
Reduced hamiltonian, 6
Reflection profiles, 295ff
Relativistic atoms, 115
Relativistic local density, 113
Restricted Hartree-Fock, 44
Restricted radial functions, 433ff
Rigid body model, 196
Rigid pseudoatom, 193, 558
Rocking curve, 185
Rotational average, 609
Rotation barriers, 398

Scandium, 486
Scattering cross section, 144, 145
Scattered intensity, 143, 147
Scattered wave, 139
Screened ions, 70
Secon order magnetic effects, 663
Secondary extinction, 223, 239, 243
Self consistent field, 31ff
Self consistent field limit, 33
Semi-classical theory, 157ff
Semi-empirical methods, 36
Series termination, 345, 365, 492
Shareholder partitioning, 387, 740
Sigma-pi symmetry, 52, 56, 794
Silicon, 67
Singlet-triplet densities, 378
Size of orbitals, 714
Slater determinant, 27
Slater type orbitals, 31, 34, 439
Sm, 509
Small orbital effects, 491
$SmCo_5$, 509
SmS, 511
$S_4N_4$, 473
Solid ammonia, 885
Spatial partitioning of density, 383, 393

Spectroscopic measurements, 731
Spherical harmonic expansion, 480ff
Spherical harmonic expansion of anharmonicity, 530
Spin coupling tensors, 654
Spin density, 16, 63, 259, 479, 650, 723
Spin density functional, 74
Spin dependent Compton profiles, 125
Spin flippers, 335
Spin Hamiltonian, 634, 639
Spin orbital, 30, 38
Spin-Orbit contribution, 636
Spin-spin coupling, 662
Statistical dynamical theory (Kato), 244
Statistical exchange, 107ff, 111
Step scanning, 297, 357
Stewart partitioning, 389, 405ff
Strong hydrogen bonds, 878
Strongly anharmonic potential, 533
Structure factors, 184, 552, 608
    C, 87, 89
    LiF, 95
    V, 126
Structure factor versus intensity averaging, 560
Substituent effect, 764
Substituted determinants, 38
Symmetry constraints, 473
Symmetry equivalents, 362

Takagi's equations, 218ff, 243
Thermal averaging of scattering, 151, 551, 556, 602
Thermal diffuse scattering, 191, 201, 205ff, 304, 315, 360, 553
Thomas Fermi Dirac theory, 64, 69, 74
Time averaged density, 521ff
Time averaging, 173ff
Time dependent perturbation, 161ff
Titanium, 488
$TiO_2$, 783
TmSb, 515
Topography of density, 21ff
Total charge, 612
Total electron scattering, 581, 589, 602
Total wavefunction, 550
Transition matrices, 647ff
Transition metal complexes, 779ff 793ff
Two centre terms, 409, 804
Two electron density, 9, 648

Unrestricted Hartree-Fock, 30
Uracil, 421
US, 510
USb, 814

Valence density, 56
Vanadium metal, 450ff
Vibration amplitude, 155
Vibrational wavefunctions, 555
Vibrations and form factors, 549
Vibrational modes, 174
Vibrations in molecular crystals, 180
Virial partitioning, 22, 385, 698

Wannier functions, 79, 123
Water polymers, 838, 867ff
Wave scattering, 137

X-Ray diffuse scattering, 188
X-Ray dynamical diffraction, 218
X-Ray total scattering, 11

Yttrium iron garnet, 808

Zeeman term, 638

MIX
Papier aus verantwortungsvollen Quellen
Paper from responsible sources
FSC® C105338

If you have any concerns about our products,
you can contact us on
**ProductSafety@springernature.com**

In case Publisher is established outside the EU,
the EU authorized representative is:
**Springer Nature Customer Service Center GmbH**
**Europaplatz 3, 69115 Heidelberg, Germany**

Printed by Libri Plureos GmbH
in Hamburg, Germany